Polymeric Dispersions: Principles and Applications

NATO ASI Series

Advanced Science Institutes Series

A Series presenting the results of activities sponsored by the NATO Science Committee, which aims at the dissemination of advanced scientific and technological knowledge, with a view to strengthening links between scientific communities.

The Series is published by an international board of publishers in conjunction with the NATO Scientific Affairs Division

A	**Life Sciences**	Plenum Publishing Corporation
B	**Physics**	London and New York
C	**Mathematical and Physical Sciences**	Kluwer Academic Publishers
D	**Behavioural and Social Sciences**	Dordrecht, Boston and London
E	**Applied Sciences**	
F	**Computer and Systems Sciences**	Springer-Verlag
G	**Ecological Sciences**	Berlin, Heidelberg, New York, London,
H	**Cell Biology**	Paris and Tokyo
I	**Global Environmental Change**	

PARTNERSHIP SUB-SERIES

1.	**Disarmament Technologies**	Kluwer Academic Publishers
2.	**Environment**	Springer-Verlag / Kluwer Academic Publishers
3.	**High Technology**	Kluwer Academic Publishers
4.	**Science and Technology Policy**	Kluwer Academic Publishers
5.	**Computer Networking**	Kluwer Academic Publishers

The Partnership Sub-Series incorporates activities undertaken in collaboration with NATO's Cooperation Partners, the countries of the CIS and Central and Eastern Europe, in Priority Areas of concern to those countries.

NATO-PCO-DATA BASE

The electronic index to the NATO ASI Series provides full bibliographical references (with keywords and/or abstracts) to more than 50000 contributions from international scientists published in all sections of the NATO ASI Series.
Access to the NATO-PCO-DATA BASE is possible in two ways:

– via online FILE 128 (NATO-PCO-DATA BASE) hosted by ESRIN,
Via Galileo Galilei, I-00044 Frascati, Italy.

– via CD-ROM "NATO-PCO-DATA BASE" with user-friendly retrieval software in English, French and German (© WTV GmbH and DATAWARE Technologies Inc. 1989).

The CD-ROM can be ordered through any member of the Board of Publishers or through NATO-PCO, Overijse, Belgium.

Series E: Applied Sciences - Vol. 335

Polymeric Dispersions: Principles and Applications

edited by

José M. Asua
Department of Applied Chemistry,
The University of the Basque Country,
San Sebastián,
Spain

Kluwer Academic Publishers

Dordrecht / Boston / London

Published in cooperation with NATO Scientific Affairs Division

Proceedings of the NATO Advanced Study Institute on
Recent Advances in Polymeric Dispersions
Elizondo, Spain
June 23–July 5, 1996

A C.I.P. Catalogue record for this book is available from the Library of Congress

ISBN 0-7923-4549-5

Published by Kluwer Academic Publishers,
P.O. Box 17, 3300 AA Dordrecht, The Netherlands.

Kluwer Academic Publishers incorporates the publishing programmes of
D. Reidel, Martinus Nijhoff, Dr W. Junk and MTP Press.

Sold and distributed in the U.S.A. and Canada
by Kluwer Academic Publishers,
101 Philip Drive, Norwell, MA 02061, U.S.A.

In all other countries, sold and distributed
by Kluwer Academic Publishers Group,
P.O. Box 322, 3300 AH Dordrecht, The Netherlands.

Printed on acid-free paper

CONTENTS

PREFACE

Polymeric dispersions are used in a wide variety of applications such as synthetic rubber, paints, adhesives, binders for non-woven fabrics, additives in paper and textiles, leather treatment, impact modifiers for plastic matrices, additives for construction materials, flocculants, and rheological modifiers. They are also used in biomedical and pharmaceutical applications such as diagnostic tests and drug delivery systems. The rapid increase of this industry is due to several reasons: i) environmental concerns and governmental regulations to substitute solvent-based systems by water-borne products, ii) polymeric dispersions have some unique properties that meet a wide range of application problems, and iii) compared to other polymerization processes, emulsion polymerization presents substantial advantages from the point of view of controllability of the operation. In addition to their commercial importance, the production, characterization, and application of dispersed polymers have aroused an increasing interest in academia because of the scientific challanges that they present.

The manuscripts included in this volume were presented at the NATO Advanced Study Institute on "Recent Advances in Polymeric Dispersions" held in Elizondo, Spain, at the Hotel Baztan during the period of June 23 - July 5, 1996. The goal of the NATO - ASI was to integrate in a single course the state of the art of the Science and Technology of Polymeric Dispersions by reviewing the fundamentals, discussing the new developments, pointing out unsolved problems, and speculating about future research directions. The areas addressed were divided into the following groups: I. Kinetics and Mechanisms in Polymerization in Dispersed Media; II. Particle Morphology; III. Characterization Methods; IV. Polymerization Reactors; and V. Applications of Dispersed Polymers. This volume reflects the above subdivision.

The Advanced Study Institute was made possible by a generous grant from the NATO Scientific Affairs Division. We were also very fortunate to obtain additional support from the following:

Comisión Interministerial de Ciencia y Tecnología, Madrid, Spain
DSM, Geelen, The Netherlands
National Science Foundation, Washington D.C., USA
Union Carbide, Cary, North Carolina, USA
Wacker Chemie GmbH, Burghausen, Germany

I express my sincere thanks to all of them for their generous support and their interest in the meeting.

In am particularly grateful to Dr. David R. Bassett, Professor Mohamed S. El-Aasser, Professor Ronald H. Ottewill, and Dr. Klaus Tauer, members of the Organizing Committee for their most valuable contributions to the scientific organization.

Many things had to be organized to guarantee the smooth running of the scientific and social programmes. In so far as the objectives have been met, this certainly has been due to the help, before, during and after the Institute, of Dr. María J. Barandiaran, Dr.

José C. de la Cal, Dr. Jacqueline Forcada, Dr. José R. Leiza, Mr. Mikel Larrañaga and Ms. Isabel Sáenz de Buruaga.

My appreciation to Dr. Philip D. Armitage for his help in the process of reviewing the papers included in this book, and to Ms. Inés Plaza who retyped many of the chapters.

I acknowledge the help of the staff of the Hotel Baztan for their expert assistance and for their warm hospitality.

The success of an Institute is ultimately determined by the interest and committment of the lecturers and participants. I want to thank all of them for their enthusiasm and collaboration.

Finally my deepest gratitude to my wife Esmeralda, and our daughter, Leire, who provided all the cooperation and inspiration needed for the success of this project.

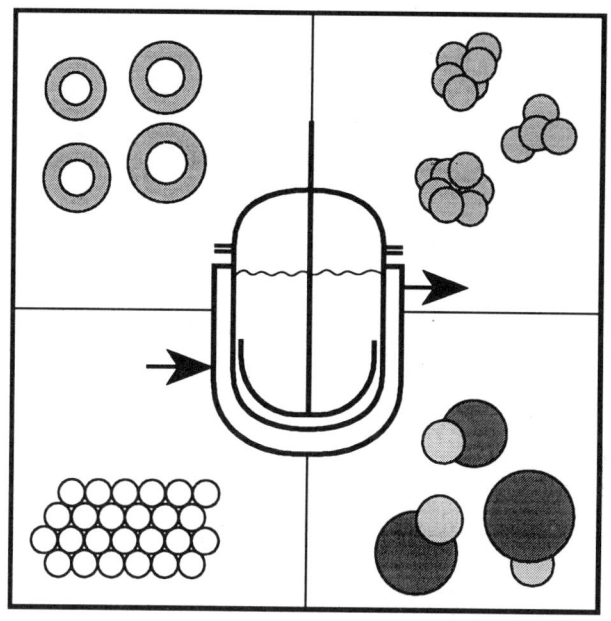

NATO Advanced Study Institute
on
Recent Advances in Polymeric Dispersions

June 23 - July 5, 1996
Hotel Baztán, Elizondo, Spain

Director

Professor José M. Asua, The University of the Basque Country, San Sebastián, Spain

Scientific Committee

Dr. David R. Bassett, UCAR Emulsion Systems, Cary, NC, USA.
Professor Mohamed S. El-Aasser, Lehigh University, Bethlehem, PA, USA.
Professor Ronald H. Ottewill, University of Bristol, Bristol, UK.
Dr. Klaus Tauer, Max Planck Institute, Teltow, Germany.

List of Contributors

Carlos Abad
3M Belgium
European Business Centre
Haven 1005, Canadastraat 11
B-2070, Zwijndrecht
Belgium

José M. Asua
The University of the Basque Country
Departamento de Química Aplicada
P.O. Box 1072
20080 San Sebastián
Spain

David R. Bassett
Ucar Emulsion Systems
410 Gregson Drive
Cary, NC 27511
USA

José C. de la Cal
The University of the Basque Country
Departamento de Química Aplicada
P.O. Box 1072
20080 San Sebastián
Spain

Françoise Candau
Institute Charles Sadron
CNRS - ULP
6 rue Boussingault
Strasbourg, Cédex 67083
France

Dominique Charmot
Rhône Poulenc Recherches
52, rue de la Haie Coq
93308 Aubervilliers
France

Emma M. Coen
The University of Sydney
School of Chemistry
Sydney, NSW 2006
Australia

Andrew J. De Fusco
Ucar Emulsion Systems
410 Gregson Drive
Cary, NC 27511
USA

Thierry Delair
U.M. CNRS-Biomerieux
ENSL
46, Allée d'Italie
69364 Lyon, Cédex 07
France

Victoria L. Dimonie
Lehigh University
Emulsion Polymers Institute
111 Research Drive, Iacocca Hall
Bethlehem, PA 18015-4732
USA

Ivon G. Durant
University of New Hampshire
Polymer Research Group
G106 Parsons Hall
105 Main Street
Durham, NH 03824-3547
USA

Mohamed S. El-Aasser
Lehigh University
Emulsion Polymers Institute
111 Research Drive, Iacocca Hall
Bethlehem, PA 18015-4732
USA

Abdelhamid Elaïssari
U.M. CNRS-Biomerieux
ENSL
46, Allée d'Italie
69364 Lyon, Cédex 07
France

Anton L. German
Eindhoven University of Technology
Den Dolech 2
P.O. Box 513
Eindhoven, 5600 MB
The Netherlands

John M. Geurts
Eindhoven University of Technology
Department of Polymer Chemistry and
Technology
Den Dolech 2
P.O. Box 513
Eindhoven, 5600 MB
The Netherlands

Robert. G. Gilbert
The University of Sydney
School of Chemistry
Sydney, NSW 2006
Australia

Archie E. Hamielec
McMaster University
Department of Chemical Engineering
Hamilton, Ontario, L85 4L7
Canada

Wolf-Dieter Hergeth
Wacker-Chemie GmbH
Werk Burghausen, ABT. 1
PF 1260
D-84489 Burghausen
Germany

Richard D. Jenkins
Ucar Emulsion Systems
410 Gregson Drive
Cary, NC 27511
USA

Ingolf Kühn
BASF AG
D-67056 Ludwigshafen
Germany

Dax Kukulj
The University of Sydney
School of Chemistry
Sydney, NSW 2006
Australia

Katharina Landfester
Max-Planck Institute for Polymer Research
Postfach 3148
D-55021 Mainz
Germany

Do Ik Lee
Dow Chemical Company
DESIGNED LATEX RESEARCH
1604 Bldg.
Midland MI 48674
USA

José R. Leiza
The University of the Basque Country
Departamento de Química Aplicada
P.O. Box 1072
20080 San Sebastián
Spain

Christopher M. Miller
Union Carbide Corporation
UCAR Emulsion Systems
410 Gregson Drive
Cary, NC 27511
USA

Massimo Morbidelli
ETH
Laboratorium für Technische Chemie LTC
Universitätstrasse 6
CH-8092 Zürich
Switzerland

Willem Norde
Wageningen Agricultural University
Department for Physical & Colloid
Chemistry
P.O. Box 8038
Wageningen, 6700 EK
The Netherlands

Ronald H. Ottewill
The University of Bristol
School of Chemistry
Cantock's Close
Bristol, BS8 1TS
United Kingdom

Christian Pichot
U.M. CNRS-Biomerieux
ENSL
46, Allée d'Italie
69364 Lyon, Cédex 07
France

Gary W. Poehlein
School of Chemical Engineering
Georgia Institute of Technology
Atlanta, GA 30332-01001
USA

Prapasri Rajatapiti
Lehigh University
Emulsion Polymers Institute
111 Research Drive, Iacocca Hall
Bethlehem, PA 18015-4732
USA

Jöel Richard
Microencapsulation Center
Parc Scientifique des Capucins
8, rue André Boquel
49100 Angers France

Krishan C. Sehgal
Ucar Emulsion Systems
410 Gregson Drive
Cary, NC 27511
USA

Andre Siani
ETH
Laboratorium für Technische Chemie LTC
Universitätstrasse 6
CH-8092 Zürich
Switzerland

Joao B.P. Soares
University of Waterloo
Dept. Chemical Engineering
Waterloo, Ontario N21 3G1
Canada

Hans Wolgang Spiess
Max-Planck Institute for Polymer Research
Postfach 3148
D-55021 Mainz
Germany

G. Storti
Dip. Ingenneria Chimica e Matematica
Universitá di Cagliari
Piazza d'Armi, 09123 Cagliari
Italy

E. David Sudol
Lehigh University
Emulsion Polymers Institute
Iacocca Hall, D330
111 Research Dr.
Betlehem, PA18015
USA

Donald C. Sundberg
University of New Hampshire
Polymer Research Group
G106 Parsons Hall
105 Main Street
Durham, NH 03824-3547
USA

Klaus Tauer
Max Planck Institute
Kantstrasse 55
Teltow-Seehof, 14513
Germany

J.J.G. Steven van Es
Eindhoven University of Technology
Department of Polymer Chemistry and
Technology
Den Dolech 2
P.O. Box 513
Eindhoven, 5600 MB
The Netherlands

Alex van Herk
Eindhoven University of Technology
Department of Polymer Chemistry and
Technology
Den Dolech 2
P.O. Box 513
Eindhoven, 5600 MB
The Netherlands

John M.G. Verstegen
Eindhoven University of Technology
Department of Polymer Chemistry and
Technology
Den Dolech 2
P.O. Box 513
Eindhoven, 5600 MB
The Netherlands

Julian A. Waters
The University of Bristol
School of Chemistry
Cantock's Close
Bristol, BS8 1TS
United Kingdom

NOMENCLATURE

The following constitutes a common nomenclature for the book. Other specific symbols are defined in each chapter. Units are indicated in terms of base physical quantities: length (L), mass (M), time (t), temperature (T), amount of substance (mol), charge (C, Coulomb), and electric current (A, ampere). In some cases, derived quantities are also used: energy ($E=ML^2/t^2$), and electric potential (V, volt).

A	heat transfer area (L^2).
A_{ijk}	Hamaker constant for the interaction between materials i and k, being both immersed in a phase j (E).
A_P	surface area of all particles in the system (L^2).
a_s	area occupied by a single surfactant molecule when the surface is saturated (L^2).
B(V,V')	rate coefficient for coagulation between particles of volume V and V' (L^3 mol^{-1} t^{-1}).
BSA	Bovine Serum Albumin.
C_{pi}	specific heat capacity (E M^{-1} T^{-1} or E mol^{-1} T^{-1}).
CPVC	critical pigment volume concentration.
CTA	chain transfer agent.
c.c.c	critical coagulation concentration (mol L^{-3}).
cmc	critical micelle concentration (mol L^{-3}).
D	impeller diameter (L).
DET	direct non radiactive energy transfer.
D_w	diffusion coefficient of the monomeric radicals in the aqueous phase (L^2 t^{-1}).
d_o	diameter of the unswollen polymer particle (L).
d_p	diameter of the monomer swollen polymer particle (L).
[E]	concentration of desorbed radicals in the aqueous phase (mol L^{-3}).
ESEM	environmental scanning electron microscopy.
ESR	electron spin resonance.
e	electron charge (C).
F_i	molar feed flow rate of compound i (mol t^{-1}).
f	efficiency factor for initiator decomposition (dimensionless).
f_I	overall initiator efficiency (dimensionless).
f_i	molar fraction of monomer i (dimensionless).
g	gravitational constant (L t^{-2})
H	reactor height (L).
HGG	Human Gamma Globulin.
HSA	Human Serum Albumin.

h	intersurface separation (L).
h_p	Planck's constant (E t).
$[I_2]$	concentration of initiator (mol L^{-3}).
$[IM_i]$	concentration of initiator derived oligomers of length i in the aqueous phase (mol L-3).
I(Q)	scattered intensity (E L^{-2} t^{-1}).
IEP	isoelectric point.
I_gG	Immuno Gamma Globulin.
j_{crit}	critical degree of polymerization for homogeneous nucleation.
K(V)	rate coefficient for propagational particle growth (L^{-3} t^{-1})
k_a	entry rate coefficient (L^3 mol^{-1} t^{-1})
k_d	exit rate coefficient (t^{-1}).
k_{dm}	first order rate coefficient for desorption of a monomeric radical (t^{-1}).
k_i	rate coefficient for initiator dissociation (t^{-1}).
$k^j_{p,w}$	propagation rate coefficient for a j-meric radical in the aqueous phase (L^3 mol^{-1} t^{-1}).
k_p	propagation rate coefficient (L^3 mol^{-1} t^{-1}).
$<k_p>$	average propagation rate constant (L^3 mol^{-1} t^{-1}).
k_p^*	rate constant for propagation to pendant double bonds (L^3 mol^{-1} t^{-1}).
k_p^1	propagation rate coefficient for a monomeric radical (L^3 mol^{-1} t^{-1}).
k_{pji}	propagation rate constant of radical j with monomer i (L^3 mol^{-1} t^{-1}).
k_t	termination rate constant (L^3 mol^{-1} t^{-1}).
$k_{t,w}$	termination rate coefficient in the aqueous phase (L^3 mol^{-1} t^{-1}).
k_{tr}, k_{trm}	rate coefficient for transfer to monomer (L^3 mol^{-1} t^{-1}).
k_{trp}	rate coefficient for transfer to polymer (L^3 mol^{-1} t^{-1}).
LCB	long chain branching.
LCST	Lower critical solution temperature.
MAS	magic angle spinning.
MFT	minimum film forming temperature (T).
M_c	molecular weigth between chemical crosslinks (M mol^{-1}).
M_e	entanglement length.
$<M_n>$	number-average molecular weight (M mol^{-1}).
M_{oi}	molecular weight of the monomer i (M mol^{-1}).

$[M]_p$	concentration of monomer in the polymer particles (mol L^3).
$[M]_{p,tot}$	total monomer concentration in the polymer particles (mol L^3).
$<M_w>$	weight-average molecular weight (M mol^{-1}).
$[M]_w$	concentration of monomer in the aqueous phase (mol L^3).
$[M]_{w,sat}$	saturation concentration of the monomer in the aqueous phase (mol L^3).
N	agitator speed (t^{-1}).
N_A	Avogrado's number (mol^{-1}).
N_P	power number ($P/\rho N^3 D^5$, dimensionless).
N_p	Number of polymer particles per unit volume (L^{-3}).
N_{Re}	Reynolds number ($D^2 N \rho/\eta$, dimensionless).
N_{We}	Weber number, ($D^3 N^2 \rho_d/\gamma$, dimensionless).
$\tilde{n}$	average number of radicals per particle.
$n(V)$	particle number distribution function (mol L^{-6}).
n_m, n_p, n_0	refractive index of the pure component (monomer), polymer and medium, respectively (dimensionless).
n_0	number of ions per unit volume (L^{-3}).
P	agitator power (E).
PDB	pendant double bond.
PEBP	pulsed electron beam polymerization.
P_i	probability of finding a radical with ultimate unit of type I.
PLP	pulsed laser polymerization.
$P(Q)$	particle shape factor (dimensionless).
PVC	pigment volume concentration
p	controlled process outputs.
Q	scattering vector.
R_{CTA}	rate of transfer to chain transfer agent (mol L^{-3} t^{-1}).
R_H	hydrodynamic radius of the particle (L).
R_{nm}	nucleation rate (mol L^{-3} t^{-1}).
R_p	polymerization rate (mol L^{-3} t^{-1}).
$[R]_w$	total concentration of radicals in the aqueous phase (mol L^{-3}).
r_i	reactivity ratio (ultimate model) (dimensionless).
r_0	radius of the unswollen polymer particle (L).
r_p	radius of the swollen polymer particle (L).
SANS	small angle neutron scattering.
SEC	size exclusion chromatography.
$S(Q)$	structure factor.
$[S_T]$	concentration that the added surfactant would have if there were no adsorption (mol L^{-3}).

$[S_W]$	concentration of emulsifier in the aqueous phase (mol L^{-3}).
TEM	transmission electron microscopy.
T_g	glass transition temperature (T).
TOF/SIMS	time-of-flight/secondary ion mass spectroscopy.
U	overall heat transfer coefficient ($EL^{-2} T^{-1} t^{-1}$).
u	manipulated variable.
u	average fluid velocity in the axial direction ($L t^{-1}$).
V_A	van der Waals' attraction energy (E).
V_B	Born repulsion energy (E).
V_m	molar volume of the monomer ($L^3 mol^{-1}$).
V_P	volume of the scattering particle (L^3).
V_R	electrostatic energy of repulsion (E).
V_T	total potential energy of interaction (E).
W_g, W_s	gel and sol fractions, respectively (dimensionless).
$W_{(r)}$	molecular weight distribution.
w_p	weight fraction of polymer (dimensionless).
X	monomer conversion (dimensionless).
XPS	X-ray photo electron spectroscopy./,
x	state variables.
y	measured variables.
z	critical degree of polymerization for entry of radicals in the polymer particles.

Greek Symbols.

α_{11}	electronic polarisability (C L).
γ	interfacial tension ($E L^{-2}$).
ε_0	permittivity of free space ($C^2 M L^{-1} t^{-2}$).
ε_r	relative permittivity of the dispersion medium (dimensionless).
ξ	zeta potential (V).
η	viscosity ($M L^{-1} t^{-1}$).
θ	angle of the scattering.
κ	Debye-Hückel reciprocal length (L^{-1}).
λ	wavelength (L).
λ_0	wavelength of light in vacuo (L).
v_i	ith moment of the polymer chain distribution.
π	osmotic pressure.
ρ	density of the reaction medium ($M L^{-3}$).
ρ_a	pseudo-first-order rate coefficient for radical entry (t^{-1}).
$\rho_{initiator}$	pseudo-first-order rate coefficient for entry radicals resulting from

	thermal radical generation (t^{-1}).
ρ_m, ρ_p	densities of monomer and polymer, respectively $(M\ L^{-3})$.
$\rho_{thermal}$	pseudo-first-order rate coefficient for entry of radicals resulting from thermal radical generation (t^{-1}).
$\rho(x), \bar{\rho}(x)$	instantaneous and cumulative crosslinking densities.
σ	surface charge density $(C\ L^{-2})$.
τ	mean residence time in the reactor (t).
φ	volume fraction of the organic phases (dimensionless).
$\phi_i{}^j$	volume fraction of component i in phase j (dimensionless).
χ_{ij}	Flory-Huggins interaction parameter between species i and j. (dimensionless).
ψ_s	electrostatic surface potential (V).

MECHANISMS FOR RADICAL ENTRY AND EXIT
– *Aqueous-phase influences on polymerization*

ROBERT G GILBERT

School of Chemistry,
University of Sydney
NSW 2006,
Australia

1. Introduction

An emulsion polymerization comprises a number of phases, as illustrated in Figure 1. Although polymerization takes place within the particle phase, the generation of free radicals usually takes place in the aqueous phase. The phase-transfer event whereby radicals go from the aqueous into the particle phase is *entry*. Free-radical activity can go from the particles back into the aqueous phase: the process denoted *exit* (or *desorption*). Further, aqueous-phase radicals arising either from initiator or from exit can have a number of fates which impinge upon radical activity within the particles, which in turn has effects on the polymerization process and products. It is these various events which form the subject of the present chapter.

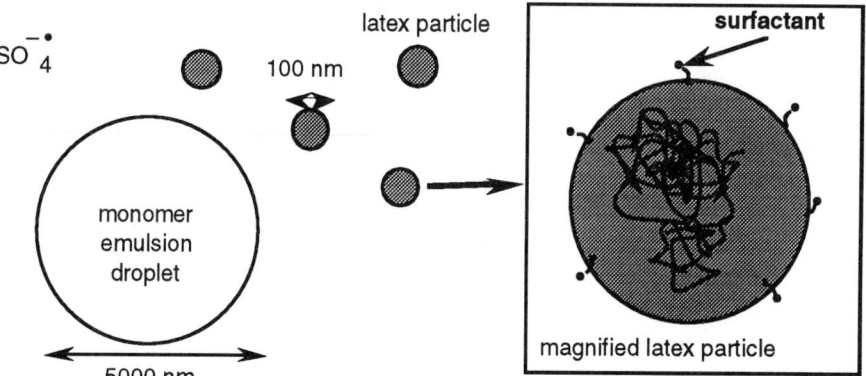

Figure 1. Illustrating the different phases in an emulsion polymerization; during particle formation, micelles are also present.

The basic equation describing the rate in an emulsion polymerization is:

$$R_p = k_p \frac{N_p}{N_A} \bar{n} [M]_p \qquad (1)$$

where k_p is the propagation rate coefficient, N_A the Avogadro constant, N_p the number of latex particles per unit volume, $\bar{n}$ the average number of propagating radicals per

1

J. M. Asua (ed.), Polymeric Dispersions: Principles and Applications, 1–15.
© 1997 *Kluwer Academic Publishers. Printed in the Netherlands.*

particle, and $[M]_p$ the monomer concentration within the particle. The value of $\bar{n}$ is in turn controlled by entry, exit and termination; the first of these two events form the subject of the present chapter.

2. Obtaining Entry and Exit Rate Coefficients from Experiment

Entry and exit rate coefficients can be obtained from the time evolution of the polymerization rate in a monodisperse seeded system, in the absence of secondary particle formation: i.e., not just the steady-state rate, but the rate at which this steady state is attained [1]. It is convenient, when considering means of obtaining experimental rate coefficients in emulsion polymerization, to consider two limiting but widely-applicable cases: a *zero-one* system, which is where entry into a latex particle which already contains a growing radical results in extremely rapid termination, and a *pseudo-bulk* system, where termination is rate-determining and where there is no effect of compartmentalization. The kinetics of a zero-one system are dominated by entry and exit, while those of a pseudo-bulk system must also take termination into account. Termination is complex to quantify because the rate coefficients depend in general on the lengths of both chains [2]. Hence unambiguous experimental results for the rate coefficients, and hence mechanisms, for entry and exit can best be obtained from zero-one systems, although this may not always be possible to achieve. It is possible to obtain entry rate coefficients from pseudo-bulk systems (e.g., [3,4]) but this requires extensive model-based assumptions and is therefore less reliable.

The kinetics of zero-one systems are complicated by having to take into account the fate of exited radicals. While equations describing these kinetics can readily be written down (e.g., [5-7]), these are not suitable for general data fitting. Fortunately, it turns out [1] that it is sufficient to consider three limiting cases for the fates of exited radicals in zero-one systems:

Limit 1: complete aqueous-phase termination:

$$\frac{d\bar{n}}{dt} = \rho_a(1 - 2\bar{n}) - k_{tr}[M]_p \frac{k_{dM}}{k_{dM} + k_p^1[M]_p} \bar{n} \qquad (2)$$

Limit 2a: complete re-entry:

$$\frac{d\bar{n}}{dt} = \rho_a(1 - 2\bar{n}) - 2 \frac{k_{tr} k_{dM}}{k_p^1} \bar{n}^2 \qquad (3)$$

Limit 2b: re-entry and re-escape:

$$\frac{d\bar{n}}{dt} = \rho_a(1 - 2\bar{n}) - 2k_{tr} [M]_p \bar{n} \qquad (4)$$

where $\rho_a = \rho_{initiator} + \rho_{thermal}$ is the pseudo-first-order rate coefficient for entry, containing components for radicals deriving directly from initiator (but not from re-entry of exited radicals) and from thermal radical generation, k_{tr} is the rate coefficient for transfer to monomer (it has been assumed for simplicity that all exit is by monomeric radicals formed by transfer to monomer), k_p^1 is the propagation rate coefficient for a monomeric radical formed by transfer, and k_{dM} is the first-order rate coefficient for desorption of a monomeric radical [8]:

$$k_{dM} = \frac{3D_W}{r_p^2} \frac{[M]_W}{[M]_p} \tag{5}$$

where D_W is the diffusion coefficient of monomeric radicals in the water phase and r_p the swollen radius of a particle. The second term in each of eqs (2–4) represents exit, and can be expressed in terms of a phenomenological exit rate coefficient k_d (specifically, the second term in these equations can be written as $-k_d\bar{n}$, $-2k_d\bar{n}^2$ and $-k_d\bar{n}$ respectively).

Solution of eqs (2), (3) or (4), depending on the particular system, in the steady state enables $\bar{n}$ to be calculated for a *zero-one* system. The computation of $\bar{n}$ in a *pseudo-bulk* system is rendered complex by the need to take chain-length-dependent termination into account (e.g., [9,10]). If a suitable value of the chain-length-averaged termination rate coefficient is known, various algorithms for finding $\bar{n}$ in a compartmentalized system are available (e.g., [11]).

Means of obtaining entry and exit rate coefficients from appropriate experimental data have been discussed in detail elsewhere (e.g., [1]). In brief, one uses the *approach* to steady state together with the steady-state value of $\bar{n}$. In turn this information is usually obtained from the experimental polymerization rate, which can be converted to $\bar{n}$ if the propagation rate coefficient is known [1,12]. The most reliable data for exit are those obtained from the time evolution after removal from a γ-radiation source, which yields $\rho_{thermal}$ and k_d directly, without the possibility of any artifact from inhibition. Such *relaxation data* are especially sensitive to k_d because it is exit which is the prime cause of loss of radical activity in a zero-one system; these values of k_d, when combined with those for the steady-state $\bar{n}$ with chemical initiator, in turn yield $\rho_{initiator}$ through equating the right-hand size of eqs (2–4) to zero. Means of processing such data have been discussed in detail elsewhere [1].

Some data are shown in Figures 2 and 3. The data for chemical initiator exhibited inhibitor artifacts: consistency tests showed that the approach to steady state in these particular data was in fact due at least in part to the consumption of inhibitor, since the value of k_d inferred from the approach to steady state in the chemically-initiated system was significantly different from that inferred from γ-relaxation data. Multiple re-insertions and removals with γ initiation/relaxation can be used to test if data are vitiated by inhibitor artifacts. Where such artifacts are absent, the approach to steady state in a chemically-initiated system can be used to find ρ_a and k_d (e.g., [12,13]).

As stated, the values of both entry and exit rate coefficients inferred from experiment depend on the assumed fate of exited radicals, i.e., which of eqs (2–4) is obeyed. Assuming different fates does not have a large effect on the value of ρ_a, but can affect k_d by 30% or more. Methods have been given [14] whereby the likely fate can be calculated by relatively straightforward means with data that are easily obtained, with sufficient accuracy that one can decide which limit is obeyed with acceptable reliability.

4

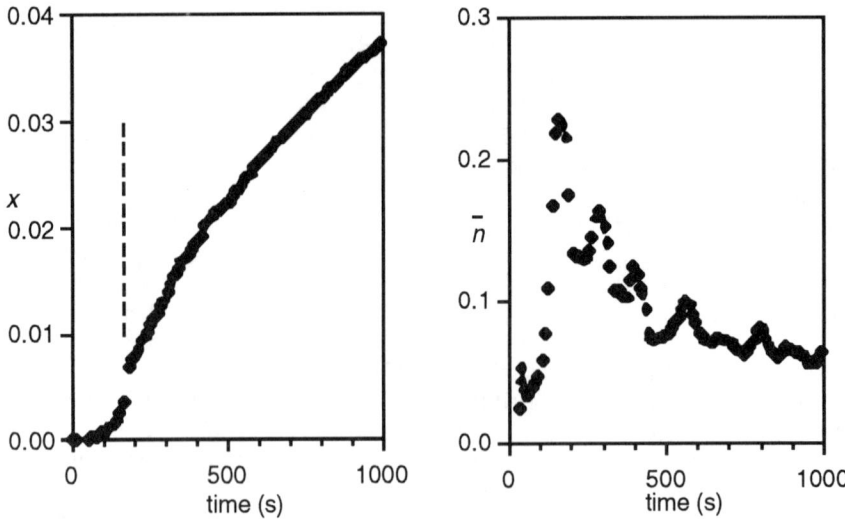

Figure 2. Conversion (x) and $\bar{n}$ as functions of time for a γ-initiated styrene seeded emulsion poly-merization system: 24 nm unswollen radius, $N_p = 5.8 \times 10^{16}$ dm^{-3}, 50°C, Interval II. The broken line indicates time of removal from radiation source; the relaxation data yield the values of the exit and thermal entry rate coefficients directly. Data obtained by Emma Coen in the author's group.

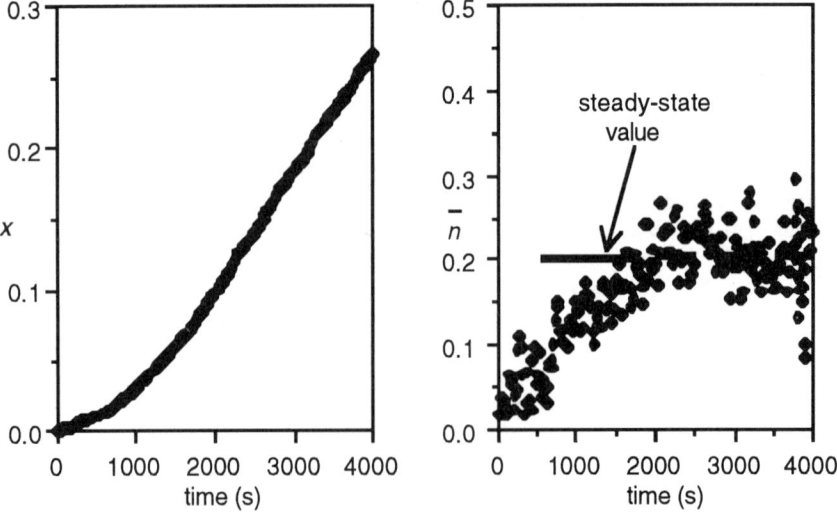

Figure 3. Conversion and $\bar{n}$ as functions of time for a chemically-initiated styrene seeded emulsion polymerization system: unswollen radius 24 nm, $N_p = 5.8 \times 10^{16}$ dm^{-3}, $[I_2] = 1.5 \times 10^{-4}$ mol dm^{-3} persulfate, 50°C, Interval II. The approach to steady state is an inhibitor artifact; together with the exit rate coefficient obtained from Figure 2, the steady-state value of $\bar{n}$ yields ρ_a. Data obtained by Emma Coen in the author's group.

In applying this methodology, one must ensure that the system is zero-one. A *necessary* but not sufficient condition for this is that $\bar{n}$ be less than $^1/_2$.

A number of suggestions have been made as to how it may be determined experimentally if the zero-one assumption is obeyed for a given system [1]. The simplest *sufficient* condition is that a true steady state is observed in Interval II in a system with $\bar{n} \leq$ $^1/_2$. This is because if termination is rate-determining (which would vitiate the zero-one assumption), then the rate at which termination occurs on entry in Interval II (when monomer concentration remains constant but the particle volume increases) must depend on particle volume (if compartmentalization is important), and thus the overall polymerization rate will change with conversion. That is, if the overall polymerization rate does *not* change with conversion in Interval II, the system must be zero-one. An example of contrary behavior is seen in MMA, when the rate and $\bar{n}$ show a steady acceleration in Interval II in small particles with low $\bar{n}$ [15], showing that the system cannot be zero-one. This is because [16] exit is rapid but the exited radical goes quickly from particle to particle until it undergoes propagation to a significant degree of polymerization; when termination takes place, it is relatively slow (i.e., rate-determining), whereas exit and re-entry are so fast that overall desorption is not rate-determining.

3. Results and Models for Entry

It is apparent that the highly hydrophilic radical arising from an initiator such as persulfate, viz., $SO_4^{-\bullet}$, cannot enter the hydrophobic environment of a latex particle, and hence the process whereby this radical activity is transferred to the interior of a particle must be complex. Some redox initiators may produce less hydrophilic radicals, perhaps (in the case of an initiator such as *tert*-butylhydroperoxide) in the interfacial region, but again this redox process will be complicated: geminate recombination must be avoided, and in cases such as *inisurfs*, this leads to very inefficient initiation [17].

The process whereby irreversible entry occurs with an initiator such as persulfate is illustrated in Figure 4. One has aqueous-phase propagation and termination until a sufficiently high degree of polymerization z is achieved so that the z-mer undergoes propagation while in the interfacial region without any other fate intervening. Now, irreversible entry may take place over a distribution of degrees of polymerization, but for simplicity one can consider entry as an all-or-nothing event, with z being an *effective* degree of polymerization. In this section, the events leading to entry in two simple systems are considered: persulfate initiator with particles stabilized by anionic and by electrosteric stabilizer.

Given the mechanism just described, the value of $\rho_{initiator}$ may be obtained by noting that z is the effective degree of polymerization at which entry occurs and no other fate is possible (a distribution of such degrees of polymerization, incorporating competition between propagation, termination and desorption, can be used to develop a more microscopic model for z in terms of the rate coefficients for the processes just described). If k_a denotes the second-order rate coefficient for entry of a z-mer (whose concentration is denoted $[IM_z]$), then:

$$\rho_{initiator} = k_a[IM_z] \tag{6}$$

$$= \frac{k_{p,w}^{z-1}[M]_w[IM_{z-1}]}{N_p/N_A} \tag{7}$$

6

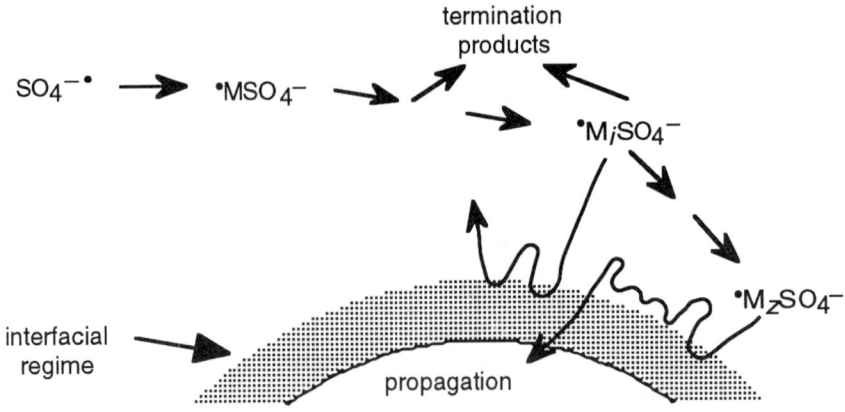

Figure 4. Illustrating the events leading to irreversible entry into the interior of a particle by a z-meric radical.

where $k_{p,w}^{z-1}$ is the propagation rate coefficient for a z-meric radical in the water phase. Eq (7) follows from the assumption that the only fate for a z-mer is irreversible entry. The value of the concentration of $(z-1)$-meric radicals is in turn found by solving, in the steady state, the rate equations for all species (taking account of the fact [18] that the propagation of a radical such as $SO_4^{-\bullet}$ is so rapid as not to be rate-determining):

$$[IM_1] = \frac{2fk_i[I_2]}{k_{p,w}^1[M]_w + k_{t,w}[R]_w} \qquad (8)$$

$$[IM_i] = \frac{k_{p,w}^{i-1}[IM_{i-1}][M]_w}{k_{p,w}^i[M]_w + k_{t,w}[R]_w} \qquad (9)$$

where $[I_2]$ is the initiator concentration, k_i dissociation rate coefficient, f the efficiency with which the initiator dissociates to form two propagating aqueous-phase radicals ($f \approx$ 0.5), $k_{t,w}$ is the termination rate coefficient in the water phase (the latter is assumed independent of the degrees of polymerization of each species; the extension to allow for chain-length dependence is trivial) and the total radical concentration is:

$$[R]_w = \sum_{i=1}^{z-1} [IM_i] \qquad (10):$$

(the contribution of desorbed radicals being neglected). Eqs (7–10) are solved iteratively. If the latex is *monodisperse* and there is no particle formation, then eqs (6) and (7) imply that the *entry rate coefficient is independent of particle size* (all other variables, such as particle number, being constant): eq (6) is not rate-determining. However, if there is a range of particle sizes, as for example during particle formation, then k_a varies with particle size (there is evidence [19] that this variation is given by the diffusion-controlled value, at least in latices stabilized by ionic surfactants, which implies $k_a \propto r_p$). Under such circumstances, eq (6) is rate-determining (e.g., entry occurs at different rates for different particle sizes provided there is a range of particle sizes present), and this size-dependence must be explicitly taken into account. This is why experimental determinations of ρ_a can only be unambiguously obtained with a monodisperse seeded system in the absence of particle formation.

If all rate coefficients are assumed independent of degree of polymerization, one obtains [18,20]:

$$\rho_{\text{initiator}} \approx \frac{2fk_i[I_2]N_A}{N_p} \left\{ \frac{\sqrt{2fk_i[I_2]k_{t,w}}}{k_{p,w}[M]_w} + 1 \right\}^{1-z} \tag{11}$$

Eq (11) shows that the entry rate coefficient can depend strongly on the concentration of monomer in the water phase, with implications for residual monomer (see [21]). Initiator efficiency varies with monomer type and with initiator and particle concentrations. While initiation with a slow-propagating, relatively water-insoluble, monomer such as styrene leads to extensive aqueous-phase termination (low overall initiator efficiency), a more rapidly propagating monomer such as MMA leads to a high overall initiator efficiency [4].

3.1. ENTRY WITH IONIC SURFACTANT

It has been found (e.g., [16,18,22]) that eqs (7–10) are able to fit extensive data on styrene particles stabilized by ionic surfactants, *provided* account is taken of the observation [23-26] that $k_{t,w}$ is extremely fast ($\approx 4\times10^9$ dm^3 mol^{-1} s^{-1}) for very short aqueous-phase radical species. The value of z so obtained is 2–3 (depending on the values assumed for $k_{t,w}$ and the $k_{p,w}^1$). Some new results are shown in Figure 5. These illustrate the accord with the predictions of eqs (7–10), as well as verifying the *prediction* of this model that the entry rate coefficient should be independent of particle size if all other variables (specifically, initiator and particle concentrations) are kept constant. The values used in the fit are $k_i = 2\times10^{-6}$ s^{-1}, $k_{t,w} = 4\times10^9$ dm^3 mol^{-1} s^{-1}, $f = 0.6$, $z = 3$, $k_{p,w}^1$ and $k_{p,w}^2 = 1200$ and 300 dm^3 mol^{-1} s^{-1} (all of which are reasonable), together with the observed values for ρ_{thermal} of 2.5×10^{-4} and 1.4×10^{-3} s^{-1} for the 24 and 44 nm latices, respectively.

It has been suggested [18] that the value of z can be estimated from an expression based on hydrophobic free-energy considerations:

$$z \approx 1 + \frac{-23 \text{ kJ mol}^{-1}}{RT \ln [M]_{w,sat}} \tag{12}$$

where $[M]_{w,sat}$ is the solubility of monomer in the water phase; this yields $z = 2$–3 for styrene. Estimates of ρ_a and hence z have also been obtained for butadiene [27] and MMA [4] which are consistent with the predictions of eq (12), but in both these cases the data are less extensive and/or less unambiguous than for styrene.

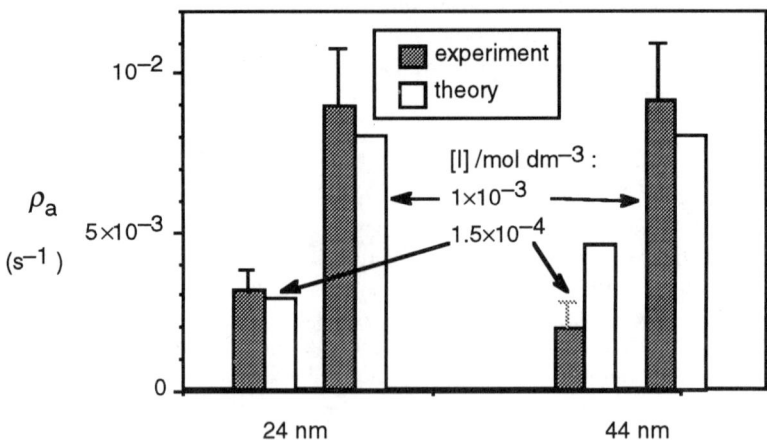

Figure 5. Experimental values for ρ_a for styrene seeded emulsion polymerizations with persulfate initiator (concentrations as indicated) at 50°C with different particles sizes; all data in Interval II, unswollen radii and uncertainties as indicated. Values of N_p (from left): 5.8, 5.2, 4.2, 5.2×10¹⁶ dm⁻³. Surfactant: 8×10⁻³ mol dm⁻³ Aerosol MA, pH = 7. Two sizes of latex were used (24 and 44 nm unswollen radii). Re-processed from data in [22]. Parameters used for simulation given in text.

3.2. ENTRY WITH POLYMERIC SURFACTANT

Experimental values have been obtained [22] for ρ_a for monodisperse polystyrene latices coated with an *electrosteric* stabilizer [28,29]. These were made with ionic surfactant (although not done here, one could add a small amount of styrene sulfonate as co-monomer [30] to improve colloidal stability); the latices were then dialyzed to remove virtually all surfactant; the latex was then swollen with styrene, acrylic acid and initiator then added and polymerization allowed to occur for a few minutes. The result is a latex stabilized by a block copolymer of acrylic acid and styrene, embedded into the latex particle, with the hydrophilic poly(acrylic acid) moiety located in the interfacial and aqueous phase regions; this is a good model for more complex systems in common industrial use.

Some results are shown in Figure 6. It is seen that putting an electrosteric stabilizer on a latex particle results in a dramatic reduction in the entry rate coefficient, and that this varies with both pH and ionic strength. This result can be rationalized in terms of the mechanism of Figure 4, if it is assumed that the electrosteric stabilizer results in a highly viscous interfacial regime (a "hairy layer") so that a z-mer now may undergo termination before it diffuses through that layer to undergo propagation in the interior of the particle. This can be quantified in terms of a higher effective value for z: e.g., the lowest ρ_a of Figure 6 corresponds to $z = 8$ rather than 3. Of course, an 8-mer would be totally insoluble in water, and this value can be seen as a measure of the average degree of polymerization which is achieved by those radicals which avoid termination in the viscous layer long enough to fully enter the interior of the particle (indeed, if (5–7)-meric species were truly present in the continuous phase, secondary nucleation would occur, since $j_{crit} \approx 5$ for styrene). It is emphasized that the effect of electrosteric stabilizer on entry is probably dependent on the degree of polymerization of the grafted polymer, and so is likely to depend on initiator and acrylic acid concentrations as well

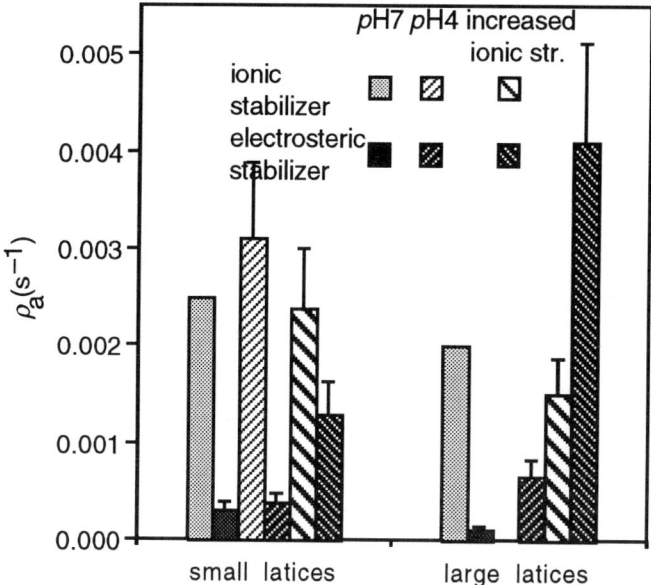

Figure 6. Showing the effect on ρ_a of electrosteric compared to ionic stabilizer; the value of ρ_a for an electrosterically stabilized particle is sensitive to changes in pH and ionic strength. Re-processed from data in [22]. Ionic strength was changed by the addition of 6×10^{-3} mol dm^{-3} NaCl. Small latices: unswollen radii 24 (ionic) and 25 (electrosteric) nm; large latices, radii 44 (ionic) and 49 (electrosteric) nm.

as on ionic strength, pH and particle size. Further quantification of this effect is seen to be an important area for the future.

3.3 VARIATION OF INITIATOR: IONIC, INISURF AND ORGANIC-PHASE INITIATORS

All of the data used above to infer the entry mechanism are for persulfate initiator. However, changing initiator type may affect entry. Some early data [31] using "V-50" (a cationic initiator which is the water-soluble equivalent of AIBN) suggested very low efficiency, but the exit rate coefficients reported in this study were also surprisingly low, suggesting that the data are suspect (e.g., there may have been inhibitor artifacts). Such experiments should be repeated with better techniques for finding ρ_a, as discussed above. It would seem reasonable that the value of z could depend on the initiating species, and thus that cationic or uncharged initiators could have higher efficiency. Testing this by appropriate studies would be useful in indicating optimal initiators.

Some studies of initiator efficiency have been reported [17] using "inisurfs", i.e., (PEO-based)-initiators which also function as surfactants [32]. It was found that these initiators had extremely low efficiency ($\approx 0.1\%$). This can be rationalized in terms of facile geminate recombination of the surface-active decomposition products from initiator when they are in the narrow confines of the interfacial region.

There is some argument as to whether that "organic-phase" initiators undergo extensive geminate recombination inside a small particle, and that under such circumstances most of the radicals causing polymerization actually arise from the initiator present in the *aqueous* phase (although AIBN is thought of as an organic-phase initiator, it has an aqueous-phase solubility comparable to styrene) [33]. However, it is more likely [34] that there is a high probability that one of the radicals formed by AIBN dissociation within a particle will desorb. Hence it is likely that many of the chains initiated by AIBN are from radicals generated in the particle phase. Naturally, AIBN can be used efficiently in large particles [35] where the more spacious confines are less conducive to geminate recombination.

4. Results and Models for Exit

4.1. TRANSFER-DIFFUSION MODEL

The transfer-diffusion model for exit quantified in eqs (2–5) is shown in Figure 7. Exit dominates the emulsion polymerization kinetics of relatively water-soluble monomers such as vinyl acetate (as discussed later), but can be important even for hydrophobic monomers such as styrene. The reason for this is even though the monomeric radical formed by transfer is relatively insoluble in water, it may be sufficiently long-lived to desorb; eqs (5) and (3) show that while the exit rate can be reduced by a low water-solubility (low $[M]_w$), this can be overcome by small particles (small r_p) and/or high transfer rate coefficient. However, eq (5) suggests that exit should be negligible for sufficiently large particles.

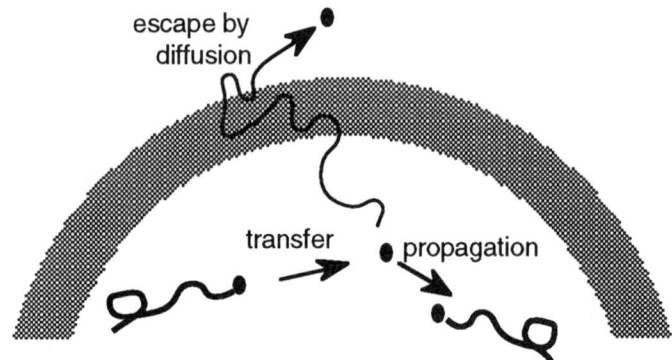

escape by diffusion

transfer

propagation

Figure 7. Illustrating the mechanism leading to exit.

4.2. SOME DATA FOR EXIT

Extensive data for the exit rate coefficient in latices with *ionic* stabilizer have been presented in the literature for styrene (e.g., [14]) and for chlorobutadiene [36]. Such data are in semi-quantitative accord with the predictions of eqs (2–5) (as illustrated in Figure 8), but not definitively so. One possible reason for this is that exit rate coefficients show strong variability from system to system (a problem which is discussed below), suggesting that there are other mechanisms operating in addition to those illustrated in Figure 7. Another cause for uncertainty is the following. In testing models against experiment, it is essential to take into account the likelihood that k_p^1 is significantly greater than k_p for a long chain, an effect predicted theoretically [37] to arise from differences in rotational entropy of activation. Moreover, a question hangs over this data interpretation, because it is not clear just to what species transfer occurs in styrene, where the hydrogen atoms would appear non-labile. While it has been suggested that transfer may be to a Diels-Alder dimer [38], such a species would be too insoluble to undergo exit. It seems likely that transfer is in fact to styrene monomer through abstraction of an H atom on the aromatic ring [39], since the observed activation energy for transfer in bulk styrene, *ca.* 50 kJ mol^{-1} [40], is not unreasonable for this process, while the frequency factor (*ca.* 10^7 dm^3 mol^{-1} s^{-1} [40]) is again of the magnitude predicted by theory [41,42].

The situation is rather different for latices with *polymeric* stabilizer. Kusters *et al.* [17] studied a latex stabilized with PEO-nonylphenyl ester and found that the exit rate coefficient was dramatically lower than that in an equivalent system with ionic stabilizer. Coen *et al.* [22] found a similar effect in electrosterically-stabilized latices (as discussed above for entry) as shown in Figure 8, with again the exit rate coefficient being sensitive to changes in pH and ionic strength. In both cases, it is reasonable to suppose that a highly viscous ("hairy") layer around the latex, originating with the polymeric stabilizer, slows down the interfacial diffusion step of Figure 7.

A final illustration of the effect of exit is in the emulsion polymerization of vinyl acetate. It has been pointed out on several occasions that plots of conversion against time for this latex show a linear region over a very wide range of x (typically up to 80% conversion). This is well beyond the transition from Interval II to Interval III [43]: i.e., the rate is apparently independent of $[M]_p$. Such a result can be easily explained [44] by noting that the rate is proportional to $[M]_p \bar{n}$, and that if the system obeys Limit 2b, eq (4), and the exit rate coefficient is large, then $\bar{n} = \rho_a/2k_{tr}[M]_p$. These results together yield a rate which is independent of $[M]_p$, consistent with observation. However, this interpretation supposes that the radical formed by transfer is sufficiently long-lived that it will be more likely to undergo re-escape than propagation when it re-enters a particle. The radical formed by transfer to monomer is likely to be the butyrolactonyl radical formed by hydrogen abstraction from the methyl group followed by cyclization. Since this is likely to be a relatively stable radical, it is likely to propagate very slowly, a presumption in accord with the postulated mechanism. Direct observation of the exit rate coefficient by γ-radiolysis relaxation [44] gives a value in quantitative accord with this mechanism and the predictions of eq (4).

4.3. ORIGINS OF VARIABILITY IN EXIT RATE COEFFICIENT

As stated, the exit rate coefficient seems to show significant variation from latex to latex, for fixed particle size[14]. There are two possible origins of this effect. The first is the interference of "hairs" (small amounts of polymeric surfactant) formed during the production of the seed latex, perhaps from peroxide or carboxylic acid species. These will in turn vary with exposure to oxygen, hydrolysis, and so on, and thus might show variability from latex to latex even though the preparations were ostensibly identical.

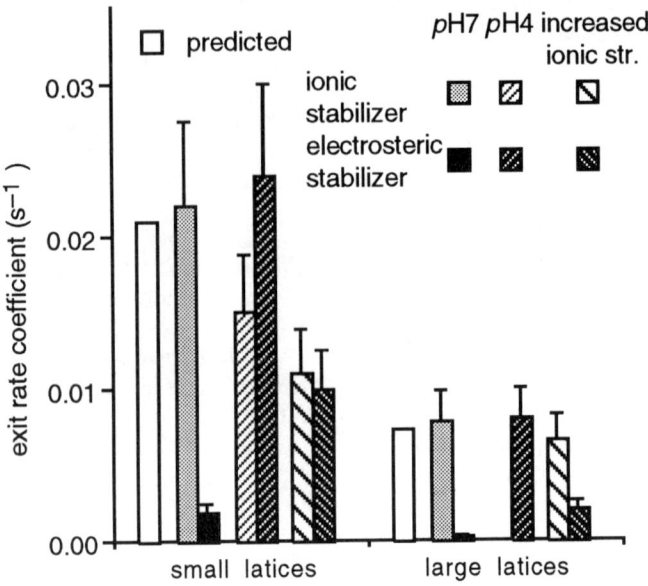

Figure 8. Effect of changes in surfactant (ionic vs. electrosteric), pH and ionic strength on exit rate coefficient in seeded emulsion polymerization of styrene; same system as in Figure 5. No data were able to be obtained for 44 nm ionically stabilized at $pH4$ because of coagulation. Re-processed from data in [22]. The theoretical predictions used eqs (3) and (5), with $k_{tr} = 2.8 \times 10^{-2}$ dm^3 mol^{-1} s^{-1}, $k_p^1 = 2.1 \times 10^3$ dm^3 mol^{-1} s^{-1}, $D_W = 1.5 \times 10^{-5}$ cm^2 s^{-1}.

Another possible cause of variation is "background thermal" polymerization, which functions as a radical loss process in a zero-one system because of "instantaneous" termination. This may arise from peroxide species formed during or after seed-latex preparation. Again, the thermal entry rate coefficient shows variability from latex to latex, consistent with variable trace amounts of peroxide species.

4.4. FATE OF EXITED RADICALS

It is emphasized that aqueous-phase radical species deriving directly from initiator (e.g., ${}^{\bullet}M_iSO_4^-$) are chemically distinct from those arising from exit (${}^{\bullet}M$), and thus have very different likelihoods of entering a particle irreversibly or terminating in the aqueous phase. Total radical concentrations in the aqueous phase are typically 10^{-9}–10^{-8} $mol\,dm^{-3}$, while particle concentration is 10^{-8}–10^{-6} $mol\,dm^{-3}$ (corresponding to $N_p \approx 10^{16}$–10^{18} dm^{-3}). Because both termination and entry seem to be diffusion-controlled (capture rate $\propto$ radius), and because radicals are much smaller than particles, it is apparent that a radical in the aqueous phase without a barrier to entry (i.e., $M^{\bullet}$) is much more likely to enter a particle than to terminate. However, ${}^{\bullet}M_iSO_4^-$ has a barrier to entry (i.e., is hydrophilic) for small i, and thus an initiator-derived radical is much more likely to terminate than entry for $i < z$.

The quite different fates of exited radicals in styrene and vinyl acetate, illustrated in Figure 9, arise principally from the very different values of k_p^1.

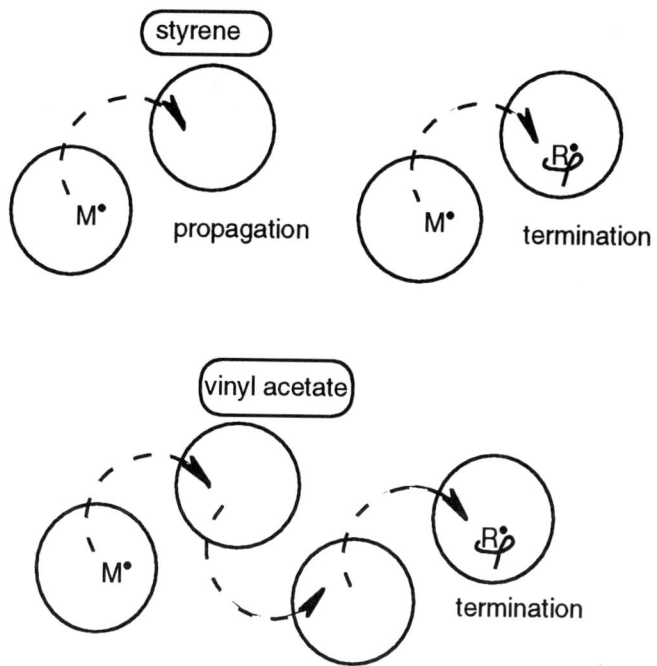

Fig 9. Illustrating the different fates dominating the behavior of exited free radicals in styrene and in vinyl acetate.

Although the studies detailed here are for homopolymer systems, the overall mechanistic precepts have been established (e.g., [45]) to apply to copolymerizations. Indeed, frequently entry or exit in a copolymer system will be dominated by the behavior of only one monomer: for example, in styrene/methyl acrylate, exit is dominated by the MA because of its higher k_{tr} and its higher water solubility.

5. Technical Implications

Some influences that exit and entry have on polymerization products and manufacturing process are as follows.

• Particle size/particle concentration is an important technical property that is controlled *inter alia* by *entry*. This is because particle formation only stops when the rate of entry into pre-existing particles sufficiently exceeds that at which an aqueous-phase radical can form a new particle [46]. This has a number of implications. For example, slow entry in a polymerically-stabilized latex means that an aqueous-phase radical has a higher probability of forming a new particle, which explains the observation that secondary nucleation is a common phenomenon in sterically or electrosterically stabilized systems (e.g., those containing acrylic or methacrylic acid). This realization in turn suggests ways around the problem, such as strategies for controlled feed of the co-monomer with regard to particle number and size so as to minimize secondary nucleation, while still being able to retain the functionality of the co-monomer.

• Overall rate, of which entry is the fundamental driving force. Depending on the kinetics, exit can be an important cause of the loss of radical activity (as is the case with vinyl acetate, discussed above) and thus can reduce the rate. In turn, rate influences optimal use of reactor time and safety (through control of the exotherm).

• Entry influences polymer products through the change in the rate of entry at high conversion and its influence on residual monomer, discussed by Kukulj and Gilbert later in this volume. The reason for this is seen in the strong dependence of ρ_a on z in eq (11).

• The many by-products of aqueous-phase kinetics include obvious termination products derived from species such as $^\bullet MSO_4^-$, and also products (such as $PhCH(OH)CH_2CHPhCH_2OSO_3^-$) from hydrolysis and transfer reactions of such species [47-49]. These may be involved in discoloration, weathering, etc. In surfactant-free and low-surfactant systems, species with initiator fragments (including termination products from aqueous-phase kinetics and long chains with an initiator-derived endgroup) are major contributors to colloidal stability.

In these and other cases, optimization of polymer product and process can be improved through mechanistic understanding of the events controlling entry and exit [1].

Acknowledgment. Support of this research by the Australian Research Council is gratefully acknowledged.

6. References

[1]   Gilbert, R.G. (1995) *Emulsion Polymerization: A Mechanistic Approach*, Academic, London .
[2]   Benson, S.W. and North, A.M. (1962) *J. Am. Chem. Soc.*, **84**, 935.
[3]   Scheren, P.A.G.M., Russell, G.T., Sangster, D.F., Gilbert, R.G. and German, A.L. (1995) *Macromolecules*, **28**, 3637.
[4]   Russell, G.T., Gilbert, R.G. and Napper, D.H. (1993) *Macromolecules*, **26**, 3538 .
[5]   Barandiaran, M.J. and Asua, J.M. (1995) *J. Polym. Sci. Part A Polym. Chem.*, **34**, 309
[6]   Asua, J.M., Sudol E.D. and El-Aasser, M.S. (1989) *J. Polym. Sci. Polym. Chem. Ed.*, **27**, 3903.
[7]   Casey, B.S., Morrison, B.R., Maxwell, I.A., Gilbert R.G. and Napper , D.H. (1994) *J. Polym. Sci. A: Polym. Chem.*, **32**, 605.
[8]   Ugelstad, J. and Hansen, F.K. (1976) *Rubber Chem. Technol.*, **49**, 536.
[9]   Russell, G.T., Gilbert, R.G. and Napper, D.H. (1992) *Macromolecules*, **25**, 2459 .
[10]   Russell, G.T. (1994) *Macromol. Theory Simul.*, **3**, 439.
[11]   Ballard, M.J., Gilbert, R.G. and Napper, D.H. (1981) *J. Polym. Sci., Polym. Letters Edn.*, **19**, 533.
[12]   Hawkett, B.S., Napper, D.H. and Gilbert, R.G. (1980) *J. Chem. Soc. Faraday Trans. 1*, **76**, 1323.
[13]   Lansdowne, S.W., Gilbert, R.G., Napper, D.H. and D.F. Sangster, (1980) *J. Chem. Soc. Faraday Trans. 1*, **76**, 1344.
[14]   Morrison, B.R., Casey, B.S., Lacík, I., Leslie, G.L., Sangster, D.F., Gilbert , R.G. and Napper, D.H., (1994) *J. Polym. Sci. A: Polym. Chem.*, **32**, 631.
[15]   Ballard, M.J., Napper, D.H. and Gilbert, R.G. (1984) *J. Polym. Sci., Polym. Chem. Edn.*, **22**, 3225.
[16]   Casey, B.S., Morrison , B.R. and Gilbert, R.G. (1993) *Prog. Polym. Sci.*, **18**, 1041.
[17]   Kusters, J.M.H., Napper, D.H., Gilbert R.G. and German, A.L. (1992) *Macromolecules*, **25**, 7043.
[18]   Maxwell, I.A., Morrison, B.R., Napper D.H. and Gilbert, R.G. (1991) *Macromolecules*, **24**, 1629.
[19]   Morrison, B.R., Maxwell, I.A., Gilbert R.G. and Napper, D.H. (1992) in: *ACS Symp. Series - Polymer Latexes - Preparation, Characterization and Applications* (eds. E.S. Daniels, E.D. Sudol and M. El-Aasser) p.28, American Chemical Society, Washington D.C.
[20]   Fitch, R.M. and Tsai, C.H. (1971) in: *Polymer Colloids* (eds. R.M. Fitch) p.73, Plenum, New York.
[21]   Kukulj, J. and Gilbert, R.G. (1997) in: *Polymeric Dispersions. Principles and Applications* (eds. J.M. Asua) Kluwer Academic, Dordrecht.
[22]   Coen, E., Lyons, R.A. and Gilbert, R.G. (1996) *Macromolecules*, **29**, 5128.
[23]   Sangster, D.F. and Davison, A. (1975) *J. Polym. Sci., Polym. Symp.*, **49**, 191
[24]   Dainton, F.S. and James, D.G.L. (1959) *J. Polym. Sci.*, **39**, 299.
[25]   Dainton, F.G. and Eaton, R.S. (1959) *J. Polym. Sci.*, **39**, 313.

[26]   Fischer, H. (1966) *Makromol. Chem.,* **98**, 179.
[27]   Verdurmen, E.M., Geurts, J.M., Vertsegen, J.M., Maxwell, I.A. and German, A.L. (1993) *Macromolecules,* **26**, 6289.
[28]   Bassett, D.R. and Hoy, K.L. (1980) in: *Polymer Colloids II* (eds. R.M. Fitch) p.1, Plenum, New York.
[29]   Napper, D.H. *Polymeric Stabilization of Colloidal Dispersions,* Academic, London.
[30]   Juang, M.S. and Krieger, I.M. (1976) *J. Polym. Sci., Polym. Chem. Edn.,* **14**, 2089.
[31]   Penboss, I.A., Napper, D.H. and Gilbert, R.G. (1983) *J. Chem. Soc. Faraday Trans. 1,* **79**, 1257.
[32]   Tauer, K., Goebel, K.-H., Kosmella, S., Neelsen, J. and Stähler, K. (1988) *Plaste und Kautschuk,* **35**, 373.
[33]   Nomura, M., Ikoma, J. and Fujita, K. (1993) *J. Polym. Sci., Part A: Polym. Chem.,* **31**, 2103.
[34]   Asua, J.M., Rodriquez, V.S. , Sudol, E.D. and El-Aasser, M.S. (1989) *J. Polym. Sci., Polym. Chem. Edn.,* **27**, 3569.
[35]   Sudol, E.D., El-Aasser, M.S. and Vanderhoff, J.W. (1986) *J. Polym. Sci., Polym. Chem. Edn.,* **24**, 3515.
[36]   Christie, D.I., Gilbert, R.G., Congalidis, J.P., Richards, J.R. and McMinn, J.H. (1995) *DECHEMA Monographs,* **131**, 513.
[37]   Heuts, J.P.A. , Radom , L. and Gilbert, R.G. (1995) *Macromolecules,* **28**, 8771.
[38]   Olaj, O.F., Kauffmann, H.F. and Breitenbach, J.W. (1977) *Makromol. Chem.,* **178**, 2707.
[39]   Moad, G. and Solomon, D.H. (1995) *The Chemistry of Free Radical Polymerization,*Pergamon, Oxford.
[40]   Tobolsky, A.V. and Offenbach, J. (1955) *J. Polym. Sci.,* **16**, 311.
[41]   Heuts, J.P.A. , Sudarko and Gilbert, R.G. *Macromol. Symp.,* (in press).
[42]   Heuts, J.P.A., Radom, L. and Gilbert, R.G. (in preparation).
[43]   Stannett, V., Klein, A. and Litt, M. (1975) *Br. Polym. J.,* **7**, 139.
[44]   De Bruyn, H., Gilbert, R.G. and Ballard, M.J. *Macromolecules,* (submitted).
[45]   Schoonbrood, H.A.S., German , A.L. and Gilbert, R.G. (1995) *Macromolecules,* **28**, 34.
[46]   Morrison, B.R. and Gilbert, R.G. (1995) *Macromol. Symp.,* **92**, 13.
[47]   Wang, S.-T. and Poehlein, G.W. (1993) *J. Appl. Polym. Sci.,* **50**, 2173.
[48]   Wang, S.-T. and Poehlein, G.W. (1994) *J. Appl. Polym. Sci.,* **51**, 593.
[49]   Morrison, B.R., Maxwell, I.A., Napper, D.H., Gilbert, R.G. , Ammerdorffer, J.L. and German, A.L. (1993) *J. Polym. Sci., Polym. Chem. Edn.,* **31**, 467.

PARTICLE GROWTH IN EMULSION POLYMERIZATION
Determination of Propagation Rate Constants and Monomer Concentration.

A.M. VAN HERK
Department of Polymer Chemistry, Eindhoven University of Technology
P.O. Box 513, 5600 MB, Eindhoven, The Netherlands

1. The Rate of Particle Growth

Particle growth in emulsion homopolymerization is determined by the following equation:

$$R_p = -d[M]/dt = k_p\,[M]_p\,\bar{n}\,N_p/N_A \tag{1}$$

where R_p is the rate of growth, k_p the (average) propagation rate constant, $[M]_p$ the monomer concentration in the particles, $\bar{n}$ the average number of radicals per particle, N_p the number of particles per unit volume of aqueous phase and N_A Avogadro's number.

In the case of an emulsion copolymerization and applying the pseudo-homopolymerization rate approach, the rate of growth is given by:

$$R_p = -d[M]/dt = \langle k_p \rangle\,[M]_{p,tot}\,\bar{n}\,N_p/N_A \tag{2}$$

where $\langle k_p \rangle$ is the average propagation rate constant as defined by eq (4) and $[M]_{p,tot}$ the total monomer concentration in the particles. In the case of more then two monomers this equation gets more complicated (see section 1.2).

In the case of an emulsion copolymerization the phenomenon of composition drift can occur. This means that the feed composition and the ratio in which the two monomers are build in the copolymer are not equal, resulting in a drift in the monomer ratio in the reactor and therefore also a drift in the composition of the formed copolymer which can result in heterogeneous copolymers. In the case of an emulsion polymerization there is an additional effect of monomer partitioning on the composition drift (see also 1.2). When the water solubility of the two monomers differs, one of the monomers is held back in the aqueous phase which can result in an enhancement of the composition drift. In some special cases (where the more reactive comonomer is also the more water soluble comonomer) the water solubility can partially compensate composition drift. So if reactivity ratios are obtained by analysis of emulsion copolymerizations one can either use the concept of apparent reactivity ratios or introduce the monomer partitioning equilibria and use the reactivity ratios obtained in homogeneous media.

J. M. Asua (ed.), Polymeric Dispersions: Principles and Applications, 17–30.
© 1997 *Kluwer Academic Publishers. Printed in the Netherlands.*

18

In this review emphasis is put on the experimental methods to obtain accurate values for k_p and $[M]_p$. The methods of obtaining ñ are only briefly discussed and were treated more extensively in the chapter of this book by Gilbert. In another part of this book the methods to obtain the particle diameter and thus N_p will be discussed. Reference is made to the initial sources as much as possible.

1.1. THE PROPAGATION RATE CONSTANT

The propagation rate constants are usually obtained from solution or bulk polymerization experiments. Can these k_p-values be transferred to an emulsion polymerization one might wonder ? In turns out that in many cases the effects of the chemical micro-environment (also solvent effects) are small. Also transferring copolymerization rate constants from e.g. bulk to emulsion systems is in general possible. Another aspect is the medium viscosity, it turns out that in general the k_p-values remain constant up till a certain polymer content , for MMA this is up to a weight fraction of polymer of 70 %. This critical conversion is dependent on the type of monomer and the temperature. Above this polymer content the propagation rate constant decreases as a consequence of the propagation becoming diffusion controlled .

1.1.1. The Pulsed Laser Polymerization/Molecular Weight Analysis Method
Determination of the propagation rate constants (k_p) is preferably performed by the pulsed laser polymerization technique (PLP) combined with size exclusion chromatography (SEC).

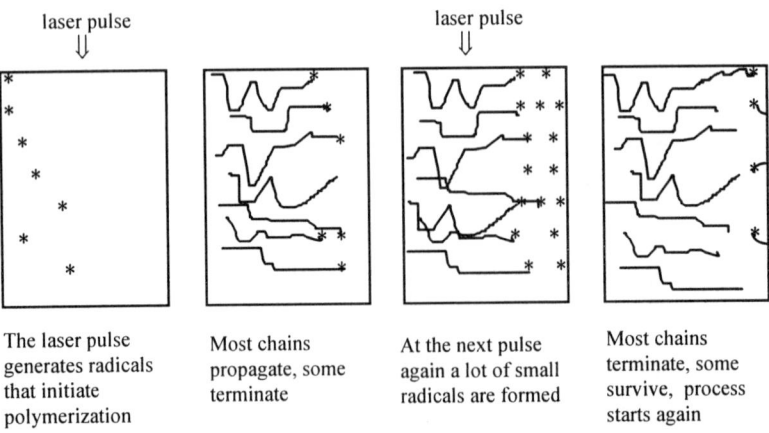

The laser pulse generates radicals that initiate polymerization

Most chains propagate, some terminate

At the next pulse again a lot of small radicals are formed

Most chains terminate, some survive, process starts again

Figure 1 Schematic representation of the pulsed laser polymerization experiment

This method is very suitable to obtain accurate propagation rate constants in homogeneous systems because there are very little assumptions made and the determination is yielding a value of k_p not coupled to the termination constant k_t in contrast to other techniques (see 1.1.2 and 1.1.3). For this reason extensive reference is made to the literature on this relatively new method. The PLP-SEC technique was described originally by Alexandrov [1] in 1977 and was further developed by Olaj in 1987 and the years after [2-7].This method comprises the generation of radicals through a photoinitiator, activated by a laser pulse. The time of growth of a polymer chain is directly determined by the time between pulses and this experiment gives direct access to k_p. The method is schematically illustrated in Figure 1. The chain length for the chains initiated and terminated by small laser induced radicals is given by the simple equation [2]:

$$L_{0,i} = i*k_p*[M]*t \qquad (3)$$

where : $L_{0,i}$ is the chain length of the polymer formed in the process of growth in the time between two laser pulses, k_p the propagation rate coefficient, [M] the monomer concentration at the site of polymerization, t is the time between two subsequent laser pulses and i=1,2,3,.... The higher order peaks (i=2,3,...) may occur when growing chains survive termination by one or more subsequent pulses. The radical profile as a function of time in a pulsed laser experiment is shown in Figure 2.

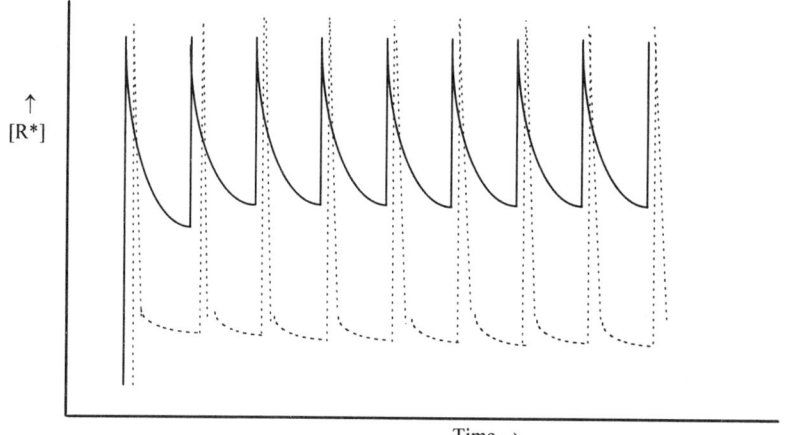

Figure 2 Radical concentration profile in a pulsed laser polymerization experiment,
——— homogeneous medium , ······ microemulsions or latex systems.

In-between two pulses normal bimolecular termination can occur which results in the so-called background polymer. Olaj [2] suggested that the inflection point at the low molecular weight side of the peaks gives a good measure for k_p. This is usually true, however in very small microemulsion droplets indications were found that the best measure for k_p shifts from the inflection point to the peak maximum (see below). On the other hand, when the k_p is known, eq (1) can also be used to obtain the monomer concentration in microemulsion droplets and in latex particles (section 1.2). In Figure 3 a typical molecular weight distribution is shown of polymer produced in a PLP experiment.

An overview of the publications of the other active groups in PLP-SEC, up to the beginning of 1996, will be given.

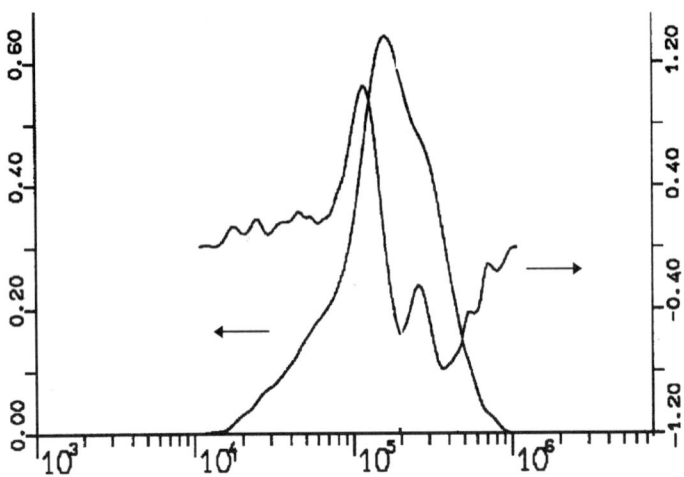

Figure 3. Molecular weight distribution (left axis) and derivative (right axis) of polymethylacrylate obtained in a PLP experiment at 1000 bar and -15°C at a frequency of 90 Hz (conversion 3.8 %). SEC measured at Röhm. Reproduced with permission of C. Kurz from her thesis, Göttingen 1995

Davis and O'Driscoll [8-17] extended the method to copolymerization, the average propagation rate constant $<k_p>$ is given by the general equation:

$$< k_p >= \frac{\overline{r_1} f_1^2 + 2 f_1 f_2 + \overline{r_2} f_2^2}{(\overline{r_1} f_1 / k_{p11}) + (\overline{r_2} f_2 / k_{p22})} \tag{4}$$

with f_i is the molar fraction of monomer i and k_{pij} is the propagation rate constant for propagation of chain end i with monomer j, where in the ultimate model

$$\overline{r_1} = r_1, \overline{r_2} = r_2, \overline{k_{p11}} = k_{p11}, \overline{k_{p22}} = k_{p22}$$

$$\text{with } r_1 = \frac{k_{p11}}{k_{p12}} \text{ and } r_2 = \frac{k_{p22}}{k_{p21}}$$

In the case that also the penultimate unit influences the reactivity, the penultimate model applies and we can also define k_{p111}, k_{p211} etc., in eq (4):

$$\overline{k_{p11}} = \frac{k_{p111}(f_1 r_1 + f_2)}{f_1 r_1 + f_2 / s_1} \quad \text{and} \quad \overline{k_{p22}} = \frac{k_{p222}(f_2 r_2 + f_1)}{f_2 r_2 + f_1 / s_2}$$

$$\overline{r_1} = \frac{r_1'(f_1 r_1 + f_2)}{f_1 r_1' + f_2} \quad \text{and} \quad \overline{r_2} = \frac{r_2'(f_2 r_2 + f_1)}{f_2 r_2' + f_1}$$

$$r_1 = \frac{k_{p111}}{k_{p112}}, \quad r_2 = \frac{k_{p222}}{k_{p221}}, \quad r_1' = \frac{k_{p211}}{k_{p212}}, \quad r_2' = \frac{k_{p122}}{k_{p121}}, \quad s_1 = \frac{k_{p211}}{k_{p111}}, \quad s_2 = \frac{k_{p122}}{k_{p222}}$$

The data in Figure 4 can be described very well by the penultimate model.

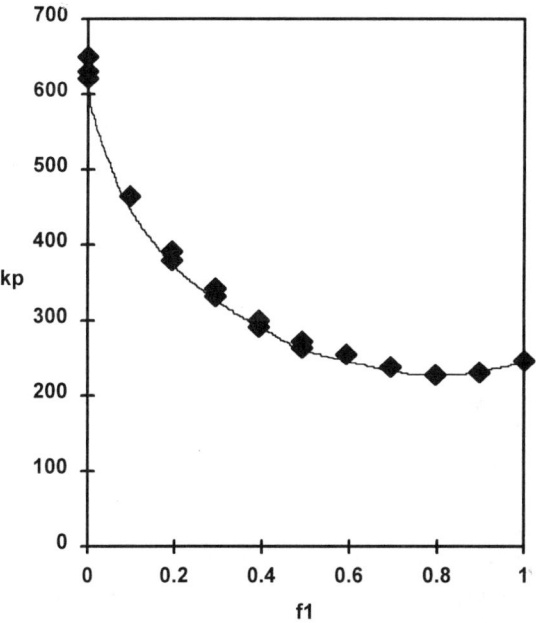

Figure 4. k_p as a function of the feed fraction f_1 for the copolymerization of styrene (M_1) and methyl methacrylate (M_2) (unpublished data, Peeters, Manders, van Herk).

O'Driscoll simulated the PLP experiments with Monte-Carlo simulations [13] and Davis investigated the molecular weight of the resulting polymer from a PLP experiment of MMA with Matrix-Assisted Laser Desorption [15]. The group of Buback [18-23] also performed PLP experiments at high pressures with subsequent SEC analysis. In 1986 a IUPAC working party entitled "Modeling of Free Radical Polymerization Kinetics and Processes" was founded which recommended the PLP-SEC method as the most reliable method for the determination of k_p [24-28]. Other active groups are: Gilbert [29-35], Hutchinson [36-41]. Holdcroft and Guillet were the first to perform PLP in transparent microemulsions [42]. This was also done by the group of van Herk and German [43-48] in collaboration with the group of Schweer.

Van Herk et al. performed pulsed electron beam polymerization in heterogeneous systems [48], following the same principles as in the PLP-SEC method. Schweer [49-54] published simulations of the PLP experiment and also used a flash light instead of a pulsed laser to do similar experiments [49]. Recently they also applied the MALDI-TOF technique to determine the molecular weight distribution [54]. Furthermore simulations were done by the groups of Moad [55,56], Yan and Zhang [57], Zhang and Yang [58, 59] and the group of McLaughlin and Hoyle [60]. The latter group published several papers on the complete molecular weight distributions formed during a PLP experiment, but not with the direct aim of determining k_p values [60, 61].

A general review on laser-initiated polymerization, including pulsed laser polymerization to obtain values for k_p, was published in 1994 by Davis [17].

Although the PLP/SEC method can result in accurate k_p values there are some problems. The method relies on accurate SEC calibration. If narrow molecular weight standards are not available, the method of universal calibration can be applied which again relies on the knowledge of the Mark-Houwink constants [40,41]. One solution to this problem is the use of an in-line viscosity detector. The values for the Mark-Houwink constants are regularly updated which is therefore also updating the values of k_p (see for example [41]).

Another problem is occurring with fast reacting monomers like vinylacetate or methylacrylate and butylacrylate [41]. The occurrence of clear peaks in the SEC trace that can be attributed to chains that are initiated and terminated by laser generated radicals can be hindered by the following processes: (1) preliminary termination of the growing chains by rapid termination or chain transfer to monomer or polymer; (2) little termination by the small laser generated radicals because of a slow termination rate or because of insufficient penetration of the laser beam in the sample.

In Table 1 a compilation of the latest (May 1996) k_p values and activation parameters are given which were obtained with the PLP-SEC method. In most cases the low molecular weight inflection point is the best measure of k_p as also confirmed by several simulations [13,57]. From the modeling studies of Schweer [52] and Manders [46] it turns out that under certain experimental conditions the maximum of the PLP peaks can be the best measure of k_p. Especially when high radical concentrations are present in the system under investigation, for example in the very small microemulsion droplets [43].

In microemulsion droplets and latex particles the radical concentration profile (Figure 2) in the pseudo-stationary state will consist of two decay curves following the rapid increase in radical concentration after the laser pulse; the first one just after the rapid increase in the radical concentration, here the rate of termination will be higher in the (micro)emulsion droplets or particles because the local radical concentration will be higher. The second decay curve comprises the stage where the number of radicals per particle is less than two, here the termination rate will be lower than in a homogeneous system because termination involves exit of a radical and entry into another particle followed by bimolecular termination.

TABLE 1. Homopropagation rate constants obtained by the PLP/SEC method

Monomer	Solvent	Arrhenius A	eq. EXP	10 °C	25 °C	30 °C	40 °C	50 °C	60 °C	70 °C	ref
Acrylamide	water, pH=1		20		16000						29
Methacrylam.	water, pH=1		20		1100						29
BA	bulk	2.51 10^7	20								33
BA	THF/toluene	1.66 10^7	17.27	10800	15600	17500	21800	26800	32500	39000	35
BA*	bulk	1.8 10^7	17.4	11100	16100	18100	22500	27700	33700	40400	41
n-BMA	bulk,1000 bar	7.28 10^6	22.9	434	708	824	1100	1450	1870	2380	20
n-BMA	bulk	3.44 10^6	23.3		274						11
n-BMA*	bulk	1.81 10^6	20.55	289	454	523	676	857	1108	1394	38
t-BMA	bulk	2.51 10^7	27.7		352			836			32
i-BMA*	bulk	2.47 10^6	21.53	252	417	496	633	798	1040	1336	38
Butadiene	chlorobenzene	8.05 10^7	35.71	20.8	44.6	56	85	138	204	295	49
Chloroprene	bulk	1.95 10^7	26.63	235	447	485	673	988	1300	1720	36
DA	bulk			17660		22113					41
DMA	bulk	3.44 10^6	21.72	339	538	622	819	1060	1350	1700	40
DMA	toluene	2.93 10^5	16.19	300	427	452	601	796	848	1010	11
EHA	bulk			13180	18030						41
EMA	bulk	1.50 10^6	20.46		258						11
EMA*	bulk	3.65 10^6	22.89	206	356	400	590	676	939	1160	38
PMOS	toluene	5.90 10^5	23.0	33.7	52	64.2	85.9	113	146	186	10
MAN	bulk/benzene	2.69 10^6	29.7	8.9	16.8	20.5	29.9	42.5	59.3	81	55
MMA	bulk	2.39 10^6	22.18	193	311	360	470	621	769	989	36
MMA**	bulk	2.65 10^6	22.34	200	323	375	497	649	833	1050	28
MMA	bulk	4.94 10^6	23.94		316			667			21
MMA	toluene/2-but.					384					19
MMA	bulk/EA/Meth.				294						8
MMA	bulk					364					18
MMA	bulk				313						4
^{1}H$_8$-MMA	bulk				270						6
^{2}H$_8$-MMA	bulk				342						6
Sty	bulk				77						4
Sty	toluene				79						4
Sty	bulk/Meth./EB				78						8
Sty	bulk	1.99 10^7	30.78	41.7	80.6	98.9	146	211	297	411	11
Sty	bulk	3.04 10^7	31.48			116		248		490	23
Sty**	bulk	4.27 10^7	32.51	42.9	85.9	107	161	237	341	480	27
Vinylacetate	bulk	2.7 10^8	27.82	1935	3420						38
Vinylacetate*	bulk	1.49 10^7	20.39	2580	3990	4570	5910	7540	9460	11700	40

* Most recent values
** IUPAC values
Italics, values calculated from the Arrhenius parameters

Besides laser initiation termination some alternative sources of initiation and termination were investigated with the same purpose of obtaining k_p-values; Olaj et al. investigated the periodic variation of initiation by dropwise addition of initiator [62] and the periodic variation of termination by addition of a radical scavenger [63]. Olaj et al. [66,67] also analyzed the molecular weight distribution of polymer obtained in a rotating sector experiment (1.1.3) which compared well with that from PLP. Holdcroft and Guillet [64] controlled the initiation and termination processes independently in a dual laser technique where two laser pulses of different wavelength were used with the aim of producing monodispers polymer.

There are several copolymerization systems investigated with the PLP/SEC method. Even more so than in the case of homopolymerization, calibration of the SEC equipment introduces problems. A compilation of the systems investigated and the observed reactivity ratios is given in Table 2.

TABLE 2. Copolymerization parameters obtained by the PLP/SEC method

Monomer 1	Monomer 2	Solvent	T (°C)	$k_{p(1)11}$	$k_{p(2)22}$	r_1	r_2	r_1'	r_2'	s_1	s_2	re
$^1H_8$-MMA	$^2H_8$-MMA	bulk	25	270	342	1	1	-	-	-	-	6
Sty	MMA	bulk	25	89.1	299.2	0.523	0.460	0.523	0.460	0.30	0.80	6
Sty	MMA	bulk	25	77.5	294	0.472	0.454	0.472	0.454	0.466	0.175	9
Sty	EMA	bulk	25	78	258	0.62	0.35	0.62	0.35	0.62	0.21	1
Sty	EMA	bulk	55	249	589	0.62	0.35	0.62	0.35	0.45	0.22	1
Sty	BMA	bulk	25	78	274	0.72	0.45	0.72	0.45	0.56	0.63	1
Sty	BMA	bulk	55	249	656	0.72	0.45	0.72	0.45	0.50	0.67	1
Sty	LMA	toluene	25	78	776	0.57	0.45	0.57	0.45	0.33	0.26	1
Sty	MA	bulk	25	78	-	0.73	0.19	0.73	0.19	1.10	0.26	1
Sty	MA	bulk	50	-	-	0.73	0.19	0.73	0.19	0.94	0.11	1
Sty	BA	bulk	25	78	-	0.95	0.18	0.95	0.18	1.89	0.21	1
Sty	BA	bulk	50	-	-	0.95	0.18	0.95	0.18	0.90	0.11	1
PMOS	Sty	toluene	25	52	86	0.82	1.12	-	-	-	-	1
PMOS	MMA	toluene	25	52	300	0.32	0.29	0.32	0.29	0.36	0.60	

A special system where one of the monomers does not homopolymerize is the system styrene/maleic anhydride where maleic anhydride does not homopolymerize[14]. The first terpolymerization system that was investigated with the PLP/SEC method is the system styrene/methyl methacrylate/methyl acrylate [45]. Recently [65] pronounced solvent effects were observed for the system Sty/MMA in benzyl alcohol using the PLP/SEC technique in combination with composition data and sequence distributions. Olaj et al. showed that the molecular weight distribution obtained in a rotating sector experiment can be used just as well to obtain k_p-values from the molecular weight distribution [66,67].

1.1.2.   The Time-resolved Pulsed Laser Polymerization Method
In 1986 [68] with the advent of powerful UV lasers a new method was introduced which is capable of determination of k_p and k_t; time resolved pulsed laser polymerization (TR-PLP). In this method the conversion of the monomer is followed as a function of time (t), for example with an infrared detector after the generation of radicals with a short laser pulse. The monomer concentration $[M]_t$ as a function of time is given by:

$$[M]_t / [M]_0 = (2k_t [R]_0 t + 1)^{-0.5k_p/k_t} \qquad (5)$$

with $[R]_0$ the radical concentration immediately after laser pulse absorption which can be calculated from the absorbed laser energy if the quantum yield is known. With known $[R]_0$ the value for k_t and thus for k_p can be inferred. A combination of the obtained ratio k_p/k_t with independently determined k_p data, for example obtained with the PLP/SEC method gives k_t values.

Another method using pulsed laser initiation is that where the overall polymerization rate is measured for a pulsed laser polymerization with a pulse frequency v [69,70]. When it is possible to determine quantitatively the end-groups introduced by the photoinitiator [71,72] the initiator efficiency can be obtained.

1.1.3.   The Rotating Sector Method and Spatially Intermittent Polymerization
Analysis of the steady state kinetic expression does allow the determination of the ratio k_p^2/k_t :

$$R_p = -\frac{d[M]}{dt} = k_p[R][M] = k_p \left(\frac{fk_d[I]}{k_t} \right)^{1/2} [M] \qquad (6)$$

The average lifetime τ of the growing chain during (pseudo)stationary-state conditions is given by:

$$\tau = \frac{[R]_s}{2k_t[R]_s^2} = \frac{1}{2k_t[R]_s} = \frac{k_p[M]}{2k_t R_p} \qquad (7)$$

A measurement of τ in conjunction with R_p will yield the ratio k_p/k_t. The oldest and most widely used method is that of the rotating sector. In this method, first applied by Burnett et al. [73], the irradiation of a photoinitiated radical polymerization is modulated by a rotating sector where illumination and dark periods are produced by cut out portions in the disk. In a typical experiment, the average rate of polymerization is measured as a function of the speed of the rotating sector, expressed in Δt, the period of irradiation. The plot of R_p as a function of $\Delta t/\tau$ can be compared with theoretical curves and the resulting value of τ, in combination with steady state experiments, can then be used to evaluate k_p and k_t . An important drawback of this method is that, because termination is diffusion controlled, k_t is a function of chain length and viscosity and it is difficult, if not impossible, to have identical reaction conditions in both experiments. For a long time in copolymerization the variations in the ratio k_p/k_t were ascribed (erroneously) to the termination rate constants, but as the termination process is not chemically controlled this is very unlikely. Fukuda [74] introduced the penultimate model for the propagation step (eq (4)). Another form of this experiment is spatially intermittent polymerization (SIP) [75].

1.1.4. Electron Spin Resonance

An elegant method to determine k_p is by the direct measurement of the radical concentration [R] in eq (8) by Electron Spin Resonance (ESR):

$$R_p = k_p [R] [M] \tag{8}$$

When the rate of polymerization is measured at the same time, with a known monomer concentration, k_p can be determined. The steady-state concentrations of propagating radicals is however very low (10^{-7}-10^{-6} mol dm^{-3}) and the accurate measurement of the radical concentration is not an easy task. Improvements to the ESR equipment has made it possible to measure this low concentrations [76-78]. Also the radical concentration directly in emulsion polymerization was measured (application of eq (1) renders k_p again), either in batch [79] or semicontinuous emulsion polymerization [80].

1.1.5. Emulsion Polymerization

In emulsion polymerization the occurrence of the so-called zero-one kinetics makes it possible to use compartmentalization to obtain values for k_p. In steady state emulsion polymerization the value of ñ is constant. In a zero-one system the value for ñ may reach the limiting value of 0.5. When a plateau is observed for the rate of polymerization (per particle) for a zero-one system as a function of initiator concentration or particle diameter, the value for ñ at that plateau is 0.5 and when $[M]_p$ in eq (1) is known this yields a value for k_p. This method was applied for butadiene [81] and shows good agreement with, for example, the PLP/SEC method [53].

1.2. THE MONOMER CONCENTRATION IN THE LATEX PARTICLES

The concentration of monomer in latex particles can be determined by gas chromatography [82], by conductimetry [83], but also (in reacting systems !) through pulsed electron beam polymerization [48]. The monomer partitioning equilibria also can be predicted from thermodynamic considerations [84] according to the Vanzo equation [85]:

$$\ln(1 - \phi_p^p) + \chi \cdot \phi_p^{p\,2} + \phi_p^p \cdot (1-1/P_n) + 2 \cdot V_m \cdot \gamma \cdot \phi_p^{p\,1/3} / r_0 = \ln ([M]_w / [M]_{w,sat}) \tag{9}$$

where ϕ_p^p is the volume fraction of polymer in the latex particles, P_n is the number average degree of polymerization, χ is the Flory-Huggins interaction parameter between the monomer and the polymer, V_m is the molar volume of the monomer, γ is the particle-water interfacial tension and r_0 is the radius of the unswollen latex particles and $[M]_{w,sat}$ the saturation concentration of monomer in the aqueous phase.

Equation (9) applies to partial swelling of latex particles (stage III of an emulsion polymerization), for saturation swelling the right hand side of eq (9) equals zero.

The monomer concentration in the latex particles (can be calculated from ϕ_p^p) depends on the interaction parameter, the particle-water interfacial tension, the particle diameter and the molar volume of the monomer. The Vanzo equation has experimentally been verified [84]. The contribution of the conformational entropy of mixing of monomer and polymer dominates the free energy of mixing at higher volume fractions of polymer in the latex particles. The interfacial free energy does not strongly contribute to the parameters that determine the degree of partial swelling of latex particles (stage III of an emulsion polymerization). This is in contrast with the results for saturation swelling of latex particles where the balance between the free energy of mixing of monomer and polymer and the interfacial free energy of the latex particles-water interface determines the degree of latex particle swelling.

Maxwell et al. [84, 86] made some assumptions that led to a major simplification of eq (9). He showed that partial swelling, especially at higher polymer fractions, can mainly be described by the combinatorial entropy of mixing, taking the interaction term and the interfacial tension term together as a constant correction term.

It was observed in emulsion copolymerization that in most systems investigated with moderately water soluble monomers the monomer ratio in the monomer droplets equals that in the polymer particles [82,84]. In Table 3 some saturation concentrations of monomers in different polymer particles are shown.

TABLE 3.   Saturation concentrations of monomers in different polymer particles at 20 °C

seed	MA (mol/l)	MMA (mol/l)	BA (mol/l)	S (mol/l)	ref
Poly(MA)	8.4/7.9*	-	3.8*	5.6	85
Poly(BA)	8.8*	-	6.1/5.5*	5.4	85
Poly(S)	6.2	-	5.4	5.4	85
Poly(S/MMA)	-	6.9	-	5.6	82

* At 35 °C

With PEBP a value of 5.8 mol/l was found for the saturation concentration of styrene in 46 nm diameter polystyrene particles at 23 °C [48]. It is important to note that the fact that the monomer concentration in the latex particles determined by gas chromatography and by PEBP are the same, means that the propagation rate constants obtained in homogeneous polymerization (used in the PEBP experiments to calculate the monomer concentration) can indeed be transferred to emulsion systems.

Equation (9) also predicts that the monomer concentration is decreasing when the particle size is decreasing which was indeed observed experimentally [87].

It turns out that it is still not possible to theoretically describe monomer partitioning in systems where one of the monomers is completely miscible with the aqueous phase, e.g. (meth)acrylic acid or hydroxy ethyl methacrylate.

When more then two monomers are present a more general description of monomer partitioning is possible and also eq (2) becomes more complicated (see for example [88]).

1.3. RADICAL CONCENTRATION

The radical concentration in the latex particles is determined by radical entry, exit and termination. Direct determination of the radical concentration in the latex particles can only be done with ESR [78,79].

Indirect determination of the radical concentration is done by the kinetic analysis through, for example, eq (1). For more information the reader should refer to the chapter in this book by Gilbert.

1.4. CONCLUSIONS

The quality of the predictions of kinetics of particle growth and of the microstructure of the formed (co)polymer has improved very much due to the improvement of the methods of obtaining kinetic and thermodynamic parameters relevant to homogeneous and heterogeneous polymerization [89]. This development will also have its impact on the possibilities of designing intelligent process strategies aimed at obtaining well defined products with better properties [90].

References

1. Aleksandrov, A.P., Genkin, V.N., Kitai, M.S., Smirnova, I.M., Sokolov, V.V. (1977) *Sov. J. Quantum Electron.* 7, 547-550 (orig. (1977) *Kvantovaya Elecktron.* (Moscow) 4, 976-979).
2. Olaj, O.F., Bitai, I., Hinkelmann, F. (1987) *Makromol. Chem.* 188, 1689-1702.
3. Olaj,O.F., Bitai, I. (1987) *Angew. Makromol. Chem.* 155, 177-181.
4. Olaj, O.F., Schnöll-Bitai, I. (1989) *Eur. Polym. J.* 25, 635-641.
5. Schnöll-Bitai, I., Olaj, O.F.(1990) *Makromol. Chem.* 191, 2491-2499.
6 Olaj, O.F., Schnöll-Bitai, I. (1990) *Makromol. Chem., Rapid Commun.* 11, 459-465.
7. Olaj, O.F., Zifferer, G. (1992) *Makromol. Chem., Theory Sim.* 1, 71-90.
8. Davis, T.P., O'Driscoll, K.F., Piton, M.C., Winnik, M.A. (1989) *Macromolecules* 22, 2785-2788.
9. Davis,T.P.,O'Driscoll,K.F., Piton, M.C.,Winnik, M.A. (1989) *J. Polym. Sci. Polym. Lett. Ed.* 27, 181-185.
10. Piton, M.C., Winnik, M.A., Davis, T.P., O'Driscoll, K.F. (1990) *J. Polym. Sci., Part A: Polym. Chem.* 28, 2097-2106.
11. Davis, T.P., O'Driscoll, K.F., Piton, M.C., Winnik, M.A. (1990) *Macromolecules* 23, 2113-2119.
12. Davis, T.P., O'Driscoll, K.F., Piton, M.C., Winnik, M.A. (1991) *Polymer Int.* 24, 65-70.
13. O'Driscoll,K.F., Kuindersma, K.F. (1994) *Makromol. Chem. Theory Sim.* 3, 469-476.
14. Sanayei, R.A., O'Driscoll, K.F., Klumperman, B. (1994) *Macromolecules* 27, 5577-5582.
15. Zammit, M.D., Davis, T.P., Haddleton, D.M (1996) *Macromolecules* 29, 492-494.
16. Coote, M.L., Zammit, M.D., Davis, T.P. (1996) *Trends in Polymer Science* 4, 189-196.
17. Davis T.P. (1994) *J. Photochem. Photobiol. A: Chem.* 77,1-7.
18. Beuermann, S., Buback, M., Russell, G.T. (1994) *Macromol., Rapid Commun.* 15, 351-355 .
19. Beuermann, S., Buback, M., Russell, G.T. (1994) *Macromol., Rapid Commun.* 15, 647-653.
20. Bergert, U., Buback, M., Heyne, J. (1995) *Macromol. Rapid Commun.* 16, 275-281.

21. Bergert, U., Beuermann, S., Buback, M., Kurz, C.H., Russell, G.T., Schmaltz C. (1995) *Makromol. Rapid Commun.* **16**, 425-434.
22. Buback, M., Busch, M., Lämmel, R. (1995) *DECHEMA Monographs* Berlin **131**, 569-577, and same authors, (1996) *Makromol. Chem. Theory and Simul.* to be published
23. Buback, M., Kuchta F.-D. (1995) *Makromol. Chem. Phys.* **196**, 1887-1898.
24. Buback, M., Garcia-Rubio, L.H., Gilbert, R.G., Napper, D.H., Guillot, J., Hamielec, A.E., Hill, D., O'Driscoll, K.F., Olaj, O.F., Shen, J., Solomon, D., Moad, G., Stickler, M., Tirell, M., Winnik, M.A.(1988) *J.Polym.Sci., Part C: Polym. Lett. Ed.* **26**, 293-297.
25. Buback, M., Gilbert, R.G., Russell, G.T., Hill, D.J.T., Moad, G., O'Driscoll, K.F., Shen, J., Winnik, M.A. (1992) *J.Polym.Sci, Part A: Polym. Chem. Ed.* **30**, 851-863.
26. Gilbert, R.G. (1992) *Pure Appl. Chem.* **64**, 1563-1567.
27. Buback, M., Gilbert, R.G., Hutchinson, R.A., Klumperman, B., Kuchta, F.D., Manders, B.G., O'Driscoll K.F., Russell, G.T, Schweer, J. (1995) *Macromol. Chem. Phys.* **196**, 3267-3280.
28. Beuermann S., Buback M., Davis, T.P.,Gilbert R.G., Hutchinson R.A., Olaj, O.F., Russell G.T., Schweer J., van Herk, A.M. (1996) *Macromol. Chem. Phys.* To be published.
29. Pascal, P., Napper,D.H,. Gilbert, R.G, Piton, M.C., Winnik, M.A. (1990) *Macromolecules* **23**, 5161-5163 and Pascal, P., Winnik, M.A., Napper, D.H., Gilbert, R.G. (1993) *Macromolecules* **26**, 4572-4576.
30. Danis, P.O., Karr, D.E., Westmoreland, D.G., Piton, M.C., Christie, D.I., Clay, P.A., Kable, S.H., Gilbert, R.G. (1993) *Macromolecules* **26**, 6684-6685.
31. Morrison, B.R., Piton, M.C., Winnik, M.A., Gilbert, R.G., Napper, D.H.(1993) *Macromolecules* **26**, 4368- 4372.
32. Pascal, P., Winnik, M.A., Napper, D.H., Gilbert, R.G. (1993) *Makromol. Chem., Rapid Commun.* **14**, 213-215.
33. Heuts, J.P.A., Clay, P.A., Christie, D.I., Piton, M.C., Hutovic, J., Kable, S.C.,. Gilbert, R.G. (1994) *Progress in Pacific Polymer Science; Proceedings* **3**, 203-216.
34. Christie D.I., Gilbert R.G. (1996) *Makromol. Chem. Phys.* **197**, 403-412.
35. Lyons, R.A., Hutovic, J., Piton, M.C., Christie, D.I., Clay, P.A., Manders, B.G., Kable, S.H., Gilbert, R.G., Shipp, D.A. (1996) *Macromolecules* **29**, 1918-1927.
36. Hutchinson, R.A., Aronson, M.T., Richards, J.R. (1993) *Macromolecules* **26**, 6410-6415.
37. Hutchinson, R.A., Richards, J.R., Aronson, M.T. (1994) *Macromolecules*, **27**, 4530-4537.
38. Hutchinson, R.A., Paquet Jr, D.A., McMinn, J.H., Fuller, R.E. (1995) *Macromolecules* **28**, 4023-4028.
39. Hutchinson, R.A., Paquet Jr, D.A., McMinn, J.H. (1995) *Macromolecules* **28**, 5655-5663.
40. Hutchinson, R.A., Paquet Jr, D.A., McMinn, J.H., Beuermann, S., Fuller, R.E., Jackson, C. (1995) *DECHEMA Monographs* **131**, 467-493 Berlin.
41. Beuermann, S., Paquet Jr, D.A., McMinn, J. H., Hutchinson, R.A. (1996) *Macromolecules* **29** , 4206-4215 .
42. Holdcroft, S., Guillet, J.E. (1990) *J.Pol.Science Part A: Polymer Chemistry* **28**, 1823-1829.
43. Manders, B. G., van Herk, A. M., German, A. L., Sarnecki, J., Schomäcker, R., Schweer, J. (1993) *Makromol. Chem. , Rapid Commun.* **14**, 693-701.
44. Schweer, J., van Herk,A.M., Pijpers,R.J., Manders,B.G.,German,A.L. (1995) *Makromol. Symp.* **92**, 31-41.
45. Schoonbrood, H.A.S., van den Reijen, B., de Kock, J.B.L., Manders B.G., van Herk, A.M., German, A.L. (1995) *Macromol. Chem. Rapid Commun.* **16**, 119-124.
46. Manders, B.G., van Herk, A.M., German, A.L. (1995) *Macromol. Theory Simul.* **4**, 325-333.
47. Manders, B.G., Chambard, G., Kingma, W.J., Klumperman, B.,. van Herk, A.M, German, A.L. (1996) *J. Polym. Sci., Polym. Chem. Ed.* To be published.
48. van Herk, A. M., de Brouwer, H., Manders, B. G., Luthjens, L. H., Hom, M. L., Hummel, A. (1996) *Macromolecules* **29**, 1027-1030.
49. Deibert,S.,Bandermann,F.,Schweer,J.,Sarnecki,J. (1992) *Makromol.Chem.,Rapid Commun.* **13**, 351-355.
50. DE 4041139 (1991), Bayer AG, inv. Schweer J.; Chem. Abstr. 116, 42230u (1992) .
51. Schweer J., Pijpers R.J.; Patent Le A 30 060 "Monomerbestimmung in Latex mit PLP/GPC", BAYER AG(1993).
52. Sarnecki, J., Schweer, J. (1995) *Macromolecules* **28**, 4080-4088.
53. Bandermann, F., Günther, C., Schweer, J. (1996) *Macromol. Chem. Phys.* **197**, 1055-1069 and *DECHEMA Monographs* **131**, 599-610 Berlin.

30

54. Schweer, J., Sarnecki, J., Mayer-Posner, J.F., Müllen, K., Räder H.J., Spickermann. J. (1996) *Macromolecules* **29**, 4536-4543.
55. Shipp, D.A., Smith , T.A., Solomon, D.H., Moad, G. (1995) *Makromol. Rapid Commun.* **16**, 837-844.
56. Deady, M., Mau, A.W.H., Moad, G., Spurling, T.H. (1993) *Macromol. Chem.* **194**, 1691-1705.
57. Yan, D., Zhang, M., Schweer, J. (1996) *Macromolecules* **29**, 793-799 .
58. Lu, J., Zhang, H., Yang, Y. (1993) *Makromol. Chem. Theory Simul.* **2**, 747-760.
59. He, J., Zhang, H., Yang, Y. (1995) *Makromol.Theory Simul.***4**, 811-819.
60. McLaughlin, K.W., Latham, D.D., Hoyle, C.E., Trapp, M.A. (1989) *J. Phys. Chem.* **93**, 3643-3647.
61. Hoyle, C.E., Trapp, M. A., Chang, C.H., Latham, D.D., McLaughlin, K.W (1989) *Macromolecules* **22**, 35-38 and 3866-3877.
62. Olaj,O.F., Schnöll-Bitai, I. (1991) *Makromol. Chem. Rapid Commun.* **12**, 373-380.
63. Schnöll-Bitai, I., Olaj,O.F. (1992) *Makromol. Chem. Rapid Commun.* **13**, 395-402.
64. Holdcroft, S., Guillet, J.E. (1991) *J. Polym. Sci., Polym. Chem. Ed.* **29**, 729-737.
65. O'Driscoll, K.F., Monteiro, M.J. (1996) *J. Polym. Sci.* To be published.
66. Olaj,. O.F., Kremminger, P., Schnöll-Bitai, I. (1988) *Makromol. Chem., Rapid Commun.* **9**, 771-779
67. Olaj, O.F.,Schöll-Bitai, Kremminger, P. (1989) *Eur. Polym. J.* **25**, 535-541
68. Buback, M., Hippler, H.,Schweer, J.,Vögele, H.-P. (1986) *Makromol. Chem. Rapid Commun.* **7**, 261-265.
69. Olaj, O.F., Bitai, I., Gleixner, G. (1985) *Makromol. Chem.***186**, 2569-2580.
70. Brackemann, H., Buback, M., Vögele, H.-P. (1986) *Makromol. Chem. Rapid Commun.* **187**, 1977-1992.
71. Buback, M., Huckestein, B., Leinhos, U. (1987) *Makromol. Chem. Rapid Commun.* **8**, 473-479.
72. Buback, M., Huckestein, B., Ludwig, B. (1992) *Makromol. Chem. Rapid Commun.* **13**, 1-7.
73. Burnett, G.M., Melville, H.W. (1947) *Proc. Roy. Soc. London Ser. A* **189**, 486-491.
74. Ma, Y-D., Fukuda, T., Inagaki, H. (1985) *Macromolecules* **18**, 26-31.
75. O'Driscoll, K.F., Mahabadi, H.K. (1976) *J. Polym. Sci. Polym. Chem. Edn.* **14**, 869-881.
76. Roth, H.-K., Wünsche, P. (1981) *Acta Polymerica* **32**, 491-494.
77. Yamada, B., Kageoka, M. Otsu, T., (1992) *Polymer Bul* **29**, 385-392.
78. Tonge, M.P., Pace, R.J., Gilbert, R.G. (1994) *Makromol. Chem. Phys.* **195**, 3159-3172.
79. Ballard, M. J., Gilbert, R.G., Napper, D. H., Pomery, P.J., O'Sullivan, P.W., O'Donnell, J.H. (1986) *Macromolecules* **19**, 1303-1308.
80. Lau, W., Westmoreland, D.G. (1992) *Macromolecules* **25**, 4448-4449.
81. Weerts, P.A., German, A.L., Gilbert, R.G. (1991) *Macromolecules* **24**, 1622-1628.
82. Aerdts, A.M., Boei, M.M.W.A., German, A.L. (1993) *Polymer* **34**, 574-580.
83. Noël, L.F.J., van Herk, A.M., Janssen, R.Q.F., van Well, W.J.M., German, A.L. (1995) *J. Colloid Interf. Sci.* **175**, 461-469.
84. Maxwell, I.A., Kurja, J., Van Doremaele, G.H.J., German, A.L., Morrison, B.R. (1992) *Makromol. Chem.* **193**, 2049-2064.
85. Vanzo, E., Marechessault, R.H., Stannett, V. (1965) *J. Colloid Sci.* **20**, 65-71.
86. Maxwell, I.A., Kurja, J., Van Doremaele, G.H.J., German, A.L. (1992) *Makromol. Chem.* **193**, 2065-2080.
87. Morton, M., Kaizerman, S., Altier, M. W. (1954) *J. Colloid Sci.* **9**, 300-312.
88. Urretabizkaia, A., Asua, J.M. (1994) *J. Polym. Sci. Chem. Ed.*, **32**, 1761-1778.
89. Gilbert, R.G. (1995) *Emulsion Polymerization, A Mechanistic Approach*, Academic Press, London
90. Schoonbrood H.A.S., Brouns H. M. G., Thijssen H.A., van Herk A.M., German A.L. (1995) *Makromol. Symp.* **92**, 133-156.

STABILITY OF POLYMER COLLOIDS

R.H. OTTEWILL
School of Chemistry, University of Bristol
Cantock's Close, Bristol BS8 1TS
U.K.

1. Introduction

The general principles of Colloid Stability, particularly as applied to aqueous systems containing electrolytes, have been treated in a number of articles including those at previous NATO Institutes [1,2]. Therefore in this Chapter a short summary of basic principles will be given and then attention will be focussed on areas where Colloid Stability, or lack of it, is particularly important in preparing and utilising polymer colloid dispersions.

2. Basic Principles of Colloid Stability in Aqueous Media

The basis of current theories of colloid stability [3] for smooth spherical particles with charged surfaces is to consider the total potential energy of interaction, V_T, as being composed of three terms, so that

$$V_T = V_R + V_A + V_B \tag{1}$$

where V_R = electrostatic energy of repulsion [3], V_A = van der Waals' attraction [4] and V_B, the Born repulsion, is a very short range repulsion which arises at close approach from molecular orbital overlap [5].

2.1. ELECTROSTATIC REPULSION

The term V_R, for spheres of radius R, can be expressed in the form,

$$V_R = 4\pi\varepsilon_r\varepsilon_0\psi_s^2 R^2 \exp(-\kappa h)/(h + 2R) \tag{2}$$

where ε_r = relative permittivity of the dispersion medium, ε_0 = the permittivity of free space, ψ_s = the electrostatic surface potential and h, the intersurface separation. κ is the Debye Hückel reciprocal length given by, $\kappa^2 = 2n_0v^2e^2/\varepsilon_r\varepsilon_0kT$, where n_0 is the number of ions of each type per unit volume, v is the magnitude of the charge on each ion, e = the fundamental electronic charge, k is the Boltzmann's constant and T the temperature. This expression is valid for $\kappa R < 3$ and for $\kappa R > 10$,

$$V_R = 2\pi\varepsilon_r\varepsilon_0\psi_s^2 R \ln[1 + \exp(-\kappa h)] \tag{3}$$

31

J. M. Asua (ed.), Polymeric Dispersions: Principles and Applications, 31–48.

At intermediate values an approximation is given by Verwey and Overbeek [3],

$$V_R = 2\pi\varepsilon_r\varepsilon_0 R(4kT/e)^2 \gamma^2 \exp(-\kappa h)/v^2 \qquad (4)$$

with $\gamma = [\exp(ve\psi_s/2kT)-1]/[\exp(ve\psi_s/2kT)+1]$.

2.2. VAN DER WAALS' ATTRACTION

The van der Waals' energy of attraction between two spheres in a vacuum is given by [4],

$$V_A = \frac{-A_{11}}{12}\left\{\frac{1}{x^2+1} + \frac{1}{x^2+2x+1} + 2Ln\frac{x^2+2x}{x^2+2x+1}\right\} \qquad (5)$$

with $x = h/2R$, or as a useful approximation

$$V_A = -A_{11}R/12h \qquad (6)$$

The Hamaker Constant A_{11}, for the material of the particles is given by

$$A_{11} = 3\pi^2 h_p v_{11}\alpha_{11}{}^2 q_{11}{}^2/4 \qquad (7)$$

with h_p = Planck's constant and v_{11} = the dispersion frequency, α_{11} = the electronic polarisability, and q_{11} = the number of atoms (or molecules) per unit volume of the particles. A similar expression can be written for the dispersion medium to give a value for A_{22}. The quantity $(\sqrt{A_{11}}-\sqrt{A_{22}})^2$ is termed the composite Hamaker Constant and will be given the symbol A_{121}; a few values of this quantity are tabulated in Table 1 using A_{11} values for the polymers [6,7] and 3.70×10^{-20} J for water [6].

TABLE 1. A_{11} and A_{121} for polymers

Polymer	$A_{11}/10^{-20}$ J	$A_{121}/10^{-20}$ J
Poly(tetrafluoroethylene)	3.80	0.33
Poly(isoprene)	5.99	0.74
Poly(styrene)	6.58	0.95
Poly(methylmethacrylate)	7.11	1.05
Poly(vinyl chloride)	7.78	1.30
Poly(vinyl acetate)	8.84	1.10

2.3. THE BORN REPULSION

It can be noted from the above equation that as $h \rightarrow 0$ then $V_A \rightarrow -\infty$. This is an unphysical answer since it suggests it would never be possible to separate surfaces. However, at short distances the situation changes. The orbitals of atoms on approaching surfaces overlap and this leads to a very strong short range repulsion which is known as the Born Repulsion. This energy of repulsion changes as approximately $1/h^{14}$. It is usually represented as a cut-off potential at a distance of about one atomic diameter from the distance origin.

3. The Potential Energy Diagram

Figure 1 shows a sketch of V_T as a function of h for an electrolyte concentration of 10^{-3} mol dm^{-3} ($1/\kappa=10nm$), R=100 nm, $\psi_s=50mV$ and a composite Hamaker Constant of 0.95×10^{-20} J.

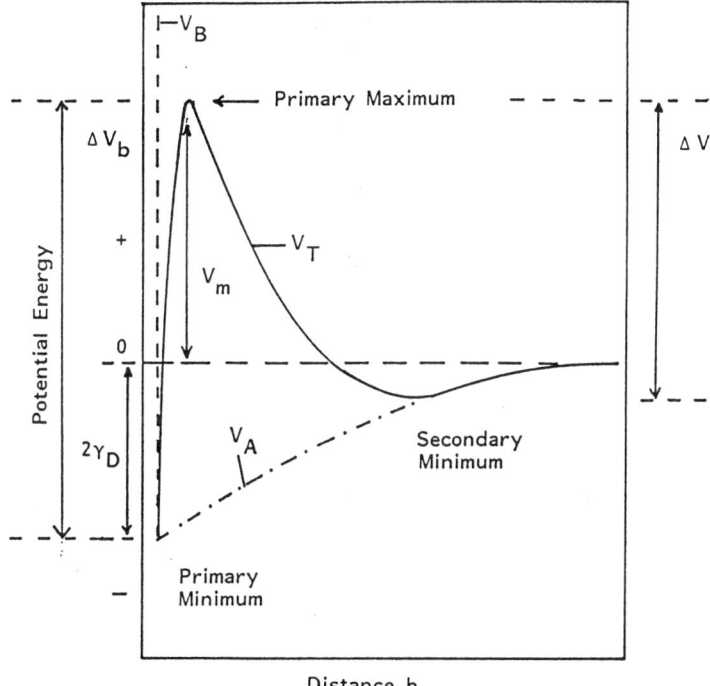

Figure 1. Potential energy diagram for particle-particle interaction.

This curve exhibits a number of characteristic features which can be summarised as follows:

a) at short distances of surface separation, a deep minimum in the potential energy curve occurs; this is termed the PRIMARY MINIMUM and determines the distance of closest approach, h_o. The depth, from $V_T = 0$, is related to twice the dispersive contribution to the surface energy [9];

b) at intermediate distances, the electrostatic repulsion, which has an exponential decay ($exp(-\kappa h)$), is larger than the attraction term (changes as $-1/h$) and hence there is a maximum in the curve. This is termed the PRIMARY MAXIMUM; the magnitude can be represented by V_m;

c) at large distances the curve is even more sensitive to the different decay rates of the repulsive and attractive contributions with distance and a minimum can occur in the

34

curve. This is known as the SECONDARY MINIMUM; the depth represented by V_{SM} tends to increase in magnitude as the particle size increases;

d) the activation energy needed to bring two particles together can be considered as $\Delta V_f/kT$;

e) the activation energy required to separate two particles located at a distance corresponding to the primary minimum position and redispersing them is ΔV_b. Typically, $\Delta V_b >> \Delta V_f$, so that once particles are brought together into a primary minimum condition considerable energy is needed to redisperse them.

Although there are a number of assumptions in this so called DLVO approach [3,10] on a semi-quantitative level it is extremely useful as a means of discussing aspects of the stability of colloidal dispersions.

3.1. EFFECT OF ELECTROLYTE CONCENTRATION AND SURFACE POTENTIAL

The potential energy curve enables an immediate idea to be obtained of the influence of electrolyte concentration, electrolyte type and surface potential on the form of the potential energy of interaction. Schematic examples are presented in Figure 2.

From these it can be observed that if the salt concentration is increased at a constant values of ψ_s, then V_m is depressed until at ca. 0.1 mol dm^{-3}, V_m tends to zero. Therefore there is no longer a substantial energy barrier keeping the particles separated and they can easily go into a primary minimum situation. Similarly, lowering the surface potential at a constant electrolyte concentration leads to a ψ_s value at which $V_m = 0$. Frequently, both ψ_s and electrolyte concentration change together.

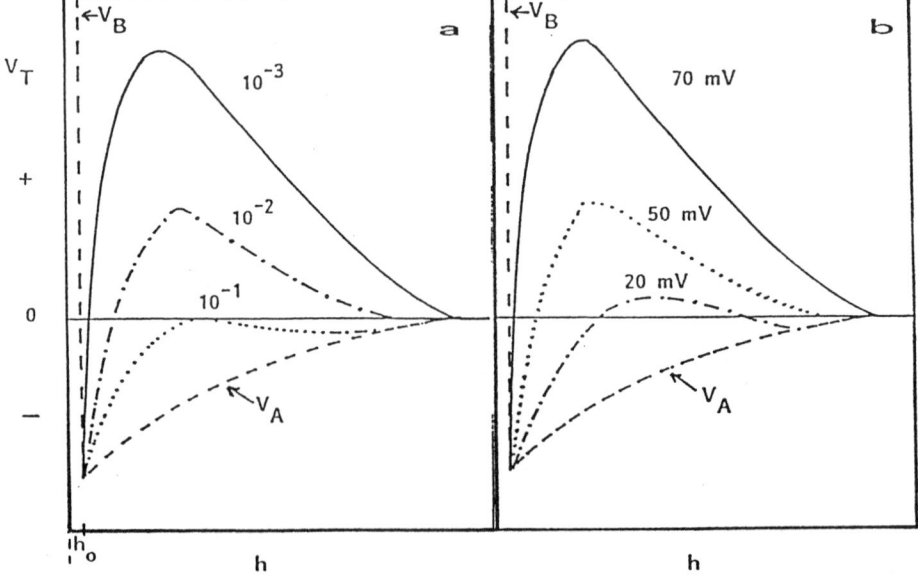

Figure 2. Effect of variation of electrolyte (mol dm^{-3}) concentration and ψ_s.

3.2. PREDICTION OF LOSS OF STABILITY

From the above arguments it can be anticipated that as $V_m \to 0$ then a dispersion of charged particles will tend to lose its colloid stability and coagulate. For a consideration of this point we can ignore V_B and write,

$$V_T = V_R + V_A \tag{8}$$

and putting for the instability condition, $V_T = V_m = 0$, we obtain

$$V_R = -V_A \tag{9}$$

and

$$dV_R/dh = -dV_A/dh \tag{10}$$

From an analysis for spherical particles using equations (4) and (6) we find the value of κ at which coagulation occurs is given by,

$$\kappa_{coag}/m^{-1} = 2.039 \times 10^{10} \, \gamma^2/Av^2 \tag{11}$$

Moreover, since for a symmetrical electrolyte κ can be directly related to the electrolyte concentration in mol dm^{-3} at which coagulation occurs, normally called the critical coagulation concentration, or c.c.c., we find

$$c.c.c./mol \; dm^{-3} = 3.853 \times 10^{-39} \, \gamma^4/A^2v^6 \tag{12}$$

with A in J, which for small values of ψ_s reduces to,

$$c.c.c./mol \; dm^{-3} = 3.451 \times 10^{-35} \, \psi_s^4/A^2v^2 \tag{13}$$

A further approximation which has some validity is to replace ψ_s by the zeta potential ζ. This is useful in the sense that ζ-potential values are often available from measurements, for example, of electrophoretic mobility.

3.3. SOME EXPERIMENTAL c.c.c. VALUES

Some values of c.c.c. for anionic and cationic polymer colloid systems are listed in Table 2.

The data given in Table 2 have been selected on the basis that the ions chosen do not chemically react with water at the pH of the coagulation experiment. The implicit assumption in eq (11) is that coagulation occurs as a consequence of compression of the diffuse double layer, i.e. a decrease of $1/\kappa$. Hence, it is found in these cases that coagulation usually occurs at a finite, but small, value of the ζ-potential.

TABLE 2. c.c.c. Values for polymer colloids

Polymer	Counterion	c.c.c./mol dm^{-3}	Reference
Poly(styrene)	H$^+$	1.3	11
Carboxyl Surface	Na$^+$	160.0	12
	Ba^{+3}	14.3	13
	La^{+3}(pH 4.6)	0.3	11
Poly(styrene)	Cl$^-$	150.0	14
Amidine Surface	Br$^-$	90.0	14
	I$^-$	43.0	14

In many cases, however, particularly with trivalent and tetravalent cations, the ions do react with water under certain pH conditions to form complex species in solution usually of higher charge [15]. For example, in the case of aluminium, at pH values below ca 3.5, the ion exists in the Al^{3+} form with six water molecules in the octahedral co-ordinate positions. As the pH is increased reaction with water molecules occurs to form the hydrolysed species [16], [Al$_{13}$ O$_4$ (OH)$_{24}$ (H$_2$O)$_{12}$]$^{7+}$. This change in the chemistry of the ion has a profound effect on its behaviour. The coagulation behaviour becomes very pH dependent, the c.c.c. decreases and because of the adsorption of the large ion charge reversal occurs. This type of behaviour is best represented by a domain diagram with the axes as salt concentration and pH [7,15].

4. Coagulation as a Kinetic Process

4.1. PERIKINETIC COAGULATION

As shown by Smoluchowski [17] coagulation can be considered as a series of rate processes starting with the single particles in a stable dispersion and then forming doublets, triplets, quadruplets etc as shown in Figure 3.

Ultimately coagulated structures can be formed which contain many particles.

In the earliest stages of coagulation the rate of disappearance of primary particles can be written as,

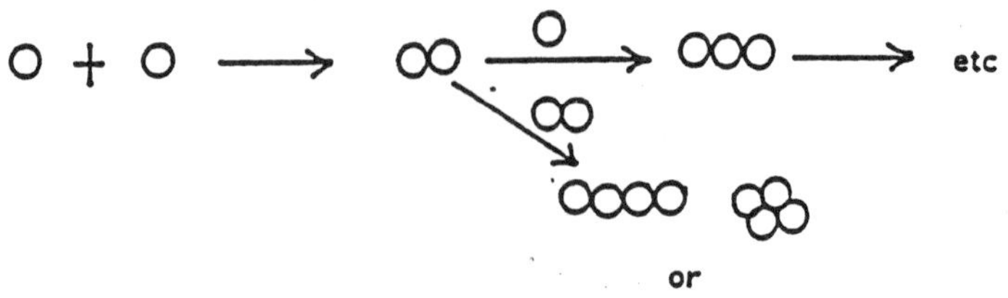

Figure 3. Kinetics of coagulation

$$-dN_1 / dt = k_o N_1^2 \qquad (14)$$

where N_1 = the number of primary particles per unit volume in the initial dispersion. In the absence of a repulsive energy barrier the rate process is diffusion controlled, so that $k_o = 8\pi R_c D$ with R_c as the collision radius of the particle and D its diffusion coefficient. As a first approximation R_c can be taken as 2R which with η = the viscosity of the medium gives,

$$k_o = 8kT / 3\eta \qquad (15)$$

In the presence of a repulsive field diffusion is modified [18,10] and eq. (14) has to be rewritten as,

$$-dN_1 = k_o N_1^2 / W = k N_1^2 \qquad (16)$$

where k = the rate constant in the presence of a potential energy barrier. W is termed the Stability Ratio and is given, assuming R_c=2R, by,

$$W = 2R \int \exp(V_T/kT) dh/(h+2R)^2 \qquad (17)$$

If the total potential energy of interaction, V_T, is reduced to zero then W = 1. Since this means that every collision should be cohesive it is spoken of as Rapid Coagulation. In the presence of a weak energy barrier the sticking probability decreases so the process is termed Slow Coagulation.

As an example of the effect of W we find that at an electrolyte concentration of 10^{-3} mol dm^{-3}, and a surface potential of 50 mV, W has a value of 10^7 so that coagulation is not perceptible on a reasonable time scale.

The value of W can be obtained from kinetic measurements [13,20] and a typical example is shown in Figure 4. The transition from rapid to slow coagulation is quite clear and the sharp break gives a value for the c.c.c.

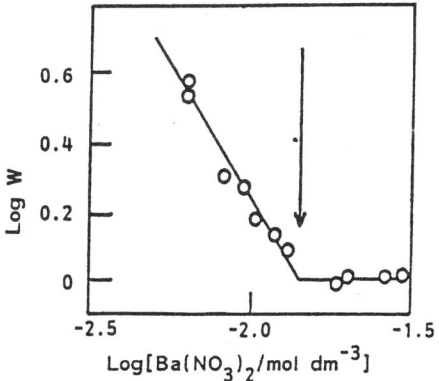

Figure 4. Log W against log concentration of barium nitrate

4.2. ORTHOKINETIC COAGULATION

An important external field is applied to the particles when the system is stirred; for simplicity the stirrer can be considered to produce a simple shear gradient, S. Hence for this situation the rate of rapid coagulation can be written as [21],

$$4R^3SN_1^2/3 \qquad (18)$$

The shear term has a third power dependence on radius so that orthokinetic coagulation for R>ca. 0.5 μm can become the dominant effect. For colloid stability to be maintained the requirement is,

$$V_m/kT > 4\eta_0 R^3 S/kT \qquad (19)$$

A helpful presentation which illustrates the effect of shear rate on dispersions is to use dimensionless quantities [22]. The colloid forces can be represented as the ratio of the electrostatic repulsive term to the attractive term, i.e. as $\varepsilon_r \varepsilon_0 \psi_s^2 R/A_{121}$ and the hydrodynamic term as the ratio of the shear term to the attractive term, that is, $6\pi\eta R^3 S/A_{121}$. This is illustrated in Figure 5. The line in the direction of the arrow shows that as shear gradient increases, the particles move out of the region of secondary minimum flocculation to a dispersion region and then into a region of primary minimum coagulation; high shear rates may also redisperse particles.

A recent examination of carboxylated ter-polymers [23] has shown that orthokinetic coagulation can be sensitive to pH when raising the pH causes an increase in particle radius and dispersion viscosity.

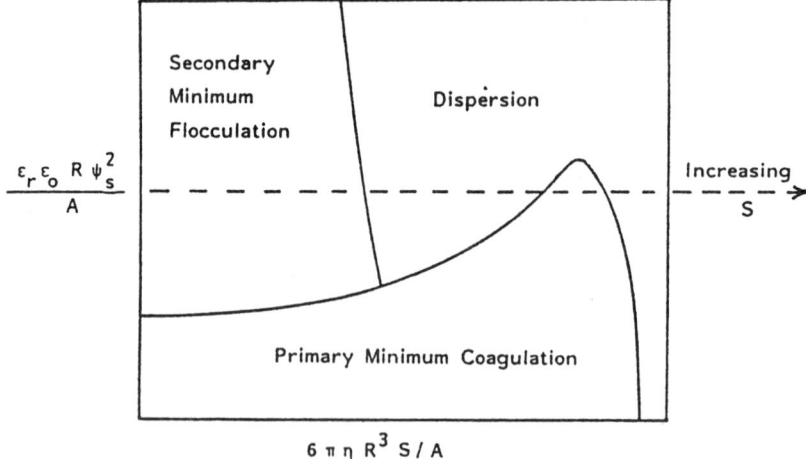

Figure 5. Orthokinetic coagulation

5. Coagulation during the Process of Particle Formation

A number of authors during the early 1970's suggested that coagulation played an important part in the process of particle formation in emulsion polymerisation. Basically the concept was that once nucleation had occurred to form the first nuclear particles, the probability was that these would be unstable in the colloid sense because of their small particle size and low surface charge; consequently, in view of the high number concentration coagulation would occur over a short period of time to form particles of a sufficient size and surface charge to provide colloid stability.

Since the size of the initial stable particles controls the number concentration of the latex during the diffusional growth period then for the same initial monomer and initiator concentration it could be anticipated that the final diameter in the medium of higher ionic strength would be larger. These predictions appear to be confirmed by experimental results. An extension of this effect has been the use of electrolyte to control the ultimate size of the particles during aqueous emulsifier-free polymerisations [24]. On the basis of a range of experiments on the preparation of polystyrene particles it was found that the final particle diameter could be related to the total ionic strength of the medium, I, by,

$$\log D = 0.238 \log \left\{ \frac{[I][M]^{1.723}}{[P]} + \frac{4929}{T} \right\} - 0.827 \qquad (20)$$

with [M] = initial monomer concentration of styrene in mol dm^{-3} based on the total volume of the system, [P] = the potassium persulphate concentration (mol dm^{-3}) and T = absolute temperature. A comparison of this relationship with experimental results is shown in Figure 6.

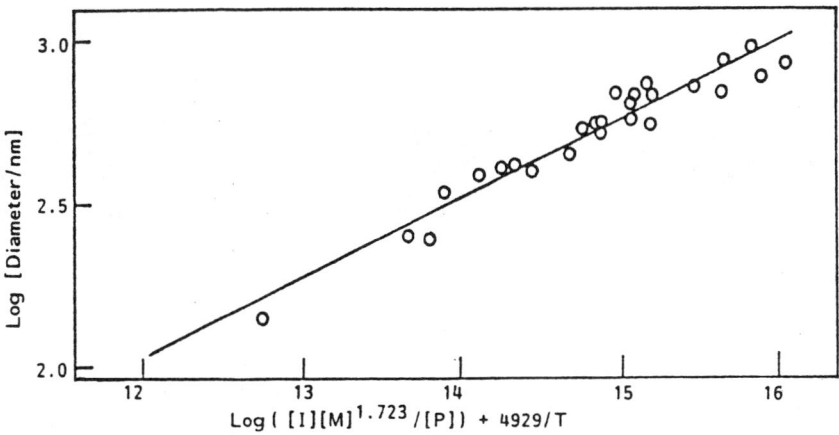

Figure 6. Emulsifier-free polymerisation, effect of salt concentration

As well as 1:1 electrolytes 2:1 and 3:1 electrolytes can be used and essentially as expected show the same effects at lower concentrations. However, because of the higher efficiency of the cation as a coagulant the range of utilisation is more restricted and there is no gain in the range of particles sizes formed [25].

40

6. Heterocoagulation

When particles of the same or different sizes, but of opposite charge, are mixed together the association of the particles can occur. The particles can be composed of the same polymer or different polymers. It is also possible for heterocoagulation to occur between large and small particles if they have the same sign of charge and the surface potentials are different.

When heterocoagulation occurs various results can be obtained and two of these are illustrated schematically in Figure 7a. This shows that extensive coagula can be formed with one type of particle causing bridging flocculation of the others. Alternatively, by careful choice of the ratios of small particles to big then surface coating can occur as shown by a scanning electron micrograph [26] in Figure 7b.

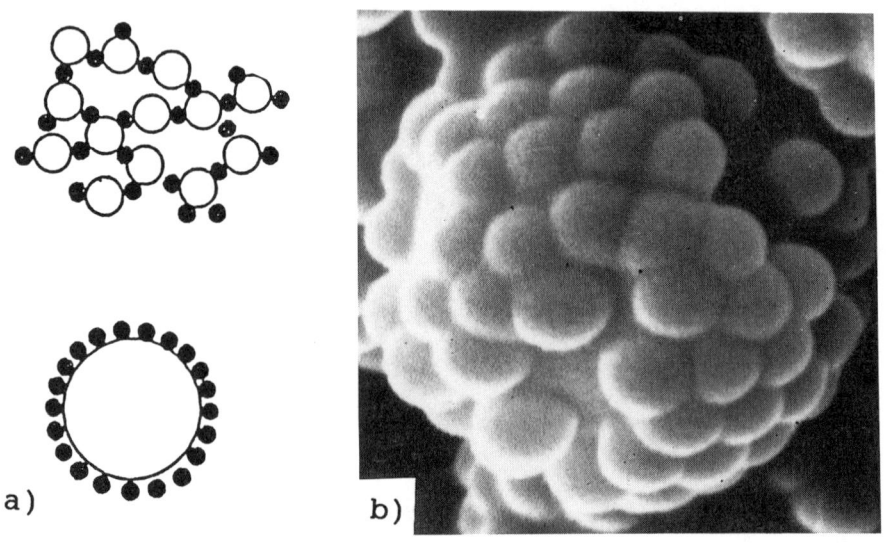

Figure 7. Heterocoagulation processes, a) schematic b) scanning electron micrograph

6.1. SURFACE COATING

For small particles of radius R_1 and large particles of radius R_2, then assuming that the particles on the larger sphere are hexagonaly close-packed then the number of small particles required to saturate the surface of the larger one with a monolayer, N_{sat}, is given by [27],

$$N_{sat} = 3.64[(R_1 + R_2)/R_2 + 1]^2 \tag{21}$$

As an example if $R_1 = 0.25$ μm and $R_2 = 0.50$ μm then $N_{sat} \approx 23$.

An alternative concept [28] is to consider the small particles of the coating as having a total surface area equivalent to that of the big particle, that is,

$$R_2^2 = N^* R_1^2 \qquad (22)$$

which for the same values of R_1 and R_2 gives N^* as 4.

Aspects of heterocoagulation are also important in the preparation of polymer colloid particles. Two examples are the growth of particles on to a seed particle in a second stage emulsion polymerisation and the formation of heteroparticles of complex morphology, e.g. core-shell particles.

6.2. COAGULATION IN SEEDED GROWTH POLYMERISATIONS

It has been found that the swelling of large particles, ca 1 μm, by monomer as a precursor to forming larger particles can be slow but if a sufficient number concentration of seed particles is present, N_L, then larger particles can be formed without secondary growth occurring [29]. As mentioned previously in the early stages of polymerisation the small nuclear particles, Ns, formed are, in the colloidal sense, unstable and consequently in the presence of the larger particles a number of kinetic processes can occur. Thus we can write the various kinetic possibilities as,

i)   small - small interaction $-dN_S/dt = k_{SS}N_S^2/W_{SS}$ (23)

ii) small - large interaction $-dN_S/dt = k_{SL}N_SN_L/W_{SL}$ (24)

iii) large - large interaction $-dN_L/dt = k_{LL}N_SN_L^2/W_{LL}$ (25)

Since the large particles are already stable and W_{LL} is large then these remain stable. In the case of the small newly formed particles these are very unstable and i) is a facile process since W_{SS} is tending to unity. This could lead to the growth of new particles as described earlier; however, provided enough seed particles are present then interaction ii) can become the dominant process leading to the small particles being scavenged by the larger ones. Moreover, since the small particles rapidly take up monomer and become monomer swollen this forms a mechanism of transfering monomer to the larger particles and hence of obtaining a particle growth process.

Figure 8 shows the variation in the amount of secondary growth obtained as ΔN = (number of particles at end of reaction - number of seeds) against the number of seed particles. It substantiates the idea that the number of seed particles present is an important factor [30].

6.3. ENGULFMENT AND NANOENCAPSULATION

The coating of a large particle by small ones as illustrated in Figure 7 can be viewed as a preliminary step in the process of forming core-shell particles or obtaining complex particles by an engulfment process. These possibilities are illustrated in Figure 9. Both

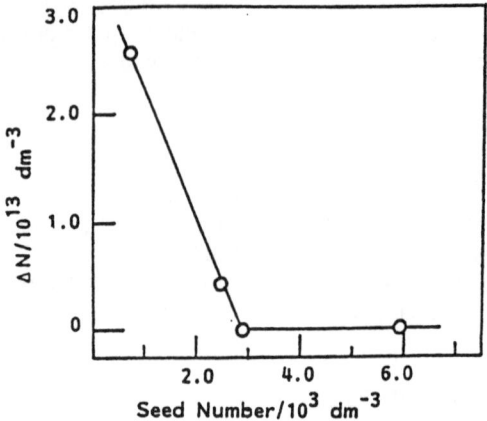

Figure 8. Secondary growth vs number of seeds added

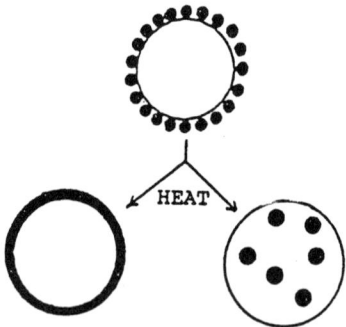

Figure 9. Particle engulfment and encapsulation

the surface energies and the glass transition temperatures of the particles also play an important role in these processes [31,32].

If the larger particle has the lower glass transition temperature, T_g, then on heating to ca. 45°C above this value, but remaining below the T_g of the smaller particle, with the appropiate interfacial energy conditions [28,33], then engulfment occurs, as shown in Figure 9.

If the smaller particles, forming the heterocoagulated coating, have the lower T_g then under similar conditions nanoencapsulation occurs to give a core-shell morphology [32].

7. Surface Coagulation

Figure 10 gives a sketch of surface coagulation, a process which takes place exclusively at the dispersion-air interface via the combination of two particles. This process can take place at electrolyte conditions which are much lower than the c.c.c. in the bulk phase. Polymer colloid particles formed from perfluorinated polymers, which have a low surface

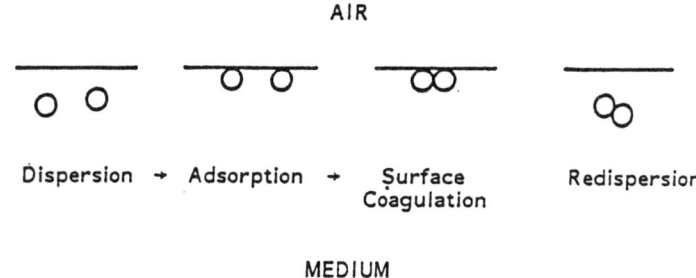

AIR

Dispersion → Adsorption → Surface   Redispersion
                          Coagulation

MEDIUM

Figure 10. Schematic illustration of surface coagulation

energy, are particularly prone to surface coagulation. A detailed investigation of this topic has been reported by Heller and co-workers [34,35].

8. Formation of Granules

In some cases the polymer colloids formed are not used as dispersions but as a dry powder in the form of granules which can then be shipped to processing plants for mechanical treatments such as extrusion. A typical example is the use of latices of poly(tetrafluoroethylene) which are coagulated by salt and then densified by shear. An added requirement is the densification of the coagula by an appropriate choice of electrolyte and mixing conditions.

9. Surfactants and Polymer Colloids

Surfactants play several roles in emulsion polymerisation and a major one is to provide colloidal stabilisation of the particles formed by adsorption to the particle surfaces.

Adsorption of the surfactant depends on a number of factors including the nature of the polymer and, in particular, the nature of the hydrophobic regions on the particle surface. In the case of anionic and nonionic latices, adsorption primarily occurs on the latter regions via the hydrocarbon chain of the surfactant. Particles of high surface charge density tend to adsorb less surfactant of the same charge as a consequence of electrostatic repulsion.

In the case of poly-(tetrafluoroethylene) particles the basic surface has a low surface energy, ca. 12 mN m^{-1} and the surface is both hydrophobic and oleophobic. Hydrocarbon chains are only very weakly adsorbed and hence perfluoroalkane surfactants are used as stabilisers. Interestingly, nonionic surfactants of the polyethylene glycol type can adsorb via the head-group leaving the hydrocarbon chain exposed to the medium, hence providing a site for further surfactant adsorption [36].

When cationic surfactants are added to a dispersion of negatively charged particles then the initial stage of adsorption arises by the positive charge of the surfactant interacting with an anionic group on the particle surface. Once all the surface charges are neutralised then $\psi_s \rightarrow 0$, and hence also $V_m \rightarrow 0$. This situation can readily be observed experimentally since coagulation occurs and the electrophoretic mobility is reduced to zero; there is thus coincidence between the c.c.c. and the concentration of surfactant at which reversal of charge occurs. This is strongly dependent on chain length [37].

Once the negative charges on the surface have been neutralised, thus rendering the surface completely hydrophobic, further adsorption on to the surface can occur via the hydrocarbon chain of the surfactant [37]. If the initial layer is sparsely populated then a monolayer is formed; if more tightly packed then a bilayer can be formed. The additional adsorption provides a substantial positive charge on the particles and colloidal restabilisation occurs with the particles in cationic form. With high additions of charged surfactants a substantial increase in the total electrolyte can occur thus reducing $1/\kappa$ and giving a second coagulation region as a consequence of electrical double layer compression [37].

With more polar polymers, e.g. polystyrene, and cationic surfactants a systematic shift of the c.c.c. occurs to lower concentrations as the chain length of the surfactant increases, a typical Traube's rule effect [37]. With poly-(tetrafluoroethylene) particles only very small differences are observed in the c.c.c. with variation of hydrocarbon chain length [36].

Nonionic surfactants also adsorb on to polymer colloid particles and often produce effective steric stabilisation of the particles (see section 10.1).

10. Steric Stabilisation

Although electrostatic stabilisation is very effective there are a number of conditions in practice where it is not appropiate or where it cannot be used. For example, it has been shown that in the presence of electrolyte charged particles coagulate at the c.c.c. Such systems therefore cannot be used in high electrolyte conditions. It is also clear that such systems are likely to coagulate on freezing, since with the separation of ice crystals the electrolyte concentration increases in the equilibrium solution phase. Although non-aqueous dispersions will not be discussed here this can be a situation of low dielectric permittivity where charge stabilisation is often not viable.

An alternative mechanism is therefore needed to prevent entry of the particles into the deep attractive well. An approach is shown schematically in Figure 11, which is to surround the core particle of radius R with a layer of thickness δ. If the Hamaker Constant of the layer is close to that of the medium the layer simply acts as a spacer and prevents the surfaces of the core particles approaching to a distance of less than 2δ, thus substantially reducing the depth of the attractive well to a value of the order of 1kT; the latter being about the kinetic energy of a colloidal particle in Brownian Motion.

A more sophisticated approach shows that the energy of steric repulsion, V_s, can be expressed in the form [11],

$$V_S = 4\pi c_\delta^2 (\psi_1 - \chi_1)(\delta - h/2)^2 (3R + 2\delta + h/2)/3V_1\rho_2^2 \qquad (26)$$

where ψ_1 is an entrophy parameter for the mixing of molecules in the ovelap region, which for ideal mixing can be taken as 0.5; χ_1 characterises the interaction of the adsorbed molecule with the solvent. It follows that if $\chi_1=0.5$, then $V_S=0$; for $\chi_1>0.5$,

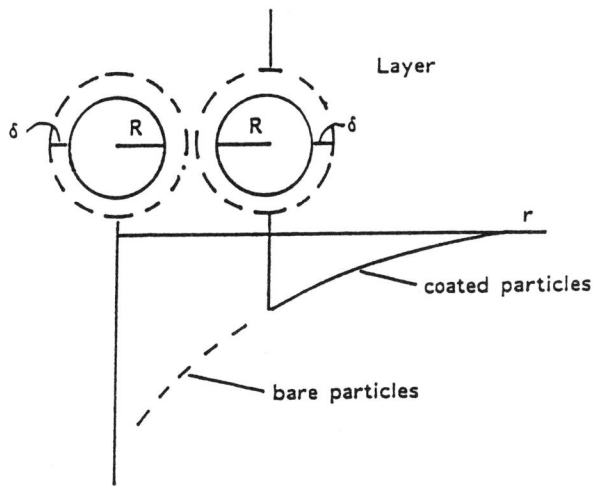

Figure 11. Steric stabilisation

then V_S is negative and hence attractive and if $\chi_1 < 0.5$, V_S is positive and hence repulsive.

10.1. NONIONIC STABILISERS

Nonionic surfactants can be quite effective as steric stabilisers. For example, addition of dodecylhexaoxyethylene glycol monoether at a concentration just above the critical micelle concentration to a polystyrene latex containing particles of radius 26nm raised the c.c.c. for lanthanum nitrate from 5.97×10^{-4} to 3.70×10^{-3} mol dm^{-3}. It was also found with this system that flocculation occurred at or very close to the Cloud Point of the nonionic surfactant. However, redispersion of the particles occurred on cooling below the Cloud Point provided that the temperature was not taken more than ca 5 to 10°C above the Cloud Point [11].

A further use of polyethylene glycol chains is to graft them directly to the particle surface using a polymerisable monomer [38]; azo initiators have also been used for this purpose [39]. For example, methoxy polyethylene glycol methacrylate, with 40 ethylene glycol units, was successfully used in the emulsion polymerisation of styrene using as the initiator system ascorbic acid and hydrogen peroxide. The particles so produced were found to be colloidally stable in 0.75 mol dm^{-3} barium chloride whereas similar charge-stabilised particles had a c.c.c. of 2.1×10^{-2} mol dm^{-3} barium chloride [40].

10.2. FREEZE-THAW STABILITY

Grafting of polyethylene oxide chains to the surface of the latex particles also produced a substantial improvement in freeze-thaw behaviour over that found with charge stabilised particles [38]. As a method of assessing this behaviour the optical absorbance was measured after thawing a dispersion which had been kept frozen for 3 days at 18°C. The result was compared with a control which had been kept at ambient conditions for the same period. Thus the freeze-thaw stability index was defined as:

$$\frac{\text{Adsorbance after thawing}}{\text{Adsorbance of Control}}$$

The sterically stabilised system had an index of 0.60 compared with 0.03 for a charge-stabilised system.

10.3. POLYMERIC DESTABILISATION - DEPLETION FLOCCULATION

The addition of polymeric species to a polymer colloid dispersion can produce a number of effects including, network formation, bridging flocculation, steric stabilisation and charge stabilisation or flocculation by reversal of charge if the polymer is a polyelectrolyte. These effects depend, in addition to the chemistry involved, also on the relative size of the particle and on the radius of gyration of the polymer, R_G.

Additional effects can be observed when the polymer does not adsorb to the particle surface, for example, when the particle is already sterically stabilised by a species of similar chemical structure. Phenomenologically what is observed is that at low dissolved polymer concentrations the particles and polymer form a stable dispersion. However, on increasing the polymer concentration a critical concentration is reached whereupon the dispersion separates into a particle-rich phase in equilibrium with a second phase dilute in particles. This effect has been termed Depletion Flocculation [41]. Although predicted in 1954 [42] it has only been investigated recently in any detail. It can be treated as a phase transition [43] and hence unlike the coagulation process discussed earlier it is reversible, the particles redispersing on removing the polymer or diluting the system.

As in the previous sections this can be expressed as an energy of interaction given by:

$$V_{dep} = \pi P_{os} \int \left[(R+R_G)^2 - r^2 \right] dr \qquad (27)$$

which for the boundary condition that $V_{dep} = 0$ when $h = 2R_G$ or $r = 2R + 2R_G = 2D$, gives,

$$V_{dep} = -4\pi P_{os} D^3 \left[1 - 3r/D + r^3/D^3 \right] / 3 \qquad (28)$$

The osmotic pressure for the polymer solution can be obtained from the virial equation,

$$P_{os} = RTc_2/M_2 + BRTc_2^2 \qquad (29)$$

with M_2 = the molecular weight of the polymer and c_2 its solution concentration; B is the second virial coefficient.

It is clear from eq. (28) that the depletion effect produces an additional attractive effect which depends on the molecular weight of the polymer in solution. The larger the molecular weight the longer the range of the attration. This is illustrated by some experimental results of Sperry [44] who used five different hydroxy ethyl celluloses which ranged in viscosity average molecular weight from 70,000 to 855,400. The flocculation concentrations for polystyrene particles of diameter 430 nm, with an

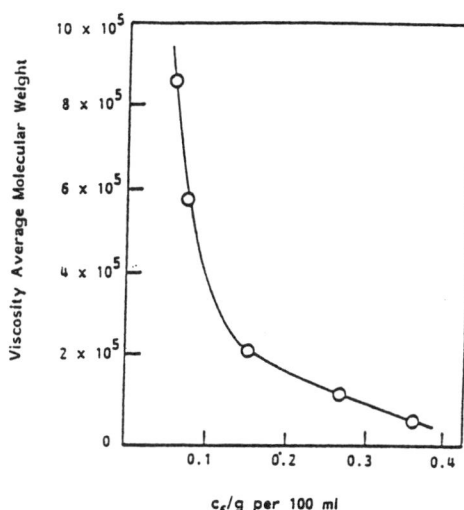

Figure 12. Depletion flocculation with ethylhydroxycellulose of various molecular weights.

adsorbed layer of a nonionic surfactant, Triton X-405, in 10^{-2} mol dm^{-3} electrolyte are shown in Figure 12. These results indicate a dependence of c_f on molecular weight.

11. References

1.   Candau, F. and Ottewill, R.H. (1990) *Scientific Methods for the Study of Polymer Colloids and their Applications*, Kluwer Academic Publishers, Dordrecht.
2.   Ottewill, R.H. (1990) in F. Candau and R.H. Ottewill (eds.), *Introduction to Polymer Colloids*, Kluwer Academic Publishers Dordrecht, pp. 129-157.
3.   Verwey, E.J.W. and Overbeek, J.Th.G.(1948) *Theory of Stability of Lyophobic Colloids*, Elsevier, Amsterdam.
4.   Hamaker H.C.(1937) Physica **4**, 1058.
5.   Israelachvili, J.N. (1985) *Intermolecular and Surface Forces*, Academic Press, London.
6.   Hough, D.B. and White, L.R. (1980) Advances Coll. Int. Sci. **14**, 3-41.
7.   Ottewill, R.H. (1982) in I. Piirma (ed) *Emulsion Polymerisation*, Academic Press, New York, pp. 1-49.
8.   Kitchener, J.A. and Schenkel, J.H. (1960) Trans. Faraday Soc. **56**, 161-173.
9.   Fowkes, F.W.(1964) Ind. Eng. Chem. **56**, 40-52.
10.   Overbeek, J.Th.G.(1980) *Pure and Appl. Chem.* **52**, 1151-1161.
11.   Ottewill, R.H. and Walker, T. (1968) Kolloid Z. u Z. Polymere, **222**, 108-116.
12.   Storer, C.C.(1968)Ph. D. Thesis, University of Bristol.
13.   Ottewill, R.H. and Shaw, J.N. (1966) Discuss. Faraday Soc. **42**, 154-163.
14.   Pelton, R. (1976) Ph. D. Thesis, University of Bristol.

48

15. Force, C.G. and Matijevic, E. (1968) Kolloid Z. u Z. Polymere, **224**, 51-62.
16. Johansson, G. (1960) Acta Chem Scand. **14**, 771-773.
17. von Smoluchowski, M. (1917) Z. Phys. **92**, 129.
18. Fuchs, N.(1934) Z. Phys. **89**, 735.
19. Overbeek, J.Th.G.(1952) in H.R. Kruyt (ed) Colloid Science, Elsevier, Amsterdam, **1**, 283-286.
20. Reerink, H. and Overbeek, J. Th.G. (1954) Discuss. Faraday Soc. **18**, 74-84.
21. Tuorila, P.(1927) Kolloidchem. Beih. **24**, 1-122.
22. Zeichner, G.R. and Schowalter, W.R.(1977) A.I. Chem. E. **23**, 243-254.
23. Husband, J.C. and Adams, J.M. (1992) Colloid Polym. Sci. **270**, 1194-1200.
24. Goodwin, J.W., Hearn, J., Ho, C.C. and Ottewill, R.H.(1974) Colloid Polym. Sci. **252**, 464-471.
25. Goodwin, J.W., Ottewill, R.H., Pelton, R., Vianello, G. and Yates, D.E. (1978) Brit. Polym. J. **10**, 173-180.
26. Goodwin, J.W. and Ottewill, R.H.(1978) Faraday Discuss. Chem. Soc. **65**, 338-339.
27. Roulstone, B.J. and Waters, J.A.(1994) European Patent 0 549 163.
28. Waters, J.A. (1994) Colloids and Surfaces, **23**, 167-174.
29. 29.Chung-Li, Y., Goodwin, J.W. and Ottewill, R.H.(1976) Prog. Colloid Polym Sci. **60**, 163-175.
30. Jayasuriya, S.(1978) Ph. D. Thesis, University of Bristol.
31. Ottewill, R.H., Schonfield, A.B. and Waters, J.A.(1996) Colloid Polym. Sci. **274**, 763-771.
32. Ottewill, R.H., Schonfield, A.B. and Waters, J.A. to be published.
33. Waters, J.A.(1995) in J.W. Goodwin and R. Buscall (eds.), Colloidal Polymer Particles, Academic Press, London, pp. 113-135.
34. Heller, W. and Peters, J.(1970) J. Colloid Int. Sci. **32**, 592-605; **33**, 578-585.
35. Heller, W. and de Lauder, W.B.(1971) J. Colloid Int. Sci., **35**, 60-65 & 308-313.
36. Bee, H., Ottewill, R.H., Rance, D.G. and Richardson, R.A.(1983) in R.H. Ottewill, C.H. Rocheser and A.L. Smith (eds.) *Adsorption from Solution*, Academic Press, London pp. 155-171.
37. Ingrma, B.T. and Ottewill, R.H.(1991) in D.N. Rubingh and P.M. Holland(eds.) *Cationic Surfactants*, Academic Press, New York pp. 87-110.
38. Ottewill, R.H., Satgurunathan, R., Waite, F.A. and Westby, M.J.(1987) Brit. Polym. J. **19**, 435-440.
39. Thompson, M. personal communication.
40. Ottewill, R.H. and Satgurunathan, R. (1987) Colloid Polym. Sci., **265**, 845-853.
41. Napper, D.H.(1983) *Polymeric Stabilisation of Colloidal Dispersions*, Academic Press, London.
42. Asakura, S. and Oosawa, F.(1954) J. Chem. Phys. **22**, 1255-1256.
43. Russel, W.B.(1987) The Dynamics of Colloidal Systems, University of Wiscosin Press, Madison, Wisconsin.
44. Sperry, P.R.(1984) J. Colloid Int. Sci., **87**, 375-384.

PARTICLE NUCLEATION AT THE BEGINNING OF EMULSION POLYMERIZATION

K. TAUER, I. KÜHN[1]
MPI für Kolloid- und Grenzflächenforschung
Kantstraße 55
D-14513 Teltow, Germany

1. Introduction

The question of how the particles are generated during an emulsion polymerization has been a matter of intensive considerations since the beginning of emulsion polymerization in the early years of this century. However, the first public scientific papers appeared after the Second World War. During this time an intensive work was done in each of the major companies to develop large scale emulsion polymerization processes. Until today every year publications appear concerning particle formation in the different kinds of heterophase polymerizations. The certainly incomplete selection of references [1 - 50] illustrates the continuous activities in this field over the last nearly 50 years. Besides these original papers, one can find nucleation chapters also in the following selection of monographs, proceedings and textbooks concerning polymer colloids and heterophase polymerization [51 - 59].

It was found very early that micelles are not a prerequisite for the preparation of latex particles [1, 6, 10]. FIKENTSCHER [1] and HARKINS [6] gave an explanation for nucleation in non-micellar systems as they proposed a reaction of dissolved monomer molecules with initiator radicals in water. The reason why the pioneers dealt mainly with micellar systems is the much higher polymerization rate as well as the better process yield [1, 6]. Especially, the work of PRIEST [9] has influenced all following nucleation models as he discussed at the beginning of the fifties already single-chain precipitation, primary particles, and interparticle combination, a phrase that is today called coagulative nucleation.

With this impressive list of research activities over the last half century one could believe that there is not anything more to do. However, at least two facts give reasons for that this not to be the case. Firstly, there is a lack with respect to nucleation kinetics to confirm one or the other nucleation mechanism. In most of the cases conclusions concerning particle nucleation are based on experimental data which are obtained when particle growth dominates compared to nucleation i.e., at already fairly high monomer conversion. Secondly, it is the state of the art to start thinking about nucleation with the

---

[1] Present address *BASF AG, D 67056 Ludwigshafen, Germany*

J. M. Asua (ed.), Polymeric Dispersions: Principles and Applications, 49–65.

peculiarities of heterophase polymerization [59] and not with the common features to other nucleation events as for instance during crystallization and condensation processes. The common features are governed by the thermodynamics and the peculiarities are determined by chemistry and kinetics of the particular process. It has to be pointed out that some approaches have been made in this direction but unfortunately, only in a very general sense without any specific conclusions [31, 52, 57] for emulsion polymerization. It should be noted that the thermodynamics of a nucleation process is basically the same as for a phase formation and precipitation whereby the more general term is phase formation.

2. Particle Nucleation in Emulsion Polymerization - a Special Case of Phase Formation

The objective of this chapter is to recognize nucleation in emulsion polymerization or in any kind of heterophase polymerization as a special case of phase separation that can be described using general principles and equations. In this sense it has to be pointed out that particle nucleation, precipitation, phase formation, and phase separation can be used as synonyms.

The understanding of particle nucleation in emulsion polymerization as phase separation or precipitation starting from a homogeneous solution becomes obvious by reconsidering experimental results obtained by FIKENTSCHER [1, 10], BAXENDALE [2], and FITCH et al. [12] for polymerization of an aqueous methyl methacrylate (MMA) solution or by Priest [9] for an aqueous vinyl acetate (VAC) solution. The reaction starts in a complete transparent aqueous solution but it ends with a latex.

At a distinct time the polymer phase precipitates in the form of latex particles i.e., particle nucleation occurs. For such a reaction the free energy function is shown in Figure 1 whereby the reaction coordinate is the concentration of polymer. The driving force for the whole reaction (affinity) Φ is given by the difference between the chemical potential of the initial (μ_1) and the final state (μ_2). With respect to stability the curve describes three different regions.

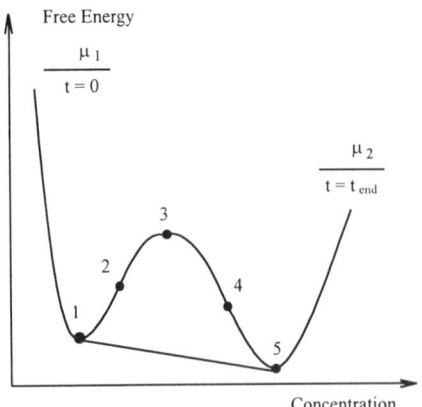

Figure 1. Change of free energy for a reaction with phase separation (schematic drawing)

The line between points 1 and 5 describes a stability curve, i.e. the system stays in one phase. Two meta-stable regions are located between points 1 and 2 and between 4 and 5, respectively. In the region behind point 3 with respect to higher concentration the system lowers its free energy by unmixing (phase separation, nucleation) into two phases. It is obvious, that if the system overcomes the instability region between points 2-3-4 the affinity for phase separation increases. This means, point 3 corresponds to a free energy of activation for the phase separation (ΔG_{max}) that strongly depends on the experimental conditions. And, as the nucleation rate is an exponential function of ΔG_{max}, it follows that nucleation is extremely sensitive to changes in experimental conditions. Therefore, a fundamental understanding of the thermodynamics, kinetics, and mechanisms of nucleation processes is of great importance for control of particle structure and morphology. This is a very general conclusion valid for any phase separation process (metal and polymer alloys, crystallization, bubble and droplet formation) in different fields of material sciences but also meteorology and medicine (formation of kidney stones or gallstones).

These considerations lead to the formulation of a general nucleation criterion that is also valid for emulsion polymerization. *Nucleation requires bringing the system into a thermodynamic unstable intermediate state.* Figure 2 illustrates this criterion by means of a phase diagram.

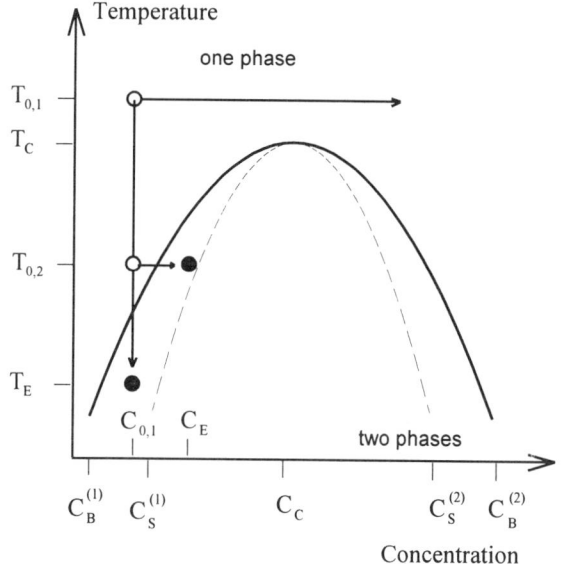

Figure 2. Schematic phase diagram illustrating nucleation

The solid line between the concentrations $C_B^{(1)}$ and $C_B^{(2)}$ describes the coexistance curve that separates the system into a single phase region and a two phase region (binodal). The dashed curves between the concentrations $C_S^{(1)}$ and $C_S^{(2)}$ represents the spinodal that controls the early initial stages of phase separation [60]. The open symbols in Figure 2 represent two different initial states. For a phase separation the system must be brought into the region between both curves indicated by the filled symbols. In the particular case depicted in Figure 2 this can be achieved either by increasing the concentration or by decreasing the temperature. For the hypothetical initial state characterized by T_{02} and C_{01} both ways are possible whereas for that characterized by T_{01} and C_{01} there remains only the temperature quench to T_E. The temperature quench has a big meaning for crystallization processes and preparation of alloys. As emulsion polymerization is carried out isothermally, the nucleation condition is reached by increasing the polymer concentration. A quantitative estimation of the free energy for a phase separation that leads to particles with a colloidal size is possible in the following general way [61].

The change in the free energy (ΔG_N) can be expressed as the difference between the free energy of a nucleus and that of the molecules forming the nucleus according to eq (1) where G(m) is the free energy of a nucleus (particle) consisting of m molecules and μ is the chemical potential of a single molecule in solution.

$$\Delta G_N = G(m) - m \cdot \mu \tag{1}$$

G(m) is given by the free energy of the m molecules in a nucleus (eq (2)) whereby due to its colloidal size it is necessary to distinguish between the molecules in the volume and at the surface. In eq (2) indexes v and s refer to volume and surface, respectively, and g stands for the free energy of one molecule.

$$G(m) = m_v \cdot g_v + m_s \cdot g_s \tag{2}$$

From eq (2) follows eq (4) if g_v is expressed as the chemical potential μ_v and if the surface tension (γ_{pw}) is defined as the surface area (A_S) based difference between the free energy of a surface molecule and a volume molecule multiplied with the number of surface molecules (eq (3)).

$$\gamma_{pw} = (g_s - g_v) \cdot m_s / A_s \tag{3}$$

$$G(m) = m \cdot \mu_v + \gamma_{pw} \cdot A_s \tag{4}$$

The chemical potential of the nucleating molecules for a diluted solution is expressed in a standard way by eq (5) where $\mu^\varnothing$ is the standard chemical potential, $k_B T$ is the thermal energy, and φ_p^w is the molar fraction of molecules in solution.

$$\mu = \mu^\varnothing + k_B T \cdot \ln \varphi_p^w \tag{5}$$

Equation (6) is valid for μ_v if the assumption is made that the molar fraction in that case corresponds to the saturation value $\varphi_{p,sat}^w$.

$$\mu_v = \mu^\varnothing + k_B T \cdot \ln \varphi_{p,sat}^w \tag{6}$$

The ratio $\varphi^w_p/\varphi^w_{p,sat}$ is the supersaturation (S) of the solution. Furthermore, it is assumed that the nuclei have always a spherical shape and hence their surface area is expressed as $c_1 \cdot (m \cdot j)^{2/3}$. c_1 is a constant and j is the chain length of the nucleating polymers. To express the surface area in such a way is reasonable as the polymers in a nucleus are randomly arranged and not ordered. So, the number of surface units is larger than the number of chains in a nucleus. Especially, this becomes obvious for only one chain per nucleus.

Combining eqs (1) to (6) follows a general relation for the change of the free energy of nucleus formation (eq (7)). This relation is well known from the classical nucleation theory (CNT) and in the special form of eq (7) adopted for a heterophase polymerization.

$$\Delta G_N = -m \cdot k_B T \cdot \ln S + c_1 \cdot (m \cdot j)^{2/3} \cdot \gamma_{pw} \tag{7}$$
$$c_1 = (4\pi)^{1/3}(3M_0/\rho_p N_A)^{2/3}$$

Figure 3 elucidates some properties of eq (7) in dependence on m and j. These are typical properties for the different kinds of nucleation processes.

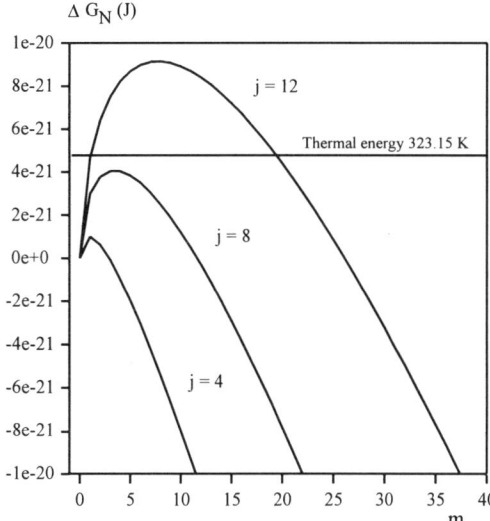

Figure 3. Free energy of nucleation dependence on m and j; values for MMA

Equation (7) describes at a critical value m_{crit} a maximum that is interpreted as the nucleation barrier. Nuclei formed with a number of molecules less than m_{crit} will dissolve again but if m is greater than m_{crit} the nuclei are stable. At this point critical conditions for the nucleation are defined. The relation to the phase diagram depicted in Figure 2 is as follows. The polymerization starts as homogeneous isothermal reaction in water with the formation of oligomers. At the very beginning the concentration of the polymer molecules in water as well as their chain length increases and the system stays homogeneous as long as it moves into a region of the phase diagram where phase separation is possible.

3. A Nucleation Model for Emulsion Polymerization Based on CNT

Equation (7) is the centrepart of a modeling framework for particle nucleation at the beginning of an emulsion polymerization [48]. To model nucleation at this very early stage of an emulsion polymerization means to predict the size of the nuclei, the number of chains per nucleus, the chain length of these chains, and the number of nuclei for that time when nucleation occurs i.e., for the critical condition (index crit). The solution of eq (7) requires expressions to calculate the supersaturation and the chain length of the oligomers. A calculation of S requires a relation for the solubility of the oligomers in the dispersion medium (corresponding to $\varphi^w_{p,sat}$) in dependence on their chain length. An approximation based on the FLORY-HUGGINS theory given by BARRETT [52] is very useful in that case. The radical polymerization kinetics gives an expression to calculate the concentration of oligomers (corresponding to φ^w_p) in dependence on chain length and time. Finally, S (eq (8)) is a function of both polymerization time and chain length of the oligomers. It combines polymer solution theory (third factor of the right hand side of eq (8) with radical polymerization kinetics (eq (9)). For details of the derivation it is necessary to refer to the original publication [48].

$$S(t,j) = c_2 \cdot \frac{\beta(I,t)}{(1+\beta(I,t))j} \cdot \left(\left[I_{2,0}\right]-\left[I_2(t)\right]\right) \cdot \frac{j(1-\exp(j\cdot(1-\chi_{pw}-1/j)))}{\exp(j\cdot(1-\chi_{pw}-1/j))} \tag{8}$$

$$c_2 = 2M_0/\rho_p$$

$$\beta(I,t) = \frac{\left(2\cdot f\cdot k_I \cdot k_t\left[I_2(t)\right]\right)^{1/2}}{\overline{k_{pi}}\cdot\left[M\right]^w} \tag{9}$$

$$\left[I_2(t)\right] = \left[I_2(0)\right]\cdot \exp(-k_I \cdot t)$$

Equations (10) to (13) result for the size of a critical nucleus (D_{crit}), for the number of chains forming a nucleus (M_{crit}), for the number of nuclei (N_{crit}), and for the free energy at the critical point (ΔG_{max}), respectively.

$$D_{crit} = c_3 \cdot \frac{j\cdot\gamma_{pw}}{k_BT\cdot \ln S} \tag{10}$$

$$c_3 = (2/3)c_1(6M_0/\pi\rho_pN_A)^{1/3}$$

$$M_{crit} = c_4 \cdot \frac{\left(j^{2/3}\cdot\gamma_{pw}\right)^3}{\left(k_BT\cdot \ln S\right)^3} \tag{11}$$

$$c_4 = ((2/3)c_1)^3$$

$$N_{crit} = c_5 \cdot \exp(-\Delta G_{max}/k_B T) \quad (12)$$
$$c_5 = 1/v_w$$

$$\Delta G_{max} = c_6 \cdot \frac{j^2 \cdot \gamma_{pw}^3}{(k_B T \cdot \ln S)^2} \quad (13)$$
$$c_6 = (4/27)c_1^3$$

The expression for N_{crit} as well as that for the rate of polymerization (eq (14)) is based on ideas coming from kinetics of precipitation [62] where k_N is the first order nucleation rate constant.

$$R_N = c_7 \cdot N_{crit} \quad (14)$$
$$c_7 = k_N$$

Besides the kinetic constants, values for χ_{pw} and γ_{pw} are needed for a solution of these equations. Swelling experiments with latex particles lead to values for the interfacial tension particle to water [63], [64] whereas value for χ_{pw} result from investigations with inverse gas chromatography [65]. The only really unknown parameter is ΔG_{max} that corresponds to an activation free energy of nucleation. Nucleation occurs when the relation $\Delta G_{max} > v k_B T$ is fulfilled. However, the value of v is unknown and its experimental determination is a problem for any nucleation process [66].

The most reliable prediction of this homogeneous nucleation model is the chain length of the nucleating oligomers. This is due to the fact that after the first particles have been formed the homogeneous model must be replaced by a heterogeneous model.

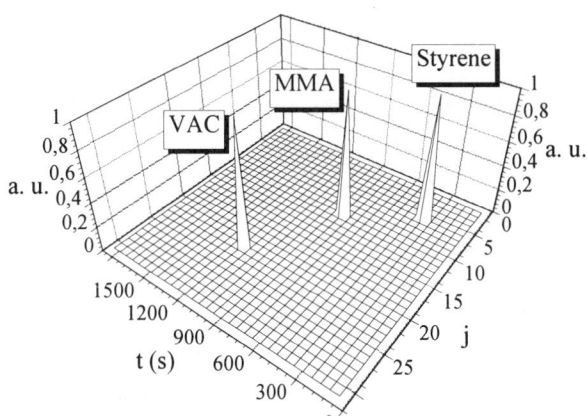

Figure 4. Calculated nucleation plane with v=10 for emulsifier free emulsion polymerization of Styrene, MMA, and VAC

Figure 4 shows a time-j plane (nucleation plane) calculated with a fairly high activation energy of $10k_B T$. Nevertheless, some important features of this model approach are visible. For different monomers, nucleation occurs at quite different times and j values. Thus , the influence of monomer properties, especially with different water solubilities is clear. The higher the water solubility the higher j_{crit} and the longer time it takes until the first particles appear.

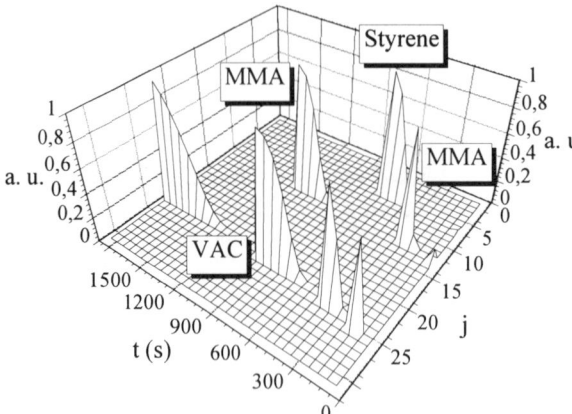

Figure 5. Calculated nucleation plane with v =1.01 for emulsifier free emulsion polymerization of Styrene, MMA, and VAC

Comparing results depicted in Figure 4 and Figure 5 it is clear that with decreasing ΔG_{max} the number of possible nucleation events increases drastically whereby the first nucleation (only to that applies the model) takes place at shorter polymerization times. However, the value for j_{crit} at which nucleation occurs changes only within a range of ± 1. Furthermore, the strong influence of the water solubility of the monomers is clearly seen. The higher the water solubility the longer the chain length of the nucleating oligomers and the more nucleation events can occur. The results summarized in Table 1 , certify the predictive power of this model with respect to j_{crit}. Furthermore, these calculations state that more than one oligomer forms a nucleus, i.e., a multi-chain precipitation takes place rather than a single-chain precipitation. Although the values for m, especially if ΔG_{max} is assumed to be 10 times $k_B T$, are very unlikely, the calculations never resulted in a value of m=1. The values for m obtained with v=1.01 seem to be likely as in that case also nucleus concentrations on the order of 10^{25} m^{-3} have been obtained.

TABLE 1. Comparison of calculated ($j_{crit,m}$) with experimental ($j_{crit,e}$) values of chain length for nucleating oligomers

Monomer	$j_{crit,m}$ (v=10)	m (v=10)	$j_{crit,m}$ (v=1.01)	m (v=1.01)	$j_{crit,e}$	Ref.
Styrene	6	887	5	3	5	[67]
MMA	11	223	12-11	4	10	[67]
VAC	22	110	23-25	4	18-20	[68]

Some important conclusions can be drawn from the CNT model approach with respect to experimental investigations of the particle nucleation in emulsion polymerization. Firstly, particle nucleation should start with a jump to a very large number of particles within a very short period of time after the critical value of the supersaturation of oligomers in the aqueous phase has reached. Secondly, the reproducibility of the experiments should be poor due to the exponential dependence of the nucleation rate on ΔG_{max}. (eq (14)). Thirdly, impurities in the water phase (dispersion medium) should have a strong influence on the nucleation behavior especially, if they influence the solubility of the nucleating polymers. Fourthly, emulsifiers and micelles should play no direct role in the nucleation step as long as they do not influence the solubility of the nucleating polymers. Fifthly, if surfactants are present they lower γ_{pw} as well as ΔG_{max}. (cf. eq (13)) and consequently the nucleation should take place earlier. Points four and five describe two possible effects of emulsifier molecules. However, the main effect of the emulsifier molecules is the stabilization of the nuclei formed. Depending on the kind and concentration of the emulsifier the number of nuclei may stay constant, increase or decrease after the first nucleation event that is described by the model.

A further increase in the predictive power of this modeling strategy is possible if it is extended to heterogeneous nucleation. This needs a modification of eq (7) as well as of the polymerization kinetics as all species dead or alive can be captured by particles. Today it is only a vision that the extended model can also be applied to treat radical entry into existing particles.

4. A Strategy for an Experimental Study of Particle Nucleation in Emulsion Polymerization

To detect and to investigate nucleation events is always a problem as at the time when one feels the rain drop on the skin or when one sees light scattering in a formerly transparent solution, it is already too late. The challenge is to be able to detect very tiny objects at low concentrations. Another problem arises from the fact that nucleation is a very fast process compared to the whole duration of an emulsion polymerization. Furthermore, if once the first particles are formed, the detection of another nucleation event is still more complicated. So, to find a proper strategy is crucial for successful experimental investigations.

It is very straightforward to use optical methods if one wants to detect the point when a solution becomes turbid. If this method is dynamic light scattering, one can determine a size without knowing anything about the particles even with dependence on the polymerization time in a simple way [45]. However, any optical technique is never a "yes-or-no" method as the response depends on at least three variables: the concentration, the path length, and the power of the light source. This means that the particular value of each of these variables could be insufficient to detect particles. Nevertheless, optical methods are very useful tools and they become still more useful if an additional method is simultaneously employed. With this second method, it should be possible to measure changes in the composition of the solute in the dispersion. Since in emulsion polymerization ionic species are present in the dispersion medium conductivity measurement is such a method.

The use of an optical method requires that the only species which contribute to turbidity or scattering are the particles. Especially, if a sparingly water soluble monomer is used, a monomer feed into the dispersion must be maintained that is high enough to ensure saturation for a certain time. On the other hand, this feed must be low enough to avoid monomer droplets that disturb the experiment. The best way to realize such a controlled monomer feed is to place a certain volume on top of the reaction mixture and confine spreading to a known area with a glass shade. Stirring has to be adjusted to avoid mixing of the monomer phase with the reaction medium.

Turbidity measurement is preferred as the optical method instead of dynamic light scattering for at least two reasons. Firstly, turbidity can be used on-line in a stirred system and secondly, it depends on both the size and the concentration of particles.

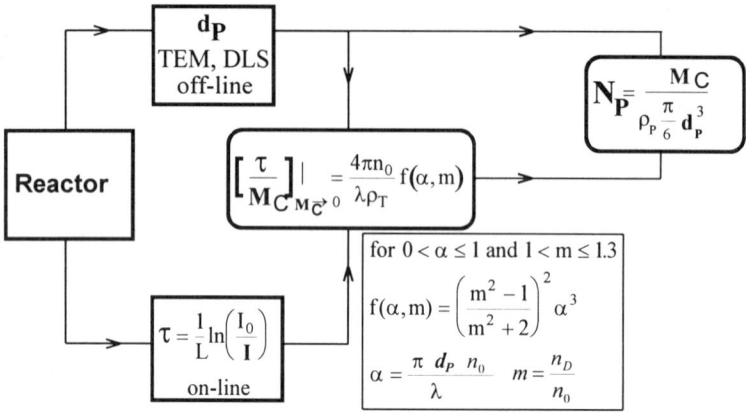

Figure 6. Clarification of the experimental strategy to determine particle number-time curves

Figure 6 shows a scheme of the experimental investigation strategy to get particle number - time curves based on these considerations. The basic idea is to measure the particle size off-line (d_p) and use that value and the on-line measured turbidity (τ) to calculate the polymer content (M_C) and subsequently the particle number (N_P). A detailed discussion of this procedure as well as results for the emulsifier-free styrene emulsion polymerization are published in [49].

5. Particle Nucleation in Styrene Emulsion Polymerization - New Experimental Results

5.1. EMULSIFIER-FREE POLYMERIZATION

During the experiments it turned out that conductivity is the only method for a detection of the onset of nucleation. Firstly, it reacts earlier than transmission on changes during the reaction [49] and secondly, it changes sharply within one second, the smallest time gap between two data points for the particular conductivity set-up.

Curve A in Figure 7 illustrates this behavior and a comparison with curve B shows the influence of styrene on the potassium persulfate (KPS) decomposition. Curve B results are in the absence of styrene and can be used to estimate the KPS decomposition rate constant. It results in the all-Teflon reactor at 60 °C a value of $6.5 \cdot 10^{-6} \, \text{s}^{-1}$ that is only slightly higher than a published value of $5.5 \ 10^{-6} \ s^{-1}$ obtained in a glass apparatus [69]. However, in the presence of styrene (the first part of curve A until the bend, Figure 7) the decomposition is much faster and it results a decomposition rate constant of $8.6 \ 10^{-5} \ s^{-1}$. This increase in the decomposition rate in the presence of organic material is typical for KPS [70].

The results depicted in Figure 7 represent a very simple case as the reaction mixture consists of only water, styrene, and KPS. The interpretation of this behavior is possible if the following results will be considered. Firstly, when additional to the standard procedure the pH is on-line recorded it turns out that the change of the proton concentration, $[H^+]$, has exactly the same shape as the conductivity curve. Even the bend occurs at the same time. Secondly, when the same experiment is conducted in the presence of a buffer (Na_2HPO_4) conductivity remains unchanged during the entire reaction. From these results it follows that under the particular experimental conditions, the conductivity is governed by protons. With this knowledge, a recalculation of the conductivity curve until the bend is possible [71].

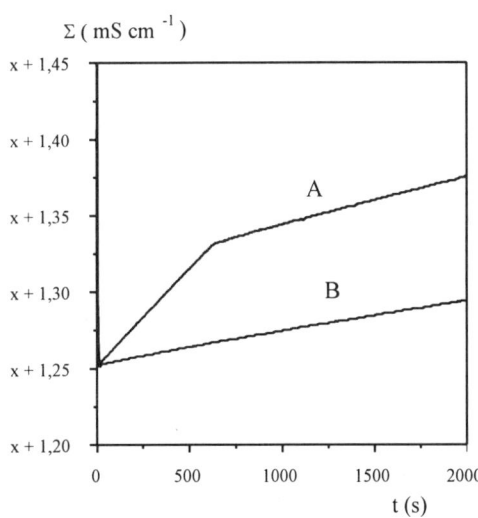

Figure 7. Conductivity change during KPS decomposition in the presence (A) and absence (B) of styrene; Recipe: Water, 25 mM KPS, styrene saturation concentration in water, 60 °C

This is a proof for the conclusion that the first part of curve A in Figure 7 corresponds to the pre-nucleation period. Since at the exact same time also a bend in the $[H^+]$-time curve occurs, one can conclude that a part of the protons produced lose their mobility. It is most likely that protons lose their mobility due to a specific adsorption or due to an incorporation in the electrical double layer of the polymer particles nucleated at this time. If this is true, an important conclusion is that a huge number of particles nucleate within a time period less than a second and hence, faster methods are necessary to investigate nucleation kinetics.

A second important conclusion is based on the reasonable assumption that the amount of protons adsorbed is proportional to the particle surface area. If this is true then the difference between the calculated (Σ_{theo}) and measured conductivity (Σ) is proportional to the proton concentration adsorbed at the particle surface ($[H^+]_s$) and hence, proportional to the total particle surface as described by eq (15).

$$\Sigma_{theo} - \Sigma \propto \left[H^+ \right]_S \propto N_P \cdot d_P^2 \tag{15}$$

Since the polymer concentration is very, low the turbidity (τ) depends on the particle number and particle diameter in a way as described by eq (16).

$$\tau \propto N_P \cdot d_P^6 \tag{16}$$

As both conductivity and turbidity are measured on-line during the whole polymerization, the combination of eqs (15) and (16) gives a possibility to calculate the particle diameter quasi on-line from turbidity and conductivity data (eq (17)).

$$\frac{\tau}{\Sigma_{theo} - \Sigma} = F \cdot d_P^4 \tag{17}$$

In eq (17) F depends on a variety of different constants and cannot be calculated in advance but, it can be fitted from measured particle diameters.

Figure 8 shows a result of such a fit with only one value for F over the entire polymerization range. The error connected with the application of eq (17) increases with increasing transmission (decreasing time). Nevertheless, particle diameter-time as well as particle number-time curves are accessible quasi on-line over the entire polymerization range. It is clearly seen that the particle number jumps from zero to 1,8 10^{19} m^{-3} in that second when the bend in the conductivity curve occurs.

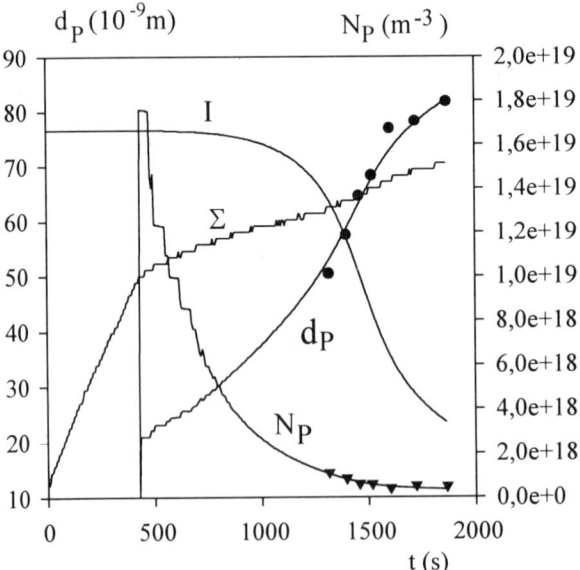

Figure 8. On-line data for particle diameter (d_P) and particle number (N_P) during an emulsifier-free emulsion polymerization of styrene; Recipe: styrene, 25 mM KPS; 60 °C; Conductivity (Σ) and transmission (I) no axis; Lines represent on-line data; Symbols represent off-line data determined according to the procedure illustrated in Figure 6

5.2. POLYMERIZATION IN THE PRESENCE OF EMULSIFIER

In this case the investigation is more complicated as the on-line approach is only possible if the initiator decomposition rate constant is known. Unfortunately, the KPS decomposition depends very strongly on the composition of the medium [70] and hence, it has to determined for each particular case. Another problem with the use if conductivity measurements arises if an ionic emulsifier is used as the absolute conductivity is increased but the sensitivity is decreased. Nevertheless, the off-line technique is applicable. Contrary to the emulsifier-free case (cf. Figure 8), the particle concentration in the presence of SDS (concentration above cmc) increases over the entire reaction [50]. Furthermore, the particle concentration is more than two orders of magnitude higher than in the emulsifier-free case.

In order to use conductivity measurements a non-ionic emulsifier (ANTA-ROX $^{®}$ CO 880, GAF Chemicals) was tested at concentrations below as well as above the cmc. Figure 9 shows the on-line conductivity data.

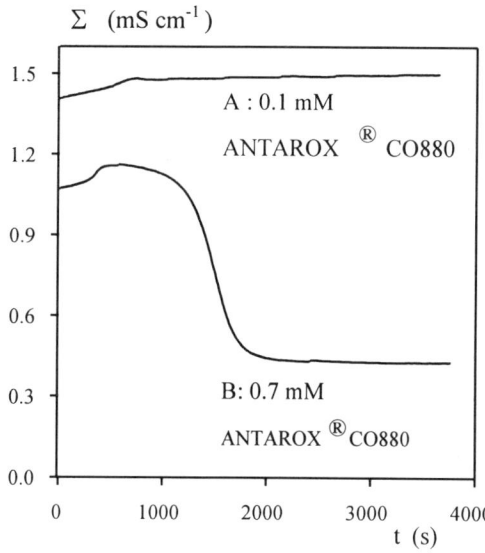

Figure 9. Change of conductivity during the emulsion polymerization of styrene in the presence of ANTAROX $^{®}$ CO 880; Recipe: styrene, 25 mM KPS, 60 °C, ANTAROX $^{®}$ CO 880; Curve A: $[S]_w = 0.1$ mM is below cmc; Curve B: $[S]_w = 0.7$ mM is above cmc

At first sight the curves look completely different compared to that of Figure 7 especially, if the emulsifier concentration is above the cmc (curve B). But, in both cases a bend occurs after the initial increase in the conductivity, as well, indicating particle nucleation.

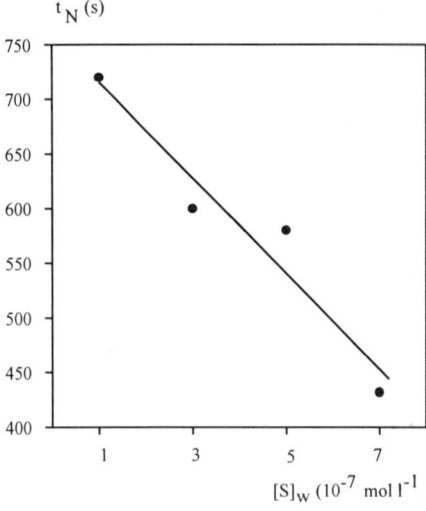

Figure 10. Dependence of the nucleation time (t_N) on the emulsifier concentration; Recipe: styrene, 25 mM KPS, 60 °C, ANTAROX ® CO 880

This interpretation seems to be likely as particles have only been detected by DLS in samples taken after the bend.

If the time of the bend is considered to be the nucleation time, the results depicted in Figure 10 are obtained for variation in the emulsifier concentration $[S]_w$. The nucleation time decreases with increasing emulsifier concentration as it is predicted (cf. chapter 3).

6. Conclusions

A summary of the experimental findings leads to the result that four of the five predictions from the model based on CNT (cf. section 3) have been clearly verified by the experimental data. The influence of additives to the dispersion medium is not clear. This is because in the presence of KPS any additive influences not only nucleation but also the initiator decomposition. The poor reproducibility of investigations during the nucleation period has been observed several times [45, 49]. Initial nucleation of particles occurs suddenly as confirmed by the conductivity results, independent of whether or not emulsifier is present even if its concentration is above the cmc. Finally, nucleation occurs the faster after the start of polymerization the higher the emulsifier concentration whereby the cmc is no special point.

A further use of CNT in emulsion polymerization may lead to models that will be able to contribute to a better understanding of heterogeneous nucleation, secondary particle nucleation, and capture of oligomers by particles.

7. References

1. Fikentscher, H. (1938) Emulsionspolymerisation und technische Auswertung, Angew. Chem. 51, 433
2. Baxendale, J.H., Evans, M.G., and Kilham, J.K. (1946) The kinetics of polymerization reactions in aqueous solutions, *Trans. Faraday Soc.* **42**, 668-684

3. Harkins, W.D. (1947) A general theory of the mechanism of emulsion polymerization, *J. Amer. Chem. Soc.* **69**, 1428-1444

4. Smith, W.V. and Ewart, R.H. (1948) Kinetics of emulsion polymerization, *J. Chem. Phys.* **16**, 592-599

5. Smith, W.V. (1948) The kinetics of styrene emulsion polymerization, *J. Amer. Chem. Soc.* **70**, 3695-3702

6. Harkins, W.D. (1950) General theory of mechanism of emulsion polymerization. II, *J. Polym. Sci.* **5**, 217-251

7. Wintgen, R. and Sinn, G. (1951) Zum Mechanismus der Emulsionspolymerisation I. Teilchenzahl und Teilchengröße bei der Emulsionspolymerization des Styrols mit Nekal, *Kolloid Zschft.* **122**, 103-109

8. Wintgen, R. and Jürgen-Lohmann, L. (1951) Zum Mechanismus der Emulsionspolymerisation II. Polystyrol-Latices mit Nekal: Methodisches, Teilchenzahl und Emulgatormenge, *Kolloid Zschft.* **122**, 111-117

9. Priest, W.J. (1952) Particle growth in the aqueous polymerization of vinyl acetate, *J. Phys. Chem.* **56**, 1077-1082

10. Fikentscher, H., Gerrens, H., und Schuller, H. (1960) Emulsionspolymerisation und Kunststoff-Latices, Angew. Chem. **72**, 856-864

11. Roe, C.P. (1968) Surface chemistry aspects of emulsion polymerization, *Ind. Engng. Chem.* **60**, 20-33

12. Fitch, R.M., Prenosil, M.B., and Sprick, K.J. (1969) The mechanism of particle formation in polymer hydrosols. I. Kinetics of aqueous polymerization of methyl methacrylate, *J. Polym. Sci.* Part C **27**, 95-118

13. Fitch, R.M. and Tsai, C-h. (1970) Polymer colloids: particle formation in nonmicellar systems, *Polymer Lett.* **8**, 703-710

14. Dunn, A.S. (1971) The role of emulsifier in emulsion polymerization, *Chemistry & Industry* **4**, 1406-1412

15. Harada, M., Nomura, M., Eguchi, W., and Nagata, S. (1972) Generation mechanisms of polymer particles and operation of reaction in emulsion polymerization, *Kobunshi Kagaku* **29**, 844-851

16. Brooks, B.W. (1973) Particle nucleation rates in continuous emulsion polymerization reactors, *Br. Polym. J.* **5**, 199-211

17. Fitch, R.M. (1973) The homogenous nucleation of polymer colloids, *Br. Polym. J.* **5**, 467-483

18. Fitch, R.M. Shih, L-b. (1975) Emulsion polymerization: kinetics of radical capture by particles, *Progr. Colloid & Polymer Sci.* **56**, 1-11

19. Sütterlin,N., Kurth, H.-J., and Markert, G. (1976) Ein Beitrag zur Teilchenbildung bei der Emulsionspolymerisation von Acrylsäure- und Methacrylsäureestern, *Makromol. Chem.* **177**, 1549-1565

20. Laaksonen, J., Nurmi, K., and Stenius, P. (1978) Emulgatorfri emulsionspolymerisation - Del I. En översikt av teorin för emulsionspolymerisation, *Kemia-Kemi* **5**, 61-64

21. Goodwin, J.W., Ottewill, R.H., Pelton, R., Vianello, G., and Yates, D.E. (1978) Control of particle size in the formation of polymer latices, *Br. Polymer J.* **10**, 173-180

22. Hansen, F.K. and Ugelstad, J. (1978) Particle nucleation in emulsion polymerization. I. A theory for homogeneous nucleation, *J. Polym. Sci.: Polym. Chem. Ed.* **16**, 1953-1979

23. Hansen, F.K. and Ugelstad, J. (1979) Particle nucleation in emulsion polymerization. II. Nucleation in emulsifier-free systems investigated by seed polymerization, *J. Polym. Sci: Polym. Chem. Ed.* **17**, 3033-3045

24. Hansen, F.K. and Ugelstad, J. (1979) Particle nucleation in emulsion polymerization. III. Nucleation in systems with anionic emulsifier investigated by seeded and unseeded polymerization, *J. Polym. Sci.: Polym. Chem. Ed.* **17**, 3048-3067

25. Chen, C-y. and Piirma, I. (1980) Emulsion polymerization of acenaphthylene. I. A study of particle nucleation in water phase and in aqueous emulsions, *J. Polym. Sci.: Polym. Chem. Ed.* **18**, 1979-1993

26. Lichti, G., Gilbert, R.G., Napper, D.H. (1983) The mechanisms of latex particle formation and growth in the emulsion polymerization of styrene using the surfactant sodium dodecyl sulfate, *J. Polym. Sci.: Polym. Chem. Ed.* **21**, 269-291

27. Fitch, R.M., Palmgren, T.H., Aoyagi, T., and Zuikov, A. (1984) Kinetics of particle nucleation and growth in the emulsion polymerization of acrylic monomers, *Die Angew. Makromol. Chem.* **123/124**, 261-283

28. Fitch, R.M., Palmgren, T.H., Aoyagi, T., and Zuikov, A. (1985) Kinetics of particle nucleation and growth in the emulsion polymerization of acrylic monomers, *J. Polym. Sci.: Polym. Sump.* **72**, 221-224

29. Bogdanova, C.V., Solovjev, J.V., Eliseeva, V.I., and Zuikov, A.V. (1985) Mechanism of particle formation in emulsifier-free polymerizations, *Kolloidn. Zhurn.* (russ.) **4**, 781-782

64

30. Adhikari, M.S., Banerjee, M., Konar, R.S. (1985) Number of latex particles formed per unit volume of the aqueous phase at 100 per cent conversions in aqueous and emulsion polymerizations of vinyl monomers, *Polym. Commun.* **26**, 181-184

31. Carrà, S., Morbidelli, M., and Storti, G. (1985) Role of surfactants in emulsion polymerization, *Physics of Amphiphiles: Micelles, Vesicles, and Microemulsions*, **XC** Corso, Soc. Italiana di Fisica, 483-512

32. Feeney, P.J., Napper, D.H., and Gilbert, R.G. (1987) The determinants of latex monodispersity in emulsion polymerizations, *J. Coll. Interface Sci.* **118**, 493-505

33. Song, Z. and Poehlein, G.W. (1988) Particle formation in emulsion polymerization: transient particle concentration, *J. Macromol. Sci. - Chem.* **A25**, 403-443

34. Song, Z. and Poehlein, G.W. (1989) Particle formation in emulsifier-free aqueous phase polymerization of styrene, *J. Coll. Interf. Sci.* **128**, 501-510

35. Dunn, A.S. (1989) Latex particle nucleation in emulsion polymerization, *Eur. Polym. J.* **25**, 691-694

36. Song, Z. and Poehlein, G.W. (1989) Particle nucleation in emulsifier-free aqueous phase polymerization: stage 1, *J. Coll. Interf. Sci.* **128**, 486-500

37. Lock, M.R., El-Aasser, M.S., Klein, A., and Vanderhoff, J.W. (1990), Role of itaconic acid in latex particle nucleation, *J. Appl. Polym. Sci.* **42**, 1065-1072

38. Whang, B.C.Y., Ballard, M.J., Napper, D.H., and Gilbert, R.C. (1991) Molecular weight distributions in emulsion polymerization: evidence for coagulative nucleation, *Aust. J. Chem.* **44**, 1133-1137

39. Boieshan, V. and Levitsch, D. (1991) Formation of latex particles during the continuous emulsion polymerization of vinyl chloride, *Acta Polymerica* **42**, 551-553

40. Guo, J.S., Sudol, E.D., Vanderhoff, J.W. and El-Aasser, M.S. (1992) Particle nucleation and monomer partitioning in styrene o/w microemulsion polymerization, *J. Polym. Sci.: Part A: Polym. Chem.* **30**, 691-702

41. Hergeth, W.-D., Lebek, W., Stettin, E., Witkowski, K. and Schmutzler, K. (1992) Particle formation in emulsion polymerization, 2 Aggregation of primary particles, *Makromol. Chem.* **193**, 1607-1621

42. Kobiakova, K.O., Gromov, V.F. and Teleshov, E.N. (1993) Formation of polymer particles in the process of polymerization of acrylamide in water-cyclohexane emulsions, *Vysokomol. Soedin.* (russ.) **35**, 10-14

43. Snuparek, J., Branda, P., Mrkvickova, L., Lednicky, F. and Quadrat, O. (1993), Effect of coagulative mechanism of particle growth on the structural heterogeneity of ethyl acrylate - methacrylic acid copolymer latex particles, *Collect. Czech. Chem. Commun.* **58**, 2451-2457

44. Schlüter, H. (1993) Theory of colloid stability and particle nucleation kinetics in emulsion polymerization, *Colloid & Polym. Sci.* **271**, 246-252

45. Shen, S., Sudol, E.D., and El-Aasser, M.S. (1994) Dispersion polymerization of methyl methacrylate: mechanism of particle formation, *J. Polym. Sci.: Part A: Polym. Chem.* **32**, 1087-1100

46. Bleger, F., Murthy A.K., Pla, F., and Kaler, E.W. (1994) Particle nucleation during microemulsion polymerization of methyl methacrylate, *Macromolecules* **27**, 2559-2565

47. Fengler, S. and Reichert, K.-H. (1995) Zur Kinetik und Teilchenbildung der Fällungspolymerization von Acrylsäure in Gegenwart von Polystyrol-block-Polyethylenoxid-Copolymeren, *Die Angew. Makromol. Chem.* **225**, 139-152

48. Tauer, K. and Kühn, I. (1995) Modeling particle formation in emulsion polymerization: An approach by means of the classical nucleation theory, *Macromolecules* **28**, 2236-2239

49. Kühn, I. and Tauer, K. (1995) Nucleation in emulsion polymerization: a new experimental study. 1. Surfactant-free emulsion polymerization of styrene, *Macromolecules* **28**, 8122-8128

50. Tauer, K., Kühn, I. and Kaspar, H. (1996) Some colloid chemical features of emulsion polymerization, *Progr. Colloid Polym. Sci.* **101**, 30-37

51. Fitch, R.M. (Ed.) (1971) *Polymer Colloids*, Plenum Press, New York-London

52. Barrett, K.E.J. (Ed.) (1975) *Dispersion Polymerization in Organic Media*, John Wiley & Sons, London, New York, Sydney, Toronto

53. Blackley, D.C. (1975) *Emulsion Polymerization Theory and Practice*, Applied Science Publishers Ltd., London

54. Becher, P. and Yudenfreund, M.N. (Eds.) (1978) *Emulsion, Latices, and Dispersions*, Marcel Dekker, Inc., New York and Basel

55. Eliseeva, V.I. (1980) *Polymer Dispersions* (russ), Chimia, Moscow

56. Fitch, R.M. (Ed.) (1980) *Polymer Colloids II*, Plenum Press, New York and London

57. Fitch, R.M. (1981) Latex Particle Nucleation and Growth in D.R. Bassett and A.E. Hamielec (eds.), *Emulsion Polymers and Emulsion Polymerization*, ACS Symposium Series No. 165, ACS, Washington, pp. 1-29
58. Piirma, I. (Ed.) (1982) *Emulsion Polymerization*, Academic Press, London, San Diego, New York, Boston, Sydney, Toronto
59. Gilbert, R.G. (1995) *Emulsion Polymerization A Mechanistic Approach*, Academic Press, London, San Diego, New York, Boston, Sydney, Toronto
60. Binder, K. (1991) Spinodal Decomposition in R.W. Cahn, P. Haasen, and E.J. Kramer (Eds.) *Material Science and Technology*, Vol. 5, VCH Verlagsgesellschaft mbH, Weinheim, pp. 405-471
61. Everett, D.H. (1988) *Basic Principles of Colloid Science*, The Royal Society of Chemistry, London
62. Nielsen, A.E. (1964) *Kinetics of precipitation*, Pergamon Press, Oxford, London, Edinburgh, New York, Paris, Frankfurt
63. Morton, M., Kaizerman, S. and Altier, M.W. (1954) Swelling of latex particles, *J. Coll. Sci.* **9**, 300-312
64. Gardon, J.L. (1968) Emulsion polymerization. VI. Concentration of monomers in latex particle, *J. Polym. Sci.* **A-1 6**, 2859-2879
65. Gündüz, S. and Dincer, S. (1980) Solubility behaviour of polystyrene: thermodynamic studies using gas chromatography, *Polymer* **21**, 1041-1045
66. Shi, F. and Seinfeld, J.H. (1994) Nucleation in the pre-coalescence stages: universal kinetic laws, *Mater. Chem. Phys.* **37**, 1-15
67. Morrison, B.R. and Gilbert, R.G. (1995) Conditions for secondary particle formation in emulsion polymerization systems, *Macromol. Symp.* **92**, 13-30
68. Schmutzler, K., Kakuschke, R., and Hergeth, W.-D. (1989) Untersuchungen zur Herausbildung der Molmassenverteilung bei der Emulsionspolymerisation von Vinylacetat, *Acta Polymerica* **40**, 238-242
69. Behrmann, E.J. and Edwards J.O. (1980) The thermal decomposition of peroxidisulfate ions, *Rev. Inorg. Chem.* **2**, 179-206
70. House, D.A. (1962) Kinetics and mechanism of oxidations by persulfate, *Chem. Rev.* **62**, 185-203
71. Kühn, I. and Tauer, K. (1996), to be published

PARTICLE SIZE DISTRIBUTIONS

EMMA M. COEN AND ROBERT G. GILBERT

School of Chemistry,
University of Sydney
NSW 2006,
Australia

1. Introduction

Particle sizes and particle size distributions are of both technical and scientific importance in emulsion polymerization systems. On the technical side, particle sizes and particle size distributions are major determinants of a number of properties of a polymer latex, such as its viscoelastic behavior. Scientifically, experimental PSDs can furnish sensitive tests of mechanistic assumptions. Given the basic mechanisms and component rate parameters, particle size distributions (PSDs) can be predicted (as discussed in section 2 of this article). The problem is to find the values of those rate parameters. The major unknowns are those involved in particle formation, and for this reason part of this chapter examines the effect of nucleation on the PSD both experimentally and theoretically. While physically reasonable parameter values can be found which can reproduce experiment (e.g., the dependence of the particle number on initiator and surfactant concentrations, and the PSD during Interval I), such models are not yet reliably predictive. For cases where these rate coefficients have been established independently, calculations of the PSD in the absence of new particle formation give good accord with experiment. Moreover, appropriate comparison between theory and experimental PSDs can be used to infer kinetic information such as the size dependence of the entry rate coefficient.

2. Models for predicting PSD

2.1. PSDS IN ZERO-ONE SYSTEMS AND IN INTERVAL I

The correct evolution equations for particle formation and growth, which governs the PSD in an *ab initio* emulsion polymerization, are one of the major unsolved fundamental problems in the field of polymer colloids. There is no shortage of evolution equations in the literature — indeed creating models for emulsion polymerizations is effectively a cottage industry. The problem is what are the *correct* equations, and what are the correct values of the parameters that should go into them. What is definitely known is that the nucleation event is a complex one.

J. M. Asua (ed.), Polymeric Dispersions: Principles and Applications, 67–78.
© 1997 *Kluwer Academic Publishers. Printed in the Netherlands.*

68

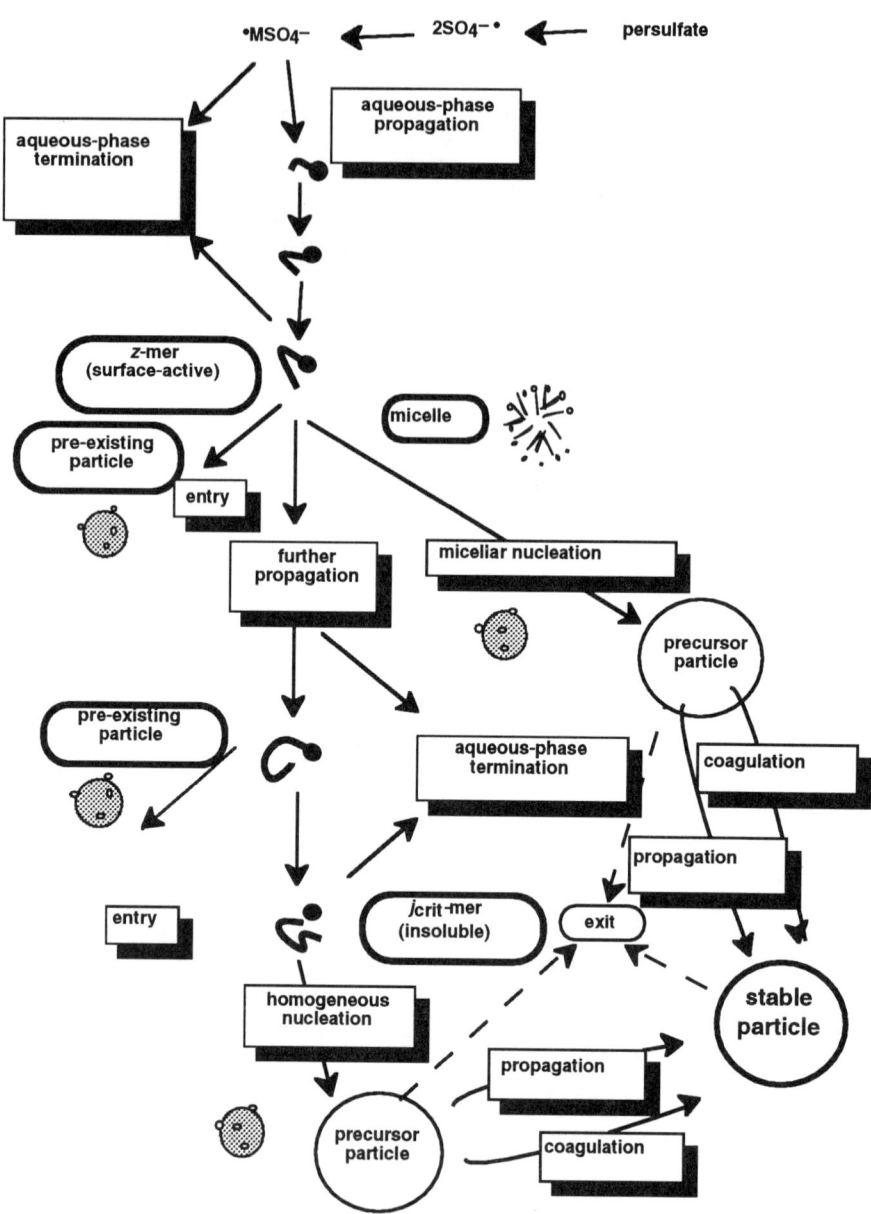

Figure 1. The events governing particle formation and the PSD in Interval 1.

Figure 1 shows one of the more complete reaction schemes for particle formation. A charged moiety formed by initiator decomposition propagates with monomer in the aqueous phase to form an oligomeric radical. This radical can have one of many fates. It may be terminated in the aqueous phase, giving a dead oligomer. When the degree of polymerization is sufficiently high that the species is surface-active (this degree of polymerization is denoted z), the radical may either enter a pre-existing particle or a micelle, if micelles are present; this creates a particle by *micellar* nucleation. Alternatively, it may grow further to a degree of polymerization j_{crit}, when it may undergo *homogeneous* nucleation. *Precursor* particles formed by either mechanism grow by propagation, but are also colloidally unstable, and so may undergo *coagulation* ("precursor particle" in this context meaning colloidally unstable). Particles grow by propagation if a radical is contained within the particle, or by coagulation. When a particle is sufficiently large, it becomes colloidally stable. Nucleation ceases when all newly-formed aqueous phase radicals undergo either entry into a pre-existing particle or aqueous-phase termination rather than forming a new particle by either mechanism. At the same time, exit (desorption) can occur. During the nucleation period, when particles are extremely small, it is reasonable to assume that the system obeys zero-one kinetics.

Previous studies [1,2] have shown that, once what has been established as reliable models for entry and exit are included, certain features of the PSD and of particle formation (including the general shape of the PSD, the nucleation time, the quantitative dependence of N_p on surfactant concentration and on ionic strength) cannot be reproduced without taking coagulation of precursors into account, *even above the cmc.*

The evolution equations for the PSDs are now given. These will be presented as *volume* distributions, $n(V,t)$, the particle distribution function at time t and *unswollen* volume V. These evolution equations are simpler in terms of unswollen volume than those in terms of unswollen radius r_0 or swollen radius r_p, because a growing particle increases uniformly in unswollen volume. However, equivalent relations in terms of the other variables exist, and indeed technically it is easier to develop numerical solutions in terms of the radius distribution. The various distributions are related by:

$$n_V(V,t) = \frac{n_r(r_0,t)}{4\pi r_0^2} \tag{1}$$

Swollen and unswollen size are related by:

$$\frac{r_P}{r_0} = \left(\frac{\rho_m}{\rho_m - [M]_p M_0}\right)^{1/3} \tag{2}$$

where ρ_m and ρ_p are the densities of monomer and polymer, and M_0 is the molecular weight of monomer. The average unswollen radius at complete conversion, $<r_0>$ and the particle number are trivially related by:

$$N_p = \frac{\text{total mass monomer}}{\frac{4}{3}\pi <r_0>^3 \rho_p} \tag{3}$$

2.2. ZERO-ONE KINETICS

The evolution of the PSD in a zero-one system are as follows (e.g., [1]). Since particles can either contain zero or one radicals, the total particle size distribution is the sum of the PSDs for particles that contain zero radicals, $n_0(V,t)$ and particles that contain one radical. Because a monomeric radical (formed by transfer) can undergo desorption, it is necessary to sub-divide the particles containing one radical into those where this radical is monomeric, n_1^M, and where it is polymeric, n_1^P:

$$n(V) = n_0(V) + n_1^M(V) + n_1^P(V) \tag{4}$$

These are normalized so that the number of particles per unit volume of the aqueous phase is given by:

$$\frac{N_p}{N_A} = \int_0^\infty \left(n_0(V,t) + n_1^P(V,t) + n_1^M(V,t) \right) dV \equiv \int_0^\infty n(V) dV \tag{5}$$

The PSD evolution equations for zero-one systems are, from Figure 1:

$$\frac{\partial n_0(V,t)}{\partial t} = \rho \left[n_1^P + n_1^M - n_0 \right] + k_{dM} n_1^M - n_0 \int_0^\infty B(V,V') \left[n_0(V') + n_1^P(V') \right] dV'$$

$$+ \int_0^\infty B(V', V-V') \left[n_0(V') n_0(V-V') + n_1^P(V') n_1^P(V-V') \right] dV' \tag{6}$$

$$\frac{\partial n_1^P(V,t)}{\partial t} = \rho_{initiator}(V) n_0 - \rho(V) n_1^P - k_{tr}[M]_p n_1^P - \frac{\partial}{\partial V} K n_1^P + k_p^1[M]_p n_1^M$$

$$+ \delta(V - V_0) \left([IM_{j-1}] k_{p,w}[M]_w + \sum_{i=z}^{j-1} [IM_i] k_{a,micelle}^i [micelle] \right)$$

$$- n_1^P \int_0^\infty B(V,V') \left[n_0(V') + n_1^P(V') \right] dV'$$

$$+ \int_0^\infty B(V', V-V') \left[n_0(V') n_1^P(V-V') + n_1^P(V') n_0(V-V') \right] dV' \tag{7}$$

$$\frac{\partial n_1^M(V,t)}{\partial t} = -\left(\rho + k_p^1[M]_p + k_{dM} \right) n_1^M + k_{aE}[E] n_0 + k_{tr}[M]_p n_1^P \tag{8}$$

The symbols are those defined elsewhere in this book [3], with some additional terms: $B(V,V')$ is the rate coefficient for coagulation between particles of volume V and V', V_0 is the volume at which particle are deemed to form, $k_{a,micelle}^i$ is the rate coefficient for capture of an i-mer by micelles, the concentration of which is [micelle], and the rate coefficient for propagational growth is given by:

$$K(V) = \frac{k_p M_0 [M]_p}{N_A \rho_p} \tag{9}$$

One also has the aqueous-phase concentrations of radicals, both desorbed (E) and arising from initiator ([IM$_i$], where i is the degree of polymerization):

$$[IM_1] = \frac{2 f k_i [I_2]}{k_{p,w}^1[M]_w + k_{t,w}[R]_w} \tag{10}$$

$$[IM_i] = \frac{k_{p,w}^{i-1}[IM_{i-1}][M]_w}{\int_0^\infty k_{aI}^i(V) n(V) dV + k_{a,micelle}^i [micelle] + k_{p,w}^i[M]_w + k_{t,w}[R]_w} \tag{11}$$

$$[R]_w = \sum_i [IM_i] + [E] \qquad (12)$$

$$[E] = \dfrac{\displaystyle\int_0^\infty k_{dM}(V)n_1^M(V)\,dV}{\displaystyle\int_0^\infty k_{aE}(V)n(V)\,dV + k_{t,w}[R]_w} \qquad (13)$$

The various entry rate coefficients are given by:

$$\rho(V) = \rho_{initiator}(V) + k_{aE}(V)[E] \; ; \; \rho_{initiator} = \sum_{i=z}^{j-1} k_{aI}^i(V)[IM_i] \qquad (14)$$

If there is no particle formation, the PSD evolution equations simplify greatly, both because the particle formation terms are absent and because the system will normally be colloidally stable, so that coagulation does not occur.

2.3. PSEUDO-BULK KINETICS

In the pseudo-bulk case, there is no limit on the number of radicals that may be present in a particle and termination is no longer instantaneous. Since the very small particle sizes that are present in nucleation mean that termination is extremely rapid, it is safe to assume that all nucleating particles will follow zero-one kinetics (although it is important to realize that systems with low $\bar{n}$ can follow pseudo-bulk kinetics: see [1,3,4]). In pseudo-bulk systems, however, it is reasonable to assume that the particles are colloidally stable, and hence both coagulation and particle formation can be ignored. The evolution equations are:

$$\dfrac{\partial n(V,t)}{\partial t} = -\dfrac{\partial\left(K\bar{n}\, n(V,t)\right)}{\partial V} \qquad (15)$$

The time evolution of $\bar{n}$ requires that, because termination is rate-determining, chain-length-dependent termination rate coefficients must be taken into account. Detailed expressions for this have been given elsewhere (e.g., [1,5]).

Numerical solutions to both the zero-one and pseudo-bulk PSD equations can be obtained by a number of techniques. Here, the finite-difference method is used, converting these relations into coupled first-order differential equations; the variable is changed from V to r_0; see [1] for details. The reason that r is used is because this greatly reduces the number of equations, since if one chooses an even volume increment, the number of equations in terms of V goes as the cube of the maximum volume.

2.4. MODELS FOR RATE PARAMETERS

Details of models and expressions for the various rate parameters in the PSD evolution equations have been given elsewhere [1]. The micellar and particle entry rate coefficient are obtained from the Smoluchowski equation:

$$k_{aI}^i(V) = \begin{cases} 0, & i < z \\ 4\pi r_P D_w N_A / i, & i \geq z \end{cases} \tag{16}$$

$$k_{aE}(V) = 4\pi r_P D_w N_A \tag{17}$$

$$k_{a,\text{micelle}}^i = \begin{cases} 0, & i < z \\ 4\pi D_w\, r_{\text{micelle}}\, N_A / i, & i \geq z \end{cases} \tag{18}$$

Micelle concentration is obtained as suggested by Giannetti [6]. The coagulation rate coefficient $B(V,V')$ is calculated using DLVO theory with the Healy-Hogg model [7] (see, e.g., [8,9]), and is dependent primarily on the surface charge densities and swollen volumes of the two particles; detailed expressions have been given elsewhere [8], with the following minor changes. The surface charge density is the sum of the contributions from adsorbed and generated surfactant. All ionic end groups that were created through the decomposition of initiator are assumed to be adsorbed onto the surface of a particle: a reasonable assumption for small particles. The expression for the surface charge density from monovalent ionic end groups is:

$$\sigma_I = e N_I / A_P \tag{19}$$

where A_P is total area of all particles in the system, e the charge on the electron and N_I is the total number of ionic moieties that have been released through initiator decomposition. The contribution from added surfactant is calculated assuming a Langmuir adsorption isotherm:

$$\sigma_s = e / A_s \tag{20}$$

where:

$$A_s = a_s \left(1 + \frac{1}{[S_w] b_s}\right) \tag{21}$$

The Langmuir adsorption isotherm parameters are a_s (the area occupied by a single surfactant molecule) and b_s (which is related to the free energy of adsorption). The amount of surfactant in the aqueous phase is given by:

$$[S_w] = [S_T] - \frac{A_P}{N_A A_s} \tag{22}$$

Here $[S_T]$ is the concentration that added surfactant would have if there were no adsorption.

$B(V,V')$ depends on the relative particle sizes, as shown in Figure 2. Large particles coagulate more readily with small particles, than with particles of the same relative size. This may account for the skewness observed in early-time particle size distributions (when expressed in terms of volume) [10], as discussed in section 4.

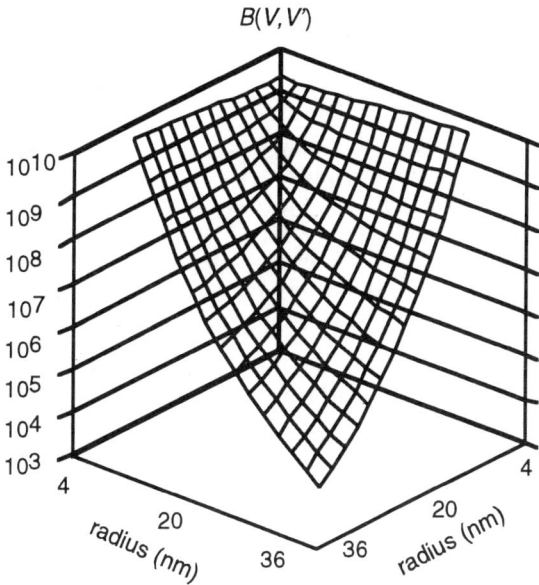

Figure 2. Calculated dependence of coagulation rate coefficient (dm^3 mol^{-1} s^{-1}) on unswollen radius of the two particles, from the Healy-Hogg model.

3. Experimental Methods of Measuring PSD

Many methods are available for measuring *average* particle size. Until recently, the only reliable method for determining particle size *distributions* was calibrated transmission electron microscopy. However, this technique is difficult, time consuming and hard to apply to some polymers which have a tendency to melt under the beam (e.g. MMA and vinyl acetate).

A more convenient method of determining particle size distributions is by Capillary Hydrodynamic Fractionation, which can give full particle size distributions for particles between 15 and 1100 nm in diameter. Values for average radius are typically within 5% of those achieved through electron microscopy [11]. Variations in the calibration curve due to temperature or variations in the conductivity of the eluent can be detected and corrected by obtaining a PSD for a sample of known size. Under typical operating conditions, CHDF is sensitive to secondary populations that differ in diameter by 10% of the diameter of the primary population. Sensitivity to small populations of secondary particles can be checked by adding small amounts of a second latex of known size. The time required to generate a full particle size distribution is usually 15 min. However, CHDF requires dilution to low solids content. Another limitation is that the effect of change in, e.g., the pH of eluent on systems that are suspected to be sensitive to such changes cannot be measured by this method, since the detection method is sensitive to changes in eluent conductivity.

4. Experiment Results

4.1. PSD IN INTERVALS 2 AND 3

It has been shown [1,12] that values of rate coefficients obtained by *kinetic* means (see elswehere in this book [3]) successfully predict the time evolution of the *particle size distribution* in Interval 2, as illustrated in Figure 3. This task is relatively straightforward, because all that is being examined is the *change* in size and polydispersity in a pre-existing PSD. Nevertheless, it is one of the few occasions in emulsion polymerization where one type of measurement is able to successfully predict a completely different measurement without any adjustable parameters!

One particularly useful aspect of the time evolution of the PSD in a zero-one Interval 2 system is furnished by *competitive growth* experiments [13,14], the data comprising the PSDs of the two components of a bimodal seed. These experiments can, with careful data interpretation [15] that takes into account the fate of exited free radicals, lead to information on the dependence of entry rate coefficients on particle size (earlier work [13,14] did not take these fates into account, and is therefore in significant error). These data support (although do not prove!) the applicability of the Smoluchowski expression for the entry rate coefficient of an oligomeric species, eq (16).

There seem to have been no systematic studies of the comparison between models and theories for the PSD in a *pseudo-bulk* system, of the type illustrated in Figure 3. Such an exercise is not likely to bring any surprises, and for that reason is certainly worth performing.

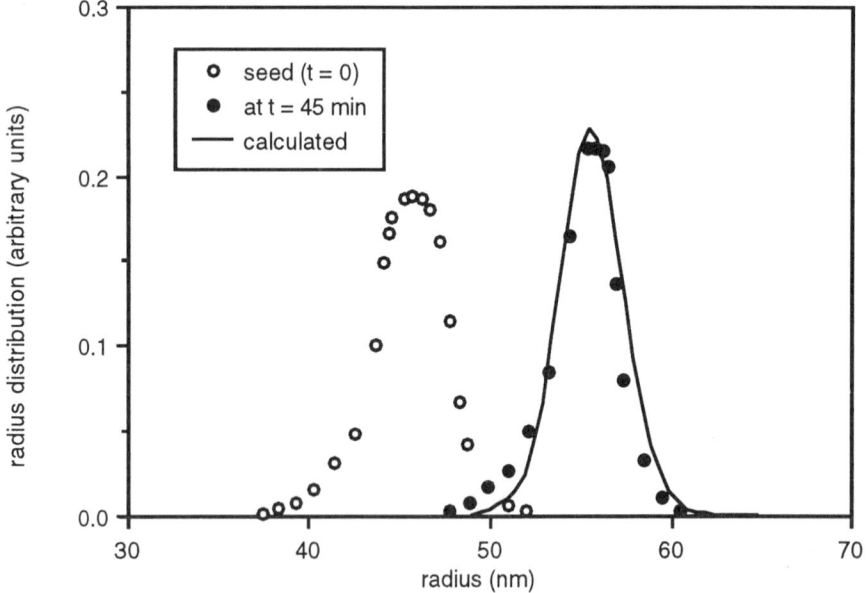

Figure 3. Comparison of measured and calculated radial PSD after 45 min growth in seed emulsion polymerization of styrene; experimental data from [12]. Calculated PSD from values of entry and exit rate coefficients measured from experimental conversion/time data.

The PSD in Interval 3 is trivially obtained from a knowledge of conversion (since virtually all monomer is contained within the particles).

4.2. SECONDARY PARTICLE FORMATION

Models for predicting particle size distributions are of great use in predicting conditions for secondary nucleation. In addition, an important test for the validity of a model is the ability to predict the initial conditions that will cause secondary nucleation.

The condition for the creation of a secondary population is that entry of newly-formed radicals into particles be slow enough so that the competing process of particle formation is significant. The PSD-based model of eqs (6–14) and section 2.4 can be used to predict the amount of secondary nucleation in a given system, as now exemplified (this illustration being an extension of the application of a simpler version given previously [16]).

The system used for this study uses data [16] for the number of new particles formed in a seeded emulsion polymerization of styrene at 50 °C over a range of N_p values, with $[I_2] = 1 \times 10^{-3}$ mol dm^{-3} persulfate, sodium dodecyl sulfate = 8×10^{-4} mol dm^{-3}, which is below the cmc, and seed particle swollen radius = 44 nm. Zero-one kinetics hold under these conditions. Results are shown in Figure 4. It can be seem that the observed number of new particles is reproduced excellently by the model, without any of the parameters being adjusted. This supports the physical assumptions and parameter values in the model, at least below the cmc.

4.3. PSD IN INTERVAL 1.

While the evolution of the PSD of a pre-formed seed is relatively easy to predict from measured or modelled entry and exit rate coefficients, predicting the size and size distribution of the particles formed in an non-seeded (*ab initio*) polymerization is much more difficult, because there are so many more rate parameters. There are two aspects to such a test: quantitative (do the model calculations accurately predict N_p and the full PSD?) and qualitative (does the model correctly predict trends such as dependences on initiator and surfactant concentrations and the qualitative shape of the PSD?). Success in qualitative prediction supports the correctness of the underlying physical assumptions, while correct quantitative prediction supports the full details of the model.

A comparison between theory and observation is for the PSD, N_p and $\bar{n}$ observed in experiments [17] for the time evolution of the PSD just after the cessation of particle formation in styrene. As such, the data should be sensitive to nucleation kinetics. The original data interpretation [17] was without the extensive mechanistic information subsequently obtained on entry and exit, and is therefore regarded as superseded. The experiment used 2.38×10^{-2} mol dm^{-3} sodium dodecyl sulfate, which is above the cmc, 1.29×10^{-2} mol dm^{-3} potassium persulfate, at 50 °C. The PSD was measured 6 min after the end of the inhibition period, when particle formation was thought to be complete (the model however predicts that particle formation ceases at *ca.* 15 min after polymerization commences). PSDs were obtained using electron microscopy.

76

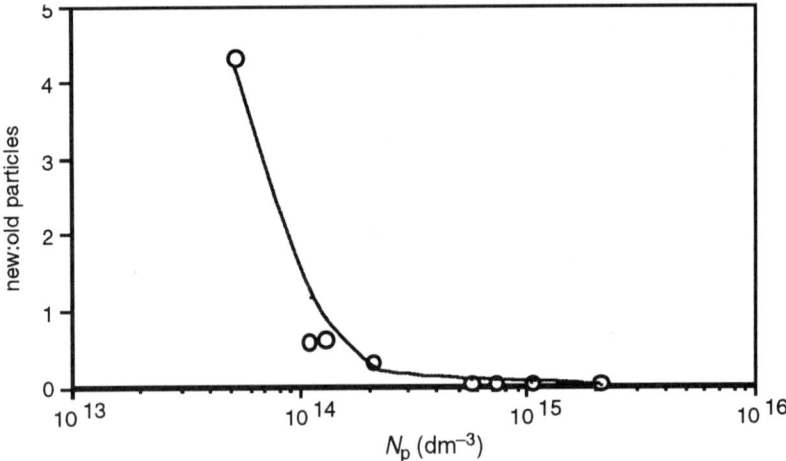

Figure 4. Predicted (lines) and observed (points) ratio of number of new to old particles in styrene emulsion polymerization; conditions given in text. Experimental data re-processed from [16].

Figures 5 and 6 show the observed and calculated radius and volume distributions. The calculations used the zero-one PSD description, eqs (6–14). Parameter values being given in the figure caption; the calculation used a simplified model for $B(V,V')$, whereby coagulation occurs at the same rate coefficient B for interactions between any particle and a particle with a swollen radius not exceeding a chosen value. The particle number was poorly predicted: an order of magnitude less than the experimental value of 3.6×10^{18} dm^{-3}. Comparison of the PSDs shows that qualitative features are reproduced, particularly the skewness in the early-time volume distribution. Accord between model and experiment for N_p could be obtained by parameter adjustment (e.g., by making the saturation value of $[M]_p$ dependent on radius, as predicted by the Morton relation [18]); this curve-fitting exercise is not undertaken here. Moreover, these experiments, being performed without thorough deoxygenation, may have had considerable inhibitor present which is known to lead to an increase in particle number (e.g., [19]), because it takes longer for precursor particles to become sufficiently large to capture all newly-formed radicals in preference to their forming new particles. Indeed, the presence of oxygen can lead to highly variable N_p unless de-gassing is carefully controlled [20]. More extensive data of the type shown in Figure 5, under carefully controlled conditions, will be required before truly reliable parameter values can be obtained.

5. Conclusions

In modelling the evolution of particle size distribution, it is critical that all of the mechanisms involved are considered. The PSD is governed by particle formation, while its

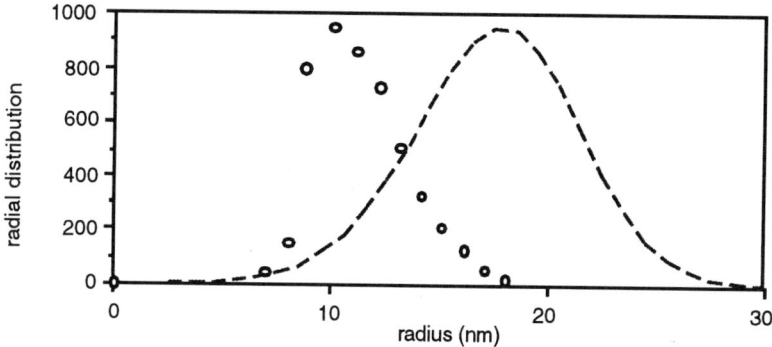

Figure 5. Calculated (lines) and observed (points) radius PSDs (arbitrary units) for styrene emulsion polymerization with SDS surfactant above the cmc, 6 min after the end of the inhibition period. Parameter values in calculations: $D_w = 1.5 \times 10^{-5}$ cm^2 s^{-1}, $z = 3$, $j_{crit} = 5$, $r_{micelle} = 2.6$ nm, $k_i = 7.2 \times 10^{-7}$ s^{-1}, $k_{p,w}^n$ (dm^3 mol^{-1} s^{-1}) = 1200 ($n = 1$), 280 ($n = 2$) and 260 ($n \geq 3$), k_p^1, k_p (dm^3 mol^{-1} s^{-1}) = 2080, 260, $k_{tr} = 2.9 \times 10^{-2}$ dm^3 mol^{-1} s^{-1}, $k_{t,w}^{1,1} = 8 \times 10^9$ dm^3 mol^{-1} s^{-1}, remaining $k_{t,w}$ from Smoluchowski equation, $a_s = 43$ Å^2, $n_{agg} = 162$, [cmc] = 3.9×10^{-3} mol dm^{-3}, [M]$_p = 5.8$ mol dm^{-3} (assumed independent of particle size), $B(V,V') = 10^9$ dm^3 mol^{-1} s^{-1} if either particles has $r_p \leq 6$ nm, = 0 otherwise. See [1,6] for explanation of parameter values.

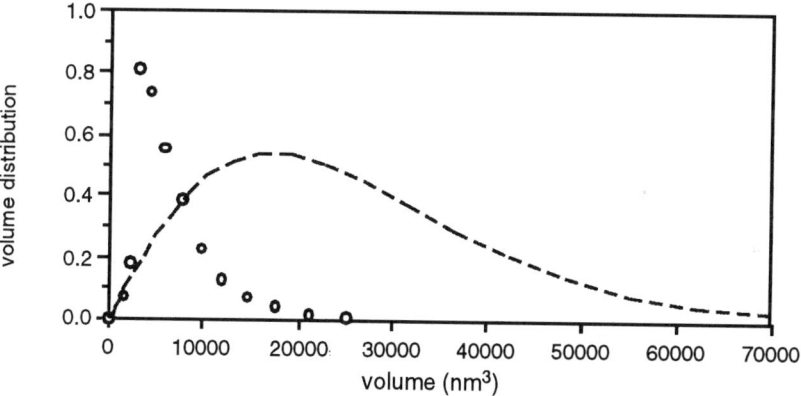

Figure 6. Data of Figure 5 plotted as volume distributions. Note positive skewness of the early-time distributions.

subsequent evolution is readily calculated from what are now known to be reliable models for entry and exit (testing of theory against experimental PSDs in termination-dominated systems is yet to be carried out). The model can be used in a wide variety of situations, including predicting where secondary particle formation occurs. Proper quantification requires coagulation of precursors to be included, even above the cmc.

There are many situations where the parameter values are uncertain, requiring further experiment. Since technologies are being developed that result in easily measured full particle distributions, a better understanding of the kinetics of particle formation and evolution can be expected. This understanding will be of great benefit in intelligent design of both experiment and industrial applications.

Acknowledgments The support of BASF, and of helpful interactions with Drs Brad Morrison and Dieter Distler, are gratefully acknowledged.

78

References

[1]   Gilbert, R.G. (1995) *Emulsion Polymerization: A Mechanistic Approach*,Academic, London.

[2]   Gilbert, R.G. (1997) in: *Emulsion Polymerization and Emulsion Polymers* (eds. P.A. Lovell and M.S. El-Aasser) Wiley, London.

[3]   Gilbert, R.G. (1997) in: *Polymeric Dispersions. Principles and Applications* (eds. J.M. Asua) p.1, Kluwer Academic, Dordrecht.

[4]   Ballard, M.J., Napper, D.H. and Gilbert, R.G. (1984)*J. Polym. Sci., Polym. Chem. Edn., **22**, 3225.

[5]   Scheren, P.A.G.M., Russell, G.T., Sangster, D.F., Gilbert , R.G.and German, A.L. (1995) *Macromolecules, **28**, 3637.

[6]   Giannetti, E. (1993) *A.I.Ch.E. Journal, **39**, 1210.

[7]   Hogg, R., Healy, T.W. and Furstenau, D.W. (1966) *Trans. Faraday Soc., **62**, 1638.

[8]   Richards, J.R., Congalidis, J.P. and Gilbert, R.G. (1989) *J. Appl. Polym. Sci., **37**, 2727.

[9]   Feeney, P.J., Napper, D.H and Gilbert, R.G. (1987) *Macromolecules, **20**, 2922.

[10]   Lichti, G., Gilbert, R.G. and Napper, D.H. (1983) *J. Polym. Sci. Polym. Chem. Edn., **21**, 269.

[11]   Miller, C.M., Sudol, E.D., Silebi, C.A. and El-Aasser, M.S. (1995) *J. Colloid Interface Sci., **172**, 49.

[12]   Lichti, G., Hawkett, B.H.,Gilbert, R.G., Napper, D.H. and Sangster, D.F. (1981) *J. Polym. Sci., Polym. Chem. Edn., **19**, 925.

[13]   Vanderhoff, J.W., Vitkuske, J.F., Bradford, E.B. and Alfrey, T. (1956) *J. Polym. Sci., **20**, 265.

[14]   Ugelstad, J., El-Aasser, M.S. and Vanderhoff, J.W. (1973) *J. Polym. Sci., Polym. Letters Edn., **11**, 503.

[15]   Morrison, B.R., Maxwell, I.A., Gilbert, R.G. and Napper, D.H. (1992) in: *ACS Symp. Series - Polymer Latexes - Preparation, Characterization and Applications* (eds. E.S. Daniels, E.D. Sudol and M. El-Aasser) p.28, American Chemical Society, Washington D.C.,

[16]   Morrison, B.R. and Gilbert, R.G. (1995) *Macromol. Symp., **92**, 13.

[17]   Feeney, P.J., Napper, D.H. and Gilbert, R.G. (1987) *J. Colloid Interface Sci., **118**, 493.

[18]   Morton, M., Kaizerman, S. and Altier, M.W. (1954) *J. Colloid Sci., **9**, 300.

[19]   Pearson, L.T., Louis, P.E.J., Gilbert, R.G. and Napper, D.H. (1991) *J. Polym. Sci., Polym. Chem. Edn., **29**, 515.

[20]   Kühn, I. and Tauer, K. (1995) *Macromolecules, **28**, 8122 .

NETWORK FORMATION IN FREE-RADICAL EMULSION POLYMERIZATION

D.CHARMOT
Rhône Poulenc Recherches
52, rue de la Haie-Coq
93308 Aubervilliers Cedex - France

1. Introduction

The share of thermoplastic polymer produced by radical polymerization in water dispersion is increasing steadily. This move towards water borne systems is obviously driven by the environmental needs to reduce the use of hydrocarbons in the manufacturing or polymer processing sites, and for the final customer to avoid exposure to solvents. Besides, emulsion polymerization technology enjoys a number of advantages : it is usually a kinetically fast reaction leading to high molecular weight polymeric materials in high solids content water dispersion ; the viscosity of which remains in the low to medium range, making temperature control during process relatively easy. Compared with homogeneous radical polymerization, the mechanisms of emulsion polymerization are still unclear as far as the reaction proceeds, and the level of predictability remains poor in real systems when the number of monomers, chain transfer agents and the range of operational variables is high. This is particularly true for the development of molecular weight distribution (MWD) in emulsion polymers produced at high conversions. The kinetic models for linear polymerization become increasingly complicated as one approaches typical industrial recipes, requiring a large number of kinetic parameters which cannot be determined independently : thus in practice these unknown rate constants are estimated by fitting measurable characteristics such as MWD to a given model. In non linear polymerization, i.e. systems where the average functionality of the monomers exceeds 2, additional events like pendant or terminal double bond polymerization and long chain branching due to chain stoppage by radical transfer to polymer, make the kinetic scheme even more difficult to handle and the mathematics so sophisticated that we may lose the actual physics which lies behind it. Non-linear polymerizations however are most encountered in commercial polymers produced in emulsions: e.g. vinyl-divinyl systems such as butadiene homo and copolymers, polychloroprene and branched polymer such as polyvinylacetate and certain polyalkylacrylates and ethylene copolymers. In a recent work, Gilbert [1] pointed out that relatively little properly characterized experimental data on MWD that enables thorough quantitative analysis, were available. They are even fewer in the field of non-linear polymers of commercial interest such as those mentioned above. Moreover important features of non-linear polymers such as crosslinks or branching densities are not obtained through direct titration of chain connection points, but rather from model dependant methods like swelling or dynamic mechanical properties. However, the measurement of the fraction of insoluble polymer, or gel content, is a powerful but

J. M. Asua (ed.), Polymeric Dispersions: Principles and Applications, 79–96.
© 1997 *Kluwer Academic Publishers. Printed in the Netherlands.*

simple method to assess the level of connectivity of the polymer. In the first part of this review, we will comment on some of the available techniques on gel fraction determination, as well as on its role on latex film formation and the mechanical properties that are derived therefrom. In the second part, the main kinetic models applied to non-linear emulsion polymerization will be introduced.

2. Molecular Weight Distribution and Crosslinking in Relation with Film Toughness.

A certain amount of recent work has clearly shown that the mechanical properties of a latex film are not reached until the polymeric chains in neighbouring particles have interdiffused across the interfaces to a depth which is equivalent to the radius of gyration of a polymer chain situated at the interface [16-19]. SANS measurements performed in parallel with stress-strain experiments on polybutylmethacrylate latexes [15] showed unambiguously that the steep increase in fracture energy corresponded to the onset of the interdiffusion process : after a 5 min annealing at 90°C the interpenetration depth is 2 nm which compares well to 3 nm, the mean molecular dimension for chain entanglement. With increasing divinyl monomer contents (Methallylmethacrylate) this interdiffusion process is more and more hindered, and when M_c, the molecular weight between chemical crosslinks is lower than M_e, the entanglement length, the films remain extremely brittle after annealing. It is highly likely that the low molecular weight fraction of the MWD is responsible for this fast rise of film toughness at short annealing time. In the latter system, surfactant was used to stabilize the polymer emulsion, wheras in commercial systems, copolymerization of hydrophilic monomers is commonly used to impart stability as well as to obtain better adhesion properties towards "difficult" substrates. It results in hydrophilic shell/hydrophobic core honeycomb-like latex film morphologies [20]. In a series of poly-butylmethacrylate-co-acrylic acid latexes, Winnik et al. [21] showed by fluorescence non radiative energy transfer, that the mean apparent diffusion coefficient of the polymer increased with (T-Tg), where T is the annealing temperature, and Tg the estimated transition temperature of the particle shell. In similar systems, Joanicot et al. (Rhône Poulenc) [22] using SANS measurements and stress-strain experiments reported that low molecular masses were able to permeate through the hydrophilic membrane ; but only at temperatures where the membrane happened to fragment, did a massive interdiffusion of the high molecular weight polymer take place, eventually yielding tough films. In all of the cited examples, the existence of crosslinked polymer and the peculiarities of the MWD of the sol polymer are crucial in the prediction of the film formation and the development of mechanical properties. Therefore, reliable methods for determination of these polymer features as well as theoretical frames to predict them are strongly needed.

3. Determination of Gel and Crosslinks Densities in Emulsion Polymers

The definition of 'gel', W_g, is for many authors a contentious issue ; diverse terms as gel, macrogel, microgel, insoluble, organogel are often met [2]. We will limit the term gel to every tridimensional covalent network produced within polymer particle during the polymerization process and formed by free radical reactions, including e.g. transfer to polymer, copolymerization of multifunctional monomers, etc... Not considered are crosslinking reactions which my take place after the polymerization process e.g. during filmification, as is the case with reactive latex systems.

3.1. PHYSICALLY AND CHEMICALLY CROSSLINKED GELS

From an experimental point of view it is necessary to distinguish between a gel produced by reversible physical interactions and a true gel consisting of covalently crosslinked polymer chains. Linear polymer particles carrying charged surface groups may give rise to large quantities of 'gel' when dispersed in a good solvent having a low dielectric constant. This ionomeric effect, resulting from the interaction of ions pair and giving rise to a three-dimensional network, is well documented [3-4] and may lead to erroneous results during the measurements of the content of 'chemical' gel. For films made from copolymer lattices and having the composition styrene/butylacrylate/acrylic acid, Cohen-Addad *et al.* [6] have shown that W_g, measured using toluene as solvent, increased rapidly with the degree of neutralization of COOH groups ; this associative effect disappeared when toluene was replaced with a solvent of high dielectric constant (THF). It is clear that the choice of a solvent is critical as it must solubilize (swell) the core polymer and disrupt the ionic bonds due to the charged groups present on the particle shell.

3.2. METHODS OF MEASUREMENTS OF CHEMICALLY CROSSLINKED GELS IN LATEX COPOLYMERS.

Crosslinked polymer molecules which occupy a substantial fraction of the particle volume are likely to have sizes of a similar scale as the particle diameter, i.e. from a few tens of nanometers to several hundreds nanometers. Hence, as opposed to crosslinked polymer formed in an homogeneous medium which have an infinite size compared with their soluble counterpart, microgels formed in latex particles will scale in size not too far from the soluble chains. We thus better understand why the method of extraction (e.g. Soxlhet apparatus) currently used for coalesced latex film have poor accuracy and reproducibility. In this case the size threshold between gel and soluble polymer is badly defined. Chromatographic techniques allow a better separation of gel and soluble polymer by adjusting the porosity of the chromatographic packing. Malihi [7] has proposed a method where the latex particles are transferred from the aqueous into the eluting solvent (THF) and the different macromolecular species separated by a size exclusion mechanism. The (micro)gel elute through the interstitial volume of the packing much faster than the soluble chains which enter the internal porosity of the stationary phase. In general the W_g values obtained from this chromatographic method (using a detector calibration procedure) tend to be higher than the extraction method. The discrepancy is explained in terms of the microgel fraction which may go through the filter and consequently may not be accounted for in the extraction test. The plugging of the SEC columns remains a major drawback of this method. Thin Layer Chromatography associated with Flame Ionisation Detection (TLC/FID) has been used to quantify the degree of crosslinking and the level of grafting in core/shell composite particles polybutadiene(-co-styrene)/styrene-co-acrylonitrile and polybutadiene(-co-styrene)/polymethylmethacrylate[8]. Uncrosslinked butadiene-co styrene material is first eluted with toluene, and a second development with methyl acetate reveals the polyacrylonitrile or polymethylmethacrylate grafted fractions. Quantification through FID was possible with the aid of calibration curves.

Ultra centrifugation techniques have been optimized and have been often used to study the sol/gel ratio of macromolecules obtained by emulsion polymerization [9]. Recently [10] a more detailed analysis has been achieved using the determination of the distribution of the sedimentation coefficients (s-distribution). A typical pattern of the s-distribution of a partially crosslinked particle exhibit a bimodal curve referring to the soluble (linear & branched species) and to the crosslinked gels respectively, and gives a direct measure of W_g. Because the force of sedimentation is independent of swelling and the force of friction is proportional to the microgel size, the rate of swelling Q can be obtained :

$$Q = \left(\frac{1}{s_{swollen}} W_g \frac{\rho_p - \rho_{medium}}{\eta} \frac{1}{18} d^2 \right)^3 \qquad (1)$$

with ρ_p and ρ_{medium}, the densities of the unswollen particles and the dispersion medium, d the unswollen particle diameter, $s_{swollen}$ the sedimentation coefficient of the swollen particles and η, the viscosity of the dispersion medium. This method allows, at the same time, the determination of the percentage of gel and the crosslinks density by applying the well known Flory-Rehner relationship [11] (high chain length approximation) :

$$\ln(1 - \Phi_p) + \Phi_p + \chi_{12}\Phi_p^2 + \frac{\overline{V_1}\rho_p}{M_c}\left(\Phi_p^{1/3} - \frac{\Phi_p}{2} \right) = 0 \qquad (2)$$

with $\Phi_p = 1/Q$, the polymer volume fraction within the particle, $\overline{V_1}$, the molar volume of the solvent, χ_{12}, the solvent/polymer interaction parameter. Some uncertainty arises however from the polymer concentration dependence of the s-distribution which leads to a considerable underestimation of the Q and M_c values [10]. As already mentioned, (§2.1) the presence of ionic groups in the latex particles gives rise, after film formation, to an ionomeric effect where the true covalent crosslinking is augmented by an additional ionic crosslinking. It is possible to remove this ionic contribution and to obtain information about the covalent gel alone, by measuring the swelling ratio, Q, of isolated particles dispersed in water where the ionomeric effect is at a minimum. A straightforward method was proposed by Vanzo et al. for the accurate determination of the solvent uptake of latex polymer particles [12]. The rate of swelling is described by (2) where the surface free energy term, $\left(4\overline{V_1}\gamma / d_0RT \right)\Phi_p^{1/3}$, is added to the left-hand side of (2), which then becomes equal to $\ln(P/P_0)$. γ is the interfacial tension between the polymer particle and serum, P the vapour pressure at a given volume fraction of polymer in the swollen polymer particle, and P_0 the vapour pressure at equilibrium swelling. Derivation of (2) shows that the extrapolation to low solvent uptake gives the interaction parameter on one hand ; on the other hand the contribution of the surface free energy term is shown to be negligible in the medium solvent activity range (e.g. : $P/P_0 < 0.7$). The pressure/solubility curve allows then a direct determination of the chemical contribution of M_c. This method was applied to a series of copolymers styrene/butadiene/acrylic acid prepared with increasing level of chain transfer agent (CTA) [13]. The M_c values thus determined

(Figure 1) , may be compared with (i) those obtained by the equilibrium swelling of coalesced films and (ii) those obtained from DMA performed on dry films [14]. It appears that chemical crosslinking, as measured on the water dispersed particles, is noticeably smaller than the total crosslinking density, as measured from the elastic modulus on the rubbery plateau, when the level of CTA approaches 1%.

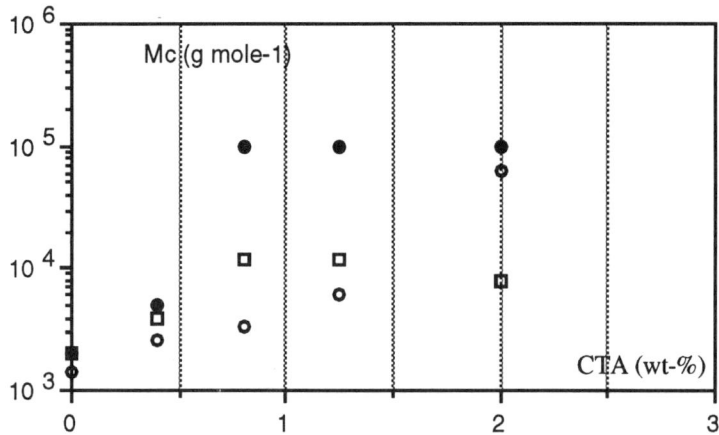

Figure 1 : Mc vs. CTA levels. (wt-% based on monomers) (1) ● : Mc values determined from pressure/ solubility curves on latex . Solvent : chloroform ; temp.: 25°C[13]. (2) ❑ : Mc values determined from films swollen at equilibrium. Solvent: chloroform ; temp:. 25°C[13]. (3) ○ : Mc values determined from DMA experiments [14].

4. Elementary Reactions in Non-linear Radical Polymerization

We will define systems as non-linear when, in the process of chain formation, connection points between primary chains are established. Primary chains are imaginary linear chains which would result if all crosslinks were severed. The most common non-linear systems are those obtained by the copolymerization of mono-ethylenic and di-ethylenic monomers, where the crosslinked units can be seen as tetrafunctional sites on a macrocospic scale (Figure 2), or, on a microscopic scale, as two trifunctional branches connected together. Actual tetrafunctional units might form if internal double bond are attacked by a macroradical : it is doubtful however that this could happen because of the their usually low reactivity.

84

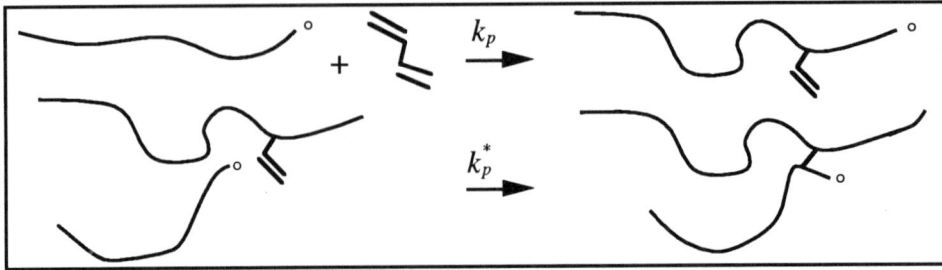

Figure 2: Tetrafunctional link in vinyl-divinyl copolymerization. k_p, k_p^* : rate propagation constant on divinyl monomer and pendant double bond , respectively.

The divinyl monomers currently used commercially include for example, divinylbenzene, ethyleneglycol and butanediol dimethacrylate, diallylphtalate, allyle methacrylate ; their rate of use is usually in the range 0-5 %. Other divinyl monomers such as butadiene and chloroprene are used as main components in the production of industrially important elastomers. Cyclization reactions are an important feature of vinyl-divinyl copolymerization : in primary cyclization the cycle forms within the primary chain whereas in secondary cyclization the cycle forms between two or more primary chains. Crosslinks formed in primary cyclization are not contributing to the elastic properties of the gel molecules ; in some instances it may even delays the onset of gelation [28]. One must keep in mind however that primary cyclization is significant only when : (i) the pendant double bond (PDB) possesses a reactivity toward radical addition of similar magnitude to that of the main monomer (for instance it is unlikely that primary cyclization would play a great role in butadiene polymerization since the vinyl 1-2 group are far less reactive than the diene) ; (ii) the monomer concentration is low.

Chain transfer to polymer and monomer gives trifunctional connection points, as sketched in Figure 3.

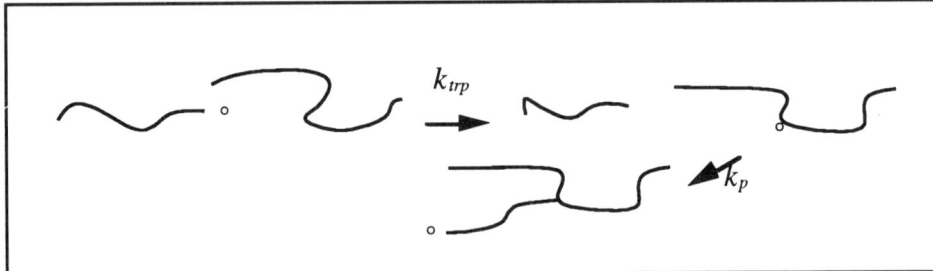

Figure 3a : Long chain branching through transfer to polymer. k_{trp} : rate constant for transfer to polymer.

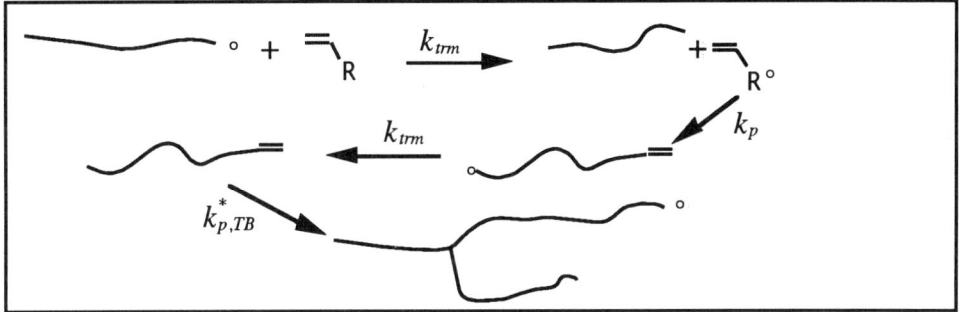

Figure 3b . Long chain branching through transfer to monomer. k_{trm} : rate contant for transfer to monomer. $k^*_{p,TB}$: rate constant for propagation on terminal double bond.

In the transfer-to-polymer reaction (Figure 3a), a radical is formed on a dead chain (generally through H-abstraction), which then adds monomer units to produce a long chain branch which eventually terminates either by transfer or radical coupling. Transfer to monomer (Figure 3b) leads to a double-bond terminated chain which can act as a macromonomer, resulting also in long chain branching (LCB). Long chain branching is well documented in vinyl acetate polymerization where transfer both to polymer and monomer are known to occur [23-24]. Transfer to poly-conjugated dienes such as polybutadiene or polychloroprene is thought to take place on the allylic protons, but it still remains difficult to establish the respective contribution of transfer to polymer and copolymerization of the PDB, to LCB and gel formation. Although not often quoted in the literature, the transfer to polymer in acrylic emulsion polymerization is substantial and leads in most cases to gelled polymer. ^{13}C NMR has been used to identify the site of grafting as the tertiary proton of the acrylic backbone : Lovell [5] *et al.* were able to measure a level of grafting as high as 2 to 4 molar % in the emulsion copolymerization of butyl acrylate and methyl methacrylate. In a comparative batch emulsion polymerization of butyl acrylate (BA) and butylmethacrylate (BMA), the gel content was 50% in the case of BA, and was virtually zero with BMA since no tertiary protons responsible for grafting are present on the polyBMA chain [15]. Radical back-biting reactions are well known in ethylene polymerization, generating short-branched structures [25]. Recently [26], intramolecular radical transfer reactions have been reported for some vinylethers bearing a carbon atom on the β-position, with a capto-dative substitution. This is the shortest branch that is possible to achieve since the back-biting occurs on the pendant group of the propagating radical. Those short branches however do not affect the MWD, nor they lead to any gelation. It is worth mentioning that some of the CTA's,which are currently used in emulsion or suspension polymerization, can lead to branched polymers : hence in the case of α-methylstyrene dimer, chain stoppage is taking place through an addition-fragmentation pathway, resulting in a terminal double bond on the dead chain. Even if the end chain α-methylstyrene group is sterically hindered towards radical addition, it will eventually react when the polymerization proceeds at high conversion. This non-linear system is virtually the same as the one resulting from transfer to monomer as described in Figure 3b.

5- Modelling Techniques in Non-linear Radical Emulsion Polymerization

Mathematical models for the prediction of the evolution of molecular weight in emulsion polymerization have been proposed (for instance [29-31] and references cited herein), based on the elementary reactions of radical polymerization, i.e. propagation, termination and chain transfer, and taking into account the kinetic events proper to emulsion polymerization : water phase initiation, radical segregation and chain desorption. Thermodynamics of monomer partitioning between water, polymer particles and droplets has been also extensively studied [32]. Classical kinetic approaches as well as statistic based models have been developed to compute instantaneous and cumulative MWD's using various numerical methods such as coupled integro-differential equations or large power probability matrix [31]. Until recently none of them actually addressed the effect of non-linear polymerization in segregated medium. On the other hand, the kinetic of network formation in homogeneous medium was first described by Flory and Stockmayer more than 50 years ago [11]. Although it has been refined by a number of authors, this mean field treatment of non-linear systems remains today a pretty good guideline for predicting main trends in crosslinking reactions. Since then, some of the limitations of the Flory-Stockmayer model have been removed by new theoretical treatments :

- Generalized Flory's approach : Hamielec and Tobita [28,33,34] took into account the history of individual chain during conversion and its role on the variance of the crosslinks density.
- Numerical fractionation : proposed by Teymour et al. [35,36], this method describes a cascade growth of chains subdivided in generations, that are constituted of branched macromolecules of increasing size and branching density
- Monte Carlo simulation : Tobita et al. [37-41] shows that in the particular case of colloidal polymer particles where the number of primary chains is finite, it is numerically feasible through a Monte Carlo algorithm to keep track of individual chains and to build the actual MWD including any branching and crosslinking reactions.

5.1. GENERALIZED FLORY APPROACH

In his pioneering work, Flory made some simplifying assumptions to account for crosslinking kinetics in vinyl-divinyl copolymerization : (i) the reactivities of all types of double bonds are equal ; (ii) all double bonds react independently of each other ; (iii) there is no intramolecular reaction in finite molecules. This system is actually a ternary polymerization composed of the vinyl monomer, the divinyl monomer (symmetrical) and the PDB. This ternary system can be eventually reduced to an homopolymerization scheme by using pseudo-kinetic rate constants [42]. In butadiene polymerization where the vinyl 1-2 pendant groups of polybutadiene show little reactivity, the instantaneous and cumulative crosslinks densities are simply computed as :

$$\rho(x) = \frac{k_p^* x}{k_p(1-x)} \quad (3a) \qquad , \qquad \overline{\rho(x)} = \frac{k_p^*}{k_p}[1+(1/x)\ln(1-x)] \quad (3b)$$

with x , the monomer conversion and k_p^* , k_p pseudo-kinetic rate constants for crosslinking and propagation respectively. Beyond the gel point the fraction of gel material is obtained according to :

$$W_g = 1 - \sum_{r=1}^{\infty} w(r)\left(1 - \overline{\rho} W_g\right)^r \qquad (4)$$

with r and $w(r)$ the length and weight distribution of the primary chains. This approach suffers however from shortcomings because it considers that all the primary chains have the same crosslinks density independent of their birth conversion (equilibrium state). This is obviously far from reality : it is known indeed that large heterogeneities in chain composition are produced with changes in operating conditions (monomer feed profiles, temperature, starved vs. flooded conditions, etc...) which is a common feature of industrial processes. A major improvement has been brought by Hamielec and Tobita [28,33] who introduced the notion of birth conversion of primary chains to consider the history of each chain in terms of composition and crosslinking during reaction time : one discriminates between crosslinks formed instantaneously during chain growth (Figure 4 : chain B, $\rho_i(x)$), and the additional crosslinks accumulated from birth conversion, θ , to conversion $x > \theta$ (Figure 4 : chain A, $\rho_a(\theta, x)$).

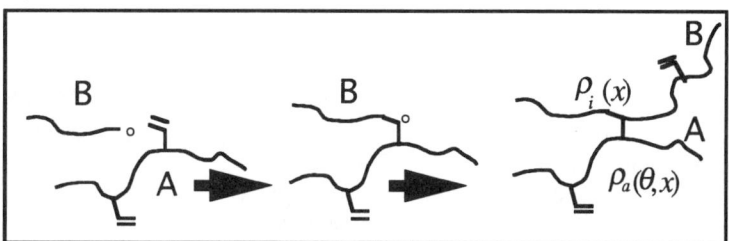

Figure 4 . Additional and instantaneous crosslinks according to [28]

When cyclization effects can be neglected, the additional crosslinks density is given by :

$$\frac{\partial \rho_a(\theta, x)}{\partial x} = \frac{k_p^*}{k_p} \frac{\left[F_2(\theta) - \rho_a(\theta, x)\right]}{1 - x} \qquad (5)$$

with $F_2(\theta)$, the rate of incorporation of the divinyl monomer at conversion θ . The instantaneous crosslinking density is computed as :

$$\rho_i(x) = \frac{k_p^*}{k_p} \frac{\left[\overline{F_2(x)} - \overline{\rho_a(x)}\right]x}{1 - x} \qquad (6)$$

Since each instantaneous crosslinked unit has a counterpart corresponding to an additional crosslinked unit on a neighbour chain, it follows that :

$$\overline{\rho}(x) = \overline{\rho}_a(x) + \overline{\rho}_i(x) = 2\overline{\rho}_a(x) = 2\overline{\rho}_i(x) \qquad (7)$$

At conversion x , the weight fraction of sol for the primary molecules formed at conversion θ is given by:

$$W_s(\theta, x) = \sum_{r=1}^{\infty} w(r, \theta)\left[1 - (\rho_a(\theta, x) + \rho_i(\theta))W_g(\theta, x)\right]^r \qquad (8)$$

If we know an analytical expression for the primary chain weight distribution $w(r, \theta)$, then $W_s(\theta, x)$ can be easily calculated numerically, and integrated over birth conversion to obtain the accumulated gel fraction :

$$W_g(x) = 1 - W_s(x) = \frac{1}{x}\int_0^x W_s(\theta, x)d\theta \qquad (9)$$

The form of $w(r, \theta)$ is straightforward when the radical compartimentalization has no effect on primary chain distribution, which is the case when radical termination is dominated by transfer. Fortunately this situation occurs quite often, for instance in vinyl acetate and vinyl chloride emulsion polymerization, where the main event for chain stoppage is transfer to monomer [23,24]. In the case of styrene-butadiene emulsion copolymerization substantial amount of CTA is used ; the chain termination regime can be estimated roughly comparing the rate of transfer to CTA (R_{CTA}), to the rate of radical flux ($R_i \approx R_{t,c}$) : there are usually several order of magnitude in favour of R_{CTA} and therefore $w(r, \theta)$ is described by the most probable distribution :

$$w(r, \theta) = \tau(x)^2 r \frac{1}{(1 + \tau(x))^r} \qquad (10)$$

with $\tau(x) = R_{CTA}(x) / R_p(x)$. This generalized Flory model was further adapted and applied to the emulsion polymerization of dienes by allowing for semi-batch operations and multi-monomer feeding [43,44]. The results below were obtained in the copolymerization of styrene and butadiene in the presence of tertiododecylmercaptan (TDM) as a CTA. The evolution of the gel content is reported in Figure 5 :

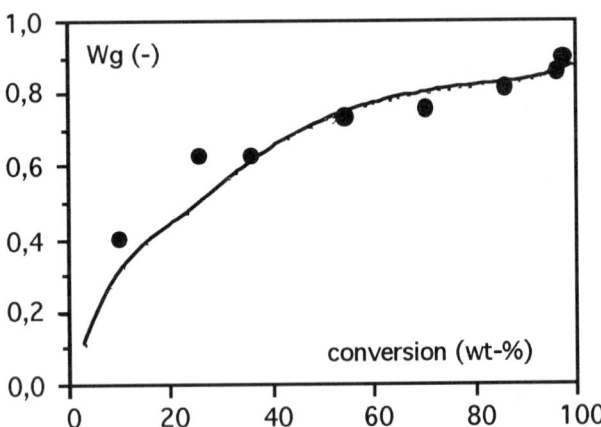

Figure 5 . Gel fraction vs. weight conversion in seeded semi-batch emulsion copolymerization of styrene and butadiene (70-30 wt-%) in the presence of TDM (0.1 wt-%). ● : experimental points. Solid curve : model (kp^*/kp set at 2.5x10^{-3}). Conditions :[43]

We observe a high insoluble fraction at the very beginning of reaction : as opposed to crosslinking reactions in homogenous medium, the polymer content within the particles, acting as micro-reactors, is relatively high even at low conversion and corresponds in batch operation to the maximum swelling monomer to polymer ratio at equilibrium (i.e. Φ_p in eq.(2)). Crosslinks density of the gel fraction obtained from (2) are compared with the model (Figure 6) : the total crosslinks density exhibits some increase all through the conversion, but hinders the actual variance in crosslinks density experienced by the chain according to their birth conversion , as depicted in Figure 7.

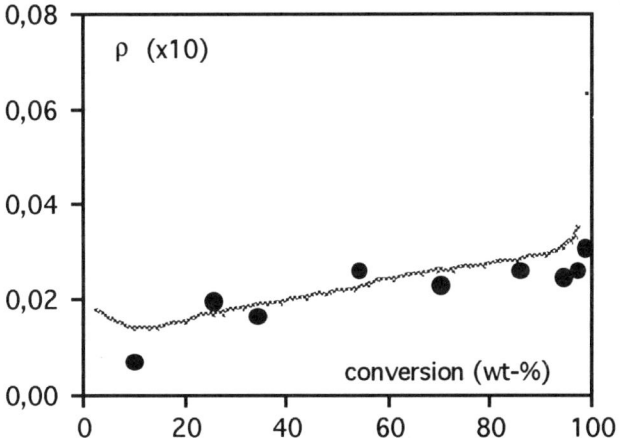

Figure 6 : Crosslinks density of the extracted gel vs. conversion. Same conditions as in Figure 5. ● : calculated data according to eq (2) from experimental swelling ratios. Dotted curve : model

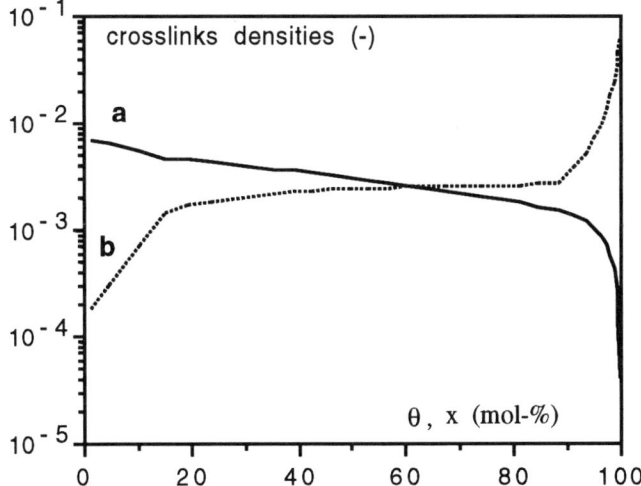

Figure 7 . Computed additional crosslinks density (**a** : $\rho_a(\theta, x)$) and instantaneous crosslinks density (**b** : $\rho_i(x)$), vs. conversion. Same conditions as in Figure 5.

The instantaneous crosslinks profile reflects the polymer to monomer ratio within the polymer particle ; (i) a first rise at early conversion (batch condition) ; (ii) a plateau during the steady-state regime and (iii) the stage III corresponding to monomer depletion. The fast increase in the PDB polymerization during stage III could give some hint as to the spatial inhomogeneities of the polymer particle ; the macroradicals are initiated in the aqueous phase and captured on the particle surface, and in addition the mean average length between crosslinks is much shorter than the kinetic length for chain transfer and it is likely that the radical are segregated on the outer layer, leading to a crosslinks gradient from the core to the shell of the particles.

5.2. NUMERICAL FRACTIONATION METHODS

An alternative approach to the mean field theory described above has been proposed, based on the distribution functions of primary chains with discrete values of crosslinking points [43]. In a vinyl-divinyl system it is shown that the instantaneous weight fraction of primary chains of length n with m crosslinks is given by :

$$w(n,m) = P_t P_p^{*m} \left[\tau P_p^{\ n} \frac{n^m}{m!} + \frac{\beta}{2} P_t P_p^{\ m} \frac{n^{m+1}}{(m+1)!} \right] \qquad (11)$$

with $\beta = R_t / R_p$, $\tau = R_f / R_p$, $P_t = (R_f + R_t)/\sum R$, $P_p = (R_p)/\sum R$ $P_p^* = (R_p^*)/\sum R$, and $\sum R = R_f + R_t + R_p + R_p^*$, i.e. the sum of the reaction rates for transfer, termination, propagation and PDB addition, respectively. The derivation of the gel content is based on the assumption that only the chains with m -values below a critical number N_{crit} remain soluble. From eq.(11) and some additional rules to account for the transitions between the different chain families with increasing level of branching ($m = 0, 1, 2$) and the gelled polymer ($m > N_{crit}$), a set of population balance equations are then formulated. The gel content is finally obtained by subtracting the amount of sol from the total mass of polymer. This model, together with the generalized model of Flory were compared in the case of styrene-butadiene emulsion copolymerization, and the evolution of the gel fraction can be seen in the Figure 8 given below :

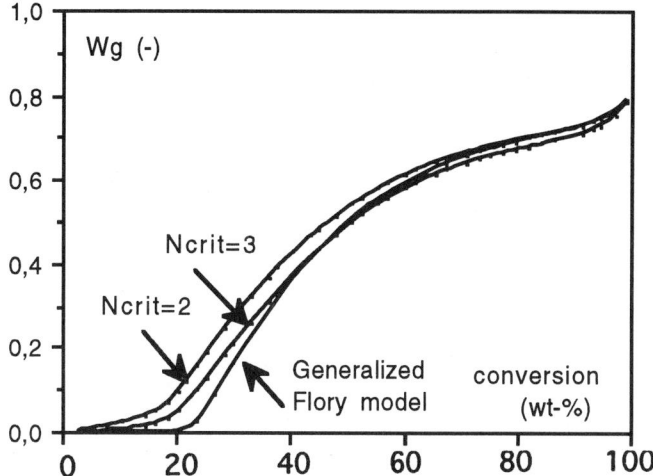

Figure 8 . Calculated gel fraction vs. conversion in a semi-batch styrene-butadiene (70-30 wt-%) with TDM as the CTA (0.8 wt-% based on monomer). Comparison between the generalized Flory model and the approach of Charmot et Guillot [43]

Contrary to the Flory approach, this model generates "gel" material at very low conversion, and interestingly as the threshold value N_{crit} increases (For clarity only N_{crit} =2 and N_{crit} =3 are represented in Figure 8) Wg converges towards the same profile for both theoretical approaches.

In the "Numerical Fractionation" method of Teymour and Campbell [35,36]the same principle is applied but generalized to a percolation theory approach, and the gel described as a fractal structure. The polymer in subdivided in linear and branched fractions called polymer generations. These generations are composed of polymer molecules of similar scale, and as one moves from one generation to the next, the average molecular size will grow geometrically leading eventually to an infinitely large cluster (gel). The cascade growth in a model system where transfer to polymer occurs and where the only termination mechanism is radical coupling is sketched in Figure 9 : As reaction starts, linear chains are being formed (L), which are subsequently reactivated through transfer to the polymer backbone leading to branched species referred to as the first branched polymer generation B1. Polymer in generation B1 will keep adding branches yet will be considered to belong to the same generation. Transfer to the next generation will occur only if a connection event takes place between 2 chains belonging to the same generation, moving to generation B2. As the reaction proceeds, higher generation will keep growing until gel appears, whereby these large sized clusters will be swept up by the gel since it is known that the rate of most crosslinking reactions depend on molecular sizes. Due to this cascade mode of growth, each generation is used to build the next higher, and its weight fraction increases up to the gel point, after which it is quickly consumed. Numerical application shows that the algorithm converges quite quickly with the number of generations n ,, and a best compromise in terms of accuracy and computing time lies somewhere between 5 to 10 generations.

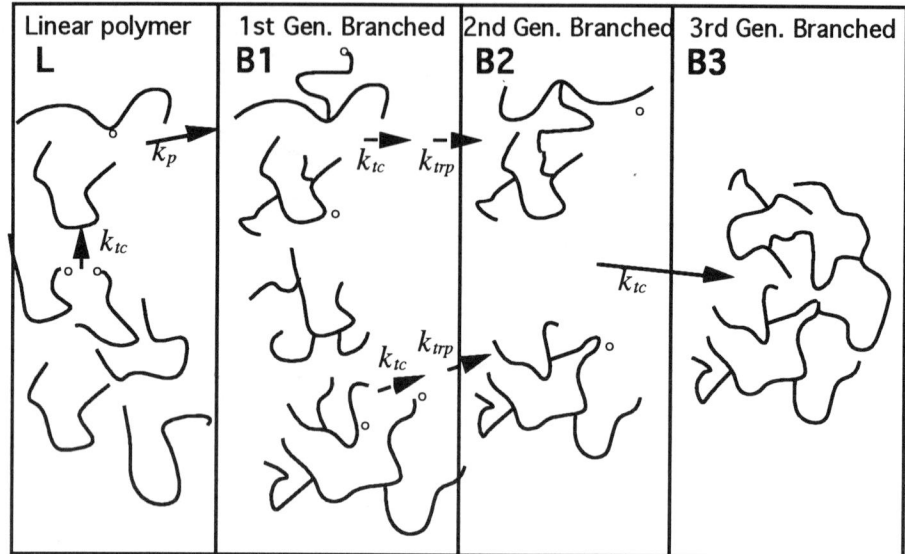

Figure 9 : Generation transfer chart . Linear (L) and branched (Bi) generations in the cascade growth as proposed by Teymour and Campbell (^{36}). k_{tc} : rate constant for termination by combination

The method of computation is based on the determination of moments of live radicals and dead chains for each generation, plus the overall moments. If n_C is the highest ranked generation before gel onset, then the model will consist in $3n_C +5$ ordinary differential equations (ODE) and $3n_C +5$ algebraic equations that can be solved with any available software package for ODE's. Thus the model enjoys a number of advantages: the mathematics remains manageable even on a personal computer and it makes possible to address non linear systems that otherwise could not be described and quantified properly as far as the gel fraction is concerned. As an example, in a commonly encountered case where transfer to polymer yields insoluble material (emulsion polymerization of acrylics) the "Numerical Fractionation" technique will give the gel fraction whereas in the Flory approach only the overall moments of the distribution are calculated up to gel point, at which the 2nd moment diverges. In the same paper [36] Teymour and Campbell proposed a method to reconstruct the MWD, wherein the molecular weight moments of an individual generation are used to generate a Shultz distribution for that species. It follows that multimodal MWD are produced. Although not physically justified, this method of MWD reconstruction is supported by some experimental evidence : for instance in the case of emulsion polymerization of chloroprene, it turned out that some multiple peaks were visible in GPC trace in the vicinity of the gel point [44].

Very recently and almost simultaneously, Arzamendi and Asua [45] and Mazzotti *et al* [46] transposed this model to emulsion polymerization to take into account the radical compartimentalization effects on gel and MWD development. As the MWD reconstruction method was still questionable, Arzamendi and Asua proposed to subdivide the first n_e branched generations into sub-generations of chains with a constant number of branches m. General balance equations were established taking into account the whole radical distribution among particles together with the transfer rules between (sub-)generations. An example of the application of this model is the emulsion polymerisation with branching arising from transfer to monomer and polymer. As expected, while the initial fractionation technique generates multimodal MWD, the new approach lead to a monomodal distribution skewed towards large Mw's, which seems in better agreement with what is usually found experimentally [23,24]. Interestingly the gel fraction was found to increase as the termination constant k_{tc} increases, but further increase in k_{tc} caused a drop of the gel content because under these circumstances the average number of radicals, $\bar{n}$, is close to 0.5 (system 0-1) and the probability of termination between large macroradicals is very low, if not zero.

5.3. MONTE CARLO TECHNIQUE

In this model, extensively studied by Tobita and co-workers, individual primary polymer molecules are generated at random and their fate is examined with regards to any connection events (branching and terminal double bond polymerization, TDBP). This technique seems particularly well suited to emulsion polymerization since the number of polymer molecules in a particle is finite (e.g.: 3×10^3 for a 0.1 μm particle size and a number-average chain length of 1000). The polymer particles are sampled at random and all the polymer molecules in each of the particle are simulated with appropriate probability functions, for instance : $N(r)$, the instantaneous number-chain length distribution of primary chain, P_b, the probability that a primary chain starts growing from a polymer backbone, P_{tdbp}, the probability of chain connection through TDBP. Thus in the simple limiting case of emulsion polymerization of vinyl acetate with chain transfer to polymer and transfer-dominated termination, a typical algorithm is outlined below [40,41] :

1-determine r of the i-th primary chain ($N(r)$)
2-start growing from a polymer backbone(P_b) ? yes goto 3 , no goto 4
3- select a polymer molecule at random on the weight basis and attach branch chain,
4- if the primary radical adds a TDB on existing polymer molecule, connect them
5- return to 1

Compared with experimental data, this method gives a good correlation with the actual MWD for moderate conversion, but tends to overestimate the high molecular weights tail of the MWD at high conversion. This approach ensures however that the actual MWD development remains within the particle boundaries, whereas in bulk models which are "forced" into segregated systems, giant molecules can be produced whose sizes are larger than the polymer particle. When termination by radical coupling is introduced in the model, a strong MWD broadening is observed which can be attributed to gel formation, although the criteria for the onset of gelation and gel quantification are not obvious [47]. Simulations were carried out in vinyl-divinyl emulsion polymerization systems with some simplifying assumptions [48] ; the gel or microgel fraction is assimilated to the weight of the largest cluster and appears qualitatively as a sharp peak on the reconstructed MWD. When the gel fraction is close to 1, the MWD resembles the particle size distribution. This narrow-shaped distribution of the so-called gel polymer can be explained qualitatively by using the same formalism as in the Numerical Fractionation scheme : the cascade growth will converge to a generation occupied by only one giant molecule with a size comparable to the particle volume, whose further growth is only allowed through transfer from lower generations. Even if this approach has not yet received a thorough experimental validation, we can note the striking resemblance between the MWD's simulated by Tobita and the GPC traces of partially crosslinked acrylic latexes that can be seen in [7].

6. Conclusions

As a conclusion to this overview in kinetic modelling of crosslinked systems we have listed in Table 1, as a guideline for selecting the appropriate model for a given situation, the different theoretical approaches together with their main capabilities.

TABLE 1. Fields of application of models in non linear free radical polymerization. (1) : a : vinyl-divinyl copolymerization (low drift in crosslinks density). b : vinyl-divinyl copolymerization (large drift in crosslinks density). c: long chain branching with radical coupling.

Model	Systems (1)	Bulk media	Segregated media	Gel content	MWD
Flory	a	●		●	
Flory/Hamielec/Tobita	a, b	●		●	
Numerical Fractionation (Teymour)	a, b, c	●		●	●
Numerical Fractionation (Teymour, Arzamendi, Mazzotti)	a, b, c		●	●	●
Monte-Carlo (Tobita)	a, b, c	●	●	●	●

As it was first stressed, the mean field treatment from Flory and its generalized form appear today as reliable tools, within the reach of any polymer chemist wishing to explore a given system. Numerical Fractionation opens a new route for unravelling complicated cases often found in the real life, such as the build-up of gel in acrylic systems. The radical segregation adds another level of complexity to the whole scheme, but recent work has shown that the compartimentalization effect can be incorporated successfully. Fortunately, many emulsion polymerization systems can be reduced to some limiting case where radical segregation plays a minimal role and the bulk or pseudo-bulk approach applies, as least for the prediction of the primary chain distribution. Last but not least, the Monte-Carlo technique has exhibited its enormous potentialities to tackle the challenging complexity of free radical emulsion polymerization, its large computing time however is so far a serious drawback. As it was noted in the introduction, experimental studies on MWD's in emulsion polymerization have been left behind compared with theoretical and modeling developments, and should be the subject for much future work.

7. References

1. Gilbert, R.G. (1995), *Emulsion Polymerization* . Academic Press Pub., 245-290
2. Murray, M.J., Snowden, M.J. (1995) The preparation, characterization and applications of colloidal microgels, *Advances in Coll. Interf. Sci.*, **54**, 73-91.
3 Weiss R.A., Turner S.R., Lundberg R.D., (1985), *J. Polym. Sci. Polym Chem. Ed.*,**53**, 525-33
4 Lundberg R.D., Makowski H.S., (1980), *J. Polym. Sci. Polym. Phys. Ed., 18*, 1821
5 Lovell P., Shah T., Heatley F., (1991), *Polymer Communications* , **32**(4), 98-103
6. Cohen-Addad, J.P., Bogonuk, C., Granier, V. (1994) Gel-like behavior of pH dependant latex films, *Macromolecules*, **27**, 5032-36.
7. Malihi, F.B., Kuo, C., Provder, T. (1983) Determination of gel content of acrylic latex by size exclusion chromatography. *J.Liquid Chrom., 6*(4), 667-83
8. Merkel, M.P. Dimonie, V.L. El Aasser, M.S. Vanderhoff, J.W. (1987), Morphology and grafting reactions in core/shell latexes, *J. Appl. Polym. Sci.*, 1219-1233
9. Shashoua, V.E., Beaman, R.G. (1958) Microgel : An idealized polymer molecule. *J. Polym. Sci.*, **33**, 101-117
10. Müller, H.G., Schmidt, A., Kranz, D. (1991) Determination of the degree of swelling and crosslinking of latex particles by analytical ultracentrifugation. *Progr. Colloid Polym. Sci.*, **86**, 70-75.
11. Flory P.J. *Principles of Polymer Chemistry*, (1953) Cornell Univ. Press, Ithaca, NY. 579
12. Vanzo, E., Marchessault, R.H., Stannett V. (1965) The solubility and swelling of latex particles, *J Coll. Sci.*, **20**, 62-71.
13. Charmot, D., Oger, N. (1989) Swelling characteristics of SB copolymers latexes and films. Unpublished data from Rhône Poulenc internal report.
14. Richard, J. (1992) Dynamic micromechanical properties of SB copolymers films., *Polymer*, **33**, 562-71
15. Zosel A. (1993), Influence of crosslinking on structure, mechanical properties, and strength of latex films.*Macromolecules*, **26**, 2222-27
16. Hahn K., Ley G., Schuller, H., Oberthür R. (1986) *Coll. Polym. Sci.*, **264**, 1092
17. Hahn, K., Ley, G., Oberthür R. (1988) *Coll. Polym. Sci.*, **266**, 631
18. Zhao, C.L., Wang,Y., Hruska, Z., Winnik M.A (1990) *Macromolecules*, **23**, 4082
19. Yoo J.N., Sperling L.H., Glinka, C.J., Klein A. (1990) *Macromolecules*, **23**, 3962
20. Joanicot M., Wong K., Richard J., Maquet J., Cabane B., (1993) *Macromolecules* **26**, 3168
21 Kim H.B., Winnik M., (1995), *Macromolecules*, **28**, 2033-2041
22. Joanicot, M., Wong, K., Cabane, B. (1996) Interdiffusion in cellular latex films *Macromolecules*, to be published
23. Friis, N., Goosney, D., Wright, J.D., Hamielec, A.E. (1974) Molecular weight and branching development in vinyl acetate emulsion polymerization, *J. Appl. Polym. Sci.*, **18**,1247-1259
24. Friis, N., Hamielec, A.E. (1975) Kinetcs of vinyl chloride and vinyl acetate emulsion polymerization, *J. Appl. Polym. Sci.*, **19**,97-113
25. Roedel, M.J. (1953) *J. Am. Chem. Soc.*, **75**, 6110
26. Sato T., Ito D., Kuki M., Tanaka H. , Ota T., *Macromolecules*, **24**, 2963-67
27. Yamada, B., Tagashira, S., Aoki, S. (1994) Preparation of polymers with allyl end group using dimer of a-methylvinyl monomer., *J. Polym. Sci. PartA : Polym. Chem.*, **32**, 2745-54
28. Tobita, H., Hamielec A.E. (1989) Crosslinking kinetics in free radical polymerization, *Polymer Reaction Engineering, Proceedings of the 3rd Berlin International Workshop,* VCH Publishers, 43-83
29. Lichti, G., Gilbert, R.G., Napper D.H. (1980) Molecular weight distribution in emulsion polymerization, *J. Polym. Sci., Polym. Chem Ed.*, **18**, 1297-1323
30. Forcada J., Asua J.M. (1991) Emulsion copolymerization of styrene and methyl methacrylate. ii. molecular weights. *J. Polym. Sci., Polym. Chem Ed.*, **29**, 1231-42
31. Storti, G., Polotti, G., Cociani, M., Morbidelli, M. (1992) Molecular weight distribution in emulsion polymerization. *J. Polym.Sci., Polym.Chem Ed.*, **30**,731-50
32. Guillot, J. (1985) *Makromol. Chem. Suppl.*, **10/11**, 235
33. Tobita, H., Hamielec, A.E. (1989) Modeling of network formation in free radical polymerization, *Macromolecules*, **22**, 3098-3015
34. Tobita, H. (1992) Cross-linking kinetics in emulsion polymerization. *Macromolecules*, **25**, 2671-78
35. Teymour, F., Campbell J.D. (1992) Numerical fractionation. a novel technique for the mathematical modelling of long chain branching and crosslinking in polymer systems. DECHEMA monograph **127**, *4th International Workshop on Polymer Reaction Engineering.* VCH Pub.,149-157
36. Teymour, F., Campbell, J.D. (1994) Analysis of the dynamic gelation in polymerization Reactors using the "numerical fractionnation" technique. *Macromolecules*, **27**, 2460-2469.
37. Tobita, H., Kimura, K., Fujita, K., Nomura, M. (1993) Crosslinking in emulsion copolymerization of methyl methacrylate and ethylene glycol dimethacrylate. *Polymer*, **34**(12), 2569-2573
38. Tobita, H., Yamamoto, K. (1994) Network formation in emulsion crosslinking copolymerization., *Macromolecules*, **27**, 3389-96
39. Tobita, H., Takada Y., Nomura M., (1994), Molecular weight distribution in emulsion polymerization, *Macromolecules*, **27**, 3804-11
40. Tobita H. (1994) Kinetics of long chain branching in emulsion polymerization : chain transfer to polymer, *Polymer*, **35**(14), 3023-31

41. Tobita, H. (1994) Kinetics of long chain branching in emulsion polymerization : vinyl acetate polymerization, *Polymer*, **35**(14), 3032-38

42. Broadhead T., Hamielec A., Mc Gregor J.F. (1985) *Makromol. Chem. suppl. 10/11*,105

43. Charmot, D., Guillot, J. (1992) Kinetic modeling of network formation in styrene-butadiene emulsion copolymers : a comparatice study with the generalized form of flory' theory of gelation, *Polymer,* **33**(2), 351-360.

44. Charmot, D., Sauterey, F. (1992) Emulsion polymerization of chloroprene. polymerization mechanisms in relation with long chain branching. DECHEMA monograph **127**, *4th* Intern.[al] *Workshop on Polymer Reaction Eng.* VCH Pub. 483-92

45. Arzamendi, G., Asua, J. (1995) Modeling gelation and sol molecular weight distribution in emulsion polymerization. *Macromolecules*, **28**, 7479-90

46. Mazzotti, M., Storti, G., Fiorentino, S., Ghielmi, A., Morbidelli, M. (1995), Kinetics of branching in emulsion polymerization. DECHEMA monograph **131**, *5th International Workshop on Polymer Reaction Engineering.* VCH Pub., 19-34.

47 Tobita H., (1993), Molecular weight distribution in free radical polymerization with long-chain branching, *J. Polym. Sci. Polym. Phys.,* **31**,1363-71

48. Tobita, H. (1995) Microgel formation in emulsion crosslinking copolymerization, DECHEMA monograph **131**, *5th International Workshop on Polymer Reaction Engineering.* VCH Pub., 3-18

POLYMERIZATION AT HIGH CONVERSION

DAX KUKULJ AND ROBERT G. GILBERT

School of Chemistry,
University of Sydney
NSW 2006,
Australia

1. Introduction

Although polymerization through to very high conversion is of great importance industrially, there has been little mechanistic work published in the area. It is necessary that there be essentially no residual monomer in a commercially viable product. The specific definition of *no* residual monomer depends on current environmental legislation requirements which are always becoming stricter, typically in the ppm range. Residual monomer is commonly removed industrially by burn-out, which usually involves addition of a redox initiator and/or temperature increase. However, the peculiarities of mechanisms at high conversion are poorly understood, and increased understanding could lead not only to better control but perhaps to new polymer properties. This paper presents some novel facets of polymerization in the high conversion region. These include the distribution of monomer between the phases, initiator efficiency, kinetics of the reaction, and the resultant molecular weight distributions.

2. Initiator Efficiency

2.1. ENTRY MODEL

In emulsion polymerization it is known that initiation of chains occurs by entry of initiator derived free radicals from the aqueous phase (ignoring transfer). Typically a hydrophilic initiator, e.g. a sulfate radical, needs to react with monomer in the water phase before it is sufficiently surface active to enter a particle irreversibly. For the case of ionically stabilized lattices the overall initiator efficiency, f_I, is approximated by [1-3]:

$$f_I \approx \left\{ \frac{\sqrt{2 f k_I [I_2] \, k_{t,w}}}{k_{p,w} [M]_w} + 1 \right\}^{1-z} \tag{1}$$

It is apparent that as the aqueous monomer concentration $[M]_w$ decreases, so does the initiator efficiency. The value for the critical degree of polymerization, z, is expected to depend on water solubility of the monomer: the (admittedly simplistic) hydrophobic free energy model [1] (see eq (11) of [3]) predicts that $z \approx 2$–3 for styrene, 4–5 for MMA and *ca.* 10 for vinyl acetate (the order follows that of the solubility of monomer in water). With the values of $[M]_w$ pertaining to "ordinary" conditions (e.g., Interval II and most

97

J. M. Asua (ed.), Polymeric Dispersions: Principles and Applications, 97–107.
© 1997 *Kluwer Academic Publishers. Printed in the Netherlands.*

of Interval III), eq (1) predicts, for example, that styrene usually has very low initiator efficiency whereas the efficiency of initiation in MMA and VAc will be very high. However, at high conversion, $[M]_w$ will become small, and f_I can decrease dramatically because of the strong dependence on $[M]_w$ for these monomers. We now further explore this possibility.

2.2. MONOMER PARTITIONING

To calculate the entry rate, the aqueous phase monomer concentration needs to be known as a function of conversion. This requires the partitioning of monomer between the particles and water phase (except for very unusual circumstances, such as monomers of very low water solubilities, the transport of monomer to particles through the aqueous phase is not expected to be rate-determining [2]).

Several monomer partitioning relationships have been used in the literature. The simplest is an empirical relation between unsaturated and saturation concentrations of monomer in the particle and water phases:

$$\frac{[M]_w}{[M]_{w,sat}} = \left(\frac{[M]_p}{[M]_{p,sat}}\right)^{0.6} \tag{2}$$

A more sophisticated approach is to use the Vanzo equation [4]:

$$\ln\left(\frac{[M]_w}{[M]_{w,sat}}\right) = \ln(1-\phi_p^P) + \phi_p^P(1-P_n^{-1}) + \chi(\phi_p^P)^2 + \frac{2V_m\gamma(\phi_p^P)^{1/3}}{r_0RT} \tag{3}$$

where ϕ_p^P = volume fraction of polymer, P_n = average degree of polymerization of the polymer, χ = Flory-Huggins interaction parameter, γ = particle-water interfacial tension, V_m = partial molar volume of monomer, and r_0 = unswollen radius of particles. This has been shown to fit data for a number of monomer systems, including those with more than one monomer present [5-7]. However, there have been no studies of this partitioning at very high ϕ_p^P.

As a start to elucidating the applicability of the Vanzo equation at high conversion, the partitioning of MMA has been measured at very high ϕ_p^P. A monomer-swollen latex was centrifuged, and the amount of monomer in the aqueous phase measured by GC. Figure 1 shows the aqueous phase monomer concentration as a function of the weight-fraction of polymer, w_p. As w_p increases the aqueous monomer concentrations drops off rapidly. Three fits are shown to the data: eqs 2 and 3, and a third-order polynomial fit. On these scales each appear to fit the data almost equally well. However when the high w_p region is emphasised by plotting against $\log w_m$ (where the weight-fraction monomer $w_m = 1-w_p$), Figure 2, the difference in the fits becomes apparent. The 0.6 power relationship deviates beyond w_p of 0.97 (i.e. $w_m > 0.03$). The Vanzo equation similarly deviates around w_p of 0.97 when physically reasonable values for χ and γ are taken (note that the Vanzo equation can be made to fit the data excellently over the whole w_p range, but this leads to unphysical values for χ and γ, e.g., negative χ) Due to the inadequacies of the two relationships a polynomial was used to fit the data in subsequent work. The temperature dependence of $[M]_w/[M]_{w,sat}$ is small [5].

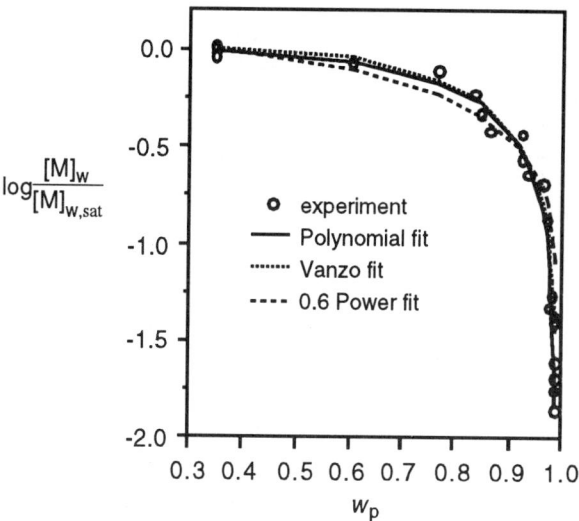

Figure 1. Data for monomer partitioning: concentration of MMA in aqueous phase (measured by GC), compared to saturation value, as function of weight-fraction polymer in 50 nm particles at 25 °C, and various fits as described in text. Vanzo fit used $\chi = 0.3$, $\gamma = 21$ dyne cm^{-1}.

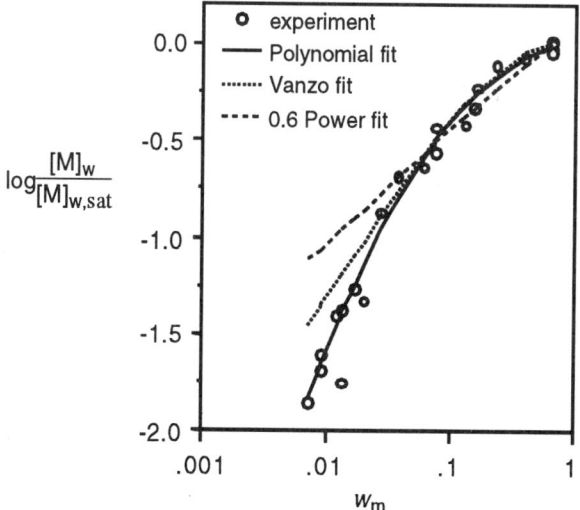

Figure 2. Data of Figure 1 re-plotted as function of monomer fraction $w_m = 1 - w_p$.

2.3. PREDICTION OF ENTRY EFFICIENCY

From a knowledge of the partitioning data the initiator entry efficiency can be estimated using eq (1). Figure 3 shows the calculated initiator entry efficiency for MMA as a function of w_p for both the 0.6 power and the polynomial forms of the partitioning data.

The calculated efficiency drops markedly in the last 10% of conversion: from 0.7 at intermediate conversion (a value which is in accord with experiment [8]) to extremely low at high conversion (note that the 0.6 power fit overestimates the entry efficiency above 98% conversion, as does the Vanzo fit; the latter is not shown). This suggests that persulfate is an ineffective initiator for the removal of residual monomer, this prediction arising from the strong dependence (fourth power, from the value of z for MMA) of the efficiency on $[M]_w$ in eq (1).

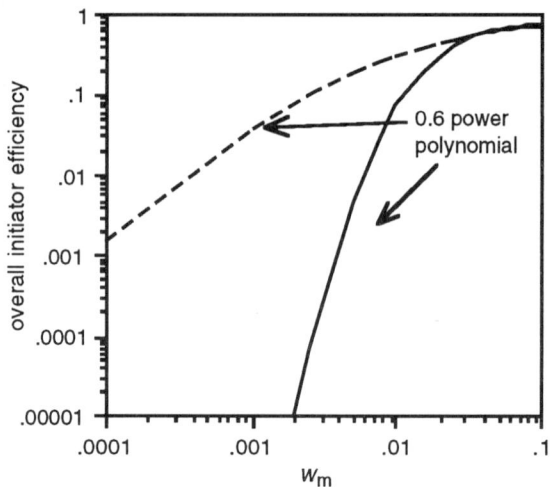

Figure 3. Dependence of initiator entry efficiency on weight-fraction monomer at high conversion for MMA, calculated using eq (1), [persulfate] = 5×10^{-3} mol dm^{-3}, $k_I = 10^{-6}$ s^{-1}, $z = 4$, for the 0.6 power and polynomial fits to the monomer partitioning data.

Figure 4 shows the calculated entry efficiency for butyl methacrylate. The partitioning of BMA has not been measured, and so the 0.6 power relationship was used for this calculation. Since BMA is much more hydrophobic than MMA the entry efficiency at $w_p = 0.9$ is much lower, nor is the predicted drop in efficiency is as sudden.

2.4. LOCATION OF MONOMER

Using the measured partitioning of MMA monomer between the particle and water phases it is straightforward to calculate the fraction of monomer in the aqueous phase as a function of conversion. Figure 5 shows this for different solids contents. As conversion increases, the *fraction* of monomer in the aqueous phase increases until a maximum at around $w_p = 0.97$, whereupon it drops off rapidly. This means that most of the residual MMA monomer at the end of a reaction is located in the particle phase. Strategies to remove residual monomer must take this into account.

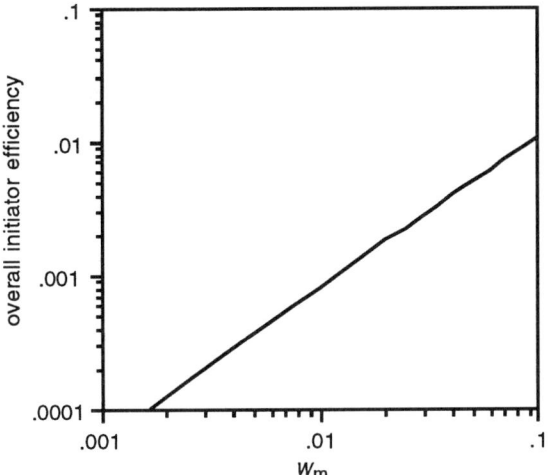

Figure 4. Predicted dependence of initiator entry efficiency on w_m for BMA, calculated using eq (1), [persulfate] = 5×10^{-3} mol dm^{-3}, $k_I = 10^{-6}$ s^{-1}, $z = 3$, assuming the 0.6 power relationship of eq (2) for monomer partitioning.

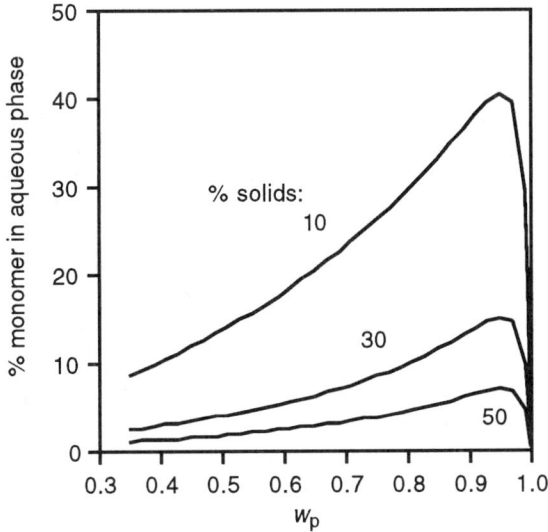

Figure 5. Percentage of MMA in the water phase as a function of w_p for different solids contents, calculated using the experimental partitioning data of Figure 1.

3. Rates at High Conversion

Conventional polymerization models predict that $\bar{n}$ should always increase with conversion. It has however been reported [9] that direct EPR measurements of $\bar{n}$ in an MMA latex at high conversion show a significant decrease in $\bar{n}$ in the last few percent conversion. The explanation suggested for this effect was that proposed [2,10] to explain the observation of an increase in the amount of low molecular weight species formed at high conversion in a styrene emulsion polymerization, viz., a spatial inhomogeneity arising from the surface-anchoring phenomenon, whereby the charged end-group of entering free radical remains attached to the surface; this mechanism will be discussed in detail in section 4. Both of these phenomena (changes in $\bar{n}$ and in the MWD) are now explored further. In each case, seeded emulsion polymerization is employed, with a pre-formed seed of the appropriate monomer.

3.1. PROPAGATION IN GLASSY SYSTEMS

In glassy polymers such as MMA and styrene, the propagation rate coefficient becomes diffusion-controlled [11-13] at high conversion, resulting in a dramatic decrease in k_p. This places inherent limitations on the rate at which residual monomer may be polymerized.

3.2. $\bar{n}$ IN RUBBERY SYSTEMS

The high conversion behavior of $\bar{n}$ in MMA is made complex by the glassy nature of the system, in addition to any complexities arising from low monomer concentration *per se*. To elucidate the reported behavior in $\bar{n}$, we carry out observation in the rate at low monomer fraction in a *rubbery* polymer, BMA. In a rubbery system, diffusion coefficients also drop as conversion increases but not as drastically as in a glassy polymer, and k_p is not expected to become diffusion-controlled. To eliminate any peculiarities which might arise from purely physical effects of low monomer concentration (as might conceivably arise form non-ideal mixing of monomer and polymer, etc.), studies were also performed where the fraction of monomer w_m was very small, but polymer fraction w_p was intermediate, by adding butyl *iso*-butyrate, the saturated analog of BMA. Figure 6 shows the measurements of $\bar{n}$ as a function of w_p for undiluted BMA at different initiator concentrations, obtained from dilatometry experiments (the relation between polymerization rate and $\bar{n}$ requires the value of k_p, which is well established for this monomer [14,15]. A decrease in $\bar{n}$ is seen for this rubbery system, as was previously reported for MMA.

Figure 7 shows the corresponding results with an inert diluent present (when of course w_m is no longer the same as $1-w_p$). Again a decrease in $\bar{n}$ was seen at high conversion. This indicates that the drop in $\bar{n}$ is a real effect caused by the low monomer concentration (consistent with the predictions of the surface-anchoring mechanism) and not caused by an anomaly at very high w_p.

It might be postulated that a side reaction of persulfate in the absence of aqueous-phase monomer at very high conversion may somehow be responsible for the observed decrease in $\bar{n}$. However, a similar decrease in $\bar{n}$ at high conversion is seen with the water-soluble azo-initiator 4,4'-azobis(4-cyanovaleric acid), showing that this effect is not a peculiarity of persulfate kinetics.

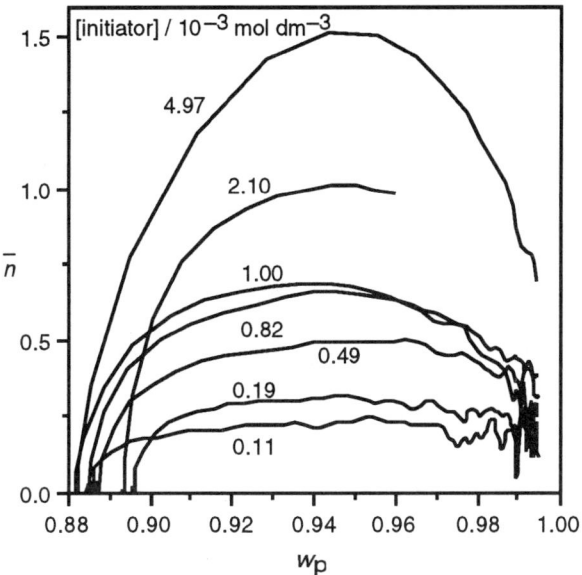

Figure 6. Experimental values of $\bar{n}$ vs. w_p for undiluted BMA at different concentrations of persulfate initiator; $N_p = 3.0 \times 10^{17}$ dm^{-3}, unswollen radius of BMA seed = 42.4 nm, 50 °C. Noise in $\bar{n}$ arises because $\bar{n}$ is obtained by numerical differentiation of dilatometric data.

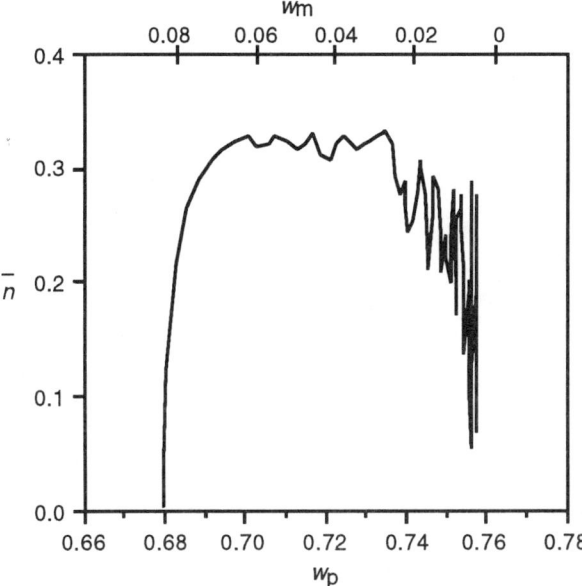

Figure 7. High-conversion $\bar{n}$ as function of w_p for BMA seeded emulsion polymerization of BMA, as in Figure 6, but with addition of inert diluent (butyl *iso*-butyrate, the saturated analog of BMA); [persulfate] $= 1.05 \times 10^{-3}$ mol dm^{-3}.

3.3. $\bar{n}$ IN GLASSY POLYMERS

The observation by Westmoreland and co-workers [9] of an increase in $\bar{n}$ in MMA as the reaction went to high conversion, followed by a drop at conversions > 95%, has also been seen by the present authors, as shown in Figure 8. The conversion at which the drop in $\bar{n}$ occurs is lowered as particle size is increased.

It is emphasized that the drop at higher conversion is not predicted by "convent-ional" emulsion polymerization models, including those which take all the complexities of chain-length-dependent termination [16] into account [8,17,18]. As seen in Figure 8, even those models which reproduce the observed behavior in these and other MMA systems (including γ-radiolysis relaxation data) at all except very high conversions, nevertheless predict a monotonic increase in $\bar{n}$. This is in contradiction to experiment, where $\bar{n}$ drops off dramatically at $w_p \approx 0.97$. The modelling includes the monomer partitioning and decrease in initiator efficiency discussed earlier, so the discrepancy is not caused by errors in the aqueous phase monomer concentration. Nor can this deficiency in model be ascribed to inaccuracies in the input dependence of k_p on w_p (that used here, obtained by early EPR studies [13], is deemed to be the best available), since there is no qualitative change if this dependence is varied within reasonable bounds. The same holds true for the possibility that the experimental behavior of $\bar{n}$ in Figure 8 is an artifact arising from calculation of this quantity from the rate obtained from dilatometric measurements (which requires both $k_p(w_p)$ and an assumed linearity in density with polymer fraction), since the same behavior is seen both in direct measurements of $\bar{n}$ by EPR [9] and in the dilute rubbery system of Figure 7.

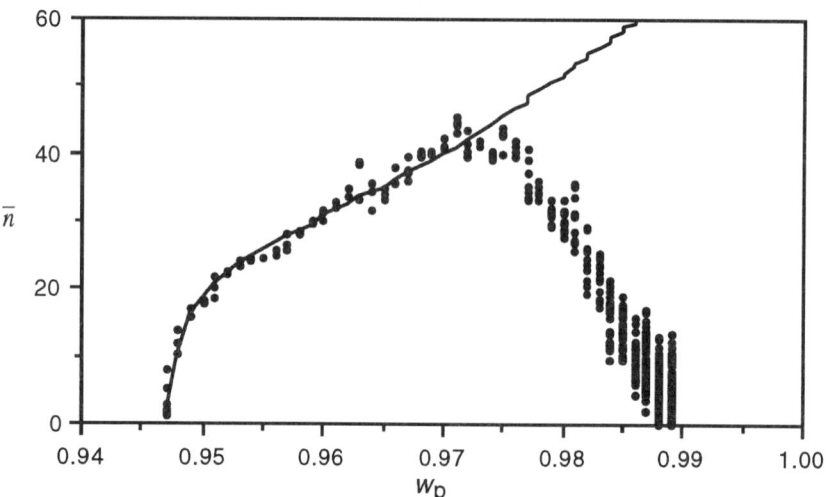

Figure 8. Observed (points) and modelled (line) dependence of $\bar{n}$ on w_p for MMA seeded emulsion poly-merization at 50 °C. Conditions: $N_p = 1.6 \times 10^{17}$ dm^{-3}, [persulfate] = 2.3×10^{-3} mol dm^{-3}, swollen particle radius = 54 nm. The simulations used models and parameter values in [8,17,18].

4. Spatial Inhomogeneities from Surface-Anchoring

The drop in $\bar{n}$ cannot be due entirely to a decrease in the entry rate of initiator-derived free radicals (see Figure 3), since this was also taken into account in the model prediction for MMA in Figure 8. During most of the duration of an emulsion polymerization, the dominant termination mechanism is between a long, relatively slow, entangled polymer radical, and a short, rapidly diffusing, radical (from transfer and/or initiator [8,17]). This should hold in a rubbery polymer even to complete conversion because the diffusion coefficients of short species do not vary drastically as w_p increases. The predicted monotonic increase in $\bar{n}$ arises from the gel-effect phenomenon of a decreasing average termination rate coefficient arising from these short radicals become longer and longer prior to termination as w_p increases [18]. There must be another mechanism occurring at high conversion which is not being accounted for in the model.

The drop in $\bar{n}$ observed can be explained by surface-anchoring: spatial inhomogeneities occurring in the latex particle at high conversion, resulting in localization of radicals in the outer volume (shell) of the particle, causing an increase in the termination rate and thereby lowering $\bar{n}$ [2,10,19]. When an oligomeric radical enters a latex particle, it is likely that the initiator-derived sulfate end group is anchored on the surface due to its hydrophilicity. At intermediate conversions, where there is plenty of available monomer, the radical end can quickly grow by propagation away from the surface of the particle. If transfer occurs (as is very likely), this spatial randomization will be increased even further by the rapid diffusion of the short species derived from transfer, prior to its propagation. However, at high conversion, the monomer concentration is much lower, and so the radical cannot grow as quickly away from the surface; the rate of transfer will be similarly reduced. This causes the radical concentration to be shifted from being evenly distributed throughout the particle to being concentrated in the shell of the particle, which can lead to an increased rate of termination by an entering radical in the shell of the particle, causing the overall drop in $\bar{n}$. The postulated mechanism is illustrated in Figure 9.

As initiator concentration is increased the rate of termination will also increased. The mechanism of Figure 9 suggests that the w_p at which the drop in $\bar{n}$ occurs should then become lower.

5. Molecular Weight Distributions

It has been pointed out above that a number of workers have observed a large increase in the amount of low ($\leq 10^5$) molecular weight material formed at very high conversion in styrene [19,20] and in MMA [9]. An example of this is given in Figure 10. This shows the molecular weight distribution as ln(number distribution) [denoted ln$P(M)$], where the number distribution $P(M)$ is the number of chains with molecular weight M, this being obtained from the GPC distribution $w(\log M)$ by means discussed elsewhere [21,22]; approximately, $P(M) = w(\log M)/M^2$. The appearance of a large excess of low molecular weight species at high conversion is most apparent. This is consistent with the surface-anchoring mechanism [2], since it is apparent from Figure 9 that this should lead to low molecular weight polymer produced in the shell of the latex particles. This could influence film-forming and water-resistance properties in the final application of the polymer.

106

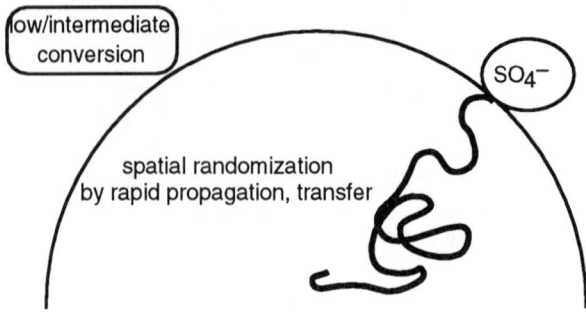

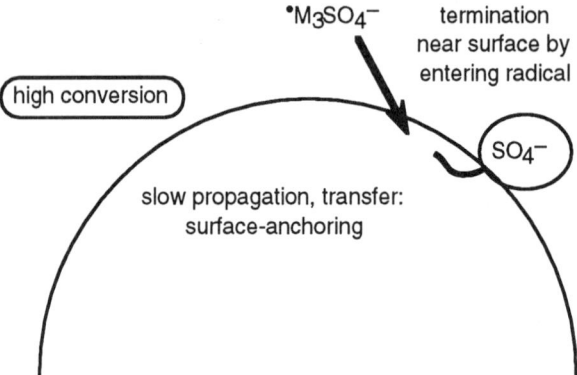

Figure 9. The suggested mechanism whereby surface-anchoring leads to a decrease in $\bar{n}$ because of increased termination in the outer shell at high conversion. This shows the expected differences in distribution of radicals between intermediate and very high conversion regimes based on chain growth rates.

6. Conclusions

Emulsion polymerization at high conversion poses the challenge of removal of residual monomer. However, the effects which operate at very high conversion also open up some challenges and possibilities of extending or shortening growth in this regime (e.g., by controlled feed) to improve control of polymer products and processes.

Acknowledgment. Support of this project by Zeneca UK, and helpful interactions with Drs John Padget and Dennis Keight of that company, and with Dr Tom Davis, are gratefully acknowledged. DK appreciates the support of an Australian Postgraduate Research Award.

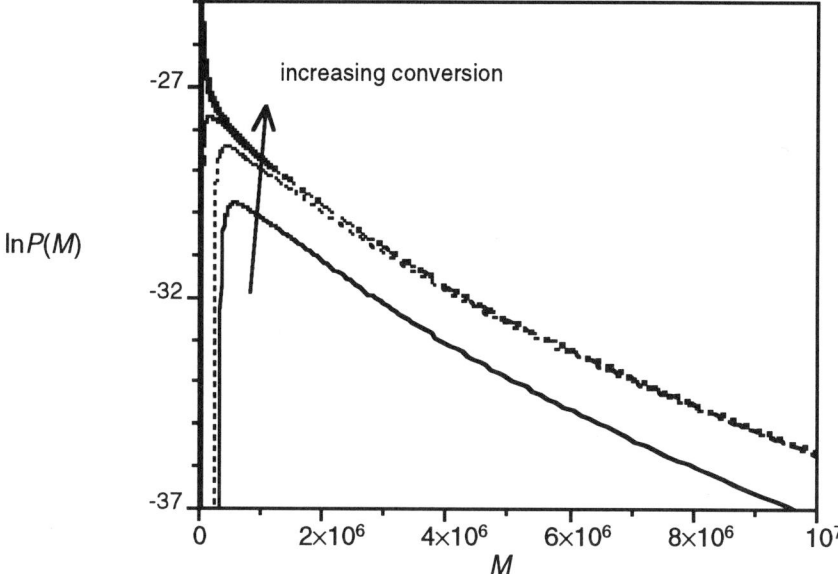

Figure 10. Cumulative number molecular weight distribution for seeded styrene emulsion polymerization of 77 nm unswollen radius particle, with the component due to the seed removed. Samples taken at 36, 53, 60 and 93 % conversion. Re-processed from data in [20,23].

References

[1] Maxwell, I.A., Morrison, B.R., Napper, D.H. and Gilbert, R.G. (1991) *Macromolecules,* **24,** 1629.
[2] Gilbert, R.G. (1995) *Emulsion Polymerization: A Mechanistic Approach,* Academic, London.
[3] Gilbert, R.G. (1997) in: *Polymeric Dispersions. Principles and Applications* (eds. J.M. Asua) p.1, Kluwer Academic, Dordrecht.
[4] Vanzo, E., Marchessault, R.H. and Stannet, V. (1965) *J. Colloid Sci.,* **20,** 62.
[5] Maxwell, I.A., Kurja, J., van Doremaele, G.H.J., German, A.L. and Morrison, B.R. (1992) *Makromol. Chem.,* **193,** 2049.
[6] Maxwell, I.A., Kurja, J., van Doremaele, G.H.J., German, A.L. and Morrison, B.R. (1992) *Makromol. Chem.,* **193,** 2065.
[7] Noël, L.F.J., Maxwell, I.A. and German, A.L. (1993) *Macromolecules,* **26,** 2911.
[8] Russell, G.T., Gilbert, R.G. and Napper, D.H. (1993) *Macromolecules,* **26,** 3538.
[9] Parker, H.-Y., Westmoreland, D.G. and Chang, H.-R. (1996) *Macromolecules,* **29,** 5119.
[10] Gilbert, R.G. (1995)*Trends in Polym. Sci.,* **3,** 222.
[11] Lau, W., Westmoreland, D.G. and Novak, R.W. (1987) *Macromolecules,* **20,** 457.
[12] Lau, W. and Westmoreland, D.G. (1992) *Macromolecules,* **25,** 4448.
[13] Ballard, M.J., Gilbert, R.G., Napper, D.H., Pomery, P.J., O'Sullivan, P.W. and O'Donnell, J.H. (1986) *Macromolecules,* **19,** 1303.
[14] Hutchinson, R.A., Paquet, D.A., McMinn, J.H. and Fuller, R.E. (1995) *Macromolecules,* **28,** 4023.
[15] Hutchinson, R.A., Paquet, D.A., McMinn, J.H., Beuermann, S., Fuller, R.E. and Jackson, C. (1995) *DECHEMA Monographs,* **131,** 467.
[16] Benson, S.W. and North, A.M. (1962) *J. Am. Chem. Soc.,* **84,** 935.
[17] Russell, G.T., Gilbert, R.G. and Napper, D.H. (1992) *Macromolecules,* **25,** 2459.
[18] Scheren, P.A.G.M., Russell, G.T., Sangster, D.F., Gilbert, R.G. and German, A.L. (1995) *Macromolecules,* **28,** 3637.
[19] Miller, C.M., Clay, P.A., Gilbert,R.G. and El-Aasser, M.S., *J. Polym. Sci., Polym. Chem. Edn.,* (in press).
[20] Clay, P.A., Christie,D.I. and Gilbert, R.G., (in preparation).
[21] Shortt, D.W. (1993) *J. Liquid Chromat.,* **16,** 3371.
[22] Clay, P.A. and Gilbert, R.G. (1995) *Macromolecules,* **28,** 552.
[23] Clay, P.A. (1995) PhD thesis, Sydney University.

PREPARATION OF LATEXES USING MINIEMULSIONS

M. S. EL-AASSER*
*Department of Chemical Engineering and Emulsion Polymers
Institute, Iacocca Hall
Lehigh University, Bethlehem, PA 18015, USA*

C. M. MILLER
*Union Carbide Corporation, UCAR Emulsion Systems
410 Gregson Drive, Cary, NC 27511, USA*

1. Introduction

The miniemulsion process is an emulsification process which involves the use of mixed emulsifier combinations, comprising a mixture of an ionic surfactant, such as sodium lauryl sulfate, and a cosurfactant, such as a long-chain alcohol, e.g. cetyl alcohol [1], or a long-chain alkane, e.g. hexadecane [2]. The product of such process is a stable oil-in-water emulsion with an average droplet diameter in the range of 100-400 nm.

Polymer latexes can be prepared by the miniemulsification process via two different routes, depending on whether the oil phase to be emulsified is a polymer solution or a monomer. The first route is by direct emulsification of a polymer solution into an aqueous phase and the subsequent removal of the solvent used in reducing the original viscosity of the polymer. The second route involves preparation of the monomer miniemulsion followed by initiation of polymerization in monomer droplets, using water-soluble or oil-soluble initiators.

In this paper, a review with references will be given of the principles of preparation of miniemulsions, polymerization of monomer miniemulsions, and applications of miniemulsions.

2. Preparation of Oil-In-Water Emulsions

2.1. PRINCIPLES OF EMULSIFICATION

The emulsification of an oil in water is the result of two competing processes: the dispersion of the bulk oil phase into droplets and the coalescence of the droplets to form the bulk phase. Coalescence is favored over dispersion from the point of view of the free energy. Thus the coalescence of droplets must be prevented or retarded if the dispersion of droplets is to be stable. The efficiency of this dynamic emulsification process is determined by the relative efficiencies of formation and stabilization of droplets, which are determined by: (i) the intensity and duration of agitation; (ii) the type and concentration of surfactants; (iii) the mode of addition of the surfactant, and the oil and water phases; (iv) the density ratio of the two phases; (v) the temperature; and (vi)

---

* Corresponding author.

J. M. Asua (ed.), Polymeric Dispersions: Principles and Applications, 109–126.
© 1997 *Kluwer Academic Publishers. Printed in the Netherlands.*

the viscosities of both phases. Much work has been done to correlate these parameters with the stability and droplet size of the emulsions [3-8].

Since preparation of an emulsion is a kinetic process, the droplet size is sensitive to the intensity of the agitation employed. Many different types of emulsification equipment are available (e.g., mixers, colloid mills, or homogenizers). Also used are ultrasonifiers, which convert electrical energy to high-frequency mechanical energy, and electric dispersers, in which the oil streaming through a capillary is subjected to a high positive potential, breaking it into droplets which are collected in an immiscible medium [9]. One interesting feature of electric dispersion is the uniformity of the emulsion droplet size [9,10].

With increasing surfactant concentration, the emulsion droplet size decreases due to the decrease in the oil/water interfacial tension to a low plateau value. There are several guidelines to the choice of the surfactant [5]: (i) it must have a specific molecular structure, with polar and nonpolar ends; (ii) it must be more soluble in the water phase so as to be readily available for adsorption on the oil droplet surface; (iii) it must adsorb strongly and not be easily displaced when two droplets collide; (iv) it must reduce the interfacial tension to 5 dynes/cm or less; (v) it must impart a sufficient electrokinetic potential to the emulsion droplets; (vi) it must work in small concentrations (vii) it shouid be relatively inexpensive, nontoxic, and safe to handle. A wide variety of commercial surfactants fulfill these requirements.

The temperature has only an indirect effect on emulsification because of its effect on viscosity, surfactant adsorption, and interfacial tension. An increase in the density difference between the oil and water phases results in a decrease in droplet size owing to the different velocities imparted to the two phases during emulsification.

2.2. EMULSIFICATION USING MIXED EMULSIFIER SYSTEMS

The emulsification of oil in water by mechanical shear using practical concentrations of conventional surfactants, such as sodium lauryl sulfate, usually gives average droplet sizes in the range of 2-5 μm, and at best, as small as 1 μm. The emulsions have broad size distribution so that an emulsion with an average droplet size of 1 μm contains droplets as small as 0.5 μm. However, it was shown that [1,2,11-14] fluid, opaque styrene-in- water emulsions of 100-200 nm average droplet size were prepared by simple stirring using 0.5-2% of the sodium lauryl sulfate- cetyl alcohol mixed emulsifier system. These important and initially unexpected results have spurred a large amount of research on the conditions under which miniemulsions may be formed and the mechanisms for their formation and stabilization. Since this initial work, many alternative cosurfactants have been proposed. In most cases these alternative "cosurfactants" have no surface activity but instead act as simple diffusion retarders (e.g., long chain alkanes, oil-soluble initiators, polymer of certain molecular weights). Regardless of the cosurfactant employed the shear device has a large influence on the droplet size obtained. The most efficient shear devices for the miniemulsification process are the Manton Gaulin Submicron Disperser (Gaulin Corp.), the Microfluidizer (Microfluidics Corp.) and Ultrasonication homogenizers. The following reviews some studies on the conditions for miniemulsion preparation and stabilization.

2.2.1. *Emulsification using Fatty Alcohols*
When a fatty alcohol is employed as a cosurfactant, the resulting miniemulsion droplet size distribution and stability are sensitive to a large number of variables including

alcohol chain length, molar ratio of alcohol to anionic surfactant, order of addition of materials, etc. In these systems it has been hypothesized that the droplet stability is attributable to two distinct mechanisms: the formation of a surfactant/cosurfactant interfacial complex preventing coalescence of droplets, and the reduction of the rate of diffusion of the dispersed phase from smaller to larger droplets. In most cases it is expected that both of these mechanisms contribute to the stabilization, but it has yet to be conclusively determined to what extent each of these are contributing. In order to characterize the effect of interfacial association between the fatty alcohol and the anionic surfactant, a large amount of research has been devoted to studying dilute aqueous solutions of these materials. The presence of liquid crystals in aqueous solutions of sodium lauryl sulfate/cetyl alcohol has been proven using birefringence [15,16,17,18] and viscosity measurements [5,18]. It has been hypothesized that these liquid crystals may continue to exist to some extent even after emulsification with an oil. Proof of this was provided by Hessel [19] who used phase-contrast light microscopy to show multilameller structures at the surface of miniemulsion droplets prepared from aqueous solutions of sodium lauryl sulfate/cetyl alcohol without homogenization. However, these droplets were approximately 1μm in diameter, considerably larger than those typically found in miniemulsion systems. The reduction of droplet size by homogenization may change the droplet/water interface. The following summarizes some additional results on the stability of miniemulsion droplet stabilized using fatty alcohol/anionic surfactant systems. It will be seen below that the nature of the interface in these systems plays an important role in not only the stability of these droplets, but also the kinetics of miniemulsion polymerization.

Several researchers have found that the order of mixing is an important variable in the preparation of stable miniemulsions using fatty alcohols [1, 20,21,22]. The common conclusion of these researchers is that in order to form stable miniemulsions, a gel phase [23] consisting of the fatty alcohol, the ionic surfactant, and the water must first be prepared by agitation at a temperature above the melting point of the fatty alcohol. When such a gel phase is prepared, spontaneous formation of miniemulsions with gentle stirring has been reported to occur [22]. This behavior is consistent with the interfacial barrier theory for stabilization of miniemulsions. On the other hand it has also been recognized that the formation of a gel phase may not be required when homogenization is used to prepare a miniemulsion. Again these differences may be due to the different mechanisms operative depending upon the droplet size distribution of the miniemulsion droplets.

The fatty alcohol chain length and the fatty alcohol-to-surfactant ratio are important variables for the formation of stable miniemulsions. Both Choi [20] and Ugelstad [24] found that the stability of miniemulsions prepared with fatty alcohols increases as the chain length of the alcohol increases (C_{12} -C_{20}). These results seem most consistent with the fatty alcohol acting as a diffusion retarder rather than forming an interfacial barrier. On the other hand, several researchers have noted that an optimum cosurfactant to surfactant ratio exists when fatty alcohols are used to stabilize miniemulsions. These data suggest some type of order or packing at the droplet/water interface preventing coalescence of droplets.

Chou et al. [25] obtained the best stability for styrene and benzene miniemulsions at a molar ratio of cetyl alcohol to ionic surfactant between 1/1 and 3/1. Choi also obtained maximum stability for his miniemulsions of styrene stabilized with sodium lauryl sulfate and cetyl alcohol in this range [20]. These researchers formed their miniemulsions by spontaneous emulsification in a pre-formed gel phase of fatty alcohol and anionic surfactant. In contrast, Goetz [21] showed using transmission electron microscopy and light scattering that a miniemulsion's average droplet size decreases with increasing cosurfactant to surfactant ratio beyond 3/1 indicating that an optimum ratio of cetyl alcohol to sodium lauryl sulfate does not exist. This result was also found by Miller et al [26]. using capillary hydrodynamic fractionation to determine the size of the miniemulsion droplets. Both of these latter researchers used a sonifier to homogenize the droplets to small size.

The apparent discrepancy between the above results may be due to the effect of droplet size reduction through homogenization on the molecular association of fatty alcohols with anionic surfactants. As mentioned previously, the stability of miniemulsions prepared using fatty alcohols may be attributed to either the formation of interfacial complexes between the surfactant and the cosurfactant preventing coalescence of droplets, the reduction of the rate of diffusion of the dispersed phase from smaller to larger droplets, or a combination of both. By this reasoning, it is possible that different mechanisms are operative depending upon the miniemulsions' droplet sizes. Specifically, for smaller droplets, (i.e., those prepared using homogenization) the stabilization may be mostly due to retardation of the rate of diffusion, while for larger droplets (those not homogenized) the stabilization may be mostly due to the presence of interfacial complexes. It follows from this mechanism that the molar ratio of cetyl alcohol to sodium lauryl sulfate should be more important for larger droplets (prepared without homogenization) than for smaller droplets prepared using a high shear homogenizer.

Finally, the water solubility of the dispersed phase has also been shown to affect a miniemulsion's droplet size and stability. Specifically, Brouwer et al. [27] showed that the ease of formation of miniemulsions increases with increasing water solubility of the dispersed phase, but the overall stability of the miniemulsions decreases with increasing water solubility. This behavior is also expected for emulsions stabilized using conventional surfactants.

2.2.2. *Emulsification using Highly Water Insoluble Compounds*
A large amount of research has also been conducted on miniemulsions prepared using water insoluble compounds such as alkanes and oil soluble initiators. Ugelstad [24,28] has suggested that the stabilization of miniemulsion using compounds such as hexadecane as a cosurfactant could be explained by the Higuchi and Misra concepts [29]. These rationalize stabilization by retardation of the interdroplet diffusion of the oil phase due to the presence of the relatively water-insoluble hexadecane (10^{-8} M) inside the droplets.

According to Higuchi and Misra, emulsion degradation by diffusion is the result of increasing water solubility of the oil with decreasing droplet radius. Accordingly, in an emulsion system the small droplets are thermodynamically unstable and tend to dissolve while the larger droplets grow at their expense. Since the rate of emulsion degradation is controlled by the diffusion rate of the compound with least water solubility, the presence of the relatively water-insoluble hexadecane or fatty alcohol in the miniemulsion causes its stabilization by retarding the diffusion rate of the oil phase. Smith et al. [30] correlated the rate of emulsion degradation to the ratio of the partition coefficients of the less water-soluble to the more water-soluble components; the higher the ratio the more efficient the compound in stabilizing the emulsion. Ugelstad emphasized that the water-insoluble compound must also have a relatively low molecular weight [24].

Ugelstad [24,28] treated the emulsification process on the basis of thermodynamic concepts where the stability of miniemulsions is considered to be the result of a semi-equilibrium which is established when the concentration potential, developed due to the transport of the more-water-soluble compound from the small to the large droplets, balances the potential energy due to the difference in droplet size. Ugelstad also explained the increased swelling capacity of the miniemulsion droplets with monomer. The large energy of mixing, mainly entropy of mixing, due to the interaction with the cosurfactant balances the increased interfacial energy due to large increase in particle size as a result of swelling with monomer

Delgado [31] studied the stability of miniemulsions prepared using hexadecane and concluded that there was no optimum ratio of co-surfactant to surfactant. This result was also found by Rodriguez [32]. Both researchers noted that as the molar ratio of hexadecane to surfactant was increased, the shelf-life and centrifugational stabilities of the emulsions increased. However, they also noted for molar ratios greater than 4/1, the stability increased only slightly. Alduncin et al. [33] showed that oil soluble initiators can also be employed for preparing stable miniemulsions. In their studies it was found that the stability is sensitive to the water solubility of the oil soluble initiator employed.

Finally, it has been reported that under certain conditions polymer predispersed into the oil phase can aid the stability of homogenized monomer droplets [34]. However, the use of polymer as a true cosurfactant seems debatable since, as pointed out by Ugelstad [24,28], the oil droplets should be unstable to diffusion degradation by the Morton-Kaizerman equation [35]. In this case it is likely that polymer is dispersed by the homogenizer forming small polymer particles at the time of polymerization initiation. Thus these systems are probably neither conventional nor miniemulsions but rather a form of seeded emulsion polymerization.

3. Preparation of Latexes by Direct Emulsification of Polymer Solutions

This section describes the preparation of artificial latexes prepared by direct emulsification of polymer solution in water. Until recently, the artificial latexes (also referred to as pseudolatexes) were the least important of the three categories outlined in the introduction section. Artificial latexes were typified by aqueous dispersions of reclaimed rubber, butyl rubber, and stereoregular rubbers such as cis-1,4-polyisoprene. Since emulsion polymerization is limited to monomers which can be polymerized by free-radical vinyl-addition polymerization, latexes of polymers which cannot be prepared by this method can only be prepared by dispersion of bulk polymer in an aqueous medium, for example, by emulsification of the fluid polymer in water.

3.1. CONVENTIONAL VS. MINIEMULSION METHODS OF PREPARING ARTIFICIAL (OR PSEUDOLATEXES)

Artificial latexes (or pseudolatexes) of polymers such as epoxy resins, polyester, ethylcellulose, stereoregular rubber, and polyurethane, may be prepared by the dispersion of the corresponding bulk polymer (or a solution of the polymer) in an aqueous medium using conventional surfactants and emulsification methods. The various methods for the preparation of latexes by emulsification of polymer solutions were reviewed by Blackley [36] and Warson [37]. Basically, there are three different methods for the preparation of artificial latexes by emulsification:

1.   Direct Emulsification [38-41]. The liquid polymer or polymer solution in a volatile water-immiscible organic solvent (or mixture of solvents) is emulsified in water containing surfactant using conventional emulsification methods, and the emulsion is steam-distilled to remove the solvent.

2. Inverse Emulsification [42-45]. The liquid polymer or polymer solution in a volatile water-immiscible organic solvent (or mixture of solvents) is compounded with a long-chain fatty acid (e.g., oleic acid) using conventional rubber mixing equipment and mixed slowly with a dilute aqueous base, to give a water-in-polymer emulsion, which then inverts to a polymer-in- water emulsion as more aqueous base is added; the emulsification is then steam-distilled to remove the solvent (if used).

3. Self-Emulsification [46-48]. The polymer molecules are modified chemically by the introduction of basic (e.g., amino) or acidic (e.g., carboxyl) groups in such concentration and location that the polymer becomes self-emulsified without surfactant upon dispersion in acidic or basic solutions.

With all three methods, the emulsification may be carried out at elevated temperatures to lower the viscosity of the polymer or polymer solution. Emulsification at temperatures above that of boiling water may be carried out under pressure.

Self-emulsification gives average particle sizes as small as 100nm, much smaller than the other two methods and fully competitive with those produced by emulsion polymerization. However, one major disadvantage is the hydrophilic functional groups of the polymer which make the coating films water- sensitive. Moreover, the concentration and location of the functional groups is critical: with too low a concentration or improper location of the functional groups, the polymer is not self-emulsifiable; with too high a concentration, the polymer forms a polymer solution upon emulsification-neutralization. Thus, although self-emulsification gives average particle sizes which are competitive with those produced by emulsion polymerization, its applications are limited by the water sensitivity of the films.

The latexes prepared by direct or inverse emulsification have average particle sizes in the range 1-10 µm with a small- particle size tail extending to about 0.5 µm, about 5-10-fold larger than the 100-300 nm average particle size of commercial coatings latexes prepared by emulsion polymerization. This 5-10 fold difference in particle size is responsible for the inferior film properties and poor shelf stability of these artificial latexes compared to the synthetic latexes prepared by emulsion polymerization. Consequently, a substantial decrease in the average particle size of latexes prepared by direct or inverse emulsification would be an important contribution to the development of water-based coatings.

3.1.1. *Critical Particle (or Droplet) Size for Settling or Creaming, and Influence of Particle Size Latex Film Formation*

The critical size for settling (or creaming) of an emulsion may be calculated from the criterion of Overbeek [49], which states that colloidal particles that settle at a rate of only 1 mm in 24 hrs according to Stokes' law will never settle in practice, because of the Brownian motion of the particles and the thermal convection currents arising from small temperature gradients in the sample. The Brownian motion, which results from the unbalanced collisions of solvent molecules with the colloidal particles, increases in intensity with decreasing particle size. The convection currents depend upon the sample size and the storage conditions.

The rate of settling (or creaming) of spherical particle according to Stokes' law is:

$$\text{Rate of settling (or creaming)} = (dp^2/18 \, \eta) \, (\rho_p - \rho_{medium})g \quad (1)$$

where dp is the particle diameter, η the viscosity of the medium, ρ_p and ρ_{medium} the densities of the particle and the medium, respectively, and g the gravitational constant. Substituting the foregoing settling rate of 1 mm in 24 hr. into eq (1) gives the critical particle size for settling. Figure 1 shows the variation of critical particle size with density difference between the particles and the medium as a function of viscosity of the medium calculated from this equation.

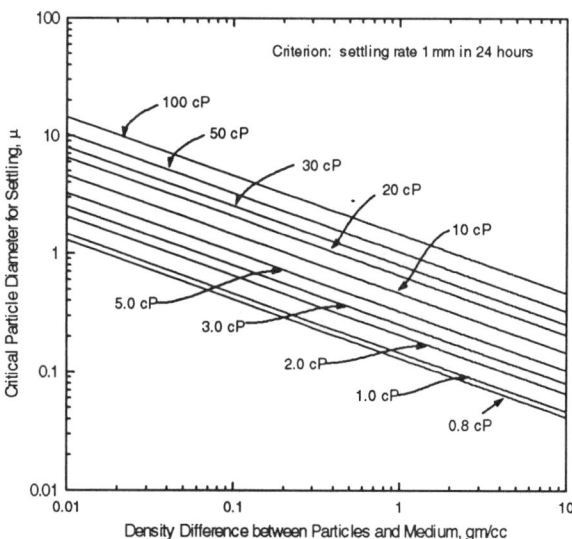

Figure 1: Variations of the critical particle diameter for settling (or creaming) with density difference between the particles and the medium as a function of viscosity of the medium.

For polystyrene latex particles in water, the density difference is 0.05 g/ml and the viscosity of the medium is about 1 cps; therefore, from Figure 1, the critical particle diameter for settling is 650 nm. This calculated size is in good agreement with the foregoing experimental observation [50] that monodisperse polystyrene latex particles of

800 nm diameter settled within 3-6 months and particles of 500 nm diameter or smaller never settled. Since most of the polymers to be emulsified to form artificial or pseudolatexes have densities in the range 1.10-1.15 g/ml, their critical particle diameters for settling would be 300 nm or smaller. Therefore, it is critical whether the emulsification process produces droplets of 1000 nm or 200 nm diameter.

The artificial or pseudolatex used for coatings must form a continuous film upon drying under given conditions. The forces exerted on the latex particles during drying are those arising from the water-air and polymer-water interfacial tensions [51,52]. The maximum shear modulus of a polymer particle that can coalesce upon drying from an aqueous latex is calculated to be about 11×10^6 Pa for a particle diameter of 100 nm at 30 dynes/cm surface tension [53]. This maximum shear modulus decreases inversely with increasing particle size, that is, the maximum shear moduli for coalescence of particle diameters of 1000 nm and 10,000 nm are 11×10^5 and 11×10^4 Pa, respectively. Thus, the larger the latex particle size, the softer must be the polymer for the particle to coalesce upon drying. If the shear modulus of the polymer is too high for the latex particle size, the coalescence will be incomplete and the film properties will be poor.

3.1.2. *Preparation of Artificial Latexes by Miniemulsification*

The miniemulsification process, as described in Section 2.2, was used in the preparation of a wide variety of polymer latexes [54-58]. Examples of such polymers include polystyrene, poly(vinyl acetate), epoxy resins, epoxy resin curing agents, ethylcellulose, cellulose acetate phthalate, polyesters, alkyd resins, rosin derivatives, synthetic natural rubbers, poly(vinyl butyral), Kraton (triblock styrene-butadiene-styrene copolymer, Shell Chemical Co.), EPDM, and silicons.

Fully cured and air-drying polyurethane latexes can also be prepared by miniemulsification [55, 56, 58, 59]. The approach used in the preparation of fully-cured polyurethane latexes involves dissolution of the urethane prepolymer (prepared by solution polymerization) in a reactive monomer system followed by the miniemulsification to form miniemulsion droplets. This is followed by polymerization of the monomers in the miniemulsion droplets using oil-soluble or water-soluble initiators to form a urethane/acrylic latex system. A similar approach was recently used to make a alkyd/acrylic latex system [60].

3.1.3. *Preparation of Core-Shell Latexes Using the Miniemulsification Process*

Latex systems with core/shell particle morphology are used in solving many practical problems in several application areas such as adhesives, sealant, self-crosslinking thermoset coatings, paper coatings and as impact modifiers for many polymer matrices. Core/shell latex particles are usually prepared by a two-stage emulsion polymerization process. The core latex is prepared in the first step, which is then used as seed in the second stage polymerization to coat the particles with the second polymer.

An alternative approach to produce core/shell latexes is to use the miniemulsification process to prepare the core latex which is then used as seed in an emulsion polymerization process to overcoat the particles with the shell polymer. This approach offers the advantage that one can select any type of polymer to be used as seed and not be limited to polymers prepared by emulsion polymerization.

The applicability of this approach was demonstrated in the preparation of a latex system with particles of rubbery core/glassy shell [61]. The rubbery core latex was prepared by miniemulsification of Kraton solution (triblock styrene- butadiene-styrene copolymer, Shell Chemical Co.). The emulsifier used in this process was a (70:30) mixture of Aerosol TR-70 and Aerosol-22 (American Cyanamid), and a molar ratio of cetyl alcohol/ emulsifier of 2:1. The polymer miniemulsion was then vacuum distilled to remove the solvent used to reduce the viscosity of the Kraton solution. The resulting Kraton latex was subjected to two ion-exchange cycles to adjust the level of the emulsifier. The Kraton latex was then used as seed in an emulsion polymerization process to coat the particles with a layer of poly(methyl methacrylate) such that the core/shell ratio was 65/35 by weight.

4. Preparation of Latexes via Polymerization of Monomer Miniemulsions

The probability of particle nucleation in monomer droplets was dismissed in conventional emulsion polymerization based on the unfavorably small surface area, and fewer number, of the monomer droplets to compete effectively with the monomer-swollen micelles. This is due to the relatively large size monomer droplets (1-10 μm in diameter), compared to the size of the monomer-swollen micelles (generally 10-30nm in-diameter). However, despite this unfavorable statistical probabilities some monomer droplets capture radicals and polymerize to form large-size microscopic particle, which can easily be seen by examination of the final latex using optical microscopy [62].

The monomer droplets could become a significant locus for particle nucleation and polymerization if their surface area were large, i.e. if the droplet size could be made small. This is the basis for polymerization in miniemulsion and microemulsion monomer systems. The reduction in the monomer droplet sizes (generally 100-500 nm in-diameter for miniemulsions, and 5-40 nm in-diameter for microemulsions) increases both the number and surface area of the droplets by several orders of magnitude, relative to conventional emulsion droplets. This results in an effective competition of initiation in monomer droplets with other particle nucleation mechanisms such as micellar and aqueous phase. Indeed, Ugelstad, El-Aasser, and Vanderhoff demonstrated experimentally that monomer droplets could become the principal locus for particle formation in styrene miniemulsions systems [1]. This initial work has since been verified for many other systems and kinetic studies of this unique nucleation mechanism have been performed to give a detailed mechanism for these polymerizations.

4.1. EARLY WORK ON MINIEMULSION POLYMERIZATION

The original work in this area started with the idea of demonstrating that monomer droplets could serve as the main locus of initiation and polymerization in an emulsion polymerization process [1]. For this purpose styrene miniemulsions were prepared by emulsifying styrene monomer into an aqueous solution of sodium lauryl sulfate-cetyl alcohol (SLS-CA) mixture. The recipe used in this early work comprised 75 parts (by weight) water, 25 parts styrene, and three different concentration of SLS-CA mixtures 0.4-0.8,0.2-0.4 and 0.1-0.4 (by weight) parts. The polymerization was carried out using 0.25 parts of potassium persulfate initiator at 70°C. Optical and transmission electron microscopy showed that the initial size of the emulsion droplets was approximately the same as the latex particles. Thus, it was concluded that monomer droplets could be the principal locus for particle formation in miniemulsion systems because of their small size and enhanced stability. The droplet size of the styrene miniemulsion prepared with the above recipe was typically in the range 100-400 nm. The small droplet size results in a relatively large number of emulsion droplets and consequently a large surface area, which leads to an increased probability of capturing free radicals generated in the aqueous phase. The ratio of the two components of the mixed emulsifier system was found to play a role in determining the particle size distribution in these miniemulsion systems. For example, a bimodal particle size distribution was obtained with 2% by weight of SLS-CA mixture in the weight ratio of 1:4; where as a uniform particle size distribution was obtained upon using 5% of the SLS-CA mixture in the weight ratio of 1:2. The analogous cationic styrene miniemulsions and the corresponding cationic latexes with similar average particle size were prepared using the mixed emulsifier system of hexadecyltrimethylammonium bromide-cetyl alcohol (HDTMAB-CA) [63,64].

Another method for preparing miniemulsions involves the use of long-chain alkane such as hexadecane instead of the fatty alcohol as a cosurfactant [65]. According to this method the hexadecane, 1-2 weight percent based on the oil phase, was homogenized into aqueous solution of the ionic emulsifier using the Manton-Gaulin Submicron Disperser followed by the addition of the monomer which diffuses through the aqueous phase to swell the hexadecane emulsion droplets. The initiation of polymerization of the resulting miniemulsions, using water-soluble or oil-soluble initiators, takes place mainly in the monomer droplets. The swelling capacity of the resulting latex particles prepared in the presence of these types of cosurfactants, or in the presence of low-molecular weight oligomers, was found to greatly exceed that of conventional polymer latex particles by several orders of magnitude [28]. Ugelstad used this method to prepare relatively uniform large-size-particles [66].

4.2. KINETIC STUDIES OF MINIEMULSION POLYMERIZATION

Kinetic studies of miniemulsion polymerization have been performed by a number of researchers. These studies have not only demonstrated many differences between miniemulsion and conventional emulsion polymerizations, but they have also demonstrated differences in the polymerization mechanism for miniemulsions prepared using different types and concentrations of surfactants and cosurfactants. As an example of these differences, Figure 2 shows the polymerization rate versus time for a styrene conventional emulsion prepared with 5mM SLS, a styrene miniemulsion prepared with 10mM SLS and 40mM hexadecane, and a styrene miniemulsion prepared with 10mM SLS and 30mM CA measured using a Mettler RC1 reaction calorimeter [67]. In this experiment, the two miniemulsions were homogenized using a Microfluidizer, and the initiator concentration was 1.33mM based on the total water. The free surfactant for all three of these polymerizations was below the critical micelle concentration. From Figure 2 it is clear that the conventional emulsion polymerization gives a different polymerization rate curve than the miniemulsions, but it is also clear that the hexadecane and cetyl alcohol stabilized miniemulsions also give unique polymerization rate curves. Studies similar to these have led to many important mechanistic postulates for miniemulsion polymerization.

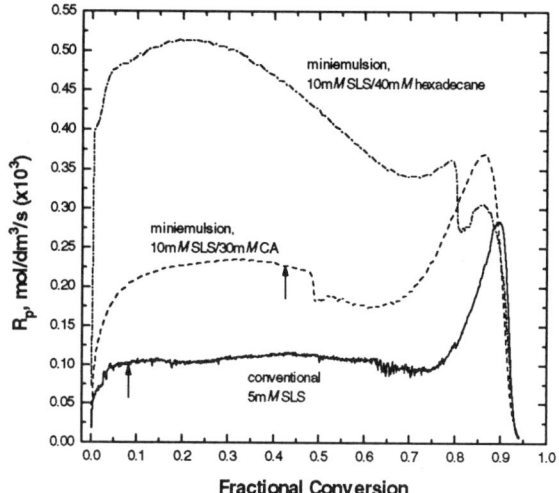

Figure 2: Polymerization rate vs. fractional conversion for styrene conventional emulsion polymerization and two styrene miniemulsion polymerizations. The polymerizations were conducted in the Mettler RC1 reaction calorimeter at 70°C. The arrows indicate the end of the nucleation period as determined by measuring the evolution of the particle size distribution [67].

4.2.1. *Mechanism for Particle Formation during Miniemulsion Polymerization using Fatty Alcohol Cosurfactants*

The polymerization kinetics of miniemulsions have been studied by a number of researchers to gain more information on the unique locus of particle formation (i.e., the droplet rather than the water phase). Several researchers have found that when cetyl alcohol is used as a cosurfactant the polymerization kinetics are characterized by a long, slow rise to a maximum rate of polymerization at about 40% conversion (see Figure 2). This increasing rate period is usually considered to correspond to the particle formation period, and its duration has spurred a number of mechanistic postulates concerning nucleation in these systems.

Choi et al. [20,68] were the first to report that styrene miniemulsions prepared using cetyl alcohol did not exhibit the interval-II characteristic of a conventional emulsion polymerization of styrene, i.e., a constant rate of polymerization. They hypothesized that the increasing rate period corresponded with the nucleation period, and the maximum polymerization rate was taken as the end of particle formation. This work also showed that the number of droplets becoming polymer particles varied with the initiator (potassium persulfate) concentration employed to the 0.3 power. Calculations were performed to estimate that approximately 20-30% of the initial droplets became polymer particles in the range of initiator concentrations investigated.

Miller et al. [69] experimentally determined the length of the nucleation period for a recipe similar to that employed by Choi [68] by withdrawing samples during the polymerization and analyzing the particle size distributions. This work showed a continuously decreasing nucleation rate ending at about 40-60% conversion, which was always past the maximum in the polymerization rate curves investigated (e.g., as illustrated in Figure 2). For most zero-one styrene polymerizations, formation of particles past the maximum polymerization rate would be quite unexpected. However,

Miller et al. [69] used thermodynamic arguments to explain how this occurs in a miniemulsion polymerization. According to their explanation, the presence of cetyl alcohol in the system prevents the disappearance of monomer droplets and results in a concentration of monomer in the polymer particles which decreases continuously with increasing conversion. As a result, the number of particles may increase but be offset by the decrease in the concentration of monomer in the polymer particles resulting in a maximum or decrease in the polymerization rate while particles are still being formed. An important implication of this work is that monomer droplets are present up to relatively high conversions. This implies that disappearance of monomer droplets by collision with polymer particles alone cannot explain the relatively small number of droplets which become polymer particles

Another common observation for miniemulsion polymerization using a fatty alcohol as the cosurfactant is that the particle size distribution obtained is quite broad, with variances on the order of 15 to 40% [1,68-71]. This result has been found even when high efficiency homogenizers such as a Microfluidizer or sonifier are used to provide a uniform initial droplet size distribution [70]. While the initial droplet size distribution obviously has a large effect on the resulting particle size distribution of the latex, Miller et al. [69] found that the length of the nucleation period was largely responsible for these broad particle size distributions. In their work, it was found that when the initiator (potassium persulfate) concentration was increased, the time until the maximum polymerization rate decreased resulting in a narrowing of the particle size distribution. It was hypothesized that given a sufficiently high initiator concentration, the final latex particle size distribution would narrow and approach the initial droplet size distribution. However, further analysis revealed that in order to approach this initial droplet size distribution a prohibitively high initiator concentration would be required.

The unusually long nucleation period obtained for miniemulsion polymerization stabilized using a fatty alcohol has been hypothesized to be caused by a "slow" entry of radicals into miniemulsion droplets. Chamberlain et al. [71] were the first to postulate that free radical entry into miniemulsion droplets stabilized using lauryl alcohol may be slow compared to similarly sized polymer particles. They proposed that an entering oligoradical must displace a surfactant molecule at the surfactant rich droplet/water interface before entering and propagating. Follow up work by Wood et al. [72] attempted to quantify the reduction in the pseudo first-order entry rate coefficient for a polystyrene seed preswollen with dodecane (not a fatty alcohol, but a swelling promoter) and styrene. Their work suggested that the entry rate coefficient was 1 to 2 orders of magnitude lower at low weight fractions of polymer and increased with increasing weight fraction of polymer. Choi et al. [68] also postulated slow radical entry into miniemulsion droplets. Their work showed that only approximately 20% of the initial droplets capture radicals, and that this fraction increased with increasing initiator concentration. Tang et al. [73] used a seeded approach similar to that used by Wood et al. [72] but using cetyl alcohol as a swelling promoter. This work also suggested that both the rate of entry and exit of radicals is reduced at low weight fractions of polymer in the seed latex.

Recently, Miller and coworkers [74-78] showed some further evidence for differences between miniemulsion droplets and latex particles formed therefrom. Their approach was to prepare miniemulsion droplets at low weight fractions of polymer by predisolving a small amount of polymer into the oil phase prior to preparing the miniemulsion. These "pre-formed polymer particles" were hypothesized to be similar to latex particles at low conversion (i.e., they looked at 0.05, 0.5, 1, and 2% polymer). It was found that the addition of a small amount of polymer had a dramatic effect on the polymerization kinetics and the number of latex particles formed. A key result of this study was that when 1% polymer was dissolved in the miniemulsion, the number of particles increased by 1 to 2 orders of magnitude and became invariant with initiator concentration. This result is reproduced in Figure 3. These results were taken as partial evidence that 100% of the droplets were present as latex particles at the end of the polymerization. Further proof of this was provided by distilling the monomer from the miniemulsion droplets prepared from polymer solutions before polymerization and measuring the pre-formed polymer particle size distribution. Using this technique it was shown that, within experimental error, the initial number of pre-formed polymer particles was the same as the final number of latex particles.

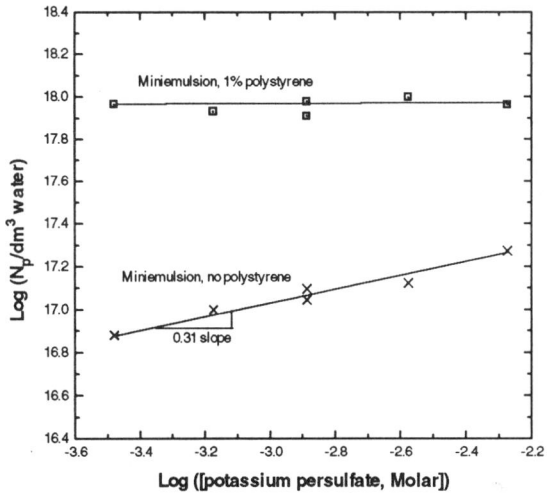

Figure 3: Effect of potassium persulfate concentration on the final number of polymer particles for styrene miniemulsions (x) and styrene miniemulsions with 1% polystyrene pre-dissolved in the oil phase (□). The miniemulsions were prepared from an aqueous gel phase consisting of 10mM SLS/30mM CA using a Microfluidizer. The polymerizations were conducted in the Mettler RC1 reaction calorimeter at 70°C. Reproduced from [76].

The results obtained by Miller et al. [74-78] appear to be the strongest to date for proving a difference between miniemulsion droplets and polymer particles. These data may ultimately provide the proof necessary to prove this important hypothesis, but as of now many questions must still be resolved about these experiments. For example, it is still unclear how the polymer added in this work is effective in increasing the number of droplets which become polymer particles. If the polymer truly is able to modify the interface allowing facile entry of radicals, some measurable property of the interface such as surface tension or surface viscosity should reflect this behavior. To date this evidence has not been conclusively provided. One would also expect the nature of the polymer end groups and the molecular weight of the polymer to effect this behavior in a predictable manner. There are also alternative explanations for the results that have not been disproved. For example, the polymer may provide additional stability for the small, uninitiated monomer droplets allowing them to more easily compete for radicals. While Miller et al. have addressed this possibility and shown it to be a less likely explanation [77], this and other explanations cannot be ignored. Finally, assuming cetyl alcohol does impede the particle nucleation process, the reason for this behavior is still unclear. Among the multitude of explanations advanced, the ability of fatty alcohols to complex with surfactants in the water phase seems to be the most probable source of an interfacial barrier preventing entry of radicals into miniemulsion droplets. As discussed above, a considerable amount of data exists suggesting some surface structure at the droplet/water interface. However, even if such a surface structure is present it has yet to be proven to effect the rate of entry into these droplets.

4.2.2. Mechanism for Particle Formation during Miniemulsion Polymerization using Highly Water Insoluble Cosurfactants

In contrast to the broad particle size distributions obtained when fatty alcohols are used as cosurfactants, Ugelstad, et al. [66,79] were the first to report that polystyrene latex with narrow particle size distribution can be prepared by initiation in monomer miniemulsion droplets by using hexadecane as a cosurfactant and adjusting the concentration of sodium dodecylsulfate and the homogenization pressure during the miniemulsification step. Tang et al. [70,73] studied the differences in the polymerization kinetics of styrene conventional emulsions and miniemulsions and stabilized with either cetyl alcohol or hexadecane. This work showed a faster initial rate of polymerization when hexadecane rather than cetyl alcohol was used as a cosurfactant which resulted in a more narrow distribution of latex particles. The faster initial polymerization rate is clearly seen in the data presented in Figure 2. In addition, Tang's work showed that the homogenization device employed has a large effect on the particle size distribution of the latex.

Recently, oil soluble initiators have been employed as cosurfactants for miniemulsion polymerization [33]. The advantage of these materials is that, unlike their long-chain alkane counterparts, they are consumed during the polymerization thus leaving no volatile organic compounds in the latex. In this work it was found lauroyl peroxide worked well for stabilizing miniemulsion droplets against degradation through molecular diffusion. Although the kinetics were not reported in this work, the latex particle size distributions were consistent with droplet nucleation when the initial droplet sizes were homogenized and then stabilized using lauroyl alcohol and sodium lauryl sulfate. More recently, dodecyl mercaptan was used as the cosurfactant in miniemulsion polymerization of methylmethacrylate [80].

4.3. MINIEMULSION COPOLYMERIZATION

Delgado et. al. [81-85] investigated the kinetics and other characteristics of miniemulsion copolymerization of vinyl acetate-n-butyl acrylate monomer mixtures and compared them to conventional copolymerization of the same system. Hexadecane was used as the cosurfactant along with sodium hexadecyl sulfate in preparing the miniemulsions. The use of hexadecane in the miniemulsions led to higher adsorption of surfactant, smaller droplet size, and higher stability of the emulsions against creaming. The copolymer composition during the initial 70% conversion was found to be rich in butyl acrylate monomer units for the miniemulsion process. The dynamic mechanical properties of the copolymer films showed less mixing between the poly(butyl acrylate)-rich core and the poly(vinyl acetate) rich shell in the miniemulsion latex films compared to the conventional latex films.

During the copolymerization process the miniemulsions gave lower overall polymerization rates and larger particles compared to conventional polymerizations.

However, the polymerization rate per particle was similar in both cases and was independent of the hexadecane concentration. The miniemulsion copolymerization showed a long nucleation stage, up to 22% conversion, characterized by an increase in the polymerization rate up to a maximum. The nucleation stage in conventional polymerization was much shorter, being completed at less than 10% conversion. Also, the dependence of the rate and number of particles on the initiator concentration was higher for the miniemulsion process. On the other hand, the surfactant concentration dependence of the number of particles for the miniemulsion process was less than for conventional process. The particle size distribution from miniemulsion polymerization was much broader, the standard deviation was 22% compared to 10% for conventional polymerization.

Rodriguez et al [86,87] investigated the miniemulsion copolymerization of 50:50 weight ratio of styrene:methyl methacrylate, using hexadecane as a cosurfactant. The effects of the hexadecane concentration and ultrasonication time on the conversion-time curves and latex particle size were studied. The results showed that an increase in hexadecane concentration and/or ultrasonication time and intensity during the miniemulsification process caused faster polymerization rate and smaller latex particle size. The mechanism of mass transport from miniemulsion droplets to polymer particles were studied by seeded emulsion copolymerization. The kinetic and particle size results from the seeded experiments suggested that collision (coalescence) between the miniemulsion droplets and polymer seed particles may play a major role in the transport of the highly water-insoluble compounds, such as hexadecane.

Aside from stabilizing the miniemulsion droplets allowing them to become the principal locus of polymerization, the cosurfactant also affects the partitioning of monomer between the different phases during miniemulsion polymerization. This is especially important in miniemulsion copolymerization where a different distribution of monomers at the reaction site will affect the copolymer sequence throughout the polymerization. A mathematical model based on thermodynamic and kinetic parameters was developed to describe the role of the hexadecane additive on the monomer distribution between the various phases, the polymerization rate, and the copolymer composition during the course of the miniemulsion copolymerization process [82,85]. It should be emphasized that the incorporation of hexadecane in the bulk monomer does not affect the kinetics of the polymerization, unless a miniemulsification process is used to form and stabilize the submicron droplets [83].

In addition to these copolymerization studies, miniemulsion terpolymerization has also been studied, as has semicontinuous miniemulsion polymerization. For example, López de Arbina and Asua [88] polymerized high solids miniemulsions comprised of styrene, 2-ethyl hexyl acrylate, and methacrylic acid using hexadecane as the cosurfactant. They found that miniemulsion polymerizations produced more stable, higher solids latexes than those prepared via conventional emulsion polymerization. Tang et al. [89] investigated semicontinuous miniemulsion polymerization of vinyl acetate and butyl acrylate. Unzué and Asua [90] investigated the semicontinuous miniemulsion terpolymerization of butyl acrylate, methyl methacrylate, and vinyl acetate. Both researchers found that feeding miniemulsions gave very broad particle size distributions through continuous droplet nucleation. Unzué and Asua [90] and Leiza et al. [91] used this approach to create high solids content latexes up to 65% solids.

5. Potential Advantages of Miniemulsion Polymerization

The major potential advantages of miniemulsion polymerization process in latex technology can be summarized as follows:

1.   It represents a novel method of introducing the monomer to the polymerization reactor with a high degree of subdivision in the submicron size range. The result of polymerization is a particle size distribution which is relatively broad, with about the same average particle size as the initial droplets. Thus, it provides an approach for controlling the particle size distribution.

2.   Due to the broad particle size distribution, some times bimodal or trimodal [1], it provides an approach for making high solids latexes [90, 91], without resorting to scheduled additions of surfactants during the course of the polymerization process.

3.   The presence of the hexadecane additive and its effect on the comonomer distribution at the site of polymerization in a miniemulsion copolymerization process provides an approach for controlling the microstructure of the polymer particles and the instantaneous copolymer composition. Thus, conceivably the addition of different types and concentrations or mixtures of cosurfactants during the miniemulsification step may lead to a novel approach for controlling polymer microstructure in the final latex particles.

4.   The different dependence of polymerization rates and particles numbers on various polymerization parameters, such as initiator concentration, emulsifier concentration, and temperature, in the miniemulsion process compared to the conventional process may allow new control strategies.

5.   Nucleation of small monomer droplets provides an unique approach for incorporating oligomers, inorganic particles, and monomers with very low water solubility into latex particles. Using conventional emulsion polymerization, these materials would have to be transported from the large monomer droplets to the growing polymer particles through the aqueous phase, resulting in precipitation and coagulation during the polymerization. On the other hand, when these materials are pre-dispersed in miniemulsion droplets, they can be more easily incorporated into the polymer particles. In order to obtain benefits from this approach, 100% of the initial droplets must be nucleated.

6.   Finally, miniemulsion polymerization provides an approach for decreasing the average latex particle size (and thus increasing the overall polymerization rate) without the need to increase the surfactant content. This has been demonstrated by decreasing the size of the initial monomer droplets via the use of more intense shear force and/or the use of a higher concentration of hexadecane concentration during the miniemulsification step [70,79].

6. References

1.   Ugelstad, J., El-Aasser, M.S., and Vanderhoff, J.W. (1973) *J. Polym. Sci. Poly. Lett.*, **111**, 503.
2.   Azad, A. R. M., Ugelstad, J., Fitch, R.M., and Hansen, F.K. (1976) *Emulsion Polymerization*, I. Piirma and J. L. Gardon (eds.), ACS Symposium Series, Vol. 24, Am. Chem. Soc., Washington, DC, p. 1.
3.   Berkman, S. and Egloff, G. (1941) *Emulsion and Foams*, Reinhold, New York.
4.   Cobb, R.M.K. (1946) *Emulsion Technology*, H. Bennett (ed.), Chemical Publishing, New York.
5.   Becher, P. (1965) *Emulsions, Theory and Practice*, 2nd ed., Reinhold, New York.
6.   Princen, L.H. (1972) *Treatise on Coatings*, R. R. Myers and J. S. Long, (eds.), Vol. I. Part III., Marcel Dekker, New York, p.77.
7.   Friberg, S. and La Force-Gillberg, G. (1973; publ. 1975)) *Recent Advances in Emulsion Technology*, Repr. Progr. Appl. Chem., **58**, 715.

8. Harusawa, F. and Mitsui, T. (1975) *Prog. Org. Coatings*, **3(2)**, 177.
9. Nawab, M.A. and Mason, S.G. (1958) *J. Colloid Sci.*, **13**, 179.
10. Watanabe, A., Higashitsuji, K., and Nishizawa, K. (1978) *J. Colloid Interface Sci.*, **64**, 278.
11. Ugelstad, J., Hansen, F.K., and Lang, S. (1974) *Makromol. Chem.*, **175**, 507.
12. Ugelstad, J., Herder Kaggerud, K., Hansen, F.K., and Berg, A. (1979) *Makromol. Chem.*, **180**, 737.
13. Ugelstad, J., Hansen, F.K., and Herder Kaggerud, K. (1977) *Faserforsch. Textilechn.-E. Polym. Forsch.*, **28**, 309.
14. Hansen, F.K., Baumann Ofstad, E., and Ugelstad, J. (1976) *Theory and Practice of Emulsion Technology*, A. L. Smith (ed.), Academic Press, London, p. 13.
15. Lack, C.D. (1985) *Emulsion Formation and Stabilization with Mixed Emulsifier Liquid Crystals*, Ph.D. Dissertation, Lehigh University.
16. Ho, C., Goetz, R.J., El-Aasser, M.S., Vanderhoff, J.W., and Fowkes, F. (1991) *Langmuir*, **7**, 56.
17. Ho, C., Goetz, R.J., and El-Aasser, M.S. (1991), *Langmuir*, **7**, 630.
18. Lack, D., El-Aasser, M.S., Silebi, C.A., Vanderhoff, J.W., and Fowkes, F.M. (1987) *Langmuir*, **3**, 1155.
19. Hessel, J. Frederick (1993) *Phase Behavior and Properties of Lamellar Surfactants*, Ph.D. Dissertation, Lehigh University.
20. Choi, Y.-T. (1986) *Polymerization of Styrene Miniemulsions*, Ph.D. Dissertation, Lehigh University.
21. Goetz, R.J. (1990) *Formation and Stabilization of Oil-in-Water Miniemulsion*", Ph.D. Dissertation, Lehigh University.
22. El-Aasser, M.S., Lack, C.D., Vanderhoff, J.W., and Fowkes, F.M. (1988) *Colloids and Surfaces*, **29**, 103.
23. Goetz, R.J. and El-Aasser, M.S. (1990) *Langmuir*, **6**, 132.
24. Ugelstad, J. and Mork, P.C. (1980) *Adv. Colloids Interface Sci.*, **13**, 101.
25. Chou, Y.J., El-Aasser, M.S., and Vanderhoff, J.W. (1980) *Polymer Colloids II*, R.M. Fitch (ed.), Plenum, New York, 599.
26. Miller, C.M., Venkatesan, J., Silebi, C.A., Sudol, E.D., and El-Aasser, M.S. (1994) *J. Colloid & Interface Science*, **162**, 11.
27. Brouwer, W.M., El-Aasser, M.S., and Vanderhoff, J.W., (1986) *Colloids and Surfaces*, **21**, 69.
28. Ugelstad, J. (1978) *Makromol. Chem.*, **179**, 815.
29. Higuchi, I. and Misra, J. (1962) *J. Pharm. Sci.*, **51**, 459.
30. Davis, S.S. and Smith, A. (1976) *Theory and Practice of Emulsion Technology*, A. Smith (ed.), Academic Press, New York, p. 325.
31. Delgado, J., El-Aasser, M.S., and Vanderhoff, J.W. (1986) *J. of Polymer Sci.: Part A; Polymer Chemistry*, **24**, 861.
32. Rodriguez, V.S. (1988) *Interparticle Monomer Transport in Miniemulsion Copolymerization*, Ph.D. Dissertation, Lehigh University.
33. Alduncin, J., Forcada, J., and Asua, J.M. (1994) *Macromolecules*, **27**, 2256.
34. Reimers, J. and Schork, F.J. (1996) *J. Appl. Polym. Sci.*, **59**, 1833.
35. Morton, M., Kaizerman, S., and Allier, M.W. (1954) *J. Colloid Sci.*, **9**, 300.
36. Blackley, D.C. (1966) *High Polymer Latices*, Maclaren, London.
37. Warson, H. (1972) *The Application of Synthetic Resin Emulsions*, Benn, London.
38. Aelony, D. and Wittcoff, H. (to General Mills, Inc.), (1959) U.S. Patent 2,899,397.
39. Miller, A.L., Robison, S.B., and Petro, A.J. (to Esso Res. & Eng. Co.), (1962) U.S. Patent 3,022,260.
40. Schnoering, H., Witte, J., and Pampus, G. (to Farbenfabriken Bayer AG), (1971) Ger Offen. 2,013,359; (1972) *Chem. Abstr.* 76:26187k.
41. Burke, Jr., O.W. (1972) U.S. Patent, 3,652,482; (1972) *Chem. Abstr.* 77:21211z.
42. Cooper, W. (to Dunlop Rubber Co., Ltd.), (1962) U.S. Patent, 3,009,891.
43. Saunders, F.L. and Pelletier, R.R. (to Dow Chemical Co.), (1972), U.S. Patent 3,642,676; (1972) *Chem. Abstr.*, 76:142492e.
44. Date, M. and Wada (to Toyobo Co. Ltd.), (1973), Japan 73 06,619; (1974) *Chem. Abstr.*, 80:38097b.
45. Suskind, S.P. (1965) *J. Appl. Polymer Sci.*, **9**, 2451.
46. Fuller, W.J. (1968) *Paint Varnish Prod.*, **58(7)**, 23.
47. Judd, P. (to W. R. Grace & Co.), (1960), British Patent 1,142,375; *Chem. Abstr.*, 70:69355g.
48. Dieterich, D., Keberle, W., and Wuest, R. (1970) *J. Oil Col. Chem. Assoc.*, **53**, 363.
49. Overbeek, J. Th. (1952) *Colloid Science*, H. R. Kruyt, (ed.), Elsevier, New York, Vol. 1, p. 80.
50. Vanderhoff, J.W., van den Hul, H.J., Tausk, R.J.M., and Overbeek, J.Th.G. (1970) *Clean Surfaces: Their Preparation and Characterization for Interfacial Studies*, G. Goldfinger (ed.), Marcel Dekker, New York, p. 15.
51. Vanderhoff, J.W., Tarkowski, H.L., Jenkins, M.C., and Bradford, E.B. (1966) *J. Macromol. Chem.*, **1**, 361.
52. Vanderhoff, J.W. (1970) *Br. Polym. J.*, **2**, 161.
53. Brown, G.L. (1956) *J. Polym. Sci.*, **22**, 423.
54. Vanderhoff, J.W., El-Aasser, M.S., and Ugelstad, J. (to Lehigh University) (1979), U.S. Patent 4,177,177.
55. El-Aasser, M.S., Hoffman, J.D., Kiefer, C., Leidheiser, Jr., H., Manson, J.A., Poehlein, G.W., Stoisits, R., and Vanderhoff, J.W. (November 1974) *Water-Base Coatings, Part I*, Final Report AFML-TR-74-208, Air Force Materials Laboratory, Wright-Patterson Air Force Base, Ohio.

126

56. Chou, Y.N., Confer, L.M., Earhart, K.A., El-Aasser, M.S., Hoffman, J.D., Manson, J.A., Misra, S.C., Poehlein, G.W., Scolare, J.P., and Vanderhoff, J.W. (November 1975) *Water-Base Coatings, Part II*, Final Report AFML-TR-74-208, Air Force Materials Laboratory, Wright-Patterson Air Force Base, Ohio.
57. El-Aasser, M.S., Pohelein, G.W., and Vanderhoff, J.W. (1977) *Preprints, ACS Org. Coat. Plast. Chem. Div.*, **37(2)**:92.
58. El-Aasser, M.S., Misra, S.C., Vanderhoff, J.W., and Manson, J.A. (1977) *J. Coatings. Technol.*, **49(635)**:71.
59. Vanderhoff, J.W., El-Aasser, M.S., and Hoffman J.D. (to Lehigh University) (1978), U.S. Patent 4,070,323.
60. Wang, S.T., Schork, F.J., Poehlein, G.W., and Wood, J.W. (1996), Accepted *J. Appl. Polym. Sci.*
61. Merkel, M.P. (1986) *Morphology of Core/Shell Latexes and Their Mechanical Properties*, Ph.D. Dissertation, Lehigh University.
62. Durbin, D.P., El-Aasser, M.S., Poehlein, G.W., and Vanderhoff, J.W. (1979) *J. Appl. Polym. Sci.*, **24**, 703.
63. Chou, Y.J., El-Aasser, M.S., and Vanderhoff, J.W. (1980) *Polymer Colloids II*, R. M. Fitch (ed.), Plenum Press, New York, p. 599; ibid. (1980) *J. Dispersion Sci. Tech.*, **1**, 129.
64. Chou, Y.J., El-Aasser, M.S., and Vanderhoff, J.W. (1982) *Computer Applications in Applied Polymer Science*, T. Provder (ed.), ACS-Symposium Ser., **197**, p. 399.
65. Hansen, F.K. and Ugelstad, J. (1982) *Emulsion Polymerization*, I. Piirma (ed.), Academic Press, New York, p. 51.
66. Ugelstad, J., Mork, P.C., Berge, A., Ellingsen, T., and Khan, A.K. (1982) *Emulsion Polymerization*, I. Piirma (ed.), Academic Press, New York, p. 383.
67. Miller, C.M. (1994) *Particle Formation and Growth During Styrene Oil-in-Water Miniemulsion Polymerization*, Ph.D. Dissertation, Lehigh University.
68. Choi, Y.T., Sudol, E.D., Vanderhoff, J.W., and El-Aasser, M.S. (1985) *J. Polymer Sci., Polym. Chem. Ed.*, **23**, 2973.
69. Miller, C.M., Sudol, E.D., Silebi, C.A., and El-Aasser, M.S. (1995) *J. Polym. Sci., Part A, Polym. Chem.*, **33**, 1391.
70. Tang, P.L., Sudol, E.D., Silebi, C.A., and El-Aasser, M.S. (1991) *J. Appl. Poly. Sci.*, **34**, 1059.
71. Chamberlain, B.J., Napper, D.H., and Gilbert, R.G. (1982) *J. Chem Soc. Faraday Trans.,I*, **78**, 591.
72. Wood, D.F., Wang, B.C.Y., Napper, D.H., Gilbert, R.G., and Lichti,G. (1983) *J. Polym. Sci., Polym. Chem. Ed.*, **21**, 985.
73. Tang, P.L., Sudol, E.D., Adams, M.E., Silebi C.A., and El-Aasser, M.S. (1992) *Polymer Latexes: Preparation, Characterization and Applications*, E.S. Daniels, E.D. Sudol and M.S. El-Aasser (eds.), ACS Symposium Series No. 492, Chapter 6, pp. 72.
74. Miller, C.M., Blythe, P.J., Sudol, E.D., Silebi, C.A., and El-Aasser, M.S. (1994) *J. Polym. Sci.: Part A: Polym. Chem.*, **32**, pp. 2365.
75. Miller, C.M., Sudol, E.D., Silebi, C.A., and El-Aasser, M.S. (1995) *Macromolecules*, **28**, 2754.
76. Miller, C.M., Sudol, E.D., Silebi, C.A., and El-Aasser, M.S. (1995) *Macromolecules*, **28**, 2765.
77. Miller, C.M., Sudol, E.D., Silebi, C.A., and El-Aasser, M.S. (1995) *Macromolecules*, **28**, 2772.
78. Kitzmiller, E.L., Miller, C.M., Sudol, E.D., and El-Aasser, M.S. (1995) *Macromol. Symp.*, **92**, 157.
79. Hansen, F.K. and Ugelstad, J. (1979) *J. Polym. Sci., Polym. Chem.*, **17**, 3069.
80. Mouran, D., Reimers, J., Schork, J. (1996), Accepted *J. Polym. Sci.*
81. Delgado, J., El-Aasser, M.S., Silebi, C.A., and Vanderhoff, J.W. (1986) *J. Polym. Sci., Polym. Chem.*, **24**, 861.
82. Delgado, J., El-Aasser, M.S., Silebi, C.A., Vanderhoff, J.W., and Guillot, J. (1988) *J. Polym. Sci., Polym. Phys.*, **26**, 1495.
83. Delgado, J., El-Aasser, M.S., Silebi, C.A., and Vanderhoff, J.W. (1989) *J. Polym. Sci., Polym. Chem.*, **27**, 193.
84. Delgado, J., El-Aasser, M.S., Silebi, C.A., and Vanderhoff, J.W. (1990) *J. Polym. Sci., Polym. Chem.*, **28**, 777.
85. Delgado, J., El-Aasser, M.S., Silebi, C.A., Vanderhoff, J.W., and Guillot, J. (1987) *Future Directions in Polymer Colloids*, M. S. El-Aasser, and R. M. Fitch (eds.), NATO-ASI Series E: Applied Sci., No. 138, Dordrecht, p. 79.
86. Rodriguez, V.S., El-Aasser, M.S., Asua, J.M., and Silebi, C.A. (1989) *J. Polym. Sci., Polym. Chem. 27*, 3659.
87. Rodriguez, V.S., Delgado, J., Silebi, C.A. and El-Aasser, M.S. (1989) *Ind. Eng. Chem. Res.*, **28**, 65.
88. López de Arbina, L.L. and Asua, J.M. (1992) *Polymer*, **33(22)**, 4832.
89. Tang, P.L., Sudol, E.D., Adams, M., El-Aasser, M.S., and Asua, J.M. (1991) *J. Appl. Polym. Sci.* , **42**, 2019.
90. Unzue, M.J. and Asua, J.M. (1993) *J. Appl. Polym. Sci.*, **49**, 81.
91. Leiza, J.R., Sudol, E.D., and El-Aasser, M.S. (1996), submitted *J. Appl. Polym. Sci.*

MICROEMULSION POLYMERIZATION

F. CANDAU
Institut Charles Sadron (CRM-EAHP), CNRS-ULP
6, rue Boussingault, 67083 Strasbourg Cédex, France

1. Introduction

In contrast with emulsion polymerization which has been extensively studied over the past 50 years and whose kinetics and mechanism are rather well understood, the concept of polymerization in microemulsions appeared only in the early eightees. Since then, the field has developed rapidly, as attested by the increasing number of papers devoted to microemulsion polymerization. The interesting features of microemulsions 1) large overall interfacial area (ca. $100m^2/ml$) ; 2) optical transparency and thermodynamic stability ; 3) small size of the domains ($\sim 10^{-2}$ μm), and 4) great variety of structures, result in a unique microenvironment which can be taken advantage of to produce novel materials with interesting morphologies or polymers with specific properties [1-3].

In this paper, we review the salient features of microemulsion polymerization at the present state of knowledge. We discuss the formulation of polymerizable microemulsions and show how the incorporation of monomers can modify the initial structure of the systems. The kinetics and mechanistics aspects are given and compared to those obtained in conventional emulsion polymerization. We also describe some recent results obtained on the formation of porous solid materials and functionalized microlatex particles, which seem quite promising for future applications.

2. Structure and Phase Diagrams of Monomer-Containing Microemulsions

Microemulsions are thermodynamically stable and isotropic dispersions containing oil, water and surfactant(s), the stability being ensured by a very low interfacial tension, capable to compensate the dispersion entropy [4]. In most cases, it is necessary to associate a cosurfactant (like an alcohol) to the surfactant in order to achieve the low interfacial tension required. Microemulsions can exist either alone or in equilibrium with water and/or oil phases in excess. Here we consider essentially the monophasic domains of the phase diagrams.

One of the most common structures encountered in microemulsions consists of water or oil droplets dispersed in an oil or water continuous phase respectively. The type of dispersion results from the preferred curvature C_0 of the surfactant layer which is by convention positive for oil-in-water (o/w) systems and negative for water-in-oil (w/o) systems.

By varying several parameters such as the w/o ratio, one can induce an inversion from an o/w to a w/o microemulsion and vice-versa. The type of structure in the inversion domain depends essentially on the bending constant K_e characteristic of the

J. M. Asua (ed.), Polymeric Dispersions: Principles and Applications, 127–140.

128

elasticity of the surfactant layer. If K_e is of the order of kT (k : Boltzmann constant, T absolute temperature), the persistence length of the film (i.e. the distance over which the film is locally flat) is microscopic. The interfacial film is flexible and is easily deformed under thermal fluctuations. The phase inversion occurs through a bicontinuous structure formed of water and oily domains randomly interconnected [5].

The incorporation of a monomer in a microemulsion either by interchanging the water by an aqueous monomer solution or the oil by a hydrophobic monomer may induce large changes in the phase diagram due to the possible cosurfactant role of the monomer and/or to the change in solubility of the surfactant in oil or water. It is therefore important to study, prior to polymerization, the phase diagram of the systems in the presence of monomer and to optimize the formulation procedure.

2.1. ROLE OF MONOMER

Most of the water-soluble monomers polymerized in microemulsion and in particular acrylamide (AM) were found to act as cosurfactants, leading to a considerable extent of the microemulsion domain in the phase diagram (Figure 1) [6-9].

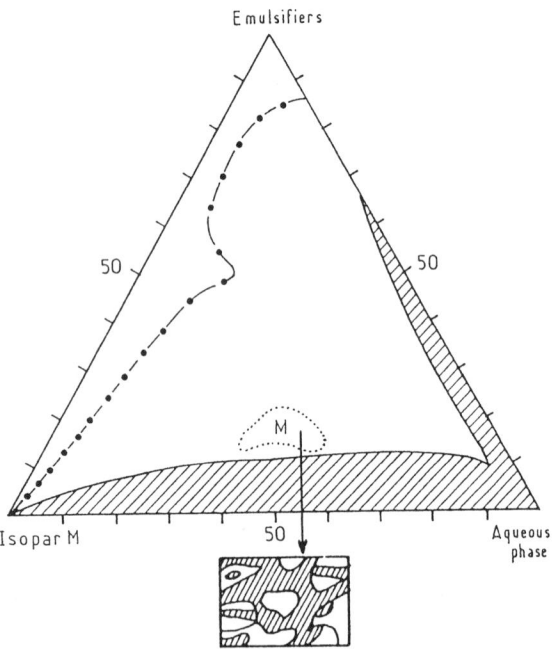

Figure 1. Pseudo-ternary phase diagram. The dotted line (-●-) is the boundary between the microemulsion (left) and emulsion (right) domains in the absence of monomers i.e. in pure water. Addition of monomers (acrylamide + sodium acrylate) to water (1.25 mass ratio) extends the microemulsion domain up to the solid line (entire white area). Polymerization reactions have been carried out in the bicontinuous M area (from ref. [7]).

The cosurfactant role of various hydrophilic monomers was confirmed by surface-tension experiments [9,10]. In the case of the widely studied AOT/water-acrylamide/toluene systems (AOT : sodium 1-4 bis (2-ethylhexyl) sulfosuccinate), the interfacial localization of the AM molecules induces attractive interactions between the droplets as seen by light scattering, small angle neutron scattering and viscometry experiments. It should be noted that without acrylamide, the systems display essentially a hard sphere behavior [6,11].

Upon further addition of acrylamide, the interaction potential becomes so attractive that transient clusters form. Above a threshold volume fraction, a large increase in the electrical conductivity is observed which is the signature of a percolation phenomenon [12]. This percolating structure was shown to affect the formation of polymer latex particles and the polymerization mechanism [12,13].

Under certain conditions, the radius of curvature can become so large that the globular configuration evolves toward a bicontinuous structure. This transition can be induced by addition of ionic monomers and/or electrolytes to microemulsions stabilized by nonionic surfactants [14,15]. In this case, the role of the monomer is twofold : as a cosurfactant, it increases the flexibility and fluidity of the interface, which favors the formation of a bicontinuous microemulsion. As an electrolyte, it decreases the solubility in water of the ethoxylated moiety of the nonionic surfactant (salting-out effect) with its progressive transfer in the oil phase via a bicontinuous phase. Here the role of the monomer is clearly demonstrated, since the corresponding systems in the absence of monomer have not a bicontinuous character but are indeed coarse emulsions. This is illustrated by the example of the pseudo-ternary phase diagram given in Figure 1.

2.2. FORMULATION RULES

The formation of a microemulsion requires far more surfactant than an emulsion, due to the necessity of stabilizing a large overall interfacial area. This drawback can restrict the potential uses of microemulsion polymerization since high solid contents and low surfactant amounts are desirable for most applications.

In the case of waters-soluble monomers, much effort has been devoted to an optimal formulation, compatible with an economical process. The reader can refer to ref. [8] for the details of the optimization procedure. The treatment is based on the so-called Cohesive Energy Ratio (CER) concept developed by Beerbower and Hill for conventional emulsions [16]. This approach lies on a perfect chemical match between the partial solubility parameters of oil (d^2_O) and lipophile tail of the surfactant (d^2_L) as well as those of water and hydrophile head. The CER concept has been improved to take into account the modification of the solubility parameters of water brought about by the presence of monomers in large proportions (~ 50% based on water) [9,15].

Another important aspect of the formulation is illustrated by the studies dealing with the preparation of porous materials. This implies to formulate systems containing large amounts of hydrophobic monomers (up to 70 wt%) either in the continuous phase of globular microemulsions, or in the oily domains of bicontinuous microemulsions [17-21]. A typical example of a phase diagram is given in Figure 2. It shows four detectable regions, three of them being monophasic microemulsions (w/o, o/w and bicontinuous). The acrylic acid here acts as a cosurfactant.

There have not been such detailed formulation studies for o/w microemulsions based on hydrophobic monomers. The main reason is that the monomer mostly investigated so far is styrene trapped in droplets stabilized by aliphatic surfactants. According to the

CER concept [16], there is a chemical mismatch between styrene (aromatic) and the hydrophobic tail of the surfactants classically used. In addition, by essence, styrene has no amphiphilic character and cannot act as a cosurfactant. As a result, the domain of existence of microemulsions is very limited.

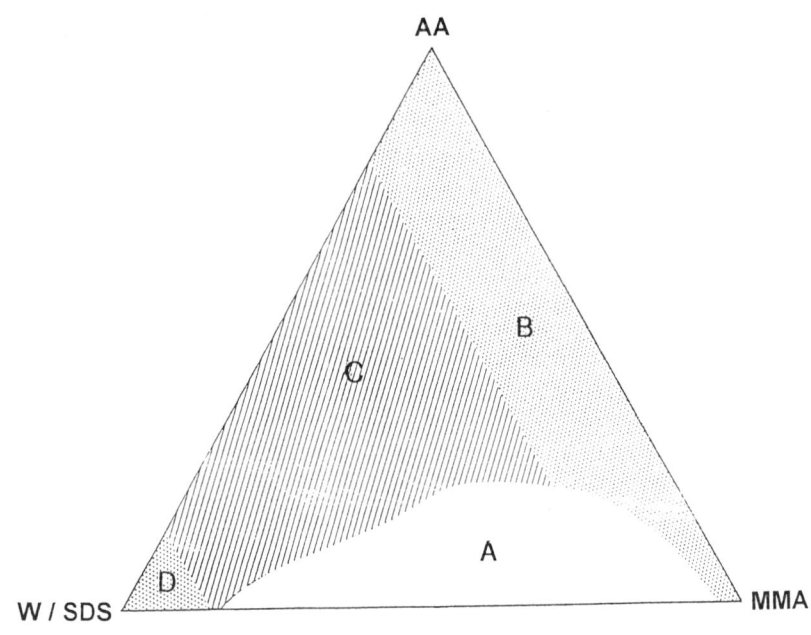

Figure 2.   Ternary phase diagram for the system, methylmethacrylate (MMA), acrylic acid (AA), 20 wt% solution of sodium dodecylsulfate in water (w/SDS) and ethyleneglycol dimethacrylate (EGDMA) at 25 ± 0.1°C and 1 atm. Compositions are on wt% basis, EGDMA content is of 4% of the combined weight of MMA and AA. Domain A : 2-phase region, Domain B : w/o microemulsion, Domain C : Bicontinuous micro- emulsion, Domain D : o/w microemulsion (from ref. [20]).

3. Polymerization in Globular Microemulsions

Most of the studies have dealt either with the free radical polymerization of hydrophobic monomers, e.g. styrene, methylmethacrylate (MMA) or derivatives [22-39] within the oily core of o/w microemulsions or with the polymerization of water-soluble monomers e.g. acrylamide (AM), within the aqueous core of w/o microemulsions [40-44]. The polymerization can be initiated thermally, photochemically or under γ-radiolysis. Beside the conventional dilatometry and gravimetry techniques, the polymerization kinetics were monitored by Raman spectroscopy [25], pulsed UV laser source [29], rotating sector technique [42], calorimetry and internal reflectance spectroscopy [39].

For both o/w and w/o systems, the amount of monomer is mostly restricted to 5-10wt% with respect to the overall mass whereas that of surfactant(s) lies within the same range or even above. The main difficulty encountered by most of the authors and which precludes the use of higher monomer concentrations lies in retaining the optical transparency and stability of the microemulsions upon polymerization. In addition to entropic factors contributing to the destabilization of microemulsions during polymerization, the compatibility between polymer and cosurfactant also influences the system. This is especially true when styrene is polymerized within o/w microemulsions containing an alcohol because the latter is a non solvent for the polymer. Conversely, the polymerization of acrylamide in alcohol-free-w/o microemulsions was already reported in 1982 to give transparent microlatexes of small particle size (d ~ 30nm) [40].

Despite these difficulties, most of the earliest studies used an alcohol in the formulation of o/w microemulsions and it is only since 1989 that the polymerization of hydrophobic monomers in three-component oil-in-water microemulsions was reported [23,24,28,31-34,35-37,39].

In order to understand the mechanism occurring in microemulsion polymerization, one has to remember that the concentration of monomer used is low (a few percent) while the concentration of surfactant is much larger than that used in an emulsion. The monomer/surfactant mass ratio is around 0.3-1 compared to 30-60 in an emulsion.

The second difference lies in the structure of the initial systems. In an emulsion, the monomer is located in large monomer droplets (d ~ 1-10 μm), in small micelles (d ~ 5-10nm) and partially solubilized in the continuous phase. In a globular microemulsion, it is solubilized within a single class of swollen-micelles (d = 5-10nm). These features added to the dynamic character of microemulsions are at the origin of the differences in the mechanisms observed in both processes.

The problem of particle nucleation was first addressed in the early eightees by Candau and coworkers in the case of water-in-oil microemulsions [6,41]. It was apprehended only in the recent years for ternary or quaternary oil-in-water microemulsions, likely due to the onset of turbidity and lack of stability with time observed by most of the authors. A scheme which is now well accepted is that of a *continuous* particle nucleation mechanism. This view is supported by several features.

i) The particle size of the final microlatex (d ~ 20-40nm) is larger than that of the initial monomer-swollen micelle. This leads to a final number of polymer particles, N_p, ca. 2 or 3 orders of magnitude smaller than that of the monomer droplets.Nucleated particles grow by addition of monomer from other inactive micelles, either by coalescence with neighboring micelles or by monomer diffusion through the continuous phase [6] (Figure 3).

ii) The number of polymer chains per particle, n_p, is in general very low, sometimes equal [6,41] or close to one [2,22,26,27,31,32,35]. In the case of o/w microemulsions, n_p was found to increase slightly with the degree of conversion [26,31]. This augmentation was accounted for by the capture of radicals by preexisting polymer particles, these competing more effectively with the monomer-swollen micelles, as the reaction proceeds. Limited flocculation at later stages of the reaction was also envisioned.

iii) The number of polymer particles was found to increase continuously with the conversion [12,26,31,37,45] (Figure 4).

132

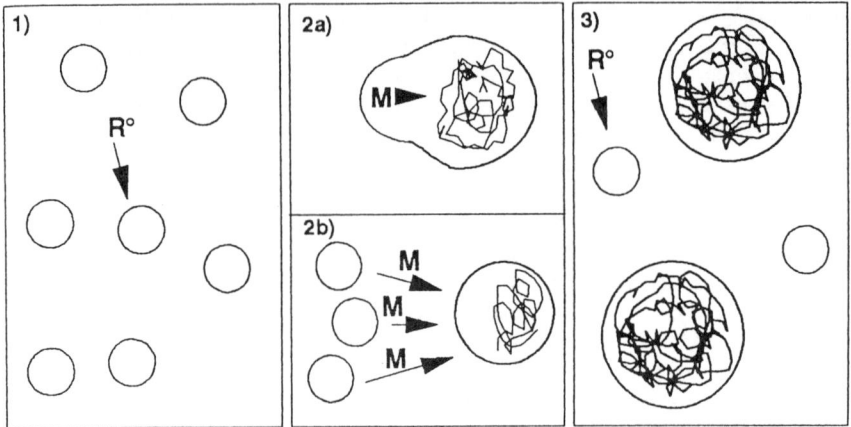

Figure 3. Microemulsion polymerization mechanism : 1) Before polymerization : monomer-swollen micelles (d ~ 5-10nm). 2) Polymer particle growth a) by collisions between particles ; b) by monomer diffusion through the continuous phase. 3. End of polymerization : polymer particles (d ~ 40nm) + small micelles (d ~ 3nm).

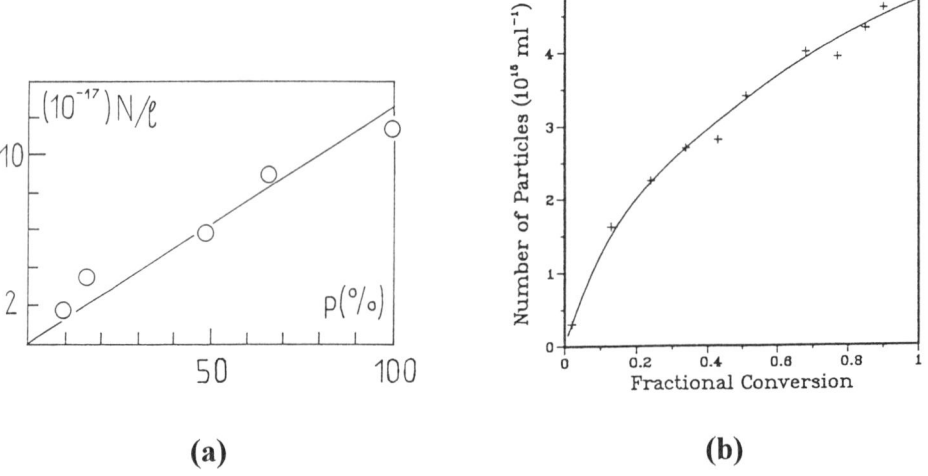

(a) (b)

Figure 4. Number of polymer particles versus conversion. a) for acrylamide polymerization in AOT w/o microemulsions (from ref. [12]) ; b) for styrene polymerization in o/w microemulsions (from ref. [26]).

As a result of the particle growth and the large amount of surfactant used in the formulation, small micelles of uniform size are always in excess throughout the reaction mixture. Their high total interfacial area relative to nucleated polymer particles implies that the micelles preferentially capture primary radicals generated in the continuous phase. This leads to a process of continuous particle nucleation with each particle formed in one single step. This mechanism was confirmed by transmission electron microscopy (TEM) experiments [12]. This behavior is in contrast with what is observed in conventional emulsion polymerization where the first nucleation stage (Interval I) is followed by a particle growth at constant particle number (Interval II). Note that these results do not follow the Smith and Ewart theory since each nucleated particle contains on average <u>one</u> collapsed macromolecule either growing or in its final form. However, all particles are not active at any given time and the average number of free radicals per particle, $\bar{n}$, averaged over the entire population, is less than one.

The phase diagrams of oil/water/surfactant systems allow one to make a direct comparison between emulsion and microemulsion polymerization processes just by varying the surfactant concentration, as shown by Gan et al. [32,35]. Figure 5 represents the polymerization rate-conversion curves for methylmethacrylate polymerization at different surfactant concentrations.

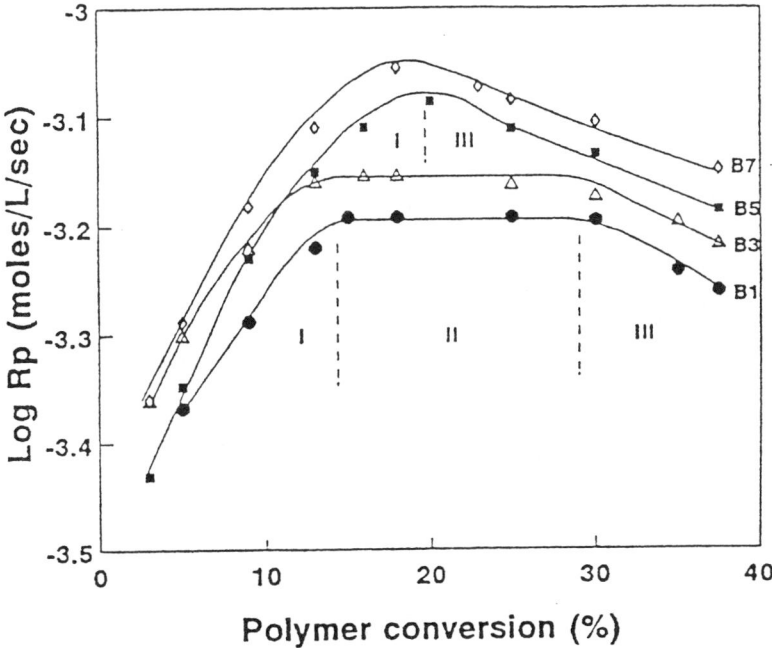

Figure 5. MMA polymerization-rate curves at 60°C for emulsions (B_1 and B_3) and micro-emulsions (B_5 and B_7) (from ref. [35]).

Three distinct intervals are observed in the emulsion regime with a rate-plateau (Interval II), as commonly observed in emulsion polymerization. Only two intervals are observed in the microemulsion regime with a maximum occurring at around 20% conversion. This behavior is very general and was also observed by several groups for microemulsion polymerization of styrene [22,28,30-32] and butylacrylate [45]. The rate decrease observed above 20% conversion is due to monomer depletion in the polymer particles, as proposed by El-Aasser and coworkers [26,27]. Termination of chain growth in polymer particles is attributed to chain transfer to monomer.

Comparative mechanistic studies on the microemulsion polymerization of styrene and methylmethacrylate have been carried out by several groups [23,25-28,31,35-37].The results could be coherently interpreted in terms of the relative monomer solubilities in water. In the case of styrene which has a very weak solubility in water (0.031%), it was postulated that initiation takes place in the microemulsion droplets. The experiments performed in MMA systems suggest that homogeneous nucleation can compete with monomer droplet initiation because of the non negligible solubility (1.56%) and the more polar and cosurfactant character of MMA.

For both monomers, the polymerization rates were generally faster with a water-soluble initiator (potassium persulfate, KPS) than with an oil-soluble (2,2'-azobisisobutyronitrile, AIBN) [25,30]. This behavior was discussed in terms of different efficiencies of the initiators in producing effective radicals for the polymerization. At an equimolar concentration of initiators, KPS generates more radicals in the aqueous phase than AIBN. These radicals are thus more effective for initiation (in the continuous phase, in the monomer droplets or at their interfaces) than AIBN radicals due to a significant autotermination of AIBN radical pairs in the small droplets (cage-effect) and the low solubility of AIBN in water. In this case, initiation is believed to occur essentially via micellar entry of single radicals arising either from the very small portion of AIBN dissolved in water or desorbed from other swollen micelles.

In the case of the photopolymerization of acrylamide in AOT reverse micelles with AIBN as the initiator [I] and toluene as the organic phase, a monoradical termination was found (polymerization rate, $R_p \propto$ [I]), caused by a degradative chain transfer to toluene [42]. A study on the locus of initiation of the same AOT/(water-AM)/toluene systems was performed by using initiators of various solubilities [43]. Initiation with AIBN was shown to take place predominantly in the water/oil interlayer where the encounter with acrylamide-cosurfactant is facilitated. With water-soluble ammonium peroxodisulfate, initiation occurs, as expected, in the micellar water-pools.

4. Polymerization in the Continuous or Bicontinuous Phases of Microemulsions

Beside polymerization in globular microemulsions, several studies have dealt with polymerization of monomers in the other phases of microemulsions. One of the main goals underlying these studies was to utilize the microstructure of microemulsions as a template to produce solid polymers with similar characteristics. For example, incorporation of large amounts of hydrophobic monomers in the continuous phase of w/o microemulsions should yield solid polymers with a swiss-cheese-like structure capable of encapsulating the disperse phase (water).

4.1. MICROEMULSIONS BASED ON HYDROPHOBIC MONOMERS

An important contribution in this field was provided by Cheung et al. who obtained porous polymeric structures by photopolymerization of monomers in single-phase microemulsions [18-20]. The systems consisted of methylmethacrylate (MMA), acrylic acid (AA), a cross-linking agent, ethylene glycol dimethacrylate (EGDMA), water and sodium dodecylsulfate (SDS) as the surfactant. Large amounts of monomers were used in the formulation (up to 70% in some cases). An example of the richness of the structures of the phase diagram is given in Figure 2. The polymeric materials were characterized by scanning electron microscopy, thermogravimetry, adsorption studies, swelling and permeability measurements and differential scanning calorimetry. Interestingly, a close correlation was found between the microstructure of the polymeric material and the nature of the initial microemulsion.

i) Polymerization in microemulsions with a water/oil droplet structure yields closed cell porous polymeric solids, having a morphology characterized by a disjointed cellular structure where the water pores are distributed as discrete pockets throughout the solid.

ii) Polymerization in microemulsions with a bicontinuous structure results in a polymer with an open-cell structure, i.e. an interconnected porous structure with water channels through the polymer. The surface area increases steadily upon increasing water content in the precursor microemulsion.

However, the length scale of the porous structure obtained (1-4μm) is considerably larger than the length scale characteristic of microemulsions (less than 0.1μm), due to phase separation effects or structural changes during polymerization. In more recent studies, Gan et al. have attempted to preserve to a greater extent the initial bicontinuous structure of microemulsions by varying the nature of the surfactant and the polymerization conditions [21]. The use of bicontinuous microemulsions based on a polymerizable surfactant like (((acryloyloxy)-undecyl)dimethylammonio) acetate (AUDMAA) seems the key parameter for obtaining transparent solid polymeric materials with an open-cell microstructure. The widths of the randomly distributed bicontinuous domains are about 50-70nm that is far below those usually obtained in the previous reports.

The preparation of novel solid materials is a huge field for applications such as microfiltration, separation membranes or their supports, microstructured polymer blends and conductive composite films [46], and porous microcarriers for the culture of living cells and enzymes. The considerable progress accomplished over the last four years permits to envision many future developments.

4.2. MICROEMULSIONS BASED ON WATER-SOLUBLE MONOMERS

The (co)polymerization of various water-soluble monomers (neutral, anionic and/or cationic) in the aqueous domains of nonionic bicontinuous microemulsions has been studied by Candau and coworkers over the last decade [7-9,47-52]. By optimizing the procedure, one can prepare microemulsions containing up to 25 wt% monomers dissolved in the same amount of water and around 8 wt% surfactant. As for polymerization of hydrophobic monomers in the bicontinuous phase of microemulsions, the initial structure is not preserved upon polymerization. However, a notable difference with the former systems is that the final system is a microlatex which is remarkably transparent (100% optical transmission) fluid and stable with a particle size remaining unchanged over years even at high volume fractions (~ 60%). The microlatex consists of water-swollen spherical polymer particles with a narrow distribution as seen from quasi-elastic light scattering (QELS) and TEM experiments. This result is of major importance with regard to inverse emulsion polymerization which is known to produce unstable latexes with a broad particle size distribution [10]. The microlatex stability was accounted for by i) reduced gravity forces ($\sim d^3$) ; ii) high entropic contribution from the droplets owing to their large number ; iii) low interfacial tension between polymer droplets and continuous phase [2].

5. Characteristics of the Final Products

The characteristics of both particle latexes and polymers formed depend critically on the formulation. In fact, the composition chosen for the system depends on whether the ultimate goal of the formulator is to prepare specific polymers or to produce small-sized latex particles. As a rule, the larger the surfactant/monomer ratio, the smaller the particle size [2,38,52,47] and the higher the molecular weight [51,52]. Therefore, high solid contents and small-sized particles can hardly be achieved simultaneously.

5.1. POLYMERS

The polymer molecular weights are high, usually ranging from 10^6-10^7, as expected for polymerization in dispersed media. When alcohols are used in the formulation, chain transfer reactions can occur which reduce the molecular weight. The distribution of molecular weights in o/w systems is usually very broad ($M_w/M_n \cong 2$-7 ,up to 12 in some cases) [32]. Note that the few polymer chains of high molecular weight (one in the limiting case) confined in the microlatex particle must be strongly collapsed in order to fit such small dimensions (d < 40nm) [2,6].

Candau and coworkers have performed a comparative study on the microstructure of copolymers prepared by polymerization in microemulsion, emulsion or solution [50,51]. An interesting finding is that microemulsion polymerization seems to improve the structural homogeneity of the copolymers with reactivity ratios close to unity. For example, microemulsion polymerization leads to almost random polyampholytes whereas those prepared in solution exhibit a strong tendancy to alternation [51]. The conformation and solution properties of these ampholytic polymers are directly related to the monomer sequence distribution : at equimolar proportions of anionic and cationic monomers, a random polyampholyte (microemulsion process) is insoluble in water whereas an alternated one (solution process) is soluble. These results are accounted for by the marked differences between the microemulsion process and others, in terms of microenvironment (charge screening and preferential orientation of the monomers at the w/o interface) and mechanism (interparticular collisions with complete mixing) [2,50].

The applications of the polymers formed by microemulsion polymerization, concern essentially porous polymers (see above) and water-soluble polymers. The latter can be used in oil-recovery processes, as flocculants in paper manufacture, mining field and water treatment. More details are given in Ref.2.

5.2. MICROLATEXES

The size of the microlatex particles has been usually determined by QELS and EM. As a general rule, the particle diameters prepared in microemulsions are much smaller than those obtained by emulsion polymerization, although they still exceed significantly those of the parental microemulsion droplets due to the particular mechanism occurring during polymerization. They are around 20-60nm when the starting microemulsions are globular (o/w or w/o). They are bigger and around 50-150nm if the microemulsions are initially bicontinuous simply due to the larger monomer incorporations (~ 25%). As can be expected, the particle size increases upon increasing the monomer content or decreasing the surfactant [22,23,32,35,38,47] and/or the initiator concentration [22,26,28]. The index of polydispersity is ca. $d_w/d_n \cong 1.05$-1.15.

Some efforts were made to control the particle size by using appropriate formulations. For example, Antonietti and Nestl reported a study using a new class of metallosurfactants (tetradecyldiethanolamine-copper) which allowed them to reduce both particle size and surfactant concentration [53]. They succeeded in getting a particle diameter as low as 14nm, with a value of the weight ratio of surfactant/monomer of 3. This results in a considerable surface area (~ 500 m^2/g) which renders these systems of particular and technical interest for subsequent functionalization.

The effect of the nature of surfactant on particle size was investigated for microlatexes of poly(acrylamide-co-sodium 2-acrylamido-2-propanesulfonate) obtained by polymerization in bicontinuous microemulsions [54]. By using different nonionic surfactant blends at the optimal conditions derived from the CER concept (section 2.2.) the authors showed a significant effect of this parameter on particle size. The results were accounted for by salting-out effects of variable importance of the ethoxylated surfactants by the anionic monomer (NaAMPS).

5.3. FUNCTIONALIZED MICROLATEX PARTICLES

The large inner surface area of microemulsions can be easily modified and functionalized by simple copolymerization reactions or embedding reactions as recently shown by Antonietti et al. [3,33,34].

Metal-complexing microlatexes were synthesized via copolymerization of styrene in microemulsions using two comonomers where a 2'2-bipyridine is coupled with or without a spacer to a methacrylic acid unit (6'-methyl-2,2'-bipyridin-6-ylmethyl methacrylate, MBM, and 4-(6'methyl-2,2'-bipyridin-6-ylmethoxy)butyl methacrylate, BMBM, [55]. The average particle size is d ~ 30nm (width of the distribution ~ 0.3). Binding experiments with Ni(II), Co(II), Cr(II) and Cu(II) ions in the aqueous disperse phase show that most of the bipyridine units are located at the latex surface.

Another way to functionalize the surface of microlatex particles is to incorporate amphiphilic block copolymers (for example polystyrene/ polyvinylpyridine) as cosurfactants together with the classical surfactants used in the formulation [33,34].

Stable microlatexes in the nano-size range (20-30nm) can be of great interest for biological applications, as for example in immuno-assays, adsorbants for proteins, immobilization of enzymes and antibodies and for control release in drug delivery. Some procedures based on inverse microemulsion polymerization have been proposed for the preparation of nanocapsules [2]. In particular, an enzyme-nanoparticle-recognition molecule was prepared to be used as a diagnostic tool for hybrization of nucleotide probes [44].

138

In immuno-assay experiments, the size of particles is a critical parameter for the detection sensitivity. In a recent patent was reported the synthesis of nanoparticles containing ca. 10wt% solid contents with good performances concerning simultaneously reduced size, reasonable polydispersity and functionalization [56]. The examples concerned the copolymerization of styrene with different polymerizable surfactants bearing various functional groups (OH, SO$_3$H, COOH ...). Typical particle sizes were around 20-30nm.

References

1. Candau, F. (1987) Microemulsion polymerization, in H.F. Mark, N.M. Bikales, C.G. Overberger and G. Menges (eds.), *Encyclopedia of Polymer Science and Engineering*, 2nd ed., Vol.9, pp. 718-724.
2. Candau F. (1992) Polymerization in microemulsions, in C.M. Paleos (ed.), *Polymerization in Organized Media*, Gordon and Breach Publishers, Philadelphia, Chap.4, pp. 215-282.
3. Antonietti, M., Basten, R. and Lohmann, S. (1995) Polymerization in microemulsions - A new approach to ultrafine, highly functionalized polymer dispersions, *Macromol. Chem. Phys.* **196**, 441-466.
4. See for example : Bourrel, M. and Schechter, R.S. (1988) *Microemulsions and Related Systems*, Marcel Dekker Publisher, New York.
5. de Gennes, P.G. and Taupin, C. (1982) Microemulsions and the flexibility of oil/water interfaces, *J. Phys. Chem.* **92**, 2294-2304.
6. Candau, F., Leong, Y.S., Pouyet, G. and Candau, S.J. (1984) Inverse microemulsions polymerization of acrylamide : characterization of the water-in-oil microemulsions and the final microlatexes, *J. Colloid Interface Sci.* **101**, 167-183.
7. Candau, F., Zekhnini, Z. and Durand, J.P. (1986) Copolymerization of water-soluble monomers in nonionic bicontinuous microemulsions, *J. Colloid Interface Sci.* **114**, 398-408.
8. Holtzscherer, C. and Candau, F. (1988) Application of the cohesive energy ratio concept (CER) to the formation of polymerizable microemulsions, *Colloids Surf.* **29**, 411-423.
9. Buchert, P. and Candau, F. (1990) Polymerization in microemulsions : 1. Formulation and structural properties of microemulsions containing a cationic monomer, *J. Colloid Interface Sci.* **136**, 527-540.
10. Candau, F. (1990) Polymerization in inverse emulsions and microemulsions, in F. Candau and R.H. Ottewill (eds.), *An Introduction to Polymer Colloids*, NATO ASI Series C N°303, Kluwer Publishers, Dordrecht, Chap.3, pp. 73-96.
11. Holtzscherer, C., Candau, F. and Ottewill, R.H. (1990) A small angle neutron scattering study on AOT/toluene/(water+acrylamide) micellar solutions, *Prog. Colloid Polym. Sci.* **81**, 81-86.
12. Carver, M.T., Hirsch, E., Wittmann, J.C., Fitch, R.M. and Candau, F. (1989) Percolation and particle nucleation in inverse microemulsion polymerization, *J. Phys. Chem.* **93**, 4867-4873.
13. Barton, J., Tino, J., Hlouskova, Z. and Stillhammerova, M. (1994) Effect of percolation on free-radical polymerization of acrylamide in inverse microemulsion, *Polym. Int.* **34**, 89-96.
14. Candau, F. (1989) Polymerization of water-soluble monomers in microemulsions in M. El-Nokaly (ed.), *Polymer Association Structures : Microemulsions and Liquids Crystals*, ACS Symposium Series n° 384, Chap.4, pp. 48-61.
15. Corpart, J.M. and Candau, F. (1993) Formulation and polymerization of microemulsions containing a mixture of cationic and anionic monomers, *Colloid Polym. Sci.* **271**, 1055-1067.
16. Beerbower, A. and Hill, M.W. (1971) The cohesive energy ratio of emulsions. A fundamental basis for the HLB concept, in McCutcheon (ed.), *Detergents and Emulsifier Annual*, Allured Publ. Co., Ridgewood, pp. 223-236.
17. Haque, E. and Qutubuddin, S. (1988) Novel polymeric materials from microemulsions, *J. Polym. Sci. Polym. Lett. Ed.* **26**, 429-432.
18. Raj, W.R.P., Sasthav, M. and Cheung, H.M. (1991) Formation of porous polymeric structures by the polymerization of single-phase microemulsions formulated with methylmethacrylate and acrylic acid, *Langmuir* **7**, 2586-2591.
19. Raj, W.R.P., Sasthav, M. and Cheung, H.M. (1992) Polymerization of microstructural aqueous systems formed using methylmethacrylate and potassium undecenoate, *Langmuir* **8**, 1931-1936.
20. Raj, W.R.P., Sasthav, M. and Cheung, H.M. (1993) Microcellular polymeric materials from microemulsions : control of microstructure and morphology, *J. Appl. Polym. Sci.* **47**, 499-511.
21. Gan, L.M., Li, T.D., Chew, C.H., Teo, W.K. and Gan, L.H. (1995) Microporous polymeric materials from microemulsions of zwitterionic microemulsions, *Langmuir* **11**, 3316-3320.
22. Guo, J.S., El-Aasser, M.S. and Vanderhoff, J.M. (1989) Microemulsion polymerization of styrene, *J. Polym. Sci. Polym. Chem. Ed.* **27**, 691-710.
23. Perez-Luna, V.H., Puig, J.-E., Castano, V.M., Rodriguez, B.E., Murthy, A.K. and Kaler, E.W. (1990) Styrene polymerization in three-component cationic microemulsions, *Langmuir* **6**, 1040-1044.

24. Larpent, C. and Tadros, Th.F. (1991) Preparation of microlatex dispersions using oil-in-water microemulsions, *Colloid Polym. Sci.* **269**, 1171-1183.
25. Feng, L. and Ng, K.Y.S. (1990) In situ kinetic studies of microemulsion polymerization of styrene and methylmethacrylate by Raman spectroscopy, *Macromolecules* **23**, 1048-1053.
26. Guo, J.S., Sudol, E.D., Vanderhoff, J.W. and El-Aasser, M.S. (1992) Particle nucleation and monomer partitioning in styrene o/w microemulsion polymerization, *J. Polym. Sci. Polym. Chem. Ed.* **30**, 691-702.
27. Guo, J.S., Sudol, E.D., Vanderhoff, J.W. and El-Aasser, M.S. (1992) Modeling of the styrene microemulsion polymerization, *J. Polym. Sci. Polym. Chem. Ed.* **30**, 703-712.
28. Puig, J.E., Perez-Luna, V.H., Perez-Gonzales, M., Macias, E.R., Rodriguez, B.E. and Kaler, E.W. (1993) Comparison of oil-soluble and water-soluble initiation of styrene polymerization in a three-component microemulsion, *Colloid Polym. Sci.* **271**, 114-123.
29. Manders, B.G., van Herk, A.M. and German, A.L., Sarnecki, J., Schomäcker, R. and Schweer, J. (1993) Pulsed-laser studies on the free-radical polymerization kinetics of styrene in microemulsion, *Makromol. Chem. Rapid Commun.* **14**, 693-701.
30. Gan, L.M., Chew, C.H. and Lye, I. (1992) Styrene polymerization in oil-in-water microemulsions : kinetics of polymerization, *Makromol. Chem.* **193**, 1249-1260.
31. Gan, L.M., Chew, C.H., Lee, K.C. and Ng, S.C. (1994) Formation of polystyrene nanoparticles in ternary cationic microemulsions, *Polymer* **35**, 2659-2664.
32. Gan, L.M., Lee, K.C., Chew, C.H. and Ng, S.C. (1995) Effect of surfactant concentration on polymerization of methylmethacrylate and styrene in emulsions and microemulsions, *Langmuir* **11**, 449-454.
33. Antonietti, M., Lohmann, S. and van Niel, C. (1992) Polymerization in microemulsions : 2. Surface control and functionalization of microparticles, *Macromolecules* **25**, 1139-1143.
34. Antonietti, M., Lohmann, S. and Bremser, W. (1992) Polymerization in microemulsion. Size and surface control of ultrafine latex particles, *Prog. Colloid Polym. Sci.* **89**, 62-65.
35. Gan, L.M., Chew, C.H., Ng, S.C. and Loh, S.E. (1993) Polymerization of methylmethacrylate in ternary systems : emulsion and microemulsion, *Langmuir* **9**, 2799-2803.
36. Rodriguez-Guadarrama, L.A., Mendizabal, E., Puig, J.E. and Kaler, E.W. (1993) Polymerization of methylmethacrylate in 3-component cationic microemulsion, *J. Appl. Polym. Sci.* **48**, 775-786.
37. Bleger, F., Murthy, A.K., Pla, F. and Kaler, E.W. (1994) Particle nucleation during microemulsion polymerization of methylmethacrylate, *Macromolecules* **27**, 2559-2565.
38. Schauber, C. and Riess, G. (1989) Préparation de microlatex acryliques par polymérisation photochimique de solutions micellaires et de microémulsions, *Makromol. Chem.* **190**, 725-735.
39. Full, A.P., Puig, J.E., Gron, L.U., Kaler, E.W., Minter, J.R., Mourey, T.H. and Texter, J. (1992) Polymerization of tetrahydrofurfuryl methacrylate in three-component anionic microemulsions, *Macromolecules* **25**, 5157-5164.
40. Leong, Y.S. and Candau, F. (1982) Inverse microemulsion polymerization, *J. Phys. Chem.* **86**, 2269-2271.
41. Candau, F., Leong, Y.S. and Fitch, R.M. (1985) Kinetic study of the polymerization of acrylamide in inverse microemulsion, *J. Polym. Sci. Polym. Chem. Ed.* **23**, 193-214.
42. Carver, M.T., Dreyer, U., Knoesel, R., Candau, F. and Fitch, R.M. (1989) Kinetics of photopolymerization of acrylamide in AOT reverse micelles, *J. Polym. Sci. Polym. Chem. Ed.* **27**, 2161-2177.
43. Vaskova, V., Hlouskova, Z., Barton, J. and Juranicova, V. (1992) Polymerization in microemulsions. 4. Locus of initiation by ammonium peroxodisulfate and 2,2'-azoisobutyronitrile, *Makromol. Chem.* **193**, 627-637.
44. Daubresse, C., Grandfils, C., Jerome, R. and Teyssie, P. (1994) Enzyme immobilization of nanoparticles produced by inverse microemulsion polymerization, *J. Colloid Interface Sci.* **168**, 222-229.
45. Capek, I. and Potisk, P. (1995) Microemulsion polymerization of butyl acrylate. IV. Effect of emulsifier concentration, *J. Polym. Sci. Polym. Chem. Ed.* **33**, 1675-1683.
46. Kaplin, D.A. and Qutubuddin, S. (1994) Electrodeposition of pyrrole into a porous film prepared by microemulsion polymerization, *Synth. Met.* **63**, 187-194.
47. Holtzscherer, C., Durand, J.P. and Candau, F. (1987) Polymerization of acrylamide in nonionic microemulsions : characterization of the microlatices and polymers formed, *Colloid Polym. Sci.* **265**, 1067-1074.
48. Corpart, J.M. and Candau, F. (1993) Aqueous solution properties of ampholytic copolymers prepared in microemulsions, *Macromolecules* **26**, 1333-1343.
49. Candau, F., Buchert, P. and Krieger, I. (1990) Rheological studies on inverse microlatices, *J. Colloid Interface Sci.* **140**, 466-473.
50. Candau, F., Zekhnini, Z. and Heatley, F. (1986) ^{13}C NMR study of the sequence distribution of poly(acrylamide-co-sodium acrylates) prepared in inverse microemulsions, *Macromolecules* **19**, 1895-1902.
51. Corpart, J.M., Selb, J. and Candau, F. (1993) Characterization of high charge density ampholytic copolymers prepared by microemulsion polymerization, *Polymer* **34**, 3873-3886.
52. Candau, F. and Buchert, P. (1990) Polymerization of trimethylaminoethylchloride methacrylate in microemulsion : formulation, characterization and rheological behavior of the

140

microlatexes, *Colloids Surf.* **48**, 107-122.
53. Antonietti, M. and Nestl, T. (1994) Polymerization in microemulsions with metallosurfactants, *Macromol. Rapid Commun.* **15**, 111-116.
54. Candau, F. and Anquetil, J.Y. (in press), New developments in polymerization in bicontinuous microemulsions, in D.O. Shah (ed.), *Micelles, Microemulsions, and Monolayers.*
55. Antonietti, M., Lohmann, S., Eisenbach, C.D. and Schubert, U.S. (1995) Synthesis of metal-complexing latices via polymerization in microemulsion, *Macromol. Rapid Commun.* **16**, 283-289.
56. Larpent, C., Richard, J. and Vaslin, S. (1992) Nanoparticules de polymères fonctionnalisées, leur procédé de préparation et leur utilisation, *French Patent* 92, 06759 to Prolabo.

DISPERSION POLYMERIZATION

E. DAVID SUDOL
Emulsion Polymers Institute
Lehigh University, Bethlehem, Pennsylvania 18015 USA

1. Introduction

The term *'dispersion polymerization'* can be considered a misnomer. The terms *bulk, solution, suspension, emulsion, miniemulsion,* and *microemulsion* all evoke mental images of how these polymerizations are carried out and are at least superficially correct if not conceptually. Generally, these refer to the initial state of the system prior to polymerization while some also describe the final state after the monomer has been converted to polymer (i.e., bulk, solution, and suspension). On the other hand, the three types of emulsions are considered to result in polymer colloids or latexes (although sometimes referred to as 'emulsion polymers'). Dispersion polymerization, as defined here, however, does not begin with a 'dispersion' of monomer in another liquid phase but rather a homogeneous solution as in solution polymerization. The end product, however, can be called a dispersion since stable polymer particles result which are indeed 'dispersed' in a continuous liquid phase. Some would say it is more akin to *precipitation polymerization* which has a similar initial state (i.e., homogeneous) but results in a 'bulk' polymer that, through polymerization, has precipitated and separated en masse from the inert liquid phase. Others would contend that it is more like an emulsion polymerization except for the absence of the initial emulsion; although this may on the surface sound absurd, the mechanisms involved in the polymerization indeed have significant similarities and much of the current knowledge of the mechanisms of dispersion polymerization has derived from that in emulsion polymerizations, an area where a great deal of fundamental research has been conducted and continues to be conducted. How the two types of polymerization came together historically is described below.

2. Historical Perspective

Dispersion polymerization as it is largely reported upon today has its roots in two developments which began in the 1950's, namely, the preparation of monodisperse latexes via emulsion and seeded emulsion polymerization in aqueous systems and the preparation of "latexes" in organic media via dispersion polymerization. Both of these initially involved the preparation of submicron particles and grew as research areas and businesses independently.

Monodisperse polystyrene (PS) latexes initially became available over 40 years ago in a variety of sizes from under 0.1 mm to just over 1.0 μm [2]. These were initially offered without charge to the scientific community, but soon their demand became so

J. M. Asua (ed.), Polymeric Dispersions: Principles and Applications, 141–154.

great that they were developed into a small business. Many applications were found ranging from use as calibration standards to biomedical supports. As a largely academic use, monodisperse latexes came to be used as seed particles for studying the fundamental kinetic mechanisms of emulsion polymerization [3, 4].

The need for larger particles in the micron size range (1 - 100 μm), however became apparent early on. Nonetheless, difficulties in producing these with narrow size distributions made them generally unavailable for more than 25 years. By the early 1980's, seeded polymerization techniques did advance sufficiently whereby monodisperse polystyrene (PS) latexes in this size range could be produced [5, 6]. However, these polymerization methods were often termed 'tedious' or 'difficult to reproduce'.

In the meantime, dispersion polymerization developed rapidly driven by the commercial needs of industry. In the 1950's, polymers for coatings applications were mostly available as low molecular weight materials in organic solvents, which were cured upon application or as high molecular weight polymers in the form of latex particles dispersed in water. Certain disadvantages inherent in each of these kinds of systems led industrial researchers to seek an alternative which combined the advantages of these systems, namely, high molecular weight polymer in the form of particles dispersed in an organic medium.

The growth in the patent literature on dispersion polymerization from the late 1950's through the early 1970's can be used as an indicator of the strong interest that companies had in the prospect of using this process to make polymers for high volume coatings applications. ICI, DuPont, and Rohm & Haas were the companies leading in patent applications over this time period. It was ICI researchers, however, that became best known for their work through the publication of "Dispersion Polymerization in Organic Media" in 1975, a book edited and written in part by K.E.J. Barrett of ICI [1]. This book describes the design and development of dispersion polymerization recipes and also the state of the mechanistic understanding of the physical and chemical processes taking place during the polymerization. In addition, results were reported whereby micron-size poly(methyl methacrylate) (PMMA) particles having a narrow size distribution could be produced [7]. This is where the two developments came together.

Soon dispersion polymerization was being investigated as an alternate method for preparing monodisperse PS and PMMA latexes in the 1 - 10 mm size range in a *single step* [8]. The earlier work had shown that indeed micron size PMMA particles could be produced by dispersion polymerization in petroleum distillates, however, little control over the particle size was found despite variations in dispersant and initiator levels [7]. This 'control' became the subject of many subsequent studies which expanded the size range and numbers of polymers that could be prepared in the form of micron-size monodisperse particles. Many of these studies have also included kinetic data plus other experimental information with which to infer something about the polymerization mechanism. As in emulsion polymerization, the heterogeneous nature of the reaction provides an increased level of complexity which for this system may be seen as being even greater, as will be shown in the following sections.

This paper will focus primarily on the developments in the understanding of dispersion polymerization as furthered by efforts to prepare micron-size monodisperse polymer particles primarily in polar media. Much of this will come by extension from the earlier works and by analogy with recent advances in the understanding of emulsion polymerization.

3. Mechanism of Free Radical Dispersion Polymerization

Stated simply, dispersion polymerization is a process by which stable colloidal polymer particles are formed in a continuous liquid medium through polymerization of a monomer (or monomers) which is *completely miscible with this medium and* in which a 'stabilizer' has been initially dissolved. Emulsion polymerization might similarly be described by replacing the expression in italics with *dispersed in this medium*. The complexity underlying this simplistic description quickly becomes evident when examining the individual physical and chemical processes which together must be taken into account to understand how the entire process is controlled. By far, nucleation is the most ill understood and yet most important process in dispersion polymerization.

The chemical reactions are those which take place in most free radical polymerizations which include decomposition of the initiator species, initiation, propagation, termination, and transfer. Initially, these will all occur in the medium (as in solution polymerization) but with the appearance of polymer particles will occur to varying extents in both phases depending on the partitioning of the various species. This partitioning is largely controlled by the nature of the medium which also controls much of the physical processes which occur in dispersion polymerization. These include those phenomena which affect both the nucleation and growth of the polymer particles: (1) formation of particle nuclei; (2) adsorption/desorption of stabilizer; (3) flocculation of nuclei; (4) adsorption/desorption of oligomers (dead or as free radicals); and (5) diffusion limited termination and propagation in the polymer particles.

The following sections will concentrate on how the choice of experimental variables affects the formation of particles in dispersion polymerization.

3.1. PARTICLE NUCLEATION

The mechanism of particle formation in dispersion polymerization is subject to uncertainty because, as in emulsion polymerization, it must be inferred from experimental data which do not directly monitor the nucleation process at the molecular level. No one has witnessed the growth of the polymer chains which lead to the formation of particles. The theories of nucleation are largely based on polymerization kinetics, particle size, and molecular weight data gained in the early stages of the reaction with additional evidence provided by more detailed analyses of the species in the medium and associated with the particles. In addition, much of the mechanistic understanding has been gained by extrapolation from its nearest neighbors, namely precipitation polymerization and emulsion polymerization (particularly for highly water-soluble monomers and systems employing surfactants below their cmc's)

Much more is known about how to make specific polymer particles in specific sizes than is understood about why certain conditions 'work' while others do not. Recipes for preparing a wide variety of monodisperse polymer particles in the micron size range by dispersion polymerization are available in the literature. These recipes have been developed largely in an empirical fashion where the choice of system components have been guided not only by the requirements of the product but the limitations of the process and the current understanding of what controls it. The particle size and its distribution are known to be affected by: (1) the monomer and its concentration; (2) the medium; (3) the amount and type of the stabilizer(s); (4) the initiator and its concentration; and (5) the temperature of the reaction. The limitations on the process include: (1) the monomer must be sufficiently soluble in the medium such that no

144

separate monomer phase exists over the range of concentrations (and temperatures) used; (2) the polymer formed must be sufficiently insoluble in the medium such that it is almost completely found in the polymer particle phase; (3) the stabilizer must be soluble in the medium and yet find its way to the surface of the particles to effect stabilization; and (4) the initiator must be soluble in the medium over the range of concentrations employed. All of these depend on the medium and what is termed its 'solvency' which can be considered to be one of the most important factors in understanding the mechanism of nucleation in dispersion polymerization.

3.1.1 Proposed Mechanisms

Various nucleation mechanisms have been proposed to account for the formation of particles in dispersion polymerization. For monodisperse particles, nucleation is considered to be complete (i.e., no new and stable particles are formed) at a low conversion (monodispersity is considered to be achieved only by having a long particle growth stage relative to the nucleation stage). By 1% conversion, particles must be already over 200 nm in diameter to achieve a final size of 1 μm, the lower limit that is being considered here. This size is already larger than the latex particles typically prepared in emulsion polymerizations.

As in any free radical polymerization, the first event is the decomposition of an initiator species to form free radicals capable of adding monomer units via propagation reactions. The possible fates of free radicals in a dispersion polymerization are represented in the schematic presented in Figure 1.

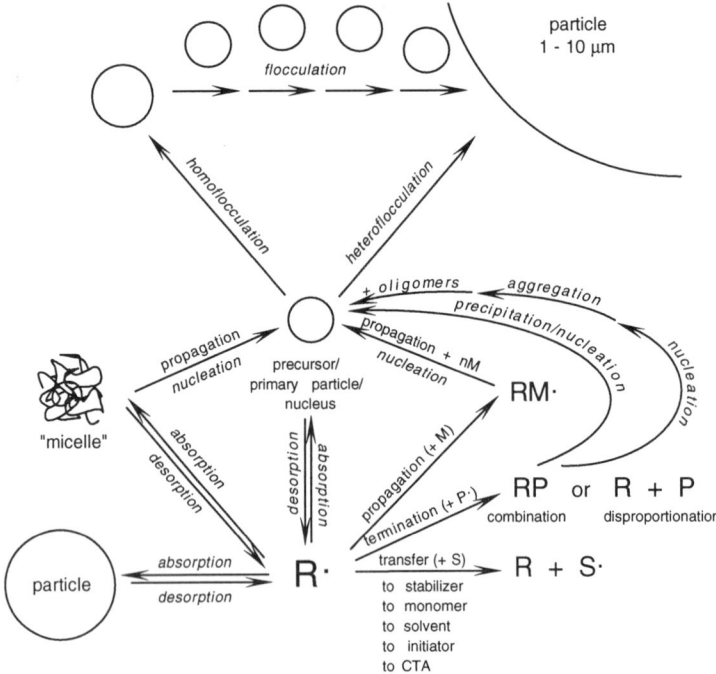

Figure 1. Schematic representation of the fate of free radicals in dispersion polymerizations; words in italics represent physical processes.

In the absence of absorption (or adsorption) sites (e.g., particles), these radicals (R•) can propagate with the monomer, terminate with other radicals or transfer to various species including the stabilizer. The latter can continue to propagate to form grafted stabilizer.

Particle nuclei can be formed by: (1) *self nucleation* where an oligomeric chain reaches a critical chain length (j_{crit}) either by propagation or termination (by combination) whereby it becomes insoluble in the medium and collapses on itself to form a nucleus which may subsequently be stabilized by adsorbed stabilizer; (2) *aggregative nucleation* where the oligomeric chains, as they grow and increase in number, associate with each other to form an insoluble aggregate or nucleus which can also be stabilized by adsorbed stabilizer; (3) *micellar nucleation* only if the stabilizer does form some type of micelle or pseudo-micelle in the medium; this mechanism has been largely discounted in dispersion polymerization since the monomer is available in high concentrations in the medium; or (4) *coagulative nucleation* where the nuclei formed by any of the preceding mechanisms coagulate and coalesce with each other (homoflocculation); after nucleation has ceased these may continue to flocculate with the existing particles (heteroflocculation). This mechanism is similar to homogeneous/coagulative nucleation as used to describe certain emulsion polymerizations (particularly for reactions without surfactant or where it is used in concentrations below its cmc, and in the case of highly water-soluble monomers) [9].

Nuclei formation is a function of all the polymerization parameters. The critical chain lengths for precipitation or aggregation are determined by the solvency of the medium (monomer plus solvent) and the nature of the polymer. The rate of nuclei formation is controlled by the number (concentration) and growth rate of the oligomeric chains which are determined by the initiation, propagation, termination, and transfer rates (in the absence of particles which can adsorb these oligomers). Stabilization of particles is determined by the rate and extent of adsorption of the stabilizer (and its efficiency in stabilization) which itself is a function of the solvency of the medium and the nature of the polymer onto which it is adsorbing. In addition, in cases where homopolymers are used as stabilizer, grafting of the stabilizer is often reported as either an additional or essential mechanism for providing stability to the particles. Particle nucleation ceases when any radicals being produced no longer result in the formation of stable particles but instead are either adsorbed by existing particles in the form of growing oligomers, form dead polymer (adsorbed or soluble in the medium), or precipitate on the particles as unstable nuclei.

3.1.2 *Effect of Solvency of the Medium*

The medium consists of a reactive component, namely the monomer(s), and a non-reactive component which itself may contain more than one component (e.g., alcohol/water). In most cases, the monomer is a good solvent for its own polymer and thus its mixture with the other components must still provide for the insolubility of the polymer. Often the choice of the solvent is optimized by trial and error to suit a specific monomer/polymer/stabilizer system in order to achieve conditions suitable for preparing micron-size monodisperse particles of controlled size and in a reproducible fashion. As an illustration, recipes used to produce monodisperse PS [10], PMMA [11], and PBA (poly(n-butyl acrylate)) [12] have employed ethanol (EtOH), methanol (MeOH), and MeOH/H_2O (90/10 wt/wt) as the non-reactive components in the medium, respectively, in combination with poly(vinyl pyrrolidone) (PVP) as stabilizer, and 2,2'-azobis (isobutryonitrile) (AIBN) as initiator. These represent extensions and expansions on previous work by Almog et al. using similar conditions for preparing monodisperse PS and PMMA [8].

Solubility Parameters. Based on much experimental evidence, the general observation is that the particle size increases with increasing solvency of the medium for the polymer. This has often been seen by simply increasing the amount of monomer in a given reaction, the increased solvency being the chief reason for the increased size. In attempts to quantify the effect of solvency on the resulting particle size, a number of authors have tried to find correlations between the particle size and the difference in solubility parameters of the polymer and the medium [8, 13]. Indeed some correlations have been found particularly within given systems (i.e., same initiator, stabilizer, and monomer), however, extensions to vastly different solvents or monomers have not correlated as well by simply using solubility parameters [14].

A second critical interaction is between the stabilizer and the medium. Paine used the three dimensional Hansen solubility parameters to correlate the particle size of polystyrene particles produced by dispersion polymerization in polar solvents using hydroxypropyl cellulose (HPC) as stabilizer [15]. Notable among his findings was that the largest particles were produced in solvents where the polarity and hydrogen bonding terms of the Hansen parameter were the closest to that of the HPC and not polystyrene. His conclusion from this was that the HPC was more the controlling factor in determining particle size in these dispersion polymerizations; the difference in the solubility properties of the polymer and the stabilizer should be sufficient to allow control of the particle size through adjustment of the solubility of the medium. Finally, Paine was also able to show that if all three Hansen solubility parameters were matched by using various solvent combinations, the same particle size resulted.

Viscosity. The viscosity of a polymer solution is not only a function of the molecular weight and concentration of the polymer but also the configuration of the polymer in solution which is determined by the interaction of the polymer with the medium (solvency). In an effort to reconcile differences found in the dispersion polymerization of n-butyl acrylate in $MeOH/H_2O$ and $EtOH/H_2O$ systems using PVP as stabilizer, solution viscosities were measured at two shear rates as a function of the percent alcohol in the alcohol/water mixture [12]. The results, illustrated in Figure 2, show a number of important differences. First, it should be noted that conditions suitable for dispersion polymerizations are indicated for each system by the horizontal arrows (i.e., limits were established by required solubilities of BA monomer and limited solubilities of PBA polymer). Two series of dispersion polymerizations were run at 70/30 $EtOH/H_2O$ and 90/10 $MeOH/H_2O$ (indicated by the vertical dashed lines). Poor stability and reproducibility were found for the former reactions with particle sizes in the range of 1 to 2 µm while good stability and reproducibility were indicated for polymerizations in the $MeOH/H_2O$ medium with monodisperse particles ranging from 1 to 5 µm being obtained by varying the stabilizer and initiator levels. These results are at least partly explained by the difference in the conformation of the stabilizer in the medium and its ability to adsorb on the flocculating nuclei to provide stability. In the $EtOH/H_2O$ system, the high viscosity and shear thinning indicate that the stabilizer molecules are well solvated (expanded) by the medium and that the molecules entangle to some extent. These would not be expected to have as strong an affinity for the resulting polymer particles as in the $MeOH/H_2O$ system where the stabilizer molecules are more coiled in solution and do not appear to interact with each other (no shear thinning). In this case, their adsorption onto the PBA particles would be expected to be even greater than in a comparable aqueous phase system. (It should be noted that PVP has also been used successfully to stabilize large-particle-size latexes in seeded emulsion polymerizations of styrene [5].) These results may also help to explain why methanol proved to be a better medium than ethanol for the dispersion polymerization of MMA [11].

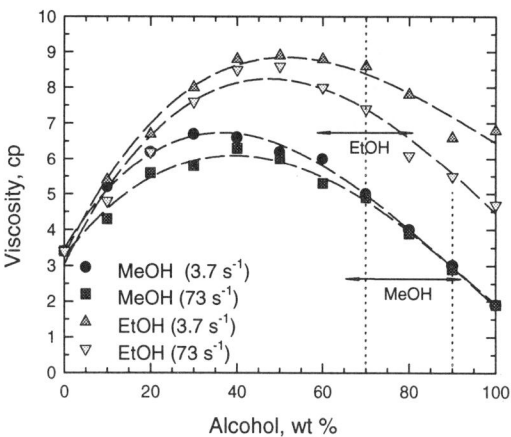

Figure 2. Solution viscosities of 1% PVP K-90 solutions in alcohol/water media.

3.1.3 *Effect of Stabilizer*

As alluded to earlier, the stabilizer plays a crucial role in the formation of stable particles in dispersion polymerization, often being used as the primary means of controlling the resulting particle size. The stabilizer is typically an amphipathic polymer which is able to adsorb onto the surface of the particles imparting stabilization via the steric mechanism. Block and graft copolymers are often used in dispersion polymerizations where, by definition, one segment favors the medium and the other the particle surface. Another category of stabilizer has been termed a graft copolymer precursor; in this case, the monomer in the dispersion polymerization grafts onto the stabilizer during the polymerization whereby it can become incorporated chemically or physically on the surface of the polymer particles.

Yet another type of stabilizer is the homopolymer which can adsorb onto the surface of the particles being in equilibrium with that dissolved in the medium; the true nature of this stabilization has been subject to debate, however. An example of this is the use of PVP as stabilizer. This homopolymer has found extensive use in stabilization systems in both aqueous [5] and non-aqueous (polar) media [8, 10-12, 17, 18]. In addition, it has been used in combination with other stabilizers such as Aerosol OT (sodium dioctyl sulfosuccinate) [10, 12], Triton N-57 (nonyl phenyl polyether alcohol) [10, 16], and Aliquat 336 (methyl tricaprylyl ammonium chloride) [8]. To illustrate the effect of the concentration of PVP on the resulting particles, a compilation of some results for the dispersion polymerization of three different monomers are presented in Figure 3.

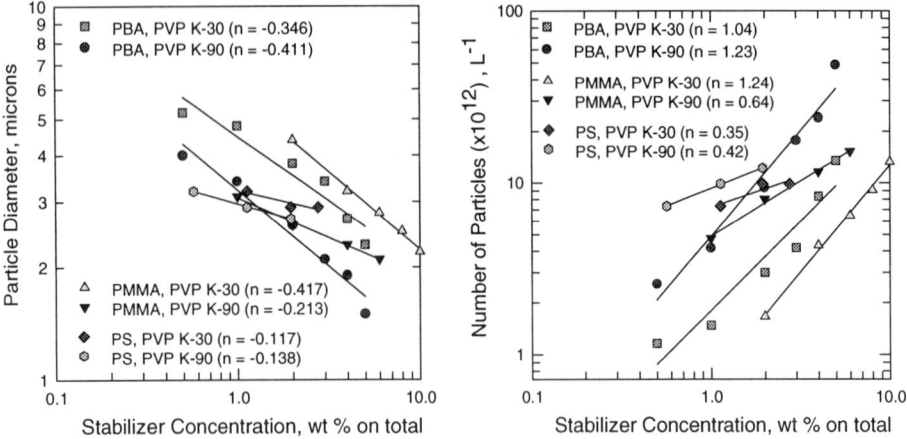

Figure 3. Log-log plots of particle diameter (left) and number of particles (right) as a function of the stabilizer concentration for three monomer systems. (n represents the slope of the regression analysis.)

The same data are represented here in two ways, as particle diameter (left) and number of particles (right) as a function of the stabilizer concentration. Both are log-log plots, the former being typical of those presented in dispersion polymerization publications while the latter is more often used to present results of emulsion polymerizations. The slopes (n), obtained by regression analysis and given in the parenthesis, represent the power dependencies ($D \propto [S]^n$, $N_p \propto [S]^n$). These results were all obtained in the same laboratory and yet illustrate the variation in recipe conditions that are used to obtain monodisperse micron-size particles from different monomers, namely, styrene [10], methyl methacrylate [11], and n-butyl acrylate [12]. The same two molecular weight PVP's were used in all three of these studies. The general results are that the particle size decreased with both increasing concentration and molecular weight of the stabilizer. Note that the concentration is represented as weight percent and not a molar concentration which indicates that fewer molecules of the higher molecular weight PVP (PVP K-90, average molecular weight of 360,000) are required to stabilize smaller and thus more numerous particles than the lower molecular weight (PVP K-30, average molecular weight of 40,000). This may be indicative of the stronger adsorption of the higher molecular weight polymer leading to greater numbers of particles. Further comparisons are difficult when one notes the other differences in the conditions of these polymerizations as described in Table 1. As can be seen, no set of conditions is duplicated in these polymerizations. Suffice it to say that optimization of recipes for preparing micron-size monodisperse particles will often preclude a direct comparison of results, at least without much more in-depth knowledge of the interaction leading to such conditions being established.

Evidence for grafting onto the stabilizer has been presented for several systems. Material balances on the stabilizer [11], the ability of the dissolved polymer recovered from cleaned dispersion particles to be redispersed as stable colloids [19, 20], and electron microscope examination of the surface and interior of the particles [20], among

TABLE 1. Comparison of reaction conditions for preparing micron-size monodisperse particles via dispersion polymerization

Monomer	Medium	Costabilizer	Initiator	Temperature	Reference
BA	MeOH/H$_2$O[1]	none	ACPA[2]	70°C	12
MMA	MeOH	none	AIBN[3]	55°C	11
Styrene	EtOH	Aerosol OT	AIBN	70°C	10

[1] 90/10 wt/wt

[2] 4,4' azobis(4-cyanopentanoic acid)

[3] 2,2' azobis(isobutyronitrile)

other methods, have all provided indirect evidence for grafting onto stabilizers which includes not only PVP but also poly(acrylic acid) (PAA) and HPC. Although this evidence is compelling, the adsorption mechanism of stabilization cannot be discounted. It is more likely that both of these mechanisms play a role in the stabilization of the particles. It should be noted that a similar mechanism has been proposed for the emulsion polymerization of vinyl acetate in the presence of poly(vinyl alcohol) polymeric stabilizer. Evidence for significant grafting has been found for this system as well [21].

3.1.4 *Effect of Initiator*
Nucleation in dispersion polymerization is expected to be affected by the rate of production of free radicals in the medium as determined by the concentration and the decomposition rate of the initiator. In emulsion polymerization, an increase in the initiator concentration typically leads to an increase in the number of particles, more radicals nucleating more particles via either micellar or homogeneous/coagulative mechanisms. However, in dispersion polymerization, the opposite trend has been reported, i.e., increased initiator producing fewer particles. This result is not universal, however, as illustrated by the results compiled in Figure 4 for the same systems reported in Figure 3. Both PMMA and PS particles increase in size (Figure 4, left) and decrease in number (Figure 4, right) with increasing AIBN concentration. However, the number of PBA particles produced has a slight but positive dependence on the initiator concentration. Again note that the polymerization conditions are not identical for each of these series as noted in Table 1 (AIBN used in all reactions); in addition, PVP K-30 was used to prepare the PMMA and PS particles while PVP K-90 was used in the BA polymerizations which can account for the greater numbers of particles but not the trend.

How can the above results be reconciled with the proposed nucleation mechanisms in dispersion polymerization? A number of authors have offered qualitative explanations for the seemingly paradoxical result of decreasing particle number with increasing initiator concentration. In fact, this phenomenon has also been reported for the emulsion polymerization of vinyl acetate using sodium persulfate as initiator and a reactive surfactant (sodium dodecyl allyl sulfosuccinate) as stabilizer where the dependency of the number of particles on the initiator concentration ranged from -0.21 to -0.41 depending on the surfactant level [22]. Although, the cause of this was not investigated, it was tentatively attributed to an electrolyte effect causing destabilization of the particles. Of course, this cannot explain the results reported above. The most popular explanation for this phenomenon is that by increasing the radical production rate, the average molecular weight of the polymer formed in the medium is decreased by termination reactions which subsequently reduces the number of chains which either aggregate or grow long enough to form nuclei thus reducing the number of particles [8, 23 - 25]. Two other variations on this reasoning have also been offered: (1) lower molecular weight grafts are formed on the stabilizer (for the same reasons) which make it more soluble in the medium and thus not as effective a stabilizer [16]; and (2) more chain transfer involving the initiator occurs with more radicals being produced, lowering the number of nuclei formed [26]. Another explanation which differs significantly from the preceding ones is that there is a competition between limited aggregation of nuclei and the adsorption (and subsequent stabilization) by the stabilizer; smaller particles are favored by faster adsorption of stabilizer and slower production of oligomers which form the nuclei [10]. Thus, lower initiator concentrations produce smaller and more numerous particles.

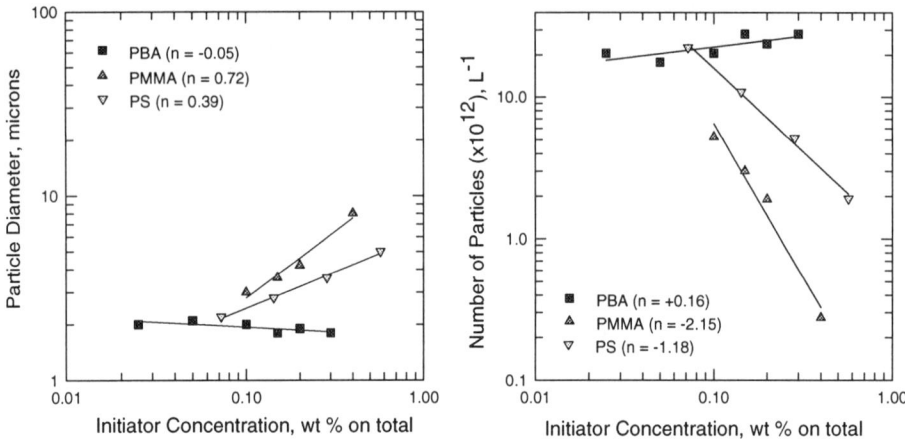

Figure 4. Log-log plots of particle diameter (left) and number of particles (right) as a function of the initiator (AIBN) concentration for three monomer systems. (n represents the slope of the regression analysis).

Additional *outside* evidence which might shed more light on this phenomenon comes from results reported for *anionic* dispersion polymerizations of styrene in hexane [27]. In this case, the particle size (3 - 5 mm) was also found to increase with increasing initiator (*sec*-butyllithium) concentration (and also decreasing stabilizer (diblock copolymer of styrene and butadiene) concentration); $D \propto [I]^{0.22}$, $N_p \propto [I]^{-0.67}$. In these polymerizations, all chains are initiated at the same time and are not terminated as in free radical polymerization. These chains grow simultaneously to nearly the same final molecular weight (10^4 - 10^5 g/mol). Chain aggregation and flocculation of nuclei are also considered to be the mechanism of nucleation operative in this system. These results would tend to support the last explanation given above since the number of chains growing beyond their j_{crit} are not reduced by termination reactions.

Still, the reason for the different dependencies in Figure 4 is not clear. The relative rates of the various physical and chemical processes must be understood more fully to characterize this behavior.

4.   Kinetics of Dispersion Polymerization

To varying degrees, polymerization takes place in both the continuous phase and particle phase in any dispersion polymerization system. The extent to which each of these occurs depends on the partitioning of the monomer and free radicals between these phases. And this partitioning is a function of variables mentioned previously, particularly the solvency of the medium and the nature of the monomer/polymer system. Therefore, the reaction kinetics is a combination of solution polymerization and polymerization inside swollen polymer particles. The latter often becomes more important for polymers in which the gel effect is notable which can increase the rate substantially over solution kinetics.

During the nucleation stage of a dispersion polymerization, the kinetics begin as in a solution polymerization, the rate depending on the number of free radicals, the monomer concentration, and the reaction temperature. As particles are formed, the partitioning described above determines the relative rates of polymerization in the two phases. This partitioning of monomer, solvent, initiator, and free radicals are largely governed by thermodynamics. Indeed, a thermodynamic model for the partitioning, based on Flory-Huggins theory [28] as extended by Morton et al. [29], was successfully applied to the dispersion polymerization of styrene in ethanol [30]. This work showed that the monomer concentration in the two phases decreased throughout the reaction, although the relative concentrations in the particles and the medium (partition coefficient) increased slightly from about 0.85 to 1.1. (It should be noted that the ethanol also partitions into the particles to a small degree.) The fraction of polymer in the particles was initially about 0.7 and increased with conversion. This high and increasing polymer content led to an increase in the polymerization rate as the termination rate decreased. These kinetics resemble the pseudo-bulk region in emulsion polymerization where the rate has little if any dependence on the particle size since they are so large and can accommodate many radicals. Therefore, the rate of reaction generally lies in between that of solution and emulsion polymerization. This effect is also seen in the molecular weights which also increase with conversion and lie between these two types of polymerization.

The phenomena described above are highly dependent on the choice of system components which can vary significantly, making it risky to generalize about the kinetics of dispersion polymerization.

5.   Variety in Dispersion Polymerization

As illustrated earlier, the difficulty in comparing results, even from the same laboratory, stems from the variability in the experimental conditions either chosen or required for preparing micron-size monodisperse particles. Although this is not unique to dispersion polymerization, the added dimension of the medium (in contrast to emulsion polymerization) and its affect on the polymerization process greatly complicates any comparison.

Over the past fifteen years, styrene has been the mostly widely used monomer for studying dispersion polymerizations, as well as emulsion polymerizations. Reactions are typically run in polar media which might comprise a pure alcohol (CH_3OH to $C_{10}OH$) or may contain either a more polar (e.g., H_2O) or less polar component, these being added to control the size and monodispersity of the particles. Monomer levels range from under 5% to 50%. Three stabilizers have been most popular: PVP, HPC, and PAA (all varying in molecular weight). Costabilizers (e.g., Aerosol OT, Aliquat 336, Triton N57) may or may not be used. Other stabilizers include macromonomers (e.g., methacryloyl terminated poly(ethylene glycol) [31]) which copolymerize with the monomer and chain transfer agents (e.g., $CH_3(OCH_2CH_2)_{113}$-)C)CH$_2$SH [32]) onto which the monomer grafts. Initiators are typically peroxides (BPO) or an azo type (AIBN, AMBN, ADVN, ACPA). Temperatures range from 55°C to 80°C. Particles typically achieve sizes up to 10 mm.

Styrene has also been popular for performing dispersion copolymerizations with monomers such as divinyl benzene [10, 17], butadiene [18], n-butyl acrylate [33], and butyl methacrylate [13, 34].

Methyl methacrylate is the next most popular monomer to study, stemming initially from the early work at ICI [1]. Block copolymers (e.g., polystyrene-poly-dimethylsiloxane [35]) and graft copolymers (e.g., poly(12-hydroxy stearic acid)-g-PMMA [36]) have continued to be used as stabilizers in hydrocarbon media (e.g., alkanes). In addition, homopolymers (e.g., polyisobutylene) in mixed solvents (e.g., CCl$_4$/2,2,4-trimethylpentane) have been used to make monodisperse particles up to 13 mm [37]. Alcohols (CH$_3$OH) have been used in a more limited way, paralleling work in polystyrene using PVP as stabilizer [8, 11].

A number of other polymers have been synthesized by this technique including: poly(N-vinyl formamide) [38], poly(1-methacryloxybenzotriazole) [39], poly(butyl-2-cyanoacrylate) [40], polyacrolein [41], polyacrylonitrile [42], polychloromethylstyrene [43], polypyrrole [44], and glycidyl methacrylate copolymers [45].

Other areas of interest include the preparation of structured particles by dispersion polymerization or seeded dispersion polymerization (this has largely been accomplished by first preparing particles via dispersion polymerization and then replacing the medium with water, followed by seeded emulsion polymerization [46, 47]). In a more 'exotic' process, dispersion polymerizations have been conducted in supercritical CO$_2$, this requiring design of different stabilizers such as poly(1,1-dihydroperfluorooctyl acrylate) [48]. All of these examples serve to illustrate the ongoing interest and potential for further developments in dispersion polymerization.

6.   Prospects

Based on the preceding, dispersion polymerization appears to be firmly entrenched as a means of preparing micron-size monodisperse polymer particles. The variety of these particles is expected to expand into many areas of specialty application. On the other hand, much work is still required on even the 'simplest' systems (e.g., styrene) to gain a more fundamental understanding of the mechanisms of particle formation and growth, and how these are affected by the various process variables. Mathematical modeling of dispersion polymerization holds promise as one avenue for increasing this understanding as illustrated by several recent efforts to model various aspects of the process [30, 49, 50].

References

1. K.E.J. Barrett (ed.) (1975) *Dispersion Polymerization in Organic Media*, Wiley, London.
2. Bradford, E.B. and Vanderhoff, J.W. (1955) *J. Appl. Phys.*, **26**(2), 864-871.

3. Sudol, E.D., El-Aasser, M.S., and Vanderhoff, J.W. (1986) *J. Polym. Sci.: Part A: Polym. Chem.,* **24**, 3499-3513.
4. Gilbert, R.G. (1995) *Emulsion Polymerization: A Mechanistic Approach,* Academic Press, London.
5. Vanderhoff, J.W., El-Aasser, M.S., Micale, F.J., Sudol. E.D., Tseng, C.M., Silwanowicz, A., Kornfeld, D.M., and Vicente, F.A. (1984) *J. Disp. Sci. Tech.,* **5**(3&4), 231-246.
6. Ugelstad, J., Mork, P.C., Berge, A., Ellingsen, T. and Khan, A. (1982) in I. Piirma (ed.) *Emulsion Polymerization,* Academic Press, New York, 383-413.
7. Barrett, K.E.J. and Thomas, H.R. (1975) in K.E.J. Barrett (ed.) *Dispersion Polymerization in Organic Media,* Wiley, London, pp. 115-199.
8. Almog, Y., Reich, S., and Levy, M. (1982) *Brit. Polym. J.,* **14**, 131-136.
9. Hansen, F.K. (1992) in E.S. Daniels, E.D. Sudol, and M.S. El-Aasser (eds.) *Polymer Latexes: Preparation, Characterization, and Applications,* ACS Symposium Series **492**, Washington D.C., pp. 12-27.
10. Tseng, C.M., Lu, Y.Y., El-Aasser, M.S., and Vanderhoff, J.W. (1986) *J. Polym. Sci.: Part A: Polym. Chem.,* **24**, 2995-3007.
11. Shen, S., Sudol, E.D., and El-Aasser, M.S. (1993) *J. Polym. Sci.: Part A: Polym. Chem.,* **31**, 1393-1402.
12. Wang, D., Dimonie, V.L., Sudol, E.D., and El-Aasser, M.S. (1996) *Graduate Research Progress Reports,* Emulsion Polymers Institute, Lehigh University, **45**, 191-202.
13. Ober, C.K. and Lok, K.P.. (1987) *Macromol.,* **20**, 268-273.
14. Lok, K.P. and Ober, C.K. (1985) *Can. J. Chem.,* **63**, 209-216.
15. Paine, A.J. (1990) *J. Polym. Sci.: Part A: Polym. Chem.,* **28**, 2485-2500.
16. Paine, A.J., Luymes, W., and McNulty, J. (1990) *Macromol.,* **23**, 3104-3109.
17. Hattori, M., Sudol, E.D., and El-Aasser, M.S. (1993) *J. Appl. Polym. Sci.,* **50**, 2027-2034.
18. Hu, R., Dimonie, V.L., Sudol, E.D., and El-Aasser, M.S. (1995) *J. Appl. Polym. Sci.,* **55**, 1411-1415.
19. Paine, A.J. (1990) *J. Coll. Int. Sci.,* **138**(1), 157-169.
20. Paine, A.J., Deslandes, Y., Gerroir, P., and Henrissat, B. (1990) *J. Coll. Int. Sci.,* **138**(1), 170-181.
21. Magallanes, G.S., Dimonie, V.L., Sudol, E.D., Yue, H.J., and El-Aasser, M.S. (1996) *J. Polym. Sci.: Part A: Polym. Chem.,* **34**, 849-862.
22. Urquiola, M.B., Dimonie, V.L., Sudol, E.D., and El-Aasser, M.S. (1992) *J. Polym. Sci.: Part A: Polym. Chem.,* **25**, 2631-2644.
23. Ober, C.K. and Hair, M.L. (1987) *J. Polym. Sci.: Part A: Polym. Chem.,* **25**, 1395-1407.
24. Kobayashi, S., Uyama, H., Yamamota, I, Matsumoto, Y. (1990) *Polym. J.,* **22**, 759-761.
25. Tuncel, A., Kahraman, R., and Piskin, E. (1990) *J. Appl. Polym. Sci.,* **50**, 303-319.
26. Uyama, H. and Kobayashi, S. (1994) *Polym. Int.,* **34**, 339-344.
27. Awan, M.A., Dimonie, V.L., and El-Aasser, M.S. (1996) *J. Polym. Sci.: Part A: Polym. Chem.,* **34**, 2651-2664.
28. Flory, P.J. (1953) *Principles of Polymer Chemistry,* Cornell University Press, Ithaca, New York, Chaps. 12 and 13.
29. Morton, M., Kaizerman, S., and Altier, M.W. (1954) *J. Coll. Sci.,* **9**, 300-312.
30. Lu, Y.Y., El-Aasser, M.S., and Vanderhoff, J.W. (1988) *J. Polym. Sci.: Part B: Polym. Phys.,* **26**, 1187-1203.
31. Capek, I., Riza, M., and Akashi, M. (1992) *Polym. J.,* **24**(9), 959-970.
32. Bourgeat-Lami, E. and Guyot, A. (1995) *Polym. Bull.,* **35**, 691-696.
33. Sáenz, J.M. and Asua, J.M. (1995) *J. Polym. Sci.: Part A: Polym. Chem.,* **33**, 1511-1521.
34. Horák, D., Svec F., and Fréchet, J.M.J. (1995) *J. Polym. Sci.: Part A: Polym. Chem.,* **33**, 2961-2968.
35. Dawkins, J.V. and Taylor, G. (1979) *Polymer,* **20**, 599-604.
36. Antl, L., Goodwin, J.W., Hill, R.D., Ottewill, R.H., Owens, S.M., Papworth, S., and Waters, J.A. (1986) *Coll. Surf.,* **17**, 67-78.
37. Williamson, B., Lukas, R., Winnik, M.A., and Croucher, M.D. (1987) *J. Coll. Int. Sci.,* **119**(2), 559-564.
38. Uyama, H., Kato, H., and Kobayashi, S. (1994) *Polym. J.,* **26**(7), 858-863.
39. Yoshida, M., Yokota, T., Asano, M., and Kumakura, M. (1989) *Colloid Polym. Sci.,* **267**(11), 986-991.
40. Douglas, S.J., Illum, L., and Davis, S.S. (1985) *J. Coll. Int. Sci.,* **103**(1), 154-163.
41. Margel, S. and Wiesel, E. (1984) *J. Polym. Sci., Polym. Chem. Ed.,* **22**, 145-158.
42. Ansarifar, M.A. and Luckham, P.F. (1988) *Colloid Polym. Sci.,* **266**(11), 1020-1023.
43. Margel, S., Nov, E., and Fisher, I. (1991) *J. Polym. Sci.: Part A: Polym. Chem.,* **29**, 347-355.
44. Digar, M.L., Bhattacharyya, S.N., and Mandal, B.M. (1994) *Polymer,* **35**(2), 377-382.
45. Hosaka, S., Murao, Y., Tamaki, H., Masuko, S., Miura, K., and Kawabata, Y. (1993) *Polym. Int.,* **30**, 505-511.
46. Shen, S., El-Aasser, M.S., Dimonie, V.L., Vanderhoff, J.W., and Sudol, E.D. (1991) *J. Polym. Sci.: Part A: Polym. Chem.,* **29**, 857-867.
47. Okubo, M. and Nakagawa, T. (1994) *Colloid Polym. Sci.,* **272**(5), 530-535.
48. Hsiao, Y.L., Maury, E.E., DeSimone, J.M., Mawson, S., and Johnston, K.P. (1995) *Macromol.,* **28**, 8159-8166.
49. Paine, A.J. (1990) *Macromol.,* **23**, 3109-3117.
50. Poehlein, G.W. and Ahmed, S.F. (1995) *Int. Polym. Coll. Newsletter,* **26**, 87-93.

METALLOCENE CATALYSTS IN DISPERSED MEDIA

J.B.P. SOARES
University of Waterloo, Department of Chemical Engineering
Waterloo, Ontario, Canada N2L 3G1

A.E. HAMIELEC
McMaster University, Department of Chemical Engineering
Hamilton, Ontario, Canada L8S 4L7

1. Introduction

Metallocene catalysts are organometallic coordination compounds in which one or two cyclopentadienyl rings or substituted cyclopentadienyl rings are bonded to a central transition metal atom, as shown in Figure 1. The nature and number of the rings and substituents (S), the type of transition metal (M) and its substituents (R), the type of the bridge, if present, and the cocatalyst type determine the catalytic behaviour of these organometallic compounds towards the polymerization of linear and cyclic olefins and diolefins.

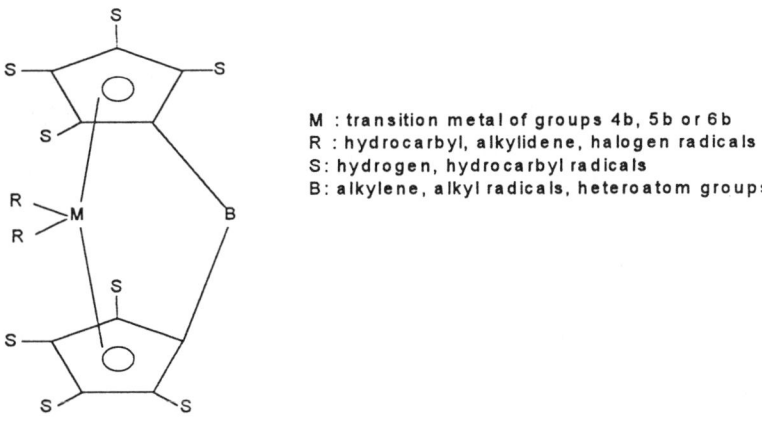

M : transition metal of groups 4b, 5b or 6b
R : hydrocarbyl, alkylidene, halogen radicals
S: hydrogen, hydrocarbyl radicals
B: alkylene, alkyl radicals, heteroatom groups

Figure 1. Generic structure of a metallocene catalyst

The importance of these new catalytic systems is revealed by the number of patents issued in this field since 1980. It has been reported that 179 metallocene patents have been granted in the U.S. in 1995 [1]. Currently, there are five review articles published in the literature on metallocene catalysts [2-6]. These review papers cover metallocene

J. M. Asua (ed.), Polymeric Dispersions: Principles and Applications, 155–176.
© 1997 *Kluwer Academic Publishers. Printed in the Netherlands.*

catalyst synthesis, nature of active sites, polymerization conditions and mechanisms, patent literature, and polymerization reactor engineering.

What is so great about metallocene catalysts? In brief, metallocene catalysts have excellent: (1) Polymer microstructure control - metallocene catalysts can produce polymer with narrow distributions of molecular weight, chemical composition, and stereoregularity. (2) Catalytic activity - metallocene catalysts have very high activities and therefore are adequate for the production of commodity polymers. (3) Versatility - metallocene catalysts can produce several different types of polymers with enhanced or entirely novel properties. (4) Compatibility - metallocene catalysts can be used in existing polymer manufacturing processes with minimal modifications.

In this article we will describe the leading features of metallocene catalysts and polymerization processes, focusing in their improved ability of polymer microstructure control.

2. Comparison of Conventional Ziegler-Natta and Metallocene Catalysts

Metallocene catalysts have been used mainly for the synthesis of polyolefins, and are therefore seen as complementary (and possible substitute) to conventional Ziegler-Natta catalysts. In its broadest definition, Ziegler-Natta catalysts are composed of a transition metal salt of metals of group IV to VIII (known as the catalyst) and a metal alkyl of a base metal of groups I to III (known as the cocatalyst or activator). For industrial use, most Ziegler-Natta catalysts are based on titanium salts and aluminum alkyls. Several industrial processes using a variety of reactor types (monomer bulk-slurry, diluent-slurry, gas-phase fluidized-bed, gas-phase stirred-bed, loop reactor, solution) exist today for the production of polyolefins using these catalysts.

The most important innovations introduced in the manufacture of polyolefins with Ziegler-Natta catalysts were the synthesis of linear high-density polyethylene (HDPE), the copolymerization of ethylene and α-olefins to produce linear low-density polyethylene (LLDPE), and the production of highly isotactic and syndiotactic polypropylene. HDPE has few or no short chain branches and no long chain branches, and because of its greater rigidity, it is used in structural applications. Copolymerization of ethylene with α-olefins disrupts the order of the linear polyethylene chain by introducing comonomer units that form short chain branches. As a consequence, the density, crystallinity, and rigidity of the polymer is decreased. By varying the amount and type of α-olefin, the type of catalyst, and the polymerization conditions, one can produce several grades of copolymers to meet specific market demands. LLDPE shares the market with high-pressure low-density polyethylene (HP-LDPE) produced by free-radical processes. Both HP-LDPE and LLDPE are used predominantly for manufacture of films.

Several types of Ziegler-Natta catalysts are stereospecific, i.e., the insertion of asymmetric monomers into the growing polymer chain in a given orientation is favoured over all other possible orientations. This characteristic of Ziegler-Natta catalysts permitted for the first time the production of highly isotactic and syndiotactic polypropylene. Isotactic polypropylene is used in several injection molding and extrusion processes due to its excellent rigidity, toughness, and temperature resistance. Only atactic polypropylene of low molecular weight, which has little commercial value, is obtained in free-radical polymerization.

Most industrial processes today utilize heterogeneous Ziegler-Natta catalysts. Conventional soluble Ziegler-Natta catalysts have not found widespread industrial applications, mainly because of insufficient catalytic stability and stereochemical control. Important exceptions are some vanadium-based systems for the production of ethylene-propylene copolymers and ethylene-propylene-diene terpolymers [7,8] and syndiotactic polypropylene [9].

Polyolefins made with heterogeneous Ziegler-Natta catalysts have a polydispersity index for molecular weight distribution significantly greater than the theoretical value of two and non-uniform stereoregularity and copolymer composition (the chemical composition of LLDPE, as measured by temperature rising elution fractionation is generally bimodal [10]).

The non-uniformity of polymer made with Ziegler-Natta catalysts has been linked to the presence of multiple site types on their surface and also to intraparticle mass and heat transfer resistances during polymerization, but it is generally accepted that the presence of multiple site types is the dominating mechanism [11] In this way, the whole polymer is considered to be a microscopic blend of chains with different average properties made on each site type. In fact, it has been claimed that the nature of these active sites can be inferred from the analysis of the molecular weight and chemical composition distributions of the produced polymer [12-14].

One of the main reasons why metallocene catalysts are considered by some to be revolutionizing the polyolefins production technology is the fact that they can make polymer with narrow distributions of molecular weight and chemical composition, as illustrated schematically in Figures 2 and 3. When compared to heterogeneous Ziegler-Natta catalysts, metallocene catalysts have a much better control over polymer microstructure, and therefore can produce polymer with well-defined mechanical and rheological properties. Additionally, it is possible to combine different types of metallocene catalysts to design the molecular weight and composition distributions of polymers [15], and from the knowledge of structure-property relationships, produce polymer with properties to match specific market demands.

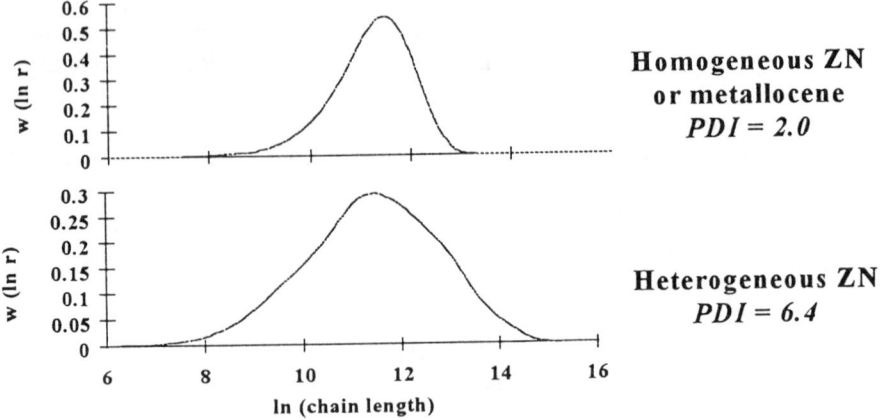

Figure 2 . Chain length distribution of polyolefins made with
metallocene *vs.* heterogeneous Ziegler-Natta catalyst

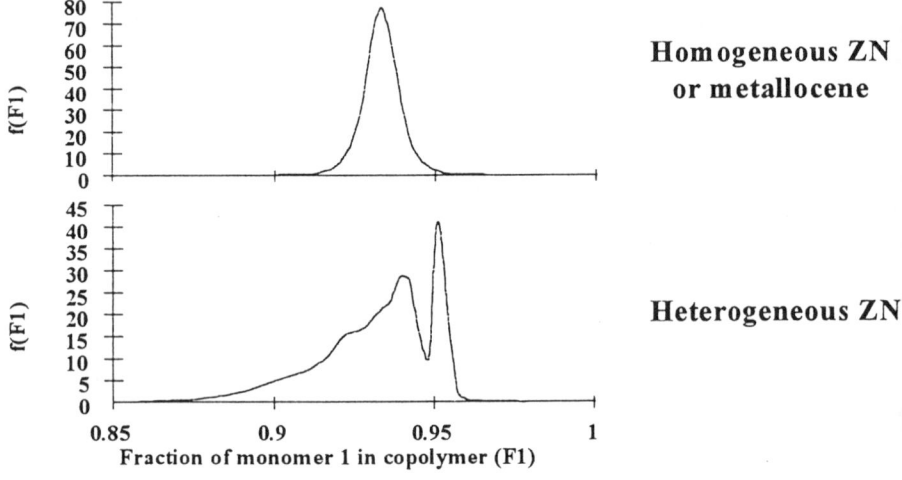

Figure 3. Copolymer composition distribution of polyolefins made with
metallocene *vs.* heterogeneous Ziegler-Natta catalyst

3. Types of Metallocene Catalysts

One of the first researchers to use metallocene catalysts for polymerization were Breslow
and Newburg [16]. They used soluble bis(cyclopentadienyl)titanium derivatives and
alkylaluminums for ethylene polymerization. Several other researchers followed this
original work, using the same catalytic system or modifications of that system,
including G. Natta [16]. However, these catalytic systems had low activity and stability

for the polymerization of ethylene and produced only low molecular weight polyethylene. Additionally, they were not active for propylene polymerization [17].

It was noticed later that the activity of metallocene/alkylaluminum catalysts could be significantly increased by the controlled addition of water to the polymerization reactor [18]. This enhanced activity was attributed to the reaction between water and alkylaluminum to form alkylaluminoxane. This single discovery led to the development of an entirely new class of soluble catalytic systems that are today the most promising branch of Ziegler-Natta catalysis.

3.1. ALUMINOXANES

The type of aluminoxane has a marked influence on the efficiency of a metallocene/aluminoxane catalytic system. Methylaluminoxane (MAO) is generally more effective as a cocatalyst than other aluminoxanes such as ethylaluminoxane (EAO) and isobutylaluminoxane (IBAO) [19]. More remarkably, the catalytic activity of the metallocene complex is directly proportional to the degree of oligomerization of the aluminoxane [18]. Additionally, for most homogeneous metallocene catalysts, a large excess of aluminoxane is required for the polymerization to reach its optimum value. Aluminum/transition metal ratios varying from 1,000 to 50,000 are commonly reported in the literature.

Despite its marked influence in catalytic performance, the exact role of the aluminoxane component is not known precisely. Experimental evidence seem to indicate that besides acting as an alkylation agent and impurity scavengers, aluminoxanes are involved in the formation of active sites and in the prevention of their deactivation by bimolecular processes. More recently, due to the discovery of aluminoxane-free cationic metallocene complexes it has been proposed that the aluminoxane may be involved in the production of the cationic active site and in the stabilization of the anion [20,21].

The exact structure of aluminoxanes is still a matter of controversy. They supposedly exist as a mixture of different cyclic or linear oligomers with degree of oligomerization commonly varying from six to twenty. Some recent experimental evident suggest that MAO can also have a three-dimensional, open cage structure. Reddy and Sivaram [4] recently published an extensive review of techniques for the synthesis of aluminoxanes.

3.2. NON-STEREOSPECIFIC METALLOCENES

The most commonly used catalysts for polyethylene production are achiral cyclopentadienyl derivatives of zirconium, titanium, and hafnium. Titanium and hafnium catalysts show a smaller activity and are less stable at temperatures above 50 °C than zirconocenes[19]. These catalysts are capable of producing polyethylene with activities as high as 40,000 kg of polyethylene (g $Zr.h)^{-1}$ for bis(cyclopentadienyl)zirconium dichloride/MAO (Cp_2ZrCl_2/MAO) at a polymerization temperature of 95 °C and ethylene pressure of $8x10^5$ Pa [22]. Catalytic activity is a strong function of the aluminum/transition metal ratio [23]. The catalytic activity of Cp_2ZrCl_2/MAO for ethylene polymerization increases steadily from 0.25 kg polyethylene (g Zr.h.Pa)$^{-1}$ to 4.8 kg polyethylene (g Zr.h.Pa)$^{-1}$ by varying the aluminum/zirconium ratio from 1,070 to 46,000.

The molecular weight of polymer made with Cp_2ZrCl_2/MAO is very sensitive to temperature, ranging from 1,000,000 at 0 °C to 1,000 at 100 °C [18]. Most soluble metallocene catalysts show the same relation between molecular weight and polymerization temperature, presumably due to an intensification of β-hydride elimination with increasing temperature. At polymerization temperatures below -20 °C, transfer reactions are so reduced that the molecular weight becomes only a function of polymerization time, thus behaving as in a living polymerization system [25].

Hydrogen is an efficient chain transfer agent when used with metallocene catalysts. However, contrary to what is observed with conventional Ziegler-Natta catalysts, only traces of hydrogen are necessary to significantly reduce the molecular weight of the polymer. The presence of hydrogen also lowers the activity of Cp_2ZrCl_2/MAO system [26], but this effect is reversible; removal of hydrogen results in a increase of the polymerization rate to its original value. Hydrogen also have a reversible effect (increase or decrease, depending on the catalytic system used) on the rate of polymerization with conventional Ziegler-Natta catalysts [27].

Achiral metallocene catalysts have also been used to produce olefin copolymers. The most remarkable property of these catalysts when used for copolymerization is their ability to produce copolymers with narrower chemical composition distribution than the ones produced with heterogeneous Ziegler-Natta catalysts. This permits an improved control of copolymer composition and it is also essential for the production of elastomers free of crystallinity. Additionally, metallocene catalysts can produce copolymers with almost random incorporation of comonomers, which results in a maximum decrease in polymer crystallinity for a given amount of comonomer incorporation, a desirable feature for the synthesis of elastomers.

Non-stereospecific metallocenes are also active for the polymerization of propylene, but only atactic polypropylene is formed.

3.3. STEREOSPECIFIC METALLOCENES

By the appropriate selection of metallocene catalysts, it is possible to produce polypropylene with different chain microstructures. Polypropylene with atactic, isotactic, isotactic-stereoblock, atactic-stereoblock and hemiisotactic configurations have been produced with metallocene catalysts (Figure 4). It is also possible to synthesize polypropylene chains that have optical activity, by using only one of the enantiomeric forms of the catalyst [25].

Designation	Fisher Projection	Examples
atactic	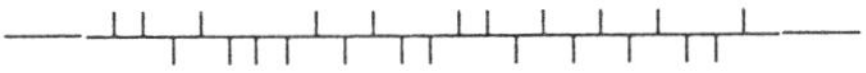	Cp_2ZrCl_2
isotactic	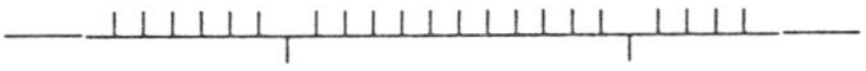	$Et(Ind)_2ZrCl_2$
syndiotactic	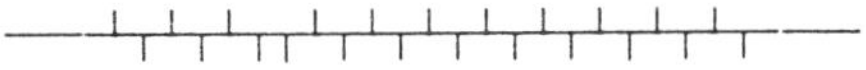	$iPr(Flu)(Cp)HfCl_2$
isotactic-stereoblock	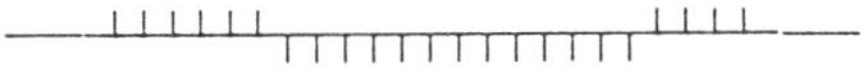	$(NMCp)_2ZrCl_2$
isotactic-atactic-stereoblock		$Et(Me_4Cp)(Ind)TiCl_2$
hemiisotactic	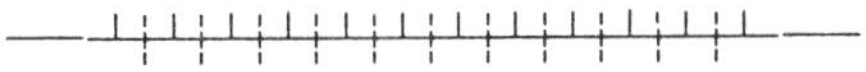	$iPr(Cp)(Ind)ZrCl_2$

Figure 4. Types of metallocene-made polypropylenes
(Cp - cyclopentadienyl, Et - ethylidene, Ind - indenyl, iPr - isopropyl, Flu - fluorenyl, NM -neomenthyl, Me - methyl)

According to Fierro *et al.* [28], C_2 symmetric precursors are necessary to obtain a catalyst for isospecific polymerization, and C_s symmetric precursors to produce a catalyst for syndiospecific polymerization. Asymmetric precursors can be used to synthesize metallocene catalysts that produce hemiisotactic and isotactic-stereoblock polypropylene.

Ewen [29] was the first to report the synthesis of isotactic polypropylene with bis(cyclopentadienyl)titanium diphenyl (Cp_2TiPh_2) and MAO, and ethylenebis(indenyl)titanium dichloride ($Et(Ind)_2TiCl_2$) and MAO. $Et(Ind)_2TiCl_2$ was produced as a mixture of 56% racemic and 44% meso forms. Of the total polypropylene produced, 63% was isotactic, and the mechanism of monomer insertion was site-controlled. The meso form of the catalyst produced the 37% atactic polymer fraction.

The bridge extending between the two indenyl rings imparts stereorigidity to the metallocene complex, preventing the rotation of the rings about their coordination axes. The spatial arrangement of the chiral racemic isomeric form favours the coordination of propylene molecules in such a way as to produce mainly isotactic chains. For the meso form, both monomer orientations are equally favoured and therefore only atactic chains are formed.

The first stereospecific metallocene catalysts could only produce polymer with low molecular weight, and although they could polymerize propylene with high degree of isotacticity (as measured by ^{13}C NMR) several regio-irregularities, such as 2-1 and 1-3 insertions, were detected in the chains. Consequently, these polypropylene resins had a melting temperature (T_m) significantly smaller than the ones of polypropylene resins made with heterogeneous Ziegler-Natta catalysts [30]. Catalysts for the production of polypropylene has evolved considerably and today it is possible to synthesize polypropylene with high molecular weight averages and high T_m. Figure 5 illustrates the evolution of metallocene catalysts for the production of polypropylene.

3.5. SUPPORTED METALLOCENE CATALYSTS

Since most of the conventional Ziegler-Natta polyolefin industrial plants are designed to use heterogeneous catalysts (with exception of EPDM plants that use soluble vanadium-based catalysts), the commercial application of soluble metallocene catalysts would require the design of new plants or the adaptation of existing ones to operate with soluble catalysts. One way of overcoming this problem is by supporting the metallocene catalyst on a "inert" carrier, hopefully without significantly losing its catalytic activity, stereochemical control, and ability to make polymer with narrow molecular weight and chemical composition distributions and, when desired, long chain branching.

Metallocenes can be effectively supported on several inorganic oxides, the most commonly used being SiO_2, $MgCl_2$, Al_2O_3, MgF_2, and CaF_2. Polyolefin particles and natural polymers such as cellulose have also been used to support metallocene catalysts [5].

The type of support as well as the technique used for supporting the metallocene and MAO have a crucial influence on catalyst behaviour. Several techniques for supporting metallocenes and MAO have been proposed, such as [34]: (1) Adsorption of MAO onto the support followed by addition of the metallocene, (2) Immobilization of the metallocene on the support, followed by contact with MAO in the polymerization reactor, (3) Immobilization of the metallocene on the support, followed by treatment with MAO.

By the appropriate choice of supporting conditions, stereo- and regioselectivity can be improved and transfer reactions can be minimized with consequent production of polymers with improved regio- and stereoregularity, and higher molecular weights.

Additionally, supported metallocenes usually require smaller aluminum/transition metal ratios than the equivalent soluble systems and some can be activated in the absence of aluminoxanes by common alkylaluminums [35-38]. This reduced dependence on the presence of aluminoxanes and on high aluminum/transition metal ratios has been related to a reduction in catalyst deactivation by bimolecular processes due to the immobility of the active sites on the surface of the support.

Aluminoxanes can be either synthesized separately and then supported on the carrier or they can be produced *in situ* by reacting an alkylaluminum directly with the water adsorbed on the support. Several patents have been issued regarding supporting technology for metallocene catalysts. For a more detailed description on supporting techniques, the reader is referred to Soares and Hamielec [5].

Supported multiple-site type catalysts can also be designed to produce polyolefins with broad molecular weight distribution. Polyolefins with broad molecular weight distributions are desired for certain applications because of their easier processability. In a series of patents, Welborn [15] claims that it is possible to produce LLDPE and HDPE with polydispersity indexes between 2.5 and 100 by combining at least one metallocene, at least one non-metallocene transition metal compound, an aluminoxane and an organometallic compound on a support.

The catalytic activity of supported metallocenes is usually inferior to the one of the equivalent soluble catalyst, probably due to deactivation of catalytic sites or inefficient production of active sites during the supporting process. Broadening of the molecular weight distribution for supported catalysts can also occur under certain supporting conditions. Although it is generally accepted that this might be caused by the formation of sites of different types due to support-metallocene interactions, recent experimental results seems to indicate that mass transfer resistances can play an important role as well [39].

3.6. CATALYSTS FOR LONG CHAIN BRANCHING FORMATION

The most suitable catalyst types for long chain branch formation appear to be those with an "open" metal active center, such as the Dow Chemical constrained geometry catalysts. The active center of these catalysts is based on group IV transition metals that are covalently bonded to a monocyclopentadienyl ring and bridged with a heteroatom, forming a constrained cyclic structure with the titanium center (Figure 6). Strong Lewis acid systems are used to activate the catalyst to a highly effective cationic form. This geometry allows the titanium center to be more "open" to the addition of ethylene and higher α-olefins, but also for the addition of vinyl-terminated polymer molecules [40]. A second and very important requirement for the efficient production of polyolefins containing long chain branches by these catalytic systems is that a high level of dead polymer chains with terminal unsaturation be produced continuously during the polymerization. Hamielec and Soares [6] discussed the reactor engineering requirements for the optimum formation of long chain branches with these catalytic systems.

Lai *et al.* [41] presented some remarkable data on the effect of polydispersity on I_{10}/I_2 (ratio of melt flow indices measured at two different conditions, generally correlated with shear thinning and breadth of molecular weight distribution for linear polyolefins) for polyolefins synthesized using classical heterogeneous titanium-based Ziegler-Natta catalysts and produced with a constrained geometry catalysts. Shear thinning, as expected, increases as the molecular weight distribution broadens for polyolefins produced with heterogeneous Ziegler-Natta catalysts. On the other hand, polyolefins synthesized with constrained geometry catalysts have narrow molecular weight distribution, with polydispersity near the theoretical value of two for single-site type catalyst. However, the I_{10}/I_2 ratio can be increased at almost constant polydispersity, by increasing the long chain branching frequency. In fact, these authors have shown how to synthesize polyolefins with narrow molecular weight distribution and sufficient degree of long chain branching that combines the excellent mechanical properties of polyolefins with narrow molecular weight distribution (impact properties, tear resistance, environmental stress cracking resistance, and tensile properties) with the good shear thinning of linear polyolefins with broad molecular weight distribution. Polyolefins with narrow molecular weight distribution and containing no long chain branches generally have poor rheological properties.

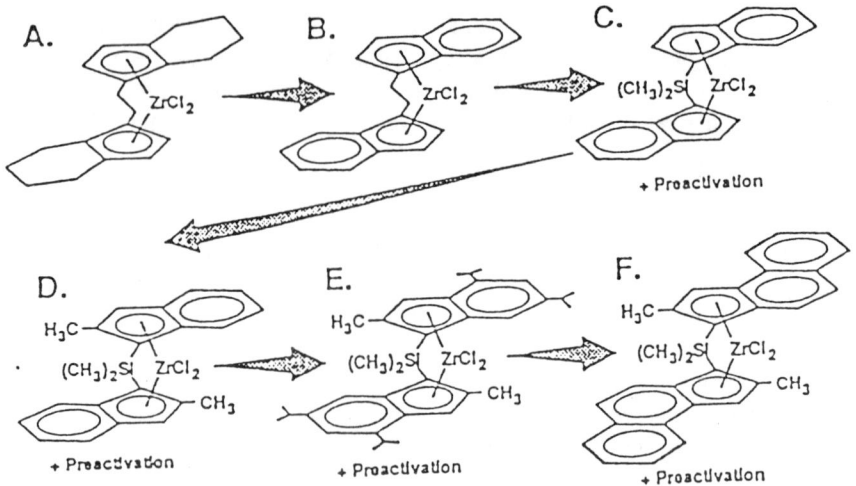

Figure 5. Evolution of metallocene catalysts for polypropylene synthesis

3.4. MAO-FREE SYSTEMS

Cationic metallocenes are catalysts in which the transition metal atom is positively charged. The metallocene complex is therefore a cation associated with a stable anion. Cationic metallocenes are prepared by combining at least two components: The first is a metallocene and the second is an ion exchange compound comprising a cation and a non-coordinating anion. The cation reacts irreversibly with at least one of the first component's ligands. The anion must be capable of stabilizing the transition metal cation complex and must be labile enough to be displaced by the polymerizing monomer. The relationship of the counterion to the bridged structure control monomer insertion and isomerization [31]. There is now enough experimental evidence to support the hypothesis that all active center types operative with metallocenes are cationic.

The hypothesis that the catalyst center is polar or ionic is further supported by the electronic effects observed in some metallocenes of the type $(X_2C_9H_5)_2ZrCl_2/MAO$, where X can be a chlorine, a hydrogen, or a fluorine atom, or a CH_3 or a OCH_3 group. It was observed that, for ethylene polymerization, electron withdrawing atoms such as fluorine significantly lowered the catalytic activity and molecular weight of the produced polymer, while electron donors such as CH_3 had little influence over the polymerization. For the case of polypropylene production, electron withdrawing groups reduced considerably the stereochemical control of the catalysts. This has been related to changes in the degree of association of the metallocene and the MAO counterion or to the increase in the strength of the metal-carbon bond between metallocene and ligands [32,33].

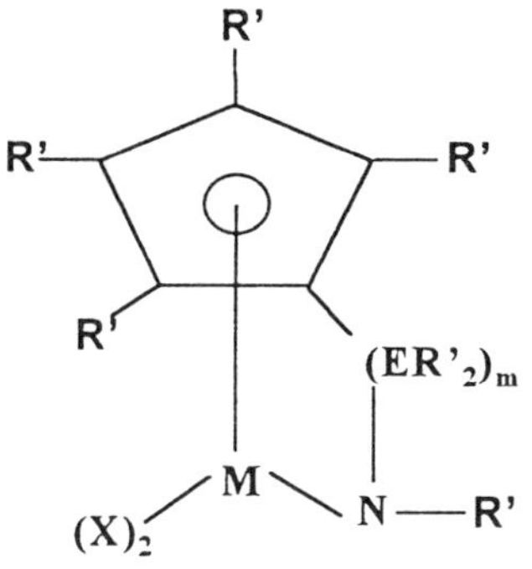

M - Ti, Zr, Hf
R' - hydorgen, silyl, alkyl, aryl
E - silicon or carbon
X - hydride, halo, alkyl, aryl
m - 1,2

Figure 6. Constrained geometry catalysts

A suitable cocatalyst specified by Lai *et al.* [41] is tris(pentafluorophenyl)borane. There is no evidence in the literature that methylaluminoxane cocatalysts are suitable for the synthesis of polyolefins containing long chain branches. It can be speculated that the presence of methylaluminoxane will promote transfer to aluminum and therefore produce dead polymer chains with saturated chain-ends, which are unavailable for long chain branch formation.

4. Mechanisms and Chain Growth Kinetics

Despite intense research activity, no definite, unequivocal polymerization mechanism has yet been defined to describe the behaviour of metallocene and Ziegler-Natta catalysts. This is hardly surprising, given the complex nature of the catalytic systems considered: the catalyst may be soluble or insoluble in the reaction medium, a cocatalyst is generally required but some catalysts are able to polymerize olefins alone, the monomers may be liquid or gaseous, electron donors may be present or not, and the polymerization can take place in gas phase, liquid monomer or suspended in a diluent with various residence-time distributions. Good reviews on polymerization mechanisms with Ziegler-Natta catalysts are available in the literature [9,42,43].

It is well established now that the two key steps in Ziegler-Natta and metallocene-catalyzed polymerizations are the complexation between the monomer and the active center, followed by insertion into the growing polymer chain. For these mechanisms, the cocatalyst acts as an alkylating and reducing agent, and polymer growth takes place *via* insertion of monomer into the transition metal-carbon bond.

One of the models with the greatest impact on the further development of monometallic polymerization mechanisms was proposed by Cossee [43]. In this model, the active site is composed of a transition metal atom having an octahedral configuration, with four chlorine ligands from the crystal lattice, an alkyl group introduced by the cocatalyst, and a coordination vacancy [3,5].

In Cossee's model, the polymer chain has to flip back to the position occupied before the monomer insertion step in order to explain isotacticity. Besides, several important phenomena, such as monomer reaction orders higher than one and copolymerization rates higher than homopolymerization rates of both comonomers can not be explained by Cossee's model [44]. Because of these shortcomings, several alternative monometallic models have been proposed based on Cossee's model. There is no agreement about the general validity of these models, but it is generally accepted that Cossee's model provides the best representation to date of the leading mechanisms governing Ziegler-Natta polymerization [45].

Recently a new mechanism has been proposed which can overcome some of the deficiencies of Cossee's mechanism [44]. This model was called trigger mechanism and involves a two-monomer transition state, where the insertion of a complexed monomer is triggered by another monomer unit. The main assumptions of this model are: (1) the monomer site is never free, since a new monomer will enter the site when the monomer that previously occupied this site is inserted in the growing chain, (2) the insertion step will not proceed, or will proceed very slowly, in the absence of another monomer unit, (3) in the transition state, two monomer units interact with each other and with the transition metal atom. The trigger mechanism is able to predict polymerization rate dependency upon monomer concentration from first to second order, and increase in polymerization rate of ethylene upon adding propylene.

Farina *et al.* [46] presented a general mechanism for polymerization with metallocenes catalysts. They pointed out that metallocene catalysts differ from conventional heterogeneous Ziegler-Natta catalysts because they have two active sites bound to the same metal atom, allowing the growing chain to shift from one site to the other. Two mechanisms are proposed for monomer insertion: in the alternating mechanism, the chain shifts positions between monomer insertions; in the retention mechanism, the chain always occupies the same position in the active site. Four statistical insertion models were proposed: (1) alternating mechanism combined with site control, (2) alternating mechanism combined with site and chain-end control, (3) alternating and retention mechanisms combined with site control, (4) alternating and retention mechanisms combined with site and chain-end control. Unfortunately no simulation results were presented.

The most likely long chain branch formation mechanism with metallocene catalyst systems is terminal branching, a mechanism which has been known in the free-radical polymerization literature for many years [47]. In free-radical polymerization, macromonomers (a long chain molecule with a reactive carbon-carbon double-bond at its end) are generated via termination by disproportionation and via chain transfer to monomer. With metallocene catalyst systems, the facile β-hydride elimination reaction appears to be responsible for *in-situ* macromonomer formation. Other reaction types, such as β-methyl elimination and trans may also generate dead polymer chains with terminal unsaturation [48].

It is generally accepted that the most effective macromonomer for addition to the active center with the generation of a long trifunctional branch is the one with terminal vinyl unsaturation, probably due to steric effects.

5. Mathematical Modelling

Mathematical models of polymerization with metallocene catalysts are similar to the ones for Ziegler-Natta-catalyzed polymerization. Molecular weight averages are conventionally estimated using the method of the moments. The number average chain length is expressed as the ratio of the first moment to the zeroth moment of the molecular weight distribution. Similarly, the mass average chain length is expressed as the ratio of the second moment to the first moment of the distribution. Higher averages (z, z+1, ...) are obtained in a analogous way. Population balances can be derived for the living and dead polymer chains and solved for the moments of the distribution. In this way, one does not need to solve the population balances directly, which, for most cases, requires enormous computational effort.

For the case of steady-state operation (or instantaneous properties), analytical solutions can be easily derived for these population balances. This approach should be used whenever possible, because it permits the calculation of the whole distribution of molecular weights at minimum computational effort. Since several rheological and mechanical properties of polymers depend upon the whole distribution of molecular weight, this approach will become increasingly more important as our knowledge of property-structure relationships increases.

Flory's most probable distribution is simply expressed as [49]:

$$w(r) = \tau^2 \, r \, \exp(-\tau \, r) \tag{1}$$

where τ is the ratio of transfer to propagation rates.

This well known expression can be used to calculate the chain length distribution of linear homopolymers produced with single-site type metallocene or Ziegler-Natta catalysts and predicts a theoretical value of two for the polydispersity index.

For the case of multiple-site-type catalysts, Flory's distribution can be applied to predict the chain length distribution of polymer molecules made on each site type (and therefore having different values of τ). The instantaneous chain length distribution of the total polymer produced with the catalyst will be a weighted average of the individual Flory's most probable distributions for each site type:

$$\overline{w}(r) = \sum_i m_i \, w_i(r) \tag{2}$$

where m_i is the weight fraction of polymer made on each site type i.

Figure 7 illustrates the predicted chain length distribution of a polymer made with a three-site-type catalyst as a superposition of individual Flory's most probable chain length distributions. This model can be used to analyze actual molecular weight distributions, as obtained by gel permeation chromatography, and to obtain information about the nature of the active sites present on the catalyst [12,13].

Soares and Hamielec [50] derived a phenomenological model for the chain length distribution of polymers produced with metallocene catalysts that allow long chain branching formation via terminal double-bond mechanism and obtained an analytical solution for the chain length distribution of the populations containing different number of long chain branches per polymer molecule. The frequency distribution of chain length for polymer populations with n long chain branches per chain is given by:

$$f(r, n) = \frac{1}{(2n)!} r^{2n} \tau^{2n+1} \exp(-\tau \, r) \tag{3}$$

where, r represents chain length and τ is given by:

$$\tau = \frac{R_\beta}{R_p} + \frac{R_{CTA}}{R_p} + \frac{R_{LCB}}{R_p} \tag{4}$$

and R_β is the rate of β-hydride elimination, R_p is the rate of monomer propagation, R_{CTA} is the rate of transfer to chain transfer agent, and R_{LCB} is the rate of macromonomer propagation or long chain branch formation. Notice that equation (3) reduces to Flory's most probable distribution for linear chains, i.e., when $n = 0$.

Molecular weight averages of copolymers can be easily calculated with the method of moments by using pseudo-kinetic rate constants, and average copolymer compositions can be obtained from the relative rate of comonomer polymerization [11,47]. However, as for the case of homopolymerization, whenever possible it is advantageous to predict the whole distribution of molecular weight and chemical composition for copolymerization. For the case of linear chains and binary copolymerization, this instantaneous bivariate distribution is given by Stockmayer's distribution [51].

Stockmayer's bivariate distribution is given by the expression:

$$w(r, y)drdy = \tau^2 r \, \exp(-\tau \, r)dr \frac{1}{\sqrt{2\pi \, \beta / r}} \exp\left(\frac{y^2 r}{2\beta}\right)dy \tag{5}$$

where,

$$\beta = \bar{F}_1\left(1 - \bar{F}_1\right)K \tag{6}$$

$$K = \left[1 + 4\bar{F}_1\left(1 - \bar{F}_1\right)\left(r_1 r_2 - 1\right)\right]^{0.5} \tag{7}$$

and y is the deviation from the average mol fraction of monomer 1 in the copolymer, $\bar{F}_1$ is the average mol fraction of monomer 1 in the copolymer, and r_1 and r_2 are the reactivity ratios.

For the case of multiple-site-type catalysts, one can assume that each active site instantaneously produces copolymer chains that follow Stockmayer's bivariate distribution. In this way, the bivariate distribution of chain length and chemical composition for the product copolymer can be obtained as a weighted average of individual Stockmayer's distributions over all site types:

$$\overline{w}(r, y) = \sum_i m_i w_i(r, y) \qquad (8)$$

Figure 8 shows the predicted chemical composition distribution for a LLDPE made with a three-site-type catalyst. Stockmayer's bivariate distribution can also be used as a mathematical model for temperature rising elution fractionation detector response for polymer made with multiple-site type catalysts [52].

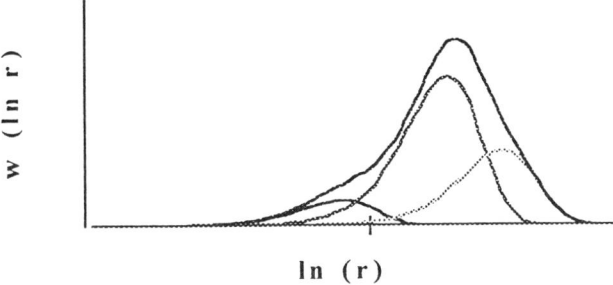

Figure 7. Chain length distribution of whole polymer as a superposition of different Flory's most probable distributions per site type

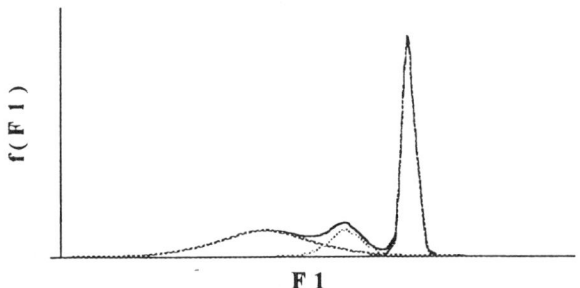

Figure 8. Copolymer composition distribution of whole polymer as a superposition of different Stockmayer's distributions per site type

For copolymerizations involving three or more monomer types, Stockmayer's bivariate distribution is no longer valid. However, Flory's most probable distribution is valid, providing a working analytical expression for the molecular weight distribution for multicomponent polymerization. The dimensionless parameter τ in Flory's equation is defined in the same general way in terms of ratios of rates. To evaluate τ for a multicomponent polymerization, one must evaluate these rates in the appropriate manner using pseudo-kinetic constants in the context of the terminal model for copolymerization or for any other copolymerization model which is applicable (e.g. the penultimate model).

For lack of any analytical expression for chemical composition distribution for a terpolymerization or higher, one can make the reasonable assumption that, for long copolymer chains, all of the chains have the same composition at an average value ($\overline{F}_1$, $\overline{F}_2$, etc.). In this manner, one can construct a multi-dimensional distribution of chain length and mole fractions of the different monomer types, for single-site and multiple-site-type catalysts.

When dealing with supported Ziegler-Natta or metallocene catalysts, it is necessary to take into account the heterogeneous nature of the catalyst, since mass and heat transfer resistances may affect the properties of the formed polymer. Heterogeneous Ziegler-Natta and supported metallocene catalysts consist of porous secondary particles, formed by loosely aggregated primary particles [53]. During polymerization, the growing polymer chains fragment these secondary particles, forming an expanding particle containing primary particles and living and dead polymer chains. This catalyst fragmentation mechanism has been documented for several types of Ziegler-Natta catalysts. One of its consequences is the well-known replication phenomenon: the shape of the particle size distribution of the polymer particles at the end of batch or semi-batch polymerization closely approximates the shape of the particle size distribution of the catalyst at the beginning of polymerization.

Based on this experimental evidence, some researchers advocate that, due to diffusion resistances, catalysts fragments at different radial positions in the polymer particle are exposed to different concentrations of monomer and chain transfer agent, and consequently produce polymer with chain length averages that differ radially inside the polymer particle. For copolymerization, monomers with different effective diffusivities and reactivities may be responsible for radial composition heterogeneity in the polymer particle. In addition, if there are appreciable heat transfer resistances, hot spots can occur inside the polymer particle, altering reaction rates and further broadening molecular weight and chemical composition distributions. Some strong experimental support for this hypothesis has been presented recently [39,54,55]. Several mathematical models accounting for intraparticle mass and heat transfer resistances have been published in the literature, especially for modelling conventional heterogeneous Ziegler-Natta catalysts. The polymeric flow model [56], the multigrain model [57], and modifications of these models have been applied extensively for different polymerization conditions. For a comprehensive review on physical models and on mathematical modelling of olefin polymerization in general, the reader is referred to Soares and Hamielec [11]. Only a few of these models have been used to simulate polymerization with supported metallocene catalysts. It is clear, however, that most of these mathematical models can be readily modified to simulate these new catalytic systems.

Soares and Hamielec [11] applied the polymeric multilayer model for supported metallocene catalysts as well as for conventional heterogeneous Ziegler-Natta catalysts. In the case of supported single-site-type metallocene catalysts, the only factors responsible for broadening of the molecular weight and chemical composition distributions are intraparticle mass and heat transfer resistances. This model can calculate the complete distributions of molecular weight and chemical composition in each model layer and for the whole polymer particle using Stockmayer's bivariate distribution. It is well known that several important mechanical and rheological properties of polyolefins depend on these distributions [58].

For copolymerization, the combined effect of different effective diffusivities and reactivities for the comonomers can generate a radial profile of chemical composition. If mass transfer resistance is significant, the mole fraction of propylene (the slower polymerizing monomer) in a ethylene-propylene copolymer increases from the surface to the center of the particle. This behaviour can be attributed to the higher reactivity of ethylene. Since ethylene is more reactive than propylene, its consumption will be affected more by mass transfer than propylene consumption (larger Thiele modulus). Consequently, the radial profile of ethylene concentration will be steeper than the one for propylene. In this way, the inner layers of the polymer-catalyst particle will produce polymer that is richer in propylene than in ethylene. For smaller diffusivities, the radial profile of copolymer composition becomes more prominent, especially for short polymerization times, in good agreement with the experimental results published by Hoel *et al.* [39] for the copolymerization of ethylene and propylene using a supported metallocene catalyst. It is important to notice that this effect is less marked for longer polymerization times due to particle expansion and consequent decrease in the concentration of catalyst sites per particle.

For single-site-type catalysts, the main conclusions that can be drawn from these models are: (1) mass transfer resistances can reduce the polymerization rate and decrease molecular weight averages, (2) particularly important for supported catalysts, increasing the concentration of highly active catalytic sites can increase the effect of mass transfer resistances and reduce catalyst performance and product quality, (3) mass transfer resistances may also be a source of composition heterogeneity for highly active and large catalyst particles, if the comonomers have reactivities that differ significantly, (4) temperature gradients in the polymeric particle are not expected to be a significant factor for reactions carried out in slurry reactors. These conclusions obtained with single-site-type models are especially important for the technology of supported metallocene catalysts, where single-site, highly active species, may be subjected to considerable resistances for mass and heat transfer.

The replication phenomenon in heterogeneous Ziegler-Natta and metallocene catalysts permits one to readily predict the particle size distribution of the polymer particles from the knowledge of the catalyst's particle size distribution. The particle size distribution of the polymer is an important variable in designing and operating polymer recovery, treatment, and processing units. Good replication is supposed to occur when there is an adequate balance between the mechanical strength of the particle and catalyst activity. If the reactivity is too high and the particle very weak, the fast growing polymer chains can rupture the catalyst particle into smaller, isolated fragments, forming undesirable fine polymer powder. On the other hand, if the particle is too strong, there will be little or no fragmentation and the polymer chains will block the catalyst pores, making the internal active sites inaccessible to monomer. Replication factors of forty to fifty (ratio of average polymer particle diameter to average catalyst particle diameter) can be obtained with third and fourth generation Ziegler-Natta catalysts.

A necessary condition to obtain a perfect replication of the catalyst particle size distribution is that the residence time of all catalyst particles in the reactor be the same. For the case of continuous operation, this requirement is only possible in plug flow reactors. In a continuous stirred tank reactor (commonly used in slurry and mechanically agitated gas-phase processes) and fluidized bed reactors (such as UNIPOL gas-phase reactor) the catalyst particles experience a distribution of residence times in the reactor which does influence the size distribution of the formed polymer particles [59].

Soares and Hamielec [59] developed a model to account for the influence of the reactor residence-time distribution on the particle size distribution of the polymer product. This model considered an average polymerization rate for all catalyst particles, and assumed that the active sites were homogeneously distributed on the catalyst particle. Two types of sites were considered: Stable sites which did not deactivate, and unstable sites which were allowed to deactivate following an exponential deactivation rate law. A numerical algorithm for predicting the polymer particle size distribution from knowledge of the catalyst particle size distribution and from the residence-time distribution of the polymerization reactors was developed. Polymerization reactors could have any residence-time distribution and could be used alone or in series. The effect of mass transfer resistance on the replication factor as calculated with the polymeric multilayer model was also accounted for. Mass transfer resistance was not found to significantly influence the growth of the polymer particle even for very low effective diffusivities. In reality, the polymeric multilayer model predicted near perfect replication for the studied polymerization conditions. However, the reactor residence-time distribution had a significant influence on polymer particle size distribution. The conclusions drawn are specially important for the case of ethylene-propylene impact copolymers. With these resins, it is necessary to produce a copolymer with an optimum ratio of isotactic polypropylene to ethylene-propylene rubber to maximize the impact properties of the product. However, a broad residence-time distribution in the polymerization reactor will produce polymer particles with varying ratios of isotactic polypropylene to ethylene-propylene rubber, consequently decreasing the quality of the product. Narrow reactor residence-time distributions are clearly beneficial for the production of impact copolymers.

In most polymerizations using soluble Ziegler-Natta catalysts in general, and metallocenes in particular, the polymer is not soluble in the reaction medium and precipitates after a critical chain length is achieved. If, after chain termination, the active site returns to solution, then intraparticle mass and heat transfer effects should not influence the polymerization. However, if the active sites are trapped inside the polymer particles, intraparticle mass and heat transfer resistances could become significant.

A mathematical model for particle growth during ethylene polymerization catalyzed with soluble metallocenes was proposed by Hermann and Böhm [60]. Unfortunately, very little detail about their model was given. They found out that the process of polymer particle formation in a slurry reactor consisted of aggregation by Brownian motion followed by diffusion controlled particle growth, leading to the formation of particles with high surface area and low bulk density.

Koivumäki et al. [61] studied the mechanism of polymer particle formation in a heat balance calorimeter for the polymerization of ethylene and 1-hexene with Cp_2ZrCl_2/MAO. For homopolymerization of ethylene, the particles formed were five times larger in size and had lower bulky density than particles formed via copolymerization. This caused a significant increase in slurry viscosity and decreased the overall heat transfer coefficient. For copolymerization, the presence of comonomer apparently favoured the formation of smaller polymer particles, causing no measurable increase in slurry viscosity. Actually, if during a homopolymerization run, comonomer is introduced in the reactor, the viscosity stops increasing after a lag time. This particle size difference can be used to explain polymerization rate enhancement during copolymerization, due to a decrease in mass transfer resistances. The authors, however, acknowledged that their model assumed Newtonian behaviour for the slurry and that this could lead to some data distortion.

6. Adaptation of Metallocene Catalysts to Existing Olefin Polymerization Processes

Metallocene catalysts have the potential of significantly changing the polyolefin market if production costs can be reduced and if new polymer grades can be implemented without significant processing difficulties.

Although metallocene-produced polyolefins can compete with commodity polyolefins synthesized with conventional Ziegler-Natta catalysts, they will not probably be restricted to the polymer commodity market. Because of the better polymer microstructure control obtained with metallocene catalysts, it will be possible to produce specialty polyolefins to compete with non-olefinic polymers, thus opening an entire new market for polyolefin applications.

The following companies are already commercializing metallocene-made polyolefins: Exxon (39 grades), Dow Chemical (39 grades), Hoechst (3 grades), BASF (3 grades), and Mitsui (3 grades). Most are polyethylene co- and terpolymers with butene, hexene and octene, but 7 grades of isotactic polypropylene are also available (4 from Exxon and 3 from Hoechst) [1,62].

Metallocene catalysts need to be supported to be used in gas-phase reactors, such as Union Carbide's fluidized-bed UNIPOL process, or BASF-NOVOLEN stirred-bed process. For these processes it is necessary to have a free-flowing catalyst powder which will form polymer particles with adequate size distribution, avoiding the formation of fine powder or particle agglomerates. In other words, good replication of the catalyst particles is essential for the efficient performance of these reactors.

Langhauser *et al.* [63] reviewed the industrial production of polypropylene (homopolymer, random copolymer, and impact copolymer) using $Me_2Si(2-MeInd)_2ZrCl_2$/MAO-supported catalyst and the NOVOLEN-BASF process. This catalyst can produce polypropylene with high molecular weight even at elevated temperatures. The polymer particles replicate well the size distribution of the catalyst particles. This catalyst can produce polypropylene with new properties, such as low extractables for food wrapping and medical applications, which is a consequence of the homogeneous microstructure of polymers produced with a single-site-type catalyst.

Impact copolymers can also be produced with this catalyst. Impact copolymers made with heterogeneous Ziegler-Natta catalysts show some crystalline domains in the amourphous elastomeric phase, while the elastomeric phase of the metallocene-produced copolymer is entirely amorphous. This new microstructure will likely enable the production of copolymers with enhanced impact properties.

According to Langhauser *et al.* [63] this catalytic system can be adapted to their existing gas-phase polymerization process without any significant technical change.

Mobil Chemical Co. is also producing LLDPE for film resins using metallocene catalysts in a gas-phase fluidized bed polymerization reactor. Minimal capital investment was necessary to adjust the processes to the new metallocene catalyst and the new resins have superior properties over corresponding Ziegler-Natta resins.

Slurry processes, either using liquid monomer or a diluent, are commonly used for laboratory-scale olefin polymerizations. Supported metallocenes and heterogeneous Ziegler-Natta catalyst will have a similar behaviour regarding macroscopic phenomena in the reactor, provided that there is no desorption of the active sites during the polymerization. It is reasonable to assume that most existing slurry polymerization reactor process can be easily adapted to use supported metallocene catalysts.

For homogeneous catalysts, the process of polymer particle formation generally leads to porous, low-density polymer particles, which can cause significant increase in slurry viscosity and reactor fouling, leading to inadequate reactor temperature control. Additionally, polymer particles with poor powder properties are undesirable for post-reactor polymer processing. These problems must be addressed before using soluble metallocene catalysts for industrial production of polyolefins in slurry reactors.

Solution processes are especially adequate for the production of polyolefins containing long chain branches. Presently, two industrial solution processes are being used to produce polyethylene: Dow Chemical's INSITE process, and Exxon's EXACT process. These processes can produce polyolefins with novel properties due to the controlled incorporation of long chain branches in a homo- or copolymer backbone.

7. References

1. Schut, J.H. (1996) Here's the latest score on single-site catalysts, *Plastic World*, **April**, 41-46.
2. Gupta, V.K., Satish, S., and Bhardwaj, I.S. (1994) Metallocene complexes of group 4 elements in the polymerization of monoolefins, *J.M.S.- Rev. Macromol. Chem. Phys.*, **C34** (3), 439-514.
3. Huang, J. and Rempel, G.L. (1995) Ziegler-Natta catalysts for olefin polymerization: mechanistic insights from metallocene systems, *Prog. Polym. Sci.*, **20**, 459-525.
4. Reddy, S. S. and Sivaram, S. (1995) Homogeneous metallocene-methylaluminoxane catalyst systems for ethylene polymerization, *Prog. Polym. Sci.*, **20**, 309-367.
5. Soares, J.B.P. and Hamielec, A.E. (1995) Metallocene/Aluminoxane Catalysts for Olefin Polymerization. A Review, *Polym. React. Eng.*, **3** (2), 131-200.
6. Hamielec, A.E. and Soares, J.B.P. (1996) Polymerization reaction engineering - Metallocene catalysts, *Prog. Polym. Sci.*, accepted.
7. Cooper, W. (1976) Kinetics of polymerization initiated by Ziegler-Natta and related catalyst, in C.H. Bamford and C.F.H. Tipper (eds.), *Chemical Kinetics*, vol. 15, Elsevier, New York, pp.133-257.
8. Tait, P.J.T. (1989) Monoalkene polymerization: Ziegler-Natta and transition metal catalysts, in Sir G. Allen (ed.), *Comprehensive Polymer Science*, vol. 4, Pergamon Press, Oxford, pp. 1-25.
9. Corradini, P., Busico, V., and Guerra, G. (1989) Monoalkene polymerization: stereospecificity, in Sir G. Allen (ed.), *Comprehensive Polymer Science*, vol. 4, Pergamon Press, Oxford, pp.29-50.
10. Soares, J.B.P. and Hamielec, A.E. (1995) Fractionation of linear polyolefins by TREF, *Polymer*, **36**, 1639-1654.
11. Soares, J.B.P. and Hamielec, A.E. (1995) General dynamic mathematical modelling of heterogeneous and homogeneous Ziegler-Natta copolymerization with multiple site types and mass and heat transfer resistances, *Polym. React. Eng.*, **3**, 261-364.
12. Vickroy, V.V., Schneider, H., and Abbott, R.F. (1993) The separation of SEC curves of HDPE into Flory distributions, *J. Appl. Polym. Sci.*, **50**, 551-554.
13. Soares, J.B.P. and Hamielec, A.E. (1995) Deconvolution of chain length distributions of linear polymers made by multiple site type catalysts, *Polymer*, **36** , 2257-2263 (1995).
14. Soares, J.B.P., Abbott, R.F., and Willis, J.N. (1996) A new methodology for studying multiple-site-type catalysts for the copolymerization of olefins, *Macromol. Rapid Commun.*, submitted.
15. Welborn, H.C., Jr. (1993) Polymerization process using a new supported polymerization catalyst, *U.S. Pat. 5,183,867*.
16. Reichert, K. H. (1983) Polymerization of α-olefins with soluble Ziegler catalysts, in R.P. Quirk (ed.), *Transition Metal Catalyzed Polymerization*, vol. 4, Harwood, New York, pp. 465-494.
17. Giannetti, E., Nicoletti, G.M., and Mazzocchi, R. (1985) Homogeneous Ziegler-Natta catalysis. II.Ethylene polymerization by IVB transition metal complexes/methyl aluminoxane catalyst, *J. Polym. Sci.: Polym. Chem. Ed.*, **23**, 2117-2133.
18. Sinn, H. and Kaminsky W. (1980) Ziegler-Natta Catalysis, *Adv. Organomet. Chem.*, **18**, 99-149.

19. Kaminsky, W. and Steiger, R. (1988) Polymerization of olefins with homogeneous zirconocene/alumoxane catalyst, *Polyhedron*, **7**, 2375-2381.
20. Jordan, R.F., Bajgur, C.S., Willett, R., and Scott, B. (1986) Ethylene polymerization by a cationic dicyclopentadienylzirconium(IV) alkyl complex, *J. Am. Chem. Soc.*, **108**, 7410-7411.
21. Jordan, R.F. (1988) Cationic metal-alkyl olefin polymerization catalysts, *J. Chem. Ed.*, **65**, 285-289.
22. Kaminsky, W. (1991) Polymerization and copolymerization of olefins with metallocene/aluminoxane catalysts, *Cat. Soc. Japan*, **33**, 536-544.
23. Chien, J.C.W. and Razavi, A. (1988) Metallocene-methylaluminoxane catalyst for olefin polymerization.II.Bis-η^5-(neomenthylcyclopentadienyl)zirconium dichloride, *J. Polym. Sci.: Part A: Polym. Chem.*, **26**, 2369-2380.
24. Chien, J.C.W.; Wang, B.P. (1988) Metallocene-methylaluminoxane catalyst for olefin polymerization.I.Trimethylaluminum as coactivator, *J. Polym. Sci.: Part A: Polym. Chem.*, **26**, 3089-3102.
25. Kaminsky, W. (1986) Preparation of special polyolefins from soluble zirconium compounds with aluminoxane as cocatalyst, in T.Keii and K. Soga (eds.), *Catalytic Polymerization of Olefins*, Kodansha-Elsevier, Tokio, pp. 293-304 (1986).
26. Kaminsky, W. and Luker, H. (1984) Influence of hydrogen on the polymerization of ethylene with homogeneous Ziegler system bis(cyclopentadienyl) zirconium dichloride/aluminoxane, *Makromol. Chem. Rapid Commun.*, **5**, 225-228.
27. Soares, J.B.P. and Hamielec, A.E. (1996) Kinetics of propylene polymerization with a non-supported heterogeneous Ziegler-Natta catalyst - Effect of hydrogen on rate of polymerization, stereoregularity, and molecular weight distribution, *Polymer*, accepted.
28. Fierro, R., Chien, J.C.W., and Rausch, M.D. (1994) Asymmetric zirconocene precursors for catalysis of propylene polymerization, *J. Polym. Sci.: Part A: Polym. Chem.*, **32**, 2817-2824.
29. Ewen, J.A. (1984) Mechanisms of stereochemical control in propylene polymerizations with soluble group 4B metallocene/methylaluminoxane catalysts, *J. Am. Chem. Soc.*, **106**, 6355-6364.
30. Cheng, H.N. and Ewen, J.A. (1989) ^{13}C nuclear magnetic resonance characterization of poly(propylene) prepared with homogeneous catalysts, *Makromol. Chem.*, **190**, 1931-1943.
31. Elder, M.J., Razavi, A., and Ewen, J.A. (1992) Process and catalyst for producing syndiotactic polyolefins, *U.S. Pat. 5,155,080.*
32. Piccolrovazzi, P., Pino, P., Consiglio, G., Sironi, A., and Moret, M. (1990) Electronic effects in homogeneous indenylzirconium Ziegler-Natta catalysts, *Organometallics*, **9**, 3098-3105.
33. Lee, I.M., Gauthier, W.J., Ball, J.M., Iyengas, B., and Collins, S. (1992) Electronic effects in Ziegler-Natta polymerization of propylene and ethylene using soluble metallocene catalysts, *Organometallics*, **11**, 2115-2122.
34. Kaminsky, W. and Renner, F. (1993) High melting polypropylenes by silica-supported zirconocene catalysts, *Makromol. Chem. Rapid Commun.*, **14**, 239-243.
35. Chien, J.C.W. and He, D. (1991) Olefin copolymerization with metallocene catalysts. III. Supported metallocene/methylaluminoxane catalyst for olefin copolymerization, *J. Polym. Sci.: Part A: Polym. Chem.*, **29**, 1603-1607.
36. Kaminaka, M. and Soga, K. (1991) Polymerization of propene with the catalyst systems composed of Al_2O_3 or $MgCl_2$-supported $Et[IndH_4]_2ZrCl_2$ and AlR_3 (R = CH_3, C_2H_5), *Makromol. Chem. Rapid Commun.*, **12**, 367-372.
37. Kaminaka, M. and Soga, K. (1992) Polymerization of propene with catalyst systems composed of Al_2O_3 or $MgCl_2$-supported zirconocene and $Al(CH_3)_3$, *Polymer*, **33**, 1105-1107.
38. Soga, K. and Kaminaka, M. (1993) Polymerization of propene with zirconocene-containing supported catalysts activated by common trialkylaluminiums, *Makromol. Chem.*, **194**, 1745-1755.
39. Hoel, E.L., Cozewith, C., and Byrne, G.D. (1994) Effect of diffusion on heterogeneous ethylene propylene copolymerization *AIChE J.*, **40** (10), 1669-1684.
40. Woo, T.K., Fan, L., and Ziegler, T. (1994) A density functional study of chain growing and chain terminating steps in olefin polymerization by metallocenes and constrained geometry catalysts, *Organometallics*, **13**, 2252-2261.
41. Lai, S.Y., Wilson, J.R., Knight, G.W., Stevens, J.C., and Chum, P.W.S. (1993) Elastic substantially linear olefins,*U.S. Patent 5,272,236.*
42. Zakharov, V.A., Bukatov, G.D., and Yermakov, Y.F. (1983) On the mechanism of olefin polymerization by Ziegler-Natta catalysts, *Adv. Polym. Sci.*, **56**, 61-100.
43. Tait, P.J.T. and Watkins, N.D. (1989) Monoalkene polymerization mechanisms, in Sir G. Allen (ed.), *Comprehensive Polymer Science*, vol. 4, Pergamon Press, Oxford, pp.533-573.
44. Ystenes, M. (1991) The trigger mechanism for polymerization of α-olefins with Ziegler-Natta catalysts: A new model based on interaction of two monomers at the transition state and monomer activation of the catalytic centers, *J. Catal.*, **129**, 383-401.
45. Dusseault, J.J.A. and IIsu, C.C. (1993) $MgCl_2$-supported Ziegler-Natta catalysts for olefin polymerization: basic structure, mechanism, and kinetic behaviour, *J.M.S.-Rev. Macromol. Sci.*, **C33**, 103-145.
46. Farina, M., Di Silvestro, G., and Terragni, A. (1995) A stereochemical and statistical analysis of metallocene-promoted polymerization, *Macromol. Chem. Phys.*, **196**, 353-367.
47. Hamielec, A.E., MacGregor, J.F., Penlidis, A. (1987) Multicomponent free-radical polymerization in batch, semi-batch and continuous reactors, *Makromol. Chem., Macromol. Symp.*, **10/11**, 521-570.

176

48. Resconi, L., Piemontesi, F., Franciscono, G., Abis, L., and Fiorani, T. (1992) Olefin polymerization at bis(pentamethylcyclopentadienyl)zirconium and -hafnium centers: chain-transfer mechanisms, *J. Am. Chem. Soc.*, **114**, 1025-1032.
49. Flory, P.J. (1953) *Principles of Polymer Chemistry*, Cornell University Press, Ithaca.
50. Soares, J.B.P. and Hamielec, A.E. (1996) Bivariate chain length and long chain branching distribution for copolymerization of olefins and polyolefin chains containing terminal double-bonds, *Macromol. Theory Simul.*, in print.
51. Stockmayer, W.H. (1945) Distribution of chain lengths and composition in copolymers, *J. Chem. Phys.*, **13**, 199-207.
52. Soares, J.B.P. and Hamielec, A.E. (1995) Analyzing TREF data by Stockmayer's bivariate distribution, *Macromol. Theory Simul.*, **4**, 305-324.
53. Noristi, L., Marchetti, E., Baruzzi, G., and Sgarzi, P. (1994) Investigation on the particle growth mechanism in propylene polymerization with $MgCl_2$-supported Ziegler-Natta catalysts, *J. Polym. Sci.: Part A: Polym. Chem.*, **32**, 3047-3059.
54. Soga, K., Yamagihara, H., and Lee, D.E. (1989) Effect of monomer diffusion in the polymerization of olefins over Ziegler-Natta catalysts, *Makromol. Chem.*, **190**, 995-1006.
55. Jaber, I.A. and Fink, G. (1994) $TiCl_4/MgH_2$-supported Ziegler-type catalyst system, 3)New findings on the concentration of active sites in ethylene/1-hexene copolymerization, *Macromol. Chem. Phys.*, **195**, 2491-2503.
56. Schmeal, W.R.; Street, J.R. (1971) Polymerization in expanding catalyst particles, *AIChE J.*, **17**, 1188-1197.
57. Ray, W.H. (1988) Practical benefits form modelling olefin polymerization reactors, in R.P. Quirk (ed.), *Transition Metal Catalyzed Polymerization*, Harwood, New York. pp.563-590.
58. Graessley, W.W. (1993) *Physical Properties of Polymers*, 2nd edition, American Chemical Society, Washington.
59. Soares, J.B.P. and Hamielec, A.E. (1996) Effect of reactor residence time distribution on the size distribution of polymer particles made with heterogeneous Ziegler-Natta and supported metallocene catalysts. A generic mathematical model, *Macromol. Theory Simul.*, **4**, 1085-1104.
60. Hermann, H.F. and Böhm, L.L. (1991) Particle forming process in slurry polymerization of ethylene with homogeneous catalysts, *Polym. Commun.*, **32**, 58-61.
61. Koivumäki, J.; Lahti, M. Seppälä, J.V (1994) Polymerization of ethylene and 1-hexene with Cp_2ZrCl_2-MAO catalyst in a heat balance reaction calorimeter,. *Angew. Makromol. Chem.*, **221**, 117-125.
62. Garbassi, F., Gila, L., and Proto, A. (1994) Metallocenes: New catalysts for new polyolefins, *Polymer News*, **19**, 367-371.
63. Langhauser, F., Kerth, J., Kersting, M., Kölle, P., Lilge, D., and Müller, P. (1994) Propylene polymerization with metallocene catalysts in industrial processes, *Makromol. Chem.*, **223**, 155-164.

THERMODYNAMIC AND KINETIC ASPECTS FOR PARTICLE MORPHOLOGY CONTROL

DONALD C. SUNDBERG AND YVON G. DURANT
Polymer Research Group
University of New Hampshe
Durham, New Hampshire 03824 USA

1. Introduction

Composite latex particles offer a wide variety of physical properties to the end user and find application in coatings, adhesives, graphic arts, and impact resistant thermoplastics, among other areas. The physical properties are achieved by a balance of polymer composition, molecular weight, and latex particle morphology. There is a wide variety of particle morphologies produced, some of which are in their most stable configuration and some which are not. Because of its importance to the final properties of latex derived polymers, particle morphology is a subject of intense interest and a great deal of effort is being expended to learn how to control the final particle structure. The frequency of articles appearing in the scientific literature over the past 15 years attests to the heightened interest in this area.

While it is not the objective of this paper to provide a complete literature review, it is important to mention some of the reports which have contributed to our understanding of the parameters and events which seem to control the morphology, both from the theoretical, or modelling, side and also from the experimental side. D.I. Lee [1,2] was one of the early contributors to this field and presented an extremely useful template of the types of morphologies that one should expect in two-stage latex particles. In a diagrammatic manner he demonstrated the types of morphologies that would likely be achieved when one varied the stage ratio (second monomer to seed polymer), the degree of monomer swelling (batch to starve feed), chain transfer agent level, polymerization temperature, and molecular weight of the seed polymer. These morphologies included completely and incompletely phase separated structures, and serves as a good qualitative guide. Strictly experimental reports during the same period of time were written by Muroi et al [3] and Okubo and co-workers [4,5]. The latter group reported what were called anomalous structures, such as "raspberry" and "void" particles. In 1985 Stutman et al [6] studied the influence of 12 process variables on latex particle morphology for the poly(butyl acrylate) seed/polystyrene second stage system by an experimental design technique. Their conclusion was that the morphology was controlled by a combination of phase separation in the monomer rich surface layer of the seed particle, and the capture of a secondary crop of polystyrene particles from the aqueous phase by the seed latex particles. For the present authors, an extremely important paper was contributed by Cho and Lee in 1985 [7]. This work demonstrated for the poly(methyl methacrylate) seed/polystyrene second stage system one could achieve a wide variety of morphologies depending upon the use of water soluble or oil soluble initiators, temperature variations,

J. M. Asua (ed.), Polymeric Dispersions: Principles and Applications, 177–188.
© 1997 *Kluwer Academic Publishers. Printed in the Netherlands.*

and choices between monomer swelling or starve feeding. These authors concluded that the anchoring effect of ionic end groups on the polymers and the local viscosity within the polymer particle were the key factors in controlling the ultimate morphology.

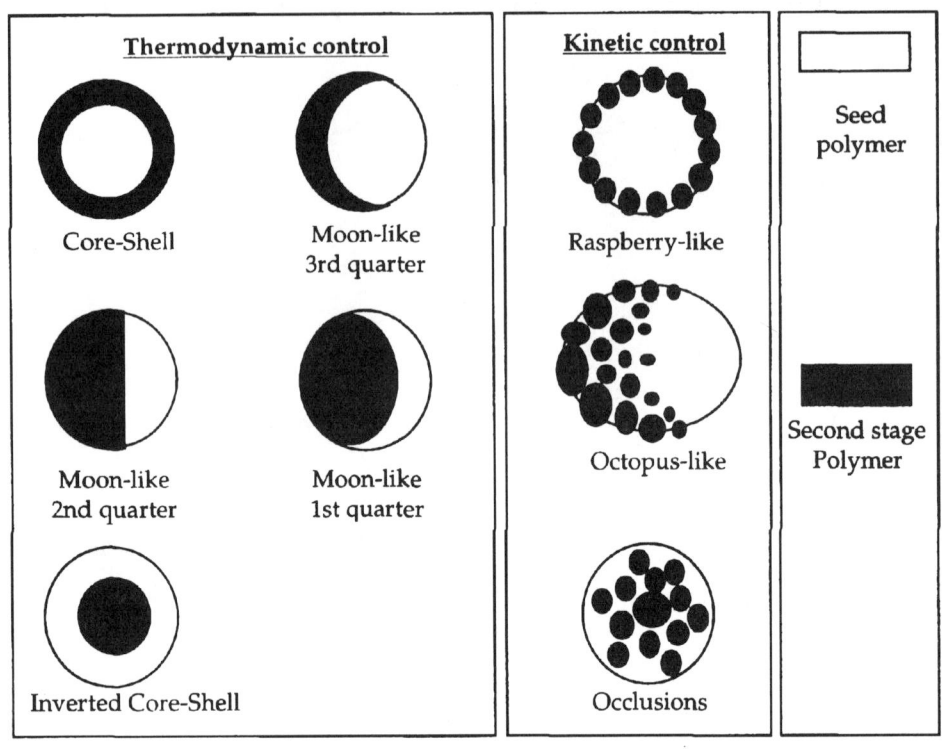

Figure 1. Some examples of morphologies of two-phase latex particles.

From the above studies it is quite obvious that a wide range of particle morphologies are possible, even within a single system, and at that time there were a number of different ideas about the parameters thought to control the particle structure. At this point is may be useful to point out that by and large, the structures of practical interest for property development are those of the classic core-shell (CS) and microdomains, or occluded structures. The former may be an equilibrium structure while the latter is most often a non-equilibrium structure. We think that it is useful to distinguish between the equilibrium and non-equilibrium morphologies and to comment on those that are fully phase separated and those that are not. Figure 1 displays the equilibrium morphologies for two phase latex particles which are fully phase separated. Here there are presented a limited number of configurations ranging from the CS to the inverted core-shell (ICS), with hemispheres and partially engulfed structures. Later we will show the continuous spectrum of structures possible for these two component systems. When we turn to three component systems, the possibilities become significantly more numerous, as demonstrated by Figure 2. Here we have shown only those arrangements which are at the extremes of their structural range (e.g. complete shells around cores rather than partial engulfments). For the three component system there are six distinctively different structural families, including core-shell-shell, cored hemisphere, hemi-core, sandwich or "snowman", tri-sectional and hemi-shell. There are 22 of these structures at the extremes, and obviously an infinite number of possibilities in between the extremes. Although we have not shown any equilibrium morphologies which are not fully phase separated, some microdomain, or occluded, structures appear to be possible when the seed polymer is crosslinked. We will comment on this again later in the paper.

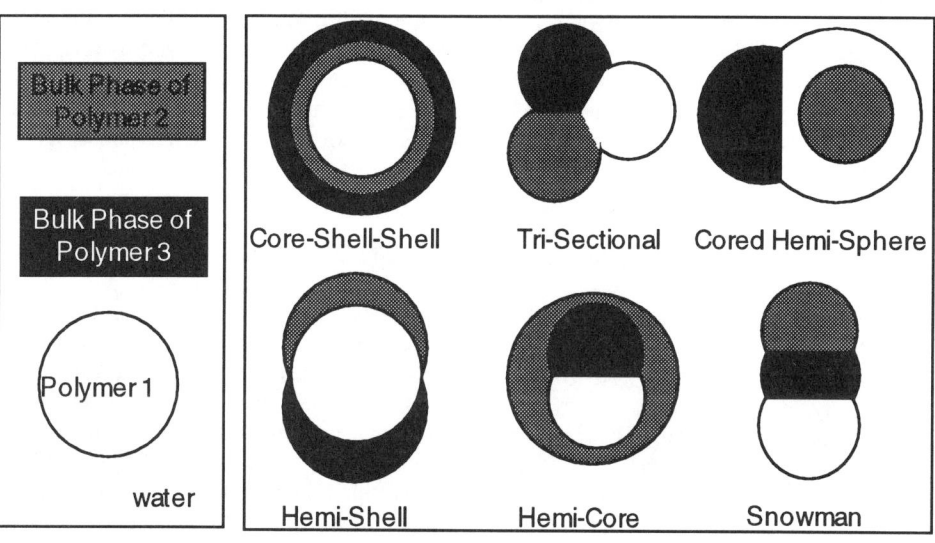

Figure 2. Some examples of morphlogies of three-phase latex particles.

When considering non-equilibrium structures, one can imagine any number of possibilities, with most of them representing incomplete phase separation, such as occluded structures. However, in addition, one often finds that the non-equilibrium structures are mixtures of arrangements such as partial engulfments and occlusions. It is possible to have fully phase separated structures such as CS which are actually non-equilibrium arrangements. These are thought to occur when the seed polymerization is operated in the starve feed mode and the seed polymer is significantly more hydrophilic than the second stage polymer. This would require that the monomer feed rate was quite slow so as to disallow the second stage monomer to penetrate very far into the seed latex particle. In this case there is the possibility of structural rearrangement after the polymerization process has been completed. This process is called latex aging and has been documented a number of years ago by Min and co-workers [8] for a poly(butyl acrylate) seed and a second stage of polystyrene. Over a period of 6-12 months it was shown that the CS morphology achieved during the starve fed polymerization process was transformed to a clearly defined hemispherical arrangement with the passage of time. Thus it would seem that any number of particle morphologies have been made by those carrying out experiments or manufacturing operations. Certainly the goal is to learn to control the structure so as to gain maximum advantage in developing physical properties for the end use application.

With all of the varieties of particle morphology cited above and the number of options in latex formulation and processing conditions, it is clear that it would be quite valuable to have a fundamental understanding of the controlling parameters for morphology development and eventually to have predictive capability. In this regard there has been a significant amount of progress for equilibrium morphology, but extremely little activity has been reported for non-equilibrium morphology. The work that has been reported has brought out the great need for information about interfacial energy at the aqueous and organic interfaces of the particle, the diffusion of monomer and polymer within the particle, and polymerization reaction in viscous media within phase separated particles. These points will be discussed to some degree in the remainder of this paper.

2. Thermodynamic Equilibrium Aspects of Morphology Development

In the context of this discussion, the use of the word "equilibrium" is meant to imply that the incompatible polymers in a two component latex particle are fully phase separated and that the particle has achieved its lowest value of free energy. Thermodynamics allows us to write the Gibbs free energy change for structural development of the particle during polymerization as a combination of terms describing the enthalpic and entropic changes in addition to the surface free energy changes. When such calculations are done for differently shaped particles at the same stage of conversion of monomer to polymer, it is clear that the enthalpic change is the same in each case. Thus in comparing the free energy between particles at the same stage of conversion, the enthalpy term may be neglected. Due to the fact that the particles are "macroscopic" compared to the size of small molecules, we may also neglect the differences in entropic free energy between the various particle shapes. This allows us to write the free energy change as

$$\Delta G = \sum_i \gamma_i S_i - \gamma_{pl/w} S^0_{pl/w} \tag{1}$$

where γ_i is the interfacial tension at the ith interface and S_i is the area of that interface. The normal procedure is to write equation (1) for any number of possible morphologies and to choose the one which displays the lowest final free energy. Before describing this approach in more detail it is important to note that more than 20 years ago Torza and Mason [9] approached a related problem in a somewhat different way. They considered the equilibrium shapes of binary particles made of incompatible oils dispersed in water in which neither oil was soluble. Their analysis of the equilibrium shape was done with spreading coefficients and showed that if one knew the various interfacial tensions, the particle morphologies (CS, ICS and hemisphere) could readily be predicted. In a series of elegant experiments they showed complete agreement between calculation and experimental results. Later in their paper they described the extension of the spreading coefficient approach to composite latex particles. That paper represents the earliest predictive approach to particle morphology that we know of. At a much later time, Hobbs et al [10] used the same approach the describe the morphology of three component bulk polymer blends produced by intensive mixing in a melt extruder. Again the results were in agreement with calculations, but not quite as obvious as with the simple liquids that Torza and Mason worked with. The spreading coefficient approach did not continue to be the choice of other investigators due to the inequalities involved in those coefficients and the very much more general approach offered by applying the well used concepts of Gibbs free energy.

In a series of reports by Berg et al [11,12], Sundberg at al [13], and Winzor at al [14,15] the basic groundwork was laid out for the application of eq. (1) to morphology development in both artificial (phase separation without reaction) and synthetic latices. It turns out that there is really no difference in approach between treating artificial and synthetic latices, when one properly accounts for the effects of the solvent in the former and the monomer in the latter. Actually when one applies eq. (1) to the very end of the process when there is no monomer or solvent present, the only interfacial tensions of importance are those of polymers against water (aqueous phase with surfactant, buffers, etc.) and polymers against polymers. The application of eq. (1) to this situation [13] results in the following equations for the free energy change for the CS and ICS morphologies,

$$\Delta\gamma_{cs} = \frac{\Delta G}{S_{p1/w}^0} = \gamma_{p1/p2} + \gamma_{p2/w}(1 - \phi_p)^{-2/3} - \gamma_{p1/w} \qquad (2)$$

$$\Delta\gamma_{ics} = \frac{\Delta G}{S_{p1/w}^0} = \gamma_{p1/w}\left[\left(1 - \phi_p\right)^{-2/3} - 1\right] + \gamma_{p1/p2}\left(\frac{\phi_p}{1 - \phi_p}\right)^{-2/3} \qquad (3)$$

where $\gamma_{P1/w}$ is the interfacial tension between the seed polymer and the aqueous phase, $\gamma_{P2/w}$ that between the second stage polymer and the aqueous phase, and $\gamma_{P1/P2}$ that between the two polymers. The stage ratio is reflected in the parameter ϕ_p which is the volume fraction of second stage polymer in the particle. Other expressions can be generated for any other fully phase separated particle structures with the only complications arising from the geometrical relationships necessary to describe them. A greatly noteworthy point about the family of equations represented by eqs. (2) and (3) is that they are independent of particle size. Thus they have application to the typically larger sized artificial latices and also to the smaller sized synthetic latices. This makes equilibrium analysis quite straightforward. The second point to make about these equations is that the difficulty lies not in the concept of how to predict morphology, but in how to determine the correct values for the various interfacial tensions. The supporting experiments described in reference [13] display the dramatic effect that a change of surfactants can have on the polymer /water interfacial tensions and the resultant changes in the particle morphology, in this instance changing from CS to hemisphere. Thus we expect to see influences on the morphology by changes in polymer polarity, surfactant type and concentration, initiator end groups, and other parameters which can influence the interfacial energy at the external surface of the particle.

182

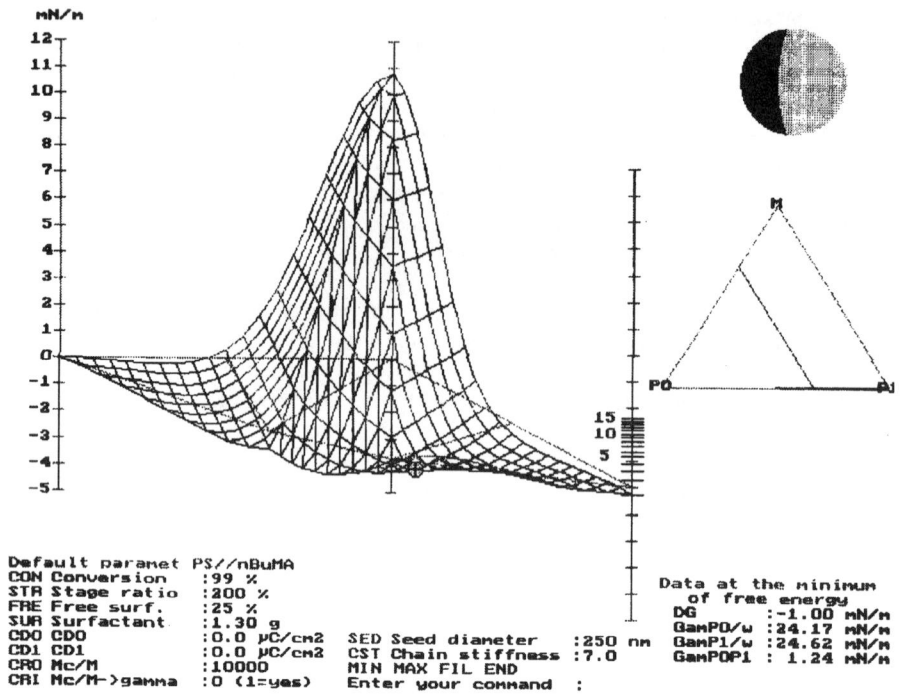

Figure 3. Surface energy map.

Extensions of the above approach have been used by other researchers to investigate conversion dependent morphologies. Chen et al (16) solved eq. (1) for a continuous spectrum of particle structures defined by the various angles at which the interfaces come into contact and made calculations as a function of conversion. The variations of interfacial tensions with conversion were estimated from experimental data and the results indicated that shifts in morphology from one shape to another may be expected during the polymerization process for those systems which remain at thermodynamic equilibrium during the entire reaction process. Throughout these calculations it is necessary to update the distribution of monomer between the polymer and aqueous phases in order to obtain a value for the interfacial tensions. As shown by Durant et al. [17], it is unfortunate that such interfacial tensions cannot be measured in-situ within the latex but must be measured in another manner. Jönsson et al [18] used another extension of eq. (1) to predict results for their experiments in PS seed/PMMA second stage particles and again find reasonable agreement between calculations and experiment. Durant and Sundberg [19] have offered another approach to utilizing eq. (1) and an extension of its solution by Chen [16], to derive a free energy *surface* which gives a clear, visual representation of the relative free energies of all possible fully phase separated (2 component) particles. Such a surface is shown in Figure 3 where the free energy is plotted on the vertical scale and the particle morphology is represented on the horizontal plane. That plane is graphically displayed in Figure 4.

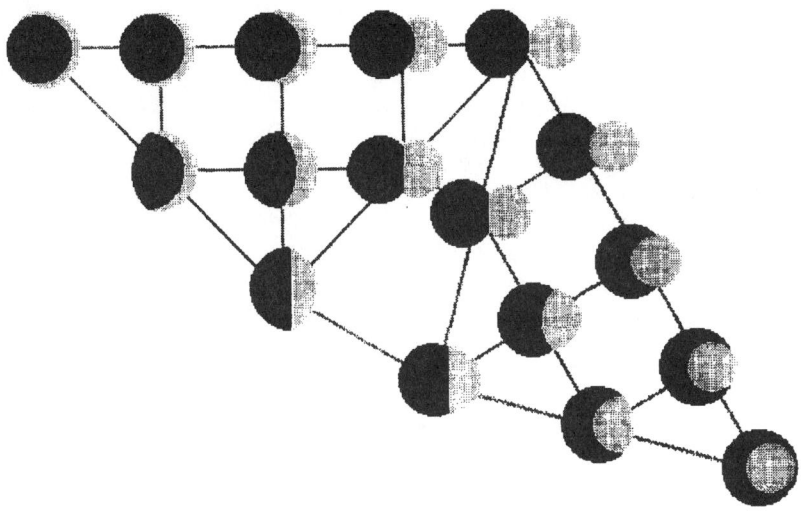

Figure 4. Topology map.

It is apparent from the above discussion that equilibrium morphologies may be predicted with reasonable reliability as long as the various interfacial tensions are known. Since these depend upon a number of variables and their interactions, the limiting factor to good predictions is the knowledge of the influence of these variables on the various interfacial tensions. Given the number of different polymers of interest to the latex community and the different surfactants and initiators used to produce them, the assembly of data sets for interfacial tensions is an important, but large task. Herein it will be useful to develop models for interfacial tensions which can be corroborated with independent data obtained in a non-latex environment. Such models for the polymer/aqueous interface have been suggested in references [14] and [15]. A model for the polymer/polymer interface (which has been much more extensively studied) is given by Broseta et al [20] and has been used by a number of the above cited references in application to latex morphology. Here again, given the variety of polymer/monomer systems of interest, there is a lot of work to do to produce data and models for use at the internal particle interfaces. This is especially true for copolymer systems, for which there has been very little information published.

Seed latex particle crosslinking and its effect on particle morphology has been treated only in a cursory way in the literature, and yet it is an important feature of many commercial latices. We have considered this effect and find that it is possible to modify eq. (1) to take crosslinking into account as follows;

$$\Delta G = \Delta G_s + \Delta G_{el} \tag{4}$$

where the s and el subscripts refer to surface and elastic forces, respectively. By considering the situation when the second stage polymer forms the core of the composite particle or when it is an occlusion within the particle, it is possible to understand how the crosslinked seed polymer would have to be elastically deformed in order for the particle to assume such a shape. Thus the free energies due to the surface forces and those due to the elastic forces may compete and yield very different equilibrium conditions than the counterpart uncrosslinked system. While the development of the elastic free energy term is beyond the scope of this paper, it can be anticipated that the elastic term will be dependent on the amount of deformation of the crosslinked polymer (a function of the amount of second stage polymer used) and the degree of crosslinking and stiffness of the seed polymer chains [21]. This creates a new and important dependency on particle size (note that in non-crosslinked systems the analysis is independent of particle size), and together with the new variable of crosslink density, makes the resultant predictions more difficult to present graphically for a given set of interfacial tensions. For the sake of this present article, we will simply say that once the interfacial tensions have been set for the system, the morphology predictions change for different seed particle sizes, stage ratios, seed polymer chain stiffness, and crosslink density. When the crosslinking is uniform within the seed polymer, the elastic free energy can be calculated from first principles. Experimental evidence [22] shows reasonable agreement with predicted results and confirms that very little crosslinking is necessary to have an important impact on the morphology. Indeed, when the crosslinking is high enough, the most favored equilibrium morphology is almost always a fully or partially engulfed structure.

The direct effect of copolymers, as compared to homopolymers, has not been reported in the literature as far as we know. The central questions that arise are related to the effects of copolymer chains on the various interfacial tensions, and the distribution of the comonomers between the polymer and aqueous phases and their attendant effect on the interfacial tensions. Because copolymer composition can drift during the polymerization of the second stage monomer, its impact on the interfacial tensions may also be a consideration. Inherent in such calculations is the need for reliable values for polymer/monomer interaction parameters and free radical polymerization rate coefficients. Although these requirements seem to be difficult to meet, our present work suggests that reasonable approximations of these parameters are useful in morphology predictions.

3. Non-Equilibrium and Kinetic Aspects of Morphology Development

It is at first important to define what is meant by "non-equilibrium" or "kinetically controlled" particle morphologies. Figure 4 shows the spectrum of possible structures for 2 component particles which are fully phase separated and at implied equilibrium. Non-equilibrium morphologies are usually incompletely phase separated structures, such as microdomains (occlusions) or mixtures of occlusions and partial or fully engulfed particles. The "raspberry" particles reported by Okubo [4] would also be considered a non-equilibrium structure. Most of the time it is sufficient to think of non-equilibrium structures as particles with incomplete phase separation. However a CS particle can be a non-equilibrium structure when the particular set of interfacial tensions would dictate otherwise under fully relaxed conditions. Thus one must be careful to distinguish the difference between the experimental or predicted structure and that expected at equilibrium.

Under all circumstances during the production of a composite latex particle there is reaction of monomer to polymer, phase separation of the second polymer within the seed polymer, and the possible diffusion of both species within the particle and perhaps at its water interface. Kinetically controlled, non-equilibrium structures develop when the rate of the polymerization reaction (and its simultaneous effect on local viscosity) is significantly faster than phase separation and polymer diffusion. These conditions may occur for any type of process conditions, not just for the expected "starve fed", or semi-batch process in which the second stage monomer(s) is fed to the seed latex over some period of time. Given that so many commercial latex processes are operated in this fashion, it is probable that most latices are produced in their non-equilibrium condition. That poses interesting questions related to latex aging (structural rearrangement after polymerization is complete) as described by Min [8]. It also suggests that it is of great value to have dynamic models of the morphology development process. This is a challenging task.

If one considers a seed latex particle swollen to some level with second stage monomer and envisions polymerization reaction taking place within the particle, one can understand that phase separation will occur through nucleation and growth processes. Both polymers may diffuse within this structured particle, and the phase separated microdomains may combine with each other, may grow by Ostwald ripening or may move within the particle due to long range van der Waals forces. This dynamic process is caused by the polymerization reaction but the particle structure development is driven by interfacial energy differences and retarded by restricted diffusion within an increasingly viscous environment. As one might imagine, the description of the reaction kinetics within such phase separated particles is perhaps not straightforward. We do not know of any publications describing such reaction kinetics, but people are working on this problem. Our group has presented a preliminary reaction rate model at a recent conference [23].

Some progress has been reported very recently regarding the description of the time dependent phase separation and growth process within structured latex particles. Gonzalez-Ortiz and Asua have described a set of models for the structural rearrangements possible during latex aging after polymerization [24] and during the polymerization process [25]. The basic concepts are that microdomains are formed through phase separation and, for the case of latex aging, these occlusions move within the particle due to long range van der Waals forces (created by interfacial tension differences). The occlusions may coalesce if they contact one another and may accumulate at the particle center or at its water interface, depending upon the nature of the interfacial tensions. Given some approximations for the interfacial tensions and for the polymer viscosity, Gonzalez-Ortiz and Asua calculated changes in particle morphology with time. Obviously the dynamics were slow in these particles which contained no monomer. For particles which contained second stage monomer, the authors had to make some assumptions about the reaction kinetics in order to compute the rate of new polymer accumulation and also assumptions as to its location within the particle at the time of phase separation. Under such assumptions they were able to calculate time dependent morphologies for a variety of chosen conditions. While it is beyond the scope of this paper to go into any more detail of these publications, the reader is recommended to consult the papers for further details. These publications represent the first reported attempts to provide computational models of this complex situation.

Reports of experimental studies concerning non-equilibrium morphology have been much more numerous. As noted earlier, Lee [1] described experimental results for the PS/P(S-co-Bd) latex system in which a variety of conditions were varied. He found that at high degrees of swelling and at low polymerization rates the morphologies were generally fully phase separated, while for the opposite conditions, incompletely phase separated conditions prevailed. For the S/PMMA (seed) system, Okubo et al [26] found that by varying the monomer feed rate the particle surface composition (judged from surfactant titration) changed from PMMA to PS as the feed rate was decreased. Raspberry shaped particles were also found. Cho and Lee [7] found similar results for the same system under similar conditions, but attributed the morphology changes to the chain anchoring effect of the SO_4 end groups at the particle surface as well as the monomer feed rate. Lee and Rudin [27] achieved making a non-equilibrium CS PMMA/PS latex by using low temperature reaction with a redox initiator. They attributed this result to the effect of temperature on the mobility of the monomer within the particle, resulting in reaction at the periphery of the particle and formation of a CS structure, while equilibrium considerations would suggest an inverted CS morphology. Jönsson et al [18,28] have also studied the popular PMMA/PS system, but did so by using each polymer as the seed and each monomer as the second stage material, while varying monomer feed rate, initiator type (water and oil soluble) and seed polymer molecular weight. Almost all of their reported morphologies were non-equilibrium and these varied with process and formulation conditions. These authors assigned the various results to variations in radical transport rates (due to restricted diffusion) within the particle during reaction. As a final comment to this part of the discussion it is interesting to note that many of the reported studies have used two glassy polymers for which polymer and monomer diffusion can become extremely restricted under certain conditions. The use of low T_G polymers within which diffusion is likely faster has not often been reported in these types of studies. However, one can speculate that if the seed is a low T_G polymer it will be more likely to have phase separation of the second stage polymer throughout the particle than might be the case for a high T_G seed. A complicating factor with low T_G seeds (e.g. PBd or PBA) is often that they are naturally or purposefully crosslinked for mechanical property reasons. The effect of crosslinking on polymer chain diffusion and occlusion migration is likely to be quite significant.

4. Confirmation of Particle Morphology

In order to test the kind of ideas that come from quantitative models, whether they be for the equilibrium or non-equilibrium cases, one of course needs specific experimental evidence. While the transmission electron microscope (TEM) provides the most obvious technique to achieve such results, it is often not all that clear what the exact particle morphology is from studying the micrographs. This is in part because the most useful micrographs are derived from microtomed and chemically stained sections of the particles, and such sections, or slices, come from randomly placed cuts through the particle. This means that for most particles one observes a lot of different looking sections on each TEM photo. Ascribing such an array of thin sections to a specifically predicted particle morphology will not always be straightforward. It is no wonder then that a number of groups are currently studying new ways to provide complementary techniques to determine the morphology of experimental latices. Among these techniques are x-ray and neutron scattering, NMR spectroscopy, and atomic force microscopy, among others.

5. Concluding Remarks

It has become clear that a knowledge of the values of the various interfacial tensions for the latex particle is essential to be able to predict what the final morphology will be. For processes run under near equilibrium conditions, these interfacial tensions and an assessment of the effect of any crosslinking present is all that is required. For non-equilibrium process conditions, a knowledge of reaction kinetics and polymer and monomer diffusivity are also required. While it is critical to make continued progress in kinetics and diffusivity, it may be that achieving an understanding of how the interfacial tensions are affected by all of the interacting variables (e.g. surfactants, initiator and carboxyl end groups, pH, salt concentration, monomer distributions, etc.) will be just as important in order to arrive at a thorough understanding of the control of latex morphology.

Acknowledgments

The authors are grateful for the financial support provided over several years from the University of New Hampshire, the Petroleum Research Fund of the American Chemical Society, the Centre Nationale Recherche Scientifique, and Rhône Poulenc.

6. References

1. Lee, D. I. (1980) Morphology and properties of two-stage latex particles, *Preprints of the 2nd Japan-Korea Joint Symposium on Polymer Science and Technology* , 97-107.
2. Lee, D. I. and Ishikawa, T. (1983) The formation of "inverted" core-shell latexes, *J. Polym. Sci., Polym. Chem. Ed.* **21**, 147-154.
3. Muroi, S., Hashimoto, H. and Hosoi, K. (1984) Morphology of core-shell particles, *J. Polym. Sci., Polym. Chem. Ed.* **22**, 1365-1372.
4. Okubo, M., Katsuta, Y. and Matsumoto, T. (1980) Rupture of anomalous composite particles prepared by seeded emulsion polymerization in aging period, *J. Polym. Sci., Polym. Ltr. Ed.* **18**, 481-486.
5. Okubo, M., Ando, M., Yamada, A., Katsuta, Y. and Matsumoto, T. (1981) Studies on suspension and emulsion. XLVII. anomalous composite polymer emulsion particles with voids produced by seeded emulsion polymerization, *J. Polym. Sci., Polym. Ltrs. Ed.* **19**, 143-147.
6. Stutman, D.R., Klein, A., El-Aasser, M.S. and Vanderhoff, J.W. (1985) Mechanism of core/shell emulsion polymerization, *I&EC Product Res. & Dev.* **24**, 404-412.
7. Cho, I. and Lee, K-W. (1985) Morphology of latex particles formed by poly(methyl methacrylate)-seeded emulsion polymerization of styrene, *J. Appl. Polym. Sci.* **30**, 1903-1926.
8. Min, T. I., Klein, A., El-Aasser, M.S. and Vanderhoff, J.W. (1983) Morphology and grafting in polybutylacrylate-polystyrene core-shell emulsion polymerization, *J. Polym. Sci., Polym. Chem. Ed.* **21**, 2845-2861.
9. Torza, S. and Mason, S. G. (1970) Three-phase interactions in shear and electrical fields, *J. of Colloid and Interfacial Sci.* **33**, 67-83.
10. Hobbs, S. Y., Dekkers, M. E. J. and Watkins, V. H. (1988) Effect of interfacial forces on polymer blend morphologies, *Polymer* **29**, 1598-1602.
11. Berg, J., Sundberg, D. C. and Kronberg, B. (1986) Microencapsulation of emulsified oil droplets by in-situ polymerization, *Polym. Materials Sci. and Eng.* **54**, 367-369.
12. Berg, J., Sundberg, D. C. and Kronberg, B. (1989) Microencapsulation of emulsified oil droplets by in-situ polymerization, *J. of Microencapsulation* **6**, 327-337.
13. Sundberg, D. C. and Muscato, M.R. (1990) Morphology development of polymeric microparticles in aqueous dispersions. 1. thermodynamic considerations, *J. Appl. Polym. Sci.* **41**, 1425-1442.

188

14. Winzor, C. L. and Sundberg, D. C. (1992) Conversion dependent morphology predictions for composite emulsion polymers: 1. synthetic latices, *Polymer* **33**, 3797-3810.
15. Winzor, C. L. and Sundberg, D. C. (1992) Conversion dependent morphology predictions for composite emulsion polymers: 2. artificial latices, *Polymer* **33**, 4269-4279.
16. Chen, Y-C, Dimonie, V. and El-Aasser (1991) Effect of interfacial phenomena on the development of particle morphology in a polymer latex system, *Macromolecules* **24**, 3779-3787.
17. Durant, Y. G., Sundberg, D. C. and Guillot, J. (1994) Determination of interfacial tensions for latex particles, *J. Appl. Polym. Sci.* **53**,1469-1476.
18. Jönsson, J-E.L. Hassander, H., Jansson, L.H. and Törnell, B.(1991) Morphology of two-phase polystyrene/poly(methyl methacrylate) latex particles prepared under different polymerization conditions, *Macromolecules* **24**, 126-131.
19. Durant, Y. G. and Sundberg, D. C. (1995) An advanced computer algorithm for determining morphology development in latex particles, *J. Appl. Polym. Sci.* **58**,1607-1618.
20. Broseta, D., Leibler, L., Kaddour, L.O. and Strazielle, C. (1987) A theoretical and experimental study of interfacial tension of immiscible polymer blends in solution, *J. of Chem. Physics* **87**, 7248-7256.
21. Durant, Y. G. and Sundberg, D. C. (in press) The effects of crosslinking on the morphology of structured latex particles. 1. theoretical considerations, *Macromolecules*
22. Durant, Y. G., Sundberg, E. J. and Sundberg, D. C. (submitted to *Macromolecules*) The effects of crosslinking on the morphology of structured latex particles. 2. experimental evidence for lightly crosslinked systems.
23. Durant, Y. G. and Sundberg, D. C. (1995) Emulsion polymerization kinetics in phase separated latex particles, Preprints of the *Fourth Pacific Polymer Conference, Pacific Polymer Federation*, page 51.
24. Gonzalez-Ortiz, L. J. and Asua, J. M. (1995) Development of particle morphology in emulsion polymerization. 1. cluster dynamics, *Macromolecules* **28**, 3135-3145.
25. Gonzalez-Ortiz, L. J. and Asua, J. M. (1996) Development of particle morphology in emulsion polymerization. 2. cluster dynamics in reacting systems, *Macromolecules* **29**, 383-389.
26. Okubo, M., Yamada, A. and Matsumoto, T. (1981) Estimation of morphology of composite polymer emulsion particles by the soap titration method, *J. Polym. Sci., Polym. Chem. Ed.* **18**, 3219-3228.
27. Lee, S. and Rudin, A. (1992) Synthesis of core-shell latexes by redox initiation at ambient temperatures, *J. Polym, Sci., Part A: Polym. Chem.* **30**, 2211-2216.
28. Jönsson, J-E, Hassander, H. and Törnell, B. (1994) Polymerization conditions and the development of a core-shell morphology in PMMA/PS latex particles. 1. influence of initiator properties and mode of monomer addition, *Macromolecules* **27**, 1932-1937.

LATEX PARTICLE MORPHOLOGY. THE ROLE OF MACROMONOMERS AS COMPATIBILIZING AGENTS

PRAPASRI RAJATAPITI[1,2], VICTORIA L. DIMONIE[1], MOHAMED S. EL-AASSER[1,3,*]

[1]Emulsion Polymers Institute, [2]Material Science and Engineering and [3]Chemical Engineering Departments, Lehigh University, Bethlehem, Pennsylvania 18015, USA

1. Introduction

Seeded emulsion polymerization is the most common preparation method for composite latexes. Unfortunately, this technique may generate particles with a variety of morphologies. The core-shell morphology, where a second polymer type totally covers the first-stage particle, is only an idealized representation based upon the sequential monomer addition. Transition morphologies, such as hemispherical particles, raspberry, sandwich, mushroom, and confetti-like structures, are frequently reported [1-3]. Design and control of latex particle morphology are often crucial to fulfill the end-use requirements for these materials.

Many polymerization parameters can affect composite particle morphology [1-15]. Basically, these factors fall into two categories: thermodynamic and kinetic. Thermodynamic factors determine the stability of the ultimate particle morphology according to the minimum surface free energy. Kinetic factors, however, control whether the particle is going to reach the thermodynamically predicted degree of phase separation. Thermodynamic parameters typically involve the compatibility between the phases in the system (e.g., hydrophilicity of each phase, particle surface polarity, and interfacial tensions). On the other hand, examples of kinetic factors are crosslinking agents, viscosity of the reaction medium, mode of monomer addition, polymer molecular weight, and polymerization temperature.

A few groups of researchers specifically examined the role of interfacial tensions in particle morphology [16-21]. According to their analyses, each particular morphology possesses a different value for free energy, based on the following equation:

$$G = \sum \gamma_{ij} S_{ij} \qquad (1)$$

where G is the Gibb's free energy of the system, γ_{ij} is the interfacial tension between phases i and j, and S_{ij} is the interfacial area between phases i and j. A seeded emulsion polymerization system tends to reach the lowest surface free energy state, that is, the one with the minimum total interfacial energy. Both the interfacial tensions between the

---

* To whom correspondence should be addressed.

J. M. Asua (ed.), Polymeric Dispersions: Principles and Applications, 189–202.
© 1997 Kluwer Academic Publishers. Printed in the Netherlands.

different polymer phases (each swollen with the second-stage monomer) and the interfacial tensions between each polymer phase and the aqueous medium are the key factors controlling the composite particle morphology. Consequently, one may control the composite particle morphology by monitoring these interfacial tensions.

Composite latex particles are similar to immiscible polymer blends in that, the phases are separated. For the polymer blends, inadequate adhesion between the phases leads to poor stress transfer across the interface [22]. Thus, it is desirable to increase the compatibility between the constituent polymers and to improve the interfacial adhesion. The most popular approach is to add a "compatibilizing agent" to the blend [22-26]. Generally, "compatibilizing agents" are block or graft copolymers that are miscible with both of the components in the blends. Their functions are to lower the polymer/polymer interfacial tension and to promote the interfacial adhesion between the polymer phases. The objective of the current studies is to apply compatibilizing agents to the composite particles. The purposes of adding the compatibilizing agents are to control the interfacial tension between the core and shell polymers and to control the morphology of the composite particles (typically, a lower polymer/polymer interfacial tension leads to a more complete coverage of the core particles by the shell polymer [16-21]).

The graft copolymers used as compatibilizing agents are prepared by the macromonomer technique [27,28]. In principle, this method offers control over the copolymer's graft length because the molecular weight of the starting macromonomer can be preselected. Additionally, the number of grafts per copolymer chain can be controlled by adjusting the macromonomer to comonomer mole ratio. Thus, through this technique, one may vary the structure of the compatibilizing agents so as to achieve control over the interfacial tension between the polymer phases and the morphology of the composite particles.

The system being studied is poly(n-butyl acrylate) (PBA)/poly(methyl methacrylate) (PMMA) composite latexes. The compatibilizing agents are graft copolymers with PBA backbones and PMMA-macromonomer side-chains. To effectively act as compatibilizing agents, graft copolymers must reside between the core and shell phases of the composite particles. Subsequently, these copolymers were incorporated onto the PBA particles in situ (during the particle preparation) prior to using these particles as seed in the second-stage polymerization. Because PMMA-macromonomer branches of the graft copolymers are hydrophilic, copolymer molecules are expected to preferentially partition close to the PBA particle/water interface. During the seeded emulsion polymerization, the presence of the graft copolymers between the core particles and the newly-formed PMMA shell is expected to lower the PBA/PMMA interfacial tension in the composite particles. The lower interfacial tension would, in turn, lead to the changes in composite particle morphology.

2. Experimental

2.1. LATEX PREPARATION

Seed latexes are PBA homopolymer prepared by miniemulsion homopolymerization of n-butyl acrylate (BA) and BA/PMMA-macromonomer copolymers prepared by miniemulsion copolymerization of BA and PMMA-macromonomers. Macromonomers are a series of linear PMMA molecules with one terminal vinyl double bond per chain

(M_w between 4.8×10^2 and 1.8×10^4 g/mol) [29-31]. Due to the extremely low water-solubility of the macromonomers, the use of miniemulsion polymerization rather than a conventional emulsion process is essential for their successful incorporation into the PBA particles [32,33]. Through this process, compatibilizing agents are incorporated into PBA particles in situ (simultaneous with the particle formation) [34].

Composite latexes were prepared with methyl methacrylate (MMA) as the second-stage monomer. Seeded emulsion polymerizations were carried out in a batch mode. Seed polymer/MMA ratio was 1:1 (w/w) [35].

Table 1 provides the miniemulsion recipe for the preparation of the seed latexes [34] and seeded emulsion polymerization recipe for the preparation of the composite latexes [35]. Reaction temperature was 70 °C in both cases. PMMA miniemulsion latex was also prepared by a recipe similar to the one described for the seed latex in Table 1, substituting the BA monomer by the MMA monomer. Table 2 gives the descriptions of the seed and composite latexes.

TABLE 1. (I) Miniemulsion polymerization recipe for preparing PBA seed latexes
(homopolymer or PBA incorporating PMMA-macromonomer)
and (II) seeded emulsion polymerization recipe for preparing PBA/PMMA composite latexes.

I. Seed Latex		II. Composite Latex	
Ingredients	Weight (g)	Ingredients	Weight (g)
BA + PMMA-macromonomer[a]	20.000	Seed latex[c]	10.000
DDI water	80.000	MMA monomer	2.000
Hexadecane (HD), 20 mM[b]	0.363	DDI water	8.000
Sodium lauryl sulfate (SLS), 5 mM[b]	0.115	Sodium lauryl sulfate (SLS), 2.6 mM[b]	0.012
Potassium persulfate (KPS), 3 mM[b]	0.065	Potassium persulfate (KPS), 3 mM[b]	0.013

(a) Weights of macromonomer and BA were varied. (See Table 2).
(b) Based on the aqueous phase.
(c) PBA homopolymer or PBA incorporating PMMA-macromonomer, solids content approximately 20%

2.2. CHARACTERIZATION OF SEED LATEXES

At the end of the polymerization, complete BA monomer conversions were confirmed from the solids contents of the latexes and by gas chromatography [34]. Results from adsorption chromatography with evaporative light-scattering detector show that macromonomers are attached to PBA backbones, and that the structures of the resulting graft copolymers vary with the amount (i.e., mole ratio or molecular weight) of macromonomer in the recipe [36]. Monomer/water phase interfacial tension decreases when macromonomer is dissolved in BA [34]. For the PBA latexes incorporating PMMA-macromonomer(s), particles are smaller with the increasing amount of macromonomer incorporated in the latex (measured by transmission electron microscopy, TEM, and capillary hydrodynamic fractionation, CHDF) [34]. These results show that PMMA-macromonomers (or PMMA-macromonomer branches of the copolymers) partition close to the monomer (or PBA particle) and water interface, and lower the interfacial tension between these phases.

TABLE 2. Descriptions of seed and composite latexes.

Sample Name	Macromonomer in Seed Latex			PBA/PMMA in Composite Latex		
	M_w (g/mol)	Macromonomer/BA $(x10^{-2})$		mol/mol	w/w	Volume Fraction of PBA[a]
		mol/mol	w/w			
BM0	---			0.781	1.000	0.530
BM253	5320	0.049	2.041	0.766	0.980	0.519
BM512	1260	0.127	1.250	0.772	0.988	0.523
BM536	3640	0.127	3.605	0.754	0.965	0.512
BM553	5320	0.127	5.263	0.742	0.950	0.504
BM596	9640	0.127	9.529	0.713	0.913	0.485
BM1053	5320	0.267	11.111	0.703	0.900	0.478
Blend	homopolymer blend			0.781	1.000	0.530

(a) Based on total volume of polymers in composite latexes; calculated based on the densities of PBA and PMMA (1.055 and 1.188 g/cm³, respectively), and of PMMA-macromonomers[33], and the weights of BA, PMMA-macromonomer, and MMA (according to the latex compositions).

The partitioning of the PMMA-macromonomer side-chains (from the graft copolymer molecules) in the PBA particles was further analyzed by the soap titration [34]. Table 3 summarizes the values of the effective area per molecule (a_s) of SLS on latex particles at surface saturation. All the copolymers possess a higher a_s value compared with the PBA homopolymer. An increase in a_s values is directly proportional to an increase in the surface polarity of the latex particles [37]. Therefore, a_s values can give a qualitative information on the partitioning of MMA units from the graft copolymer at the particle interface.

By increasing the macromonomer/BA weight ratio up to 3-4%, based on BA (independent of the molecular weights of macromonomers), one observes an increase in a_s values of SLS adsorbed on the particle surface. Above this concentration, a_s values remained almost unchanged (between 56 and 57 Å²). These values suggest a continuous enrichment of the particle/water interphase zone with the MMA units. The constant values of a_s suggest that, at the particle/water interface, the maximum packing density of the PMMA side-chains is reached. By further increasing the concentration of the macromonomer, most of the additional MMA units will settle inside the PBA particles. At the maximum packing density of the MMA units on the copolymer particles, however, these particles are still not as hydrophilic as the PMMA homopolymer particles ($a_s = 92$ Å²). The lower a_s values for the copolymer particles compared with the value for the PMMA homopolymer particles is possibly due to the attachment of the MMA units to the PBA main-chains. Because of this attachment, the surface of the copolymer particles can never be free of the BA units from the copolymer backbones. The second possible reason for the low a_s values could be the inhomogeneous composition of the particles. During the miniemulsion polymerization of BA in the presence of macromonomers, because of the nonuniform size distribution of the monomer droplets or some homogeneous nucleation of BA, a number of PBA particles may be formed with little or no macromonomer in the particles .

TABLE 3. Effective surface area per molecule of SLS (at the cmc of SLS) on latex particles.

Latex	Macromonomer			Area/ molecule of SLS $(a_s, Å^2)^a$
	M_w (g/mol)	x10^{-2} (based on BA)		
		mol/mol	w/w	
PBA homopolymer				34 ± 2
BA/PMMA-macromonomer copolymer				
M248	4.8×10^2	0.049	0.19	43 ± 3
M212	1.3×10^3	0.049	0.49	36 ± 4
M236	3.6×10^3	0.049	1.40	44 ± 1
M253	5.3×10^3	0.049	2.04	51 ± 1
M296	9.6×10^3	0.049	3.56	57 ± 3
M1036	3.6×10^3	0.267	7.61	56 ± 3
M1053	5.3×10^3	0.267	11.11	56 ± 3
PMMA homopolymer				92 ± 10

(a) Values averaged from at least 3 measurements.

2.3. CHARACTERIZATION OF COMPOSITE LATEXES

2.3.1. *Morphological Observations of Composite Latex Particles*
Morphologies of the composite latex particles were studied using the TEM [35]. Figure 1 shows micrographs of composite latex particles prepared using (A) PBA homopolymer seed latex (BM0) and (B) - (D) BA/PMMA-macromonomer copolymer seed latexes (BM253, BM553, and BM1053, see Table 2 for latex compositions). In samples B, C, and D, macromonomer/BA mole ratio was varied, while M_w of macromonomer was constant (5.3×10^3 g/mol).

As seen from the micrographs, the morphologies of composite particles are strongly dependent on the amount of PMMA-macromonomer used in the preparation of the seed latexes. The composite particles prepared from PBA homopolymer seed latex (BM0) show a mainly hemispherical morphology (Figure 1A). This type of morphology (high degree of phase separation) is expected because the two polymers are incompatible. Despite their smaller sizes, composite particles prepared from PBA seed incorporating the lowest macromonomer/BA mole ratio (BM253, Figure 1B) has a similar morphology. However, composite latexes prepared from PBA seed incorporating higher amounts of macromonomer (BM553 and BM1053; Figures 1C and 1D) form particles with a mixture of morphologies (i.e., some large particles with a PMMA-rich surface and smaller particles which have either hemispherical or multiphase morphology). The uniform core-shell morphology of the larger particles may be attributed to the effect of the compatibilizing agents (BA/PMMA-macromonomer graft copolymers) partitioning on the surface of the seed particles. During the second-stage polymerization, the presence of the graft copolymer layer on the surface of these seed particles prevents the newly formed PMMA chains from segregating. Thus, PMMA resides on the surface of the particles as a continuous shell. The hemispherical and multiphase particles result from

194

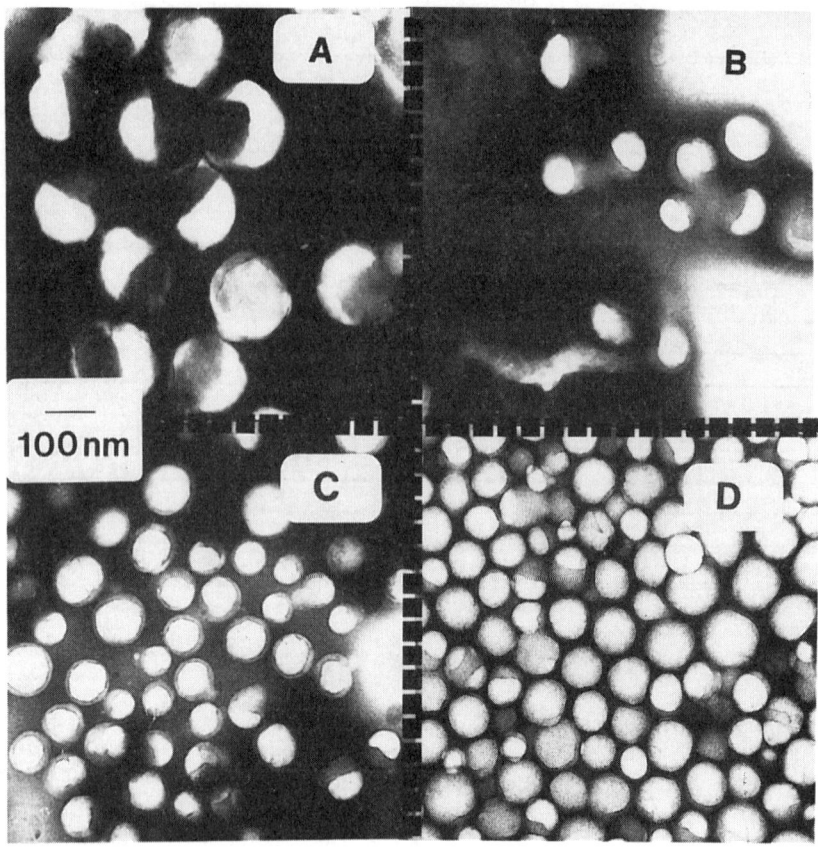

Figure 1. TEM micrographs of PBA/PMMA composite latex particles stained with PTA and RuO$_4$. (A) Prepared from PBA homopolymer seed latex particles (BM0), (B) - (D): prepared from BA/PMMA-macromonomer copolymer seed latex particles (BM253, BM553, and BM1053). (See Table 2 for descriptions of these samples). In these micrographs, PBA is stained dark, whereas PMMA appears light.

the inherent composition inhomogeneity of the seed latex. (In the miniemulsion polymerization of BA in the presence of macromonomer, some PBA seed particles were formed with little or no graft copolymer on their surfaces. The evidence for the inhomogeneous seed particle composition was already shown in the soap titration results.)

In the samples shown in Figure 1, the mole ratio of macromonomer to BA in the seed latexes was varied. However, the molecular weight of the macromonomer remained the same. Figure 2 shows TEM micrographs of another set of samples in which the molecular weight of the macromonomer in the seed latex was varied while the macromonomer/BA mole ratio was kept constant (0.127×10^{-2}) to maintain a constant number of grafted PMMA chains (samples BM512, BM553, and BM596). Therefore, the weight ratio of macromonomer to BA was increased with the increasing molecular weight of the macromonomer (see Table 2).

The composite latex particles prepared from PBA seed incorporating low molecular weight macromonomer (BM512) show multiple white domains of second-stage PMMA on the PBA phase (Figure 2A). Per unit volume, the total interphase area between the two polymer phases in these particles is larger than in those with a hemispherical morphology. This type of morphology suggests that, by adding a relatively hydrophilic macromonomer to the seed latex preparation, the PBA/PMMA interfacial tension is decreased (to some extent). The lower interfacial tension allows a more uniform coverage of the seed polymer by the shell PMMA in the composite particles. However, the amount of the graft copolymer in this sample may not be sufficient or its structure does not favor its partitioning at the seed particle surface. Thus, they are not as effective in lowering the interfacial tension between the two polymer phases as when macromonomers of higher molecular weights are used to prepare the seed latexes.

When the higher molecular weight macromonomers are used in the seed latex preparation at the same mole ratio, there are larger numbers of MMA units and longer PMMA-macromonomer grafts in the seed particles. Both of these lead to a higher concentration of MMA units at the particle interface. Consequently, the interfacial tension is reduced to a lower value. As a result, more uniform coverage of the seed by the shell polymer is observed. When the PBA seed latex incorporating the macromonomer of the highest molecular weight is used as a seed latex (BM596), the micrograph exhibits all composite particles having the PBA core completely covered by the PMMA shell (Figure 2C).

2.3.2. Dynamic Mechanical Analysis (DMA)

To supplement the morphological characteristics observed for the composite latex particles, the dynamic mechanical properties of the films prepared from freeze-dried composite latexes and homopolymer latex blends by compression molding were measured [35]. These properties are related to the morphology of the composite particles as well as the presence of an interphase layer between the two polymers [35,38,39]. A comparison between the experimentally measured moduli and the values predicted based on Dickie's model [40-42] also provides estimated quantity of the interfacial polymer in the films [35].

The G'-temperature curves in Figure 3 (top) show that all the composite samples behave similarly to the blends, i.e., having two transitions corresponding to the T_g's of the PBA and PMMA. These transitions correspond to the phase separation of PBA and PMMA within the composite particles. The G' curves between the two transitions are

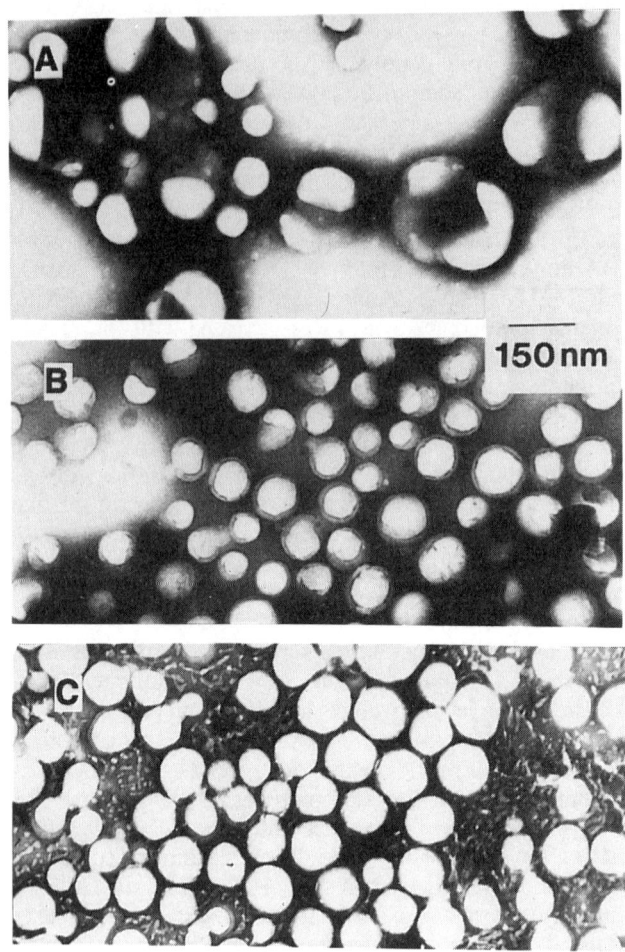

Figure2. TEM micrographs of PBA/PMMA composite latex particles stained with PTA and RuO_4. Macromonomer/BA mole ratio in seed latex = $0.127x10^{-2}$. (A) - (C): samples BM512, BM553,and BM596, respectively. (See Table 2 for descriptions of these latexes). In these micrographs, PBA phase is stained dark by RuO_4, whereas PMMA remains light.

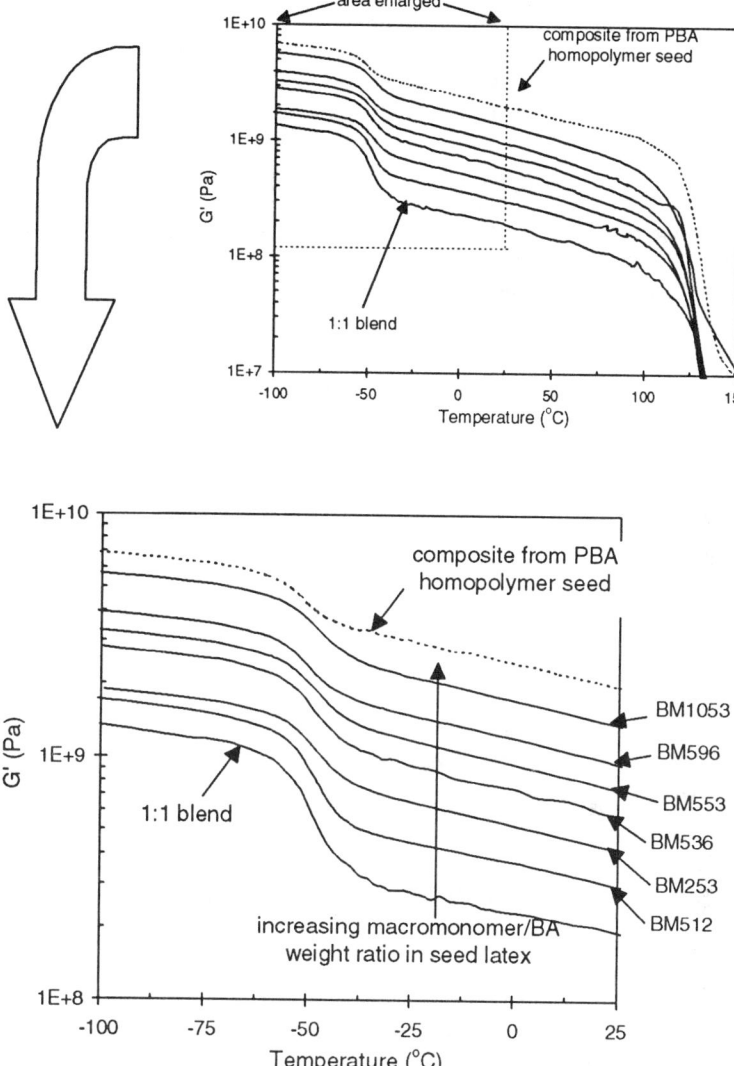

Figure 3. Temperature dependence of G' for films prepared from PBA/PMMA homopolymer blend (1:1 w/w PBA/PMMA) or composite polymer latexes (seed polymer/MMA in composite latex recipe = 1:1 w/w), showing the influence of seed latex compositions. (The curves were shifted vertically by multiplying the G' values by 1.00, 1.75, 1.25, 1.80, 4.00, 2.00 and 4.50, respectively; for the blend, BM512, BM253, BM536, BM553, BM596, BM1053 and BM0). See Table 2 for details of the latexes.

parallel to that of the PBA/PMMA homopolymer latex blend. The slopes of the curves indicate that, as in the blend, composite latex polymers form films which consist of PBA domains in a PMMA matrix [35]. However, Figure 3 (bottom) shows that, although the compositions of all these samples are nearly identical (Table 2), their G' curves are different. (These curves were vertically shifted for easier identification of the transitions.) The values of G' in the region between the PBA and PMMA transitions are the lowest for the homopolymer blend. For the case of composite latexes prepared from PBA incorporating macromonomer seed, the values increase with the amount of macromonomer (macromonomer/BA weight ratio) in the seed latexes. However, the values of G' are the highest when the composite latex was prepared from the PBA homopolymer seed particles (BM0). The difference in the G' value reflects the difference in the volume fraction of the dispersed phase in the films [35,40-42].

Previously, Cavaille et al. [43] used a parameter relating modulus of films in their relaxed and unrelaxed states to explain the interactions between inclusions and matrix in films prepared from latexes consisting of polystyrene (PS) and PBA. Here, a similar parameter, the change in the modulus before and after the T_g of PBA (G'_{20}/G'_{-100}, G' at 20 °C divided by G' at -100 °C), is used to approximate the volume fractions of the films' dispersed phase [35]. Considering that compression-molded films of freeze-dried latexes consist of discrete PBA domains dispersed in a PMMA matrix, one may calculate the values of G'_{20}/G'_{-100} for PBA/PMMA systems of given polymer compositions, based on Dickie's model [40-42] and the experimentally measured moduli of PBA and PMMA homopolymers. By comparing the model-predicted G'_{20}/G'_{-100} values for PBA/PMMA at various compositions to the values for G'_{20}/G'_{-100} obtained experimentally for each sample, one can estimate the volume fractions of the dispersed phase in the compression-molded films [35]. The results are given in Table 4. Figure 4 shows the schematic diagrams of films prepared from a homopolymer latex blend or from the composite latexes.

TABLE 4 Volume fractions of different phases in the latexes and in the compression-molded films.

Sample Name	Volume Fractions				% Interfacial polymer (based on PBA)[a]
	(A) Latex PBA	(B) Seed polymer in latex	(C) Film dispersed phase	DV (C-A)	
Blend	0.530	0.530	0.670	0.140	b
BM0	0.530	0.530	0.535	0.005	1.00
BM512	0.523	0.529	0.640	0.117	b
BM253	0.519	0.529	0.590	0.071	13.68
BM536	0.512	0.529	0.590	0.078	15.23
BM553	0.504	0.529	0.583	0.079	15.67
BM596	0.485	0.528	0.569	0.084	17.32
BM1053	0.478	0.528	0.578	0.100	20.92

(a) Approximated from DV divided by volume fraction of PBA in the composite latex (column A).
(b) Cannot be determined by this method because of the interconnected domains (PMMA occlusions inside the PBA dispersed phase).

The numbers in Table 4 show that the dispersed phase volume fraction in the film made from the homopolymer latex blend (0.670) is much higher than the volume fraction of PBA in the original latex blend (0.530). Since the dispersed phase is essentially the PBA homopolymer (no grafting between PBA and PMMA), this unexpectedly high volume fraction of the dispersed phase possibly arise from the interconnection of the PBA domains (Figure 4A) [35,40-46]. In the melt-state during the compression molding, individual PBA particles in the blend randomly come into contact with each other. This process enables the occlusions of PMMA domains within the PBA dispersed phase. This film morphology makes the volume fraction of the dispersed phase in the film higher than the actual volume of PBA in the latex.

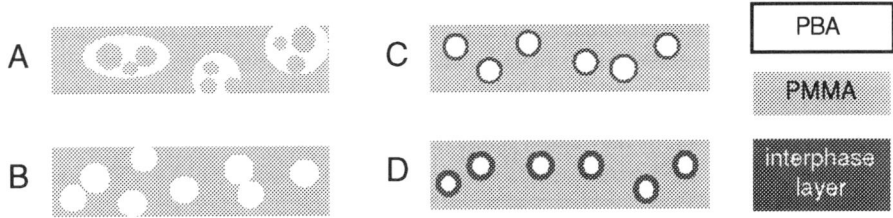

Figure 4. Schematic representations of compression-molded films: (A) showing interconnected PBA domains in PMMA matrix (blend and BM512), (B) showing individual PBA domains with no interphase region (BM0), (C) and (D) showing individual PBA domains with thicker interphase.

In spite of their similar composition (same volume fraction of PBA in the original latex, see Table 2), the volume fraction of the dispersed phase in a film derived from the composite latex prepared from the PBA homopolymer seed (BM0, 0.535) is much lower than in the one derived from the homopolymer blend (0.670). The only difference between these two samples is that, unlike individual PBA particles in the blend, the PBA phase in the composite particles is attached to the PMMA shell. This attachment slows the migration and diffusion of the PBA chains. The resulting PBA domains are thus smaller and more separated by the PMMA phase (i.e., minimum or no interconnection of PBA domains). The film morphology shown in Figure 4B for BM0 also represents the dispersed phase volume fraction in the film being almost the same as the PBA fraction in the original composite latex.

Although the volume fractions of seed polymer (i.e., PBA plus macromonomer) in all composite latex samples are nearly the same (between 0.528 and 0.530, column B, Table 4), the DMA-estimated dispersed phase volume fractions in these films (column C) are quite different. However, all of the films prepared from the composite latexes show volume fractions of the dispersed phase higher than the PBA fraction in the original composite latexes (DV values between 0.071 and 0.117).

For the composite samples prepared from the PBA seed latexes incorporating the lowest molecular weight PMMA macromonomer (BM512, see Table 2 for latex composition), the volume fraction of the dispersed phase appears similar to that of the blend (0.640 vs. 0.670). For this sample, particle agglomeration and interconnected PBA domains (Figure 4A) are anticipated because of the multiphase morphology of the particles (Figure 2A) and the higher mobility of the rubbery phase (T_g of this macromonomer is lower than room temperature [33]).

For all other composite samples, the film dispersed phase volume fractions (Table 4, column C) decrease with decreasing latex PBA volume fraction (or the increase in the latex macromonomer content, Tables 2 and 4). Nevertheless, the difference between the latex's PBA volume fraction (column A) and the film's dispersed phase volume fraction (column C), DV, increases with the macromonomer content (Table 4). Based on the DMA results of sample BM0, little domain interconnection occurs in films prepared from composite latexes. Thus, DV may only be explained by the presence of an interphase zone between PBA and PMMA in the composite sample when macromonomer is present in the seed particles [35,45,46].

Previously, Nelliappan and co-workers [47,48], using ^{13}C NMR, clearly showed that an interphase layer exists between the PBA core and PMMA shell of the composite latex particles. They also reported a thicker interphase region for the composite latex particles formed from seed PBA particles incorporating PMMA-macromonomer compared with the one prepared from PBA homopolymer seed particles. Additionally, seed particle surface analysis already showed that macromonomer side-chains of the BA/PMMA-macromonomer graft copolymers partition close to the seed particles/water interface [33,34,36]. Subsequently, in the composite films, the macromonomer side-chains of the graft copolymers would reside at the seed particle/PMMA interface. This arrangement would allow a higher fraction of the rubbery inclusions to extend into the PMMA matrix and compatibilize these two polymer phases, forming an interphase layer. The difference between the PBA volume fraction in the original latexes and the dispersed phase volume fraction as measured in the films (DV) is indicative of the volume of this interphase layer. The DV correlates well with the increase in the amount of macromonomer (weight ratio of macromonomer to BA) used in the seed latex preparation. The positive values of DV could be explained by two possible reasons: (1) the PBA volume in the sample is lower than the number calculated based on weight of BA used to prepare the original seed latex (column A), because a large fraction of the PBA chains is copolymerized with macromonomer and become a part of the interphase layer, and (2) the model used for the determination of the dispersed phase volume fraction does not account for the presence of the interphase zone, but assumes discrete PBA domains in the PMMA matrix.

In conclusion, the numbers in Table 4 show that, by using PBA seed incorporating PMMA-macromonomer to prepare the composite latexes, the PBA homopolymer volume fraction decreases, and the volume fraction of the interphase region between the two polymer phases (DV) increases. By increasing the number of MMA units in the BA/PMMA-macromonomer graft copolymer (either by increasing the mole ratio of macromonomer to BA or by increasing the molecular weights of the macromonomers), the volume of the interfacial layer increases. Both the decrease in the rubber phase volume fraction and the higher volume of the interphase layer correlate with the better compatibility of PBA and PMMA phases in the composite particles when the graft copolymers were present in the seed particles (as previously observed from the TEM micrographs in Figures 1 and 2, and by ^{13}C NMR[47,48]).

3. Conclusions

Graft copolymers with PBA backbones and PMMA-macromonomer side-chains were used as compatibilizing agents for PBA core-PMMA shell latexes. These graft copolymers were incorporated in situ into the PBA particles via miniemulsion (co)polymerization of BA in the presence of PMMA-macromonomer(s), prior to using these particles as seed in the second-stage emulsion polymerization of MMA [34]. Structures and compositions of the graft copolymers were varied by changing the mole ratio between macromonomer and BA (to control the number of grafts per chain) or varying the molecular weights of the macromonomers (to control the graft length) [36]. In the seed latexes, PMMA-macromonomer branches of the copolymers preferentially partition close to the seed particle/water interface because of their hydrophilic nature. The presence of the MMA units (from the copolymers) on the seed PBA particles' surface increases the particle polarity and lowers the particle/water interfacial tension [33,34,36].

TEM observation combined with the preferential staining showed that the degree of phase separation between the two polymers in the composite particles is affected by the amount of macromonomers used in the seed latex preparation (i.e., mole ratio of macromonomer to BA and molecular weight of the macromonomers) [35]. The observed morphologies show quite a good agreement with the decrease in the $polymer_1$/aqueous phase interfacial tensions observed when PMMA macromonomer was incorporated into the PBA seed particles [34]. The dynamic mechanical analysis interpreted according to Dickie's model for phase-separated polymer blends [40-42] enable the quantification of the interphase region [35]. It shows a decrease in the PBA volume fraction and an increase in the volume fraction of the interface layer when the BA/PMMA-macromonomer copolymer seed particles were used. These results also agree with the observed particle morphology [35] and Nelliappan's ^{13}C-NMR studies on PBA core-PMMA shell latexes [47,48]. All these results suggest that the decrease in polymer/polymer interfacial tension enhance the seed coverage by the shell polymer and increase the volume fraction of the interphase polymer. Thus, interfacial tension is considered one of the main parameters controlling particle morphology in composite latexes.

4. References

1. Okubo, M., Katsuta, Y., and Matsumoto, Y. (1980) *J. Polym. Sci., Polym. Lett. Ed.*, **18**, 481.
2. Okubo, M. (1990) *Makromol. Chem., Macromol. Symp.*, **35**, 307.
3. Rudin, A. (1995) *Makromol. Symp.*, **92**, 53.
4. Cho, J. and Lee, K.W. (1985) *J. Appl. Polym. Sci.*, **30**, 1903.
5. Lee, S. and Rudin, A. (1989) *Makromol. Chem., Rapid Commun.*, **10**, 655.
6. Lee, S. and Rudin, A. (1992) *Polymer Latexes*, E. S. Daniels, E. D. Sudol, and M. S. El-Aasser (eds.), ACS Symposium Series, **492**, American Chemical Society, Washington D.C., Chapter 15 234.
7. Lee, S. and Rudin, A, (1992) *J. Polym. Sci., Polym. Chem. Ed.*, **30**, 2211.
8. Vandezande, G.A. and Rudin, A. (1994) *J. Coating Tech.*, **66**, 99.
9. Mills, M.F., Gilbert, R.G. and Napper, D.H. (1990) *Macromolecules*, **23**, 4247.
10. Durant, Y.G.J. and Guillot, J. (1993) *Colloid Polym. Sci.*, **271**, 607.
11. M. Okubo, Y. Katsuta, and T. Matsumoto, *J. Polym. Sci., Polym. Lett. Ed.*, **20**, 45, (1982).
12. Min, T.I., Klein, A., El-Aasser, M.S., and Vanderhoff, J.W. (1983) *J. Polym. Sci., Polym. Chem. Ed.*, **21**, 2845.
13. Dimonie, V.L., El-Aasser, M.S., Klein, A. and Vanderhoff, J.W. (1984) *J. Polym. Sci., Polym. Chem. Ed.*, **22**, 2197.
14. Eckersley S. and Rudin, A. (1994) *J. Appl. Polym. Sci.*, **53**, 1139.

202

15. Jonsson, J.E., Hassander, H. and Tornell, B. (1994) *Macromolecules*, **27**, 1932.
16. Torza, S. and Mason, S.G. (1970) *J. Colloid Inter. Sci.*, **33**, 67.
17. Sundberg, D.C., Casassa, A.P., Pantazopoulos, J., Muscato, M.R., Kronberg, B., and Berg, J. (1990) *J. Appl. Polym. Sci.*, **41**, 1425.
18. Dimonie, V.L., El-Aasser, M.S., and Vanderhoff, J.W. (1988) *Polym. Mat. Sci. Eng.*, **58**, 821.
19. Chen, Y.C., Dimonie, V.L., and El-Aasser, M.S. (1991) *J. Appl. Polym. Sci.*, **41**, 1049.
20. Chen, Y.C., Dimonie, V.L., and El-Aasser, M.S. (1991) *Macromolecules*, **24**, 3779.
21. Chen, Y.C., Dimonie, V.L., and El-Aasser, M.S. (1992) *J. Appl. Polym. Sci.*, **45**, 487.
22. Paul, D.R. (1978) *Polymer Blends*, D. R. Paul and J. W. Barlow (eds.), Academic, Vol. 2, Chapter 12.
23. Gaylord, N.G. (1989) *J. Macromol. Sci., Chem.*, **26**, 1211.
24. Paul , D.R. and Barlow, J.W. (1980) *J. Macromol. Sci., Chem.*, **18**, 109.
25. Barlow, J.W. and Paul, D.R. (1984) *Polym. Eng. Sci.*, **24**, 525.
26. Fayt, R., Jerome, R., and Teyssie, P. (1989) *Multiphase Polymers: Blends and Ionomers*, L. A. Utracki and R. A. Weiss (eds.), American Chemical Society, Washington, D.C., Chapter 2.
27. Meijs, G.F. and Rizzardo, E. (1990) *J. Macromol. Sci., Rev. Macromol. Chem. Phys.*, **30**, 305.
28. M.K. Mishra (ed.) (1994) *Macromolecular Design: Concept and Practice*, Polymer Frontiers International, New York, Part 1.
29. Enikolopyan, N.S., Smirnov, B.R., Ponomarev, G.V., and Belgovskii, I.M. (1981) *J. Polym. Sci., Polym. Chem. Ed.*, **19**, 879.
30. O'Driscoll, A.F., O'Driscoll, K.F., and Rempel, G.L. (1984) *J. Polym. Sci., Polym. Chem. Ed.*, **22**, 3255.
31. Sanayei, R.A. and O'Driscoll, K.F. (1989) *J. Macromol. Sci., Chem.*, **26**, 1137.
32. Chotirotsukon, W. (1991) *M.S. Thesis*, Lehigh University.
33. Rajatapiti, P. (1996) *Ph.D. Dissertation*, Lehigh University.
34. Rajatapiti, P., Dimonie, V.L. and El-Aasser, M.S. (1995) *J. Macromol. Sci., Chem.*, **32**, 1445.
35. Rajatapiti, P., Dimonie, V.L., El-Aasser, M.S., and Vratsanos, M.S., accepted.
36. Rajatapiti, P., Dimonie, V.L., and El-Aasser, M.S., (1996) *J. Appl. Polym. Sci.*, **61**, 891.
37. Vijayendran, B.R. (1979) *J. Appl. Polym. Sci.*, **23**, 733.
38. Zosel, A. (1995) *Polym. Adv. Tech.*, **6**, 263.
39. Gauthier C. and Perez, J. (1995) *Polym. Adv. Tech.*, **6**, 1042.
40. Dickie, R.A. (1973) *J. Appl. Polym. Sci.*, **17**, 45.
41. Dickie, R.A., Cheung, M.F., and Newman, S. (1973) *J. Appl. Polym. Sci.*, **17**, 65.
42. Dickie R.A. and Cheung, M.F. (1973) *J. Appl. Polym. Sci.*, **17**, 79.
43. Cavaille, J.Y., Vassoille, R., Thollet, G., Rios, L., and Pichot, C. (1991) *Colloid Polym. Sci.*, **269**, 248,.
44. Cavaille, J.Y., Jourdan, C., Kong, X.Z., Perez, J. and Pichot, C. (1986) *Polymer*, **27**, 693.
45. Jourdan, C., Cavaille, J.Y., and Perez, J. (1988) *Polym. Eng. Sci.*, **28**, 3218.
46. Pichot, C., Kong, X.Z., Guillot, J., and Cavaille, J.Y. (1991) *Polym. Mat. Sci. Eng.*, **64**, 276.
47. Nelliappan, V., El-Aasser, M.S., Klein, A., Daniels, E.S., and Roberts, J.E. (1996) *J. Polym. Sci., Polym. Chem. Ed.*, **34**, 3173.
48. Nelliappan, V., El-Aasser, M.S., Klein, A., Daniels, E.S., and Roberts, J.E. (1996) *J. Polym. Sci., Polym. Chem. Ed.*, **34**, 3183.

CHARACTERIZATION OF PARTICLE MORPHOLOGY BY SOLID-STATE NMR

KATHARINA LANDFESTER, HANS WOLFGANG SPIESS
Max-Planck-Institut für Polymerforschung
Postfach 3148, D-55021 Mainz, Germany

1. Introduction

Nuclear magnetic resonance (NMR) today probably is the most valuable tool for elucidation of molecular structure, order, and dynamics. For the characterization of macromolecules [1,2], NMR can be divided into two major areas. High-resolution NMR provides detailed information about the chain microstructure in solution. Solid-state NMR allows to characterize the molecular structure and the organization of the macromolecules in the bulk. Special techniques offer to study dynamic aspects over a large range of characteristic rates. Solid-state NMR allows to characterize morphology properties, e.g. chain alignment in oriented polymers [3]. Multidimensional techniques offer fundamental advantages [4] by introducing an additional frequency dimension to increase the spectral resolution and by providing routes to new information, unavailable from 1D spectra even in the limit of high resolution [5]. One aspect to establish structure-property relationships for polymer materials is the characterization of domain sizes in heterogeneous materials. Advanced polymer materials with more than one component are phase separated in most cases. Small domains of only a few nanometers, as well as interfaces between the different phases, are particular difficult to characterize. Here, new solid-state NMR techniques nicely supplement well-established scattering and microscopic methods.

This article thus briefly outlines the basics of solid-state NMR and then illustrates the information available from solid-state NMR methods about the particle morphology in latex particles.

2. Solid-State NMR

NMR spectra are site-selective because the magnetic fields that the nuclei experience are slightly different from the external field B_0 due to the shielding by surrounding electrons. The effect is known as the chemical shift, which spans about 10 ppm for ^1H and 200 ppm for ^{13}C in different functional groups. This allows the structural characterization of liquids or components in solution.

Owing to the presence of angular-dependent anisotropic interactions, the spectral resolution of solid-state NMR spectra is orders of magnitude lower than that of high-

J. M. Asua (ed.), Polymeric Dispersions: Principles and Applications, 203–216.

resolution NMR in liquids [6,7]. The interaction of nuclear spins with their surrounding involves, e.g., a magnetic dipole-dipole coupling of nuclei among themselves. This leads to broad NMR lines covering approximately 50 kHz for ^{1}H-^{1}H homonuclear coupling and approximately 25 kHz for ^{1}H-^{13}C heteronuclear coupling. The anisotropy of the chemical shift results in powder patterns covering approximately 15 kHz at a field strength of 7 T. Another coupling for nuclei only with spin I > ½ is the nuclear coupling to the electric field gradient at the nuclear site. For ^{2}H (I = 1) in C-^{2}H bonds, this leads to spectral splittings of about 250 kHz. The anisotropic interactions can be removed under magic angle spinning (MAS). This high-speed mechanical rotation at an angle of 54.7 ° to the magnetic field B_0 yields liquid-like spectra if the spinning speed ω_r is significantly larger than the width Δ of the anisotropic powder pattern ($\omega_r > \Delta$). In the case of $\omega_r < \Delta$, the centerband at the isotropic shift is flanked by sidebands at multiples of ω_r, thus retaining information about the anisotropic coupling.

Further improvements for high-resolution in solid-state NMR spectra were achieved in the 1970s by combining MAS with ingenious manipulations of the nuclear spins, such as multiple pulse irradiation, high-power decoupling and cross polarization [8]. Cross polarization is a transfer of magnetization from ^{1}H to ^{13}C to enhance the ^{13}C signal up to the ratio of the gyromagnetic ratios $\gamma_H / \gamma_C \approx 4$ and due to the higher permissible repetition rate of experiment by the very much faster relaxation of ^{1}H compared to that of ^{13}C. For cross polarization almost a Hartmann-Hahn contact is applied where two radio-frequency fields are simultaneously in resonance with ^{13}C and ^{1}H respectively with the same frequency of the nuclei in the rotating coordinate system.

For the characterization of latexes two different kinds of experiments are applied: relaxation methods and spin-diffusion techniques combined with 2D separation experiments. In the following the different methods are introduced. Then the transfer and the application of these techniques to the characterization of latexes is described.

2.1. RELAXATION

Relaxation times play an important role in the NMR of bulk matter. The relaxation times are sensitive to the spectral density of molecular dynamics with rates in the range of the characteristic frequency: $\omega_R = \gamma B_R$ [9]. B_R denotes the strength of the relevant magnetic field B. For the longitudinal relaxation time in the laboratory frame (T_1), this is the B_0 field; for the longitudinal relaxation time in the rotating frame ($T_{1\rho}$), B_R is the B_1 field of the coil that 'locks' the magnetization.

The miscibility and the interfacial region can be characterized by the various relaxation times [10]. $T_{1\rho}$ (^{1}H) and $T_{1\rho}$ (^{13}C) measurements yield information about molecular dynamics in the range of kilohertz and about spatial relationships. The relaxation time T_1 is sensitive to the short-range spatial proximity of the interacting dipole moments; the detected motions are in the range of megahertz. The ^{13}C relaxation times T_1 (^{13}C), $T_{1\rho}$ (^{13}C) and the cross relaxation T_{CH} can be determined separately for each position of the molecule. In contrast, the $T_{1\rho}$ of the protons is a volume property averaged over a distance of about 2 nm due to the small mean distance between the abundant ^{1}H nuclei compared to the rare ^{13}C nuclei. For the measurement of $T_{1\rho}$ (^{1}H) two different measurement processes are available. In most cases the length of the contact time in a cross polarization experiment is varied. The signal enhancement by cross polarization, described above, implies that the signal intensities are influenced by the kinetics of this transfer of magnetization and are not a priori quantitative. The mechanism of coupling ^{13}C to ^1H is the dipole-dipole interaction with a $(1/r^3)^2$ dependence resulting in a slower cross polarization of quaternary ^{13}C nuclei without ^1H than ^1H-substituted ^{13}C nuclei. It has to be noted that MAS also reduced the dipole-dipole coupling. Thus, this experiment allows to determine the magnetization rate $1/T_{CH}$ at short contact times, while at longer contact times the relaxation rate $T_{1\rho}$ (^1H) is measured.

As an alternative to the $T_{1\rho}$ measurement via variable length of contact time, it is possible to carry out the ^1H spin-lock experiment, and only in the end of the ^1H spin-lock time to cross polarize with a fixed contact time as usual [11].

2.2. SPIN DIFFUSION

The spin-diffusion experiments allow the characterization of heterogeneities on the length scale of one monomer unit up to 150 nm [12]. Thus, spin diffusion is particularly suited for characterization of small domains, nanoheterogeneities, or concentration fluctuations on length scales of a few nanometers, where other methods often fail owing to limitations in resolution or contrast. Thus, spin-diffusion measurements are well established in solid-state NMR of polymers [3]. Moreover, several advanced approaches combining ^1H spin diffusion with highly selective ^{13}C detection have been introduced [5,13].

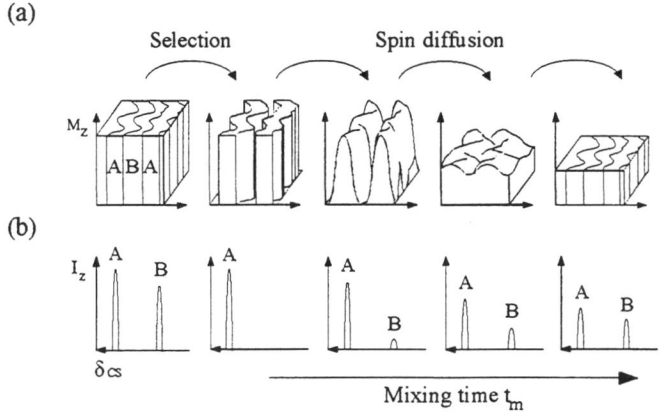

Figure 1. Schematic representation of ^1H spin diffusion in a two-phase system (A,B) with spatially constant proton density: (a) spatial distribution of ^1H magnetization; (b) NMR spectra for different mixing times t_m.

A typical ^1H spin-diffusion experiment consists of three steps: First the proton magnetization of one component is selected by a suitable filter to generate a non-equilibrium distribution of proton magnetization (Figure 1 a); second, ^1H spin diffusion, i.e. a spatial diffusion of nuclear magnetization without material transport, occurs during a mixing time t_m which is varied systematically; third, the resulting distribution of proton magnetization after the mixing time is detected in a ^1H

spectrum or after cross polarization in a ^{13}C spectrum (Figure 1 b). For short times, the mean-square distance $<x^2>$ that the magnetization moves within the time t_m is given by $<x^2> = aDt$, where D is the spin-diffusion constant of the system under study. The factor a depends on the geometry of the packing (e.g. lamellar, cylindrical, or spherical). The spin-diffusion constant is related to the strength of the dipolar coupling as reflected in the linewidth of the ^{1}H NMR spectrum, and has been calibrated for polymers of different molecular mobility by comparing spin-diffusion data with direct measurements of domain sizes by X-ray scattering and electron microscopy [14,15].

Spin diffusion requires that the different components contained in the sample can be distinguished in their NMR parameters. This is particularly straightforward if the components differ in mobility. The dipolar filter technique selects regions with differences in mobility of the components. It is based on the application of multiple pulse homonuclear decoupling. Although this multiple pulse sequence in principle is capable of averaging all dipolar couplings, it is applied here in such a way that only weak dipolar couplings are averaged and the corresponding signals are retained, whereas strong dipolar couplings lead to an irreversible decay. If there are two polymers with an ideal phase separation only two different components in mobility are found. Generally, there is a region of gradual change in structure and molecular mobility between different phases, the interface. In the case of a mobility gradient dipolar couplings also exhibit a gradual change between the values of the pure phases. Without any filter the whole particle is detected. With increasing filter strength regions with different mobilities due to different transverse relaxation times T_2 can be selected. The strength of the filter can be increased by prolongation of the delay time t_d between the pulses or by increasing the number of cycles n_{cycle} (Figure 2). After one filter cycle most of the rigid components with strong dipolar couplings are suppressed. Since the mobility of the soft components is reduced in the interface, this amount is also reduced by the dipolar filter. In reverse, the mobilized portion of the rigid components immersed in the mobile phase is still detected after applying a weak filter (e. g. $n_{cycle} = 1$). The remaining magnetization can be detected in ^{1}H spectra or after transfer to ^{13}C through cross polarization in ^{13}C CP/MAS spectra. In the ^{1}H spectra the mobile component detected as a narrow line can be quantified easily. The evaluation of the rigid component from these spectra is inaccurate because of the broad lines with line widths up to about 50 kHz.

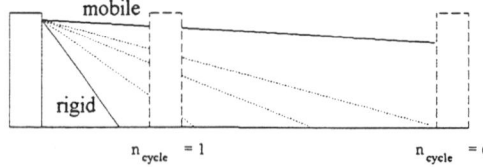

Figure 2. At weak filter strength (e.g. $n_{cycle} = 1$) only the magnetization of rigid components is suppressed; mobilized and mobile components with longer T_2 relaxation times are detected. With increasing number of filter cycles n_{cycle} (e.g. $n_{cycle} = 6$) only the highly mobile components are detected.

2.3. 2D SEPARATION EXPERIMENT

The 2D-WISE (Wideline Separation) experiment allows the combination of structural and dynamic information obtained from the isotropic chemical shift in the ^{13}C dimension and the proton line shape in the ^{1}H dimension, respectively [16]. These 2D-NMR spectra reveal changes of mobility in different chemical surroundings making use of the ^{1}H NMR line widths. The power of this technique can be demonstrated on a 50:50 wt.-% blend of polystyrene (PS) and poly(vinyl methyl ether) (PVME), cast from toluene. This blend appears homogeneous by most classical techniques. The ^{1}H wideline spectrum consists of a rather featureless superposition of components with different dipolar linewidths, which are nicely separated in the second frequency dimension and related to their ^{13}C chemical shift. About 60 K above the caloric glass transition of the blend, it is possible to detect substantial motional heterogeneities of mobility in the blend, i.e. the PVME is more mobile than the PS (Figure 3 a). By introducing a mixing time to allow for spin diffusion, the ^{1}H line shapes equilibrate after only 5 ms indicating domain sizes of different mobility in the range of 3.5 ± 1.5 nm (Figure 3 b).

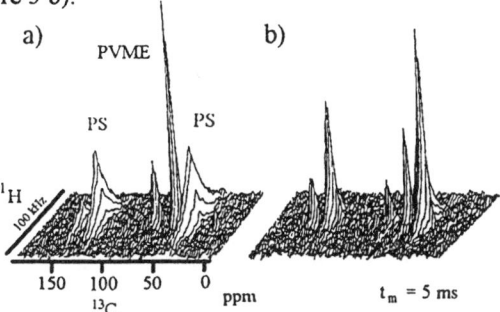

Figure 3. 2D WISE NMR spectrum at T = 320 K, (a) without spin diffusion, note different linewidths for PS and PVME; (b) with spin diffusion over a time t_m = 5 ms, note that now all lines have equal ^{1}H linewidths.

3. Characterization of Latexes

Emulsion polymerization is a well-known technique for preparing latex polymers with defined structures. Depending on the polymerization parameters and conditions, the reaction can selectively yield a variety of particles with different morphologies [17,18,19,20], e.g. core-shell, sandwich structures, hemispheres, and raspberry- or confetti-like structures.

The synthesis of core-shell latexes usually does not lead to an ideal core-shell morphology with a complete phase separation [21]. Figure 4 displays possible substructures of such core-shell latexes with different interfaces. These interfaces consist of mixed phases which are composed of the core component and the shell component. Depending on the compatibility of the two polymers and the reaction conditions, the components in the interface can be mixed on a molecular level with a continuous concentration gradient or microdomains can be formed.

The morphology of the entire particle and the interface between the two components in core-shell polymers can sensitively change the macroscopic properties of materials over a wide range. Therefore, the investigation of the internal particle morphology is an important task for further applications of core-shell systems such as paints, adhesives, coatings, or impact resistance plastics.

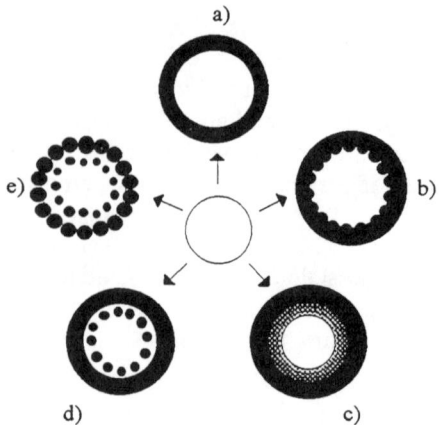

Figure 4. Morphologies deviating from the (a) ideal core shell: (b) interface with a wavy structure; (c) interface with a gradient of both components; (d) interface with microdomains; (e) microdomains in the interface and an island structure as shell.

3.1. INVESTIGATION OF MOLECULAR MISCIBILITY IN COPOLYMER LATEXES

The molecular miscibility in a latex of poly(butyl acrylate-co-acrylic acid) was investigated by $T_{1\rho}$ (^{1}H) measurements [10]. The films obtained from the dispersion shows spatial heterogeneities with one phase predominantly containing the hydrophobic butyl acrylate blocks, and other regions which consist principally of the hydrophilic acrylic acid. The evaluation of the signal amplitude as a function of the proton spin lock leads to an exponential decay and gives unitary $T_{1\rho}$ (^{1}H) values for all signals. The measured $T_{1\rho}$ (^{1}H) is between the values for the homopolymers poly(butyl acrylate) and poly(acrylic acid), and shows that acrylic acid and butyl acrylate are present alongside one another at spacing of less than 1 nm.

3.2. INFLUENCE OF SURFACE COVERAGE ON THE INTERFACE FORMATION OF PBUT/PMMA LATEXES

It was shown that the interface thickness and the average composition can be obtained using a combination of differential scanning calorimetry (DSC) and solid-state NMR relaxation experiments [22,23]. The influence of core-shell ratio and surface coverage of the seed latex by the surfactant on the interface characteristics of polybutadiene / poly(methyl methacrylate) (PBut/PMMA) latexes was examined. In contrast to DSC which can not be used successfully at low PMMA content (<30 wt.-%) as well as at high PMMA content (>75 wt.-%) solid-state NMR is a suitable method to examine the interface in the complete range of composition. It was demonstrated that the surface coverage of the PBut seed latex by the surfactant and the core-shell ratio dramatically change the relaxation time $T_{1\rho}$ (^{1}H) of PMMA, the glass transition temperature, and the interface of the core-shell latexes. The relaxation time $T_{1\rho}$ (^{1}H) are 13.8 ms and 13.1 ms for the C_α and the C_β of pure PMMA, respectively, (in the cited papers denoted as C_5 and C_6). By varying the surface coverage of the PBut seed latex at a constant PMMA content of 50 wt.-% the $T_{1\rho}$ (^{1}H) can be decreased to values lower than 10 ms indicating an interaction of the PMMA protons with neighboring PBut nuclei by direct dipole-dipole interactions, or indicating a changing of molecular motion of the PMMA chain by sufficient PBut neighbors. With increasing surface coverage from 0.17 to 0.36 the values of the relaxation times $T_{1\rho}$ (^{1}H) and the glass transition temperatures T_g decrease due to a decreasing agglomeration. A lower agglomeration at the early stages of polymerization leads to a decrease of the contact area between the two phases, PBut and PMMA. An optimum surface coverage of PBut seed latex particles with a value of 0.36 was found to introduce the greatest amount of interface during the emulsion polymerization. In this case the lowest values of $T_{1\rho}$ (^{1}H) and T_g are obtained. At higher surface coverage the formation of secondary PMMA homopolymer latexes was found. The measurements show that there is an interaction of PBut and PMMA in the interfacial region with an increase of PMMA mobility; the relaxation measurements allow to determine the amount of interfacial PMMA which is formed probably of grafted PMMA and physically interacting PMMA. The thickness of the interface was estimated to be in the range of 7 nm.

3.3. INFLUENCE OF PROCESS PARAMETERS ON THE INTERFACE REGION OF PDVB/PBUA LATEXES

The interfacial region was determined by relaxation measurements in poly(divinylbenzene) / poly(n-butyl acrylate) (PDVB/PBuA) core-shell latexes synthesized with different process parameters, including the mode of addition of the second stage monomer, the rate of addition, and the extent of conversion of the PDVB seed latex at the time of addition of the second stage monomer [24]. In $T_{1\rho}$ (^{1}H) measurements of PBuA added at two different conversion levels (75 % and 100 %) the existence of two different slopes representing two population of PBuA with regimes of different mobilities was observed. The population with the faster relaxation time probably detects PBuA crosslinked by divinylbenzene and PBuA is also grafted to the PDVB seed. The population of PBuA with a relaxation time closer to the homopolymer represents more mobile PBuA which is probably only grafted to the PDVB seed, but is not crosslinked, and has only minimal physical interaction with the PDVB seed. In the case of semicontinuous addition of PBuA, the $T_{1\rho}$ (^{1}H) of PBuA is significantly lower compared to the relaxation time of PBuA added in a batch mode relating to the higher incidence of grafting reactions in the case of semicontinuous addition. In the batch process, a crop of secondary particles of PBuA homopolymer determined by electron microscopy leads also to an increase of $T_{1\rho}$ (^{1}H). The $T_{1\rho}$ (^{1}H) values of the PBuA phase are sensitive to the composition of the core-shell latex. The PBuA component of a core-shell latex can be thought to be composed of interfacial PBuA perturbed by interaction with the PDVB and of unperturbed PBuA. The change of $T_{1\rho}$ (^{1}H) of the perturbed portion of PBuA as a function of the core-shell ratio allows to calculate the thickness of the interface region using a model proposed by McBrierty [25]. The length scale of the interface was in the range of 5-7 nm for a core-shell latex in the case of 100 % conversion of the PDVB seed latex in a semicontinuous process under starved-feed conditions.

3.4. INFLUENCE OF DIFFERENT PARAMETERS ON THE INTERFACE MORPHOLOGY AND THICKNESS OF PBUA/PMMA LATEXES

The morphology and thickness of the interface of PBuA/PMMA latexes depend on the synthesis conditions and the sizes of the particles. To elucidate the relationship between morphology, interface structure, and preparation conditions, advanced solid-state NMR techniques, involving WISE experiments [26] and spin diffusion [27], are used for the characterization of the particles [28,29,30]. The core-shell latexes presented here are composed of mobile PBuA with a low T_g of -45°C and the high T_g component PMMA (T_g = 120 °C). At the interface a contact region of the two components is built up. This leads to a partial immobilization of the soft component and a partial mobilization of the rigid component. This gradient of mobility can be characterized by the dipolar filter while the WISE experiment detects the quality of the phase separation. The PMMA mobilized by the PBuA can be quantified by filter experiments with ^{13}C detection. ^{1}H spin diffusion allows the quantification of the immobilized PBuA. This experiment allows also the determination of the interface thickness. The individual measurements are presented in the following for a low-temperature latex as an example, the results of the characterization of different latexes are summarized.

3.4.1. WISE experiment for detection of the quality of phase separation

As an example the 2D WISE spectrum of a latex synthesized at low temperature (20 °C) is shown in Figure 5.

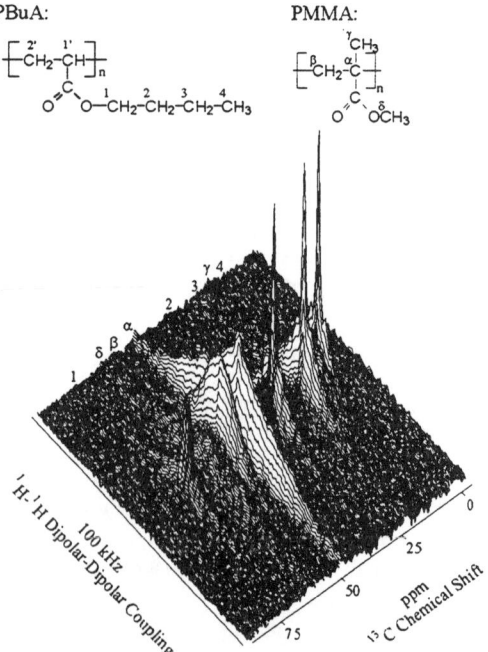

Figure 5. 2D WISE spectrum of low-temperature latex with assignment of ^{13}C chemical shifts.

As expected for PMMA with its T_g well above room temperature the slices are broadened in the ^1H dimension due to the strong dipolar couplings. The superposed narrow lines result from mobilized components in the interface. PBuA also shows a superposition of narrow and broad lines. However, the narrow components are much stronger indicating higher chain mobility.

Thus, the WISE spectrum reveals the existence of a pure PBuA phase, a pure PMMA phase, and a region where the two components are mixed. In the interface the dynamics of both the rigid and the mobile components are different from the respective dynamical behavior in the pure phases.

3.4.2. Filter experiments for the quantification of mobilized PMMA

In Figure 6 a) the effect of the dipolar filter on a low-temperature latex detected in ^{13}C spectra after magnetization transfer via cross polarization is demonstrated. The ^{13}C spectrum without any filter ($n_{cycle} = 0$) detects the whole particle. With increasing number of cycles, PMMA and immobilized PBuA are suppressed successively. The PMMA magnetization is evaluated from the signal of C_α at 45 ppm, indicated by the arrow, because there is no overlap with signals of PBuA (Figure 6b). The PMMA peaks are only partially eliminated for weak filter strengths. After one filter cycle 63.2 % of the PMMA are suppressed, while 36.8 % are still detected in the spectrum. In view of the differences in cross-polarization efficiencies this may be interpreted as indicating that at least 36.8 % of PMMA are mobilized. A total suppression is reached with $n_{cycle} = 7$ where only PBuA signals remain.

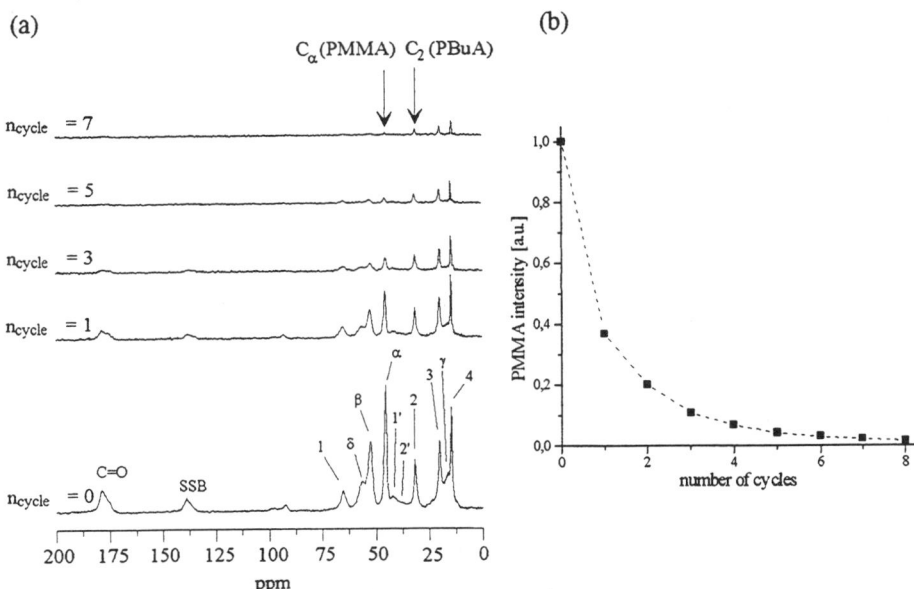

Figure 6. (a) Filter experiments on a low-temperature latex with ^{13}C detection. With increasing filter strength the rigid PMMA is suppressed (see arrow at 45 ppm for C_α). Simultaneously immobilized portions of the PBuA are suppressed as visible from the decreasing C_2 signal of PBuA; (b) The PMMA intensity of the C_α signal is plotted versus the number of filter cycles. The intensity of the first spectrum without any filter is scaled to 1. After one filter cycle 36.8 % of PMMA can still be detected as mobilized PMMA. A number of cycles $n_{cycle} = 7$ is necessary to eliminate the entire PMMA.

3.4.3. Spin-diffusion experiments for the quantification of immobilized PBuA and the interface thickness

In order to quantify the thickness of the interface, ^1H spin-diffusion experiments with ^1H detection were carried out with varying filter strengths. As an example the results for a low-temperature latex are shown in Figure 7. The number of cycles was varied between 1 and 6. The initial value is always normalized to 1.0.

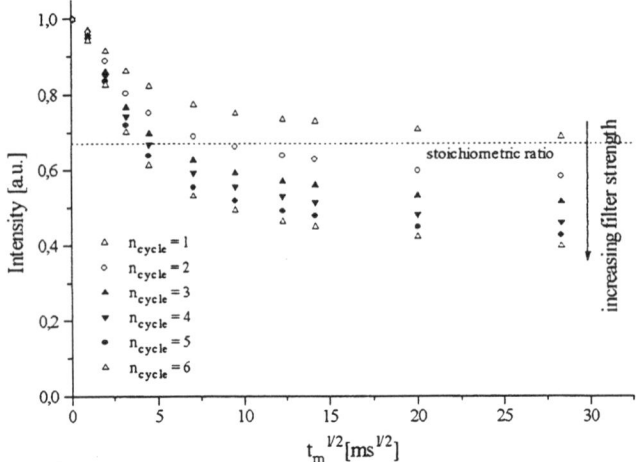

Figure 7. Spin-diffusion experiment for the low-temperature latex with ^1H detection and varying filter strength. The intensity of the PBuA signal is plotted versus the square root of the mixing time. Note the decrease of the final value with increasing filter strength and the differences in decay at small mixing times indicating the interface of the particles.

The signal decay occurs in two steps. At short mixing times the magnetization of PBuA decreases fast followed by a slower diffusion process at longer mixing times. This indicates two different structures in the particles (Figure 8). The fast decay detects the spin diffusion in the interface. This spin-diffusion process is sensitive to the thickness and the structure of the interface in the particle. The slow decay is related to the spin diffusion in the whole particle. With increasing filter strength the first decay becomes faster. The final value for long mixing times corresponds to the amount of mobile component selected in the experiment. After the weakest filter $n_{cycle} = 1$, a fraction of 69 % for the low-temperature latex presented here is observed which is nearly that expected from the proton ratio of the two polymers PBuA and PMMA (67:33). With increasing filter strength the final values in the spin-diffusion experiments are lower, reaching 40 % for $n_{cycle} = 6$ because magnetization of PBuA in the interface is also suppressed. The fast decay at short mixing times detects the thickness of the interface region. The data at short mixing times can be fitted with a one dimensional diffusion model. In the spin-diffusion time of about 100 ms the magnetization reaches only the interface occurring as a lamellar structure because of the large diameter of the particle. The simulation yields a size of about 10 nm for the interface thickness.

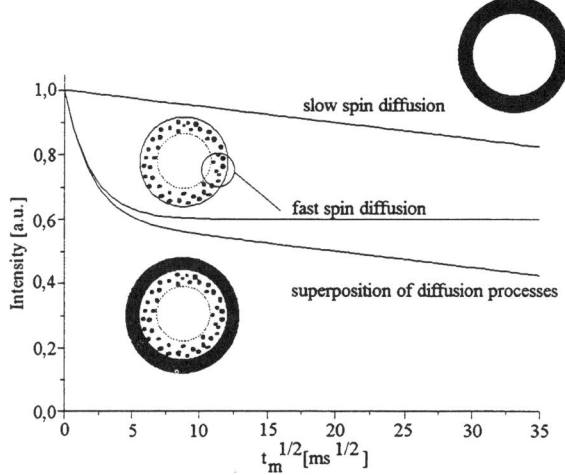

Figure 8. The entire core-shell structure leads to a slow spin diffusion; substructures in the interface allow a fast spin diffusion. Characterizing core-shell particles a superposition of diffusion processes is observed.

The combination of transmission electron microscopy and advanced solid-state NMR methods allows the characterization of the morphology and the interface structure of PBuA/PMMA polymers obtained by a two-step emulsion polymerization. The results of filter experiments, spin diffusion, and WISE allow to characterize particles with different interface morphologies shown in Figure 9. The PBuA/PMMA latexes have a core-shell morphology with an interface region between the two components which depends on the synthesis conditions. The low-temperature PBuA/PMMA latex consists of core-shell particles with an interface formed by the two components mixed on a molecular level with a continuous concentration gradient of the components. The high-temperature PBuA/PMMA latex forms small microdomains in an interface which is as thick as in the low-temperature latexes. The differences of the interfaces revealed from the spin-diffusion experiments for the high- and the low-temperature latexes can be explained with different material diffusion lengths of the oligomers during the synthesis. Three effects should be considered: First, at high temperature the oligomers can diffuse easier than at low temperature where the higher viscosity results in a reduced material diffusion coefficient. Therefore, the oligomer chains in the high-temperature latexes can diffuse towards each other and form microdomains. Second, at low temperature microphase separation of polymerized chains may be hindered due to the lower material diffusion coefficient. And third, the χ-parameter may change with temperature resulting in different compatibilities of the polymers. The interface of particles with a shell grafted on the core is only slightly different compared to the interface of the high-temperature latex. The interface thickness decreases indicating that the grafting hinders the monomer to diffuse into the core. With increasing crosslinking density in the core the thickness of interface decreases significantly. With varying ratio of core to shell polymer, it was found that the interfacial PBuA content grows until the PBuA content reaches 33 %. Up to this ratio a material diffusion into the interfacial region and a growing of the shell which is still incomplete can be detected at the same time. The further addition of PBuA does not result in an increase of the quantity of interfacial PBuA (see also [22]). For the investigated range of particle sizes between 100 and 400 nm the interface thickness does not change, only the size of the microdomains is growing with increasing particle size. For smaller particles the interface thickness decreases. It is possible to characterize even larger interface of particles with an extended interfacial region built up by a copolymer which has a composition gradient of both components.

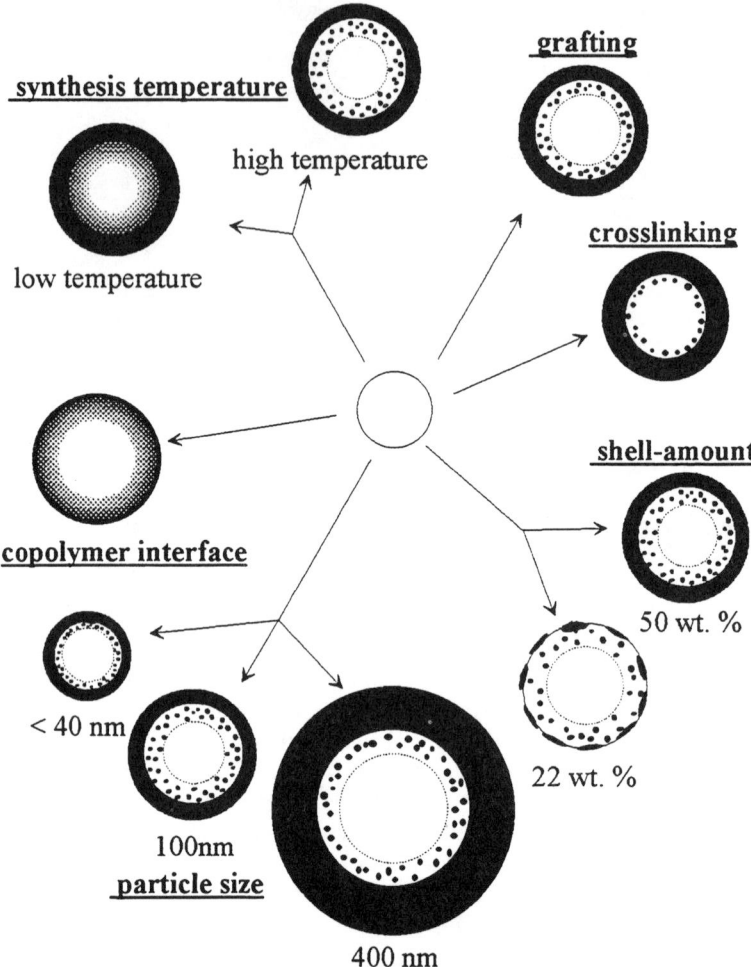

Figure 9. Overview of morphologies of PBuA/PMMA core-shell latexes due to different parameters of synthesis.

4. Conclusions

Solid-state NMR is a powerful technique for the characterization of heterogeneities in polymers and polymer blends and can be transferred to the characterization of latex systems. The techniques of solid-state NMR based on relaxation measurements and spin diffusion allow to characterize the morphology including substructures in the interface and its thickness. It is possible to detect fractions with different mobilities. A particular emphasis is put on the characterization of the interface between the two components in core-shell latexes. The measurements allow to correlate synthesis parameters to the morphologies of latexes and give information about the structure-property relationships. Recently developed experimental approaches will allow to characterize changes of the particle structure during the film formation or annealing of the particles.

5. References

1. Flory, P.J. (1953) *Principles of Polymer Chemistry*, Cornell University Press, Ithaca, NY.
2. Kroschwith, J.I. (ed.) (1990) *Concise Encyclopedia of Polymer Science and Technology*, Wiley, New York.
3. McBrierty, V.J. and Packer, K.J. (1993) *Nuclear Magnetic Resonance of Solid Polymers*, Cambridge University Press, Cambridge.
4. Ernst, E.E., Bodenhausen, G. and Wokaun, A. (1987) *Principles of Nuclear Magnetic Resonance in One and Two Dimensions*, Oxford University Press, Oxford.
5. Schmidt-Rohr, K. and Spiess, H.W. (1994) *Multidimensional Solid-State NMR and Polymers*, Academic Press, London.
6. Komoroski, R.A. (ed.) (1986) *High Resolution NMR of Synthetic Polymers in Bulk*, VCH, Deerfield Beach, FL.
7. Bovey, F.A. (1982) *Chain Structure and Conformation of Macromolecules*, Academic Press, New York.
8. Mehring, M. (1983) *High Resolution NMR in Solids*, Springer Verlag Berlin, Berlin.
9. Abragam, A. (1961) *The Principles of Nuclear Magnetism*, Oxford University Press, London.
10. Voelkel, R. (1988) High-Resolution Solid-State ^{13}C-NMR Spectroscopy of Polymers, *Angew. Chem. Int. Ed. Engl.* 27, 1468-1483.
11. Aulja, R.S., Harris, R.K., Packer, K.J., Prarmeswaran, M., Say, B.J., Bunn, A., and Cudby, M.E.A. (1982) Discriminatory Experiments in High-Resolution ^{13}C NMR of Solid Polymers, *Polymer Bulletin*, 8, 253-259.
12. VanderHart, D.L. (1990) Proton Spin Diffusion as a Tool for Characterizing Polymer Blends, *Makromol. Chem., Macromol. Symp.* 34, 124-159.
13. Schmidt-Rohr, K., Clauss, J., Blümich, B., and Spiess, H.W. (1990) Miscibility of Polymer Blends Investigated by ^{1}H Spin Diffusion and ^{13}C NMR Detection, *Magn. Reson. Chem.* 28, 3-9.
14. Clauss, J. Schmidt-Rohr, K., and Spiess, H.W. (1993) Determination of domain sizes in heterogeneous polymers by solid state NMR, *Acta Polym.* 44, 1-17.
15. Spiegel, S., Schmidt-Rohr, K., and Spiess, H.W. (1993) ^{1}H spin diffusion coefficients of highly mobile polymers, *Polymer* 34, 4566-4569.
16. Schmidt-Rohr, K., Clauss, J., and Spiess, H.W. (1992) Correlation of Structure, Mobility, and Morphological Information in Heterogeneous Polymer Materials by Two-Dimensional Wideline-Separation NMR Spectrsocopy, *Macromolecules* 25, 3273-3277.
17. Daniel, J.C. (1985) Latex de particules structurées, *Macromol. Chem., Suppl.* 10/11, 359-390.
18. Okubo, M., Katsuta, Y., and Matsumoto, T.J. (1980) Rupture of anomalous composite particles prepared by seeded emulsion polymerization in aging period, *Polym. Sci., Polym. Lett. Ed.* 18, 481-4486.
19. Okubo, M. (1990) Control of Particle Morphology in Emulsion Polymerization, *Makromol. Chem., Macromol. Symp.* 35/36, 307-325.
20. Okubo, M., Kanaida, K., and Matsumoto, T. (1987) Production of anomalously shaped carboxylated polymer particles by seeded emulsion polymerization, *Colloid Polym. Sci.* 265, 876-881.
21. Shen, S., El-Aasser, M.S., Dimonie, V.L., Vanderhoff, J.W., and Sudol, E.D. (1991) Preparation and Morphological Characterization of Microscopic Composite Particles, *J. Polym. Sci., Polym. Chem. Ed.* 29, 857-867.
22. Tembou Nzudie, D., Delmotte, L., and Riess, G. (1991) Magic-angle carbon-13 nuclear magnetic resonance analysis of polybutadiene-poly(methyl methacrylate) core shell latex, *Makromol. Chem., Rapid Commun.* 12, 254-458.
23. Tembou Nzudie, D., Delmotte, L., and Riess, G. (1994) Polybutadiene-poly(methyl methacrylate) core-shell latexes studied by high-resolution solid-state ^{13}C NMR and DSC: Influence of the surface coverage of

the polybutadien seed latex and the latex composition on on the interphase formation, *Macromol. Chem. Phys.* **195**, 2723-2737.

24. Nelliappan, V., El-Aasser, M.S., Klein, A., Daniels, E.S., and Roberts, J.E. (1995) Characterization of the Interphase in Poly(divinylbenzene) / Poly(n-butyl acrylate) Core/Shell Latexes by High-Resolution Solid-State NMR Relaxation Studies, *J. Appl. Polym. Sci.* **58**, 323-330.

25. McBrierty, V.J. and Douglass, D.C. (1981), *J. Polym. Sci., Macromol. Rev.* **16**, 295-366.

26. Schmidt-Rohr, K., Clauss, J., and Spiess, H.W. (1992) Correlation of Structure, Mobility, and Morphological Information in Heterogeneous Polymer Materials by 2D Wideline-Separation NMR Spectroscopy, *Macromolecules*, **25**, 3273-3277.

27. Cai, W.Z., Schmidt-Rohr, K., Egger, N., Gerharz, B., and Spiess, H.W. (1993) A solid-state NMR study of microphase structure and segmental dynamics of poly(styrene-b-methylphenylsiloxane) diblock copolymers, *Polymer*, **34**, 267-276.

28. Landfester, K., Boeffel, C., Lambla, M., and Spiess, H.W. (1995) Synthesis and characterization of core-shell latexes with microscopic and solid-state NMR methods, *Macromol. Symp.* **92**, 109-116.

29. Spiegel, S., Landfester, K., Lieser, G., Boeffel, C., Spiess, H.W., and Eidam, N. (1995) Microheterogeneities of core-shell latexes probed by ^{1}H spin diffusion and transmission electron microscopy, *Macromol. Chem. Phys.* **196**, 985-993.

30. Landfester, K., Boeffel, C., Lambla, M., and Spiess, H.W. (1996) Characterization of interfaces in core-shell polymers by advanced solid-state NMR methods, *Macromolecules*, in press.

SCATTERING TECHNIQUES-FUNDAMENTALS

R.H. OTTEWILL
School of Chemistry, University of Bristol
Cantock's Close, Bristol BS8 1TS
U.K.

1. Introduction

There are three types of radiation which are commonly used for scattering experiments. These are shown in Table 1.

TABLE 1. Types of radiation

Radiation	Source	Wavelength/Å
Light	Mercury Arc	ca. 4000-6500
	Laser	" "
X-rays	Laboratory Generators	0.4-1.5
	Synchroton Radiation	1->5
Neutrons	Thermal	ca. 1.0-5.0
	Cold	5.0-20.0

1.1. LIGHT SCATTERING

The theory of light scattering was first investigated by Lord Rayleigh in 1871 [1] and then subsequently developed in the early part of the 20th Century by Mie [2] and Debye[3]. Experimental investigations were greatly enhanced by the invention of the photomultiplier in the 1940's and its subsequent development in recent years to rapid response photon detectors, ca 1 nsec. Together with development of coherent light sources, lasers, and computer based correlators it has become a powerful technique [4].

The scattering of light depends on the electronic polarisability of the atoms composing the molecule/particle and hence is dependent on the refractive indices of the scattering object and the medium.

1.2. X-RAY SCATTERING

X-ray crystallography was quite well developed by the 1930's [5] and has since become a very sophisticated subject. However, although the theory of small angle X-ray scattering was clearly presented in a classic book by Guinier and Fournet [6] in 1955 the experimental aspects only developed in a few specialised centres. At the present time, however, small laboratory machines are becoming more readily available and there are a number of special centres, including the ESRF (European Synchroton Radiation Facility) at Grenoble [7].

J. M. Asua (ed.), Polymeric Dispersions: Principles and Applications, 217–228.
© 1997 *Kluwer Academic Publishers. Printed in the Netherlands.*

X-ray scattering depends on the electron density of the atoms in the scattering object and hence scales with atomic number. Hydrogen is a very weak X-ray scattering element and platinum a very strong one.

1.3. SMALL ANGLE NEUTRON SCATTERING

The neutron was discovered by Chadwick in 1932 [8] and although it was shown soon after this that neutrons could be diffracted it was not until the 1970's with the building of high flux reactors that neutron scattering facilities became widely available and the subject rapidly developed.

Neutrons are scattered by the nuclei of atoms and hence the scattering ability is isotope specific to the extent that different isotopes of the same element, e.g. H^1 and H^2 (deuterium), can scatter very differently. This makes isotopic labelling a valuable experimental tool.

The neutron can be considered as a particle of mass m travelling at a velocity v or as waveform having a wavelength λ, since by the de Broglie relationship,

$$\lambda = h_p/mv \qquad (1)$$

where h_p = Planck's constant.

2. Angular Scattering

2.1. THE SCATTERING ANGLE, θ

θ is defined as the angle of the scattering measurement with respect to the direction of the incident beam as shown in Figure 1.

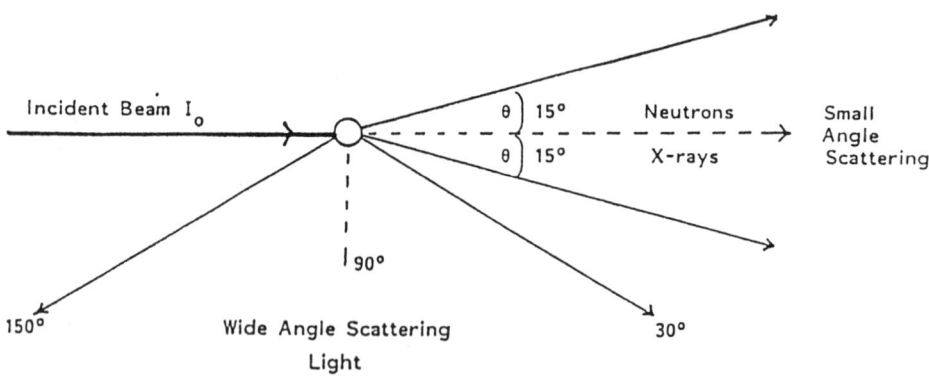

Figure 1. Usual angular regions of wide angle and small angle scattering

Both X-ray and neutron scattering measurements are made at small angles whereas light scattering with a rotatable detector is usually carried out at angles from ca 30° to 150°.

2.2. THE SCATTERING VECTOR, Q

For elastic scattering, Q, is defined as,

$$Q = 4\pi n_o \sin(\theta/2)/\lambda_o \qquad (2)$$

for light scattering, with n_o = the refractive index of the medium and λ_o = the wavelength of light in vacuo; in the medium $\lambda = \lambda_o/n_o$.
    For neutrons,

$$Q = 4\pi \sin(\theta/2)/\lambda \qquad (3)$$

The dimensions of Q are $(Length)^{-1}$. Thus measurements are related to behaviour in reciprocal space.

2.3. RANGE OF Q VALUES

Table 2 indicates the range of Q space which can be probed by the various techniques, and since the spatial distance probed can be regarded as of the order of $2\pi/Q$ this is also included.

TABLE 2. Spatial distances probed by scattering

Method	Q, Range/ Å^{-1}	Spatial Distance/ Å
Light Scattering	0.001-0.002	ca. 6300-3100
Neutron Scattering	0.001-0.25	ca. 6300- 25
X-ray Scattering	0.005-0.25	ca. 1250- 25

3. Intensity

All examinations of scattering have to be related to a fundamental measurement, that of intensity, which has to be converted by calibration into absolute units, for example, in the case of neutrons, to the number of neutrons per unit solid angle per unit of time per unit incident intensity. Thus if this quantity is called I(Q), i.e. intensity, we obtain a basic equation in the form,

$$I(Q) = AV_p^2 N_p[\quad]P(Q) \qquad (4)$$

for dilute noninteracting systems, where A = an instrument calibration factor, V_p = the volume of the scattering particle and N_p = the number of particles per unit volume; it should be noted that for unit volume, $N_P V_P = \phi$, the volume fraction of the system.
    The quantity to be inserted in the square brackets is a scattering parameter which is expressed in terms of:
        Refractive Index, n, for light scattering
        Electron Density, ρ_c, for X-rays
        Scattering Length Density, ρ_{sc}, for neutrons.

The term P(Q) is a particle shape factor and, as an example, for spheres of radius, R, is given by [6],

$$P(Q) = \left\{ \frac{3(\sin QR - QR \sin QR)}{Q^3 R^3} \right\}^2 \tag{5}$$

The formulae for P(Q) are the same whether used in light scattering, X-ray scattering or neutron scattering.

4. Light Scattering

Three different regions of light scattering are usually considered. Briefly these are:

4.1. RAYLEIGH SCATTERING

This occurs for the condition that $R << \lambda_o$; frequently it is taken as $R < \lambda_o/20$. For this situation P(Q) is unity and eq. (4) can be written as

$$I(Q)_R = A N_p V_p^2 \frac{8\pi^4 n_o^4}{\lambda_o^4} \left\{ \frac{n^2 - n_o^2}{n^2 + 2n_o^2} \right\}^2 (1 + \cos^2 \theta) \tag{6}$$

if unpolarised light is used as the incident radiation, where n = the refractive index of the particle and n_o that of the medium. For incident length with the electric vector polarised perpendicular to the scattering plane i.e. for laser radiation the scattered intensity is given by [10,11]

$$I(Q) = A N_p V_p^2 \frac{16\pi^4 n_o^4}{\lambda_o^4} \left\{ \frac{n^2 - n_o^2}{n^2 + 2n_o^2} \right\}^2 \tag{7}$$

and is independent of scattering angle, proportional to the 6th power of the radius and inversely dependent on the 4th power of the wavelength.

4.2. RAYLEIGH-GANS-DEBYE SCATTERING

This occurs for the condition that $(n-n_o)R/\lambda << 1$, essentially meaning that $(n-n_o)$ must be small so that $n \approx n_o$. Phase shifts occur between the electric fields scattered from different parts of a particle and the scattered intensity is relative to that of Rayleigh particle scattering, so that [12]

$$I(Q)_{RGD} = I_R P(Q) \tag{8}$$

with I_R = the intensity of Rayleigh scattering.

4.3 MIE SCATTERING

Mie theory provides a complete solution for spheres of all sizes and refractive index. The solutions although complex are precise in the region when $(n-n_o)R/\lambda > 1$. This approach has been discussed by Rowell at a previous NATO Institute [12].

5. Small Angle Neutron Scattering

The basis of small angle neutron scattering and small angle X-ray scattering have much in common and hence this article will primarily deal with the former. Moreover, since the phase shifts are usually small, and inherently the refractive indices are close to unity, then a number of aspects are similar to those of the Rayleigh-Gans-Debye region of light scattering.

5.1. SCATTERING LENGTH DENSITY

The coherent scattering length density is an important quantity for a molecule which is defined by the equation,

$$\rho_{sc} = \Sigma\, b_i\, /\, V \qquad (9)$$

where b_i is the coherent scattering length of the ith isotope of which the molecule is composed. The values for isotopes are tabulated, for example, in Bacon [13]; they are independent of λ and θ.

Some values of ρ_{sc} are tabulated in Table 3. The difference in values between H_2O and D_2O and between hydrogenated and deuterated molecules should be noted.

TABLE 3. Neutron scattering length densities for various molecules

Material	Formula	Coherent Neutron Scattering Length, $\rho_{sc}/10^{10}$ cm^{-2}
Water	H_2O	-0.56
Deuterium Oxide	D_2O	6.35
h_{14}-Hexane	C_6H_{14}	-0.58
h_{26}-Dodecane	$C_{12}H_{26}$	-0.46
d_{26}-Dodecane	$C_{12}D_{26}$	6.43
Polystyrene	$[C_8H_8]_n$	1.42
d-Polystyrene	$[C_8D_8]_n$	6.47
Polyacrylonitrile	$[C_3H_3N]_n$	2.28

6. The Neutron Scattering Equation

For a monodisperse dispersion of non-interacting homogeneous spherical particles eq (4) can be rewritten for small angle neutron scattering in the form,

$$I(Q) = A(\rho_p - \rho_m)^2 N_p V_p^2 P(Q) \qquad (10)$$

where ρ_p and ρ_m are respectively the coherent neutron scattering lengths of the particles and the medium. Figure 2 illustrates with experimental data the form of ln P(Q) for dispersions of polystyrene particles of increasing sizes [14,15].

An interesting point is that data are obtained over a much wider Q range than with light scattering and for small particles there is a much greater range of intensity variation.

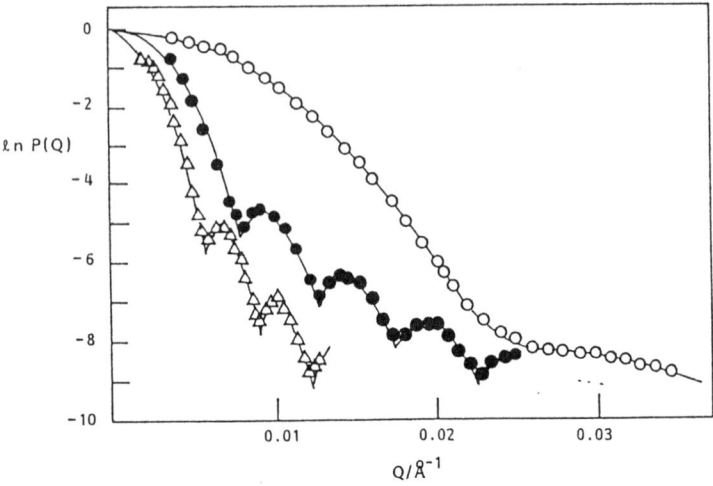

Figure 2. In P(Q) against Q for polystyrene latex particles of various diameter.
○ diameter 346 Å; ● diameter 1320 Å; △ 2020 Å.

6.1. SCATTERING AT Q = O

Although the intensity of scattering at zero θ, I(0), cannot be measured it can be obtained by extrapolation. Moreover, since P(Q) = 1.0 at θ = 0 then,

$$I(0) = A(\rho_p - \rho_m)^2 N_p V_p^2 \qquad (11)$$

and for the condition $\rho_p = \rho_m$, I(0) becomes zero. Thus by measuring the intensity of scattering from the particles as a function of ρ_m, for example by using H_2O-D_2O mixtures, then ρ_p can be obtained experimentally. This is illustrated in Figure 3.

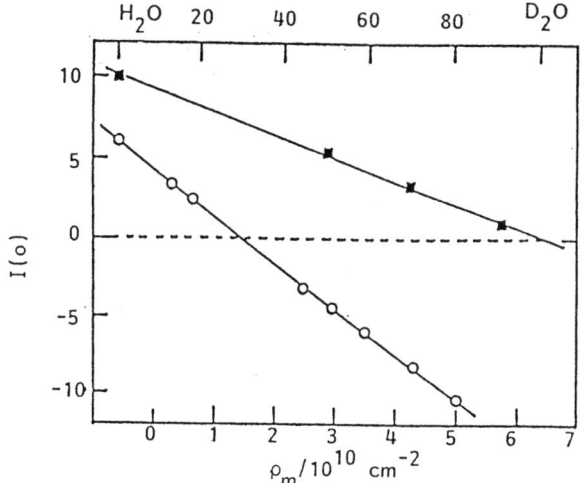

Figure 3. $\sqrt{I(0)}$ against scattering length density of dispersion medium, ρ_m
●, Polystyrene latex, ■ polydeuterostyrene latex

6.2. SCATTERING AT SMALL QR

For values of QR<1.0 the function P(Q) can be expanded to give,

$$I(Q) = I(0)\exp(-Q^2 R_g^2 / 3) \qquad (12)$$

or in alternative form

$$\ln I(Q) = \ln I(0) - Q^2 R_g^2 / 3 \qquad (13)$$

and hence from a plot of ln I(Q) against Q^2, which should be linear, a value can be obtained for the radius of gyration, R_g. Expressed in this form as a limiting law, eq. (13) is usually known as the Guinier equation.

6.3. POLYDISPERSITY

Even so-called monodisperse latices are rarely, if ever, completely monodisperse and so a correction has to be applied for this. A polydispersity function also has to be included in eq. (10) to allow for the fact that there is a distribution of particle sizes. Since many colloidal dispersions are found to have a logarithmic distribution it is convenient to use as a functional form [16].

$$p(R) = \frac{\exp\left\{-\left[\ln R - \ln R_m\right]^2 / 2\sigma_o^2\right\}}{(2\pi)^{1/2}\sigma_o R_m \exp(\sigma_o^2)} \qquad (14)$$

where p(R) is the fraction of particles in a particular size range, R_m the modal radius and σ_o a parameter giving a measure of the width of the distribution; for a narrow distribution the standard deviation, σ, is given by $\sigma = \sigma_o R_m$. In this form it was found convenient in combination with eq. (3) for computer programming. Fits can be made using σ_o and R_m as variables; these are shown as the full lines in Figure 2.

A comparison of a particle size distribution obtained by transmission electron microscopy and one derived from small angle neutron scattering is shown in Figure 4.

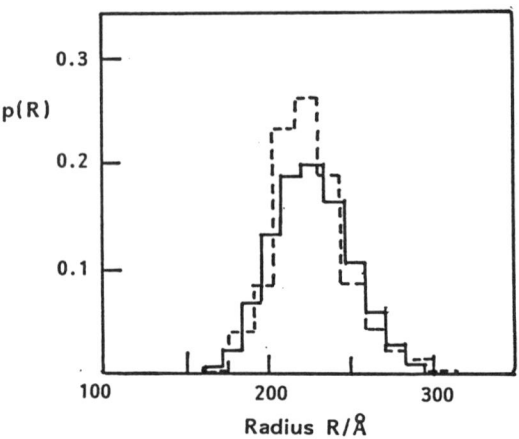

Figure 4. Particle size distribution. _____ , small angle neutron scattering; - - - - , electron microscopy.

7. Particle Morphology

A useful function to obtain information about the internal structure of a particle is the radial density distribution, P(r). This can be obtained by Fourier transformation of good scattering data as a function of Q, namely,

$$P(r) = \int_0^\infty QrI(Q)\sin(Qr)dQ \qquad (15)$$

For a homogeneous sphere P(r) is given by the expression [6],

$$P(r) = \text{Constant} \left[r^2/R - 3r^3/4R^2 + r^3/16R^4 \right] \qquad (16)$$

Figure 5 shows a comparison between a curve calculated using this equation for R = 165 è and that obtained by the transformation of the scattering data. This indicates that this polystyrene particle, diameter 160 Å, prepared by emulsion polymerisation and cleaned by mixed-bed ion-exchange resin was essentially homogeneous.

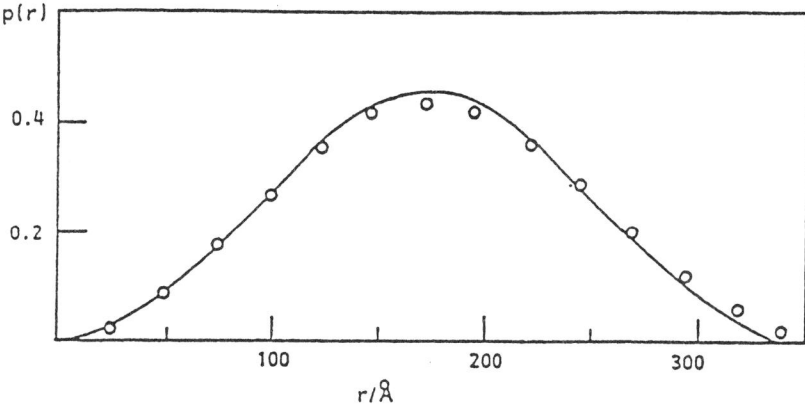

Figure 5. P(r) against r for a polystyrene particle diameter 320 Å.

8. Dynamic Light Scattering

8.1 MONODISPERSE SYSTEMS

The invention of the laser in 1960 by Maiman [17] produced a coherent light source with a fine beam which was very suitable for light scattering. This was followed by the development of electronic computers which could access electronic signals rapidly and store large amounts of data for subsequent processing. Also development of photon detectors reduced the response time down to the region of 0.1 µsec to 10 nsec.

The particles in a dispersion can be regarded as forming a random three-dimensional diffracting array which gives rise to a "speckle" pattern consisting of small bright spots, where constructive interference occurs between light scattered form the individual particles and dark areas, where destructive interference occurs. Since colloidal particles are in constant Brownian motion as the particles move the phase relationship changes and the pattern also changes through a series of random configurations. Thus the temporal fluctuations in scattered intensity provide information on the particle motions.

The dynamic light scattering optical arrangement is in many ways similar to that used for conventional time-average light scattering. However, it does need a coherent light source and the observed volume in the sample needs to be small; this is achieved by using fine (30 µm or so) adjustable apertures so that essentially the intensity of one fluctuating speckle is measured as a temporal signal. Hence, in principle the photon counting device measures the fluctuation in intensity of one speckle at short time intervals, e.g. of the order of 1 µsec. The average of the fluctuations taken over a long time interval gives the conventional time average intensity, Figure 6.

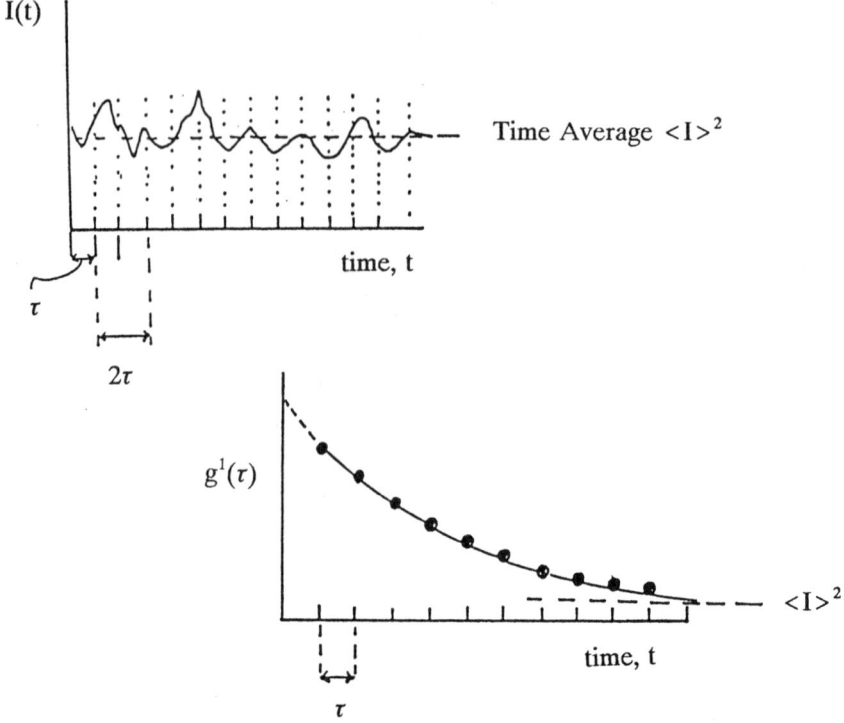

Figure 6. Intensity fluctuation against time and $g^1(\tau)$ against time

However, by taking the intensity in time steps of 1,2,3 etc intervals, τ, a normalised time correlation function is obtained which is defined as:

$$g^2(Q,\tau) = \frac{< I(0)I(\tau) >}{< I >^2} = 1 + C\left[g^1(Q,\tau)\right]^2 \tag{17}$$

where the angular brackets denote a time-averaged result. In eq. (17) C is a constant of order unity and $g^1(Q,\tau)$ is the temporal correlation function of the scattered light field which is given by,

$$g^1(Q,\tau) = \exp(-DQ^2\tau) \tag{18}$$

where D is the diffusion coefficient and τ is the correlation delay time. Since D is given by,

$$D = kT / 6\pi\eta R_H \tag{19}$$

where R_H is the "hydrodynamic radius" of the particle and η the viscosity of the medium this provides a means of estimating a hydrodynamic particle size.

9.2. POLYDISPERSITY

Polydispersity results in the correlation function becoming an average of many exponential decays and in a Q dependence of the correlation function for particles large enough to have maxima and minima in their particle form factors in the accessible Q range [18,19]. For polydisperse systems, $g^1(Q,\tau)$ consits of the sum of a distribution of exponential terms. This gives a mathematical problem in the analysis to give a particle size distribution since the sum of exponential terms is still essentially a single exponential. However, for reasonably narrow size distributions, the method of cumulants is frequently used to obtain an average particle size [20,21]. In this procedure the term $g^1(Q,\tau)$ is expanded and written in the form,

$$\ln g^1(Q,\tau) = -AQ^2\tau + BQ^4\tau^2/2 + \ldots\ldots\ldots \tag{20}$$

and the experimental data fitted by the method of least squares to obtain the coefficients A and B of the quadratic.

A then gives the average hydrodynamic radius, $<R_H>$, in the form $\overline{R}_H^6 / \overline{R}_H^5$ from,

$$A = kT / 6\pi\eta(\overline{R}_H^6 / \overline{R}_H^5) \tag{21}$$

The ratio A/B^2 has the form,

$$A/B^2 = (\overline{R}_H^6\overline{R}_H^5)/(\overline{R}_H^5)^2 - 1.0 \tag{22}$$

which provides a measure of polydispersity [21].

9.3. MICROELECTROPHORESIS

Dynamic light scattering since it measures movement can also be used to measure the velocity of particles under an applied electric field and provides a method of obtaining the electrophoretic mobility and hence the zeta-potential of polymer colloid particles. A number of commercial instruments are available which use this approach.

10. References

1. Rayleigh, Lord (1871) Phil. Mag. **41**, 107-120, 274-279 and 447-454.
2. Mie, G. (1908) Ann. Physik, **25**, 377.
3. Debye, P. (1915) Ann. Physik, **46**, 809.
4. Berne, B.J. and Pecora, R. (1975) *Dynamic Light Scattering*, John Wiley and Sons Inc., New York.
5. Bragg, W.H. and Bragg, W.L. (1933) The Crystalline State, Bell, London.
6. Guinier, A. and Fournet, G. (1955). *Small Angle Scattering of X-rays*, John Wiley and Sons Inc., New York.
7. Daresbury Laboratory Report, 1993/94.
8. Chadwick, G. (1932) Nature, **129**, 132; Proc. Roy. Soc. **A136**, 692-708.
9. Bohren, C.F. and Huffman, D.R. (1983) *Absorption and Scattering of Light by Small Particles*, John Wiley and Sons Inc., pp. 130-135.
10. Pusey, P.N. (1982) in J.W. Goodwin *Colloidal Dispersions*, Royal Society of Chemistry, London, pp. 128-142.
11. Kerker, M. (1969) *The Scattering of Light and Other Electromagnetic Radiation*, Academic Press, New York.
12. Rowell, R.L. (1990) in F. Candau and R.H. Ottewill (eds) *Scientific Methods for the Study of Polymer Colloids and their Applications*, Kluwer Academic Publishers, Dordrecht, pp. 187-208.

228

13. Bacon, G.E. (1977) *Neutron Scattering in Chemistry*, Butterwoths, London.
14. Ottewill, R.H. (1990) in F. Candau and R.H. Ottewill (eds) *Scientific Methods for the Study of Polymer Colloids and their Applications*, Kluwer Academic Publishers, Dordrecht, pp. 349-372.
15. Ottewill, R.H. (1991) J. Appl. Cryst., **24**, 436-443.
16. Espenscheid, W.F., Kerker, M. and Matijevic, E. (1964) J. Phys. Chem., **68**, 3093-3097.
17. Maiman, T.H. (1960) Nature, 33.
18. Pusey, P.N. and van Megen, W. (1984) J. Chem. Phys. **80**, 3513-3520.
19. Candau, S.J. (1990) in F. Candau and R.H. Ottewill (eds) *Scientific Methods for the Study of Polymer Colloids and their Applications*, Kluwer Academic Publishers, Dordrecht, pp. 329-347.
20. Koppel, D.E. (1972) J. Chem. Phys. **57**, 4814-4820.
21. Pusey, P.N., Koppel, D.E., Schaefer, D.W., Camerini-Otero, R.D. and Koenig, S.H. (1974) Biochemistry, **13**, 952.

APPLICATION OF SCATTERING TECHNIQUES TO POLYMER COLLOID DISPERSIONS

R.H. OTTEWILL
School of Chemistry, University of Bristol
Cantock's Close, Bristol BS8 1TS
U.K.

1. Introduction

During the 1960's interest in Polymer Colloids developed very rapidly both for their intrinsic commercial value and also, because of their spherical shape and very narrow range of particle sizes, as model colloidal dispersions for fundamental investigations. During this period also there were substantial developments in the technology of scattering equipment and from the early 1970's onward small angle neutron scattering and photon correlation spectroscopy gave a real impetus to the use of scattering methods. The first structure factor for a strongly interacting polymer colloid dispersion was published in 1975 [1,2] and this indicated the way in which concentrated colloidal dispersions could be examined both experimentally and theoretically.

2. Concentrated Charge-Stabilised Latices

2.1. THE STRUCTURE FACTOR

In the last chapter examples were shown of the spectra obtained by small angle neutron scattering from dilute dispersions of polymer colloids. In such systems the number concentration is such that only a few particles interact per unit of time, in binary collisions, as a consequence of Brownian motion. In concentrated dispersions, however, the particles are in constant interaction and an ordering occurs which is dependent on the number concentration and the strength of the repulsive interactions. The spatial correlations produced by the interaction lead to interparticle interference effects which can be clearly recognised in the spectra [3] as shown in Figure 1. This means that the intensity of scattering as shown in reference [4], has to include an additional term which is called the structure factor, $S(Q)$, to give

$$I(Q) = AN_pV_p^2(\rho_p - \rho_m)^2 P(Q)S(Q) \qquad (1)$$

or rewriting in terms of volume fraction

$$I(Q)_{con} = A\phi_{con}(\rho_p - \rho_m)^2 P(Q)S(Q) \qquad (2)$$

229

J. M. Asua (ed.), Polymeric Dispersions: Principles and Applications, 229–242.
© 1997 *Kluwer Academic Publishers. Printed in the Netherlands.*

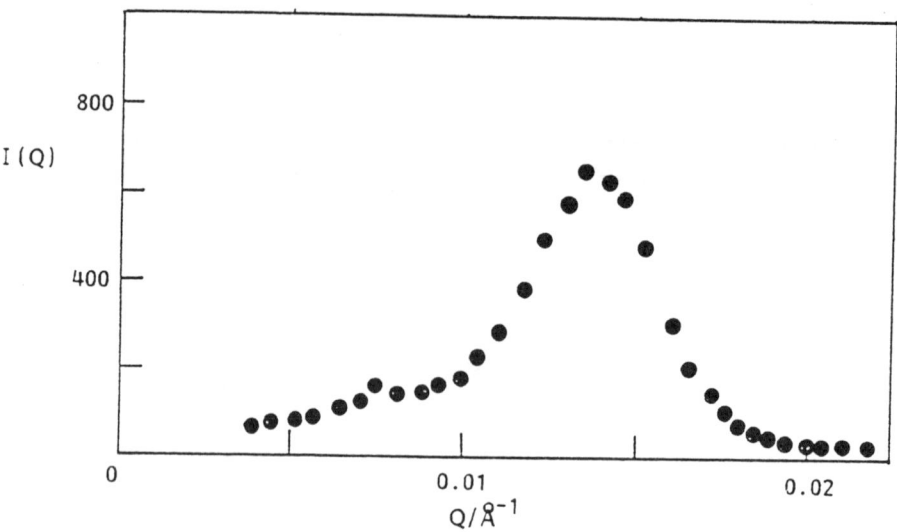

Figure 1. Intensity against Q. Polystyrene latex, volume fraction, 0.13 in 10^{-4} mol dm^{-3} NaCl

where S(Q) is given by,

$$S(Q) = 1 + \left(4\pi N_p / Q\right)\int_0^\infty [g(r) - 1]\sin Qr \, dr \tag{3}$$

and as in [4] r = the centre to centre interparticle separation and g(r) the pair correlation function. Since, for the same type of system examined in a dilute noninteracting mode, we have

$$I(Q)_{dil} = A\phi_{dil}(\rho_p - \rho_m)^2 P(Q) \tag{4}$$

hence

$$S(Q) = \frac{I(Q)_{con}\phi_{dil}}{I(Q)_{dil}\phi_{con}} \tag{5}$$

and S(Q) becomes experimentally accessible (3). The excellent sphericity of polymer colloid particles and their narrow distribution of particle sizes makes them excellent samples for this type of work.

Several curves obtained by this technique are shown in Figure 2 which illustrate the effect of volume fraction at constant salt concentration and the effect of varying salt concentration at constant volume fraction. These curves clearly indicate that a structure is built up which depends both on volume fraction (ϕ) and ionic strength. This is indicated by the increase in magnitude of the first peak in S(Q) and its decrease at low Q values as ϕ increases at constant ionic strength;

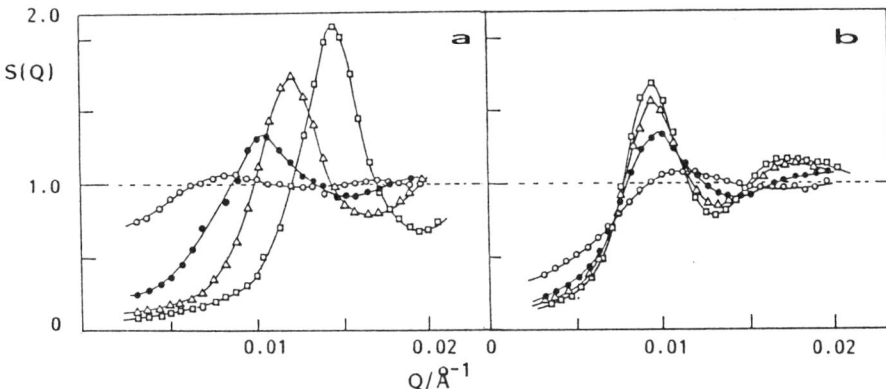

Figure 2. S(Q) against Q for polystyrene latices a) Various volume fractions at 10^{-3} mol dm^{-3} NaCl:-$^\circ$, 0.01; l, ● 0.04; Δ, 0.08; ❑ 0.13, b) various NaCl concentrations (mol dm^{-3}), at a volume fraction = 0.04; ❑, ion-exchanged; Δ, 10^{-4}; ● 10^{-3}; O, 5x10^{-3}.

the first peak also moves to higher Q as ϕ increases. Figure 2b indicates that with increase in ionic strength at constant ϕ a broadening of the peak occurs, which indicates a decrease in strength of the interparticle interaction and a greater motion of the particles.

2.2. CORRELATION WITH V_R

In a previous chapter in this volume [4] the electrostatic repulsive pair potential was written in the form,

$$V_R = 4\pi\varepsilon_r\varepsilon_o R^2\psi_s^2 \exp(2\kappa R)\exp(-\kappa r)/r \qquad (6)$$

with r = h+2R. It was shown by Hayter and Penfold [5] and by Hansen and Hayter [6] that by using this potential form in the Ornstein-Zernicke equation then the form of S(Q) against Q could be modelled. This meant that by using their approach in a fitting program, it was possible to compare the model with the experimental data. Since R was known from experiments with dilute systems and κ was known from the amount of salt added, the only variable used for the fits was the surface potential, ψ_s. A comparison between the experimental points and the theoretical fits [7] is given in Figure 3.

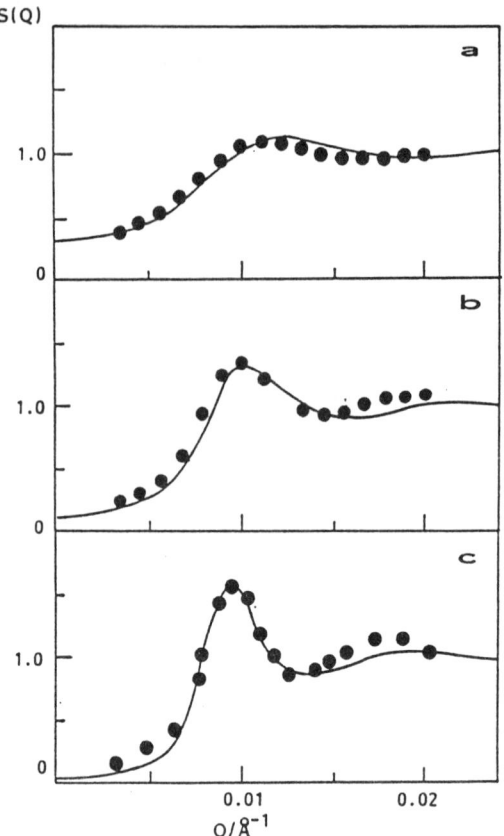

Figure 3. S(Q) against Q for polystyrene latex, volume fraction = 0.04 in various NaCl concentrations (mol dm^{-3}) a) 5 x 10^{-3}, |ψ_s| = 47 mV; b) 10^{-3}, |ψ_s| = 49 mV; 10^{-4}, |ψ_s| = 51 mV; ●, experimental points; _____ , fitted curve.

2.3. A COMPARISON BETWEEN ζ- POTENTIAL AND |ψ_s|

The values of ζ- potential were obtained using electrophoresis on the same particles as used for small angle scattering [7]. The results are compared in Table 1.

TABLE 1. A Comparison of ζ- Potential and |ψ_s|

NaCl/mol dm^{-3}	ζ- Potential/mV	|ψ_s|/mV
1 x 10^{-4}	-62 ± 5	51
3 x 10^{-4}	-54 ± 5	
1 x 10^{-3}	-54 ± 5	49
3 x 10^{-3}	-62 ± 5	
5 x 10^{-3}		47

Within the experimental error the agreement is not unreasonable although the ζ-potentials appear to be consistently slightly lower than |ψ_s|. An important point from these results is that it appears to be the diffuse electrical double layer which determines the interaction potential.

2.4. THE PAIR CORRELATION FUNCTION g(r)

Fourier transformation from Q-space (reciprocal space) to real space gives g(r) in the form,

$$g(r) = 1 + \frac{1}{2\pi^2 r N_p} \int_o^\infty [S(Q) - 1]\, Q \sin Qr\, dQ \qquad (7)$$

The parameter g(r) is an important one since it is the probability of finding the centre of another particle at a distance r, in real space, from the centre of the reference particle. It can also be written

$$g(r) = N_p(r) / N_p \qquad (8)$$

where $N_p(r)$ is the radial distribution of particle density, a microscopic quantity; $4\pi r^2 N_p(r)$ is the radial distribution function.

Figure 4 shows the results [7] obtained for polymer colloid particles of diameter 310 Å in 10^{-4} mol dm^{-3} sodium chloride solution at volume fractions of 0.01, 0.04 and 0.13. For the lowest volume fraction the curve indicates that the particles have an excluded volume region and g(r) rapidly approaches unity. At $\phi = 0.04$ a clear peak is apparent indicating a degree of short-range order. At $\phi = 0.13$ the peak has moved to a lower r, indicating a closer centre to centre spacing, and a second and a third peak. On a long-range scale, however, little order is indicated; a typical indication of fluid-like behaviour.

These experiments indicate that the pair potential given by eq. (6) can predict the behaviour of dispersions of charged polymer colloid particles with quite reasonable accuracy.

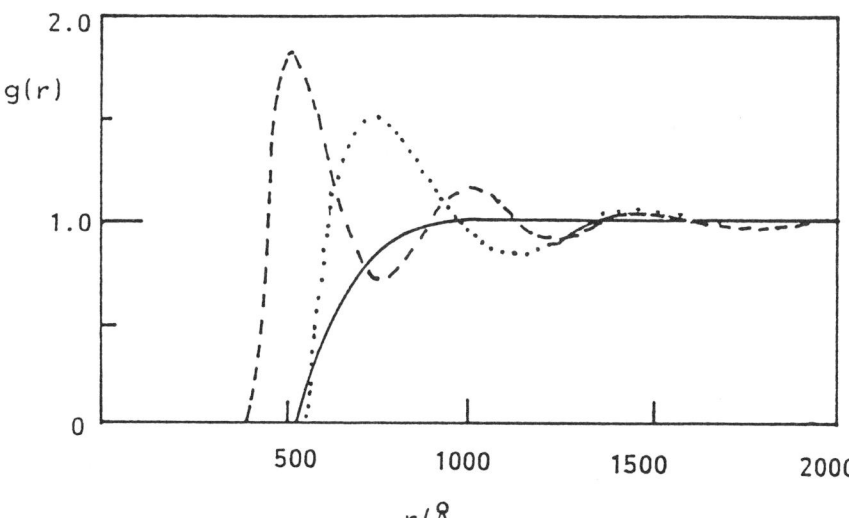

Figure 4. g(r) against r for a polystyrene latex in 10^{-4} mol dm^{-3} sodium chloride at various volume fractions, _____ , 0.01; ..... , 0.04; - - - - - , 0.13.

2.5. OSMOTIC PRESSURE

In the limit of zero scattering vector, the structure factor S(0) is directy related to the osmotic compressibility of the dispersion by,

$$\frac{d\pi}{dN_p} = \frac{kT}{S(0)} \tag{9}$$

where π is the osmotic pressure, so that from a determination of S(0) at various particle number concentrations the osmotic pressure can be obtained from

$$\pi = kT \int_o^{N_p} dN_p / S(0) \tag{10}$$

π is also related to g(r) by the expression [8,9,10]

$$\pi = N_p kT - \frac{2\pi N_p^2}{3} \int_o^\infty g(r) r^3 \frac{dV_R}{dr} dr \tag{11}$$

giving an alternate route to osmotic pressure via the scattering data; strictly V_R should be the potential of mean force [11].

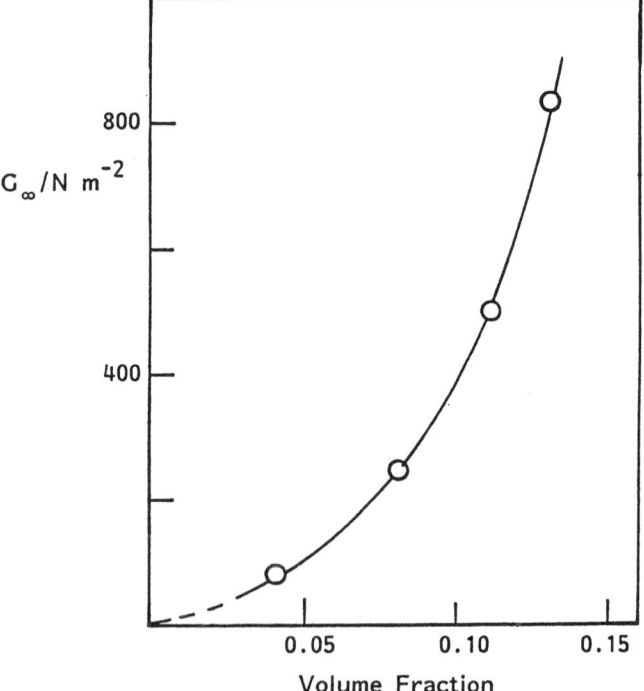

Figure 5. Shear modulus, G_∞ against volume fraction of polystyrene latex, in 10^{-4} mol dm^{-3} NaCl

2.6. SHEAR MODULUS

The shear modulus, G_∞, is also related to V_R and $g(r)$ by the equation [8,11],

$$G_\infty = N_p kT + \frac{2\pi N_p^2}{15} \int_o^\infty g(r) \frac{d}{dr} \left(r^4 \frac{\partial V_R}{\partial r} \right) dr \qquad (12)$$

This means that values of G_∞ can be obtained from small angle scattering measurements as well as by direct measurements, e.g. by the use of a shearometer [12]. Some results obtained on a polystyrene latex are shown in Figure 5.

3. Nonaqueous Dispersions of Polymer Colloids

Similar studies of $S(Q)$ and $g(r)$ have been carried out on dispersions, of poly(methyl methacrylate) stabilised by poly-12-hydroxystearic acid in dodecane (13). These demonstrate very clearly the much harder nature of the interaction which occurs between sterically stabilised particles. In fact these interacts as very nearly "hard spheres" (14).

These systems also have the advantage that it is possible to change the refractive index of the medium by addition of another medium in order to get either to, or close to, the match point between the particle and the medium.

3.1. DIFFUSION IN CONCENTRATED DISPERSIONS

The refractive index matching technique can be employed to great advantage in studying, by photon correlation spectroscopy, the diffusion processes which occur in concentrated systems since one set of particles can be taken at a high volume fraction and refractive index matched and then a second set of particles added, of different refractive index, as a tracer.

Studies of this sort have been carried out with particles having a core of poly(vinyl acetate) and sterically stabilised by poly-12-hydroxystearic acid (PHS); these particles had a refractive index of 1.471 and could be optically matched by using a mixture of cis-decalin (refractive index 1.481) and trans-decalin (1.469). Particles with a core of poly(methyl methacrylate) and also stabilised by PHS were used as tracer particles; these had a refractive index of 1.483. The hydrodynamic radius of the PMMA particles was 83 $\pm$ 2 nm and that of the PVA particles 85 $\pm$ 2 nm. Since the same stabiliser molecule was used for both sets of particles and the sizes were the same within experimental error it was assumed that the interactions between all particles were identical (15).

Photon correlation spectroscopy was used to measure the self-diffusion of the tracer particles. By making measurements on a short time scale, such that the particles had moved a distance corresponding to only a small fraction of their diameter, the short time self-diffusion coefficient D^S was obtained. Then by also making measurements on a timescale such that the particle diffused over several diameters and thus acted colloidally and hydrodynamically with other particles, the long-time self-diffusion coefficient was obtained. The diffusion coefficient D_0 for the freely diffusing particles was obtained from measurements on very dilute dispersions. The results are plotted in Figure 6.

236

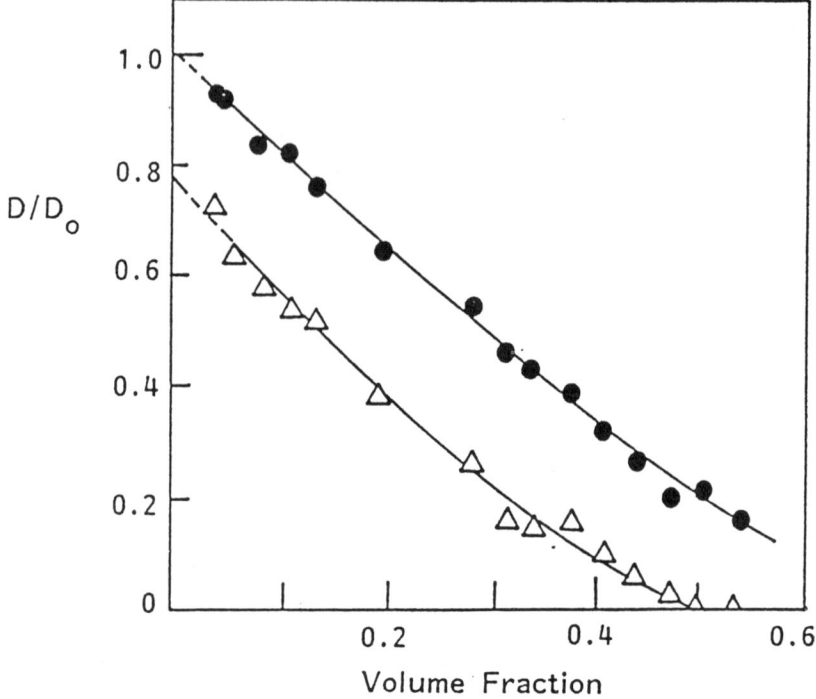

Figure 6. Ratio of self-diffusion coefficient/single particle diffusion coefficient, D_o against volume fractionl, ● short time self-diffusion; Δ , long-time self-diffusion

The normalised long-time self-diffusion coefficient, D^L/D_0 decreases with increase in volume fraction, ϕ, and tends to a very small value as $\phi \rightarrow 0.5$. This is an interesting result since it was predicted [16] that the freezing transition for hard-sphere systems should occur at a volume fraction of 0.495 and the melting transition at 0.545. It therefore appears that for nearly hard spheres long-time self-diffusion, which can be interpreted as motion through the particle array, ceases at the freezing transition. These results correlate very well with the onset of a crystalline structure in PMMA dispersions [7,14]. Short-time self-diffusion is still occurring above $\phi = 0.5$ and thus indicates motion of particles within the cage of other particles; this motion appears to continue up to a ϕ of ca. 0.64, (random close packing).

4. Crystallisation of Polymer Colloids

The fact that monodisperse polymer colloid particles can form crystalline arrays has been well investigated by a variety of techniques including, scanning electron microscopy [16], dark field microscopy [17], optical diffraction [18,19], small angle neutron scattering [7], confocal microscopy [20] etc.

However, there are considerable advantages for the visual observation of the crystallisation process in closely matching the refractive index of the particle to that of the medium. This has been clearly demonstrated for PMMA-PHS particles in an organic medium [7,14] and for fluorinated polymer colloid particles in an aqueous medium [21]. An additional advantage is that scattering techniques can then be used in order to confirm the crystal structure from the light diffraction peaks observed. This was demonstrated recently in a study of binary mixtures of PMMA-PHS particles which crystallised as an AB$_2$ type crystal [22]. Scanning electron micrographs are shown in Figure 7 and the corresponding optical diffraction patterns in Figure 8, with indexing of the peak positions.

Recent simulations [23] on mixtures of hard spheres have shown regions of super lattice structures and fluid phases that are entirely consistent with those experimental findings. Moreover, Frenkel has pointed out that at high particle densities the gain in entropy due to the increase in free volume exceeds the loss in configurational entropy and hence favours crystallisation [24].

5. The Glassy State with Polymer Colloids

In the PMMA-PHS nonaqueous systems it was found that the particles formed a glassy state, at a volume fraction above ca. 0.55. This seems to be a consequence of the fact that diffusional processes have virtually ceased at this concentration and hence nucleation is inhibited. Consequently, this leads to a high volume fraction disordered state [7,14].

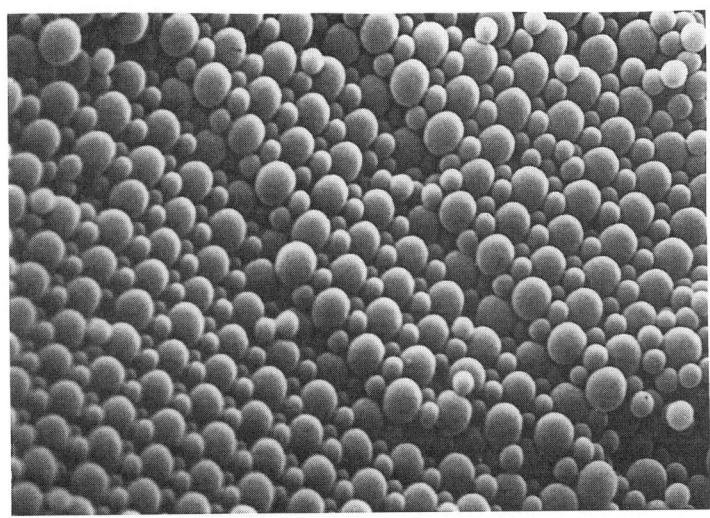

Figure 7. Scanning Electron Micrograph. Polymer Colloid AB$_2$ Crystal

238

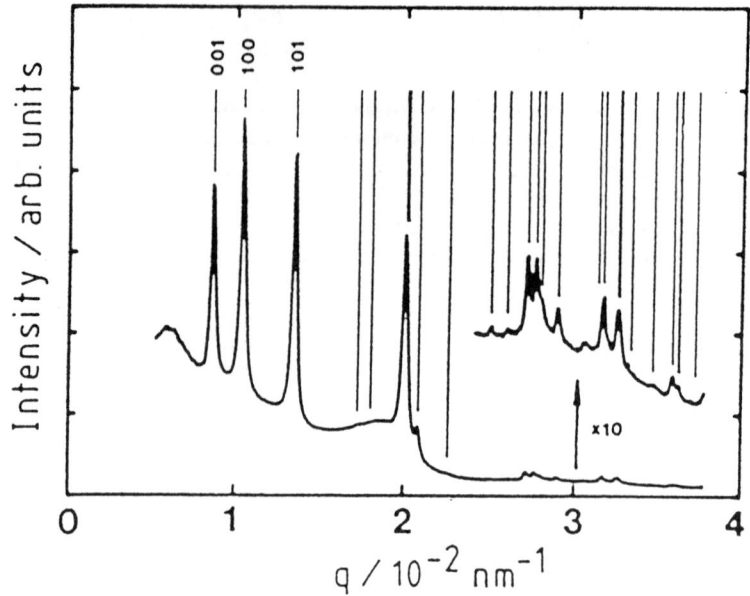

Figure 8. Optical diffraction pattern from AB_2 crystal

The same phenomena was observed in polymer colloid systems of perfluorinated polymers except that it occurred at low volume fractions at low salt concentrations, the strong electrostatic interactions again apparently inhibiting diffusion. The glassy state is more easily recognisable from dynamic light scattering studies than from structure factor studies, in that the time-delay curve deviates from the exponential form suggesting a non-Gaussian diffusional process in this state [25,26].

6. Bimodal Polymer Colloid Dispersions

In recent studies a small angle neutron scattering study has been made of binary mixtures of monodisperse polymer colloid dispersions [27,28]. As well as the formation of binary crystals other effects are also possible, including particle segregation and network formation by heterocoagulation [28].

The experiments were carried out using hydrogenated particles of radius 168 Å (A particles) and deuterated particles of radius 510 Å (B particles) giving a ratio of R_A/R_B of 0.33. Measurements were made at number concentration ratios of N_A/N_B of 9 and 15. By exploiting the variation of scattering length available from H_2O and D_2O and their mixtures it was possible to match out the hydrogenated polystyrene in one experiment, the deuterated polystyrene in another, and then to carry out a third experiment in which neither particle was matched. Since in mixed systems there are three partial structure factors $S(Q)_{AA}$, $S(Q)_{BB}$ and $S(Q)_{AB}$ enough experimental information was available to determine all three quantities.

The results for the three structure factors are shown in Figure 9 for $N_A/N_B = 15$.

1) The form of $S(Q)_{BB}$ against Q suggests that the B particles are colloidally stable in the presence of the A particles and the position of the peak indicates an average correlation distance for short range order of ca. 2000 Å. This peak was in a position close to that observed in the absence of the A particles.

2) The form of $S(Q)_{AA}$ obtained in the presence of the B particles is profoundly different from that with the B particles absent. The peak at a Q value of ca. 0.009 Å^{-1} indicates that there is some weak ordering of the A particles in the presence of B but the upturn at low Q indicates that some clustering of the A particles has occurred which is not entirely random.

3) The form of the $S(Q)_{AB}$ against Q curve indicates a negative correlation between the A and B particles suggesting separation of the two species in the overall structure which are uncorrelated with the structures formed by the A-A and B-B interactions. In essence the presence of the B particles and their excluded volume as a consequence of electrostatic repulsion means that the A particles are excluded from these regions.

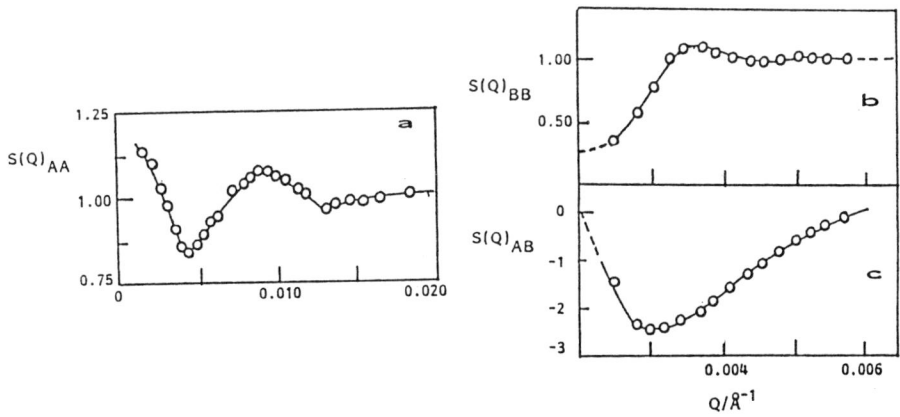

Figure 9. Partial structure factors against A for a binary mixture, $R_A/R_B = 0.33$; $N_A/N_B = 15.0$

7. Particle Morphology

7.1. INTRODUCTION

As well as examining structure in dispersions scattering measurements are also capable of providing information on the internal structure of particles and an example of this was given in reference [4].

7.2. CORE-SHELL PARTICLES

For particles with a core particle of radius R and a shell of thickness δ the intensity of scattering is given by,

$$I(Q) = AN_p\left[(\rho_s - \rho_m)(F_2 - F_1) + (\rho_c - \rho_m)F_1\right]^2 \tag{13}$$

with terms as defined previously. F_1 and F_2 are the particle form factors for the core and the core-shell particle as given by,

$$F_1 = 3V_c\left[(\sin\ QR\ -\ QR\cos\ QR)/(QR)^3\right] \tag{14}$$

$$F_2 = 3V_T\left[(\sin\ Q(R+\delta) - Q(R+\delta))/\left(Q(R+\delta)^3\right)\right] \tag{15}$$

with $Vc = 4p\ R^3/3$ and $V_T = 4p\ (R+\delta)^3$; when $\delta = 0$ equation (13) reverts to that of a homogeneous sphere. The equations only apply to dilute non-interacting systems and thus experiments need to be carried out in salt concentrations of ca. 10^{-2} mol dm^{-3} 1:1 electrolyte and at concentrations of 1% or less to minimise multiple scattering effects.

Figure 10 shows some simulated curves for a core-shell particle, at values of ρ_m corresponding to pure H_2O, and 73% and 81% D_2O v/v respectively; for convenience the curves have been normalised to unit intensity on the ordinate.

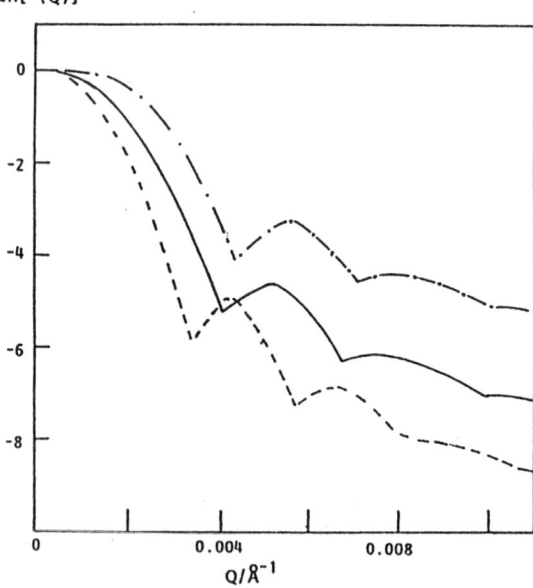

Figure 10. Simulated curves for a core shell particle at various contrasts, -----, H_2O; -.-.-, 73% D_2O; _____, 81% D_2O

As can be seen both the intensity vs Q and the positions of the maxima and minima change substantially with the values of ρ_m; for homogeneous spheres the curves would all superimpose on each other when plotted in this manner.

Figure 11 shows some small angle neutron scattering results on polymer colloid polymer particles prepared from deuterated styrene and hydrogenated methyl methacrylate. The continuous lines represent the fits obtained using eq. (13). These results indicate quite clearly that the particles have a core-shell structure with a core radius of 126 nm and a shell thickness of 9.4 nm. However, the values of ρ_s and ρ_c obtained from the fit also indicate that both the core and the shell appeared to be mixtures of poly(styrene) and poly(methylmethacrylate) but it could not be deduced whether this was a physical blend or a consequence of copolymerisation. What is clear from the results is that the shell is rich in hydrogenated methyl methacrylate [29].

As shown in the recent work of Ballauff [30-32] similar studies can be carried out using small angle X-ray scattering. In this case, solutions of sucrose at various concentrations can be used to provide the constrast variation required for morphology analysis.

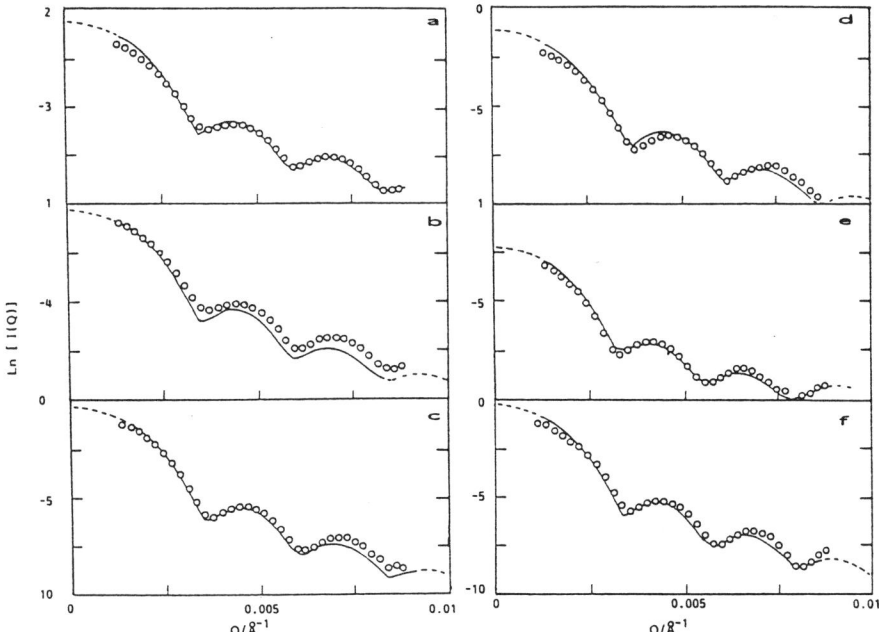

Figure 11. Neutron scattering results for a deuterated polystyrene - polymethacrylate latex: in a) H_2O b) 20% D_2O c) 40% D_2O d) 50% D_2O e) 80% D_2O f)D_2O. _____ , fitted curve

8. References

1. Brown, J.C., Pusey, P.N., Goodwin, J.W., and Ottewill, R.H. (1975) J.Phys. A. Math. Gen., **8**, 664-682.
2. Brown, J.C., Goodwin, J.W., Ottewill, R.H. and Pusey, P.N. (1976) Colloid and Interface Science, IV, 59-72.
3. Cebula, D.J., Goodwin, J.W., Jeffrey, C.G, Ottewill, R.H., Parentich, A. and Richardson, R.A. (1983) Faraday Discuss. Chem. Soc., **76**, 37-52.
4. Ottewill, R.H. *Scattering Techniques - Fundamentals*, this volume.
5. Hayter, J.B. and Penfold, J. (1981). Mol. Phys., **42**,109-118.

242

6. Hansen, J.P. and Hayter, J.B. (1982). Mol. Phys., **46**, 651-656.
7. Ottewill, R.H. (1989) Langmuir, **5**, 4-11.
8. Zwanzig, R. and Mountain, R.D. (1965), J. Chem. Phys., **43**, 4464-4471.
9. McQuarrie, D.A. (1973) *Statistical Mechanics*, Harper and Row, London.
10. Ottewill, R.H. (1981). In *Colloidal Dispersions* (ed). J.W. Goodwin, Royal Society of Chemistry, London, 143-163; 197-217.
11. Ottewill, R.H. (1985) Ber. Bunsenges. Phys. Chem. **89**, 517-525.
12. Buscall, R., Goodwin, J.W., Hawkins, M.W. and Ottewill, R.H. (1982) J. Chem. Soc. Faraday Trans. I, **78**, 2873-2887.
13. Markovic, I., Ottewill, R.H., Underwood, S.M., Tadros, Th. F. (1986) Langmuir, **2**, 625-630.
14. Pusey, P.N. and van Megen, W. (1986) Nature, **320**, 340-342.
15. Ottewill, R.H. and Williams, N. St. J. (1987) Nature, **325**, 232-234.
16. Ottewill, R.H. (1977) J. Coll. Int. Sci., 58, 357-373.
17. Hachisu, S., Kobayashi, Y. and Kose, A. (1973), J. Coll. Int. Sci. **42**, 342-348.
18. Hiltner, P.A. and Krieger, I.M. (1969) J. Phys. Chem., **73**, 2386-2389.
19. Goodwin, J.W., Ottewill, R.H. and Parentich, A. (1980).J. Phys. Chem., **84**, 1580-1586.
20. van Blaaderen, A. (1992). Doctoral thesis, de Rijksuniversiteit Utrecht.
21. Ashdown, S., (1990) Ph.D. thesis, University of Bristol.
22. Bartlett, P., Ottewill, R.H. and Pusey, P.N. (1992) Phys. Rev. Letters, **68**, 3801-3804.
23. Eldridge, M.D., Madden, P.A. and Frankel, D. (1993) Nature, **365**, 35-37.
24. Frankel, D. (1993) Physics World, **Feb**. 24.
25. Megen, W., Underwood, S.M., Ottewill, R.H., Williams, N.St.J. and Pusey, P.N. (1987) Faraday Discuss. Chem. Soc., **83**, 47-57.
26. Ottewill, R.H. (1990) Faraday Discuss. Chem. Soc., **90**, 1-15.
27. Ottewill, R.H., Hanley, H.J.M., Rennie, A.R. and Straty, G.C. (1995) Langmuir, **11**, 3757-3765.
28. Ottewill, R.H. and Rennie, A.R. (1996) Prog. Colloid Polym. Sci., **100**, 60-63.
29. Ottewill, R.H., Cole, S.J. and Waters, J.A. (1955) Macromol. Symp. **92**, 97-107.
30. Dingenouts, N., Kim, Y.S. and Ballauff, M. (1994) Colloid Polym. Sci., **272**, 1380-1387.
31. Dingenouts, N., Pulina, T. and Ballauff, M. (1994) Macromolecules, **27**, 6133-6136.
32. Dingenouts, N. and Ballauff, M. (1993) Acta Polymer., **44**, 178-183.

OPTICAL SPECTROSCOPY ON POLYMERIC DISPERSIONS

W.-D. HERGETH
Wacker-Chemie GmbH
LKE, P.O. Box 12 60
D-84480 Burghausen, Germany

1. Introduction

Optical spectroscopy covers the "middle" range of the electromagnetic spectrum from the ultraviolet (UV, $\lambda \gg 10$ nm) to the infrared (IR, $\lambda < 1$ mm). Experimental equipment for optical spectroscopy in this wavelength range is built with typical optical elements like mirrors, lenses, gratings, optical filters etc. The incident energy is delivered by "light" sources. The long-wavelength end of the electromagnetic spectrum (i.e. λ in the mm range) requires a completely different experimental techniques (e.g. hollow conductors for microwave spectrometry) as the short-wavelength end does ($\lambda \ll 100$ nm) with special X-ray and γ-ray equipment.

Electronic absorptions determine the spectral features observable in the UV/VIS wavelength range. Molecular vibrations, rotations and combined rotation-vibration transitions of molecules can be detected in the near (NIR) to the far (FIR) infrared wavelength range. Spectra can either be recorded in absorption (UV, VIS, NIR, IR, FIR) or emission (IR, Raman, fluorescence).

Infrared and Raman spectroscopy are the most important methods of vibrational spectroscopy to analyze polymers. It should be noted, that the interaction probabilities of light with the molecule for the two techniques are quite different. Infrared absorption is favored when a molecule has a permanent dipole which is modulated by the vibration. Raman scattering (emission) occurs when the molecule is polarizable, with the polarizability modulated by the vibration. Thus, both methods provide us with complementary information about the molecule.

The main focus of the present paper is on the application of Raman spectroscopy on polymeric dispersions. Raman spectroscopy is one of the fastest growing areas in analytical chemistry today. The multiplex and throughput advantages of Fourier-transform spectrometers in combination with NIR excitation drove the rediscovery of this technique for polymer analysis. The detection limit has been greatly improved. CCD (Charged-Coupled Device) based instruments pave new ways for reaction monitoring especially in aqueous systems. The application of fiber optics allows remote sensing. Thus Raman spectroscopy has the greatest potential for application on colloids in the future.

Several other spectroscopic techniques with relevance for colloids will be discussed briefly.

J. M. Asua (ed.), Polymeric Dispersions: Principles and Applications, 243–256.

2. Raman Spectroscopy

2.1. GENERAL

The basic condition for a molecular vibration (or rotation) to be Raman active requires that the polarizability $\alpha_\parallel$ of the molecule (or that of its vibrating parts) at equilibrium distance ($q = 0$) changes during the vibration:

$$\left(\frac{\partial \alpha_\parallel}{\partial q} \right)_{q=0} \neq 0 \tag{1}$$

where q is the normal coordinate of the vibration. Thus Raman spectroscopy is sensitive to non-polar molecular vibrations. Hence, double or triple bonds in monomeric or polymeric molecules are very strong Raman scatteres. Vinyl or diene monomers can easily be identified, and their concentrations determined because of their double bonds. Therefore, Raman spectroscopy is an ideal tool for monitoring the disappearance of monomers during the course of polymerization (conversion) or for detecting residual monomers in the final products (VOC's).

The Raman line intensity is directly proportional to the number of oscillators in the scattering volume, and the intensity of the illuminating radiation. Therefore, line intensities changes primarily reflect concentration changes within the sample. However, the actual number of scatteres is usually unknown, and the scattering volume may change during the reaction. In addition, there are light intensity losses at optical interfaces, losses due to absorption or diffraction, and relative differences caused by detector/spectrograph characteristics. Hence, Raman line intensities are relative intensities, requiring a calibration standard for quantitative interpretation. The change in optical system properties during the course of the reaction requires the use of an internal standard rather than an external one. However, the addition of substances as internal standards may influence reaction kinetics, partition of components, stability of the system, or the properties of the end product. Therefore, the internal calibration standard should be an inherent ingredient of the reacting system itself.

The Raman effect is an extremely weak, inelastic scattering process. For excitation frequencies sufficiently away from electronic absorptions, the Raman scattering intensity is in the order of 10^{-6} -10^{-9} of the exciting beam intensity. However, in Raman spectroscopy there is always a competitive mechanism for light emission present, sample fluorescence. The quantum yield of fluorescence, i.e. the ratio of emitted to absorbed energy quanta, is several orders of magnitude higher than the Raman scattering efficiency. Hence, in samples with small amounts of impurities, additives, or degradation products, the sample fluorescence may completely obscure the Raman spectrum. The most convenient way to avoid fluorescence interference is excitation at longer wavelengths. However, there is a price to pay for this benefit. The scattering process has an inverse dependence to the wavelength raised to the forth power. Therefore, as the wavelength is lengthened, the scattering intensity decreases severely.

Water, with a strong dipole that is very sensitive to interatomic distance, has an intense infrared absorption and a very weak Raman response. Thus water obscures infrared spectra, whereas Raman scattering is more or less oblivious to the presence of water. Hence, Raman spectroscopy can be applied to aqueous solutions and dispersions.

Raman spectroscopy does not require extensive sample preparation. Raman spectra can be taken from almost all samples even when they are colored or black. Optically transparent containers can be used as sample holders. With a confocal optical arrangement, it is even possible to obtain spectra from samples through colored or opaque bottle walls.

In the literature, a variety of linear and non-linear Raman techniques has been described. Here, we focus on conventional non-resonant Raman spectroscopy. Some applications of Surface Enhanced Raman Spectroscopy (SERS) and Stimulated Raman Spectroscopy (SRS) will be discussed in a later subsection (2.4.). The reader should be referred to excellent textbooks for theory, instrumentation, special techniques and application of infrared and Raman spectroscopy [1 - 3].

2.2. INSTRUMENTATION

In Raman spectroscopy, the interaction of the incident light with the vibrating molecule leads to a frequency shift of the scattered light with respect to the incident frequency. Thus, a Raman spectrum can be obtained by measuring the intensity of the scattered photons as a function of this frequency difference. Hence, the breadth of a Raman spectrum on the wavelength scale depends on the excitation wavelength.

In Figure 1, the wavelength ranges of Raman spectra are shown as bars relative to the excitation at 514 nm (Argon), 632 nm (HeNe), 752 nm (Krypton), and 1064 nm (Nd:YAG). The Rayleigh line of these spectra (0 cm^{-1} Raman shift) is at the left hand end of each bar, and the 3600 cm^{-1} wavenumber of the Stokes Raman shift is on the longer wavelength end. It should be noted that all spectra have the same length on the wavenumber scale (0 - 3600 cm^{-1}). For comparison, the bar indicating the IR spectrum covers only the wavenumber range from 3600 cm^{-1} (2.778 µm) to 2500 cm^{-1} (4 µm).

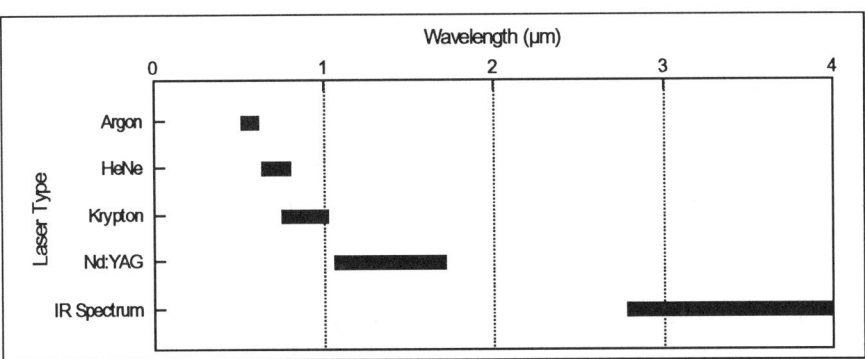

Figure 1. Vibrational spectra on the wavelength scale (see text)

Early Raman spectrometers were of the scanning dispersive type equipped with double or triple grating monochromators and operating in the visible wavelength range. The well-known λ^{-4} dependence of the Raman (as well as the Rayleigh) scattering cross-section favors this shorter wavelength excitation. However, most real life samples show very strong fluorescence with visible-light excitation. Hence, Raman spectroscopy on polymers was limited to a small number of applications in the early days.

The development and application of Raman spectroscopy experienced some sort of a renaissance after Hirschfeld's and Chase's recommendation to use Nd:YAG laser radiation at 1064 nm for the excitation of Raman spectra, interferometers to record interferograms, and fast Fourier transformation to convert them into spectra [4]. Advantages and disadvantages of FT-Raman spectroscopy have been discussed in several textbooks and papers [1 - 8]. FT Raman instruments are quite sensitive, easy to use, and relatively compact.

With NIR excitation, sample fluorescence is very much reduced. Since absorption processes are minimized in the near infrared, thermal degradation of the samples and background emission are also lessend but may still be the cause of problems. Aqueous solutions or dispersions can show effects of sample heating since the O-H stretch overtones of water (Figure 2) can absorb some of the incident light intensity.

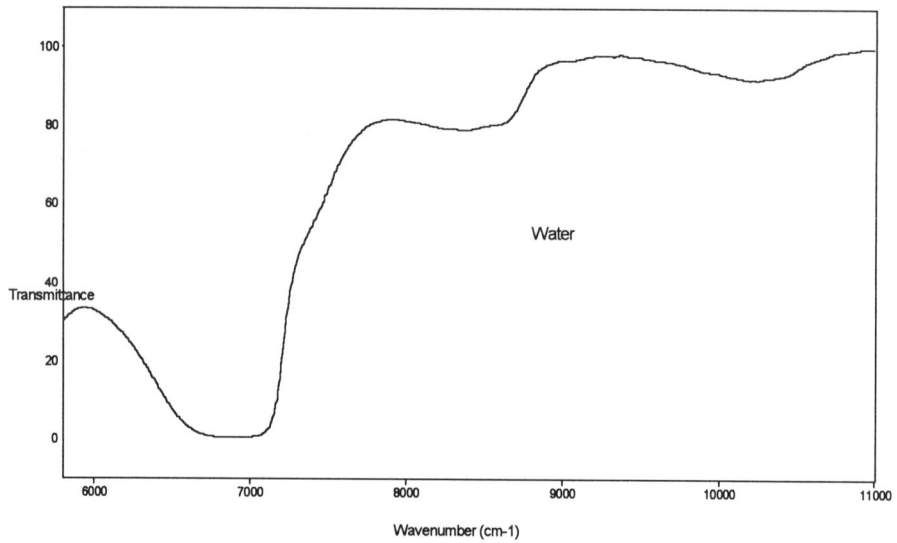

Figure 2. NIR transmittance of water

The reduction of scattered light intensity in NIR Raman spectroscopy with respect to the visible region is over-compensated by the higher optical throughput of FT instruments compared to conventional dispersive instruments.

A unique feature of FT Raman instruments is the distributed noise. The detector views both signal and noise components simultaneously. The Fourier transformation process distributes the noise component over the entire spectrum. Even light intensity that shows up at a certain wavenumber can in fact originate at a different frequency, and the Fourier transformation shifts it.

FT-Raman spectroscopy has also been carried out with Ti:Sapphire excitation at 780 nm [9] and Ar^+ excitation at 488 nm [46]. The main advantage of excitation at shorter wavelength is in the field of water-based samples due to the avoidance of self-absorption of the Raman spectrum. However, the sensitivity of CCD based dispersive instruments is superior to FT Raman in the shorter wavelength range.

Over the last several years, CCD array detectors have contributed to major advances in Raman spectroscopy [8, 10]. Current CCD's have a very high quantum efficiency, fast response and a spectral range from 350 to about 1000 nm. The two-dimensional nature of the CCD can be exploited for multichannel detection. Array detectors suffer from either limited resolution or limited band width. However, the two-dimensionality of the CCD in combination with cross-dispersion using echelle systems or holographic transmission split gratings enables one to detect complete spectra with high resolution [8].

The insensitivity of a CCD for wavelengths > 1000 nm leads to limitations for the wavelength of the incident light. The excitation with a CCD based instrument has to be done at wavelengths well below 800 nm in order to obtain the entire Raman spectrum up to 3600 cm^{-1} (see Figure 1).

Important developments to improve the performance of dispersive instruments include i) holographic notch filters to remove the Rayleigh line and/or Raman signals from other optical parts (e.g. glass fibers), and ii) holographic transmission gratings to increase the optical throughput [8]. These monochromators are suitable for industrial applications because no moving parts guarantee greater reliability.

The application of telecommunication fiber optic cables enables remote sampling and requires no aligning from one sample to another. Fiber optics connect the remotely located instrument with the reactor for on-line reaction monitoring [11].

Near infrared diode lasers are small enough to allow the Raman instrument to be compact, and they have low power requirements, making the spectrometer portable. They are relatively inexpensive and have a potentially longer operating lifetime compared to near infrared gas lasers [8, 10, 12]. Early problems with mode hopping are now eliminated [12 - 14].

2.3. APPLICATIONS

There have been only a few literature papers, through the beginning of the nineties, dealing with Raman reaction monitoring. The validity of this method has been demonstrated for the suspension polymerization of styrene [15] and vinyl chloride [16], the thermal polymerization of styrene [17 - 20] and methyl methacrylate [20], the solution polymerization of methyl methacrylate [21], the γ-initiated diacetylene polymerization [22], and the microemulsion polymerization of styrene and methyl methacrylate [23]. In these publications, the decreasing intensity of the ν(C=C) monomer Raman lines during the course of the reaction was monitored as an indicator for the extent of monomer conversion to polymer. The positions of the double bond stretching vibration in the Raman spectra of several monomers are listed in Table 1.

TABLE 1: Double bond Raman lines

monomer	$\nu(C=C)$ Raman line
vinyl chloride	1607 cm^{-1}
acrylonitrile	1610 cm^{-1}
styrene	1631 cm^{-1}
methyl acrylate	1635 cm^{-1}
2 ethyl hexyl acrylate	1637 cm^{-1}
butadiene	1639 cm^{-1}
methyl methacrylate	1641 cm^{-1}
vinyl acetate	1648 cm^{-1}

In Figure 3, Raman spectra of styrene monomer and polystyrene Latex are shown. It is obvious, that the vinyl C=C double bond of the monomer at 1631 cm^{-1} disappears during the course of emulsion polymerization. Additionally, there are other spectral changes over the entire spectrum that can be exploited to monitor the reaction. However, this requires more elaborate chemometric analysis techniques.

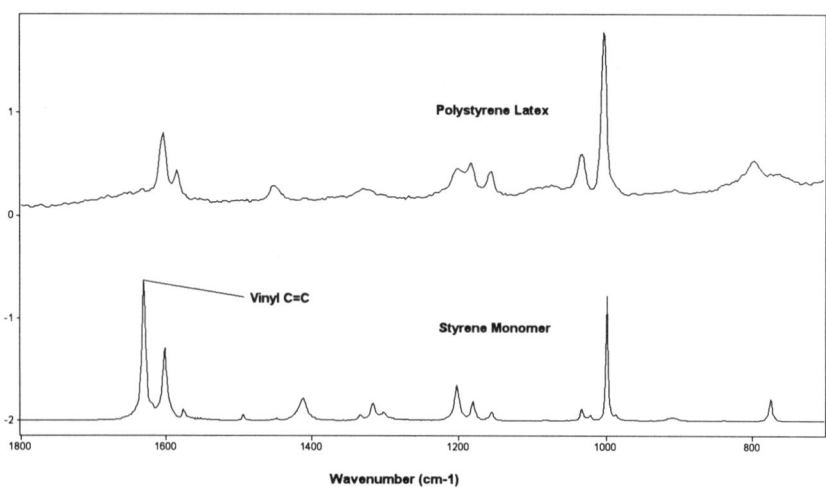

Figure 3. Raman spectra of styrene monomer and polystyrene latex

Residual monomer levels in latexes [24] and solid polymers [25] as well as the monomer partition between polymer and aqueous phase [26] have been detected by means of Raman spectroscopy. The feasibility of poly butadiene rubber conformation studies (i.e. the *vinyl - cis - trans* microstructure of residual double bonds) has also been demonstrated [27 - 29].

Recently, there has been a growing interest in the application of fiber-optic based Raman spectroscopy to monitor the kinetics of emulsion polymerization [11, 30 - 32]. Trends and future directions for on-line monitoring by means of Raman spectroscopy, and for other industrial Raman applications have been discussed in detail elsewhere [31 - 38].

The direct monitoring of methyl methacrylate emulsion polymerization by fiber-optic Raman spectroscopy using water as an internal standard has been described by Wang et al. [32]. This method of internal calibration worked very well with excitation in the visible at 514 nm (Ar-ion laser). However, with excitation in the NIR water does no longer serve as a simple internal standard. During the course of the reaction, the growing number of growing particles change the scattering volume where the light is collected from, as well as alter the light path through the disperse system of the exciting and the emitted light due to multiple scattering. With excitation at 1064 nm, the Raman spectrum of an aqueous sample partially overlaps with the near infrared absorption of water (compare Figures 1 and 2). The change of the light path in combination with the sample's self-absorption influences the spectra in a specific, non-linear manner [39]. This effect is illustrated in Figure 4.

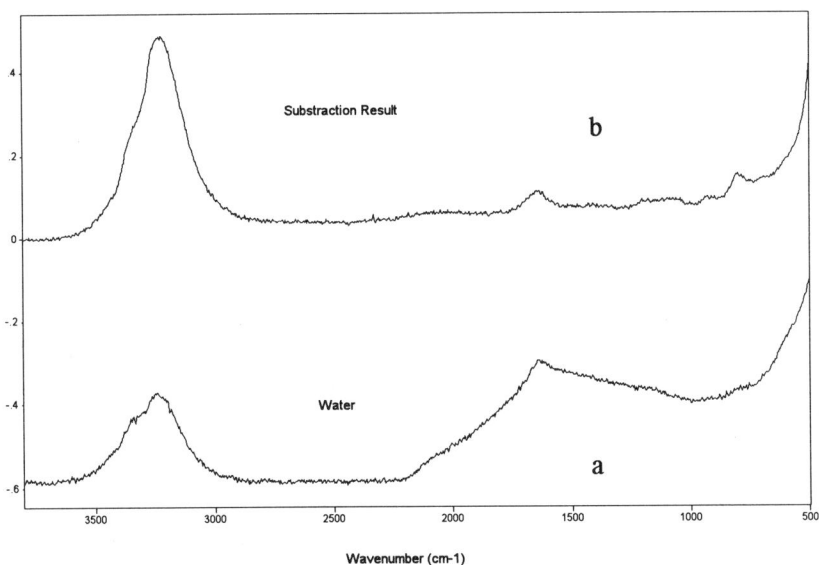

Figure 4. Raman spectrum of pure water (a) and difference spectrum of two latexes with different solids content (b, see text)

The water Raman spectrum (a) was obtained using a FT-Raman spectrometer with excitation at 1064 nm. Two band complexes of water show up in this spectral region: i) the rather weak and broad bending mode around 1640 cm^{-1}, and ii) several O-H stretching vibrations between 3000 and 3800 cm^{-1}. Spectrum (b) is a substraction result: As a first step, the Raman spectrum of a 34% total solids rubber latex was taken. Then, the latex was diluted to about 5% total solids, and another Raman spectrum was taken. The spectra were substracted from each other in such a manner that the spectral features of the rubber were gone. The resulting difference spectrum is the contribution of the water to the total Raman spectrum of the latex. Obviously, it looks completely different compared to spectrum (a). This effect complicates the quantitative analysis of Raman spectra obtained with NIR excitation.

Even in a clear solution, NIR water absorption alters spectra of dissolved substances considerably. The spectra shown in Figure 5 were obtained with NIR 1064 nm excitation. The relative band intensities in the range below about 2000 cm^{-1} do not change when acrylonitrile is dissolved in water. However, water NIR absorption becomes dominant for wavenumbers above about 2000 cm^{-1} at this excitation. Thus, the intensities of the acrylonitrile C≡N stretching band at 2237 cm^{-1} and the C-H stretching bands around 3000 cm^{-1} are reduced with respect to e.g. the double bond stertching vibration at 1610 cm^{-1}. This self-absorption effect has been addressed by several authors [39 - 42].

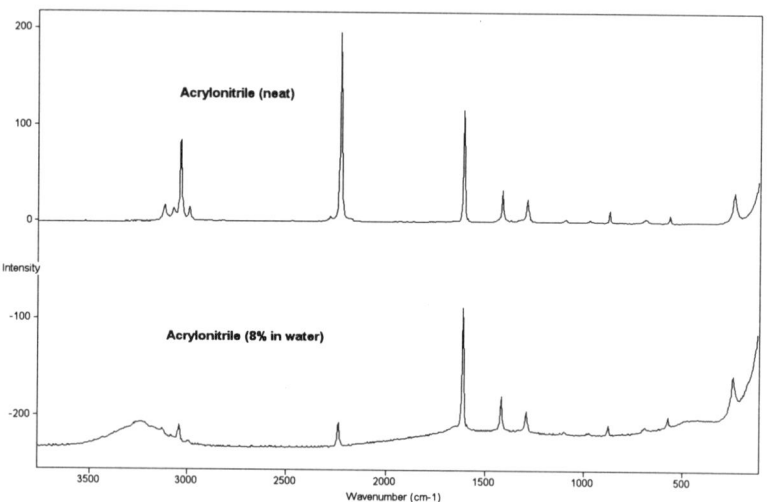

Figure 5. Raman spectra of pure acrylonitrile and an aqueous solution of acrylonitrile

A result of this effect is, that relative band intensities in Raman spectra may depend on sample path length and sample position relative to the focal point of the collection optics. This is illustrated in Figure 6. Due to the optical design of the FT-Raman instrument, a maximum of the Raman signal can be reached by moving the sample along the optical axis of the excitation laser. In order to observe the effect of water absorption, acrylonitrile-water solutions were placed in a tube and examined as a function of location along the laser optical axis. In Figure 6, the integrated peak intensities of several acrylonitrile stretching vibrations in dependence on sample position are shown. As can be seen on this plot, the nitrile peak at 2237 cm^{-1}, with the strongest water absorption, reaches its maximum first, followed by the C-H (3036 cm^{-1}, weaker absorption), and lastly the C=C at 1610 cm^{-1}, which is almost free of water absorption.

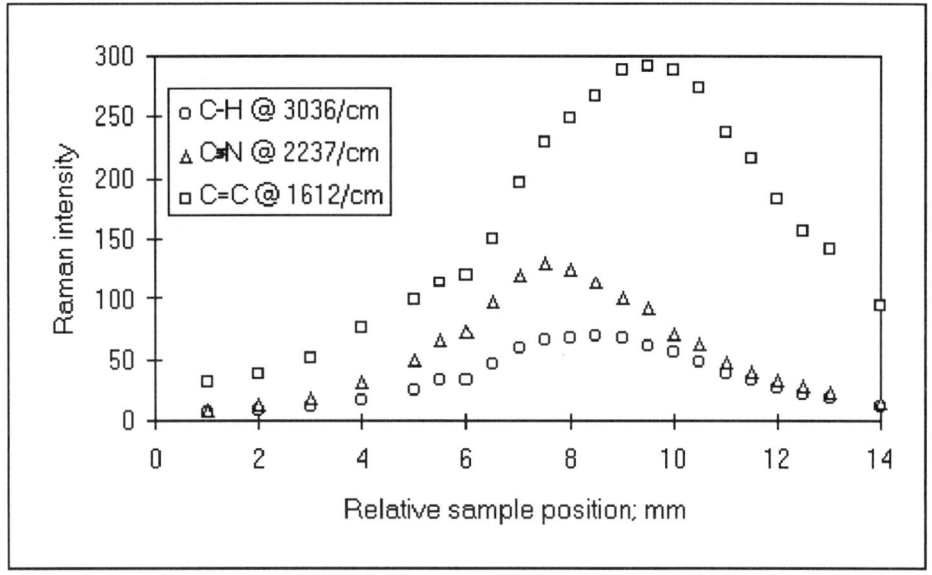

Figure 6. Integrated peak intensities as a function of sample location
for acrylonitrile water solution

In disperse systems, the partition of various components (e.g. partially water-soluble monomers) between aqueous phase and water has to be considered in order to quantify the Raman spectra of the mixture. Intensity corrections must be made for contributions to the Raman peaks arising from components dissolved in media having different refractive indexes (particle and water). According to the analysis of Sidorov [43], the corresponding internal field effect on the intensity of the Raman lines can be taken into account by means of a Mirone type correction [44, 45]. The correction factor, B_C, which describes the ratio of the scattering coefficients of a solute in the solution and in the pure liquid, is defined as follows:

$$B_C = \frac{n_m}{n} \left[\frac{3n^2}{n_m^2 + 2n^2} \right]^4 \tag{2}$$

where n_m and n are the refractive index of the pure component (monomer) and the mixture (monomer - water or monomer - particle), respectively. The refractive index n of the monomer containing subsystems of the latex (water and particle) can either be measured independently or calculated according to the Lorenz-Lorentz equation

$$\frac{n^2 - 1}{n^2 + 2} = \left(1 - \phi_P^P\right)\frac{V_m}{V} \cdot \frac{n_m^2 - 1}{n_m^2 + 2} + \phi_P^P \frac{V_p}{V} \cdot \frac{n_p^2 - 1}{n_p^2 + 2} \tag{3}$$

where ϕ_p^P is the weight fraction of the polymer in the swollen particles, V_m and V_p are the molar volume of the monomer and the polymer, respectively, and n_p is the refractive index of the polymer. The mixture rule relates the molar volumes and weight fractions as follows

$$V = \phi_m^P \cdot V_m + \phi_p^P \cdot V_p \tag{4}$$

Similar equations hold for the water-monomer subsystem. This concept has been successfully applied to determine the partition of acrylonitrile in rubber latex by means of Raman spectroscopy [26].

An interesting application of Raman spectroscopy has been published by Vorsina et al. [47]. The authors used Raman spectroscopy to examine the reaction stages and specific features of structural changes in the persulphate ion during the thermal decomposition of ammonium persulphate. Berger et al. [48] describe the measurement of aqueous dissolved gases using NIR Raman spectroscopy. The structure of water in polymer systems revealed by Raman spectroscopy has been reviewed by Meada and Kitano [49]. The sorption of water by polymers [50] as well as phase transitions in aqueous surfactant systems [51] have been studied using Raman spectroscopy. A Raman study investigating the origin of a number of weak features in the spectra of organic compounds in the wavenumber range 2800 - 2630 cm^{-1} has been reported by Lawson et al. [52]. The bands have been found to be characteristic for certain CH_3-C structural moieties which provides new information on the extent of methyl branching in organic substances. Davie et al. [53] describe the application of FT-Raman spectroscopy to analyze pigmented acrylic latex films. A new windowless cell for FT-Raman spectra of liquids and aqueous solutions has been developed by Brooker and coworkers [54].

2.4. OTHER RAMAN TECHNIQUES

Classical Raman spectroscopy as discussed above has some limitations because of the poor efficiency of the Raman effect, the overlap with high quantum yield sample fluorescence, or the inability to detect some "silent" vibrational modes. In the literature, several Raman techniques have been developed to overcome these problems.

In a conventional Raman experiment, the laser wavelength is chosen to avoid light absorption by the sample. In contrast, resonance Raman scattering is based on the concept that the excitation is close to an electronic (or even a vibrational) transition of the molecule. As a result, Raman scattering cross section increases dramatically by four or six orders of magnitude, and thus increasing the sensitivity of the measurement. One of the most useful advantages of resonance Raman spectroscopy is its abililty to obtain spectra at extremely low concentrations. So far, most of the applications described in the literature are in biological or biochemical systems. In polymer science, resonance Raman spectroscopy has been shown to be useful to study conjugated macromolecules.

The study of Raman spectra of adsorbed molecules on surfaces is one of the most promising areas in Raman spectroscopy. Molecules adsorbed on metal surfaces show a giant enhancement of the scattering efficiency by up to seven orders of magnitude. This Surface Enhanced Raman Scattering (SERS) has been shown to be ultrasensitive for detecting numerous compounds which adsorb to silver surfaces.

SERS delivers vibrational spectra of adsorbed molecules. Thus SERS has the potential of dual selectivity arising from both adsorption and detection. Originally, SERS experiments have been carried out with either bare metal surfaces or metal colloids mixed in solution with the sample. Recent developments have shown that covering the substrates with extremely thin modification layers enables one to construct chemical sensors [54 - 58]. The determination of the pH in aqueous samples with surface enhanced Raman fiber optic probes has been reported by Mullen [59].

A non-linear Raman technique, Stimulated Raman Scattering (SRS), has been applied to monitor the bulk polymerization of styrene and methyl methacrylate [60]. In contrast to spontaneous Raman scattering, most compounds show only a few Stokes lines in the SRS spectrum. As a result, spectral interference of SRS-active compounds in a reaction mixture is reduced to a minimum. The structure of water within the Gouy-Chapman diffuse electric double layer of water aerosols has been studied by morphology-dependent SRS [61].

3. UV - VIS - NIR - IR Spectroscopy

UV-VIS spectroscopy is well established in quantitative analytical chemistry because of its high sensitivity. In latex systems, light scattering by polymer particles accounts for much of the attenuation of the incident light. Thus particle sizing by UV-VIS methods is common practice in colloid laboratories. Theory and applications have been described in detail elsewhere (see e.g. [62, 63]).

Gossen et al. [64] describe the application of UV and NIR spectroscopy in combination with multivariate calibration of the optical spectra to determine both composition and particle diameter for a series of styrene/methyl methacrylate copolymer latexes. By UV spectroscopy of the highly diluted latexes, they have been able to determine the weight fractions of styrene and polystyrene in the latex as well as the mean particle diameter of samples with narrow particle size distributions. Water and methyl methacrylate compositions were not well predicted by the UV spectra. The NIR spectra were taken with a transflectance fiber-optic probe. Up to about 30% total solids, the multivariate calibration models of the NIR spectra yielded very good predictions of mean particle diameters and of the concentrations of all components.

Near-infrared spectroscopy (NIR) has become a rapid and powerful method, and is now used in many industrial applications. Absorptions in the NIR are overtones and combinations of the fundamental mid-IR vibration bands. Vibrational intensities in the NIR are considerably lower than those of corresponding infrared bands. Hence, optical layers in the order of millimeters may be transmitted. Both transmission/ transflectance and diffuse reflectance NIR spectroscopy enables one to monitor directly chemical reactions [3, 65, 66]. The most prominent NIR bands in polymeric material are those related to O-H, C-H, or N-H groups. Thus the detection of water has been one of the oldest applications of NIR spectroscopy (see Fig. 2). On the other hand, water interference may complicate NIR spectroscopy of aqueous samples [67]. A typical problem with NIR measurements is the need for instrument calibration because the instrumental response may change after a certain period of time [68].

Infrared (IR) spectroscopy has a long tradition of use in polymer science. For aqueous polymer colloids, there are some limitations owing to the strong absorption bands of water in the mid-IR. Attenuated Total Reflection (ATR) techniques help to overcome difficulties with high absorption samples. ATR probes have also been used to monitor chemical reactions (see e.g. [69, 70]. Various applications of Fourier Transform-IR spectroscopy in colloid and interface science have been reported in the literature [71].

4. Fluorescence Spectroscopy

Fluorescence techniques have been demonstrated to be useful for in-situ monitoring of cure and polymerization [72 - 76]. Pyrenyl fluorescent probes have been used to observe the formation and growth of polymer particles during the emulsion polymerization of styrene [75]. Lacik et al. [76] reported steady-state fluorescence measurements to study the inverse microemulsion polymerization of acrylamide.

The characterization of polymer colloids by fluorescence quenching techniques has been described by Winnik et al. [77]. The authors have been able to draw conclusions about the internal particle structure, transport phenomena across boundaries, the conformation of stabilizers, and particle flocculation in latexes.

A traditional field for the application of various fluorescence techniques are the micellization behavior as well as clouding phenomena of surfactants in solution [78 - 83]. Fluorescence measurements have been shown to be useful to study various aspects of latex coalescence and film formation [84 - 88].

5. Data Analysis

Quantitative spectroscopy usually requires calibration functions which reflect the relation between the measured quantity and the concentration of the components. For multicomponent analysis, a multitude of mathematical procedures has been developed. Multivariate techniques (chemometrics) are now widely applied to design measurement procedures and provide maximum chemical information by analyzing spectral data. Additionally, a basic condition for high-quality quantitative spectroscopy is the (frequency) calibration of the instruments.

Instrument calibration and quantitative data analysis are beyond the scope of this paper. The reader should be referred to textbooks (e.g. [3, 89]) and journals [90].

Acknowledgements: The author is indebted to P. Codella (GE CRD) and M. Hetem (GEP Europe) for experimental support and collaboration. Helpful discussions with M. Krell and H. Zecha (Wacker Chemie) are greatly appreciated. The author would like to thank Wacker-Chemie GmbH for the permission to publish this paper.

6. References

1. Hendra, P.J., Jones, C., and Warnes, G. (1991) *Fourier Transform Raman Spectroscopy - Instrumentation and Chemical Applications*, Ellis Horwood, Chichester
2. Chase, D.B. and Rabolt, J.F. (1994) *Fourier Transform Raman Spectroscopy*, Academic Press, San Diego
3. Schrader, B. (1995) *Infrared and Raman Spectroscopy - Methods and Applications*, VCH, Weinheim
4. Hirschfeld, T. and Chase, D.B. (1986) *Appl. Spectrosc.* **40**, 133
5. Schrader, B., Hoffmann, A., Simon, A., Podschadlowski, R. and Tischer, M. (1990) *J. Molec. Struct.* **217**, 207
6. Crookell, A., Hendra, P.J., Mould, H.M., and Turner, A.J. (1990) *J. Raman Spectrosc.* **21**, 85
7. Schrader, B., Hoffmann, A., Simon, A., and Sawatzki, J. (1991) *Vib. Spectrosc.* **1**, 239
8. Chase, B. (1994) *Appl. Spectrosc.* **48**, 14A
9. Hendra, P.J., Pellow-Jarman, M.V., and Bennett, R. (1993) *Vib. Spectrosc.* **5**, 311
10. Denton, M.B. and Gilmore, D.A. (1995)) *SPIE Proceedings* **2388**, 121
11. Vickers, T.J. and Mann, C.K. (1992) *SPIE Proceedings* **1637**, 62
12. Angel, S.M., Carrabba, M., and Cooney, T.F. (1995) *Spectrochim. Acta* **A51**, 1779
13. Cooper, J.B., Flecher, P.E., Albin, S., Vess, T.M., and Welch, W.T. (1995) *Appl. Spectrosc.* **49**, 1692
14. Cooney, T.F., Skinner, H.T., and Angel, S.M. (1995) *Appl. Spectrosc.* **49**, 1846
15. Witke, K. and Kimmer, W. (1976) *Plaste & Kautschuk* **23**, 799
16. Witke, K., Buge, H.-G., Brzezinka, K.-W., and Kimmer, W. (1983) *Acta. Polym.* **34**, 627
17. Carius, W., Palm, K., and Schröter, O. (1977) *Wiss. Z. Päd. Hochschule Erfurt-Mühlhausen* **13**, 133
18. Chu, B., Fytas, G., and Zalczer, G. (1981) *Macromolecules* **14**, 395
19. Sears, W.M., Hunt, J.L., and Stevens, J.R. (1981) *J. Chem. Phys.* **75**, 1589 and 1599
20. Gulari, E., McKeigue, K., and Ng, K.Y.S. (1984) *Macromolecules* **17**, 1822
21. Damoun, S., Papin, R., Ripault, G., Rousseau, M., Rabadeux, J.C., and Durand, D. (1992) *J. Raman Spectrosc.* **23**, 385
22. Kamath, M., Kim, W.H., Li, L., Kumar, J., Tripathy, S., Babu, K.N., and Talwar, S.S. (1993) *Macromolecules* **26**, 5954
23. Feng, L. and Ng, K.Y.S. (1990) *Macromolecules* **23**, 1048
24. Wanchek, P.L. and Wolfram, L.E. (1976) *Appl. Spectrosc.* **30**, 542
25. Dywan, F., Hartmann, B., Klauer, S., Lechner, M.D., Rupp, R.A., and Wöhlecke, M. (1993) *Makromol. Chem.* **194**, 1527
26. Hergeth, W.-D. and Codella, P.J. (1994) *Appl. Spectrosc.* **48**, 900
27. Cornell, S.W. and Koenig, J.L. (1969) *Macromolecules* **2**, 540
28. Sloane, H.J. and Bramston-Cook, R. (1973) *Appl. Spectrosc.* **27**, 217
29. Poshyachinda, S., Edwards, H.G.M., and Johnson, A.F. (1991) *Polymer* **32**, 338
30. Wang, C., Vickers, T.J., Schlenoff, J.B., and Mann, C.K. (1992) *Appl. Spectrosc.* **46**, 1729
31. Chong, C.K., Shen, C., Fong, Y., Zhu, J., Yan, F.-X., Brush, S., Mann, C.K., and Vickers, T.J. (1992) *Vib. Spectrosc.* **3**, 35
32. Wang, C., Vickers, T.J., and Mann, C.K. (1993) *Appl. Spectrosc.* **47**, 928
33. Williams, K.P.J. and Mason, S.M. (1990) *Spectrochim. Acta* **2**, 187
34. Ben-Amotz, D., LaPlant, F., Jiang, Y., and Biermann, T. (1994) *Proceedings 35th Ann. Polyurethane Technical/Marketing Conference*, 375
35. Vickers, T.J. and Mann, C.K. (1995) *SPIE Proceedings* **2367**, 219
36. Vickers, T.J. and Mann, C.K. (1995) *SPIE Proceedings* **2504**, 310
37. Owen, H. and Pelletier, M. (1995) *Laser Focus World* , 95
38. Chalmers, J.M. and Everall, N.J. (1996) *Trends Anal. Chem.* **15**, 18
39. Everall, N. (1994) *J. Raman Spectrosc.* **25**, 813
40. Schrader, B., Hoffmann, A., and Keller, S. (1991) *Spectrochim. Acta* **47A**, 1135
41. Everall, N. and Lumsdon, J. (1991) *Vib. Spectrosc.* **2**, 257

256

42. Petty, C.J. (1991) *Vib. Spectrosc.* **2**, 257
43. Sidorov, N.K., Stalmakhova, L.S., and Bogachyov (1971) *Opt. Spectrosc.* **30**, 375
44. Mirone, P. (1966) *Spectrochim. Acta* **22**, 1897
45. Fini, G., Mirone, P., and Patella, P. (1968) *J. Mol. Spectrosc.* **28**, 144
46. Brenan, C.J.H. and Hunter, I.W. (1995) *Appl. Spectrosc.* **49**, 1086
47. Vorsina, I.A., Grishakova, T.E., and Mikhailov, Yu.I. (1995) *Inorg. Mater.* **31**, 1321
48. Berger, A.J., Wang, Y., Sammeth, D.M., Itzkan, I., Kneipp, K., and Feld, M.S. (1995) *Appl. Spectrosc.* **49**, 1164
49. Maeda, Y. and Kitano, H (1995) *Spectrochim. Acta* **A51**, 2433
50. Stuart, B.H. (1995) *Polym. Bull.* **35**, 727
51. Nickolov, Zh.S. and Earnshaw, J.C. (1995) *J. Molec. Struct.* **348**, 273
52. Lawson, E.E., Edwards, H.G.M., and Johnson, A.F. (1995) *Spectrochim. Acta* **A51**, 2057
53. Davie, A.S., Kavanagh, P.E., and Leong, W.-H. (1995) *J. Coat. Technol.* **67**, 63
54. Nishikawa, Y., Fujiwara, K., Ataka, K.-I., and Osawa, M. (1993) *Anal. Chem.* **65**, 556
55. Hill, W., Wehling, B., Gibbs, C.G., Gutsche, C.D., and Klockow, D. (1995) *Anal. Chem.* **67**, 3187
56. Hill, W., Wehling, B., Fallourd, V., and Klockow, D. (1995) *Spectrosc. Europe* **7**, 20
57. Stokes, D.L., Alarie, J.P., and Vo-Dinh, T. (1995) *SPIE Proceedings* **2504**, 552
58. Pal, A., Stokes, D.L., Alarie, J.P., and Vo-Dinh, T. (1995) *Anal. Chem.* **67**, 3154
59. Mullen, K.I., Wang, D.X., Crane, L.G., and Carron, K.T. (1992) *Anal. Chem.* **64**, 930
60. Lai, E.P.C. and Ghaziaskar, H.S. (1994) *Appl. Spectrosc.* **48**, 1011
61. Aker, P.M., Moortgat, P.A., and Zhang, J.-X. (1995) *SPIE Proceedings* **2547**, 110
62. Rowell, R.L. and Ford, J.R. (1981), in D.R. Bassett and A.E. Hamielec (eds.), *Emulsion Polymers and Emulsion Polymerization*, ACS Symposium Series vol. 165, ACS, Washington, 85
63. Rowell, R.L. (1990), in F. Candau and R.H. Ottewill (eds.), *An Introduction to Polymer Colloids*, Kluwer Academic Publ., Dordrecht, 187
64. Gossen, P.D., MacGregor, J.F., and Pelton, R.H. (1993) *Appl. Spectrosc.* **47**, 1852
65. Mijovic, J. and Andjelic, S. (1995) *Polymer* **36**, 3783
66. Mijovic, J., Andjelic, S., and Kenny, J.M. (1996) *Polym. Adv. Technol.* **7**, 1
67. Reeves, J.B. (1994) *J. Near Infrared Spectrosc.* **2**, 199
68. Bouveresse, E., Hartmann, C., Massart, D.L., Last, I.R., and Prebble, K.A. (1996) *Anal. Chem.* **68**, 982
69. Dietz, J.E., Elliott, B.J., and Peppas, N.A. (1995) *Macromolecules* **28**, 5163
70. MacLaurin, P., Crabb, N.C., Wells, I., Worsfold, P.J., and Coombs, D. (1996) *Anal. Chem.* **68**, 1116
71. Scheunig, D.R. (1990) *Fourier Transform Infrared Spectroscopy in Colloid and Interface Science*, ACS Symposium Series vol. 447, ACS, Washington
72. Dousa, P., Konak, C., Fidler, V., and Dusek, K. (1989) *Polym. Bull.* **22**, 585
73. Pekcan, Ö., Egan, L.S., Winnik, M.A., and Croucher, M.D. (1990) *Macromolecules* **23**, 2210
74. Wang, Z.J., Song, J.C., Bao, R., and Neckers, D.C. (1996) *J. Polym. Sci. B Polym. Phys.* **34**, 325
75. Rudschuk, S., Adams, J., and Fuhrmann, J. (1995) *DECHEMA Monographs* **131**, 159
76. Lacik, I., Barton, J., and Warr, G.G. (1995) *Macromol. Chem. Phys.* **196**, 2223
77. Winnik, M.A. and Croucher, M.D. (1987), in M.S. El-Aasser and R.M. Fitch (eds.) *Future Directions in Polymer Colloids*, Martinus Nijhoff, Dordrecht, Boston, Lancaster
78. Yan, Y.D. and Clarke, J.H.R. (1989) *Adv. Colloid Interface Sci.* **29**, 277
79. Winnik, F.M. (1989) *J. Phys. Chem.* **93**, 7452
80. Wilhelm, M., Zhao, C.-L., Wang, Y., Xu, R., Winnik, M.A., Mura, J.-L., Riess, G., and Croucher, M.D. (1991) *Macromolecules* **24**, 1033
81. Astafieva, I., Khougaz, K., and Eisenberg, A. (1995) *Macromolecules* **28**, 7127
82. Holland, R.J., Parker, E.J., Guiney, K., and Zeld, F.R. (1995) *J. Phys. Chem.* **99**, 11981
83. Komaromy-Hiller, G. and von Wandruszka, R. (1996) *J. Colloid Interface Sci.* **177**, 156
84. Pekcan, Ö., Winnik, M.A., and Croucher, M.D. (1990) *Macromolecules* **23**, 2673
85. Zhao, C.-L., Wang, Y., Hruska, Z., and Winnik, M.A. (1990) *Macromolecules* **23**, 4082
86. Pekcan, Ö. and Canpolat, M. (1996) *J. Appl. Polym. Sci.* **59**, 277
87. Canpolat, M. and Pekcan, Ö. (1996) *J. Appl. Polym. Sci.* **59**, 1699
88. Canpolat, M. and Pekcan, Ö. (1996) *J. Polym. Sci.B Polym. Phys.* **34**, 691
89. Martens, H. and Næs, T. (1989) *Multivariate Calibration*, Wiley, Chichester
90. *Chemometrics and Intelligent Laboratory Systems*, Elsevier

MONITORING POLYMERIZATION REACTORS BY ULTRASOUND SENSORS

M. MORBIDELLI, G. STORTI [1], A. SIANI
Chemical Engineering Department /LTC
Universitätsstrasse 6, ETH-Zentrum
CH-8092 Zürich, Switzerland

1. Introduction

The propagation of small pressure pulses or sound waves in a fluid is a source of informations for single and multiphase dipersed systems. A sensing instrument based on the measurement of the ultrasound speed (frequencies between 20 kHz and 100 MHz) represents a cheap, rapid, low energy and non-invasive apparatus, able to monitor on-line the evolution of some physical properties in both single and two-phase dispersed systems.

In this chapter, after a brief summary of the main aspects characterizing the ultrasound propagation velocity (upv) in a single fluid phase, the case of dispersed systems is examined. Then, applications to polymerizing mixtures and, in particular, to the emulsion process, are discussed; details concerning sensor calibration and its use in conversion monitoring of homo- and copolymeric systems are presented.

2. Ultrasound Speed in Single Phase Fluid Systems

In general, the upv in a single phase is related to some properties of the medium through the following equation:

$$c^2 = \frac{1}{\rho\beta} = -\frac{1}{\rho V}\left(\frac{\partial V}{\partial P}\right)_s \tag{1}$$

where c indicates the upv, ρ the density and β the so-called adiabatic compressibility; it is defined as in the right hand side of the previous equation and is related to the volume variation imposed by the forcing pressure wave to the fluid phase.

When an ideal gas phase is considered, the compressibility reduces to $1/\gamma P$, where γ, the ratio between the constant pressure and constant volume heat capacities, is a function of temperature only. In the case of real gases, if molecular weight, specific heats and an equation of state are known, the upv can be calculated a priori. Thus, the upv in gas phase is essentially function of pressure and temperature, being substantially independent of the frequency of the ultrasound wave at least in the typical accessible range of values.

[1]permanent address: Dip. Ingegneria Chimica e Matematica, Università di Cagliari, Piazza d'Armi, 09123 Cagliari, Italy

J. M. Asua (ed.), Polymeric Dispersions: Principles and Applications, 257–266.

In the case of liquids, the situation is practically the same as that of the real gases but with sound speed values independent of both pressure and frequency. Through eq (1), the measured values of upv allow us to estimate one between density and compressibility, being the other known. A typical application to pure components is the measurement of the β dependence upon the temperature, being the same dependence for the density available.

In the case of mixtures, the relationships between density and compressibility and composition are required. Then, the same methodologies usually applied to the evaluation of the volumetric behavior of fluid mixtures, may be used. Namely, an equation of state for the mixture or some mixing rule, properly accounting for the interactions among the various components in the mixture, is an essential prerequisite to any data reconciliation. As an example, if an empirical mixing rule for the mixture quantities and the corresponding pure component values is available, the measured upv values may be used to estimate the mixture composition. This is a common application in the food industry, where a concentration sensor based on ultrasonic measurements is particularly attractive because of its non-invasive nature.

3. Ultrasound Speed in Two-Phase Fluid Systems

3.1. GENERAL ASPECTS

When two-phase systems are examined, the wave speed is known to be dependent on the properties of the two constituents as well as their relative amounts. Moreover, in the case of dispersed systems such as suspensions, slurries and emulsions, informations concerning the size of the dispersed phase may be extracted and this aspect is very attractive from the applicative viewpoint. Examples of applications in both situations are the morphological characterization of polymeric blends [1], the probing of marine sediments [2], and the monitoring of sedimentation of colloidal dispersions [3].

In this case, an essential aspect is the "degree" of dispersion of one phase in the other, ranging from the completely segregated case to a truly dispersed situation, according to a dimensionless quantity, called acoustic wavenumber, which is defined as $kd_p = \pi d_p / \lambda$,

being λ the sonic wavelength and d_p the particle diameter.

For large wavenumber values $\left(\geq 10^{-2} \right)$, the dispersed system behaves as two segregated phases and the corresponding upv is evaluated by a suitable average of the two values corresponding to the two separated phases. This means that, from the viewpoint of sound propagation, the system behaves as a "series" of two different sound paths. The same relationships discussed in the previous section apply to each phase, being the resulting upv value given by [4]:

$$c = \left(\frac{\varphi_1}{c_1} + \frac{1 - \varphi_1}{c_2} \right)^{-1} \tag{2}$$

where φ_1 indicates the volume fraction of phase 1 and c_1 and c_2 the upv values corresponding to phase 1 and 2, respectively. Obviously, no information about the particle size can be obtained in this case.

For low values of the acoustic wavenumber, the dispersed nature of the system has to be accounted for and different theoretical formulations are available to evaluate the ultrasonic propagation in such systems (cf. [5]).

The first proposed approach is phenomenological [6]. It is based on an extension of the relationship valid for a single phase, eq (1), to suspensions by the introduction of some "effective" values of both density and compressibility, i.e.:

$$c^2 = \frac{1}{\rho_{eff} \, \beta_{eff}} \tag{3}$$

If the index 1 indicates the suspending phase and 2 the suspended one, the simplest rules for evaluating these effective properties are the following volumetric averages:

$$\rho_{eff} = \rho_1 \varphi_1 + \rho_2 \varphi_2 \tag{4}$$

$$\beta_{eff} = \beta_1 \varphi_1 + \beta_2 \varphi_2 \tag{5}$$

Note that these equations represent the so-called volume additivity rule when calculating the average density of a two-phase liquid system and the only averaging rule in common use when calculating the average compressibility. By more comprehensive treatments, such as those briefly discussed below, the previous equations have been found accurate when dispersions of very fine particles with viscosity at least one order of magnitude larger than that of the dispersing medium are considered $\left(kd_p \cong 10^{-5} \div 10^{-4} [7]\right)$ eventhough a systematic analysis of their applicability limits is still lacking.

More fundamental approaches are available in the literature for evaluating the propagation of pressure waves through emulsions and suspensions, with particular emphasis on the case of particles with viscosity larger than that of the suspending medium (solid or polymeric particles). They may be classified into two classes [5]: coupled-phase models and multiple scattering treatments.

In the first case, the conservation equations of mass, momentum and energy, to be combined with a stress-strain relation and equations of state, are written for each separate phase. Some "coupling" terms are then included, i.e. the equation for the drag on one phase by the other. Examples are the treatments by Ahuja [8] [9]. In the second case, the ultrasonic scattering theory was initially applied to single, isolated particles [10] [11] and more recently extended to the case of interacting particles through the multiple scattering theory [12].

The results of the two approaches in terms of upv and attenuation (loss of energy experienced by the travelling pressure wave due to the conversion of organized, systematic motions of the particles into uncoordinated, random motions of thermal agitation) have been occasionally compared, but no systematic test has been reported. Notably, in the work by Harker and Temple [5] some requirements to be fulfilled so as to guarantee that the simple Urick's formula (eqs (3)-(5)) may be safely used are reported (the suspended material must have acoustic impedance close to that of the suspending phase, being this quantity given by the product ρc, density times the ultrasound velocity characterizing the material). Moreover, in the case of solid particles, parametric calculations of upv for different systems, particle sizes and ultrasound frequencies are reported, as obtained through the detailed model. One among the graphs in the paper mentioned above is reproduced in Figure 1: solid particles (Fe_3O_4) in water are considered

260

and the upv calculated for three values of the particle radius (1, 10, 100 µm), ultrasound frequency ranging from 10 kHz to 10 MHz and different volume fractions of the suspended material are shown (the previoulsy defined acoustical wavenumber is reported on the horizontal axis, using the symbol $k_p a$, instead of $kd_p/2$). By inspection of the calculated curves, it may be concluded that both particle size and volume fraction play a role in determining the actual upv value. However, by tuning the ultrasound frequency, operating conditions where the first dependence becomes negligible may be identified, being the upv affected by the solid content only (the case of the smallest particles in the figure).

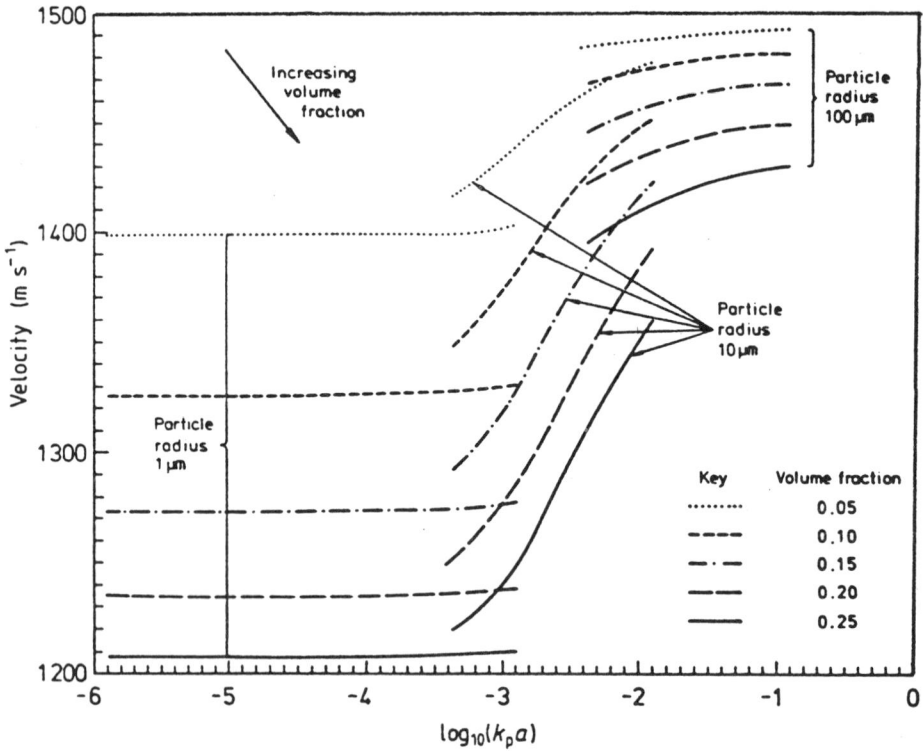

Figure 1. Ultrasound speed as a function of frequency, particle size and volume fraction of the dispersed phase (Fe_3O_4 in water; after Harker and Temple [5]).

Thus, in conclusion, when the applicative interest is mainly focused on the measurement of the dispersed phase content, a judicious selection of the ultrasound frequency is required, being the upv affected by relative amounts of the two phases, densities, compressibilities, viscosities and particle sizes.

3.2. APPLICATION TO EMULSION POLYMERIZATION

The application of an ultrasonic sensor to latex reactors for on-line monitoring of the polymer content is now examined. First, it has to be noted that two dispersed phases may be present in the system, the monomer droplets (about 10 µm of initial diameter, d_D) and polymer particles (about 0.1 µm of final diameter, d_p). As already noted, it is essential to compare these sizes with the wavelength of the ultrasound wave, i.e. to evaluate the corresponding wavenumbers. With reference to a frequency of 1 MHz, the following values can be estimated: $kd_D \cong 5 \ 10^{-2}$ and $kd_p \cong 3 \ 10^{-4}$. These results indicate that the two dispersed phases scale very differently with respect to the selected wavelength and, according to the discussion in section 3.1, the original three-phase system may be "lumped" in a system constituted of two segregated phases: the oil droplet phase, D, and a pseudo-phase water with polymer particles, WP. The second one is a true two-phase dispersion of polymer particles in water, as sketched in Figure 2. From a different viewpoint, the same schematization is naturally obtained when the numbers of particles of the two dispersed phases contained in a generic elemental volume of the emulsion with cubical shape and size equal to one tenth of the wavelength λ are estimated. While no more than a few tens of monomer droplets are obtained, the number of polymer particles is of the order of tens of millions. Then, the upv in the system may be evaluated through eq (2), i.e. through a formula which applies to the case of completely segregated phases.

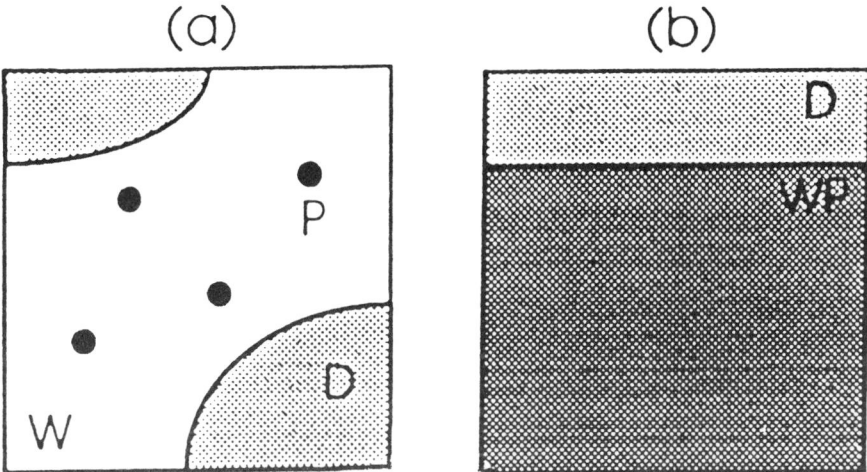

Figure 2. Schematic representation of the emulsion. (a) the real three-phase system and (b) the equivalent two-phase one. D = oil droplets, P = polymer particles, W = aqueous phase, WP = pseudo-phase of particles dispersed in water.

About the pseudo-phase WP, it is a dispersion of very fine polymer particles in water and one among the models discussed in the last previous section can be applied. In particular, due to the relevant viscosity difference between the dispersed and the continuous phase (polymer particles and water, respectively), the model proposed by Ahuja has been selected in the class of the coupled-phase models. Thus, under the assumption of small particles with viscosity at least one order of magnitude larger than that of the suspending fluid, the following explicit and relatively compact expression is obtained for the sound speed in phase WP (as in Figure 2, indices W and P characterize water and particle phase properties, respectively):

$$c_{WP}^2 = c_W^2 \frac{1 - \varphi L \cos \varepsilon}{\left[1 - \varphi\left(1 - \beta_p / \beta_w\right)\right]\left[1 + \varphi L(\tau \cos \varepsilon + s \sin \varepsilon)\right]} \tag{6}$$

$$\tau = \frac{1}{2} + \frac{\delta}{2d_p} \frac{(2\eta_w + 3\eta_P)^2}{(\eta_w + \eta_P + d_p\eta_w/6\delta)^2 + (d_p\eta_w/6\delta)^2} \tag{7}$$

$$L = \frac{\rho_P/\rho_w - 1}{[(\rho_P/\rho_w + \tau)^2 + s^2]^{1/2}} \; ; \quad \varepsilon = \tan^{-1}\frac{s}{\rho_P/\rho_w + \tau} \tag{8}$$

$$s = \frac{3\delta^2}{d_p^2}(2\eta_w + 3\eta_P) \frac{\left(1 + \frac{d_p}{2\delta}\right)(\eta_w + d_p\eta_w/d_p\delta) + \eta_w d_p^2/12\delta^2}{(\eta_w + \eta_P + d_p\eta_w/6\delta)^2 + (\eta_w d_p/6\delta)^2} \tag{9}$$

being the parameter δ defined as $(2\eta_W/\rho_W\omega)^{0.5}$, with ω the angular frequency of the sonic wave $(= 2\pi f)$ and φ the fractional volume of particles in the WP pseudo-phase. Note that the particle compressibility, β_P, is evaluated through eq (5) as adapted to a monomer-polymer homogeneous mixture, i.e. $\beta_P = \beta_m\phi_m + \beta_{pol}(1 - \phi_m)$, where ϕ_m is the volume fraction of monomer in particle and β_m, β_{pol} the compressibilities of monomer and polymer, respectively. Being the polymer particle viscosity much larger than ten times the water viscosity, the dimensionless quantity L approaches zero, with the ratio d_p/δ always less than one [13]. Therefore, the previous equations may be further reduced and the oversimplified, phenomenological Urick's equation (eqs (3)-(5)) is obtained. Thus, for the particular application under examination here and the selected frequency value, the ultrasonic technique results to be ideally suited to monitor the solid content, i.e. the conversion, wathever is the polymer particle size.

Thus summarizing, the upv in a reacting latex may be evaluated through eq (2) and eqs (3)-(5) during Intervals I and II and the last three equations alone during Interval III. The volume fractions of all phases can be evaluated as a function of conversion combining these equations with standard material balances for the organic (monomer and polymer) and aqueous phases, together with suitable monomer interphase partitioning laws (cf. [13]).

4. Illustrative Applications

The capabilities of an on-line sensor of conversion based on upv measurements during the polymerization in latex reactors have been assessed by experimental analysis of both homo- and copolymerization systems, in batch and semibatch operating modes. In all cases, a commercial sensor manufactured by Nusonics has been used. A detailed description of the sensor characteristics and of its application to a well stirred reactor may be found elsewhere [14]. The sound speed is measured through the "pulse travelling" technique, that is by measuring the time needed by an ultrasonic pulse to travel between two piezoelectric transducers at a fixed distance. The sensor provides on-line measurements (about one upv value per second) and can be directly plugged into the reacting mixture, without any sampling circuit.

Before examining the experimental results, few additional comments are required concerning the sensor calibration. When the conversion has to be estimated from upv measurements through the equations of the last previous section, the values of all involved parameters (densities and compressibilities) have to be available. In the case of density, literature values have been directly used for water, monomer and polymer. About compressibility, due to the lack of literature data for the monomers and to some scattering in the reported values for the polymers, the direct measurement by the sensor resulted to be the most effective way to operate with satisfactory accuracy. In particular, these measurements may be readily performed in the case of liquid monomers, while in the case of polymers, measurements on spent latexes at decreasing water-to-polymer ratios, W/P, are suitable to estimate the corresponding compressibility by extrapolating the measured values at zero W/P value. Moreover, the following check of the sensor calibration just at the beginning of each reaction is advisable. From the reaction recipe (and in absence of polymer), a theoretical value of the sound speed is readily estimated and compared with the actual experimental value as obtained on-line: in the case of limited discrepancies, a minor adjustement of the water phase compressibility is made and justified in terms of the effect of solubilized monomer.

4.1. HOMOPOLYMERIZATION

In Figure 3, the upv behavior as a function of conversion for the emulsion polymerization of methylmethacrylate (MMA) at three different values of the monomer-to-water ratio is reported. The experimental data are the on-line measured upv values together with the corresponding conversion values estimated by interpolating data obtained off-line by gravimetry. When comparing these data with the curves calculated through the equations of the previous section, a positive indication about the instrument capabilities as on-line sensor of conversion is obtained.

The calculated curves exhibit two slope changes, the first one between 20 and 30% of conversion and the second one close to 60%. While the first one is expected and related to the oil droplets disappearance, the second one has been obtained by an empirical modification of the expression for evaluating the polymer particle compressibility. In particular, the right hand side of eq (5) has been multiplied by the corrective term, G, defined as:

$$G = 1 \qquad\qquad \text{if} \quad \phi_m \geq \phi_{cr}$$

$$G = \frac{2.9 - \phi_m \left(\dfrac{1 - \phi_m}{1 - \phi_{cr}} \right)^b}{2.9 - \phi_{cr}} \qquad\qquad \text{if} \quad \phi_m < \phi_{cr} \qquad (10)$$

where ϕ_m indicates the volume fraction of monomer in the particles and b and ϕ_{cr} are two adjustable quantities. Their evaluation may be performed by fitting the model predictions to the upv data during interval III in reacting systems or, better, by an independent experiment of monomer addition to a spent latex, so as to simulate the same reaction interval in a non-reacting system. Usual values of b are between 4 and 6, being ϕ_{cr} always less than the saturation value and probably related to some spatial rearrangement of the macromolecules in the particles when a critical value of the monomer-to-polymer ratio is reached in the reaction locus.

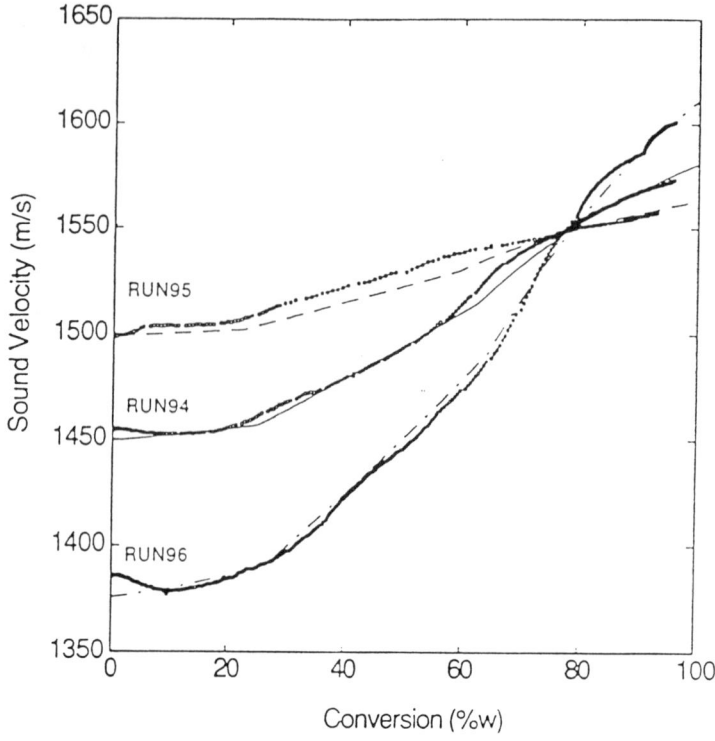

Figure 3. Ultrasound speed vs. conversion for the MMA homopolymerization at three different values of the monomer-to-water ratio, M/W: run 95 = 0.072; run 94 = 0.144; run 96 = 0.288.
• = experimental data; curves = calculated values.

Similar results have been obtained for different monomers such as styrene, vinyl acetate and butyl acrylate [13] [15].

4.2. COPOLYMERIZATION

When copolymerization systems are considered, a role of the composition of both the reacting monomer mixture and the produced polymer is expected. This behavior is well illustrated in Figure 4 where the upv curves as a function of conversion are reported for two homopolymers (styrene and butyl acrylate) and the corresponding 50/50w% copolymer. In this case, the copolymer curve is intermediate between those of the two homopolymers, thus confirming the effect of the system composition.

The same equations previously used in the homopolymer case have been adopted. However, in this case the knowledge of the corresponding composition values of both residual monomer mixture and produced copolymer is required, so as to properly evaluate the average values of density and compressibility of oil droplets and copolymer particles. These values have been estimated combining the equations for evaluating the upv with a model relating composition to conversion described elsewhere [16]. The main advantage of this model is that the required parameter values (reactivity ratios and interphase partitioning laws of the monomer species) are usually found in the literature, thus allowing one to use the model in a genuinely predictive way, without introducing new adjustable quantities. An additional difficulty arises when the evaluation of the compressibility of the copolymer particle at high conversion is examined, i.e. when b and ϕ_{cr} in eq (12) have to be estimated. Composition depent expressions for both these quantities are in fact not available and the same parameter estimation procedure previously discussed in the homopolymer case should be applied for each particular copolymer composition, thus reducing the sensor flexibility.

The conversion evolution in various copolymeric systems has been successfully monitored on-line through this type of sensor in batch and semibatch reactors [15] [17]. In particular, when combined with the theoretical monomer feed policies discussed elsewhere in this book, constant composition copolymers have been obtained without the typical problems of time irreproducibilities encountered when implementing the same monomer feed strategy using time instead of conversion as evolutionary coordinate.

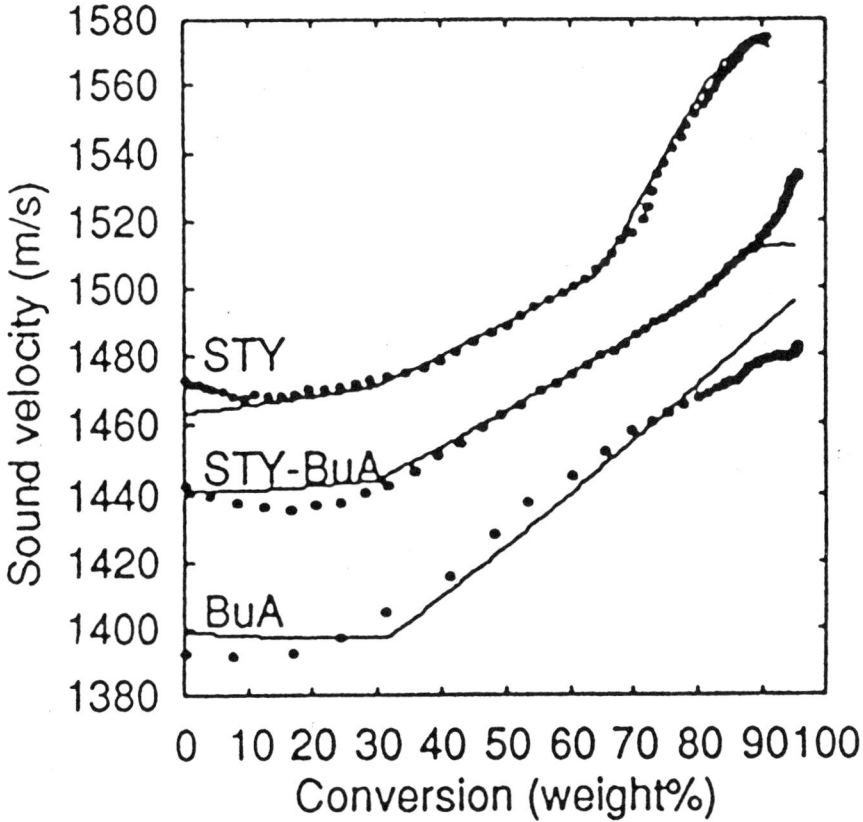

Figure 4. Ultrasound speed vs. conversion for the 50/50%w copolymerization of styrene and butyl acrylate and the two corresponding homopolymerization reactions.
• = experimental data; curves = calculated values.

266

5. References

1. Hourston, D.J., and Hughes, I.D. (1978) *Polymer*, **19**, 1181.
2. Hovem, J.M. (1980) *J. Acoust. Soc. Am.*, **67**, 1559.
3. Howe, A.M., Mackie, A.R., and Robins, M.M. (1986) *J. Dispersion Sci. Technol.*, **7**, 231.
4. Bonnet, J.B., and Tavlarides, L.L. (1987), *Ind. Eng. Chem. Res.*, **26**, 811.
5. Harker, A.H., and Temple, J.A.G. (1988), *J. Phys. D: Appl. Phys.*, **21**, 1576.
6. Urick, R.J. (1947) *J. Appl. Phys.*, **18**, 983.
7. Atkinson, C.M., and Kytomaa, H.K. (1992) *Int. J. Multiphase Flow*, **18**, 577.
8. Ahuja, A.S. (1972) *J. Acoust. Soc. Am.*, **51**, 916.
9. Ahuja, A.S. (1973) *J. Appl. Phys.*, **44**, 4863.
10. Epstein, P.S., and Carhart, R.R. (1953) *J. Acoust. Soc. Amer.*, **25**, 553.
11. Allegra, J.R., and Hawley, S.A. (1972) *J. Acoust. Soc. Amer.*, **51**, 1545.
12. McClements, D.J., and Povey, M.J.W. (1989) *J. Phys. D: Appl. Phys.*, **22**, 38.
13. Canegallo, S., Apostolo, M., Storti, G., and Morbidelli, M. (1995) *J. Applied Polym. Sci.*, **57**, 1333.
14. Apostolo, M., Canegallo, S., Siani, A., and Morbidelli, M. (1995) *Makromol. Chem., Macromol. Symp.*, **92**, 205.
15. Siani, A., Apostolo, M., Morbidelli, M., and Storti, G. (1995) in K.-H.Reichert and H.-U.Moritz (eds), *5th International Workshop on Polymer Reaction Engineering*, DECHEMA Monographs, Frankfurt, Vol. 131, p.149.
16. Canu, P., Canegallo, S., Morbidelli, M., and Storti, G. (1994) *J. Applied Polym. Sci.*, **54**, 1899.
17. Galimberti, F., Siani, A., Morbidelli, M., and Storti, G. (1997) *Chem. Eng. Comm.*, submitted.

ON-LINE CHARACTERIZATION METHODS

W.-D. HERGETH
Wacker-Chemie GmbH
LKE, P.O. Box 12 60
D-84480 Burghausen, Germany

1. Introduction

Increasing demands on product quality and constancy (i.e. the repeatability of product properties within narrow specifications for certain applications) require increasing efforts to characterize the final products as well as to control the processes for making them. Reaction control strategies rely on both efficient monitoring methods and state estimation and filtering techniques. The latter techniques as well as multivariable statistics and expert systems are beyond the scope of the present paper. The main focus is on the physical aspects of instrumentation for reaction monitoring.

In addition to product quality, the safe operation of a reactor is based on continuously monitoring some of the fundamental parameters of the process and the reactor itself. In a typical industrial reactor for emulsion polymerization, only a few parameters are measured continuously or on-line. Those parameters are temperature, pressure, agitation (stirring rate), feed rates of ingredients or withdrawing rates of reaction products, i.e. mainly physical parameters that describe the reactor operation. Parameters that describe the state of the (reacting) system are typically measured, if at all, off-line or only at the end of the reaction. As a general observation, industrial on-line applications are far away from technical opportunities.

On-line monitoring of physical and chemical properties of the reaction mixture contributes to i) increase the efficiency of the process via control strategies, ii) reduce labor costs by avoiding manual work, and iii) save time for analysis and reduce emissions by avoiding the transport of samples.

An excellent overview of on-line methods for polymerization monitoring has been published by Chien and Penlidis [1]. They pointed out that "some of today's off-line techniques may become tomorrow's on-line techniques." In the past, several off-line methods have been converted into on-line methods by automated, robotic assisted semi-continuous withdrawing samples out of the reactor or from a sampling loop, and feeding them into off-line instruments. Thus, a number of typical polymer as well as colloidal characterization methods have become available for on-line monitoring, and more methods will become available in the future.

Other techniques have been applied for on-line monitoring by integrating them into a sampling loop or by-pass. Reaction mixtures that are circulated in sampling loops or by-passes are prone to demixing and/or flocculation owing to their shear

J. M. Asua (ed.), Polymeric Dispersions: Principles and Applications, 267–288.
© 1997 *Kluwer Academic Publishers. Printed in the Netherlands.*

sensitivity. Choosing the right valves and the right pumping system (peristaltic vs. piston vs. membrane) is very important to avoid clogging of the pipes. The chemical and swelling resistance of the pipe material to monomers has also to be taken into consideration.

Recent tendencies of sensor development are i) miniaturization and modular assembly of the parts, ii) the integration of the sensor with signal transmission and analysis, iii) faster response times and increased sensitivity, and iii) remote sensing e.g. by the application of fiber optics.

In general, on-line characterization does not only include on-line monitoring of the monomer transformation to polymer but also monitoring of colloidal parameters (e.g. particle sizes or charges) during the course of the reaction. Even on-line monitoring of some specific applications of polymeric dispersions like film formation is possible. The present paper is divided into sections that describe methods to monitor i) the fractional conversion (i.e. the overall monomer conversion as well as the molecular chemical composition of the reacting mixture and of the polymer formed), and ii) the colloidal properties of the dispersion. We will discuss well-established methods as well as a couple of recent developments with potential for future application as on-line monitoring techniques. No attempt has been made to achieve completeness in describing methods and reviewing the literature.

2. Fractional Conversion

The most important technique to determine monomer conversion is still off-line gravimetry. It serves as the standard method for the calibration of other techniques such as the on-line techniques discussed below.

2.1. DENSITOMETRY AND DILATOMETRY

The physical basis for both methods, densitometry and dilatometry, is the density difference between monomer and polymer (Table 1). As the polymerization proceeds, this density difference leads to an increase of the latex density, and to an overall shrinkage of the reaction mixture volume.

Both changes can be measured continuously, and therefore, the reactor conversion can be calculated at any time knowing the amount of ingredients initially charged into the reactor and / or continuously fed to the reaction mixture. There are a couple of issues to be considered by applying these techniques:

i) It has to be checked separately whether the monomer-polymer mixtures and, in the case of water soluble monomers, the monomer-water solutions are ideal mixtures or not (i.e. check if the volume of the mixture is the sum of the volumes of the components).

ii) In the case of homopolymerizations, the density or volume change during a batch reaction is directly related to the fractional conversion $x(t)$ of a single monomer [2 - 4]:

$$x(t) = \frac{1/\rho_{emul} - 1/\rho(t)}{1/\rho_{emul} - 1/\rho_{end}} \tag{1}$$

where ρ_{emul}, ρ_{end} and $\rho(t)$ are the density of the monomer emulsion, the polymer latex and the reaction mixture, resp. For copolymerizations, the density (volume) of the mixture depends on both the overall monomer conversion and the copolymer composition. Therefore, one single integral variable (density, volume, temperature or pressure) is insufficient to describe multi-monomeric reactions. One either has to measure as many independent variables as reacting monomers are in the system, or additional information is required about the chemical composition of the formed copolymer or of the residual monomer phase (chromatography, spectroscopy).

iii) Accurate densitometric and dilatometric measurements rely substantially on the accuracy of temperature control to be better than ±0.1 K. This is, of course, also valid for ultrasonic measurements, pressure, etc.

TABLE 1: Densities of monomers and polymers at 20°C

monomer	density @ 20°C (g/cm^3)	
	$\rho_{monomer}$	$\rho_{polymer}$
acrylic acid	1.051	1.37
acrylonitrile	0.806	1.17
butyl acrylate	0.899	1.08
ethyl acrylate	0.924	1.12
2-ethyl hexyl acrylate	0.887	0.99
methyl acrylate	0.954	1.22
methyl methacrylate	0.944	1.19
styrene	0.906	1.05
vinyl acetate	0.932	1.18

There are several principles known to measure the density within a reactor. Probably the most robust method is to measure the resonance frequency of a U-shaped pipe placed in the sampling loop. This frequency is a very sensitive function of the mass of fluid within the tube, and, hence, of its density. After calibration of the instrument with a fluid of known density ρ_s (period of U-tube oscillation T_s), the density of the sample $\rho(t)$ can be derived from the change of the period of oscillation $T(t)$ [2 - 4]:

$$(\rho(t) - \rho_s) = const.(T^2(t) - T_s^2) \tag{2}$$

Clogging of the pipe and gas bubbles in the latex may cause problems with this method. They can partially be avoided by the inverse arrangement: a tuning fork inserted into the latex. Here again, the change of the resonance frequency of the fork is related to the density of the surrounding fluid. However, this fork is a dual sensor. The amplitude with which the fork is oscillating is a function of the medium's viscosity.

Difficulties with this method may occur because of coagulum formation at the tip of the fork.

The modified maximum bubble pressure method developed by Schork and Ray [2 - 4] both for surface tension and density measurements encounters trouble in industrial applications because of capillary clogging and coagulum formation. The method is based on the pressure p in a gas bubble of radius r at the capillary tip immersed in the latex (height h)

$$p = \rho \cdot g \cdot h + \frac{2 \cdot \gamma}{r} \tag{3}$$

γ is the surface tension of the bubble. For density measurements, two identical capillaries are inserted into the latex at different heights. The surface tension of the latex can be determined by using two capillaries with different diameter inserted into the latex at the same height.

Since pressure measurements are easy to carry out, the barometric difference of the hydrostatic pressure Δp measured at different heights Δh within the reactor can be used to calculate the latex density according to

$$\rho = \frac{\Delta p}{\Delta h \cdot g} \tag{4}$$

However, pressure meter readings are often modified by the hydrodynamics of the stirred polymeric dispersion.

Recent papers on the application of densitometry to reaction monitoring have been published by Morbidelli et al. [5 - 8]. The authors also combined densitometry with ultrasound velocity measurements to develop optimal monomer feeding policies in order to control the composition in multimonomer emulsion polymerization (see section 2.2.).

Recently, the application of dilatometry to reaction monitoring has been discussed in detail by Gilbert [9]. In general, dilatometry is a very useful tool for academic research, but only of limited value for reaction monitoring in continuously stirred industrial polymerization reactors.

2.2. ULTRASOUND

A physical method that is receiving increasing attention for on-line application is monitoring the propagation of ultrasound in the medium. The planar elastic ultrasonic wave (frequency f with $\omega = 2\pi f$) can be described by its complex pressure amplitude p(x,t)

$$p(x,t) = p_0 \cdot \exp\left[i\omega\left(t - \frac{x}{c}\right)\right] \cdot \exp[-\alpha \cdot x] \qquad (5)$$

with the ultrasonic velocity c which is defined in liquid system as

$$c_{liq}^2 = \frac{1}{\rho \cdot \beta}, \qquad (6)$$

and α being the attenuation of the wave in the liquid

$$\frac{\alpha_{liq}}{f^2} = \frac{2 \cdot \pi^2}{\rho \cdot c^3}\left(\frac{4}{3}\eta_s + \eta_v\right) \qquad (7)$$

where η_s and η_v are the shear and the volume viscosity of the liquid, respectively. It is obvious, that the measurable quantities c and α are determined by the system's density, viscosity and adiabatic compressibility β. Therefore, measuring e.g. the pulse travelling time between two ultrasonic transducers, and/or the attenuation of the sound pressure amplitude over the same distance enables one to draw conclusions about the state of the medium that is in between the transducers. This ultrasonic technique has been already widely applied to on-line monitor the concentration of solutions, emulsions and suspensions in the concentration range between 0.01 % and 100 % (e.g. sugar in water, alcohol in beer, fat in milk, etc.). It is a very easy-to-use, very fast (response time in the millisecond range), safe and nondestructive method. Reviews on the propagation of ultrasound in suspensions were published by Harker [10], Anson [11], and Farrow [12].

In dispersed polymeric systems, several loss mechanisms contribute to the sound attenuation α_{disp} [13]:

$$\alpha_{disp} = \alpha_0 + \alpha_{vis} + \alpha_{therm} + \alpha_{scatt} + \alpha_{relax} \qquad (8)$$

α_0 is the attenuation of the dispersion medium; α_{vis} summarizes viscous losses within the particles and at the interface between particles and continuous phase; α_{therm} describes thermal losses in dependence on heat conductivity and capacity; α_{scatt} is the contribution due to sound scattering in dispersed media; α_{relax} is caused by the dynamic relaxation of the polymeric material in dependence on frequency and temperature. All of those contributions are complicated functions of temperature, frequency, particle size and number. Thus, attenuation measurements are not very reliable for on-line reaction monitoring because of i) the complex nature of α_{disp}, and ii) technical problems measuring the attenuation at high solids and in the presence of gas bubbles in stirred reactors over an extended frequency range. Additionally, attenuation measurements are less accurate than velocity measurements

For reaction monitoring, the sound propagation velocity c is the the quantity of choice because it is easy to measure with high accuracy even at high conversion and solid contents of the latex. Both the density and the adiabatic compressibility change when monomer is transformed into polymer, i.e. the sound velocity is closely related to the conversion (see Table 2).

TABLE 2: Ultrasound velocities of monomers
and polymers at 20°C

monomer	sound velocity @ 20°C (m/s)	
	$c_{monomer}$	$c_{polymer}$
butyl acrylate	1233	1375
styrene	1354	2120
vinyl acetate	1150	1853
water	1483	

Early work on the application of ultrasound for reaction monitoring was reported in the 1980's by Hauptmann and Dinger [14, 15]. Very recently, Morbidelli and coworkers published a series of papers describing sound velocity measurements during the course of homo- and copolymerizations in combination with semi-empirical models to relate sound velocity and conversion [16 - 18]. The quantitative analysis of the sound velocity change during the reaction to calculate the monomer conversion will be discussed in detail in the paper by Morbidelli. In this analysis, it has to be considered that the latex is a dispersed system where the ratio between particle size and wavelength of the sound wave determines whether the system has to be treated as a homogeneous or as a particulate phase. Additionally, the non-ideal mixing rules both for the density and the compressibility in binary, ternary, etc. mixtures (e.g. monomer-polymer, monomer-surfactant-initiator-water) leads to a distinct non-ideal dependence of the sound velocity on the concentration of the ingredients. However, the sound velocity can either be i) calibrated by an independent (off-line) conversion measurement for a certain recipe and reaction procedure, or ii) calculated on the basis of theoretical models with a couple of assumptions, and by fitting of the unknown parameters, or iii) by a combination of both, i.e. a simplified linearized model which includes some pre-determined parameters of the components and of binary/ternary mixtures (velocity-temperature-coefficients, velocity-concentration-coefficients, etc.).

In industrial emulsion copolymerization processes, the theoretical evaluation of the sound velocity in the reacting system is more or less impossible. However, the sound velocity can be calibrated with off-line conversion measurements in subsequent runs under constant conditions. The most important aspect of this approach is its sensitivity to irreproducibilities of the polymerization process.

Ultrasonic sensors based on pulse travelling time measurements (emitter-receiver arrangement of the transducers) can either be mounted in a sampling loop or they can directly be plugged into the reaction vessel. Care should be taken in the latter case, because plugged-in transducers may cause coagulum formation at their edges.

The partition of monomers between aqueous phase and polymer particles can easily be derived from ultrasound velocity measurements. It is even possible to calculate sorption rates of monomers into polymer particles from the time dependence of the sound velocity after the addition of monomer to the latex [19, 20].

The frequency of ultrasonic sensors is typically in the 1 - 10 MHz range. In principle, the accuracy of the sound velocity measurement is poor for very low frequencies (< 100 kHz or so), and it increases with increasing frequency. However, the attenuation does also increase for f >> 10 MHz, i.e. the distance between the transducers has to be shortened in order to get a signal from the emitting to the receiving transducer. Shorter distances between the transducers reduce the accuracy of the pulse travelling time measurement. In highly concentrated systems, it might be difficult do circulate the reaction mixture between the transducers. Additionally, gas bubbles can get stuck and distort the measurement.

The disturbing influence of gas bubbles can be reduced by ring sensors. The formation of coagulum with immersion-type sensors can be avoided by sensors built-in flat into the reactor wall. There are some recent developments to measure the acoustic impedance Z via reflexion coefficient measurements with very high accuracy (10^{-5} to 10^{-6}). The advantage of this type is sensor is the high accuracy of density measurements. The reflexion coefficient R is defined as

$$R = \frac{Z2 - Z1}{Z1 - Z2} \tag{9}$$

with

$$Z = \rho \cdot c \tag{10}$$

Ultrasonic sensors can also be used to measure the viscosity with high precision and very fast response time, and the volume flux in tubes even when they are not completely filled. In combination with an ultrasonic density sensor, the mass transport in a tube is measurable [21, 22]. The filling level in reactors is accessible with ultrasonic distance sensors [23].

Sensors based on the Quartz Crystal Microbalance (QCM) principle are very sensitive for a wide variety of chemical applications. The high precision and ease of measuring the QCM resonance frequency has made it a useful tool for measuring slight changes in mass on the QCM electrode surface [24]. By coating the surface with a sensitive layer, the QCM can be used as selective gas or even under-liquid sensor. In a recent publication, Lin and Ward [25] describe the determination of contact angles and surface tensions with the QCM.

2.3. CALORIMETRY

Polymerizations are exothermic reactions. Hence, temperature or heat flux changes during the course of the polymerization reflect directly the transformation of monomer to polymer. Calorimetry seems to be the simplest on-line technique for conversion measurements. On-line calorimetry to monitor reactions has been used by chemical companies for several decades (see e.g. [26 - 29]).

In a calorimeter, the rate of polymerization R_p is related to the heat generated by the reaction per unit time dQ_r/dt [30]:

$$Rp = \frac{d\,Q_r\,/\,dt}{\Delta H_p \cdot V_r} \qquad (11)$$

where ΔH_p is the enthalpy of polymerization, and V_r the volume of material in the reactor (usually the amount of water). From this, the fractional conversion $x(t)$ can be obtained by

$$x(t) = \frac{\int_0^t Q_r\,dt}{\Delta H_p \cdot M_0} \qquad (12)$$

where M_0 is the initial molar amount of monomer.

The heat flux owing to the reaction dQ_r/dt is the quantity of interest. In copolymerizations, it is the sum of the heat of reaction of the components. However, it is not the only quantity that influences the overall temperature-(or heat flux)-time characteristic of a stirred tank reactor. The energy balance of the reactor is also determined by heat transfer between reactor interior and jacket dQ_{trans}/dt, by losses owing to heat conduction and radiation of the reactor mantle dQ_{loss}/dt, by dissipation of mechanical energy due to stirring the viscous liquid dQ_{stirr}/dt, and by the accumulated heat flow dQ_{acc}/dt [31].

The accumulated heat accounts for temperature changes of the reactor because of different rates of heat generation and heat flow out-of (into) the reactor. This quantity is a major cause of uncertainty and error because it depends on the derivative of the reactor temperature T_r

$$\frac{d\,Q_{acc}}{dt} = \frac{d\,T_r}{dt} \cdot C_{p,tot} \qquad (13)$$

where $C_{p,tot}$ is the heat capacity of reactor plus latex, (i.e. sum of specific heat capacity times mass for reactor and latex). dQ_{loss}/dt can be experimentally determined, and kept almost constant by insulation. dQ_{stirr}/dt has also to be determined by separate experiments. The heat transfer between reactor interior and jacket depends on both their temperatures T_r and T_j, respectively,

$$\frac{d\,Q_{trans}}{dt} = U \cdot A \cdot \left(T_j - T_r\right) \qquad (14)$$

The reactor area for heat exchange with the jacket A is almost constant. The heat transfer coefficient U typically changes during the course of the reaction because it depends on e.g. the latex viscosity and density, the heat conductivity through the reactor-jacket interlayer, the hydrodynamics of the reactor. Therefore, precalibration of U by separate experiments may lead to errenous results. Despite changing U, correct heat balance calorimetry can be performed by measuring the jacket inlet and outlet temperatures [32]. Reichert [33] developed another elegant method to overcome the difficulty of changing U: temperature oscillation calorimetry. A small sinusoidal temperature change is added to the overall temperature-time performance of the reactor. This allows the simultaneous on-line calculation of the heat transfer value of the reactor as well as the rate of reaction.

In semicontinuous polymerizations, a mixing contribution dQ_{mix}/dt to the overall heat flow has to be taken into account if there is a temperature difference between feeding material and reactor ΔT[31]

$$\frac{d\,Q_{mix}}{dt} = \frac{dm}{dt} \cdot c_{p,feed} \cdot \Delta T \qquad (15)$$

where $c_{p,feed}$ is the specific heat capacity of the feed material.

Reaction calorimetry is a powerful tool to on-line monitor industrial polymerizations as well as to study the kinetics of polymerizations or to develop control strategies for reactions [34 - 37].

2.4. CHROMATOGRAPHY AND SPECTROSCOPY

Spectroscopic techniques can in principle be used to determine the average composition of the latex and to give additional information on microstructure and morphology. The application of optical spectroscopy (i.e. UV, VIS, NIR, IR, Raman, fluorescence) is subject of a separate paper. Applications of dielectric spectroscopy will be discussed in Section 3.1.

Nuclear Magnetic Resonance (NMR) is probably the most important spectroscopic technique to characterize latex or copolymer composition and polymer microstructure, since the signals of the monomeric units and the signals of their various spatial arrangements and configurations are much more pronounced in NMR than in any other spectroscopic technique. However, NMR is still an off-line technique because of elaborate sample preparation requirements, and the time involved to run spectra [38]. Jones and Stronks [39] published a note on quasi on-line application of pulsed NMR in latexes. They have shown that the solids content of rubber latex can be determined without any special sample preparation with a total measurement time of approximately 10 seconds.

Chromatographic methods have shown to be very valuble for on-line sensing of monomer / copolymer composition, particles sizes and size distribution, and even molecular weight distribution. Several experimental setups have been reported in the literature. In most cases, the samples were taken from a circulation loop of the reaction mixture by means of an injection valve, then automatically diluted, injected, eluted, and detected.

Gas chromatography (GC) both of the emulsion (e.g. [40 - 42]) and the head space of the reactor [43, 44] can be applied for on-line determination of residual monomer composition. It takes several minutes for the GC instrument to analyze the reaction mixture. Typical reaction times for emulsion polymerization runs are in the order of hours, in some industrial processes even more than one day. Hence, there is a sufficient number of monomer composition results per time to efficiently control the reaction based on the GC analysis.

Head space GC is not straightforward for direct reaction monitoring because it requires knowledge of the partition of each monomer between the gas phase above the reaction mixture, the aqueous phase, the monomer droplets, and polymer particles. Additionally, the head space should be in thermodynamical equilibrium with the fluid reaction mixture.

With emulsion GC, the absolute monomer concentration can be determined if either one of two conditions is fulfilled: i) use of an internal standard, or ii) inject always exactly constant sample volumes. Internal standards may influence the reaction kinetics or monomer partitioning. A very small and constant injection volume is not easy to guarantee. However, it is in general sufficient to know the monomer ratio, and to combine GC with another on-line technique (e.g. densitometry) in order to calculate the absolute monomer concentrations. Emulsion GC may suffer from demixing or flocculation of the mixture in the sampling loop, in valves or pumps. Some authors report clogging in the pipes at higher solids. The sampling loop, pumps and valves have to be cleaned carefully on a regular basis.

In principle, liquid chromatography techniques (High Performance Liquid Chromatography, Gel Permeation Chromatography) can also be carried out as on-line characterization methods [1, 45]. The emulsion samples withdrawn from the reaction mixture can be injected directly on the column. GPC is somewhat complementary to GC, because it provides directly the chemical composition of the copolymer formed rather than the composition of the non-transformed monomer.

Size exclusion chromatography, hydrodynamic chromatography and fractionation techniques for particle size analysis will be discussed in a later subsection (3.1.).

3. Colloidal Characterization

3.1. PARTICLE SIZE AND PARTICLE SIZE DISTRIBUTION

An extremely important colloidal characteristic of polymer latexes is their particle size distribution. Almost all properties of the latex (electrical, optical, rheological, etc.) as well as its stability are affected by the particle size distribution. A wide variety of experimental techniques is available for off-line determination of particle sizes and their distribution functions, e.g. microscopy based techniques (SEM, STEM, TEM, ESEM or AFM), scattering methods (light, X-rays, neutrons) or chromatographic methods (SEC, HDC, FFF). The extensive sample preparation requirements for microscopical determination of particle sizes, and the operating conditions of the instruments are prohibitive for the on-line application of these techniques.

3.1.1. Scattering Techniques

The theoretical background, basic principles and applications of scattering techniques are reviewed in the paper by Ottewill. Neutron scattering is a very powerful tool to study e.g. particle sizes, internal particle morphologies, particle dynamics (even under shear), the structure and dynamics of concentrated dispersions, or film formation. The main advantage of neutron scattering is the very elegant way to alter the scattering contrast by partial or total deuteration of the polymeric material. With this contrast variation it is possible to verify structures that will not be detectable by other techniques. The main drawback is that neutron scattering relies on the availability of a nuclear reactor which makes the method somewhat exotic for on-line application. The same holds for X-ray scattering: i) The beam intensity of conventional X-ray tubes is relatively low. Therefore, typical collection times for scattering intensity profiles are comparable to or longer than the characteristic time scale of the chemical reaction. ii) The use of the highly intense synchrotron radiaton for on-line particle monitoring is possible, but it depends, of course, on the availability of a synchrotron. Additionally, it has to be taken into consideration that intense X-rays may interfere with the reaction ingredients itself.

Particle size determination (include. size distribution) by light scattering techniques is a well established broad field of experimental methods. All of them are applicable in on-line mode either by mounting the instrument in a sampling loop or by "looking" through a window of the reactor wall. The latter methods have become available with the development of fiber optic devices. Most of the light scattering techniques require extensive dilution of the latex to avoid multiple scattering effects.

Turbidity measurement is a technique which is easy to perform. The wavelength-dependent turbidity τ of diluted latex sample will provide information on particle size and concentration. Several papers in the literature describe the on-line determination of particle size during emulsion polymerization by specific turbidity τ/φ or turbidity ratio measurements [46, 47]. The turbidity is related to the polymer volume fraction φ according to

$$\tau = \frac{3}{2}\varphi \frac{\int_0^\infty d_p^2 \cdot K\left(\frac{d_p}{\lambda_m}, \frac{n_p}{n_m}\right) f(d_p)\, d\,d_p}{\int_0^\infty d_p^3 f(d_p)\, d\,d_p} \tag{16}$$

where K, f, λ_m, and d_p are the scattering coefficient, the size distribution function, the wavelength of light in the medium, and the particle diameter, respectively. It is obvious, that K and τ are functions of both the refractive indices of the particles n_p and the medium n_m, and of the relative size of particles to the wavelength of the light d_p/λ_m. If the size distribution function f is known, the particle size distribution can be estimated from specific turbidity measurements at several wavelengths. Despite the complicated dependence of τ on K, and its direct dependence on the distributional form f (which is sometimes unknown), the specific turbidity can yield i) the turbidity average particle diameter and the volume-surface average diameter for small and large particles, resp., for any value of $m = n_p/n_m$, ii) the weight average diameter for $m < 1.15$ and particles that are smaller than the wavelength of the light, and iii) an estimate of the weight average particle diameter in all other (monomodal) cases if a log-normal particle size distribution is assumed [47].

Recently, efforts were reported to extend the theoretical basis of turbidity to higher concentrations which provides e.g. information about particle interactions [48, 49].

Colloidal refractometry is a light scattering method where one obtains the refractive index n of a latex, and analyzes it using Mie theory. This method allows the probing of the structure of concentrated dispersions in their undiluted state. The measured values of n can provide a measure of the volume average particle size [50].

Several instruments are on the market which automatically dilute the latex to a desired (extremely low) concentration and perform static multiple angular light scattering. Several modes of light scattering (incl. Frauenhofer diffraction) are typically combined within one instrument to enable particle size analysis over a wide range of sizes (e.g. "MICROTRAC" by Leeds and Northrup, or "COULTER LS" by Coulter Electronics).

In single particle counting techniques (e.g. Flow Ultramicroscopy, Light-Scattering Counters, Phase-Doppler Analysers), the sufficiently diluted sample is passed through a laser beam such that light is scattered by only one particle at a time [51]. Very high resolution can be achieved by Aerosol Spectroscopy either with Ar-laser or Xe-light illumination [52, 53]. In several instruments, hydrodynamic (or aerodynamic) focusing of particle trajectories in relation to the measurement zone is used to reduce errors when measuring particle sizes [54]. So far, on-line applications of these techniques have not been reported in the literature.

A fast-response multi-channel photometer capable for on-line monitoring even at moderate concentrations has been described by Moser [55]. This static light scattering instrument has a response time of 100 ms and an angular resolution of 1°. The scattered light is simultaneously measured at 168 angles.

Photon Correlation Spectroscopy (PCS) is a well known technique to study the dynamic behavior and the structure of colloidal systems. In highly diluted dispersions, it is relatively easy to determine the (hydrodynamic) particle size distribution as well as electrokinetic properties of the latex. With the development of fiber optics it is even possible to apply this method to highly concentrated systems (FOQELS - Fiber Optic Quasi Elastic Light Scattering [56, 57]). Several probe designs have been developed which focus the light through a window into the latex, or the tip of the sensor is directly immersed into the dispersion. A serious problem of the FOQELS method is the influence of multiple scattering of light on the results. Even with excellent fiber optics, the application of this method is restricted to particle concentrations well below the range of industrial importance.

The effect of multiple light scattering in moderately concentrated dispersions can partially be overcome by i) using extremely thin cells (i.e. decreasing the number of particles in a very small scattering volume), or ii) decorrelation techniques. Thin cells are not suitable for on-line measurements at high concentration. Decorrelation experiments have been developed for 90° single-color scattering, and as two-color scattering for arbitrary scattering angle. With Two-Color Dynamic Light Scattering (TC-DLS), the cross-correlation between the scattered light at two different wavelengths is determined rather than the autocorrelation of light at one wavelength as in conventional PCS [58 - 60]. This method allows the elimination of single-scattered light at both wavelengths. The multiple-scattered light is highly un-correlated due to the scattering geometry.

On the other hand, multiple scattering of light can also be useful to study highly concentrated dispersions. The path of multiple scattered light in the latex is similar to the path of particles or molecules under the influence of Brownian diffusion. The concept of Diffusing Wave Spectroscopy (DWS) has already been applied to particle size determination in the industry [60 - 66].

3.1.2. *Chromatographic Methods (Fractionation)*
In the past decade, particle chromatographic methods (particle fractionation) have been developed that are capable for particle size on-line monitoring. Size Exclusion Chromatography (SEC) has been shown to be applicable with porous (Liquid Exclusion Chromatography LEC) and non-porous packing of the column (Hydrodynamic Chromatography HDC). The main advantage of these methods is that the particle size distribution can be obtained directly (i.e. without any assumption about the particle size distribution) after calibration of the instrument with particle size standards. Disadvantages are the limited resolution because of radial dispersion, and relatively long elution times up to half an hour [45].

In disc centrifugation, particle separation is caused by differences in particle velocity in an applied centrifugal field. This makes the results dependent on particle density as well as on their size. The typical time for centrifuge runs is in the range of half an hour up to several hours. Hence, disc centrifugation is not suitable for on-line monitoring despite the high resolution of this method.

Much higher resolution than with SEC can be achieved by Capillary Hydrodynamic Fractionation (CHDF [67 - 70]). The fluid in a capillary has a parabolic velocity profile with the greatest fluid velocity at the center of the tube, and zero velocity at the wall. Particles in this laminar flow will be moving radially due to their Brownian motion. Larger particles are unable to approach the wall as closely as smaller particles. Hence, larger particles will travel through the tube faster than smaller particles. This separation effect is exclusively a function of particle size, it is independent of particle density. The efficiency of capillaries to separate particles depends on the eluent viscosity, the flow rate, and the capillary diameter. The optimum particle size for CHDF analysis is $\ll 1\mu m$. The main disadvantage of this method is the possibility of capillary clogging. Recently, the CHDF method has been appplied to monitor the evolution of the particle size distribution during emulsion [71] and miniemulsion polymerization [72]. Industrial applications are also known.

Field-Flow Fractionation (FFF) is a family of chromatographic-like elution techniques based on influencing the rate of particle flow through a narrow channel by applying an external field perpendicular to the flow direction. FFF can be classified according to the type of applied field into sedimentation, thermal, electrical, cross-flow and steric. The most suitable types of external fields for particle size analysis are cross-flow and sedimentation field. The particle separation in cross-flow occurs according to the hydrodynamic radius of the particles, whereas their effective weight separates particles in a sedimentation field. The particle size ranges for cross-flow and sedimentation field FFF are 10 nm to 100 µm and 50 nm to 100 µm, resp. Overviews on FFF have been published by Giddings [73, 74]. Recently, FFF (incl. thermal FFF) has been applied to characterize size and composition of core-shell latices [75]. On-line coupling of FFF with multi-angle laser light scattering has been described by Roessner [76] and Wyatt [77].

3.1.3. *Spectroscopy*

Dielectric spectroscopy (in particular dielectric relaxation spectroscopy) has been rarely applied to disperse systems in the past [78, 79]. With this technique, the real or storage component $\varepsilon'(\omega)$ of the complex dielectric permittivity $\varepsilon^*(\omega)$ and the imaginary or loss component $\varepsilon''(\omega)$ are measured over several decades of the angular frequency ω in a relatively simple experimental set-up: two or four electrodes immersed into the liquid:

$$\varepsilon'(\omega) = \varepsilon'_0 - \frac{\left(\varepsilon'_0 - \varepsilon'_\infty\right)\omega^2 \tau^2}{1 + \omega^2 \tau^2} \tag{17}$$

$$\varepsilon''(\omega) = \frac{\left(\varepsilon'_0 - \varepsilon'_\infty\right)\omega\tau}{1 + \omega^2 \tau^2} \tag{18}$$

where ε'_0, ε'_∞ and τ are the the low and high frequency limits of ε' and the relaxation time, respectively, and $\varepsilon^*(\omega) = \varepsilon'(\omega) + i\varepsilon''(\omega)$. The sigmoidal change in $\varepsilon'(\omega)$ and the appearance of a peak in $\varepsilon''(\omega)$ is characteristic of a relaxation process. The processes involved are i) orientation of permanent dipols, ii) distortion of electrical double layers, iii) charge transfers at the electrodes, and iv) diffusion or displacement of ions. In the low frequency region, dielectric spectroscopy has been shown to provide information on particle size as well as on electrical double layer properties [80]. For a latex exhibiting a low-frequency relaxation process, the relaxation time at the loss maximum $\tau = 1/\omega_{max}$ is proportional to the square of the particle radius [81]. Within a couple of minutes measuring time it is possible to determine the particle size (and an estimate of size distribution), surface charge density and ionic strength of the system [79]. These characteristics change also during the reaction, and on-line monitoring is possible. Moreover, dielectric spectroscopy is especially valuable in looking at concentrated and opaque systems.

A large number of applications of dielectric spectroscopy published so far are for cure monitoring of resins (see e.g. [82]). Just as ultrasonic, optical, NMR and mechanical spectroscopy, dielectric spectroscopy of the polymer itself is capable to detect changes of the polymer crosslink density during the course of polymerization [83].

For the application of electroacoustic spectroscopy see chapter 3.2.

3.2. ZETA-POTENTIAL AND SURFACE CHARGE

In emulsion polymerization, one normally wants to produce a colloidally stable latex. The only quantity that allows indirect conclusions with respect to colloidal stability during the reaction is the particle size distribution. However, the particle size distribution is determined by several independent processes (growth, nucleation, agglomeration/flocculation, coagulation, swelling). Therefore, a direct measure of latex stability (either particle potential or surface charge) would be helpful in understanding the influence of colloidal stability on particle size distribution during the course of the reaction.

ζ-Potential measurements offer an opportunity to obtain information on stability parameters of the reacting system. Most of the experimental ζ-potential techniques described in the literature are not suitabe for on-line analysis of reaction mixtures. They typically need extensive dilution of the latex for e.g. laser doppler velocimetry or off-line observation of single particle motion in the dark field of a microscope. Mass transport of particles under the influence of an electric field in the original, undilutes state of the latex is difficult to determine because of the small density differences between polymer particles and and surrounding water. However, recently developed, electro-acoustic methods allow the determination of ζ-potentials in the original latex.

Electro-acoustic ζ-potentials are determined by applying an alternating field to the latex. This could be either an electric or an acoustic field. If one applies an alternating electric field to the latex the particles will move along the field lines according to their surface charge. This oscillatory particle motion leads to a net momentum transfer to the liquid because of density differences between particles and dispersion medium. In the vicinity of the electrodes, the particle motion generates a sound wave within the liquid that is characterized by a phase and an Electrokinetic Sonic Amplitude ESA [84]. ESA is the pressure amplitude normalized to the applied field, and, therefore, the direct analogy to the electrophoretic mobility (particle velocity with respect to the applied field). The ESA effect was discovered in the mid-1980s [85].

The inverse effect occurs when an alternating pressure field (sound) is applied to a colloidal dispersion. The density difference between dispersed matter and dispersion fluid causes a relative motion between the particles and the surrounding medium. This relative motion leads to a periodic displacement between the charged particles and the oppositely charged counterions of their electric double layers, i.e. an oscillating electric dipole is generated with the frequency of the sound wave. In electrolytes, this effect is known as the Debye-potential for about 60 years [86]. Nowadays, it is described as Ultrasonic Vibration Potential (UVP) or Colloid Vibration Potential (CVP) in the literature.

It was shown by the theoretical treatment of O'Brien et al. [87 - 90] that the ESA signal is directly proportional to the electrophoretic mobility of the particles μ_d (ω) at the frequency ω, and related to the CVP signal for the case of parallel plate cell geometry:

$$ESA(\omega) = \frac{P}{E} = c \cdot \Delta\rho \cdot \varphi \cdot G_f \cdot \mu_d(\omega) \tag{19}$$

$$CVP(\omega) = \frac{\Delta\psi}{U_0} = \frac{c \cdot \Delta\rho \cdot \varphi \cdot G_f \cdot \mu_d(\omega)}{K*} = \frac{ESA(\omega)}{K*} \tag{20}$$

where P and E are the pressure and electric field amplitude, respectively, c is the sound velocity in the latex, $\Delta\rho$ is the density difference between particle and medium, $\Delta\psi$ is the potential difference at the electrodes, φ is the particle volume fraction, U_0 is is the velocity amplitude of the applied sound wave, and K* is the complex high-frequency conductivity of the latex. G_f is a factor for electrode geometry. ESA can be applied like CVP. However, in order to use the CVP one must know the conductivity of the dispersion as a function of frequency over the whole measuring range.

In the case of diluted latexes ($\varphi < 10\%$) with particles having thin electric double layers ($\kappa r_p > 50$; κ - Debye-Hückel parameter, r_p - particle radius), the relation between particle mobility and ζ-potential in an ESA experiment is similar to Smoluchowski's equation:

$$\mu_d(\omega) = \frac{2 \cdot \varepsilon \cdot \zeta}{3 \cdot \eta}(1 + f)G(\alpha) \tag{21}$$

with

$$\alpha = \frac{\omega \cdot r_p^2}{v} = \frac{\omega \cdot r_p^2 \cdot \rho}{\eta} \tag{22}$$

where ε, η and v are the dielectric constant, viscosity and kinematic viscosity of the liquid, respectively. The factor f is for correcting the electric field strength near the particle surface, i.e. it depends on both the frequency and the surface conductance of the particles. $G(\alpha)$ takes inertia forces on the particles into account. It is obvious from eq. (19), (21) and (22) that the ESA signal is related to both the particle size and the ζ-potential of the particles. Hence, it is possible to calculate particle size distributions as well as effective electrical potentials at the particle surface from measurements of the mobility spectrum over a certain frequency range. The linear relationship between ESA/CVP signal and particle volume fraction holds for $\varphi < 10\%$, it is nonlinear for higher particle concentrations. Albeit the lack of a theory for $\varphi \gg 10\%$, it is possible to determine the ESA/CVP signal and use it as a measure of the effective particle charge.

The sizing range of commercial ESA instruments (AcoustoSizer from Matec Applied Sciences) is about 100 nm to 10 µm for frequencies between 200 kHz and 15 MHz. ζ-potentials (either ESA or CVP) can be measured for even smaller particles. CVP-ζ-potentials and particle sizes in the same order of magnitude can also be measured by acoustic attenuation spectroscopy (1 to 100 MHz) with the PenKem AcoustoPhor instrument.

The electroacoustic method has been applied to particle size and potential measurements of several inorganic suspensions (e.g. TiO_2, SiO_2, Al_2O_3, $CaCO_3$ [87 - 90]) as well as to concentrated emulsions [91]. So far, there is no paper in the literature dealing with on-line electroacoustic measurements of particle potentials or sizes during the course of emulsion polymerization (e.g. in a by-pass). In Figure 1, it is shown that effective ESA particle potentials in poly (vinyl acetate) latexes can easily be measured up to polymer fractions of about 50 wt.-%. Differences arise from different stabilization systems.

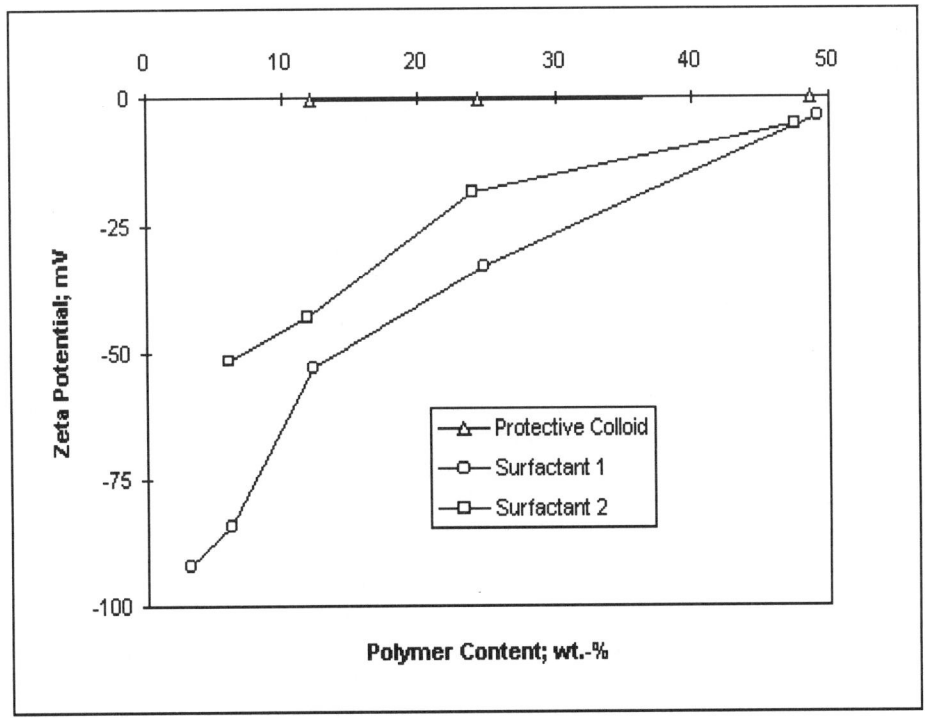

Figure 1. Effective ESA potentials of several poly (vinyl acetate) latexes
(Matec ESA 8000)

Methods other than electroacoustics or dielectric spectroscopy capable for on-line particle charge detection have not been reported in the literature. A technique of potential interest is the particle charge detector based on streaming potential measurements (when connected to a robot or on-line sampler [53, 92].

3.3. CONDUCTIVITY AND pH

The electrical conductivity Λ_d of a disperse system is a function of both the volume fraction of the dispersed phase φ ($\varphi \ll 1$) and the electrical double layer properties of the particles (expressed as surface conductivity Λ_s)

$$\Lambda d = \frac{1}{F} \Lambda 0 + \Lambda s \tag{23}$$

where Λ_0 is the conductivity of the dispersion medium. The formation factor F accounts for the shape and the volume of the non-conducting particles [93].

The electrical conductivity of an emulsion changes substantially during the course of monomer conversion. Processes that lead to a change of conductivity are e.g. the disappearance of monomer droplets and monomer-swollen micelles, the redistribution of monomer and surfactant in the period of particle growth, the change of ionic conductivity of the dispersion medium, changes of surface conductivity of the growing particles.

It has been shown by Janssen [94] that there are three major contributions that explain the behavior of the conductivity signal during emulsion polymerization: i) the free surfactant concentration in the aqueous phase, ii) the surfactant incorporated into micelles (surfactant concentration above the critical micelle concentration CMC), and iii) the initiator concentration. In emulsion homopolymerization, the conductivity will decrease as a function of conversion in intervals I and II as a result of growing particle surface area. The conductivity reaches a minimum at the interval II-III transition. In interval III, the surfactant concentration will be below the CMC in most cases depending on the monomer concentration in the aqueous phase. Generally, the conductivity will increase either by an increase of the CMC, or by surfactant desorption from the particle surface during interval III.

Noël et al. [95] describe the determination of maximum swellability of particles with monomer during the reaction by conductivity measurements. They have shown, that the maximum swellability of styrene - methyl methacrylate copolymer particles with styrene and methyl methacrylate is independent of both the temperature (over the range 20 - 50°C) and the copolymer composition.

Emulsion polymerization in the presence of nonionic surfactants is often complicated by phase inversions from oil-in-water to water-in-oil type emulsions. Electrical conductivity measurements are very sensitive for those phase inversions [96, 97].

The surfactant partition between particle surface, aqueous phase, and micelles is a determining factor for the electrical conductivity of latices. This influence can be used to determine the adsorption of ionic surfactants on latex particles from conductivity measurements [98].

According to eq. (23), the electrical conductivity of dispersions depends on the surface conductivity of the latex particles. Thus, the conductivity is also related to the Zeta-potential. However, this relation is rather complicated. In a typical reaction mixture, the contributions of all other ingredients (initiator, buffer, surfactant) outweigh the influence of the surface conductivity. The latex has to be cleaned carefully for this application of conductivity measurements.

In several industrial polymerization reactions it is essential to monitor the pH-value of the reaction mixture because of its dramatic influence on the colloidal stability of the latex. pH measurements are easily to perform with standard glass electrodes. Glass electrodes with counter pressure are applicable for on-line monitoring of reactions up to temperatures of about 80°C and pressures up to about 60 bar. Problems may arise with glass electrodes due to film formation on the glass surface, and because of the mechanical instability of the glass itself.

Solid-state pH sensors are based on the semiconductor ISFET technology (Ion Sensitive Field Effect Transistor). Their applicability is limited to pressures of about 2 bar, and temperatures up to about 85°C. They also suffer from latex film formation.

In some cases, the pressure of reactions in the presence of e.g. ethylene is in the 80 bar range. There is no sensor on the market that would be applicable for those pressures.

Niedrach [99] describes a ZrO_2 sensor for pH measurement under high pressure (83 bar) and high temperature (285°C). However, this sensor is still not commercially available.

4. Concluding Remarks

The techniques discussed in the preceding sections are either applied successfully to monitor polymerization reactions, or their potential applicability in the future has been shown. The use of on-line sensors and the combination of on-line with off-line techniques will be increasing in polymerization plants.

The applicability of several monitoring methods for on-line detection of reactor operation, for monomer conversion and latex properties is shortly summarized in Table 3. This table is by far not complete. "(X)" means that there are some limitations, "X" denotes successful applications demonstrated in the literature.

Acknowledgements: Helpful discussions with P. Hauptmann (University of Magdeburg), M. Krell and H. Zecha (Wacker Chemie) are gratefully acknowledged. The author would like to thank Wacker-Chemie GmbH for the permission to publish this paper.

TABLE 3: Methods for reaction monitoring

method		on-line monitoring	disperse systems	technical application
gravimetry			X	X
energy balance	calorimetry	X	X	(X)
	heat transport	X	X	X
	temperature	X	X	X
density	dilatometry	X	X	
	densitometry	X	X	X
	mechanical vibrations	X	(X)	X
	radioactivity	X	X	X
ultrasound	attenuation	X	X	
	velocity	X	X	X
	impedance	X		
	electro-acoustics	X	X	
electrical properties	conductivity	X	X	
	pH	X	X	X
	dielectric prop.	X	X	
optical properties	refractive index	X	X	(X)
	light scattering	X	X	(X)
	IR/NIR	X	(X)	(X)
	UV/VIS	X	X	(X)
	Raman	X	X	(X)
	Fluorescence	X	X	
chromatography	GC, HPLC	X	X	
	SEC	X	X	
	CHDF, FFF	X	X	
NMR			X	
viscosity		X	X	X
surface tension		X	X	

5. References

1. Chien, D.C.H. and Penlidis, A. (1990) *J. Macromol. Sci. - Rev. Macromol. Chem. Phys.* **C30**, 1
2. Schork, F.J. and Ray, W.H (1981), in D.R. Bassett and A.E. Hamielec (eds.), *Emulsion Polymers and Emulsion Polymerization*, ACS Symposium Series vol. 165, ACS, Washington, 505
3. Schork, F.J. and Ray, W.H (1983) *J. Appl. Polym. Sci.* **28**, 407
4. Schork, F.J. and Ray, W.H. (1987*) J. Appl. Polym. Sci.* **34**, 1259
5. Storti, G., Canegallo, S., Canu, P., and Morbidelli, M. (1992) *DECHEMA Monogr.* **127**, 379
6. Canegallo, S., Storti, G., Morbidell, M., and Carra, S. (1993) *J. Appl. Polym. Sci.* **47**, 961
7. Canu, P., Canegallo, S., Morbidelli, M., and Storti, G. (1994) *J. Appl. Polym. Sci.* **54**, 1899
8. Canegallo, S., Canu, P., and Morbidelli, M., and Storti, G. (1994) *J. Appl. Polym. Sci.* **54**, 1919
9. Gilbert, R.G. (1995) *Emulsion Polymerization - A Mechanistic Approach*, Academic Press, San Diego.
10. Harker, A.H. and Temple, J.A.G. (1988*) J. Phys. D: Appl. Phys.* **21**, 1576
11. Anson, L.W. and Chivers, R.C. (1984) *Ultrasonic Report 8401*, University of Surrey
12. Farrow, C.A., Anson, L.W., and Chivers, R.C. (1995) *Acustica* **81**, 402
13. Hauptmann, P. (1996) *VDI Berichte* **1255**, 199
14. Dinger, F. (1984) *Thesis*, Merseburg (Germany)
15. Hauptmann, P., Dinger, F., and Säuberlich, R. *Polymer* **26**, 1741
16. Siani, A., Apostolo, M., and Morbidelli, M. (1995) *DECHEMA Monogr.* **131**, 149
17. Carra, S. , Canegallo, S., and Morbidelli, M. (1995) *Chim. Ind. (Milan)* 77, 485
18. Canegallo, S., Apostolo, M., Storti, G., and Morbidelli, M. (1995) *J. Appl. Polym. Sci.* **57**, 1333
19. Hörnig, K., Hergeth, W.-D., and Wartewig, S. (1989) *Acta Polym.* **40**, 257
20. Hörnig, K., Hergeth, W.-D., Häusler, K.-G., and Wartewig, S. (1991) *Acta Polym.* **42**, 174

21. v. Jena, A. and Magori, V. (1992) *VDI Berichte* Nr. 939, 165
22. Fischer, B., Magori, V., and v. Jena, A. (1995) *Patent* 0 483 491 B1
23. Hauptmann, P. (1991) *Sensoren - Prinzipien und Anwendungen,* Hanser Verlag, München
24. Ward, M.D. and Buttry, D.A. (1990) *Science* **249**, 1000
25. Lin, Z. and Ward, M.D. (1996) *Anal. Chem.* **68**, 1285
26. Tolin, E.D., Fluegel, D.A. (1966) *Patent* US 3 275 809
27. Beckingham, B.F., Hendy, B.N., and Simons, J.V. (1979) *Patent* GB 1 549 841
28. Hendy, B.N. (1970) *Patent* GB 1 217 325
29. Walker, L.C. (1979) *Patent* DE 28 34 569
30. Miller, C.M., Sudol, E.D., Silebi, C.A., and El-Aasser, M.S. (1995) *J. Polym. Sci. A Polym. Chem.* **33**, 1391
31. Riesen, R. (1987) *Thermochim. Acta* **119**, 219
32. Schmidt, C.-U. and Reichert, K.-H. (1987) *Chem.-Ing.-Tech.* **59**, 739
33. Tietze, A., Proß, A., and Reichert, K.-H. (1995) *DECHEMA Monogr.* **131**, 673
34. Asua, J.M., Saenz de Buruaga, I., Arotcarena, M., Urretabizkaia, A., Armitage, P.D., Gugliotta, L.M., and Leiza, J.R. (1995) *DECHEMA Monogr.* **131**, 655
35. Urretabizkaia, A., Sudol, E.D., El-Aasser, M.S., and Asua, J.M. (1993) *J. Polym. Sci. A Polym. Chem.* **31**, 2907
36. Gugliotta,, L.M., Arotcarena, M., Leiza, J.R., and Asua, J.M. (1995) *Polymer* **36**, 2019
37. Varela de la Rosa, L., Sudol, E.D., El-Aasser, M.S., and Klein, A. (1996) *J. Polym. Sci. A Polym. Chem.* **34**, 461
38. Ibbett, R.N. (1993) *NMR Spectroscopy of Polymers,* Chapman and Hall, New York
39. Jones, S.A. and Stronks, H.J., BRUKER minispec application note # 20
40. Guyot, A., Guillot, J., Pichot, C., and Rios Guerrero, L. (1981), in D.R. Bassett and A.E. Hamielec (eds.), *Emulsion Polymers and Emulsion Polymerization,* ACS Symposium Series vol. 165, ACS, Washington, 415
41. Urretabizkaia, A., Leiza, J.R., and Asua, J.M. (1994) *AIChE Journal* **40**, 1850
42. Verdurmen-Noel, E.F.J. (1994), *Thesis,* Eindhoven (The Netherlands)
43. Alonso, M., Recasens, M., and Puigjaner, L. (1986) *Chem. Eng. Sci.* **41**, 1039
44. Alonso, M., Alivers, M., Puigjaner, L. and Recasens, M., (1987) *Ind. Eng. Chem. Res.* **26**, 65
45. Schork, F.J., (1992), in M.S. El-Aasser (ed.) *Advances in Emulsion Polymerization and Latex Technology,* 23rd Annual Short Course, Lehigh, Bethlehem, PA (USA)
46. Kiparissides, C., MacGregor, J.F., Singh, S., and Hamielec, A.E. (1980) *Can. J. Chem. Eng.* **58**, 65
47. Kourti, T., MacGregor, J.F., and Hamielec, A.E. (1990) *Polym. Mat. Sci. Eng* **62**, 301
48. Apfel, U., Grunder, R., and Ballauff, M. (1994) *Colloid Polym. Sci.* **272**, 820
49. Apfel, U., Hörner, K.D., and Ballauff, M. (1995) *Langmuir* **11**, 3401
50. Mohammadi, M. (1995) *Adv. Colloid Interface Sci.* **62**, 17
51. Nicoli, D.F., Wu, J.S., Chang, Y.J., McKenzie, D.C., and Hasapidis, K. (1995) *American Laboratory*
52. Löhr, G. and Reinecke, R. (1980) *Angew. Makromol. Chem.* **85**, 181
53. Fischer, J.P. and Nölken, E. (1988) *Progr. Colloid Polym. Sci.* **77**, 180
54. Ovod, V.I. (1995) *Part. Part. Syst. Charact.* **12**, 207
55. Moser, A.O., Fromheim, O., Hermann, F., and Versmold, H. (1988) *J. Phys. Chem.* **92**, 6723
56. Horn, D., Auweter, H, Ditter, W., and Eisenlauer, J. (1986) *Org. Coat.*, 251
57. Lilge, D. and Horn, D. (1991) *Colloid Polym. Sci.* **269**, 704
58. Drewel, M., Ahrens, J., and Podschus, U. (1989) *J. Opt. Soc. Am.* **A7**, 206
59. Schätzel, K. (1991) *J. Mod. Optics* **38**, 1849
60. Müller, J. (1993) *Thesis,* Kiel (Germany)
61. Maret, G. and Wolf, P.E. (1987) *Z. Phys. B* **65**, 409
62. Wolf, P.E., Maret, G., Akkermans, E., and Maynard, R. (1988) *J. Phys. (France)* **49**, 63
63. Fraden, S. and Maret, G. (1990) *Phys. Rev. Lett.* **65**, 515
64. Maret, G. (1992) *Phys. Bl.* **48**, 161
65. Bicout, D. and Maret, G. (1994) *Physica A* **210**, 87
66. Wiese, H. and Horn, D. (1992) *Ber. Bunsenges. Phys. Chem.* **96**, 1818
67. Dos Ramos, J.G. (1988) *Thesis,* Bethlehem, PA (USA)
68. Silebi, C.A. and Dos Ramos, J.G. (1989) *J. Colloid Interface Sci.* **130**, 14
69. Dos Ramos, J.G. and Silebi, C.A. (1989) *J. Colloid Interface Sci.* **133**, 302
70. Dos Ramos, J.G. and Silebi, C.A. (1990) *J. Colloid Interface Sci.* **135**, 165
71. Venkatesan, J. (1992) *Thesis,* Bethlehem, PA (USA)
72. Miller, C.M., Sudol, E.D., Silebi, C.A., and El-Aasser, M.S. (1995) *J. Colloid Interface Sci.* **172**, 249

288

73. Giddings, J.C. (1993) *Science* **260**, 1456
74. Giddings, J.C. (1995) *Anal. Chem.* **67**, 592A
75. Ratanathanawongs, S.K., Shiundu, P.M., and Giddings, J.C. (1995) *Colloids Surf. A* **105**, 243
76. Roessner, D. (1994) *Thesis*, Hamburg (Germany)
77. Wyatt, P.J. (1993) *Anal. Chim. Acta* **272**, 1
78. Becher, P. and Yudenfreund, M.N. (1978) *Emulsions, Latices, and Dispersions*, Marcel Dekker, New York and Basel, 221 and 257
79. Fitch, R.M., Su, L.S., and Tsaur, S.L. (1987), in M.S. El-Aasser and R.M. Fitch (eds.) *Future Directions in Polymer Colloids*, Martinus Nijhoff, Dordrecht, Boston, Lancaster
80. Su, L.S., Jayasuriya, S., and Fitch, R.M. (1995), in J.W. Goodwin and R. Buscall (eds.) *Colloidal Polymer Particles*, Academic Press, London, 101
81. Sauer, B.B., Stock, R.S., Lim, K.-H., and Ray, W.H. (1990) *J. Appl. Polym. Sci.* **39**, 2419
82. Sheppard, N.F. and Senturia, S.D. (1986) *Adv. Polym. Sci.* **80**, 1
83. Harrison, D.J.P., Yates, W.R., and Johnson, J.F. (1985) *J. Macromol. Sci. - Rev. Macromol. Chem. Phys.* **C25**, 481
84. O'Brien, R.W., Cannon, D.W., and Rowlands, W.N. (1995) *J. Colloid Interface Sci.* **173**, 406
85. Oja, T., Petersen, G.L., and Cannon, D.W. (1985) *US-Patent # 4,497,207*
86. Debye, P. (1933) *J. Chem. Phys.* **1**, 13
87. O'Brien, R.W. (1988) *J. Fluid Mech.* **190**, 71
88. O'Brien, R.W., Midmore, B.R., Lamb, A., and Hunter, R.J. (1990) *Faraday Discuss. Chem. Soc.* **90**, 301
89. Loewenberg, M. and O'Brien, R.W. (1992) *J. Colloid Interface Sci.* **150**, 158
90. Rider, P.F. and O'Brien, R.W. (1993) *J. Fluid Mech.* **257**, 607
91. Carasso, M.L., Rowland, W.N., and Kennedy, R.A. (1995) *J. Colloid Interface Sci.* **174**, 405
92. Fischer, J.P. and Löhr, G. (1986) *Org. Coat. Sci. & Technol.* **8**, 227
93. Wright, M.H. and James, A.M. (1973) *Kolloid-Z. Z. Polym.* **251**, 745
94. Janssen, R.Q.F. (1995) *Thesis*, Eindhoven (The Netherlands)
95. Noël, L.F.J., Janssen, R.Q.F., van Well, W.J.M., van Herk, A.M., and German, A.L. (1995) *J. Colloid Interface Sci.* **175**, 461
96. Hergeth, W.-D., Bloß, P., Biedenweg, F., Abendroth, P., Schmutzler, K., and Wartewig, S. (1990) *Makromol. Chem.* **191**, 2949
97. Bloß, P. (1990) *Thesis*, Merseburg (Germany)
98. Zwetsloot, J.P.H. and Leyte, J.C. (1995) *J. Colloid Interface Sci.* **175**, 1
99. Niedrach, L.W. (1987) *Angew. Chem.* **99**, 183

TRANSPORT PHENOMENA IN EMULSION POLYMERIZATION REACTORS

J.B.P. SOARES
University of Waterloo, Department of Chemical Engineering
Waterloo, Ontario, Canada N2L 3G1

A.E. HAMIELEC
McMaster University, Department of Chemical Engineering
Hamilton, Ontario, Canada L8S 4L7

1. Introduction

Polymerization reactor type and operation conditions have a marked influence on polymer properties such as distribution of molecular weight, chemical composition, and particle size. Due to the very nature of polymeric chains, if the synthesized polymer does not have the desired properties when exiting the reactor, it is very difficult and costly to improve its properties by further processing and purification, since most fractionation methods that are economically viable for small molecule compounds will fail for macromolecules [1,2].

Some of the most important operational and design features of polymerization reactors are its agitation and heat removal systems. In this article we will review some aspects of polymer reactor engineering that have important effects on the performance of polymerization reactors. Some of these design considerations will be exemplified with a case-study for the emulsion copolymerization of ethylene and vinyl acetate under conditions of mass transfer-controlled polymerization rate.

2. Types of Polymerization Reactors

Emulsion polymerization can be performed in several different types of reactors. Stirred-tank reactors are generally preferred, especially under semi-batch operation, because of the flexibility in controlling polymerization conditions and polymer properties. However, tubular reactors, either single-pass or loop, are also used for emulsion polymerizations. Some of the advantages and disadvantages of each reactor type will be discussed in the following sections.

2.1. STIRRED-TANK

Stirred-tank reactors (STR) are widely used in the polymerization industry because of their simple design and operational flexibility. Independently of their mode of operation (batch, semi-batch, or continuous), the main characteristic of STRs is that, ideally, all fluid elements are intimately mixed due to vigorous stirring. The good mixing

J. M. Asua (ed.), Polymeric Dispersions: Principles and Applications, 289–304.
© 1997 *Kluwer Academic Publishers. Printed in the Netherlands.*

characteristics of STRs result in concentration and temperature spatial uniformity, and consequently allows the production of polymer with well-defined properties.

2.1.1. Batch Operation

In a batch stirred-tank reactor (BSTR), all reactants are placed in the reactor at the start of the polymerization and the polymerization is allowed to proceed to its completion. This is a very simple operation mode and requires minimal equipment and operational costs. Another advantage of BSTRs is the precise control of the initial conditions in the reactor. On the other hand, it is not possible to control the polymerization by selective addition of reactants or initiators, as done in semi-batch reactors. For the case of copolymerization of comonomers with different reactivity ratios, composition drift will occur and this might have an undesirable effect on polymer properties. Additionally, the temperature control of batch reactors can be difficult, since it is not possible to control the rate of polymerization by slow addition of one of the components or by feeding reactants at lower temperatures [3]. This mode of operation is generally, but not always, restricted to exploratory polymerizations in laboratory-scale reactors.

2.1.2. Semi-batch Operation

Semi-batch stirred-tank reactors (SBSTR) are probably the most common emulsion polymerization reactors because of their improved process control characteristics and relative ease of operation. Only part of the recipe is initially fed to the reactor; the remaining charge is fed throughout the rest of the polymerization. In this way, it is possible to manipulate the rate of polymerization and the properties of the produced polymer by an adequate strategy of monomer, initiator and emulsifier transfer. This controlled addition of reactants is also a convenient way to prevent reaction runaway [4]. Since charge feed strategy can significantly alter the properties of the produced polymer, it is clear that this mode of operation requires a detailed knowledge of polymerization mechanisms and requires sophisticated control schemes to be used to its maximum capability [5].

2.1.3. Continuous Operation

In continuous stirred-tank reactors (CSTR), all components of the polymerization recipe are fed continuously to the polymerization reactor by means of common or separate feed lines. Continuous operation offers some advantages over batch and semi-batch operation, especially for high production rates. Costs of emptying, cleaning and recharging the reactor are evidently minimized, since CSTRs can be designed to operate for a long time without requiring maintenance shut-downs. Continuous operation can also produce polymer with more uniform properties since the CSTRs operate at steady-state during most of the polymerization run time, therefore eliminating inter-batch variations of polymer properties. However, sustained oscillations in monomer conversion and particle concentration, as well as multiple steady-states have been observed for the continuous operation of CSTRs in emulsion polymerization [6,8].

The use of CSTRs might be disadvantageous if changes between product grades take place frequently, since off-specification products will always be produced during grade transitions, although proper control schemes can be designed to minimize the amount of off-specification material between grade changes.

A negative consequence of the complete backmixing of CSTRs is that the conversion in a CSTR will always be smaller than the one obtained in a tubular (plug flow) reactor operating under the same conditions for reaction orders higher than one, as prevalent in polymerization reactors. Additionally, the exponential residence-time distribution of CSTRs will lead to the production of latexes of broader particle size distribution, which might be undesirable for certain film applications. This effect can be minimized if several CSTRs are combined in series to approximate the plug flow characteristics of a tubular reactor, as done in some industrial processes for the manufacture of SBR [8].

2.2. TUBULAR REACTORS

The narrow residence-time distribution of tubular reactors enable the production of latexes with narrow particle size distribution. Particle size distribution has a strong influence on the film-forming properties of latex and can also affect its molecular weight distribution.

Some additional advantages of tubular reactors over stirred-tank reactors for emulsion polymerization are a larger ratio of heat transfer surface to reactor volume, and stable operation due to reduced backmixing. However, there is a significant risk of reactor fouling and plugging by the latex. It has been suggested that fouling and plugging problems could be minimized by adding a pulsation source to the reactor feed, without significantly increasing backmixing [6,7].

Loop reactors can also be used for emulsion polymerization, provided care is taken to avoid fouling and plugging. Since the circulation rates in these reactors is very high, they generally behave as a CSTR with the added advantage of large ratio of heat transfer surface to reactor volume of a tubular reactor [8]. Lower capital cost for high pressure reactors, such as required for vinyl acetate-ethylene emulsion copolymerization may be another advantage of this reactor type.

3. Mixing

Mixing equipment (impeller type, size and location) promote fluid circulation and generate velocity gradients (shear rate) within a tank [9]. Power input into a mixing tank, P, is proportional to the pumping capacity of the impeller, Q, and to the velocity head, h:

$$P \propto Q \, h \tag{1}$$

The velocity head is directly related to the shear rates in the tank. The choice of impeller type and location is crucial for the proper operation of polymerization reactors.

3.1. IMPELLER TYPE

Impellers can be classified as turbine impellers and close-clearance impellers. Turbine impellers are generally small compared to reactor dimensions and are used in a vast range of applications in the chemical industry. Close-clearance impellers are only used for stirring high viscosity systems, such as the ones that can occur in solution and bulk polymerization.

Turbine impellers can be subdivided according to the main direction of the flow they generate into axial- or radial-flow. Pitched-blade turbines and high efficiency turbines (also called fluidfoil, airfoil, hydrofoil, or laserfoil) are common types of axial-flow impellers. The angled blades of these impellers provide mainly axial flow. Radial flow impellers, such as the straight-blade turbine and the disk turbine produce both radial and rotational flow. For the same pumping capacity, the pitched-turbine needs about one-quarter of the power requirement of a straight-blade turbine. This power requirement can be decreased by a factor of 2 or 3 if a high-efficiency turbine is used. Evidently the shear rate is also lowered by going from the straight-blade turbine to the high-efficiency axial turbine. A classification of impellers according to their fluid shear rate and pumping capacities is presented in Figure 1 [9].

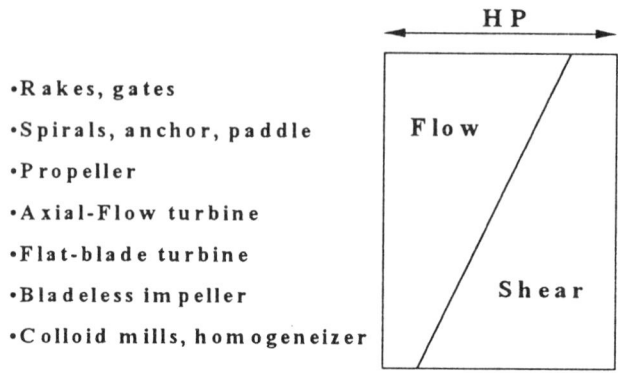

Figure 1. Flow capacity and shear rate of different impeller types

Combinations of axial and radial flow turbines are used frequently to provide high shear and turbulence, while maintaining good fluid circulation in the reactor. For reactors with high H/T ratio (Figure 2), several impellers at various liquid levels might be used simultaneously to ensure appropriate mixing.

If the main objective of mixing is to produce circulation of particles without interparticle mixing, low shear rate impellers should be used. This can either be done by selection of impeller type or by altering the D/T ratio of an existing impeller. At a given power level, a higher diameter impeller operating at a slower speed will have its shear rate decreased according to the relationship:

$$(Q / h)_P \propto D^{8/3} \tag{2}$$

As a general guideline, processes requiring flow rate rather than turbulence uses D/T ratios of 0.4 to 0.6; to enhance turbulence and micromixing (section 3.5), D/T ratios of 0.25 to 0.35 are selected [10]. For emulsion polymerization high shear rates can be detrimental because of coagulum formation. However, sufficient agitation is required to avoid monomer pooling and guarantee a uniform concentration of reactants throughout the polymerization reactor.

Close-clearance impellers, such as the helical and anchor impellers, reach the fluid near the reactor wall and minimize stagnant regions and vortex formation that might occur when high viscosity fluids are used [11]. Helical impellers have the additional advantage of providing a top to bottom turnover, which is absent in anchor impellers.

Since these agitators operate with highly viscous and non-Newtonian fluids, the rheological properties of the medium should be taken into account for proper impeller design [12]. Close-clearance impellers can be designed to come very close to or actually to scrape the reactor walls to improve heat transfer coefficients [13]. Close-clearance impellers are not efficient for low viscosity liquids.

For small to medium scale reactors, the agitator shaft generally enter the reactor from the top. However, for large polymerization reactors this might be impractical because that would require very long and thick, and consequently very costly, shafts. Large reactors generally use bottom entrance agitators. Special care should be taken when specifying seals for bottom entrance agitators to avoid leakage and permit seal replacement during a polymerization run [14].

Draft tube circulators are not commonly used in polymerization reactors, likely because of fouling of the fluid-conveying section, but they allow efficient flow circulation and can be advantageous for increasing the rates of gas-liquid mass transfer [15].

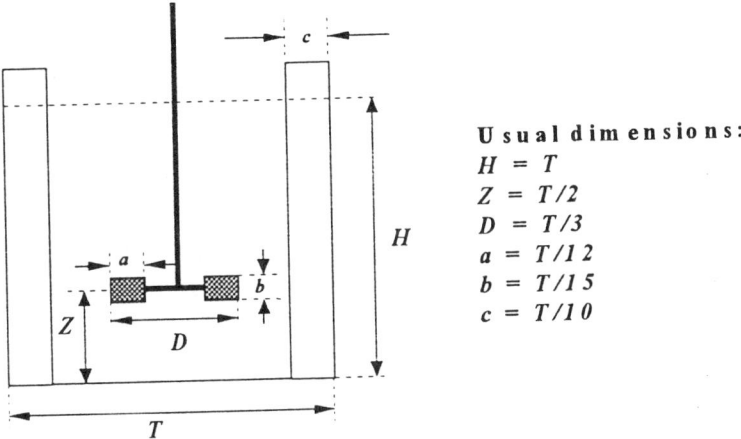

Usual dimensions:
$H = T$
$Z = T/2$
$D = T/3$
$a = T/12$
$b = T/15$
$c = T/10$

Figure 2. Standard stirred-tank geometry

There are also several designs of static mixers available. These mixers are claimed to provide excellent radial mixing coupled with axial plug flow (reduced backmixing), thus increasing the heat transfer efficiency as compared to empty tubes. Some new designs make use of internal heat exchanger tubes, therefore considerably increasing the heat removal capability of these systems and allowing the use of large diameter units [16]. The main limitation of these systems as polymerization reactors is again the increased risk of fouling due to the presence of internal packing.

3.2. MIXER DESIGN

It is important to remember that optimal operation conditions for polymerization reactors can vary with monomer conversion for the case of batch or semi-batch operation, and steady-state polymer concentration for the case of continuous operation, due to the increase in viscosity for higher polymer concentrations at higher monomer conversions. At low viscosities, turbine impellers and baffled reactors might be adequate to ensure optimal heat transfer and mixing characteristics, but at higher viscosities it might be necessary to use close-clearance impellers, such as helical or anchor [17]. However, for most practical applications of emulsion polymerization reactors, such abrupt variations of viscosity do not take place, which simplifies considerably the design of the mixing equipment.

Mixing requirements are also a function of reactor operation regime. For batch reactors, intense agitation might be needed in the initial stages of polymerization to avoid monomer segregation and promote good dispersion of the several components of the recipe, but lower agitation intensity is necessary for the later stages of polymerization to avoid shear-induced coagulum formation.

Three important factors should be considered when dealing with impeller design: impeller power, impeller flow, and heat transfer. Heat transfer will be discussed in section 4 of this paper.

3.2.1. Impeller Power

The basic relationship between agitator power, P, fluid density, ρ, shaft speed, N, and impeller diameter, D, is given by a dimensionless group called the power number, N_P:

$$N_P = \frac{P}{\rho N^3 D^5} \tag{3}$$

Under turbulent conditions (i.e., $N_{Re} > 1000$), the power number is nearly constant for turbine impellers operating in a baffled tank. Typical values of for N_P are 0.2-0.5 for high-efficiency turbines, 1.2-1.5 for pitched-blade turbines, 3.8-5.6 for straight-blade turbines, and 5.8-6.2 for disk turbines [18].

The Reynolds number for stirred-tanks is given by:

$$N_{Re} = \frac{D^2 N \rho}{\eta} \tag{4}$$

Figure 3 illustrates the general relationship between N_P and N_{Re} for two impeller types. It is very important to realize that these curves are affected by impeller location and design. It is not possible to obtain generic curves for power number. N_P has to be measured using actual process conditions or obtained directly from the mixer manufacturer.

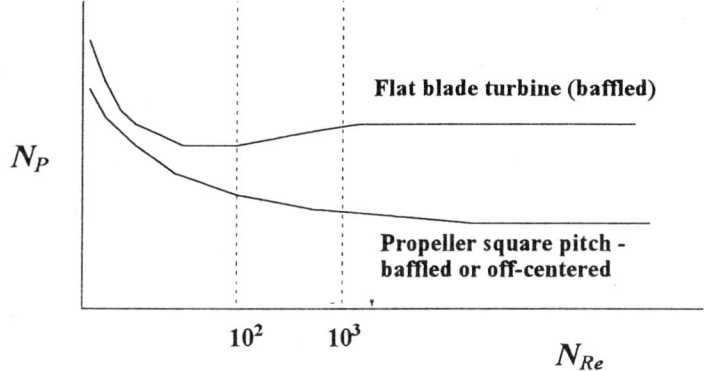

Figure 3. Reynolds number versus Power number curve for two different impeller types

3.2.2. *Impeller Flow*

Accurate predictions of fluid motion are much more difficult to obtain than those for impeller power. One of the main difficulties is the accurate prediction of the flow patterns in a stirred-tank for different impeller configurations and vessel accessories such as baffles and cooling coils. Actually, there is experimental evidence that the circulation pattern in stirred-tanks is subject to time variations [19]. Additionally, due to recirculation patterns inside a stirred-tank, it is virtually impossible to distinguish between direct flow and induced flow [18].

3.3. EFFECT OF BAFFLES

Baffles are used to alter flow patterns in agitated vessels and eliminate vortex formation near the wall. The usual design consists of four baffles 90^0 apart with a width of T/10 to T/12, and located T/72 from the reactor wall. However, the presence of suspended solids or non-Newtonian fluids might cause formation of stagnant regions around the baffles.

Full length baffles can also eliminate surface vortex, which can be detrimental in polymerizations involving gaseous monomers. This can be corrected by placing the baffles below the liquid surface; in this case the presence of baffles will lead to vortex formation and increase gas-liquid transfer rates.

3.4. NON-IDEAL MIXING / MICROMIXING AND MACROMIXING

Fluid uniformity is achieved through a combination of intense local mixing (high shear rates) and bulk fluid circulation. Even when gross fluid uniformity is achieved, minor fluctuations in temperature and concentration might be present in the reactor and this may have a significant impact on the rate of polymerization and on polymer properties. For instance, it has been shown that non-ideal mixing can cause broadening of the molecular weight distribution of polymer produced in a semi-batch stirred-tank reactor [20].

It is important to realize that the reactor residence time distribution does not uniquely define the state of the mixing. The residence time distribution in a reactor only defines the level of macromixing, i.e., the bulk flow patterns that cause different fluid elements to stay in the reactor for different periods of time. However, it could be that these fluid elements are partially or completely segregated, and therefore do not "mix" in the conventional sense. Micromixing characterizes the degree of mixing between these fluid elements. For a CSTR, a maximum amount of molecular mixing (maximum mixedness) gives the lowest possible conversion for reaction orders greater than one and the highest possible conversion for reaction orders lower than one. First order reactions do not depend on the level of micromixing [21,22]. Different levels of macro- and micromixing can have a definite effect on polymer properties such as molecular weight and composition distribution [22].

Mixing non-idealities can be easily modelled by subdividing the stirred-tank reactor in several homogeneous regions that have different temperatures and concentrations, interconnected in series or in parallel [21]. Each of these regions can be treated as ideal stirred-tank reactors to provide a model for the segregated fluid flow in the actual reactor. Effects such as by-pass regions can be modelled as a combination of a continuous stirred-tank reactor and a plug flow reactor [23]. It should be noted that the polymer in polymer particles is segregated, while the rapidly diffusing monomer is micromixed during emulsion polymerization.

4. Heat Transfer

Adequate heat removal from polymerization reactors is essential for their satisfactory operation. Temperature oscillations can cause broadening of molecular weight and chemical composition and have undesirable effects on the final polymer properties. For commercial scale polymerization reactors, heat removal is often the productivity limiting factor [8].

Since the synthesis of a macromolecule from monomer molecules will cause a significant decrease in entropy, a negative enthalpy of propagation (exothermic reaction) is necessary to make polymerization reactions thermodynamically favourable:

$$\Delta G_r = \Delta H_r - T_r \Delta S_r \qquad (5)$$

Heat of polymerization is commonly removed from laboratory-scale reactors by circulating cooling fluids through an external jacket or cooling coils. For large reactors, several techniques can be used to increase heat removal rates: [8,14] (1) Addition of cooling coils, baffles, and tube bundles. These additional pieces of equipment interfere with mixing patterns, might increase fouling, and are an obstacle for reactor cleaning; (2) Use of external loop heat exchangers. This might provide considerable additional heat transfer surface, but can only be used for systems were fouling is not severe; (3) Reflux cooling by means of a condenser mounted on top of the reactor. This permits efficient heat removal at a reduced operational cost [24,25], but care must be taken to avoid accumulation of inert non-condensable gases in the condenser. Additionally, the initiator must be non-volatile to avoid polymerization in the condenser, causing fouling and eventual plugging; (4) non-geometrical scale-up of the reactor, to increase the ratio of heat transfer surface to reactor volume, i.e., larger reactors benefit from a larger H/T ratio; (5) use of jacket designs that increase the heat transfer coefficient on the jacket side, such as dimple and half-tube jackets; (6) substitution of glass lined reactors by reactors with a polished-metal internal surface; (7) use of cold feed streams; (8) decreasing the temperature or increasing the flow rate of the cooling liquid.

Fouling on the reactor walls can significantly decrease the internal heat transfer coefficient [26]. The selection of proper emulsifier and stabilizers should minimize this undesirable phenomenon. However, some fouling is likely to occur under most circumstances, and it is particularly important to take fouling into account for the design of heat transfer equipment for continuous polymerization reactors in order to avoid frequent maintenance shut-downs.

4.1. HEAT TRANSFER COEFFICIENTS

The overall heat coefficient, U, of a jacket-cooled stirred-tank reactor (when the inner and outer heat transfer surfaces are approximately the same) is given by the expression:

$$\frac{1}{U} = \frac{1}{h_i} + ff_i + \frac{x}{k_w} + ff_j + \frac{1}{h_j} \qquad (6)$$

where h_i and h_j are the inside and outside heat transfer coefficients, ff_i and ff_j the inside and outside fouling factors, x the thickness of the reactor wall, and k_w the thermal conductivity of the reactor wall.

For internal cooling coils, the following equivalent expression can be used for the outer (process side) overall heat coefficient U_0:

$$\frac{1}{U_0} = \frac{1}{h_i} + ff_i + \left(\frac{x}{k_c}\right)\left(\frac{d_{co}}{d_{cm}}\right) + ff_{co} + \left(\frac{1}{h_j}\right)\left(\frac{d_{co}}{d_{ci}}\right) \qquad (7)$$

where k_c is the thermal conductivity of the coil, and d_{ci} and d_{co} are the inside and outer coil diameters.

The inside heat coefficient, h_i , has the following functional relationship:

$$N_{Nu} = A\left(N_{Re}\right)^a \left(N_{Pr}\right)^b \left(\eta/\eta_w\right)^c \qquad (8)$$

where Nusselt number, N_{Nu} ,and Prandtl number, N_{Pr} , are defined as usual:

$$N_{Nu} = \frac{h_i T}{k} \qquad (9)$$

$$N_{Pr} = \frac{c_p \eta}{k} \qquad (10)$$

where k is the thermal conductivity of the reaction mixture. For jacket vessels, convective heat transfer, and turbulent flow, $a = 2/3$, $b = 1/3$, and $c = 0.14$. in most correlations published in the literature. Some authors suggest that the value of the parameter c should be set to 0.25 instead of 0.14 [27]. The parameter A is a function of impeller type and reactor design. Several correlations for different reactor and impeller configurations are available in the literature [28,29]. For non-Newtonian liquids N_{Nu} depends also on its rheological behaviour but the literature on heat transfer coefficient for these systems is still scarce [27].

The effect of power on heat transfer coefficients is weak, $h_i \propto P^{0.22}$, therefore it is not practical, especially for large reactors, to use mixer power to increase heat transfer coefficients [9]. In reality, for very viscous fluids, higher impeller power could actually increase the reactor temperature because of viscous dissipation effects.

Several correlations are available for the outside (coolant side) heat transfer coefficient for jackets, coils, and baffles , h_j, and depend on the design of the heat removal apparatus being used [28,30].

Inside and outside fouling factors, ff_i and ff_j , are important design parameters, but there are no correlations available for their *a priori* estimation; one has to rely on previous operation experience to obtain these parameters. Some general estimates can be obtained in well-established handbooks on process equipment sizing and design [31]. A method for designing heat transfer exchangers taking into account time-dependent fouling thermal resistance have been proposed recently [32], but unfortunately the kinetics of fouling has to be determined experimentally *a priori*.

4.2. JACKET SELECTION AND DESIGN

Simple, unbaffled annular jackets are commonly used for small laboratory-scale reactors, but they are inefficient for controlling the temperature of large-scale reactors [3], since their heat transfer coefficients are limited to 20 to 320 Btu/h.ft^2.°F (126 to 2020 kJ/h.m^2.°C) [30].

The heat transfer coefficient of heat transfer jackets can be increased by the use of agitation nozzles to increase turbulence and modify flow patterns in the jacket, causing a two- to three-fold increase in the heat transfer coefficient as compared to simple jackets, at the cost of a higher pressure drop in the jacket. More efficient jacket designs can also be applied, such as spirally-baffled jackets, dimple-jackets, and partial-pipe jackets, leading to higher heat transfer coefficients, but also to increasing installation costs [30,33]. Several correlations are available to predict heat transfer coefficients using these improved jacket designs [34].

5. Scale-up Considerations

A considerable complication in scaling-up polymerization reactors is that different phenomena may dominate reactor behaviour at different scales. For instance, for small-scale, laboratory reactors, the ratio of heat transfer surface to reactor volume might be enough to justify the use of a simple jacketed stirred-tank reactor. However, upon scaling-up this ratio will be much smaller and it might be necessary to use internal cooling coils or other heat removal devices, which in their turn will also affect flow patterns inside the reactor. Baffles are not generally required for laboratory scale reactors but are generally needed to overcome the tendency of vortex formation for the scaled-up reactor [35].

Scale-up has to be done with considerable care, since the conditions that prevail at the laboratory scale can be rather distinct from the ones present in the large scale reactor. Special care should be taken when dealing with gas-liquid systems, since the presence of gas can significantly alter impeller performance. High gas flow rates can indeed cause the impeller to cavitate, drastically reducing its pumping capacity [36]. Another important consideration for gas-liquid systems is sparger design. The most common types are ring spargers and porous gas diffusers. Both have been reported to be more efficient than single orifice dip-tubes, clearly because of the formation of smaller gas bubbles with a higher superficial area to bubble volume ratio, which favours gas-liquid transfer. The optimal ring diameter/impeller diameter ratio is around 0.8, and the best location is always close to the bottom of the reactor, above or bellow the impeller, depending on the flow pattern and reaction characteristics of the system under investigation [37].

In order to maintain geometrically similar agitation patterns during scale-up, it is necessary to keep constant the Froude number, the Reynolds number, and the Power number. Froude number, N_F, is defined as:

$$N_F = \frac{DN}{g} \tag{11}$$

Since there are three equations and only two variables, N and P, the system is overdeterminated. Beckmann [14] suggests that the viscosity in the large reactor should be transformed into a variable as well, according to:

$$\eta_l = \eta_s \left(\frac{D_l}{D_s}\right)^{1.5} \tag{12}$$

where the subscripts l and s refer to the large and small reactors, respectively. In this way, in order to ensure geometric similarity during the scale-up, it is necessary that the viscosity in the large reactor be higher than the one in the small reactor.

It is important to point out, however, that geometric similarity of flow patterns is only one of the several possible criteria that can be used during reactor scale-up [38]. There is no easy recipe for scaling-up polymerization reactors; each process should be examined individually and the effect of scale-up on the main process parameters carefully evaluated. Leng [10] suggests some useful guidelines for a successful scale-up of mixing processes.

One of the major scale-up difficulties of mixing systems is the fact that several process parameters do not scale-up equally [35], as shown in Table 1. Therefore, the designer should determine which process parameters to keep constant and which to be allowed to vary, and this requires detailed knowledge of the process under consideration. Leng's [10] very interesting publication describes several problems and solutions that can occur during scale-up of several mixing processes with different process requirements.

TABLE 1 - Relationship between different design parameters during scale-up

Parameters	Pilot Scale (76 L)		Plant Scale (4,320 L)		
P	1.0	216.0	7776.0	36.0	0.16
P/V	1.0	1.0	36.0	0.16	0.0007
N	1.0	0.3	1.0	0.16	0.03
D	1.0	6.0	6.0	6.0	6.0
Q	1.0	65.0	216.0	36.0	6.0
Q/V	1.0	0.3	1.0	0.16	0.03
ND	1.0	1.8	6.0	1.0	0.16
N_{Re}	1.0	10.8	36.0	5.8	1.0

6. Case Study - Emulsion Copolymerization of Ethylene and Vinyl Acetate

The semi-batch emulsion copolymerization of ethylene and vinyl acetate was extensively studied by Scott *et al.* [39-44] through a series of carefully designed factorial experiments. One of the primary objectives of their work was to increase the amount of ethylene content in the copolymer at reduced polymerization pressures and temperatures. Twelve process variables were investigated, including pressure, temperature, emulsifier type and concentration, the addition of stabilizer, the addition of co-solvent, agitation, buffer, vinyl acetate feed rate and reactor configuration. Reactions conditions under which ethylene mass transfer was the rate controlling step were identified and correlated to the effects of impeller design, gas sparging, and agitation. This review will concentrate on the reactor engineering aspects of their work to illustrate some of the points covered in sections 1 to 5.

In emulsion copolymerization, the concentration of monomers in the particles control copolymer composition, molecular weight, and rate of polymerization. For gas-liquid emulsion polymerizations, the diffusion of ethylene from the liquid phase to the polymer particles must be considered because of the relatively low gas-water interfacial areas.

The reactor set-up used is shown in Figure 4. The polymerization reactor is a 2 litre Parr stand autoclave with $H = 26$ cm and $T = 10$ cm. The outlets for the ethylene, vinyl acetate, and initiator solution feeds are located below the liquid level of the emulsion, close to the agitator blades, to ensure good mixing. The agitator shaft is provided with two impellers (either radial or axial) located 4 and 9 cm from the bottom of the reactor. Other equipment details and polymerization procedures is described elsewhere [39].

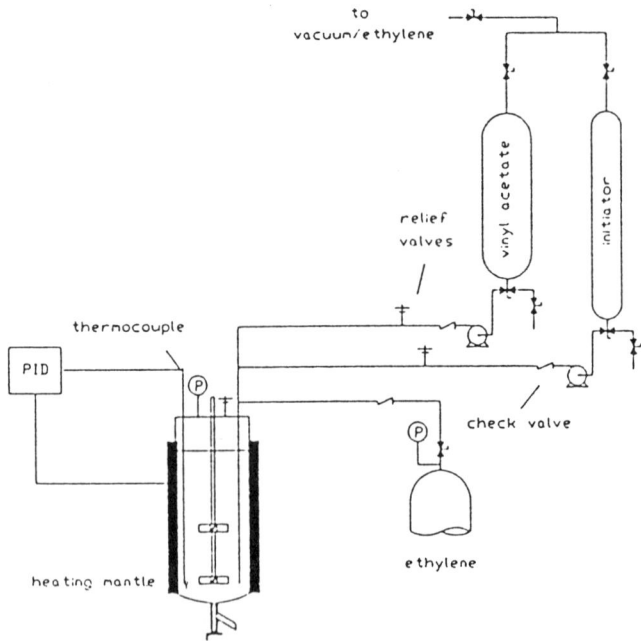

Figure 4. Schematic representation of semi-batch reactor for ethylene-vinyl acetate emulsion polymerization

TABLE 2 - Reactor configurations and results for the reactor design
experiments for ethylene - vinyl acetate copolymerization

Experiment	Reactor configuration	Final solids (wt%)	Comments
E1	axial-flow impellers single-orifice dip-tube 200 rpm	47.1	some fouling some free floating coagulum
E2	axial-flow impellers porous gas sparger 200 rpm	48.2	some coagulum and fouling of impeller blades much more foaming
E3	radial-flow impellers porous gas sparger 200 rpm	47.5	no coagulum
E4	radial-flow impeller porous gas sparger 400 rpm	50.9	no coagulum sampling problems due to foaming
E5	axial-flow impeller porous gas sparger 400 rpm	48.1	no coagulum sampling problems due to foaming

Figure 5 shows the copolymer composition as a function of polymerization time. Reactor conditions for runs E1 to E5 are shown in Table 2. Experiment E1 was considered the base case. The weight fraction of ethylene incorporated in the copolymer decreases as the polymerization advances. As the latex viscosity increases, so does the mass transfer resistance for ethylene transport from the gas phase through the water to the polymer particles Throughout the polymerization, vinyl acetate conversion remained constant, after an initial start-up period, and therefore a change in the concentration of ethylene was responsible for the decrease in ethylene content in the copolymer, not a change in the partitioning behaviour as the polymerization progressed.

In experiment E2, ethylene was fed through a porous sparger, instead of the single-orifice dip-tube used in experiment E1. The smaller diameter gas bubbles and much larger gas-water interfacial area generated by the porous sparger helped compensate for the increasing latex viscosity, resulting in a less steep copolymer composition drift than the one observed in experiment E1.

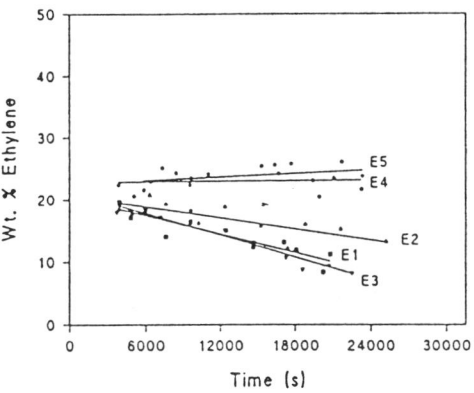

Figure 5. Copolymer composition as a function of polymerization time [42]

302

For experiment E3, the two axial-flow impellers of experiment E2 were substituted by two radial-flow impellers. As can be noticed in Figure 5, the use of radial-flow impellers in place of axial-flow impellers increases the copolymer compositional drift to the point of canceling the advantage of using the porous sparger. The use of radial-flow impellers will lead to a higher average residence time of gas bubbles in the liquid phase, but may also decrease surface vortex formation and therefore decrease the rate of gas-liquid transfer from the reactor headspace (gas headspace-water interfacial area is much smaller than gas bubble-water interfacial areas). This clearly indicates that gas transfer from the reactor head-space is important for establishing the equilibrium concentration of ethylene in the liquid phase for this reactor configuration. This also illustrates the risks of relying on strict rules for reactor design, since in a superficial analysis one could easily assume that increasing the contact time of gas bubbles-liquid by changing to radial-flow impellers would certainly increase the concentration of ethylene in the water phase.

In experiment E4, the impeller speed was increased from 200 rpm to 400 rpm, still using radial-flow impellers. Composition drift is practically eliminated at the higher impeller speed. This suggests that any mass transfer limitations present at lower agitation rates were completely eliminated and that the concentration of ethylene in the polymer phase is probably close to its thermodynamic equilibrium value throughout the polymerization.

Axial-flow impellers were used for experiment E5 at an agitation rate of 400 rpm. There is no marked difference between the copolymer composition obtained in experiments E4 and E5, which indicates that at this agitation level the impeller design plays a minor role. However, high impeller speed can be responsible for coagulum formation and will result in increase energy consumption (a major consideration for large scale reactors), therefore under some circumstances it might be useful to operate the impeller at lower speeds using a porous sparger and axial-flow impellers.

It is interesting to notice that extrapolating the curves for experiments E1 to E3 to time zero, a copolymer composition close to the ones in experiments E4 and E5 is obtained, which indicates that under those reaction conditions mass transfer limitations were present from the very beginning of the polymerization.

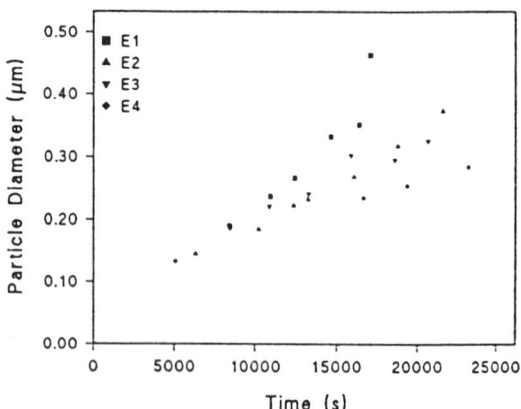

Figure 6. Particle size distribution as a function of polymerization time [42]

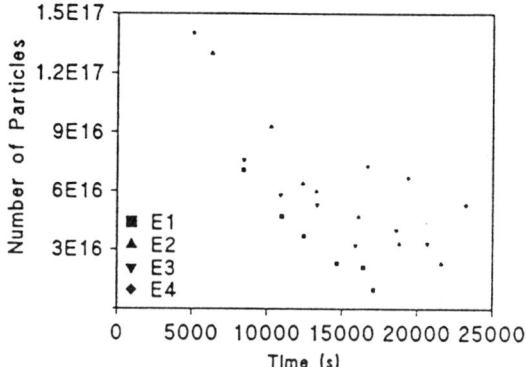

Figure 7. Number of particles as a function of polymerization time [42]

Particle size and number of particles were also affected by these different mixing and sparging configurations, as shown in Figures 6 and 7. Experiment E1 (base case) shows the fastest increase in particle size and decrease in particle number, which was attributed to particle flocculation. Coagulum was present at the end of the polymerization and the impeller blades were subjected to mild fouling. Adding a porous gas sparger for experiments E2 and E3 decreased but did not eliminate completely the formation of coagulum. A change of the flow patterns caused by the smaller size bubbles was considered to be responsible by this decrease in particle flocculation. The amount of coagulum present in experiment E3 (radial-flow impeller) was less than the one in experiment E2 (axial-flow impeller), therefore the flow patterns in the reactor can significantly affect particle flocculation for the adopted polymerization recipe. The higher impeller speeds used in experiment E4 reduced coagulum formation to a minimum. This might indicate that a more effective emulsifier or higher surface coverage of polymer particles by emulsifier were required, since flow patterns could affect particle size and number.

Acknowledgments. The authors would like to express their gratitude to Professor Alex Penlidis from the University of Waterloo, Department of Chemical Engineering, for his helpful discussions during the preparation of this article.

7. References

1. Gerrens, H. (1982) How to select polymerization reactors - Part I, *Chemtech* **June**, 380-383.
2. Gerrens, H. (1982) How to select polymerization reactors - Part II, *Chemtech* **July**, 434-443.
3. Dimitratos, J., Eliçabe, G., and Georgakis, C. (1994) Control of emulsion polymerization reactors, *AIChE J.* **40**, 1993-2021.
4. Stoessel, F. (1995) Design thermally safe semibatch reactors, *Chem. Eng. Prog.* **September**, 46-53.
5. Hamielec, A.E., MacGregor, J.F., and Penlidis, A. (1987) Multicomponent free-radical polymerization in batch, semi-batch and continuous reactors, *Makromol. Chem., Macromol. Symp.* **10/11**, 521-570.
6. Paquet, D.A. and Ray, W.H. (1994a) Tubular reactors for emulsion polymerization: I.Experimental investigation, *AIChE J.* **40**, 73-87.
7. Paquet, D.A. and Ray, W.H. (1994b) Tubular reactors for emulsion polymerization: II. Model comparisons with experiments, *AIChE J.* **40**, 88-96.
8. Poehlein, G.W. (1986) Emulsion Polymerization, in H.F. Mark and J.I. Kroschwitz (eds.), *Encyclopedia of Polymer Science and Engineering*, John Wiley & Sons, New York, pp. 1-51.

304

9. Oldshue, J.Y. (1987) Mixing and blending, , in J.J. McKetta and W.A. Cunningham (eds.), *Encyclopedia of Chemical Processing and Design*, Marcel Dekker, New York, pp. 274-309.
10. Leng, D.E. (1991) Succeed at scale up, *Chem. Eng. Prog.* **June**, 23-31.
11. Coyle, C.K., Hirshland, H.E., Michel, B.J., and Oldshue, J.Y. (1970) Heat transfer to jackets with close clearance impellers in viscous materials, *Can. J. Chem. Eng.* **48**, 275-278.
12. Carreau, P.J., Chhabra, R.P., and Cheng, J. (1993) Effect of rheological properties on power consumption with helical ribbon agitators, *AIChE J.* **39**, 1421-1429.
13. Uhl, V.W. and Voznick, H.P. (1960) The anchor agitator, *Chem. Eng. Prog.* **56**, 72-77.
14. Beckmann, G. (1973) Design of large polymerization reactors, *Chemtech* **May**, 304-310.
15. Litz, L.M. (1985) A novel gas-liquid stirred tank reactor, *Chem. Eng. Prog.* **November**, 36-39.
16. Mutsakis, M., Streiff, F.A., and Schneider, G. (1986) Advances in static mixing technology, *Chem. Eng. Prog.* **July**, 42-48.
17. Dickey, D.S. and Hill, R.S. (1993) Use the right specialty-polymer pilot plant, *Chem. Eng. Prog.* **June**, 22-29.
18. Dickey, D.S. (1991) Succeed at stirred-tank-reactor design, *Chem. Eng. Prog.* **December**, 22-31.
19. Chapple, D. and Kresta, S. (1994) The effect of geometry on the stability of flow patterns in stirred tanks, *Chem. Eng. Sci.* **49**, 3651-3660.
20. Tosun, G. (1992) A mathematical model of mixing and polymerization in a semibatch stirred-tank reactor, *AIChE J.* **38**, 425-437.
21. Levenspiel, O. (1972) *Chemical Reaction Engineering*, John Wiley & Sons, New York, pp. 326-345.
22. Nauman, E.B. (1974) Mixing in polymer reactors, *J. Macromol. Sci.-Revs. Macromol. Chem.* **C10**, 75-112.
23. Soares, J.B.P. and Hamielec, A.E. (1995) Effect of reactor residence time distribution on the size distribution of polymer particles made with heterogeneous Ziegler-Natta and supported metallocene catalysts. A generic mathematical model, *Macromol. Theory Simul.* **4**, 1085-1104.
24. Soni, Y. and Albright, L.F. (1981) Optimum design and operation of batch reactors for polymerization, *J. Appl. Polym. Sci.: Appl. Polym. Symp.* **36**, 113-132.
25. Albright, L.F. and Soni, Y. (1982) Designing and operation of reactors for suspension polymerization of vinyl chloride, *J. Macromol. Sci.-Chem.* **A17**, 1065-1080.
26. Matejicek, A., Snuparek, J., Jr., Vladyka, J., Kastanek, A., and Kleckova, Z. (1980) Heat transfer in emulsion polymerization products. Influence of contamination of the reactor wall, *Int. Polym. Sci. Tech.* **7**, T/107-T/112.
27. Wichterle, K. (1994) Heat transfer in agitated vessels, *Chem. Eng. Sci.* **49**, 1480-1483.
28. Bondy, F. and Lippa, S. (1987) Heat transfer, agitated vessels, in J.J. McKetta and W.A. Cunningham (eds.), *Encyclopedia of Chemical Processing and Design*, Marcel Dekker, New York, pp. 431-453.
29. Cheremisinoff, N.P. (1994) *Guidebook to mixing and compounding practices*, Prentice Hall, Englewood Cliffs, New Jersey.
30. Bolliger, D.H. (1987) Heat transfer, jacketed vessels, in J.J. McKetta and W.A. Cunningham (eds.), *Encyclopedia of Chemical Processing and Design*, Marcel Dekker, New York, pp. 69-82.
31. Ludwig, E.E. (1965) *Applied Process Design for Chemical and Petrochemical Plants*, vol. 3, Gulf Publishing, Houston, Texas, pp. 57-58.
32. Sanatgar, H. and Somerscales, E.F.C. (1991) Account for fouling in heat exchanger design, *Chem. Eng. Prog.* **December**, 53-59.
33. Fogg, R.M. and Uhl, V.W. (1973) Heat transfer resistance in half-tube and dimpled jackets, *Chem. Eng. Prog.* **69**, 76-80.
34. Markovitz, R.E. (1987) Heat transfer, jacketed vessels selection and design, in J.J. McKetta and W.A. Cunningham (eds.), *Encyclopedia of Chemical Processing and Design*, Marcel Dekker, New York, pp. 82-104.
35. Oldshue, J.Y., Mechler, D.O., and Grinnell, D.W. (1982) Fluid mixing variables in suspension and emulsion polymerization, *Chem. Eng. Prog.* **May**, 68-74.
36. Lakin, M.B. (1988) How to mix a reactor, *Chemtech* **May**, 300-303.
37. Oldshue, J.Y. (1983) *Fluid Mixing Technology*, McGraw-Hill, New York.
38. Zanker, A. (1987) Mixing and blending, in J.J. McKetta and W.A. Cunningham (eds.), *Encyclopedia of Chemical Processing and Design*, Marcel Dekker, New York, pp. 309-317.
39. Scott, P.J., Penlidis, A., and Rempel, G.L. (1993a) Semi-batch emulsion polymerization with gaseous comonomer: Ethylene-vinyl acetate case study, *Trends Chem. Eng.* **1**, 409-429.
40. Scott, P.J., Penlidis, A., and Rempel, G.L. (1993b) Ethylene-vinyl acetate semi-batch emulsion copolymerization: Experimental design and preliminary screening experiments, *J. Polym. Sci.: Part A: Polym. Chem.* **31**, 403-426.
41. Scott, P.J., Penlidis, A., and Rempel, G.L. (1993c) Ethylene-vinyl acetate semi-batch emulsion copolymerization: Use of factorial experiments for improved process understanding, *J. Polym. Sci.: Part A: Polym. Chem.* **31**, 2205-2230.
42. Scott, P.J., Penlidis, A., and Rempel, G.L. (1994a) Ethylene-vinyl acetate semi-batch emulsion copolymerization: Use of factorial experiments for process optimization, *J. Polym. Sci.: Part A: Polym. Chem.* **32**, 539-555.
43. Scott, P.J., Penlidis, A., and Rempel, G.L. (1994b) Reactor design considerations for gas-liquid emulsion polymerizations: The ethylene-vinyl acetate example, *Chem. Eng. Sci.* **49**, 1573-1583.
44. Scott, P.J., Penlidis, A., and Rempel, G.L. (1995) Ethylene-vinyl acetate emulsion copolymerization: Monomer partition and preliminary modelling, *Polym. React. Eng.* **3**, 93-130.

REACTION ENGINEERING FOR EMULSION POLYMERIZATION

GARY W. POEHLEIN
School of Chemical Engineering
Georgia Institute of Technology
Atlanta, GA 30332-0100, USA

1. Introduction

The mechanisms and kinetics of emulsion polymerization reactions have been presented in earlier papers. Although some uncertainties remain, especially for complex commercial recipes, significant progress has been made in our understanding of the physical and chemical phenomena involved in these reactions. This expanded knowledge base has proven to be useful for the design of latex products and manufacturing processes.

Reaction engineering of emulsion polymerization systems comprises design of the reactors, specification of the operational procedures and establishing methods for measuring and controlling important operating parameters and product characteristics. Reaction engineers require kinetic data and models but they also must be concerned with heat and mass transfer, mixing, control mechanisms, how streams are added and removed from the reactors and, last but not least, how all of these factors influence the quality of the product and its application performance.

This paper will be divided into four sections. The first will focus on reaction engineering fundamentals such as mass energy and particle population balances, some issues related to mixing and reactor configurations. The remaining three sections will be concerned with batch, semi-batch (sometimes called semi-continuous) and continous reactor processes.

The knowledge of emulsion polymerization kinetics presented in earlier papers is necessary to develop appropriate reaction models. The major focus of this paper, however, will examine how different reactor configurations influence process productivity and product properties.

2. Reaction Engineering Fundamentals

Reaction engineering textbooks generally deal with two ideal and quite different reactor types: the well-mixed tank (STR) and the plug-flow tube (PFT). A wide variety of impellers are used for emulsion polymerization in stirred tanks -- swept-back curved turbines, marine-type propellers and axial-flow turbines.

Heat transfer jackets are often used with both STRs and PRTs. Other heat transfer methods are also employed; especially with stirred-tank reactors. These include reflux condensers if recipe ingredient volatilities are adequate; internal heating/cooling surfaces such as coils and hollow baffles; and heat exchangers in external recirculation flow loops. Removal of the heat of polymerization and temperature control can be a major design issue with large reactors.

305

J. M. Asua (ed.), Polymeric Dispersions: Principles and Applications, 305–331.

Multiple feed points and heat transfer sections are possible with tubular reactors. Tubular reactors configured in the form of a recirculation loop are also used to produce latex products. These loop reactors are intermediate between the PFT and the STR in terms of a number of behavior characteristics. Loop reactors are treated extensively in another paper in this volume and will not be considered further here. The remaining parts of this section will include the balance equations and discussions of mixing and feed point considerations.

2.1. MASS BALANCES

The balance equations for individual species can be written as shown by eq. (1) for species 'i'.

$$F_{io} - F_{ie} - \int R_i dV = dN_i/dt \tag{1}$$

where F_{io} and F_{ie} are molar flow rates for the feed and effluent streams, R_i is the rate of disappearance of species 'i' per unit volume of reacting fluid, N_i is the total number of moles of 'i' in the reactor, V is reactor working volume and t is time. The units on each complete term are moles/time. F_{io} may be comprised of several streams. The parameters F_{io}, F_{ie}, V and R_i can vary with time. Hence eq. (1) is a general relationship which is valid for any reactor type. The units on the reaction term, however, are usually different for heterogeneous systems. Emulsion polymerization reaction rates, for example, are normally expressed as the molar time rate of monomer polymerizing per unit volume of aqueous phase. In this case the reaction term in eq. (1) would be written as $(1-\varphi)\int R_{pi}dV$. R_{pi} is the rate of polymerization of monomer 'i' and φ is the volume fraction of the organic phases which may also vary with time.

Equation (1) is an integral form of the mole balance which is not convenient for tubular reactor calculations. The differential form is given by eq. (2) for a steady-state tubular reactor.

$$dF_i/dV = -R_{pi}(1-\varphi) \tag{2}$$

Here F_i represents the molar flow rate of species 'i' at a particular point, V, in the reactor and R_{pi} is the reaction rate at the same point. One must be sure to use the correct units for R_{pi} (i.e. per volume of aqueous phase) when eq. (2) is applied to emulsion polymerization reactions. Also note that φ can be a function of V.

The mole balance equations contain more than one dependent variable. Thus these equations must be combined with other relationships such as feed-time profiles and energy balances which are presented in the following sub-section 2.3.

2.2. POPULATION BALANCES

Application performance of latex products can depend strongly on particle concentration and size distribution. These particle characteristics can be modeled by population balances. General models are given by Min and Ray [1] for the particle volume distribution. Population balance models include terms for the following phenomena.

- Change of the total number of particles with time.
- Change of particle size by internal growth due to polymerization.
- Formation and disappearance of particles by coalescence.
- Formation of particles by various nucleation mechanisms (i.e. micelle and droplet radical entry and oligomer precipitation).
- Population change due to flow in and out of the reactor.

The general models are complex partial-differential-integral equations which will not be presented here. These general equation can be simplified by calculating only the first several moments of the distribution, for steady-state continuous reactors and for seeded reactions in which nucleation and coalescence are avoided.

The single steady-state stirred tank reactor yields, for example, the particle number model given by eq. (3).

$$N_p = R_{nuc}\tau \tag{3}$$

where N_p is the number of particles per unit volume of aqueous phase, R_{nuc} is the rate of particle formation and τ is the mean residence time in the reactor.

Particle size distributions can be calculated from age distributions and a knowledge of particle growth rates. The basic relationship is given by eq. (4).

$$U(v)=f(t)/(dv/dt) \tag{4}$$

where $U(v)$ is the particle volume distribution, $f(t)$ is the particle age distribution and dv/dt is the growth rate determined from kinetic models. Equation (4) can be quite easy to use with seeded reactions in any reaction system in which the residence time distributions can be determined.

2.3. ENERGY BALANCES

The basic form of the overall energy balance is the same as that of the mass balances, namely:

INPUT - OUTPUT + GENERATION = ACCUMULATION

One can think of these separate contributions in terms of the reactor process which may involve feed and effluent streams, heat transfer to heating or cooling systems and to the surroundings, work input by agitators or pumps, heat generated by the polymerization reactions, and changes in the energy content of the system caused, for example, by temperature transients. If the contributions due to changes in kinetic, potential and pressure-volume energies are neglected the overall transient energy balance can be written as shown by eq. (5).

$$W_s - Q + \Sigma F_{io}H_{io} - \Sigma F_{ie}H_{ie} = \Sigma d(\int H_i C_i dV)/dt \tag{5}$$

where the H_i's are enthalpies of species 'i' in the feed and effluent streams and in the reactor, C_i is the species 'i' concentration (moles/volume of reaction media) which may (like H_i) vary from point-to-point in the reaction volume, W_S is the work done on the system and Q is the rate of heat loss from the system. Another term, $d\{M_R C_{pR}\}/dt$, should be added to account for energy changes due to temperature transients in the mass of the reactor, agitator and other equipment that can transfer heat to or from the reaction. The magnitude of this term would need to be estimated via calculation or calibration runs.

Equation (5) can be put in more useful forms for plug-flow and continuous stirred-tank reactors (PFR & CSTR). The transient balance for a PFR is obtained from a balance over a differential volume, dV, and is given by eq. 6 with shaft work and heat losses to the surroundings neglected.

$$Ua(T_c\text{-}T) - \partial\{\Sigma F_i C_{pi}T\}/\partial V + \Sigma R_{pi}[-\Delta H_{Ri}(T)] = \partial \Sigma C_i C_{pi}\{T\}/\partial t \qquad (6)$$

U is the overall heat-transfer coefficient between the cooling fluid at T_c and the reacting mixture at T, 'a' is the heat transfer area per unit volume of reactor, C_{pi} is the specific heat capacity of 'i' and $-\Delta H_{Ri}(T)$ is the heat of polymerization of monomer 'i' at temperature T. The equivalent relationship for a well mixed, transient CSTR is given by eq. (7).

$$W_S - Q - \Sigma F_{io} C_{pi}(T\text{-}T_0) - \Sigma \Delta H_{Ri}(T) R_{pi} V = d \Sigma\{C_i C_{pi} VT\}/dt \qquad (7)$$

The factor (1-φ) would need to be added to the reaction terms in eqs. (6) and (7) if the rate units were per unit volume of water. The heat loss term, Q, can be expressed as the sum of two terms.

$$Q = UA\Delta T_{lm} + Q_S \qquad (8)$$

A is the total heat transfer area, ΔT_{lm} is the log-mean temperature difference between the cooling fluid and the reactor contents and Q_S is the heat lost to the surroundings. If more than one type of heat transfer method is used (jacket, internal tubes, external heat exchanger loop, reflux condenser) several quantities will comprise the Q equation.

Many of the parameters in eqs. (6) - (8) are known, can be easily measured, or can be evaluated by combining these equations with the mass balance and reaction rate relationships for numerical modeling. The rate equations can certainly add some uncertainty to the model simulations. Mechanisms and kinetics, however, are reviewed in detail in previous papers and will not be discussed further here. Other parameters which can be troublesome include the heat-transfer coefficient (U), the heat loss to the surroundings (Q_S) and the shaft work (W_S). Fortunately W_S is usually small and it can be measured. Q_S can be estimated from heat-transfer theory or, preferably, measured with non-reaction calibration experiments. This term is often small for large-scale commercial reactors but it can be very significant for bench-scale and pilot-plant reactors.

The overall heat-transfer coefficient (U) is determined by four resistances in series: the films on the coolant and reagent sides of the reactor surface, the wall and last, but sometimes not least, the resistance due the fouling (polymer buildup) on the wall. Changes of the viscosity of the reacting media and wall fouling will result in changes in U. Hence, mathematical models and control algorithms must be updated as U changes.

Another approach to this problem, however, is to use measurements of flows and temperatures of coolant streams to determine heat transfer rates. If, for example, a jacketed, insulated reactor is used the Q term in eq. (8) can calculated from the following relationship.

$$Q = M_s C_{pw}(T_{out} - T_{in}) \qquad (9)$$

M_s is the mass flow rate of the coolant fluid, C_{pw} is its specific heat capacity and T_{out} and T_{in} are outlet and inlet coolant stream temperatures.

All of the terms on the right side of eq. (9) can be easily measured. If eq. (9) is used in eq. (7) and the other flow and thermodynamic parameters are known, or can be measured, one has a direct determination of polymerization rate. This is most easily done for isothermal polymerizations with only one monomer. If multiple monomers are reacting only the heat-of-reaction summation term in is Equation 5 is determined. The relative reaction rates of the different monomers need to be known in order to determine the rates for the individual monomers.

Heat transfer can be a major scaleup problem as the heat flow terms in eqs. (6) - (8) clearly indicate. The parameter 'a' in eq. (6) is the area for transfer per unit volume of tubular reactor. This area is equal to 4/D for a cylindrical tube of diameter D. Thus 'a' decreases as tube diameter is increased. Tubular reactors have relatively large surface areas so this fact should not be an issue for most emulsion polymerization reactions. Such is not the case, however, for jacket-cooled stirred-tank reactors. The ratio 'A/V' for jacketed tanks also changes inversely with diameter if geometric similarity is part of the scaleup criteria. Hence, reactions are often heat-transfer limited in large reactors. Small pilot-plant and bench-scale reactors can be run fast in a 'reaction limited' mode and this often done. When the results of such R&D efforts are moved to large commercial reactors the products can be quite different. **Good practice would be to duplicate the cycle time and operation procedures of the projected commercial reactor in the R&D program.**

Heat transfer limitations can be overcome by the use of heat transfer surfaces other than, or in addition to jackets around the reactor walls -- internal tubes, external heat exchangers and possibly reflux condensers. These options can, however, have significant disadvantages such as increased fouling and more costly cleaning procedures. External heat exchangers can also be used to subcool the entering feed streams in semi-batch and continuous processes. Fouling and cleaning concerns are less with these streams but the amount of energy removal is limited to the sensible heats and cooling below normal water temperatures is expensive.

2.4. MIXING

Scaleup and mixing can be a troublesome issue in emulsion polymerization processes because fluid dynamics and mixing normally serve a number of functions: emulsification, blending, and promotion of heat and mass transfer. These functions need to be handled well without causing excessive coagulation due to fluid shear. Optimum design of an agitation system for any one of these functions is not likely to be best for the others. Hence, most scaleups will involve compromise in design of the reaction process and the mixer.

Mixing has been important since the beginning of the process industries. Much of the early study of mixing phenomena resulted in the development of empirical correlations for macroscopic characteristics such as power consumption, heat transfer coefficients, blend times, interfacial areas in dispersions and minimum impeller speeds for suspension of solids. The Foreword in the Proceedings of the 6th European Conference on Mixing (Pavia, Italy, May 1988) indicates, however, that "in the last two decades the study of mixing has undergone a dramatic change in that empirical methods have more and more been replaced by scientifically based methods". Advances in theoretical fluid dynamics due to expanded computing power and new experimental measurement techniques have contributed to this change. Unfortunately much more research is needed in order to place the design and operation of mixers for complex multi-phase processes on a firm fundamental foundation. The basic principles of mixing and scaleup have been covered in the previous chapter. Hence the remainder of this section will only include some issues related to blending time, emulsification, coagulum formation and physical reactor design.

2.4.1. Blending Times

The time required to blend miscible liquids at the start or before a reaction and/or to mix entering flow streams is important in many chemical processes; including batch, semi-batch and continuous emulsion polymerization reactions. Blend time, θ, as would be expected, is a function of the ratio between the volume of the reactor (V) and the pumping capacity of the impeller (Q) and perhaps other variables. The dimensionless Blend Number, θN (being N the agitator speed), has been correlated with Reynolds Number to yield plots with shapes similar to the N_p-N_{Re} graphs. Hence, θN is relatively constant in the turbulent region where the ratio $V/Q \approx 1/N \approx \theta$. An examination of Table 1 in the previous chapter of this volume shows, however, that to achieve constant blend time (i.e. constant Q/V) for a 216:1 scaleup would require a 36-fold increase in the power to volume ratio (P/V). This large change in power is rarely feasible. Hence blend times are often reduced considerably upon scaleup as would be predicted by the inverse of the numbers in the Q/V line in that table.

2.4.2. Emulsification

Dispersion of monomers and other hydrophobic recipe components in the continuous aqueous phase is an important part of emulsion and miniemulsion polymerization processes. The importance of the degree of dispersion (i.e. droplet size) in emulsion polymerization will depend on the type of process; batch, semi-batch, continuous, seeded or unseeded, monomer-starved or flooded, mixing in the reactor or in other equipment, and how the feed streams are added. In general, one wants the dispersion of monomers to be adequate to facilitate mass transfer to the reaction sites in the monomer-swollen polymer particles and to avoid separation which can lead to bulk polymerization and the formation of large particles and excessive fouling.

Miniemulsion reactions, in contrast to conventional emulsion polymerization, are based on particle nucleation and polymerization in very small (>1μm) monomer droplets. Hence emulsification prior to miniemulsion polymerization is a key process step which almost always involves dispersion in equipment which generates high fluid deformation rates and stresses. Some monomer-droplet polymerization undoubtedly takes place in conventional emulsion polymerization reactions (see Durbin et al. [2]) and the extent of monomer emulsification as well as the presence of any highly water-insoluble ingredients will influence the amount. Monomer emulsification is clearly an important function of mixing, either in the reactor or in separate equipment. Hence, some of the fundamental aspects of emulsification and selected literature will be reviewed in the remainder of this section.

Liquid-liquid dispersions are important in many process applications involving reactions and separations. Hence there is an extensive literature on the influence of mixing on the formation of dispersions in stirred tanks and other equipment. Dimensionless variables, such as those discussed earlier, are often used to correlate experimental dispersion data. Relevant new dimensionless groups include the Weber number ($N_{We} = D^3N^2\rho_d/\gamma$), the volume fraction of the dispersed phase, ϕ, and ratios of physical properties (viscosities and densities) of the two phases. Important new parameters to correlate are various droplet diameters and interfacial area per unit volume of dispersion (a). D is the diameter of the impeller, ρ_d is the density of the dispersed phase and γ is the interfacial tension between the two phases. Much of the published work in this field is for systems that do not involve added emulsifiers because rapid phase separation is desired after many contacting operations. Eckert et al. [8], for example, provide a good review of the literature and report on an extensive experimental study of liquid-liquid interfacial areas formed by turbine impellers in baffled tanks. They studied the dispersion of a number of organic fluids (without emulsifiers) in water in order to vary interfacial tensions and disperse-phase viscosities and densities. Three tank sizes were employed with different size Rushton turbines operating over a range of speeds. Volume fractions of the dispersed phase, however, were 8% or less. Eckert et al. give a number of correlations for interfacial area, 'a', and recommend the following relationship for scaleup for the same fluids with the same volume fractions.

$$a \approx N^{1.11}D^{1.23}T^{-0.18} \tag{10}$$

where T is the diameter of the reactor. Since power per unit volume is proportional to N^3D^2 one can see that scale-up for constant 'a' requires less specific power (P/V).

Considerable work has also been published on the minimum (critical) impeller speed necessary for dispersion and to achieve target droplet sizes (maximum and average) and size distributions. Nagata [4] provides a review of some of this work. Examples of correlations are given by eqs. (11) and (12) for critical impeller speed (N_{crit}) and Sauter-mean diameter ($d_{SM}=1/a$) respectively.

$$N_{crit} = KT^{-2/3}(\eta_c/\rho_c)^{1/9}([\rho_c-\rho_d]/\rho_c)^{0.26} \tag{11}$$
$$d_{SM} = const \times N_{We}^{-0.36}(D/T)^{-k} \tag{12}$$

η is viscosity, ρ is density, K and k are constants and the subscripts c and d represent the continuous and dispersed phases.

Almost all of this work was based systems that did not contain emulsifiers. Emulsifiers are important ingredients in emulsion and miniemulsion polymerization reactions. They lower the equilibrium interfacial tension and serve to help stabilize the emulsions against coalescence. The break up of the disperse phase to form small droplets is usually a very rapid process (often of the order of 1ms) and equilibrium conditions will not exist at the interface. In fact van den Temple [5] indicated that the major function of an emulsifier in the dispersion process was to produce gradients in interfacial tension which allow the interface to resist tangential stresses. Lucassen-Reynders and Kuijpers [6] examined the role of interfacial viscosity and elasticity on the emulsification process. Their theory was based on simple shear flow but the experimental flow conditions were not well characterized. They showed that interfacial viscoelasticity is capable of increasing the effective viscosity of the disperse phase by more than an order of magnitude. They also pointed out that the interfacial tension, γ, would vary considerably over the surface during droplet breakup and the average value would be higher than the equilibrium value. As a consequence emulsion droplet size would be less sensitive to interfacial tension than expected.

Davies [7,8] studied the influence of turbulent energy dissipation rates on drop sizes. Dispersion in isotropic turbulent flows is relatively simple to treat theoretically. One assumes that the final stage of drop breakage is caused by dynamic pressure fluctuations rather than viscous shear. The magnitude of these fluctuations is given by $Dp = \rho_c(v')^2$, where v' is the time-average value of the velocity fluctuations. Davies [7] indicates that v' is proportional to $(d_{max}/P_M)^{1/3}$, where d_{max} is the diameter of the largest drop that will not be further divided by the flow field and P_M is the power dissipation in the flow per unit of mass. The pressure holding the drop together is $4\gamma/d$. These relationships can be combined to yield eq. (13) for the maximum diameter.

$$d_{max} = \text{const.} \times (\gamma/\rho_c)^{0.6}P_M{}^{-0.4} = \text{const.} \times \gamma^{0.6}\rho_c{}^{-0.2}(P/V)^{-0.4} \qquad (13)$$

The minimum drop size is estimated by noting that when the drop Reynolds number $(v'd\rho_c/\eta_c)$ is less than 5 no further breakup will occur. If v' from this relationship is used to estimate the pressure fluctuations which are then set proportional to the drop interfacial pressure, eq. (14) is obtained for minimum drop diameter, d_{min}.

$$d_{min} = \text{const.} \times (\eta_c{}^2/\rho_c\gamma) \qquad (14)$$

Davies [8] considered emulsification in four types of equipment: (1) valve homogenizers, (2) colloid mills, (3) liquid whistles and (4) turbine impellers. He estimated typical P_M and v' values for the high-energy emulsifiers, for the impeller volume of the turbine and computed d_{max} and the length scales, l_k, of the Kolmogoroff eddies. These parameters are compared with experimental d_{max} and d_{min} values for emulsions of low viscosity oils in water. Table 1 shows the results of this work.

TABLE 1. Fluid dynamics parameters and drop sizes (μm) for different types of emulsification equipment.

Type of equipment	Typical local P_M (W/kg)	Typical Local v'(m/s)	l_k	d_{max} Eq 16	d_{max} Exp	d_{min} Exp
Fine clearance valves	400×10^6	12	0.22	0.7	~	~.05
Colloid mills	0.44×10^6	1.6	1.3	10.5	6	0.1
Liquid whistles	12×10^6	3	0.5	2	2	<0.1
Turbine impellers	6×10^3	0.2	3.6	70	50	12

Davies [12] units on l_k and d's are microns.

Davies points out that the minimum size drops that are smaller than l_k could be 'satellite' drops formed during the break-up of larger drops. He used interfacial tension values obtained without added emulsifiers in his calculations and he concludes that "the principle effect of added emulsion stabilizers is to prevent re-coalescence of the droplets". A slightly modified equation was presented for the emulsification of higher viscosity liquids.

$$d_{max} = const. \ x \ (\gamma + \eta \ _d v'/4)^{0.6} \rho_c^{-0.6} P_M^{-0.4} \qquad (15)$$

McManamey [9] used a relationship equivalent to eq. (13) with P_M calculated as the total power input to the impeller divided by the fluid mass in the volume swept out by the impeller. He was able to correlate data for Sauter-mean diameters(μm) for a number of systems (again without emulsifier) with prefactor constants in the range of 0.18 to 0.265, (P_M in W/kg, γ in N/m & ρ_c in kg/m^3). The data of Brown and Pitt [10] for three different organic-water systems and two turbine sizes fit eq. (13) very well when the prefactor was 0.192. McManamey also indicated that eq. (13) could be written as shown by eq. (16) for geometrically similar equipment.

$$d = const. \ x \ (\gamma/\rho_c)^{0.6} N^{-1.2} D^{-0.8} \qquad (16)$$

A comparison of eqs. (16) and (10) (for 'a' by Eckert et al.), noting that 'd' and 'a' should be inversely related, shows that the exponents on N and D are close. This is quite satisfying since eq. (10) is based on empirical fitting of data and eq. (16) comes from rather simple turbulence theory. One must be aware, however, that much of the data in the literature are based on systems without emulsifier and at lower dispersed phase content than is the case for emulsion polymerization systems. In addition, the theory presented here has not addressed the formation of satellite drops, re-coalescence, or the influence of time in the turbulent field, i.e. the process kinetics, on the drop size characteristics.

The presence of emulsifier would help to stabilize small droplets if Oswald ripening is not significant and re-coalescence will not be considered further. Theory and experiments related to the kinetics of emulsification has been published by Braginsky and Kokotov [11]. Their theory considers residence time in the emulsification zone and the influence of phase viscosities. Experiments were carried out in four different size baffled tanks with turbine and paddle impellers of varying size containing 2-6 blades. Six liquid pairs with differing viscosities and interfacial tensions were studied. Surface active agents were added to prevent re-coalescence. Specific power input and volume fraction of the dispersed phase (1 to 4%) were also varied. They found that drop size decreased rapidly at first and approached the final values in 15 to 30 minutes. The drop size distributions at steady-state were approximately Gaussian.

In summary, although emulsification is certainly not completely understood, empirical correlations and turbulence-based theories do provide potential scale-up criteria. Uncertainty remains concerning systems with high disperse-phase loading and as to the effective interfacial tension when emulsifiers are used.

2.4.3 Coagulum Formation and Surface Fouling
Coagulation and reactor fouling can represent a major problem for emulsion and other heterogeneous polymerization processes. Fluid dynamics, both during the reaction and in post-reaction processing, can significantly influence coagulation and fouling. Hence,

this brief discussion is included in the 'Mixing' section. Other factors, however, can also contribute to this problem area. Vanderhoff [12] proposes two general mechanisms for the formation of coagulum: (1) a failure of the colloidal stability of the latex during or after the polymerization to cause flocculation of the particles and eventually form microscopic coagulum; (2) polymerization of the monomer by a mechanism other than that of emulsion polymerization, to give polymer of different form than latex particles. Mixing can influence both of these mechanisms.

If, for example, the reactor contents are not well mixed coagulum can be formed for one or more of the following reasons.

- Poor emulsification of the organic phase can lead to large monomer drops and even monomer pools. Polymerization in these loci will form masses larger than the latex particles which can remain in the latex or deposit on internal surfaces.
- Slow blending of streams that enter during the reaction can result in high local concentrations of electrolyte which can cause flocculation; of emulsifier which can cause excessive particle nucleation; of monomer which can form large drops and pools as mentioned above; and temperature variations which can influence reaction and colloidal phenomena.

The literature on flocculation of colloids is extensive. Three mechanisms are known to be important in emulsion polymerization reactions. (1) Small particles are subject to Brownian motion which causes frequent collisions between particles. If the energy level of a collision is sufficient to overcome the repulsive energy barrier flocculation will occur. Brownian motion is inversely proportional to particle size. Hence it is most important for submicron particles. (2) Local fluid motion increases particle collision frequency and, in some cases, collision energy. Shear flocculation is more important with larger particles and with mixtures of large and small particles. Theory predicts that the flocculation rate will be proportional to the fluid shear rate, the third power of the sum of the particle diameters and the second power of the particle concentration. von Smoluchowski [13,14] has developed theories for both of these mechanisms. (3) Surface flocculation is a third mechanism that could be important in emulsion polymer systems. Heller and coworkers [15-18] demonstrated that surface flocculation rates can be more important than bulk rates if significant gas-liquid interface is present. They attribute this to a higher electrolyte concentration at the interface, a lower dielectric constant near the surface and asymmetry of the particle double layer at the interface. Lowry et al. [19,20] also suggested that there would be a higher concentration of particles at the interface.

Extensive research in colloid interactions and flocculation have increased our understanding of the fundamental mechanisms and kinetics. Extrapolation of this knowledge to commercial processes and scaleup, however, remains as a major problem. The influence of fluid dynamics and mixing is especially important in manufacturing processes. Hence, some recent work related to this issue will be reviewed in the remainder of this discussion followed by some recommendations for reducing the magnitude of the problem.

Lowry et al. [19,20] and Hoedemakers [21] have studied the influence of mixing on flocculation in systems which are quite different. Lowry et al. considered the influence of agitator speed on coagulum formation during the emulsion polymerization of styrene in a small (500-ml) reactor at Reynolds numbers in the laminar flow regime. The shear rate in this regime should be directly proportional to impeller speed (N) and theory predicts that -ln(1-c) should increase linearly with N at constant shear time (where c is the fraction of particles coagulated). The experimental results for a 3.8 cm diameter pitch-blade impeller fit this model at rotational speeds between 2 and 8 rps. Lowry et al. also analyzed the results of Rubens' [22] study of coagulation in vinyl chloride/ethyl acrylate (65/35) emulsion copolymerization under turbulent conditions at higher Reynolds numbers. Rubens used a 7.6 liter reactor with three different-diameter flat-bladed impellers (5.1, 7.6 & 10.2 cm) and he varied the solids content from 40 to 50 wt%. Total reaction time was maintained constant at 9 hours. Turbulence theory predicts that average or maximum shear intensity should be proportional to $(P/V)^{1/2}$ and volume fraction of the dispersed phase, φ. The linear relationship between $P^{1/2}$ and $-V^{1/2}\ln(1-c)/\varphi$ expected from turbulence theory does fit the data for each solids content except at lower power inputs. Lowry et al. [19] suggest that this might be because some minimum shear is needed to initiate flocculation.

Lowry et al. [20] also studied the influence of surface flocculation in stirred tanks (with and without a gas-liquid interface) and in a bottle polymerizer. Equation (17) was proposed to account for both shear and surface coagulation.

$$dc/dt = P_1 + P_2c + P_3c^2 \qquad (17)$$

where the P's are constants to be determined by fitting experimental data. They were successful in fitting %coagulum vs time data for non-reacting, shear-sensitive, modest-solids ($\approx$ 35wt%) latexes in both the stirred tank and the tumbling bottles. The stirred-tank results showed that coagulation was less when the gas-liquid interface was eliminated and that increasing impeller speeds from 695 to 852 rpm significantly increased the coagulation rate. The amount of coagulum was linear with time for the bottle experiments (except for high acid concentrations) because surface coagulation was expected to be most important in this low-shear system.

Hoedemakers [22] studied the emulsion polymerization of styrene with a rosin acid soap as the emulsifier in stirred tanks (batch and continuous) and in a continuous pulsed-packed column. He measured conversion, polymerization rate, particle numbers, weight-average particle diameters and fraction of surface coverage by the soap. Monomer, initiator, electrolyte and emulsifier concentrations were varied as were temperature and shear rate. Four soap levels were used in the stirred-tank experiments. The results, for all but the highest soap level, showed that the number of particles peaked early in the reaction and then decreased to a relatively constant level after 20 to 40% conversion. Surface coverage of the particles in the final latexes for these runs was nearly the same at 71 to 75%. This limited flocculation was controlled by surface coverage and the particles formed were still in the submicron size range. Hence, these particles would not be considered to be coagulum as that term is normally used.

Hoedemakers also studied the influence of stirring rate on the number of particles in the final latex. Particle number was plotted against both power input per mass and impeller speed. The initial emulsions were all produced at an impeller speed of 500 rpm. Particle numbers were relatively independent of energy input at both low and high values. The transition from higher to lower numbers of particles occurred at the same rotational speed and energy for all three levels of soap. Unfortunately only one impeller size was used so the effects of speed and energy input could not be separated. Good agreement was obtained, however, when soap surface coverage was plotted against maximum energy dissipation for both the batch stirred-tank reactions and those carried out in the pulsed-packed column.

Vanderhoff [23] reported on coagulum formation caused by dissolving polymer in the monomer prior to reaction in a bottle polymerizer. Two types of coagulum were formed; soft-powdery and hard-glassy. The amount of hard-glassy coagulum formed was directly proportional to the amount of polymer added and electron micrographs showed that this material was comprised of particles formed from the monomer drops with flocculated latex particles. The soft-powdery material appeared to be hollow and comprised of the smaller latex particles. Surface flocculation around vapor bubbles was suggested as a possible mechanism.

In summary, coagulum and surface fouling remains a problem for emulsion and miniemulsion polymerization processes. Both recipe ingredients and reactor design and operation can influence coagulum formation or the lack thereof. Vanderhoff [12,23] recommends that "the most effective approach is to determine the mechanism by which it is formed and the approximate conversion at which it is formed". He also suggests the following methods for potentially reducing coagulum formation.

Modification of the polymerization recipe and/or technique by:
- use of a seed to eliminate nucleation,
- addition of a stabilizing emulsifier at the appropriate conversion,
- rigorous temperature control,
- varying the mode of monomer addition e.g. continuous addition,
- variation of the agitation rate during the reaction,
- addition of a chain transfer agent which may form non-surface-active oligomers by transfer in the water phase, which may flocculate and reduce the overall polymer-water interface.
- developing a better understanding of the reaction system.

Modification of the reaction system design by:
- use of a semi-batch or continuous process,
- use of a different reactor configuration,
   modification of the agitator and baffle system to ensure uniform agitation and complete but mild mixing of the ingredients,
- use of a different mode of addition of ingredients, e.g. addition below the surface rather than by dropping through the reactor top onto the upper surface of the latex.

2.4.4. Physical Reactor Design

A large variety of reactors and agitators have been used for the manufacture of emulsion polymers. This last part of the 'Mixing' section will only describe some typical reactor characteristics. Reactor vessels are normally cylindrical with dished tops and bottoms. The ratio of the tank height(H) to diameter(T) is usually 1.0 to 1.3 with an impeller diameter of about 1/2 of the tank diameter. Larger H/T ratios and multiple impellers are used in some reactors. The internal surfaces of the reactors are smooth to minimize fouling. Glass lined reactors were once the industry standard but polished stainless steel vessels have become more popular because they are less expensive and have less wall resistance to heat transfer.

Baffles will enhance mixing in most reactor designs but their use in latex reactors is quite variable. The mixing industry standard would be four radially mounted baffles with a width of about 1/10th of the tank diameter. Latex reactor baffles are often mounted slightly away from the wall in an effort to avoid low-flow regions where fouling could be more severe. Pipe baffles are sometimes used and they can serve a dual role as heat transfer surfaces. Likewise anything that protrudes into the reaction mixture can produce some baffle-like effects (e.g. reagent feed tubes and instrument sensors). Twist-element baffles, a relatively new design, have been studied by Lehtola et al. [24]. The purpose of the twist construction is to increase mixing in zones where it is poorest. Their work involved the use of KCl-solution injections followed by concentration measurements in various parts of the tank. A homogeneity index was calculated as a function of time and the power input needed to achieve a specific level of mixing was measured for both twist-element and conventional baffles. The twist-element baffles were shown to require less power at higher levels of P/V. No data were presented on total flows or local turbulence near the baffles. The design would suggest, however, that the both dead spaces and local turbulence could be less with the new baffles. Hence, they may offer some advantages for emulsion polymerization systems.

Reactor design and mixing requirements are likely to be different for batch, semi-batch and continuous reactors. These issues will be addressed in the individual sections on these reaction processes which comprise the remainder of this paper.

3. Batch Reaction Processes

3.1 GENERAL CONSIDERATIONS

Batch reactors are widely used in academic research and for preliminary product and process screening tests. Some commercial production is carried out in batch reactors but other processes are much more common. Batch reactors are relatively simple to operate and they can be used for small production runs of multiple products. Feed streams are not added after the reaction starts and so maintaining a homogeneous reaction volume is less troublesome than with other processes. The level of the reacting mixture is nearly constant which generally reduces the formation of wall polymer and permits utilization of all of the heat transfer area during the entire reaction cycle.

The major drawbacks of batch processes are their lack of flexibility in varying and/or controlling important product characteristics and their heat transfer limitations; especially with larger reactors that are only equipped with jacket cooling. Control options are restricted once a batch reaction is started with a full charge. Temperature can be controlled if the heat-transfer surfaces are adequate to remove the reaction energy release. The polymerization can be stopped by injection of inhibitor or accelerated by adding more initiator if problems are detected early enough. Reaction kinetics and transport processes will generally lead to drifts in copolymer composition and perhaps in molecular and particle architecture (e.g. branching, crosslinking and particle morphology). In addition, batch-to-batch variations can be caused by changes in the purity of ingredients, reactor operation and particle nucleation which occurs early in the reaction.

3.2. SEEDED REACTIONS

The use of a small-particle seed latex is a common technique for significantly reducing batch-to-batch variations in both batch and semi-batch processes. Nucleation is a rapid phenomenon which can be sensitive to the reaction ingredients, their impurities, mixing, temperature and perhaps the phase of the moon. It is **THE PART** of an emulsion polymerization reaction that is most likely to vary. The effective use of seed latexes can remove or significantly reduce particle number uncertainties and, thereby, reduce product variability in the final reaction process.

Nucleation variations will remain in seed preparation. The fundamental concept of processes that utilize seed, however, is to accept these variations and then to reformulate the final-product recipe with each new batch of seed. This requires careful characterization of the number concentration and size parameter(s) of the seed. When this information is known the calculation of the amount of seed needed to yield a target latex particle concentration and size in the final product is straightforward. One batch of small-particle-size seed can be used for many batches of final product.

Seed latexes are made by use of higher concentrations of emulsifier as would be expected. Seed stability is a potential problem because small particles are prone to flocculate due to high Brownian motion if they are not adequately protected. Most emulsion polymerization reactions produce latex particles that are not surface-saturated with emulsifier. Hence, some form of post-reaction stabilization may be necessary to ensure storage stability and batch-to-batch uniformity with each lot of seed. Seed stability can also be reduced via emulsifier desorption when the seed is mixed with the other ingredients in the recipe. Ideally one would like enough emulsifier to nearly saturate the seed in the total recipe without leaving any free emulsifier to support the nucleation of new particles when the reaction is initiated. A properly formulated seeded process will operate with a constant particle number, avoiding both nucleation and flocculation. Emulsifier may need to be added during the course of the reaction to maintain this balance. An exception to the constant-particle-number criteria occurs when one wishes to produce a multimodal particle size distribution. This will be discussed in the section on semi-batch reactions.

In-situ seed formation at the beginning of a batch reaction represents an alternate seeding process. In-situ seed is formed by adding all or a major part of the emulsifier to a portion of the other ingredients, forming the seed latex and then adding the remaining reagents. Hence, in a strict sense, this could be considered semi-batch operation. Batch-to-batch uniformity is less than with the use of premade and well characterized seeds but usually better than unseeded batch reactions.

3.3. DESIGN EQUATIONS

3.3.1. *Mass Balance*
The mass-balance relationship presented earlier eq. (1) reduces to eq. (18) for a well-mixed batch reaction.

$$(1-\varphi)VR_{pi} = -dN_i/dt = -(1-\varphi)VdC_{iw}/dt \tag{18}$$

where C_{iw} is the molar concentration of species 'i' per unit volume of aqueous phase. Both V and φ can vary with time in a batch reactor but the product $(1-\varphi)V$ is the total volume of aqueous phase which will be constant in the absence of large temperature changes. Hence the simplest form of the mass balance for a batch emulsion homopolymerization reaction is given by:

$$R_p = k_p[M]_p\bar{n}N_p/N_A = -dC_{Mw}/dt \tag{19}$$

where k_p is the propagation rate coefficient, $[M]_p$ is the monomer concentration in the polymer particles, $\bar{n}$ is the average number of free radicals per particles, N_p is the particle concentration (number per volume of water), N_A is Avogadro's number and C_{Mw} is the monomer concentration per volume of water. Unfortunately $[M]_p$, $\bar{n}$, N_p, C_{Mw}, and perhaps k_p, all vary with time over the course of an unseeded batch reaction. $[M]_p$ is often treated as a constant during Intervals 1 and 2 and it is equal to $\varphi [M]_p/(1-$

φ) during Interval 3. Np is constant for seeded reactions without new nucleation. C_{Mw} is equal to $C_{Mwo}(1-X)$ where C_{Mwo} is the initial moles of monomer per volume of water and X is fractional monomer conversion. Even with these simplifications, however, eq. (19) cannot be used for designing batch reactors without experimental data and/or empirical correlations which relate all variables to either time or conversion.

Equation (19) can, however, be used to fit experimental conversion-time curves. These X-t profiles are normally sigmoidal with an increasing slope during the particle nucleation period (Interval 1) and a relatively constant slope in mid-conversion ranges (through and sometimes beyond Interval 2). After transport of monomer from the drops stops, $[M]_p$ will decrease but ñ may increase due slow termination. These counter effects can result in either increases or decreases in rate. Eventually, however, the rate will slow as the monomer is depleted and/or the glass point is reached - reducing k_p. In addition the free radical flux into the particles can slow because of decreasing initiator concentration and a lack of monomer that is needed in the continuous phase to form the oligomers that will enter an organic phase. Increasing the temperature and post addition of an initiator that partitions more strongly into the particles are common techniques for reducing residual monomer.

The volume fraction of the dispersed phases in a batch reaction is relatively high over the entire cycle because all ingredients are added at the beginning. Conversion of monomer to more dense polymer results in a continuing decline in the volume fraction of the organic phases, with the volume of the monomer droplets decreasing and the volume of the latex particles increasing up to the point where monomer transport from the drops stops at the transition between Intervals 2 and 3. This transition is likely to be the point of highest viscosity due to the maximum apparent volume fraction of the dispersed phases which includes a contribution of the double-layers around the particles and drops. Flocculation, mixing and heat transfer problems can occur during this part of the reaction.

3.3.2 *Energy Balance*
A form of the energy balance for a batch reactor can be obtained by combining eqs. (7) and (9) and adding a term, $\psi dT/dt$, to account for sensible heat transients in the reactor mass.

$$W_s - M_s C_{pw}(T_{out} - T_{in}) - \Delta H_{Ri}(T) R_{pi} V(1-\varphi) = \qquad (20)$$
$$d\Sigma\{C_i C_{pi} T\}/dt + \psi dT/dt$$

where ψ is a constant. If heat losses to the surroundings are small or can be measured, and the reactor operates at a constant temperature, the right side of eq. (20) is zero. In this case relatively simple measurements of W_s and the coolant mass flow rate and temperatures can be used for calculation of reaction rate. Such rate measurements can be more precise than rates determination by differentiation of conversion-time data.

Combination of eqs. (7) and (8) for a constant-temperature batch reactor results in the following energy balance relationship.

$$-\Delta H_{Ri}(T) R_{pi}(1-\varphi) = U\Delta T_{lm}\{A/V\} + \{Q_s - W_s\}/V \qquad (21)$$

If heat transfer is accomplished only through the reactor walls, the ratio A/V decreases with reactor size when geometric scale-up is used. This can be a serious problem for batch reactors and is one of the reasons they are not often used for commercial production. This problem can be overcome by adding heat transfer area inside the reactor or in an external flow loop -- surfaces that can foul and require cleaning. Another method for increasing the productivity of batch reactors is via nonisothermal operation. The reaction can be initiated at a lower temperature and the heat capacities of the reactor and reacting mixture used to absorb part of the heat of reaction. Common practice is isothermal or near isothermal reactions but there are examples of commercial batch reactions with temperature changes as large as 40°C. I have seen two such cases. One polymerization was run nearly adiabatically with a 35 - 40°C temperature change. The recipe contained only about 20 wt% solids so the water served as a major heat sink. The second was a high solids (50%) acrylic latex carried out in a reactor that could only remove about 1/2 of the reaction heat. The temperature changed about 30°C before the jacket cooling was able to reduce the temperature late in the cycle.

3.4. SELECTED OTHER ISSUES

The relationships presented above are relatively general and they can be applied to many reaction systems. The assumption that the polymerization takes place completely in the particle phase is inherent in the way the rate components are written with the $(1-\varphi)$ factor and especially with the product $\bar{n}N[M]_p$ in eq. (19). An additional term would be needed if significant reaction occurred in the continuous phase. Likewise, if the monomer droplets were significant loci for polymerization, separate rate expressions would be required for both the droplets and the monomer-swollen latex particles. These issues are not important for most conventional recipes. Comonomers with high water solubility, however, can react mostly in the aqueous phase. In this case partition coefficients and the kinetics in both phases need to be quantified.

At the other end of the spectrum, comonomers and other ingredients (e.g. chain transfer agents) with very low water solubility may be slow to transfer through the water phase to the latex particles. They will help to stabilize the monomer drops against Oswald ripening and thereby promote droplet polymerization. This will also result in polymerization in multiple loci and changes in the batch reactor equations.

Recipes with multiple monomers, which is often the case with commercial products, will produce copolymers of different compositions during the course of a batch reaction. This compositional drift, as mentioned earlier, will result in nonuniform particle morphologies which can influence application performance. Semi-batch and continuous processes, which will be discussed in more detail in the following two sections, can be used to overcome many of these limitations.

4. Semi-Batch Reaction Processes

4.1. GENERAL CONSIDERATIONS

Most commercial products are manufactured in semi-batch processes. These reactors can be used effectively for multiple products and the operation procedures and recipes can be specified to control important latex characteristics. Latex products can be classified as **Products-by-Process** and semi-batch (also called semi-continuous) processes represent a classic example of a process in which the design and operation strongly influences product properties and application performance. Li and Brooks [25] provide a concise review of the literature on semi-batch processes for emulsion polymerization.

Some of the more important concepts and process alternatives will be outlined in the remainder of this section.

Semi-batch reactions involve the initial charging of a portion of the recipe ingredients followed, usually after a fixed time or extent of reaction, by the controlled addition of the remaining ingredients. Important process variables include:

- Composition and preparation of the initial charge.
- Time or reaction condition (e.g. conversion) when the flow of the remaining recipe ingredients is started.
- Rate of ingredient additions.
- Composition of the flow stream(s) as a function of time or
- reaction condition.
- Reaction temperature profile.
- Mixing.

These variable can be manipulated to influence one or more of the following reaction and/or product characteristics.

- Reaction rate and therefore the heat load on the cooling system.
- Copolymer composition parameters.
- Particle concentration (number) and size distribution.
- Solids content.
- Particle morphologies and surface characteristics.
- Molecular weights and molecular structure.

Process flexibility afforded by the large number of operational options is the reason that semi-batch reactors are so widely used in the industry. This large number of process options, however, does add complexity to the task of designing an optimum and robust process for a specific product. The remainder of this section will focus on some of the issues involved - excluding recipe selection.

4.2. INITIAL CHARGE

Semi-batch processes can be divided into two classifications: monomer-addition and emulsion-addition. Monomer-addition systems start with part of the monomers and all of the other ingredients in the initial charge; except perhaps some oil-soluble minor ingredients such as chain transfer agents which can be dissolved in the monomer stream to be added later. In the absence of a seed latex, the first part of the reaction includes particle nucleation and some growth. If a seed latex is used some mixing time is often employed to permit monomer swelling of the seed particles before the initiator is added.

Part of the water, emulsifier, monomer(s) and sometimes other ingredients, are held back for later addition in emulsion-addition processes. The initial charge in these systems, therefore, contains a higher volume fraction of dispersed phase and less emulsifier - i.e. for the same total recipe. If a seed latex is used and one does not wish to nucleate new particles the emulsifier level in the initial charge must be lower than what would be required to saturate the system surface area. The initial charge in either monomer-add or emulsion-add processes may include less than the total amount of initiator. Later additions are used to maintain the free radical flux during the entire reaction cycle; especially near the end to reduce the level of residual monomer.

4.3. MIXING

Mixing requirements for semi-batch reactions are clearly different than for batch reactions for two major reasons. First, the level of the mixture in the reactor can vary significantly. Second, feed streams are added to the reacting mixture. Variations in the height of the mixture in the reactor can influence fluid flow and all of the important mixing phenomena; power input, blending time, emulsification, heat transfer, coagulation, surface fouling and the desired physical design of the tank and agitator.

If emulsification is carried out in the reactor, for example, one impeller must extend deep into the reactor and be designed for the emulsifying task - i.e. generate regions of high shear. The placement and design of this impeller may not allow adequate mixing as the reactor level is increased. Hence multiple impellers may be required to meet all of the mixing demands during the course of the reaction cycle. These impellers may be of different design and size to accommodate the different mixing requirements during the reaction. Agitator speed changes may also be employed to change fluid dynamics in the reactor during the semi-batch cycle.

Level changes can clearly effect heat transfer and therefore reaction cycle time since many processes are heat-transfer limited. Surfaces above the liquid level are not effective for heat transfer and they are more likely to foul due to coating by repeated splashing and draining of the emulsion. Internal coiling coils are sometimes used to increase heat removal capacity and frequent cleaning is employed to prevent excessive polymer buildup.

Mixing or blend times for the delayed-addition steams can also be important in semi-batch processes. Fluid dynamics in the reactor and the method and location of stream injection can both be important. Some recipe ingredients are electrolytes which can cause local flocculation if concentrations are too high. Emulsifiers that are not mixed quickly can nucleate new particles, even in unsaturated systems. Water-soluble polymers are known to have the potential to stabilize or coagulate colloids, sometimes depending on how they are mixed with the bulk system. Three factors need to be considered with regard to these issues: the mixing environment, how and where the streams are injected, and the concentrations of the streams. One would normally want to inject dilute feed streams at reasonably high velocities into regions of good mixing and flow.

4.4. DELAYED FEED STREAMS

Manipulation of the delayed feed streams represents one key to controlling the course of the reaction and the properties of the latex product. Establishing the particle number concentration by controlled nucleation or by the use of a seed latex is a second important factor. Reaction rate, and therefore heat release, is the most common process parameter to be controlled via feed stream addition. Many semi-batch processes operate in a 'monomer-starved' regime in which nearly all of the monomer is in the submicron polymer particles with a smaller amount in the aqueous phase. Equilibrium between the two phases is generally assumed for the monomer. The polymerization rate, after an initial adjustment, clearly cannot be greater than the rate of monomer addition. Copolymer composition will also mirror the feed composition if the monomer feed rate is slow enough.

Operation under monomer-starved conditions also means that the polymer will be formed in an environment of high polymer and low monomer concentrations. Such conditions can exert a significant influence on the molecular weights and architecture of the polymer molecules formed. The gel-effect will be more important, leading to higher numbers of active radicals per particle and longer radical lives. The combination of longer radical growth times and lower monomer concentrations can change the average size of the kinetic chain length in either direction when compared to polymerization in a monomer-saturated system. If reactions with polymer (chain transfer or propagation via residual double bonds) are important, monomer-starved conditions can result in the formation of significantly branched and/or crosslinked molecules.

The mobility of molecules, especially oligomers and polymers, is severely reduced in monomer-starved reactions. Polymers will, therefore, tend to remain where they are formed and this will effect particle morphology. If water-soluble initiators are employed the radicals will form new polymer on or near the particle surfaces. Initiators that are partitioned inside the particles will form new polymer throughout the particles if the monomer diffusion rates are sufficient to reach the internal radicals.

Copolymer composition, as mentioned above, will be also be influenced by semi-batch feed stream policies. Monomer-starved reactions offer a straightforward method for producing either constant or varying composition copolymer. A single premixed monomer stream can be converted to copolymer of the same composition or the monomer mixture can be varied during the addition period. Bassett and Hoy [26] describe a rather simple system for achieving a wide range of composition profiles by pumping different monomer mixtures from and through several feed tanks. Guyot et al. [27], in contrast, used on-line gas chromatographic analysis of the reacting monomer mixture to control the feed rate of the more reactive monomer to produce constant-composition copolymer. Their reactions did not operate in the monomer-starved regime. Asua and coworkers [28,29] describe a semi-empirical approach for determining monomer feed policies which minimize reaction time and control copolymer composition profiles. A series of semi-batch reactions were carried out to determine kinetic parameters which were then used to determine optimal feed policies. Methyl methacrylate-ethyl acrylate and styrene-butyl acrylate copolymerizations were studied.

Another, and sometimes subtle, difference between batch and semi-batch reactions involves the influence of inhibitors in the raw materials. Inhibitors are effective free radical scavengers and they delay the start of a batch reaction, after which the reaction proceeds in a normal manner. In semi-batch reactions inhibitors in the feed stream reduce the rate of generation of effective free radicals. Hence the initiator concentration or feed rate must be high enough to generate a radical flux that reacts with the inhibitor and sustains the polymerization. When inhibitor concentrations vary the amount of initiator in the recipe may need to be changed. If the inhibitor concentrations are high the effective initiation rate can increase significantly at the end of the delayed feed addition. This can result in higher polymerization rates and temperature excursions; especially with reactions that are not operated in the monomer-starved regime.

Particle size and number characteristics can be manipulated with seed latexes and feed stream policies. If an adequate amount of seed latex (probably determined by surface area) is used and the emulsifier concentration is low enough very few or no new particles will be formed and the seed will determine the size and total number of particles in the final latex. Emulsifier may be added during the course of the reaction to prevent coagulation but the system must be maintained in the unsaturated state if particle nucleation is to be avoided. Particle concentration and the final size distribution can, of course, be altered by adding more seed and/or emulsifier sufficient to cause further nucleation during the reaction. Products with multimodal or broad size distributions can be produced in this manner.

4.5. DESIGN EQUATIONS

4.5.1 *Mass Balance*

The general mass balance [eq. (1)] can be simplified slightly for semi-batch reactors because there is no effluent stream. Hence the semi-batch mass balance for monomer 'i' is given by eq. (22) for a well mixed reactor.

$$F_{io} - R_{pi}V(1-\varphi) = dN_i/dt \qquad (22)$$

N_i is the total number of moles of monomer 'i' in the reactor. V is the volume of the reacting mixture and $(1-\varphi)$ is the volume fraction which is aqueous phase. The product $R_{pi}V(1-\varphi)$ represents the total rate of conversion of monomer to polymer in the reactor. Wessling [30] points out that experimental observations often show a constant rate period if the total number of particles in the reactor remains constant. The total rate of conversion depends on the feed rate, F_{io}, in a monomer-starved reaction but is independent of feed rate when the system contains enough monomer to form a droplet phase. If a constant monomer feed rate is employed in a starved system and if, as Wessling indicates, the monomer conversion rate is constant the accumulation term, dN_i/dt is a constant (or zero) and N_i can increase during the feed period. The rate of monomer conversion is given by eq. (23).

$$R_{pi}V(1-\varphi) = k_p[M_i]_p \bar{n} N_p V(1-\varphi)/N_A \qquad (23)$$

If the total number of particles $[N_p V(1-\varphi)]$ remains constant during the reaction, changes in rate will be determined by the value of the product $[M_i]_p \bar{n}$, unless k_p varies due to temperature changes or a decrease in monomer mobility at high conversion. If all the water is added with the initial charge, as in a monomer-add process, the term $V(1-\varphi)$ will be constant during the remainder of the cycle. A combination of eqs. (21) and (22) along with a knowledge of Np, an appropriate relationship for the average number of radicals per particle, $\bar{n}$, monomer partition coefficients and values for the rate coefficients can be used to model semi-batch reactions. Examples are given for specific systems by Wessling [30], Guyot et al. [27], Asua & coworkers [28,29] and Dimitratos et al. [31].

4.5.2 *Energy Balances*

The energy balances presented earlier (eqs. (6) and (7) with (9)) are applicable to semi-batch reactions. The variation of volume of the reacting media can, however, lead to problems in quantifying some of the terms in the energy balances. Changes in the amount of immersed heat transfer area and fluid flows can, for example, add uncertainties to reactor performance prediction. The energy content of feed streams is reflected in eqs. (6) and (7) and these streams present an opportunity for enhancing reaction heat removal. Feed streams can be cooled in external heat exchangers and serve as a heat sink as they are added to the reactor.

On-line measurements of temperature and stream flow rates (e.g. eq. (9)) can be used with the energy balance equations to determine how much reaction has taken place. Feed stream flow policies, starting points and flow rates, should ideally be determined by reactor conditions such as conversion, emulsifier surface coverage and residual monomer concentrations rather than time. Energy balance equations coupled with kinetic relationships and on-line measurements can be used to achieve better process control and less batch-to-batch variation in cycle time and products properties.

5. Continuous Reaction Processes

5.1. GENERAL CONSIDERATIONS

Continuous emulsion polymerization reactors are important for two reasons. First, they can be used for the economical manufacture of commercial products. Second, continuous stirred-tank reactors (CSTRs) operated at steady-state can be useful for study of fundamental mechanisms and kinetics. A very wide range of reactor types have been used for continuous reactions. These include CSTRs, usually several connected in series, single-pass and recirculation tubular reactors, packed beds, concentric-cylinder couette flow systems and tubular or plug-flow devices connected in series with CSTRs. The recirculation tube or loop reactor is the subject of the next chapter of this book and will not be discussed in detail here. If the loop recirculation rate is significantly greater than the flow-through rate, however, this reactor has a distribution of residence times which is nearly the same as that of single CSTR. Hence, some of the CSTR discussion in this paper will apply to the loop reactor.

Continuous reactors have been used to manufacture latex products for about 50 years. Styrene-butadiene elastomers (SBR) were manufactured in the United States beginning in the World War II era in processes comprised of 10-15 equal-size CSTRs connected in series. A number of these processes, with modifications, are still operating and SBR remains as perhaps the largest volume product made in continuous reactors. The patent literature on continuous emulsion polymerization is fairly active and a number of patents will be reviewed in the remainder of this paper as will some of the basic concepts of continuous reactors; especially CSTRs. More comprehensive review papers have been published by the author [32,33].

5.2. SINGLE STEADY-STATE CSTR

A single, steady-state CSTR can be a useful research tool but it is not likely to be a viable commercial reactor for reasons that will become apparent in this discussion. Understanding the fundamentals of this ideal reactor, however, will serve as a good starting point for examining other reactor configurations and the processes described in the patent literature.

5.2.1. *Distribution of Residence Times*
Perhaps the most significant difference between an ideal (i.e. perfectly mixed) CSTR and batch and semi-batch reactors is the broad distribution of residence times as given by eq. (24).

$$f(t) = (1/\tau)\exp(-t/\tau) \qquad (24)$$

Where t is residence time of an individual part of the effluent stream and τ is the mean residence time of the effluent in the reactor. This relationship also describes the distribution of ages of particles in the effluent stream from a CSTR. Hence, products with narrow particle size distributions cannot be produced in such a reactor. The broad range of particle ages is an asset, however, if one wishes to test a kinetic model for the relative growth rates of different size particles. Prediction of the particle size distribution (PSD) in the product from a CSTR is a stern test for particle growth models. Lee and Poehlein [34] used this concept to examine the influence of chain transfer agents on reaction rate, PSD, and radical desorption from the particles in the emulsion polymerization of styrene.

5.2.2. Polymerization Rate and Particle Number

Direct measurement of reaction rate is another potential advantage of the steady-state CSTR as a research tool. The mass balance for such a reactor is given by eq. (25).

$$F_{io} - F_{ie} = R_{pi}V(1-\varphi) \tag{25}$$

Hence, one only needs to measure the characteristics of the feed and effluent streams and the reactor working volume to determine the rate. That rate will be associated with the unchanging conditions in the reactor so the problems associated with sampling a transient reaction environment and measuring rates of change are avoided. Multiple runs are necessary, however, to establish the influence of reaction parameters such as concentrations.

The rate of polymerization in an unseeded CSTR can also be quite different than would be observed with the same recipe in a batch or semi-batch reaction. This is most dramatically illustrated by contrasting rate and particle number for Smith-Ewart Case II type kinetics in batch and continuous reactors. Since ñ=1/2 for such systems, rate is directly proportional to particle concentration and eqs. (26) and (27) show the theoretical predictions for a batch reactor and for a CSTR at modest to large values of mean residence time.

$$N_p \approx R_p \approx R_i^{0.4}S^{0.6} \qquad \text{Batch} \tag{26}$$

$$N_p \approx R_p \approx R_i^{0}S^{1.0}\tau^{-0.67} \qquad \text{CSTR} \tag{27}$$

The rate and particle number depend on the initiation rate, R_i, in the batch reaction but not in the unseeded CSTR. The exponents on the surfactant concentration, S, are quite different and the value of the mean residence time influences the performance of the CSTR.

Equation (27), as mentioned above, is only valid for modest to high values of τ. The complete relationship for particle number in an unseeded CSTR is more complex and Np displays a maximum as a function of mean residence time. If τ is small the number of free radicals formed in a typical sample of the emulsion will be limited and Np will be small and directly proportional to R_i. If τ is large the average size of the particles will be larger and the surfactant charged will only be able to stabilize a limited number of particles. A maximum number of particles will be formed at some intermediate value of τ. The maximum number of particles formed for Smith-Ewart Case II kinetics will only be about 50 to 60% of the number that would be formed with the same recipe in a batch reactor. Typical values of mean residence time are often higher than the point of maximum N_p and one can easily nucleate less than 10% of the particles that could be produced in a batch reaction. Obviously, the size of these particles, with the same amount of monomer conversion, will be larger. Shoaf and Poehlein [35] examined this problem in a project aimed at developing a continuous process to replace a commercial batch process.

5.2.3. *Copolymer Composition*
Copolymer composition changes with conversion in a batch reactor. The copolymer formed first is rich in the more reactive monomer and that formed last can be nearly a homopolymer of the least reactive monomer. The composition of copolymer formed in a steady-state CSTR will be uniform but not the same composition as the monomer feed mixture unless the conversion is high. If multiple CSTRs are used the composition of the copolymer formed in the different CSTRs will vary unless intermediate feed streams are used to adjust the monomer ratios in the reactors. The use of multiple reactors is, therefore, a potential method for producing structured particles which cannot be made in a single CSTR.

5.2.4. *Influence of Inhibitors*
Inhibitors that are present in recipe ingredients flow into the CSTR and reduce the effective rate of initiation. This is quite different from the effect of inhibitor on a batch reaction but analogous to inhibitors entering a semi-batch reactor during the delayed feed part of the reaction cycle. One can adjust for this effect by increasing the rate of initiator feed to the CSTR. This will increase the electrolyte concentration in the reactor and may lead to higher initiation rates in downstream reactors. Pettelkau and Ehrig [36] report, for a multistage polychloroprene process, that additional inhibitor was sometimes added to the second CSTR in order to lower the rate of polymerization and heat release in that reactor.

5.2.5. *Unstable Behavior and Use of Seed*
Steady-state operation is sometimes difficult to achieve in a CSTR because particle nucleation and growth phenomena can cause sustained oscillations in monomer conversion, particle concentration, PSD and surface tension. These oscillations are caused by alternating periods of particle nucleation and growth. Large numbers of particles are formed when the surfactant concentration exceeds that required to saturate the particle and droplet surfaces. When these particles grow the interfacial area increases and the system becomes unsaturated or surfactant-starved. Very few or no new particles are formed under such conditions. The effluent stream, however, carries unsaturated surface out of the reactor and emulsifier is added via the feed streams. When the saturation point is passed the particle nucleation rate becomes high again and the cycle is repeated. Typical cycles are long (6-10 mean residence times) but they are triggered by relatively short periods of rapid particle nucleation. This problem can be solved by using a pre-made seed latex in the feed stream or by generating a seed latex in a pre-tubular reactor [32-36]. The use of a seed can also set the particle number and overcome the particle number limitations of an unseeded CSTR which were discussed earlier.

5.2.6. *Start-Up and Product Changeover*
Off-spec material can be produced before a continuous reactor reaches steady-state upon start-up or when product changes are made without shutting down. Careful consideration of potential operational policies and modeling of transient conditions can minimize this problem. Product blending can also sometimes reduce the amount of waste material. These issues need to be examined in detail in the process development program for any product that may be manufactured in continuous reactors.

5.2.7. *Polymerization, Nucleation and Multiplicities*
Polymerization rates, particle concentrations and the potential for steady-state multiplicities depend on operational variables and receipt ingredients. Important

monomer characteristics are water solubility or hydrophilic-hydrophobic nature; reactivity in propagation, transfer and termination reactions and the strength of the Trommsdorf or Gel Effect.

More hydrophilic monomers tend to enhance particle nucleation and stabilization. Chain transfer to small molecules (e.g. monomer) increases free radical flux to and from the particles. This effect can lead to lower valves of ñ in the particles but higher nucleation rates early in the reaction. In contrast to the prediction of eq. (27), initiation rate has a positive influence on polymerization rate when chain transfer to monomer or other small molecules is important.

Both nucleation phenomenon and a gel effect can contribute to the possibility for multiple steady-state operating points for a CSTR. Essentially the mass balance equation -- (IN-OUT= AMOUNT REACTED) can be satisfied at more than one conversion level. The gel effect at high polymer concentration increases the rate enough to sustain the high conversion steady state.

5.3. CONTINUOUS TUBULAR REACTORS

Continuous tubular reactors have been used by a number of industrial and academic workers. A good review of this work has been published by Paquet and Ray [37]. Tubular reactors have the advantage of providing large heat-transfer area and they can produce latexes with narrow particle size distributions. Wall fouling, inadequate mixing, including phase separation, and problems with the introduction of intermediate streams are potential disadvantages. Small-scale tubular reactors are likely to operate in the laminar flow regime and have a relatively large distribution of residence times (RTD). Paquet and Ray studied a single-pass tubular reactor in the form of a helical coil. They found that the RTD was relatively narrow with both steady-flow and pulsed-flow operation. The pulsed-flow system, however, did reduce the formation of wall polymer and tube plugging.

The material of construction of the reactor tube can also influence fouling and plugging. Tubes made from polymeric materials, especially fluoropolymers, have been more successful than those made of glass and metals. Plastic tubing can be replaced at little cost but heat transfer is reduced because of low thermal conductivity. Fouling and plugging can also be minimized by only carrying out the early part of the polymerization in the tube as is done with tube-CSTR systems.

5.4. SELECTED PATENTS AND OTHER REACTOR SYSTEMS

A number of continuous reactor processes have been proposed. Most configurations are aimed at eliminating conversion oscillations, narrowing particle size distributions, controlling polymer composition and/or particle morphologies and reducing coagulation and the formation of fouled surfaces. Examples of different continuous reactor systems will be presented in the remainder of this paper.

Devana and Shay [38] describe a hybrid semi-continuous process in which the polymerization is started like a semi-batch reaction but the delayed feed stream is continued after an effluent stream begins to flow from the reactor. The feed stream is continued until 1.5 to 4 reactor volumes have been charged. This reactor would not achieve a steady-state but productivity would be increased. The final mixed product would have a narrower PSD than latex prepared in a single steady-state CSTR.

Hoedemakers and Thoenes [39] and Meuldij et al. [40] used a packed column with a pulsed-flow system. Relatively narrow residence time distributions (RDT) and high conversions were achieved. Reactor models were developed and compared with experimental results for styrene and vinyl acetate polymerizations. Imamura et al. [41] achieved narrow RDTs with a concentric cylinder reactor. The axial flow rate in the annulus and the rotational speed were adjusted to generate flows with Taylor vortices which yield narrow RTDs.

Stone [42,43] employed a spiral-flow vessel as a prereactor for CSTRs. The purpose of this prereactor was to produce a seed particle stream for the CSTR. Greene et al. [44] and Berens [45] achieved the same result by using a tubular prereactor and premade seed, respectively. Operation without conversion oscillations and high particle concentrations are the advantages of such systems. Korte and Suling [46] also used a tube-CSTR system for AN/VAc/Styrene latex production. Some of the recipe was fed directly into the CSTR, bypassing the tubular seed reactor.

A number of processes use seeds in multiple reactor systems which will generally produce latexes with narrower PSDs than a single CSTR. Heil et al. [47] fed seed latex to a 3-CSTR train. Monomer streams of different composition were fed into the first two reactors to produce soft-core, hard-shell particles. Daniels and Lenney [48] provide a detailed flow diagram for a seed-fed 2-CSTR process used to produce EVAc copolymer latexes. Achille and Bucci [49] developed a 3-reactor process with two CSTRs followed by a tubular-type finishing reactor with internal baffles. A plug-flow final stage will help reduce residual monomer.

Single CSTRs, in spite of some potential problems, are described in some patents. Pettelkau and Ehrig [36] used an up-flow stirred reactor for polymerization of chloroprene; a very reactive monomer. Kastner and Heinze [50] also describe an up-flow reactor with two impellers for vinyl chloride emulsion polymerization. Their reactor had a large height-to-diameter ratio to promote plug flow. Scott and Feast [51] used a single CSTR for a high-temperature (80-100°C) adiabatic polymerization. Cold feed streams were used to help with the reaction heat load. Sutterlin et al. [52] used tubular reactors whose internal surface was a saturated polyolefin or fluorinated polyolefin for low-cost reactors that are less likely to foul and are easier to clean if they do foul. Such reactors will also have relatively narrow RTDs.

6. Summary

Effective engineering of emulsion polymerization reactions requires an understanding of a number of important issues: fluid dynamics and mixing, heat transfer, process monitoring and control, mass transfer, phase equilibria, colloid science and, last but not least, the mechanisms and kinetics of the multiple chemical and physical reaction steps. The utilization of this knowledge in an integrated product-process development effort will help to reduce costs, decrease time-to-market and increase product quality. Some of the important aspects of batch, semi-batch and continuous reactors have been reviewed in this paper. The equally important areas of reaction mechanisms and kinetics, on-line measurements, control methods and some alternate reaction systems are the subject of other papers in this book.

330

Acknowledgments

The support of the National Science Foundation under Grant CTS-9417306 and Georgia Institute of Technology is gratefully acknowledged.

7. References

1. Min. K.W. and Ray, W.H. (1974) Modelling emulsion polymerization reactors, *J. Macromol., Sci-Revs. Macromol. Chem.*, **C11 (2)**, 177-255.
2. Durbin, D.P., Poehlein, G.W., El-Aasser, M.S. and Vanderhoff, J.W. (1979) Influence of monomer pre-emulsification on formation of particles from monomer drops in emulsion polymerization, *J. Appl. Polym. Sci.* **24**, 703-707.
3. Eckert, R.E., McLaughlin, C.M. and Rushton, J.H. (1985) Liquid-liquid interfacial areas formed by turbine impellers in baffled, cylindrical mixing tanks, *AIChE J.* **31:11**, 1811-1820.
4. Nagata, S. (1975) Mixing: Principles and Applications, John Wiley & Sons, New York.
5. van den Temple, M. (1960) The function of stabilizers during emulsification, *Proc. 3rd Int. Congress of Surface Activity, Cologne, Verlag der Universitatsdruckerei Mainz GmbH* **2**, 573.
6. Lucassen-Reynders, E.H. and Kuijpers, K.A. (1992) The role of interfacial properties in emulsification, *Colloids and Surfaces* **65**, 175-184.
7. Davies, J.T. (1972) *Turbulence Phenomena*, Academic Press, New York/London.
8. Davies, J.T. (1985) Drop sizes of emulsions related to turbulent energy dissipation rates, *Chem. Engr. Sci.*. **40:5**, 839-842.
9. McManamey, W.J. (1979) Sauter mean and maximum drop diameters of liquid-liquid dispersions in turbulent agitated vessels at low dispersed phase hold-up, *Chem. Engr. Sci.*, **34**, 432-434.
10. Brown, D.E. and Pitt, K. (1974) Effect of impeller geometry on drop break-up in a stirred liquid-liquid contactor, *Chem. Engr. Sci.*, **29**, 345-348.
11. Braginsky, L.N. and Kokotov, I.V. (1993) Influence of turbulence viscosities on the kinetics of drop breaking, *J. Dispersion Sci. & Technol.*, **14:3**, 373-394.
12. Vanderhoff, J.W. (1981) The formation of coagulum in emulsion polymerization, in D.R. Bassett and A.E. Hamielec (eds.), *Emulsion Polymers and Emulsion Polymerization*, ACS Sym. Series 165, ACS, Washington, D.C., pp. 199-208.
13. von Smoluchowski, M. (1916) Drie vortrage uber diffusion Brownsche molekularbegung und koagulation von kolloidteilchen I & II, *Physik. Z.*, **17**, 557-571, 585-599.
14. von Smoluchowski, M. (1918) Versuch einer mathematischen theorie der koagulationskinetik kolloider losungen, *Z. Phys. Chem.*, **92**, 129-168.
15. Heller, W. and Peters, J. (1970) Mechanical and surface coagulation I, *J. Coll. & Interface Sci.*, **32:4**, 592-605.
16. Heller, W. and Peters, J. (1970) Mechanical and surface coagulation II, *J. Coll. & Interface Sci.*, **33:4**, 578-585.
17. Heller, W. and DeLauder, W.B. (1971) Mechanical and surface coagulation III, *J. Coll. & Interface Sci.*, **35:1**, 60-65.
18. Heller, W. and Peters, J. (1971) Mechanical and surface coagulation IV, *J. Coll. & Interface Sci.*, **35:2**, 300-307.
19. Lowry, V., El-Aasser, M.S., Vanderhoff, J.W. and Klein, A. (1984) Mechanical coagulation in emulsion polymerization, *J. Appl. Polym. Sci.*, **29**, 3925-3935.
20. Lowry, V., El-Aasser, M.S., Vanderhoff, J.W., Klein, A. and Silebi, C. A. (1986) Kinetics of agitation-induced coagulation of high-solid latexes, *J. Coll. & Interface Sci.*, **112:2**, 521-529.
21. Hoedemakers, G.F.M. (1990) Continuous emulsion polymerization in a pulsed packed column, *PhD Thesis*, Technische Universiteit Eindhoven, The Netherlands, 127-189.
22. Rubens, R.W. (1958) The effect of agitation on the polymerization of a vinyl chloride-ethyl acrylate latex, *M.S. Thesis*, Newark College of Engineering, Newark, NJ.
23. Vanderhoff, J.W. (1990) Autoclave buildup, in G.W. Poehlein, R.H. Ottewill and J.W. Goodwin (eds.), *Science and Technology of Polymer Colloids*, I, Martinus Nijhoff Publishers, The Hague, 167-187.
24. Lehtola, T., Söderman, J. and Laine, J. (1988) Twist-element baffles, *Proceedings of the 6th European Conf. on Mixing*, Pavia, Italy, 85-90.
25. Li, B. and Brooks, B.W. (1992) Semi-batch processes for emulsion polymerization, *Polymer International*, **29**, 41-46.
26. Bassett, D.R. and Hoy. K.L. (1981) Nonuniform emulsion polymers: process description and polymer properties, in D.R. Bassett and A. E. Hamielec (eds.), *Emulsion Polymers and Emulsion Polymerization*, ACS Sym. Series 165, ACS, Washington, D.C., 371-387.
27. Guyot, A., Guillot, J., Pichot, C. and Guerrero, L. (1981) New design for producing constant-composition copolymers in emulsion polymerization: comparison with other processes, *Ibid.*, 415-436.
28. Leiza, J.R., Arzamendi, G. and Asua J.A. (1993) Copolymer composition control in emulsion polymerization using technical grade monomers, *Polymer International*, **30**, 455-460.

29. Echevarria, A., de la Cal , J. C. and Asua, J.M. (1995) Minimum-time strategy to produce nonuniform emulsion copolymers: II Open-loop control, *J. Appl. Polym. Sci.*, **57**, 1217-1226.
30. Wessling, R.A. (1968) Kinetics of continuous addition emulsion polymerization, *J. Appl. Polym. Sci.*, **12**, 309-319.
31. Dimitratos, J., El-Aasser, M.S., Georgakis, C. and Klein, A. (1990) Pseudosteady states in semicontinous emulsion polymerization, *J. Appl. Polym. Sci.*, **40**, 1005-21.
32. Poehlein, G.W. (1993) Emulsion polymerization and copolymerization in continuous reactor systems, *Polymer International*, **30**, 243-251.
33. Poehlein, G.W. (In press 1996) Continuous processes for emulsion polymerization, in M.S. El-Aasser and P.A. Lovell (eds.), *Emulsion Polymerization and Emulsion Polymers*, John Wiley & Sons, Ltd., Sussex, U.K., Chapter 8.
34. Lee, H.-C. and Poehlein, G.W. (1987) Emulsion polymerization in a seed-fed CSTR: Influence of transfer reactions, *Polym. Process Engr.*, **5(1)**, 37-74.
35. Shoaf, G.L. and Poehlein, G.W. (1989) Batch and continuous emulsion copolymerization of ethyl acylate and methacrylic acid, *Polym.-Plast. Technol. Engr.*, **28**, 289-327.
36. Pettelkau, P. and Ehrig. H. (1976) Reactors for continuous emulsion polymerization of chloroprene, *German Patent No. 2,520,891*, assigned to Bayer AG.
37. Paquet, D.A.Jr. and Ray, W.H.Jr. (1994) Tubular reactors for emulsion polymerization: I. experimental investigation, II. model comparisons with experiments, *AIChE J.*, **40:1**, 73-87 & 88-96.
38. Devona, J.E. and Shay, G.D. (1990) Semi-continuous emulsion polymerization process, *US Patent 4.946,891*, assigned to DeSoto Inc.
39. Hoedemakers, G.F. and Thoenes, D. (1990) Continuous emulsion polymerization in a pulsed-packed column, in L.A. Kliertjens (ed), *Integration of Fundamental Polymer Science and Technology*, Elsevier, London.
40. Meuldijk, J., van Strien, C., van Doormalen, F. and Thoenes, D. (1992) A novel reactor for continuous emulsion polymerization, *Chem. Engr. Sci.*, **47**, 2603-2608.
41. Imamura, T., Mizuguchi, K., Shabai, Y., Ishii, K., Ishikura, S. and Saito, K. (1994) Continuous polymerization method and process, *US Patent 5,340,891*, assigned to Nippon Paint Co. Ltd.
42. Stone, J.M. (1971) Method for controlling reaction rate in aqueous emulsion polymerization to form elastomeric polymers, *US Patent 3,600,349*, assigned to Copolymer Rubber and Chemical Corp.
43. Stone, J.M. (1973) Method and apparatus for controlling reaction rate, *US Patent 3,730,928*, assigned to Copolymer Rubber and Chemical Corp.
44. Greene, R.K., Gonzalez, R.A. and Poehlein, G.W. (1976) Continuous emulsion polymerization - steady state and transient experiments with vinyl acetate and methyl methacrylate, in I. Piirma and J.L. Gardon (eds), *Emulsion Polymerization*, ACS Sym. Series 24, ACS, Washington, D.C.
45. Berens, A.R. (1974) Continuous emulsion polymerization of vinyl chloride, *J. Appl. Polym. Sci.,*, **18**, 2379-2390.
46. Korte, S. and Suling, C. (1978) Acrylonitrile containing copolymers blends and process for their preparation, *US Patent 4,122,136*, assigned to Bayer AG.
47. Heil, E., Wenzel, F., Arndt, P.J. and Schellhaas, W. (1985) Impact resistant resins and method for making the same, *US Patent 4,543,383*, assigned to Rohm GmbH.
48. Daniels, W.E. and Lenney, W.E. (1979) Continuous emulsion polymerization of vinyl acetate and ethylene, *US Patent 4,164,489*, assigned to Air Products and Chemicals, Inc.
49. Achille, M.D. and Bucci, M. (1977) Continuous polymerization process in emulsion, *US Patent 4,022,744*, assigned to Montedison Fibre S.p.A.
50. Kastner, P and Heinze, C. (1978) Process and apparatus for the continuous production of vinyl chloride polymers in aqueous emulsion, *US Patent 4,125,574*, assigned to Hoechst AG.
51. Scott, C.M. and Feast, A.A.J. (1980) Aqueous emulsion polymerization of vinyl monomers using mixtures of emulsifiers, *US Patent 4,239,669*, assigned to I.R.S. Holding SARL.
52. Sutterlin, N., Blitz, H.-D., Mager, T., Jagsch, K.-H. and Tessmer, D. (1987) Continuous emulsion polymerization process, *US Patent 4,713,434*, assigned to Rohm GmbH Chemische Fabrik.

THE LOOP PROCESS

C. ABAD[1], J.C. DE LA CAL, J.M. ASUA
Grupo de Ingeniería Química, Departamento de Química Aplicada,
Facultad de Ciencias Químicas, Universidad del País Vasco, Apdo 1072,
20080, San Sebastián, Spain.
email: qppasgoj@sq.ehu.es

1. Introduction

Continuous reactors, specially continuous stirred tank reactors (CSTRs), are used to manufacture high production emulsion polymers [1]. The use of CSTRs provides constant quality products and easiness of on-line control. However, in these reactors the heat removal is not effective due to their low heat transfer area/volume ratio, therefore only small conversion increments can be achieved in a single CSTR, and a series of CSTRs has to be used. Continuous loop reactors, CLRs, are becoming and attractive alternative for the production of emulsion polymers. A CLR, whose employment for the manufacture of emulsion polymers was first time described by Lanthier [2,3], consists of a tubular loop that connects the inlet and the outlet of a recycle pump. Reactants are continuously fed into the reactor and the product is also continuously withdrawn from the reactor. The history of this development can be found in ref. [4]. The CLR presents several advantages for the emulsion polymerization process. Its large heat transfer area/reactor volume ratio allows high conversions (98%) in short residence times (8 min.) to be achieved [5]. This results in a substantial reduction of the reactor volume. Geddes [6] reported that a 5,000 liter batch tank reactor can be replaced in terms of production rate by a 50 liter CLR. Because of the small volume and the short residence time at which this reactor is operated, the CLR can be used with great flexibility and minimum losses in the manufacture of different emulsion polymers. The small volume and the absence of head space make the process intrinsically safe. The loop reactor presents also some disadvantages for the manufacture of emulsion polymers: i) core-shell particles cannot be produced, ii) the particle size distribution is difficult to modify and small particle sizes require substantial amounts of surfactants, iii) recipes with high mechanical stability are required to prevent shear induced coagulation; iv) It may suffer from cyclic behavior due to intermittent nucleations, also found in CSTRs; and v) very little has been published on this process in both open and patent literature. In this paper, a review will be given of the designs and flow pattern of the CLRs, polymerization studies carried out in these reactors, and mathematical modeling of CLRs.

---

[1] Current adress: 3M Belgium N.V. Chemical Group, European Business Centre, Haven 1005, Canadastraat 11, B-2070, Zwijndrecht, Belgium.

J. M. Asua (ed.), Polymeric Dispersions: Principles and Applications, 333–347.

2. Continuous Loop Reactors: Designs and Flow Patterns

A scheme of an industrial unit is presented in Figure 1. The tube is usually enclosed by a cooling jacket and coiled into a trombone shape for compactness and accessibility for maintenance. For a given reactor volume, production is limited by the heat removal rate per volume unit of the reactor. Therefore, there is a strong incentive to increase the length/diameter ratio. Values of this ratio as high as 1000/1 have been employed [7]. This ratio is limited by the pressure drop through the loop that can not exceed the design pressure differential of the circulation pump. Cooling area can also be increased by using a multitube loop reactor [Figure 2]. Geddes and Khan [8] reported that the rate of production of a 4-tube loop reactor was five times greater than a single tube loop reactor of the same volume. They also claimed that unforeseeable improvement of the product quality was achieved. This improvement seems to be due to the lower particle size obtained in the multitube loop which gave higher latex viscosity and better film properties. The lower particle size may result from the lower residence times used in the multitube loop rather than from the special flow pattern of this reactor.

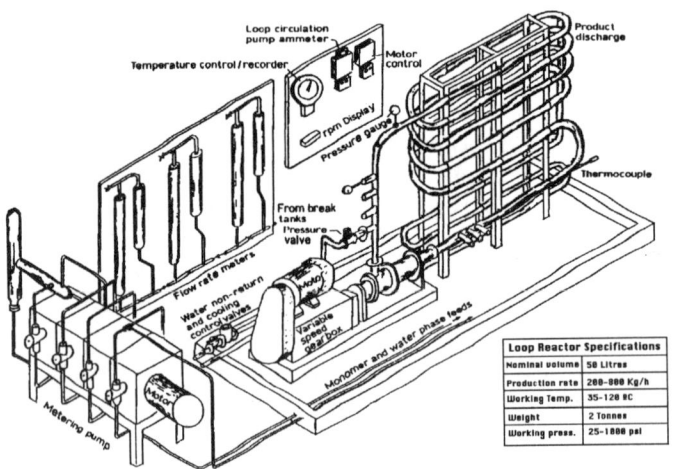

Figure 1. Industrial unit [7].

According to Geddes [6] in the most common form of the industrial scale loop reactor, the stabilizers are feed into the reactor through, or just before, the circulating pump. The rationale behind this is to provide extra mechanical stability of the polymer particles in the pump, i.e., when they are subjected to most gear. Monomers enter the reactor immediately after the pump and the product is taken from a point immediately prior to the stabilizer feed entry.

Adams [9] has proposed to add a tubular reactor after the loop reactor, the length of the tubular reactor being larger than that of the loop section by a factor of at least two. It is claimed that core-shell particles can be produced in this system by introducing a further monomer or mixture of several monomers between the loop and the tubular sections. The tubular section can also be used to reduce the level of residual monomers. Other claimed advantages of this arrangements are the ability to produce polymer particles smaller than in a loop reactor without tubular section and the improvement of production.

The economics of the loop process as compared with the tank process has been extensively discussed by Geddes [4,6]. The cost of a low pressure loop reactor is only about 5-10% less than a batch reactor plant of similar capacity, although further savings are made on building costs. A more startling comparison is for a plant capable to process ethylene. In this case, the batch plant would be at least four time more expensive than the equivalent loop plant. It has to be pointed out that although patents for the production of vinyl acetate (VAc)/ethylene copolymers in CLRs have been filed [10] the process remains to be commercialy implemented. The running costs are not detailed although it is reported that the loop process is more energy efficient.

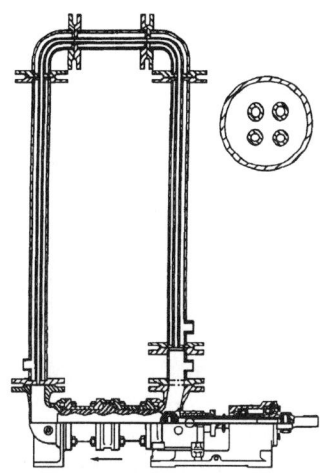

Figure 2. Multitubular loop reactor [8].

The number of industrial units is still limited. Currently, nine commercial reactors are in operation [11]. Most of them are single pipe 0.05 m^3 reactors with a yearly production of 2,500 tonnes. Polymers manufactured include p-vinyl acetate, and copolymers of VAc and Veova 10 (vinyl ester of a highly branched decanoic acid, Shell Trade Mark) and VAc - 2 ethyl hexyl acrylate. Two of the reactors can work at 1.1×10^7 Pa, and might be used for the production of VAc-ethylene copolymers.

Although the flow patterns of loop reactors have been extensively studied [12] no studies of the macromixing characteristics of industrial loop reactors for production of emulsion polymers have been reported. Lee et al. [13] determined the residence time distribution (RTD) of a 1.4 m long lab scale loop reactor made of 0.046 m diameter glass tubes. The recirculation of the reaction mixture was provided by a blade impeller. The authors reported that at recycle ratios (R_r=flowrate in the reactor/feed flow rate) as high as 124 the flow model deviated significantly from that of a perfectly backmixed CSTR. This is surprising because theoretical calculations show that for this range of recycle ratios the behavior of the loop reactor is almost that of a perfectly backmixed CSTR [14]. Abad et al. [15,16] characterized the flow pattern of the lab scale reactor presented in Figure 3a by means of tracer experiments. The RTDs obtained at low values of R_r initially presented large oscillations followed by an exponential decay. The oscillations vanished when the recycle ratio increased, and for R_r=55 a RTD typical of a perfectly backmixed CSTR was obtained. The macromixing of this system can be represented by a loop formed by two tubular reactors with axial dispersion [Figure 3.b]. The dispersion coefficients were calculated by means of the approach described by Warnecke et al. [12].

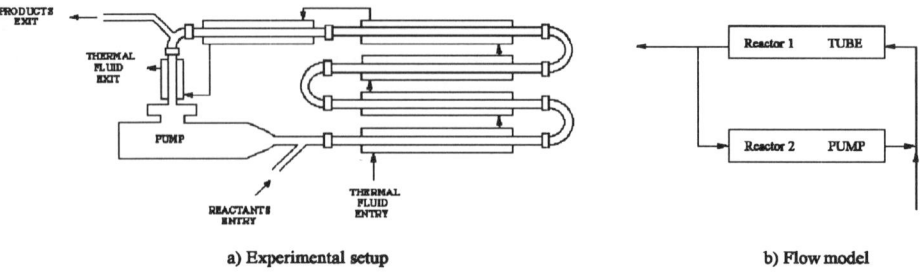

a) Experimental setup b) Flow model

Figure 3. Lab scale continuous loop reactor and flow model.

The fact that, at recycle ratios typical of process conditions, the RTD is close to that of a perfectly backmixed CSTR does not mean that, under polymerization conditions, the reaction mixture is homogeneous through the CLR. Actually, there are indications of the opposite. Lanthier [2,3] disclosed that a critical feature of the CLR is that the initiator system forms free radicals in quantity sufficient to initiate polymerization within the time required for the reaction mixture to circulate once through the reactor. Geddes [6] reported that it is customary that the monomers enter the loop immediately after the pump where the free surfactant concentration is at its highest. The product is taken from a point immediately prior to the stabilizer feed entry, where conversion is greatest. Abad et al. [16], for the redox initiated emulsion copolymerization of VAc and Veova 10 carried out in the CLR of Figure 3a, have demonstrated that the steady state polymerization rate obtained by using a preemulsified feed was higher than that obtained when the monomers and the aqueous phase were fed into the reactor without any mixing. In addition the number of polymer particles was found to be independent of the type of feed used. This means that the variations of the polymerization rates were due to monomer mass transfer limitations.

3. Emulsion Polymerization Studies in Continuous Loop Reactors

Only a few emulsion polymerization studies in CLRs have been reported in the open literature. Lee et al. [13,17,18] studied the emulsion polymerization of styrene in a 2.35×10^{-3} m^3 rectangular loop reactor made of 0.046 m diameter glass tubes. Polymerizations were carried out using a very low solids content (mostly 8 wt%) using sodium lauryl sulfate (SLS) as emulsifier and potassium persulfate (KPS) as initiator.

Polymerizations varying the emulsifier and initiator concentrations, the mean residence time, the polymerization temperature and the solids content were carried out. It was reported that the evolution of the monomer conversion presented overshoots at low emulsifier concentrations (the lowest [SLS] used was close to the critical micelle concentration, cmc), whereas a smooth evolution with a steady state conversion independent of the emulsifier concentration was obtained at high emulsifier concentrations. The dependences of steady state values of the polymerization rate and the number of polymer particles on the SLS concentration were [18]:

$$Rp \div [SLS] \qquad Np \div [SLS] \qquad \text{for} \qquad [SLS] \leq 8 \text{ kg/m}^3 \text{ water}$$

$$Rp \div [SLS]^0 \qquad Np \div [SLS]^{0.67} \qquad \text{for} \qquad [SLS] > 12 \text{ kg/m}^3 \text{ water}$$

The results for Np agree with the model predictions reported in literature for the emulsion polymerization of styrene in perfectly backmixed CSTRs [19]. However, the dependence of Np on [SLS] is not consistent with that of Rp because at low emulsifier concentrations when large particle sizes were obtained $Rp \div Np$ whereas for small values of dp, $Rp \div Np^0$. On the other hand, Lee et al. [17,18] reported that for polymerizations carried out using emulsifier concentrations in the range of the cmc the number of particles was independent of both the residence time and the solids content (in the range 8-15 wt %).

Geddes [20, 21] studied the 55 wt % solids content emulsion copolymerization of VAc and Veova 10 in a 2.7×10^{-3} m^3 CLR. Pipe length was 0.38 m and internal diameter 0.022 m. Circulation was provided by a modified progressing cavity pump. For each run the CLR was initially filled with the "water phase" that was a solution of a mixture of anionic surfactant, hydroxyethyl cellulose, sodium acetate and sodium metabisulfite. The process was started by feeding a stream of water phase and other of a mixture of VAc and Veova 10 (82/10 wt/wt) and a quantity of t-butyl hydroperoxide. Polymerizations started immediately and temperature rised from ambient to 50°C in 5 min. At this point cooling was applied to the jackets and the reactor temperature controlled at 60°C. A very high conversion (97-98.5 %) was achieved after 30-40 min (4-5 residence times). Conversion reached 99% by the time the emulsion reached the cooling tank. Conversion was quite stable but closer examination to the unreacted vinyl acetate levels revealed some cycling. Cycling was more obvious for particle size distribution. Single peaks were found in the early samples, then two peaks and finally three. The diameter of the polymer particles of each peak was found to grow linearly with time. This means that the particle volume growth rate was proportional to dp^2 but no mechanistic interpretation was given. These results also mean that the CLR shares with the CSTR one of its disadvantages: cyclic behavior due to intermittent nucleations.

Abad et al. [15,16,22-24] studied the emulsion copolymerization of VAc and Veova 10 in the CLR presented in Figure 3a. The following aspects were addressed; i) The extent of the shear-induced coagulation; ii) The effect of the start-up procedure; iii) Comparison between the CLR and the CSTR, and iv) The effect of several operation variables. Technical monomers were used and the stability of the latex was provided by a mixture of emulsifiers including Alipal CO430 (ammonium salt of sulfated nonylphenol poly(ethylenoxy) ethanol (4 ethylenoxy), Rhône-Poulenc), Arkopal N230 (nonylphenol poly(ethylenoxy) ethanol (23 ethylenoxy), Hoechst), and hydroxyethyl cellulose. $K_2S_2O_8$ and $Na_2S_2O_5$ were used as a redox pair.

Shear-induced coagulation can be critical in the performance of CLRs because coagulation will affect both the quality of the product and the duration of the continuous operation. Figure 4 presents the steady state particle size distributions (PSD) measured by transmission electron microscopy in polymerizations carried out varying the speed of the circulation pump. It can be seen that no differences were observed within the experimental error ruling out the shear-induced coagulation.

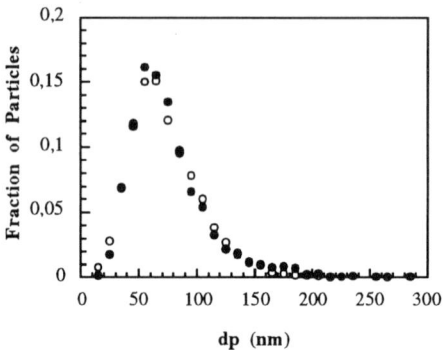

Figure 4. Particle size distributions obtained at different pump speeds. 100 rpm ($\bigcirc$) and 200 rpm ($\bullet$)

The effect of the start-up strategy on the smoothness of the operation and the amount of out-of-specifications product produced during the start-up was investigated by Abad et al. [22] for a high solids content system (55 wt%). The strategies studied were as follows:

- Strategy A: Reactor initially filled with water and heated to the reaction temperature (60°C)
- Strategy B: Reactor initially filled with water, emulsifiers, protective colloid, and $Na_2S_2O_5$ and heated to the reaction temperature (60°C).
- Strategy C: Reactor initially filled with a preemulsion of water, emulsifiers, protective colloid, $Na_2S_2O_5$ and the monomer mixture in the same proportions as in the overall recipe, with a temperature profile starting at room temperature and reaching 60°C in 90 min.
- Strategy D: Reactor initially filled with a latex from a previous reaction and heated to the reaction temperature (60°C).

The space time in the reactor was 26.5 min and the recycle ratio was 55.

Abad et al. [22] reported that the steady-state conversions were almost independent of the start-up procedure. From the point of view of the smoothness of the operation, the start-up strategies showed widely different behaviors. The polymerization carried out following strategy A proceeded smoothly with an easy control of the reactor temperature and a smooth increase of the viscosity of the reaction mixture. A substantial increase of the viscosity was observed in strategy B. This increase caused some problems in the stability of the feed flow rates. Later the viscosity decreased towards a steady-state value. A huge increase of the viscosity was observed for the start-up C making this experiment difficult to control. After some time the viscosity of the latex decreased and reached a steady-state value. When the reactor was initially heated to 60°C the start-up C could not be completed because there was a dangerous increase of the reactor pressure and a thermal runaway. When strategy D was used no significant change of the viscosity was observed during the start-up.

The latex viscosity depends on both the volume fraction of the dispersed phase and the PSD, the higher the volume fraction and the smaller the particle size the higher the viscosity. According to Abad el al. [22], the different latex viscosities were due to the evolution of the number of polymer particles in those experiments because the maximum solids content was limited by the recipe ($\cong$ 55 wt%) and, at the steady state, was the same for all the start-up procedures. Figure 5 shows that a large number of particles, which are responsible for the viscosity increases, were nucleated during start-ups A,B and C.

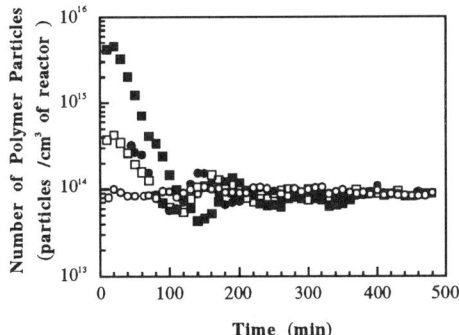

Figure 5. Evolution of the number of polymer particles for the different start-ups. strategy A ($\square$), strategy B ($\blacksquare$), strategy C ($\bullet$) and strategy D ($\bigcirc$)

The oscillations of N_p may be avoided by using the right start-up procedure which also minimizes the amount of out-of-specification product. Nevertheless, for the range of experimental conditions used in this study, the steady-state values of the conversion, particle size, number of polymer particles and molecular weight distributions were reported to be independent of the start-up procedure [22]. However, Abad [24] presented some cases in which the start-up procedure affected the properties of the polymer produced under steady-state conditions (see below).

Tracer experiments show that the macromixing in the CLR is determined by the recycle ratio, R_r. For large values of the recycle ratio ($R_r > 40$), the residence time distribution of the CLR is almost identical to that of a CSTR [15]. On the other hand, there are indications of concentration profiles along the loop [2, 3, 6, 16]. In order to further elucidate the similarities and differences between the loop reactor and the CSTR, Abad et al. [23] studied the redox-initiated emulsion copolymerization of VAc and Veova

10 in both types of reactors under comparable industrial-like conditions, namely, similar macromixing, same feed compositions and residence times, and using high solids content (55 wt%) recipes. To somehow counteract the differences in heat transfer area/reactor volume ratio due to the geometry, the CSTR was smaller (0.47×10^{-3} m^3) than the CLR (2.8×10^{-3} m^3). The start-up strategies B, C and D described above were used. These strategies lead to different evolutions of the reaction mixture viscosity which in turn affects the heat removal capacity of the reactor. It is reported [23] that under the operation conditions in which the heat removal capacity of the reactor exceeded the heat generation rate (strategies B and D) there were no differences between the CLR and the CSTR in terms of the steady state values of conversion, particle size, number of polymer particles and molecular weight distributions (Figure 6). However, under conditions in which the heat generation rate is high (strategy C) a thermal runaway occurred in the CSTR whereas the temperature of the CLR was easily controlled. This difference was due to the different geometry of the reactors and is critical from the point of view of the production rate, but, at least for the case studied, this seemed to be the only difference between both reactors.

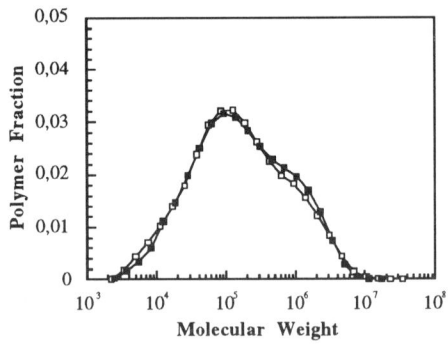

Figure 6. Molecular weight distributions for strategy D.
CLR (□), and CSTR (■)

Abad [24] carried out experiments varying the residence time (8.8 min, 17.7 min and 26.5 min) using the start-up procedure B. Figure 7a shows that the steady state conversion decreased as τ decreased. The 8.8 min residence time run was reported to be difficult to handle because the high viscosity of the latex and the high internal flow rate increased the pressure at the discharge of the circulation pump and the feeding system was unable to feed the required high flow rates smoothly.

Abad et al. [24] also reported an experiment in which the polymerization was started with the CLR initially filled with a latex, after about 10 residence times (265 min) the feed flow rate was increased to reduce τ to 17.7 min, and after about 10 more residence times (177 min) the feed flow rate was further increased to achieve $\tau=8.8$ min. The whole run proceeded smoothly, but surprisingly, the conversion was independent of the space time (Figure 7b). Similar results were reported by Lanthier [2,3]. The results reported in Figure 7 suggest steady state multiplicity. Multiple steady states for emulsion polymerization systems in CSTRs have been reported by Schork and Ray [25]. The practical importance of these results is that high monomer conversion at low residence times, i.e., maximum production, can be achieved if the adequate start-up procedure is used, namely, starting the reaction with a large space time and reducing it when the steady state is reached.

The improvement of the product quality encourages the reduction of the polymerization temperature, because the lower this temperature the higher the molecular weight, which has a beneficial effect on the scrub resistance of the films produced from the latex. Abad [24] studied the effect of the polymerization temperature on the reactor performance reporting that runs carried out at 50°C, 60°C and 70°C, all yielded a 92% conversion, namely, no effect of the reaction temperature was observed. This is surprising because a marked effect of the temperature on the conversion was observed in the emulsion copolymerization of these monomers in a semicontinuous reactor [26]. The authors explained this behavior as terms of the interplay between the rate of formation of radicals and the geometric characteristics of the loop reactor. The redox initiation is known to be very rapid, and according to Adams [9] and Lanthier [2,3] the redox pair reacts completely before completing a pass through the loop. Under these circumstances, temperature has two counteracting effects. The higher the temperature the higher the rates of radical formation and polymerization, but the shorter the time in which a particular element of fluid reacts because the initiators are rapidly consumed. These counteracting effects, together with the low sensitivity of the system when it operates at high conversions might explain the results obtained.

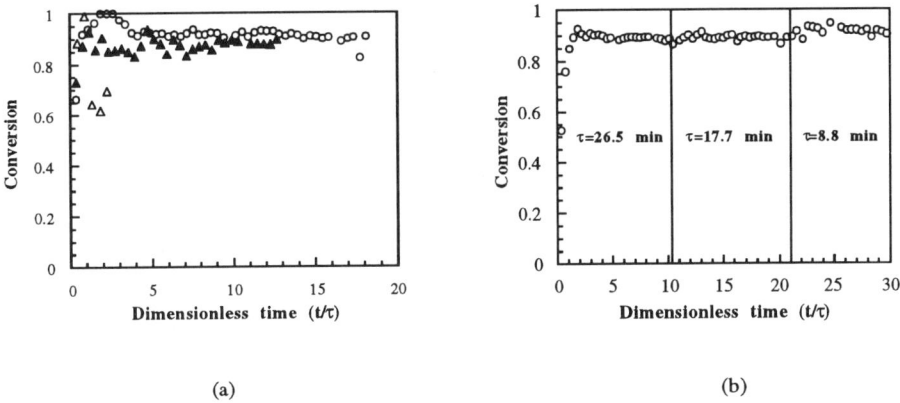

(a)   (b)

Figure 7. Effect of the space time on the evolution of the conversion for different start-up procedures (see text). Legend for figure 7a: τ= 26.5 min (○), τ= 17.7 min (▲),
τ= 8.8 min (△).

Economic reasons related with the latex cost and the improvement of the product quality encourages the reduction of the amount of emulsifier used in the recipe. In an attempt to reduce the amount of emulsifier, Abad [24] carried out several experiments varying the concentration of the emulsifiers from 2.4 wt% to 7.2 wt%. It was found that with 3.6 wt% and 7.2 wt% of stabilizers a conversion of about 90% was achieved whereas in the reaction carried out with the lowest stabilizer concentration (2.4 wt%) the latex coagulated. On the other hand, the steady state conversion was reported to be independent of the emulsifier concentration.

Abad [24] reported that no effect of the initiator concentration on the conversion was observed at high initiator concentrations whereas coagulation occurred when 0.15 wt% of initiator was used. Experiments carried out in a semicontinuous reactor show that, because of the decrease in the ionic strength enhances the efficiency of the stabilizer system, the lower the initiator concentration the higher the number of polymer particles [26]. During the start-up of the polymerization carried out with 0.15 wt% of initiator a large number of polymer particles was initially produced and the amount of stabilizer was not enough to efficiently cover these particles when they grew. The problem was overcome by starting the process with a higher amount of initiator, and later reducing this amount to the desired level as is shown in Figure 8. The whole experiment proceeded smoothly and no coagulation was observed. On the other hand, a null effect of the initiator concentration on the conversion was found. The practical significance of these results is that highly stable latexes of a high molecular weight polymer can be obtained at high polymerization rates if the polymerization is started using a higher initiator concentration that is later reduced to the desired value.

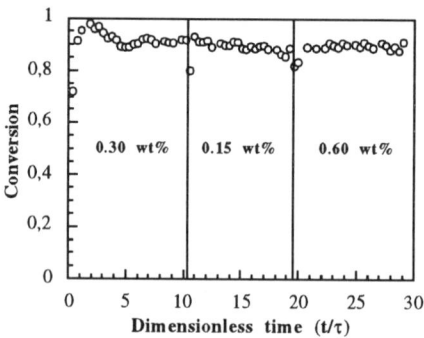

Figure 8. Effect of the initiator concentration.

4. Modelling Emulsion Polymerization in Continuous Loop Reactors

The mathematical model should combine the kinetics characteristic of emulsion polymerization with the flow model of the reactor. Based on the fact that at the large values of R_r usually employed in the operation of CLRs, the macromixing of these reactors corresponds to perfectly backmixed continuous reactors, and that for reactors with heat removal capacities larger than the heat generation rate there are no differences between the CLR and the CSTR [23], Abad et al. [15] employed the CSTR model to describe the loop reactor. However, a more general model is needed to include all the characteristics of a CLR. Abad et al. [27] developed a mathematical model for a CLR by considering that the macromixing is described by a loop formed by two tubular reactors with axial dispersion (Figure 3b). The mathematical model includes the population, material and energy balances in the reactors, plus the corresponding balances in the intersection points.

4.1. POLYMER PARTICLES POPULATION BALANCE

In the absence of coagulation, the population balance in each reactor is as follows:

$$\frac{\partial n(V^*)}{\partial t} = D_i \frac{\partial^2 n(V^*)}{\partial z^2} - \frac{\partial u n(V^*)}{\partial z} - \frac{\partial r_v(V^*)n(V^*)}{\partial v} \tag{1}$$

where $n(V^*)$ is the number density distribution function, D_i is the dispersion coefficient in the reactor i that was assumed to be the same for all the components in the reaction mixture, u the average fluid velocity in the axial direction, and $r_v(V^*)$ the rate of volumetric growth of unswollen latex particles of volume V^*.

$$r_v(V^*) = \sum_{h=A,B} (k_{P_{Ah}} P_A^P(V^*) + k_{P_{Bh}} P_B^P(V^*))[h]_p(V^*) \frac{\bar{n}(V^*)}{N_A} \frac{M_h}{\rho_p} \tag{2}$$

$$n(V_0^*) = R_{nuc} / r_v(V_0^*) \tag{3}$$

where $k_{P_{ih}}$ are the propagation rate constants, $[h]_p(V^*)$ the concentration of monomer h in polymer particles of unswollen volume V^*, N_A Avogadro's number, $P_h^P(V^*)$ the time averaged probability of finding an active chain with ultimate unit of type h in the polymer particles [26], M_h the molecular weight of monomer h, $\bar{n}(V^*)$ the average number of free radicals per latex particle of unswollen volume V^*, ρ_p the density of the copolymer, V_0^* the unswollen volume of the polymer particles when they are nucleated, and R_{nuc} the nucleation rate. In eq. (3) the contribution of the nucleation rate was included to have the same number of free radicals, namely, the stochastic broadening effect was neglected. This is a reasonable assumption in redox initiated systems because of the high entry rate of radicals into the polymer particles.

Nucleation was modelled by means of a variation of the pragmatic approach proposed by Urretabizkaia et al. [29]. The rate of nucleation was defined as follows:

$$R_{nuc} = k_1 (N^*) P_N \tag{4}$$

where N^* is the concentration of particle precursors, k_1 an adjustable parameter, and P_N the probability that a precursor becomes a polymer particle given by the following empirical equation:

$$P_N = k_2 \frac{k_3([S]_w - cmc)/cmc}{1 + k_3([S]_w - cmc)/cmc} exp[-k_4 N_P] \qquad for \ [S]_w > cmc \tag{5}$$

$$P_N = 0 \qquad for \ [S]_w < cmc \tag{6}$$

where $[S]_w$ the concentration of emulsifier in the aqueous phase (including the micelles), and N_p the concentration of polymer particles given by:

$$N_p = \int_{V_o^*}^{V_1^*} n(V^*)dV^* \tag{7}$$

where V_1^* is chosen in such a way that the whole PSD is included between V_o^* and V_1^*.

Equations (5) and (6) account for the fact that nucleation ceases when the number of polymer particles reaches a given value or the concentration of emulsifier is lower than the cmc.

4.2. MATERIAL BALANCES

The material balance for the component j in the reactor i is as follows:

$$\frac{\partial C_j}{\partial t} = D_i \frac{\partial^2 C_j}{\partial z^2} - \frac{\partial u C_j}{\partial z} + R_j \tag{8}$$

where C_j is the concentration of compound j, D_i the dispersion coefficient that was assumed to be the same for all the components in the reaction mixture and R_j the rate of generation of component j.

This material balance applies to the monomers, initiators, precursors, emulsifier and water.

4.3. TOTAL VOLUME BALANCE

Because of the higher density of the polymer as compared with the monomer, the fluid velocity decreases with conversion. Neglecting the changes in density due to temperature variation, the change of the velocity is only due to the conversion of monomer into polymer [14].

$$\frac{\partial u}{\partial z} = \sum_{h=A,B} \frac{R_h M_h}{\rho_h} - \sum_{h=A,B} \frac{R_h M_h}{\rho_p} \tag{9}$$

4.4. ENERGY BALANCE

The energy balance for each reactor is as follows:

$$\frac{\partial T}{\partial t} = \frac{\lambda_i}{\rho C_p} \frac{\partial^2 T}{\partial z^2} - u \frac{\partial T}{\partial z} - \frac{1}{\rho C_p} \left[\frac{4U_i}{d_i}(T - T_w) + \sum_{h=A,B} R_h(-\Delta H_r)_h \right] \tag{10}$$

where T is the temperature, λ_i the thermal conductivity in the axial direction in reactor i, U_i and d_i the overall heat transfer coefficient and the reactor diameter, respectively, in reactor i, ρ and C_P the density and specific heat capacity, respectively, of the reaction mixture, T_w the temperature of the thermal fluid, and $(-\Delta H_r)_h$ the heat of polymerization of monomer h.

4.5. MONOMER CONCENTRATIONS

The concentrations of the monomers in the different phases were calculated by means of the approach developed by Armitage et al. [30] for the calculation of the monomer partitioning in polydisperse emulsion copolymerization systems. In this approach it was assumed that the monomer mass transfer rates between the different phases of the system were much larger that the polymerization rate, and hence the monomer concentrations were at the thermodynamic equilibrium values.

4.6. FREE RADICAL BALANCES AND LATEX VISCOSITY

It was assumed that the pseudo-steady state assumption applied for the free radicals in both polymer particles and aqueous phase. The average number of radicals in polymer particles of unswollen volume V^* was calculated using a variation of the approach proposed by Ugelstad and Hansen [31].

The viscosity of the latex was estimated using the approach proposed by Sudduth [32].

4.7. SOLUTION OF THE MODEL

The mathematical model is a system of non linear partial differential and algebraic equations. The partial differential equations were transformed into a set of ordinary differential equations by means of the orthogonal collocation method [33]. Classical orthogonal collocation was used for the axial dispersion, whereas orthogonal collocation on moving infinite elements [34] was used for the particles size distribution. Appropiate initial and boundary conditions were considered [27].

Figure 9 presents the evolution of the conversion and PSD during the emulsion copolymerization of VAc and Veova 10 carried out using the start up strategy A. It can be seen that a fairly good agreement between experimental results and model predictions was achieved. However, the limits of the model remain to be investigated.

5. Conclusions and Future Challenges

Continuous loop reactors are an attractive alternative for the production of emulsion polymers because their large heat area/reactor volume ration allows to conduct the process at high polymerization rates. In addition, these reactors are intrinsically safe and the installation and running costs are lower that for CSTRs. Their current use is limited to the manufacturing of VAc homopolymers and copolymers and it seems that their future use will be limited to the production of homogeneous copolymers because core-

346

shell particles cannot be produced. The scarce literature available suggests that , at the high recycle ratios employed in the commercial units, the CLRs share many of the characteristics of the CSTRs (oscillations due to intermittent nucleation, steady-state multiplicity, ...) but much better thermal behavior. The start-up strategy is critical not only for both the smoothness of the operation and the minimization of the amount of out-of-specification product, but for the feasibility of the process and the properties of the polymer produced under steady state conditions. A mathematical model for the process has been reported but only limited use of the model has been carried out.

Future challenges include a better understanding of the micromixing; improvement of the mathematical model of the process by including detailed models for particle nucleation and radical formation in redox initiation; increasing of the reactor production by maximizing the heat removal capacity; production of a wider variety of polymers including all acrylics latexes; control of the particle size distribution; and commercial implementation of the production of vinyl acetate-ethylene copolymers in high pressure reactors.

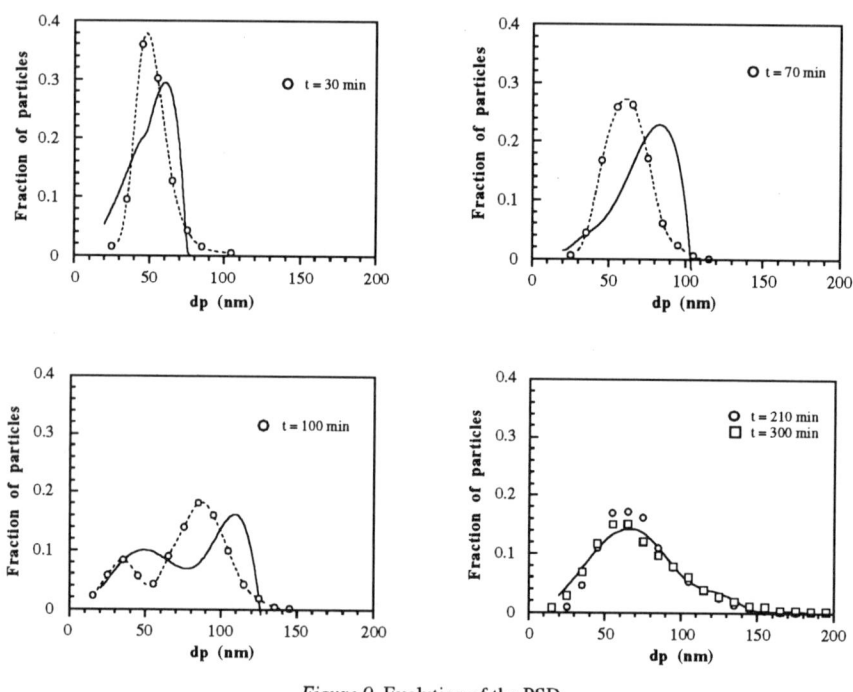

Figure 9. Evolution of the PSD.
(- - -) Experimental results and (——) Model Predictions.

Acknowledgements. The financial support by the CICYT (Grant MAT 91-0195) and the Diputación Foral de Gipuzkoa, and the scholarship of C. Abad from the Basque Government are gratefully appreciated.

6. References

1.- Poehlein, G.W. (1996), Continuous reactors in J.M. Asua (ed.), *Polymeric Dispersions. Principles and Applications*, Kluwer Academic Publishers, Dordrecht.
2.- Lanthier, R., (1970). Continuous vinyl polymerization process, US Patent 3, 551, 396.
3.- Lanthier, R. (1992). Continuous vinyl polymerization process, Canadian Patent 907795.
4.- Geddes, K., (1983), The loop process, *Chemistry and Industry* 21, 223-227.
5.- Geddes, K. (1986), Paint emulsions by the loop process 176, 494-500.
6.- Geddes, K. (1983), The loop reactor process, *JOCCA*, 76, 330-339.
7.- Geddes, K. (1990), The loop reactor process in, A.D. Wilson, J.W. Nicholson and H.J. Prosser (Ed.) *Coatings Surface Coatings-3*, Elsevier Applied Science, London, 199-228.
8.- Geddes, K. and Khan, M.B. (1990), Polymerization processes and reactors, European Patent 0417893 Al.
9.- Adams, D.C. (1984), Production of emulsion polymers, European Patent 0145325 A2.
10.- Lin, J.K. and Douek, M. (1976), Interpolymerization of ethylene with other monomers in aqueous emulsions, Canadian Patent 985844.
11.- Tonge, C. Personal Communication.
12.- Warnecke, H.J., Prüss, J. and Langemann (1985), On the mathematical model for loop reactors I and II, *Chem. Eng. Sci.* 40, 2321-2331.
13.- Lee, D.Y., Kuo, J.F., Wang, J.H. and Chen, C.Y. (1990). Study of the continuous loop tubular reactor for emulsion polymerization of styrene. *Polym. Eng. Sci.* 30, 187-192.
14.- Zacca, J.J. and Ray, W.H. (1993). Modelling of the liquid phase polymerization of olefins in loop reactors, *Chem. Eng. Sci.* 48, 3743-3765.
15.- Abad, C., de la Cal, J.C. and Asua, J.M., (1994). Emulsion copolymerization in continuous loop reactors, *Chem. Eng. Sci.* 49, 5025-5037.
16.- Abad,C., de la Cal, J.C. and Asua, J.M. (1995), Macromixing and micromixing in continuous loop reactors, in Reichert K.H. and Moritz H.U. (eds.), 5th International Workshop on Polymer Reaction Engineering, *DECHEMA Monographs*, 131, 87-94.
17.- Lee, D.Y., Kuo, J.F., Wang, J.H. and Chen, C.Y. (1990), The Performance of a continuous loop tubular reactor for emulsion polymerization of styrene, *J. Chem. Eng. of Japan*, 23, 290-296.
18.- Lee, D.Y., Wang, J.H. and Kuo, J.F. (1992), The performance of the emulsion polymerization of styrene in a continuous loop tubular reactor, *Polym. Eng. Sci.*, 32, 198-205.
19.- Poehlein, G.W. (1992). Emulsion polymerization in continuous reactors, in I. Piirma (ed.), *Emulsion Polymerization*, Academic Press, 357-382.
20.- Geddes, K.R. (1987), The loop reactor process, *Advances in Organic Coatings Science and Technology Series*, Technomic Publishing Co., Lancaster, 9, 30-44.
21.- Geddes, K.R. (1989), Start-up and growth mechanisms in the loop continuous reactor, *Br. Polym. J.* 21, 433-441.
22.- Abad, C, de la Cal, J.C. and Asua, J.M. (1995), Start-up procedures in emulsion copolymerization of vinyl esters in a continuous loop reactor, *Polymer*, 36, 4293-4299.
23.- Abad, C, de la Cal, J.C. and Asua, J.M. (1995), Emulsion copolymerization of vinyl esters in continuous reactors: comparison between loop and continuous Stirred tank reactors, *J. Appl. Polym. Sci.*, 56, 419-424.
24.- Abad, C. (1995), Estudio del proceso de obtención en un reactor loop de copolímeros de características comerciales, PhD Dissertation, Universidad del País Vasco, San Sebastián, Spain, 113-129.
25.- Schork, F.J. and Ray, W.H. (1987), The dynamics of the continuous emulsion polymerization of methylmethacrylate, *J. Appl. Polym. Sci.*, 34, 1259-1276.
26.- Urretabizkaia, A and Asua, J.M. Unpublished results.
27.- Abad, C, de la Cal, J.C. and Asua, J.M. (1995), Modelling nucleation and particle growth in emulsion copolymerization in continuous loop reactors, *Macromol. Symp.*, 92, 195-204.
28.- Forcada, J. and Asua, J.M., (1985). Modelling the microstructure of emulsion copolymers, J. Polym. Sci., Polym. Chem. Ed., 23, 1955-1962.
29.- Urretabizkaia, A., Arzamendi, G. and Asua, J.M. (1992), Modelling semicontinuous emulsion terpolymerization, *Chem. Eng. Sci.*, 47, 2579-2584.
30.- Armitage, P.D., de la Cal, J.C. and Asua, J.M. (1994). Improved methods for solving monomer partitioning in emulsion copolymer Systems. J. Appl. Polym. Sci., 51, 1985-1990.
31.- Ugelstad, J. and Hansen, F.K. (1976). Kinetics and mechanims of emulsion polymerization, Rubber Chem. Technol. 49, 536-609.
32.- Sudduth, R.D. (1993). A generalized model to predict the viscosity of solutions with suspended particles. J. Appl. Polym. Sci., 48, 25-36.
33.- Villadsen, J.V. and Stewart, W.E. (1967). Solution of boundary-value problems by orthogonal collocation, 22, 1483-1501.
34.- Finlayson, B.A. (1980). Non Linear Analysis in Chemical Engineering, McGraw Hill, New York.

OPEN-LOOP CONTROL OF POLYMERIZATION REACTORS

M. MORBIDELLI, G. STORTI[1]
Chemical Engineering Department /LTC
Universitätsstrasse 6, ETH-Zentrum
CH-8092 Zürich, Switzerland

1. Introduction

The end-use properties of polymeric materials are determined by the characteristics of the macromolecules such as length, composition, sequence of monomer units, end groups, and so on. Moreover, due to the statistical nature of the population of macromolecules, these properties are not necessarily the same for all individuals but rather they change following specific distributions. As a consequence, in order to determine the end-use properties of a polymer one has to control the molecular weight distribution, the chain composition distribution, the monomer sequence distribution and so on.

When dealing with radical polymerization an additional complication arises because of the short life time of the macromolecules, which is of the order of 1 sec. In the case of a batch polymerization reactor, the duration of the entire process needed to achieve the desired monomer conversion is typically of the order of 10^3 sec. This indicates that the final product in a batch reactor is constituted by a collection of macromolecules, each of which experienced in his life rather different growth conditions. In other words, each chain started and ended in a relatively short period of time located at different points of the polymerization batch where the reaction "environment" was rather different.

For example, in a system where the chain transfer to monomer is not the dominant termination mechanism, it is apparent that the chain propagation reaction is more favored for the macromolecules synthetized at the beginning of the process then for those synthetized towards its end, due to the monomer concentration decrease during the batch reaction. This has obviously a strong effect on the distributions of the macromolecules properties mentioned above. One of the main objectives of reactor operation is precisely to control the polymerization "environment" so as to obtain a final polymer with desired molecular properties.

2. The Polymerization Environment

By polymerization environment we mean all those characteristics of the reacting medium inside the reactor (such as temperature, composition and so on) which affect the properties of the final product. Before discussing different control techniques, it is then convenient to identify in detail which are the characteristics which affect each specific

---

[1]permanent address: Dip. Ingegneria Chimica e Materiali, Università di Cagliari, Piazza d'Armi, 09123 Cagliari, Italy.

J. M. Asua (ed.), Polymeric Dispersions: Principles and Applications, 349–361.

molecular property. We will do this with reference to radical homogeneous polymerizations. The extension to heterogeneous polymerizations is straightforward when considering that the characteristics of interest are those of the reaction locus, e.g. the polymer particles in suspension and emulsion polymerization. Moreover, only three specific molecular properties are considered in the following.

2.1. CHAIN COMPOSITION DISTRIBUTION, CCD

For the case of a copolymer involving monomers A and B, the instantaneous polymer composition is given in terms of the mole fraction of A, y_A by the well known Mayo and Lewis relation [1]:

$$y_A = \frac{(r_1 x_A + x_B)x_A}{(r_2 x_B + x_A)x_B + (r_1 x_A + x_B)x_A} \tag{1}$$

where x_A and x_B are the mole fractions of the two monomers in the reacting mixture (polymer free), while r_1 and r_2 represent the reactivity ratios of active radicals of type A and B, respectively. These are defined as the ratio between the direct and the cross propagation rate constants of the specific radical.

From the observation of the relation above, we see that the instantaneous polymer composition is determined solely by the monomer phase composition and the rate constants of the propagation reactions, which in turn are functions only of temperature and, in the case of large conversion values, also of conversion due to the occurrence of the polymer glass transition (the so-called glass effect). Thus, besides this last aspect, we can conclude that in order to control the chain composition distribution of the final polymer one should control temperature and the monomer phase composition. For example, by keeping constant the two quantities above during the entire polymerization process it is possible to produce a polymer where each individual chain has the same composition.

The result above has been obtained in the case of two monomer species. However, it can be readily generalized to situations involving any number of monomer species.

2.2. CHAIN SEQUENCE DISTRIBUTION, CSD

The sequence of different monomer units along a polymer chain is determined, again in the case of a copolymer, by the probability of having say monomer A, given by [3]:

$$p_A = \frac{\alpha}{1+\alpha}, \quad \text{where } \alpha = \frac{r_1 x_A}{x_B} \tag{2}$$

and similarly for monomer B. Thus, for example the probability of having along the polymer chain a sequence of n units of monomer A, is given by:

$$p_n = \frac{\alpha^{n-1}}{(1+\alpha)^n} \tag{3}$$

From these expressions we can conclude that also for the chain sequence distribution, the only variables which are of interest are temperature and monomer phase composition, which then define in this case the polymerization "environment". Again, this result applies to cases involving any number of monomer species.

2.3. MOLECULAR WEIGHT DISTRIBUTION, MWD

The length of macromolecular chains is determined by the competition between the chain propagation and the various chain termination reactions present in the polymerization system. It can be shown that only two dimensionless parameters ultimately define the molecular weight distribution (for example, see [4]). These are the ratio between the characteristic times for propagation and termination by bimolecular combination defined by:

$$\beta = \frac{\tau_p}{\tau_{tc}} = \frac{k_{tc}\left[R^*\right]}{k_p[M]} \qquad (4)$$

and the sum of the ratios between the characteristic times for propagation and all the other termination events, i.e. chain transfer to monomer, transfer to the chain transfer agent, S, and bimolecular termination by disproportionation:

$$\gamma = \frac{\tau_p}{\tau_{trm}} + \frac{\tau_p}{\tau_{trs}} + \frac{\tau_p}{\tau_{td}}$$
$$= \frac{k_{trm}}{k_p} + \frac{k_{trs}[S]}{k_p[M]} + \frac{k_{td}\left[R^*\right]}{k_p[M]} \qquad (5)$$

In all relations above the total concentration of active radicals can be computed through the pseudo-steady state approximation as follows:

$$\left[R^*\right] = \left[R_I /(k_{tc} + k_{td})\right]^{1/2} \qquad (6)$$

where R_I represents the rate of production of active radicals, say by initiator decomposition, per unit volume of the reacting medium. Using the above parameters the instantaneous MWD can be computed in terms of number-average molecular weight and polydispersity index as follows:

$$<M_n> = \frac{1}{(\gamma + \beta/2)}; \quad PDI = \frac{2(\gamma + 3\beta/2)(\gamma + \beta/2)}{(\gamma + \beta)^2} \qquad (7)$$

This result allows to readily identify the variables which have to be controlled in order to achieve a desired MWD. For example, let us assume that we desire to produce a polymer with the narrowest possible MWD. This means that our goal is to obtain a cumulated polydispersity index equal to the instantaneous one, which is intrinsic of the specific monomer under examination. This is obtained by running the polymerization so that the number-average molecular weight remains constant during the entire reaction, i.e. we should maintain constant the values of the two parameters β and γ. Of course this is achieved through different strategies depending upon the specific termination mechanism which is dominating in the particular polymerization system under examination. Thus, in general the variables to be kept under control include: the reaction rate constant of the dominating reactions (which reduces to temperature at least as long as a significant gel effect is not present), the rate of radical production R_I, the monomer concentration [M] and the ratio between chain transfer agent and monomer concentrations [S]/[M].

In the important case where more than one monomer species are present in the system, the relations above remain approximately correct if the true rate constants are replaced by pseudo-homopolymerization or pseudo-kinetic rate constants. These are given by appropriate averages, weighted on the composition of the monomer phase, of the corresponding true rate constants in the copolymerization system [5] [6]. For example, the pseudo-homopolymerization propagation reaction of the i-th monomer species is given by:

$$\bar{k}_{pi} = \sum_j k_{pji} p_j \tag{8}$$

where k_{pji} is the propagation rate constant of radical j with monomer i and p_j is the probability of having a growing radical ending with a j monomer unit, which is a function of the monomer phase composition.

Thus, we can conclude that in the case where several monomer species are involved, in order to control the values of the parameters β and γ we should control the values of the pseudo-homopolymerization rate constants, which includes controlling not only temperature but also the monomer phase composition. This becomes then a requirement not only for composition control but also for the control of the molecular weight distribution.

3. Open-loop Control Strategies

The open-loop control of a polymerization reactor implies to operate the reactor without on-line monitoring of the polymer characteristics, such as CCD, CSD and MWD, which have to be controlled. This is instead done in closed-loop control strategies, where a comparison between measured and desired polymer characteristics allows to produce a signal which corrects the current operation of the reactor (cf. [7]). Thus, a good open-loop control strategy must rely on the close control of all the variables, i.e. the polymerization "environment", which are known to affect the desired polymer characteristics. In the previous section we have identified which are such variables in the various instances of interest and we will analyse them individually in the following.

It should be mentioned that in principle we could be interested in obtaining polymers with many different characteristics and we could in fact obtain them by properly changing the controlling variables during the polymerization process. However, in most cases of practical interest the aim is to obtain a product with uniform characteristics. For this we are interested in keeping constant during the polymerization the corresponding controlling variables.

With this aim we will now review the control strategy for the different variables identified in the previous section.

Temperature is usually controlled by standard heat transfer devices. For this it is convenient to have reacting media with low viscosity, as it is best obtained using heterogeneous polymerization in suspension or in emulsion. Care should be taken to prevent excessive fouling of the heat transfer surfaces.

For the rate of active radicals production, R_I, it is usually sufficient to select initiators whose half-decomposition time at the polymerization conditions is low compared to the duration of the polymerization process.

As it can seen from eqs (4) and (5), monomer concentration is of relevance when controlling the MWD in systems where the bimolecular terminations by combination or disproportionation or the termination by transfer to a chain transfer agent are dominant. Apparently, keeping constant the monomer concentration in a polymerization reactor is in general rather difficult. A possibility would be to play with temperature or with the rate of active radicals production so as to maintain constant the parameters β or γ. It should be mentioned that emulsion polymerization offers in this case an attractive alternative. In this system in fact the overall monomer concentration in the reaction locus, i.e. the polymer particles, remain automatically constant in the first two intervals of the Smith-Ewart model, i.e. as long as monomer droplets are present in the reaction medium. Since this is often the period of the process where a significant amount of polymer is produced (usually from 20 to 40% of conversion), emulsion polymerization may offer a solution to this problem in many instances of practical interest. Note that it is possible to control the monomer concentration also in the case where the reactor is operated under starved conditions, as discussed in the following section. However, here the monomer concentration is generally rather low, thus leading to a significant gel effect which may strongly affect the properties of the produced polymer.

Finally, we have seen in the previous section that a key aspect is the control of the monomer phase composition. Since in general different monomer species exhibit different reactivities, it follows that the monomer phase composition tends to change during the polymerization process. This is in fact illustrated by the Mayo and Lewis relation reported above. Thus, in order to maintain constant the monomer phase composition we need to add some of the monomers (generally the most reactive ones) during the process so as to replace the quantities consumed in excess by the polymerization reactions. This aspect is discussed in detail in the next section, where semibatch reactors will be considered.

4. Semibatch Reactors for Composition Control

A closed-loop strategy for controlling the monomer phase composition in the reactor is based on the comparison of the on-line measured value of this composition with the desired value and then on a consequent action on the addition flowrates of the various monomers. In the open-loop approach we aim at the same result but without using on-line composition measurements. There are basically two strategies for solving this problem which, although fully general, will be discussed in the following with reference to a binary copolymer (cf. [8-10] and [11-16], respectively).

For the generic monomer A, we can write the mass balance in a semibatch reactor, where F_A is the molar feed flowrate of A, as follows:

$$\frac{dM_A}{dt} = F_A - R_{PA} \qquad (9)$$

where M_A indicates the overall number of moles in the reactor. The rate of polymerization of monomer A can be expressed in terms of a first-order reaction kinetics:

$$R_{pA} = k'_{pA} M_A \qquad (10)$$

where k'_{pA} is a pseudo-first order rate constant, which is given by the product between the overall concentration of active chains, $[R^*]$, and the average propagation rate constant, k_{pA} estimated through the pseudo-homopolymerization approach as a function of composition. By introducing the dimensionless time $\tau = t/\bar{t}$, where $\bar{t}$ represents the duration of the polymerization process, the above mass balance reduces to

$$\frac{dM_A}{d\tau} = \bar{t}\left(F_A - k'_{pA} M_A\right) \qquad (11)$$

An analysis of the dynamics of this process indicates that if we take the duration of the process $\bar{t}$ long enough compared to the characteristic time of the polymerization reaction, given by $1/k'_{pA}$, so that:

$$\bar{t}k'_{pA} \gg 1 \qquad (12)$$

then the reactor approximates pseudo-steady state operating conditions. That is the term in parenthesis in eq (11) vanishes and the polymerization rate becomes:

$$R_{pA} = k'_{pA} M_A = F_A \qquad (13)$$

This result can be readily confirmed by solving eq (11) in the particular case where both F_A and k'_{pA} are assumed constant, leading to:

$$M_A = \left(M_A^0 - \frac{F_A}{k'_{pA}}\right)\exp\left(-k'_{pA}\,\bar{t}\tau\right) + \frac{F_A}{k'_{pA}} \qquad (14)$$

It is apparent that in the case where $k'_{pA}\,\bar{t}, \gg 1$ after a small initial time interval, the amount of monomer in the reactor remains constant and equal to (F_A/k'_{pA}) as indicated in eq (13). In other words, by feeding the reactor slowly enough compared to the potential polymerization rate, i.e. $\bar{t} \gg 1/k'_{pA}$, we force the reaction rate to become equal to the feed rate, thus keeping constant the total amount of each monomer in the reactor. This procedure, which is usually referred to as starved polymerization, leads to a situation where the composition inside the reactor is self-controlled and equal to that in the feed stream. It is remarkable that this result is obtained without any specific knowledge of the system kinetics, by simply elongating enough the batch duration. The price to be paid of course is the low productivity of the reactor. Moreover, it should be reiterated that due to the low monomer concentration, in the starved procedure the gel effect is usually rather relevant thus complicating the system behavior.

An alternative procedure is based on the idea of feeding the monomers so as to maintain constant the monomer phase composition in the reactor. In general, we can think of having four degrees of freedom in operating a semibatch copolymerization reactor: the amount of each monomer initially charged to the reactor, i.e. M_A^0 and M_B^0, and the feed flowrates of the two monomers, i.e. $F_A(t)$ and $F_B(t)$. The constraints arise as follows:

- the initial monomer phase composition should be such as to produce the polymer with desired composition (according to the Mayo and Lewis model):

$$\frac{M_A^0}{M_A^0 + M_B^0} = \text{fixed}; \tag{15}$$

- the total amount of monomers introduced in the reactor should be equal to the total amount of polymer that we whish to produce in a single batch;
- we should feed the monomers so that the monomer phase composition remains constant:

$$\frac{d}{dt}\left(\frac{M_A}{M_A + M_B}\right) = 0 \tag{16}$$

which using eq (9) for M_A and the corresponding one for M_B, leads to:

$$M_B\left(F_A - k'_{pA} M_A\right) = M_A\left(F_B - k'_{pB} M_B\right) \tag{17}$$

The constraints above saturate only three of the four degrees of freedom, so that the last one can be used as an optimization parameter with respect to some specific performance index.

A natural choice is to require the batch time to be minimum, so as to maximize production. In this case it can be shown that the optimal policy is to charge all the less reactive monomer at the beginning of the process, while the second one should be fed with a flowrate F_A given by eq (17), where F_B is set equal to zero. This procedure obviously maximizes the polymerization rate, which in many instances may lead to rates of heat production which exceed the capacity of the cooling system of the reactor. In these cases it is convenient to relax the requirement of minimum batch time and solve the problem above by introducing the constraint that in any moment of the polymerization the heat production rate should not exceed the heat removal rate of the cooling system.

Whatever is the specific performance index selected, this operation mode presents a major difficulty. In particular, when using eq (17) to evaluate a priori the feed flowrate of the monomers, one should know the pseudo-homopolymerization first-order rate constants, k'_{pi} and the amounts of all unreacted monomers currently present in the reactor. Since there is no feedback, as in closed-loop control strategies, to correct possible errors in the estimation of these quantities, the entire procedure is subject to failure. This is particularly serious in radical polymerization, since the behavior of the radical species is particularly subject to irreproducibilities.

A convenient way to overcome this problem is to replace time as the independent variable with overall monomer conversion. In this case the behavior of the total amount of monomer A in the reactor, M_A as a function of the total mass of polymer produced, M_C is described by the relation

$$\frac{dM_A}{dM_C} = \frac{\bar{k}_{pA}M_A}{\bar{k}_{pA}M_A + \bar{k}_{pB}M_B} \tag{18}$$

It can be seen that the right hand side in this case does not involve the radical concentration and therefore becomes much less sensitive to the irreproducibilities typical of these systems. The price to be paid is that this approach requires the application of some sensor to monitor on-line the total amount of polymer produced as a function of time. Sensors of this type have been developed as discussed elsewhere in this volume [17] [18].

5. Applications of Open-Loop Control Strategies

The concepts discussed above in the context of homogeneous polymerizations can be extended to heterogeneous polymerizations. Of course the relevant equations become more complicated because of the partitioning of monomers among the various phases present in the system and, in the case of emulsion polymerization, also because of the compartmentalization effect in the polymer particles. Nevertheless, the underlying concepts remain the same, so that in the following we will discuss some examples of application of the open-loop control strategies discussed above with reference to emulsion polymerization systems [19] [20].

5.1. POLYMER COMPOSITION CONTROL

Let us consider the case of a ternary system methyl methacrylate-vinyl acetate-butyl acrylate (MMA-VAc-BuA) where the objective is to produce a polymer with molar composition 0.40, 0.30 and 0.30, respectively. Three experimental runs are considered: run 72 is conducted in a batch reactor with no composition control, while runs 80 and 74 are both conducted in a semibatch reactor. In both cases conversion is monitored on-line through a densitometer, and the feed flowrate of the two more reactive monomers, i.e. MMA and BuA, is calculated a priori as a function of conversion (not time) through an appropriate model.

The difference between runs 80 and 74 is that in the first one a little amount of $NaHCO_3$ has been added to the reacting mixture. As shown in Figure 1, representing polymer conversion as a function of time, this has a significant effect on the kinetics of the polymerization. However, since the presence of this additive is ignored in the model, it can be regarded as a source of irreproducibility for the process with respect to run 74. The aim in considering this experiment is in fact to test the robustness of the control procedure with respect to possible process irreproducibilities.

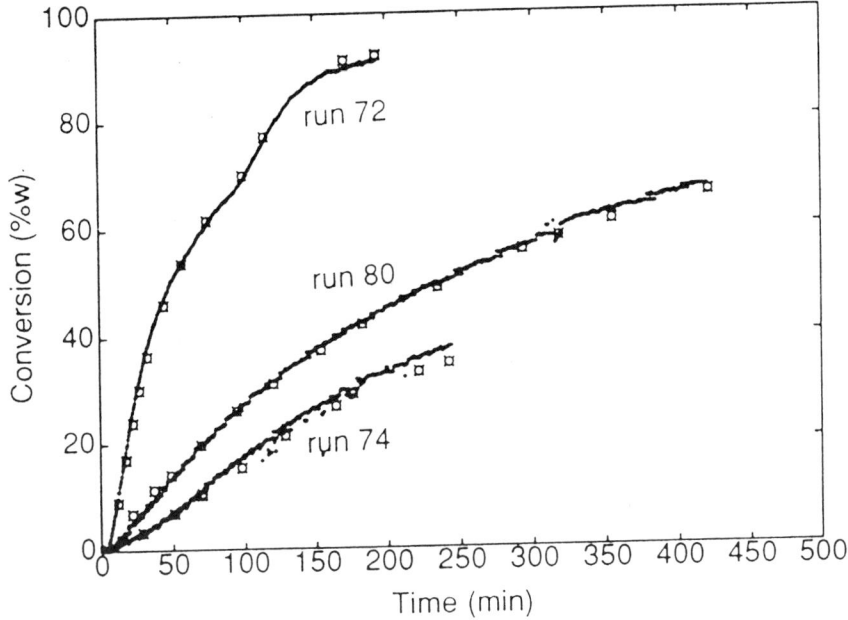

Figure 1. Conversion vs. time curves for the ternary system MMA-VAc-BuA.
• = densimetric data; o = gravimetric data; run 72 = batch; runs 74 and 80 = semibatch.

This is well elucidated in Figures 2a and 2b representing the amount of BuA added to the reactor as a function of time and of conversion, respectively. As expected the curves for runs 80 and 74 are rather different when time is used as the independent variable, while they become coincident when conversion is used instead: this proves that the second strategy has removed the effect of the process irreproducibilies. A similar behavior is obtained when considering the addition rate of MMA.

Finally, in Figure 3 the overall composition of the monomer mixture in the reactor as a function of conversion is shown. It appears that in both runs 80 and 74, as opposed to the batch run 72, the objective of producing a polymer with constant composition has been achieved.

358

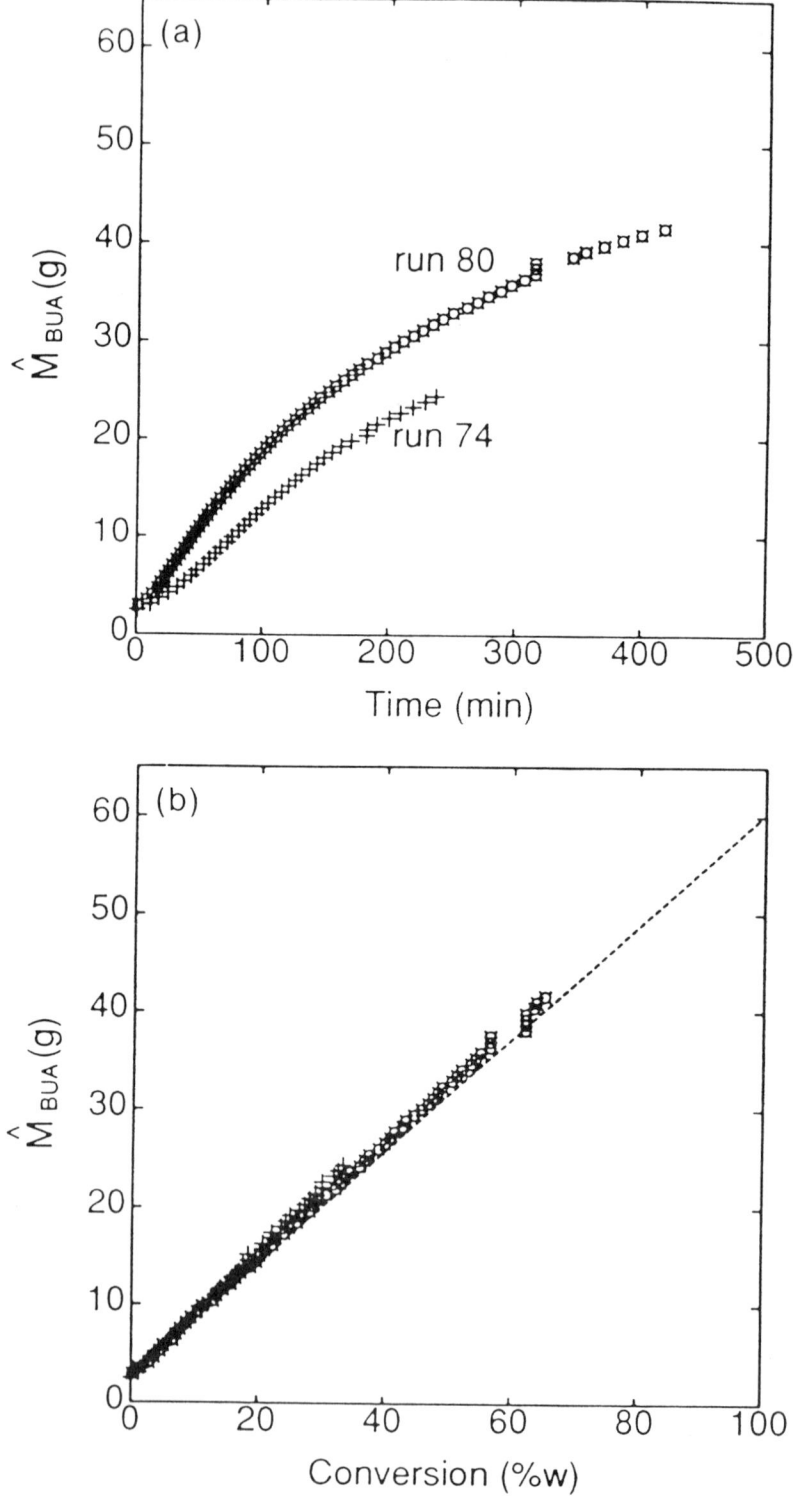

Figure 2. Amount of BuA charged to the reactor in runs 74 and 80 as a function of
time (a) and of conversion (b). +,o = experimental data; --- = calculated values

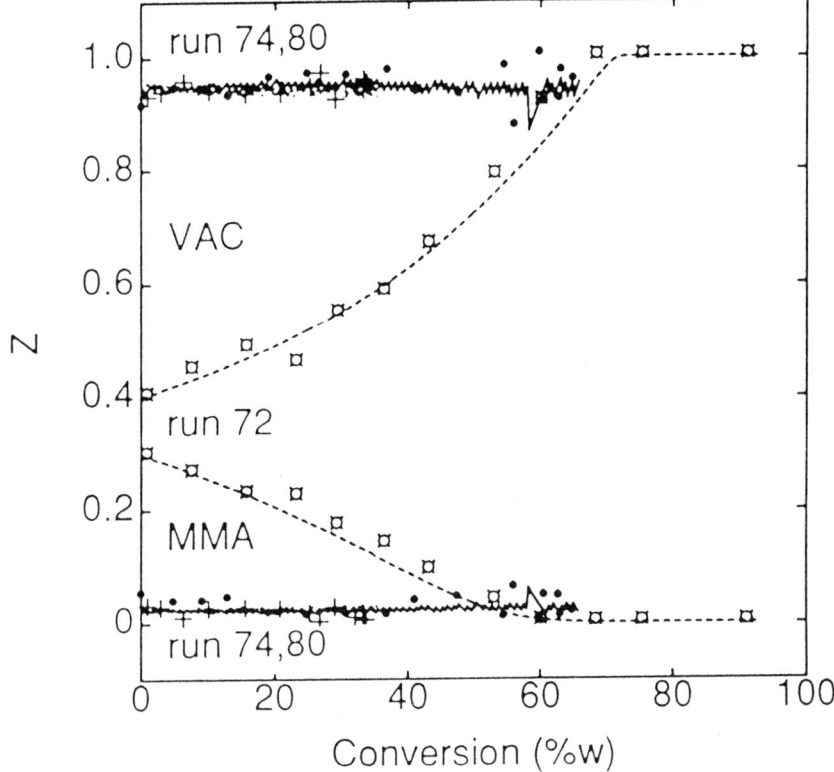

Figure 3. Mole fraction of the residual monomer mixture, Z_{MMA} and Z_{VAc}, as a function of conversion
Experimental data: o = run 72; + = run 74; • = run 80.
Calculated curves: --- = run 72; -.-.-. = run 74; —— = run 80.

5.2 POLYMER MOLECULAR WEIGHT CONTROL

We consider the homopolymerization of styrene, St, in the case where the MWD is entirely controlled by the presence of a chain transfer agent (CCl₄). Since termination is controlled by chain transfer to CCl₄, the parameter which determines the MWD is the parameter γ defined earlier, which in this case reduces to:

$$\gamma = \frac{k_{trs}[CCl_4]}{k_p[St]} \tag{19}$$

In order to obtain the narrowest cumulated MWD, we need to keep constant the instantaneous number-average molecular weight, which in this case reduces to maintain constant the parameter γ, i.e. the reactor temperature and the concentration ratio [CCl₄]/[St]. The latter objective is achieved following the same composition control

360

procedure discussed above for the monomer phase. In particular, in this case all the transfer agent is introduced in the reactor at the beginning of the process, while styrene is added during the reaction so as to keep constant the concentration ratio above. The final result is shown in Figure 4, where the GPC chromatograms obtained at various conversion values are compared for a batch run and for the semibatch run with composition control. It is apparent that, as opposed to the batch run, in the second case the number-average molecular weight remains constant during the polymerization, thus resulting in a narrower molecular weight distribution.

6. References

1. Mayo, F.R., and Lewis, F.M. (1944) *J. Am. Chem. Soc.*, **66**, 1594.
2. Trommsdorff, E., Kohle, E., and Lagally, P. (1948) *Makromol. Chem.*, **1**, 169.
3. Ham, G.E. (1964) in *Copolymerization*, G.E. Ham (ed.), *High Polymers* Vol. XVIII, J. Wiley, p. 42.
4. Hamielec, A.E. and MacGregor, J.F. (1983) in *Polymer Reaction Engineering*, K.H. Reichert, W. Geisheler (eds.), Hansen Pub., p. 21.
5. Broadhead, T.O., Hamielec, A.E., and MacGregor, J.F. (1985) *Makromol. Chem. Suppl.*, **10/11**, 105.
6. Tobita, H., and Hamielec, A.E. (1988) *Makromol. Chem., Macromol. Symp.*, **20/21**, 501
7. Stephanopoulos, G. (1984) in *Chemical Process Control*, Prentice-Hall, Inc. N.J.
8. Chujo, K. Harada, Y., Tokuhara, S. and Tanaka, A.R. (1969) *J. Polym. Sci.; Part C*, **27**, 321.
9. Snuparek, J., and Krska, F. (1977) *J. Applied Polym. Sci.*, **21**, 2253.
10. Misra, S.C., Pichot, C., El-Aasser, M.S., and Vanderhoff, J.W. (1983) *J. Polym. Sci. Polym. Chem. Ed.*, **21**, 2383.
11. Rios, L., and Guillot, J. (1982) *Makromol. Chem.*, **183**, 531.
12. Guillot, J., and Rios, L. (1982) *Makromol. Chem.*, **183**, 1979.
13. Dimitratos, J., Georgakis, C., El-Aasser, M.S., and Klein, A. (1989) in *Polymer Reaction Engineering*, K.H. Reichert, W. Geiseler (eds.) VCH, Berlin, p. 33.
14. Dimitratos, J., Georgakis, C., El-Aasser, M.S., and Klein, A. (1991) *Chem. Eng. Sci.*, **46**, 3203.
15. Arzamendi, G., and Asua, J.M. (1989) *J. Applied Polym. Sci.*, **38**, 2019.
16. Arzamendi, G., and Asua, J.M. (1990) *Makromol. Chem., Macromol. Symp.*, **35/36**, 249.
17. Hergeth, W.-D. (1997) in *Polymeric Dispersions. Principles and Applications.*, J.M. Asua (ed.), Kluwer Academic Publishers, Dordrecht.
18. Morbidelli, M., Storti, G., and Siani, A. (1997) in *Polymeric Dispersions. Principles and Applications.*, J.M.Asua (ed.), Kluwer Academic Publishers, Dordrecht.
19. Canu, P., Canegallo, S., Morbidelli, M. and Storti G. (1994) *J. Applied Polym. Sci.*, **54**, 1899.
20. Canegallo, S., Canu, P., Morbidelli, M. and Storti G. (1994) *J. Applied Polym. Sci.*, **54**, 1919.

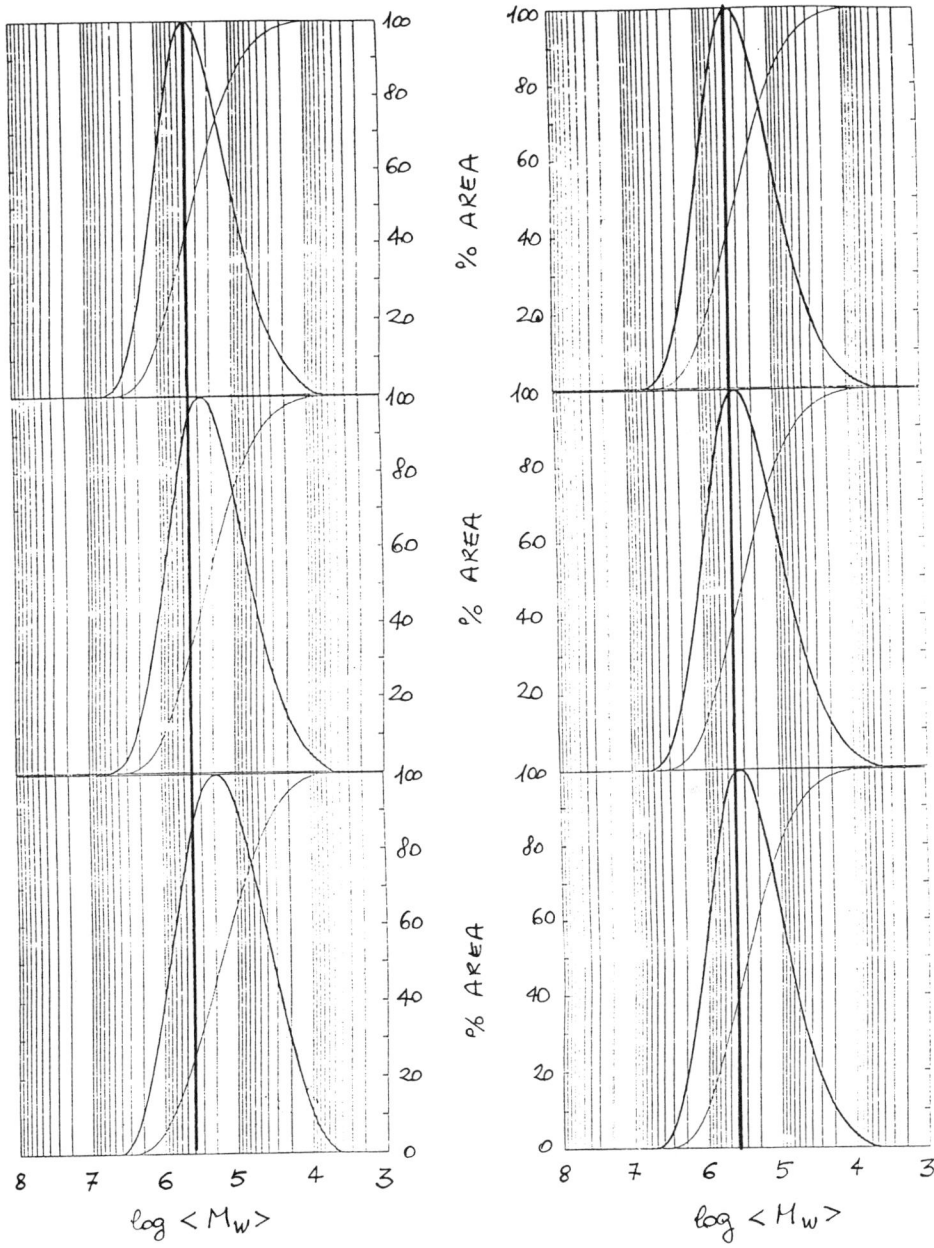

Figure 4. GPC traces at various conversion values for batch and semibatch emulsion polymerizations of styrene. From top: conversion = 22, 65 and 90% w. Left side: batch reaction; right side: semibatch reaction.

FEEDBACK CONTROL OF EMULSION POLYMERIZATION REACTORS

A Critical Review and Future Directions

J. R. LEIZA and J. M. ASUA
Grupo de Ingeniería Química, Dpto. Química Aplicada,
Facultad de Ciencias Químicas, Universidad del Pais Vasco.
Apdo. 1072, 20080 Donostia-San Sebastián, Spain

1. Introduction

In the production of dispersed polymers, the main objectives to be fulfilled are:

a) *Safety*. The reactor temperature must be kept under safe limits to avoid thermal runaways. In addition, violation of environmental regulations both in the plant environment and in the finished products must be avoided.

b) *Production rate*. The amount of product output required of a plant at any time is usually dictated by market specifications. Thus, this specifications must be met and maintained as much as possible in order to have a profitable process.

c) *Product quality*. The required quality is given by the end-use properties such as viscosity, film forming, tensile strenght, flexibility, elasticity, toughness, and opacity among others. Finished products not meeting the required specifications must be discarded as waste or whenever possible reprocessed at extra cost.

In order to implement the process conditions that lead to the required specifications of safety, production rate and product quality, it is necessary to develop suitable control strategies.

Different control configurations have been developed and implemented in chemical processes to achieve at least partially the above mentioned goals. Feedback control, feedforward control and open-loop control being the most usual configurations. In this work, we will focus on feedback control strategies for emulsion polymerization processes carried out in batch, semi-batch and CSTR reactors. A simple feedback control strategy is presented in Figure 1. The feedback control strategy requires the following steps:

a) To measure the value of the outputs to be controlled.

b) To compare the measured outputs with the desired values (set-points).

c) To decide which actions must be taken (values of the manipulated inputs) in order to track the set-points as close as possible.

When formulating the problem, it has been stated that the quality of a polymer latex is given by its end-use properties. However, the direct feedback control of these properties is usually impossible because such a properties are rarely on-line measurable. On the other hand, the current understanding of the emulsion polymerization does not allow us to establish the relationships between the operational variables of the reactor

J. M. Asua (ed.), Polymeric Dispersions: Principles and Applications, 363–378.
© 1997 *Kluwer Academic Publishers. Printed in the Netherlands.*

and the end-use properties, but only the relationships between the reactor operational variables and fundamental molecular and morphological characteristics of the polymer latex such as monomer conversion, copolymer composition distribution (CCD), molecular weight and molecular weight distribution (MWD), particle size and particle size distribution (PSD), degree of branching and crosslinking, and particle morphology.

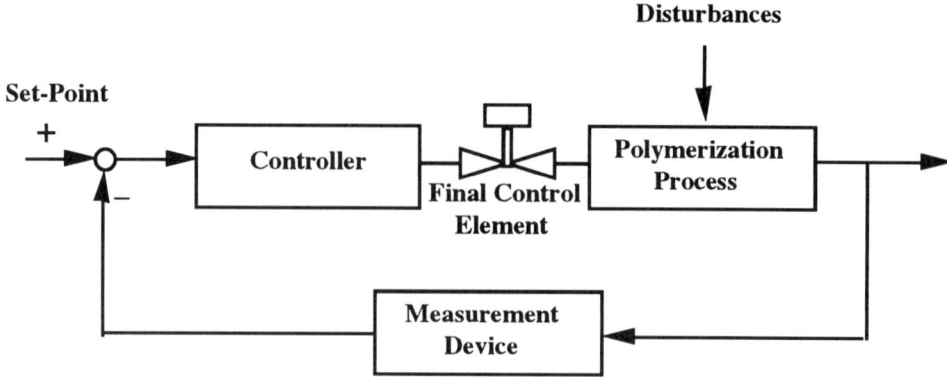

Figure 1. The feedback control configuration.

Therefore, the goal of the feedback control is reformulated to produce a latex with well defined molecular and morphological characteristics. Such characteristics, so-called controlled process outputs, are variables that directly affect the end-use properties of a polymer latex. Only some of the controlled process outputs are measurable and the approach is to use the ones which are measurable to estimate the others and apply feedback control strategies to these variables too.

A complete control objective would require to control all the controlled process outputs that affect the end-use properties of a polymer latex having as constraints safety and maximum production rate considerations. Alternatively, one can use maximum production considerations as the objective function and the controlled process outputs and safety as constraints.

A general mathematical formulation is

$$\underset{u}{\text{Max}} \quad J = f(p, t) \tag{1}$$

subject to

$$\frac{dx}{dt} = h(x, u, k, t) \tag{2}$$

$$y = h(x, u, t) \tag{3}$$

$$p = c(x, u, t) \tag{4}$$

$$u_{min} \le u \le u_{max} \tag{5}$$

$$x_{min} \le x \le x_{max} \tag{6}$$

$$y_{min} \le y \le y_{max} \tag{7}$$

$$p_{min} \leq p \leq p_{max} \qquad (8)$$
$$x(t_0) = x_0 \qquad (9)$$

where J is the objective function, x the state variables, u the manipulated variables, k the parameters, y the measured variables and p the controlled process outputs. Notice that some components of the vectors y and p may coincide, i.e., some of the molecular characteristics may be measured. In addition, the constraints on the manipulated state, measured and controlled process variables are explicitly included in the formulation. Alarms for stirring and cooling failures might also be included.

The solution of the multivariable control objective posed above is not straightforward and to our knowledge it has not been yet fully formulated and solved for emulsion polymerization systems. The main difficulties encountered in developing such general or even more simple feedback control structures can be summarized in the following points: i) there is a lack of robust, reliable and rapid on-line sensors to monitor either the end-use properties of the polymer latexes or the related controlled process variables, and ii) emulsion polymerization is a complex non-linear process not fully understood yet that can only be described by extremely non-linear models that involve a great number of usually unknown parameters.

Nevertheless, many attempts to achieve simpler control objectives have been carried out. In what follows, we present an overview of the works appeared in the literature where different feedback control strategies were used to control latex properties such as monomer conversion, polymer composition, molecular weight, particle size distribution and particle morphology by both computer simulation and real-time implementation. In addition, the potential application to the control of emulsion polymerization processes of techniques successfully applied to other polymerization processes such as solution and bulk polymerizations will be discussed. The first section will deal with state estimation techniques useful when on-line measurements of the controlled process variables are not available or the error associated with them is important. In the subsequent sections and separately, a critical literature review of feedback control strategies for maximum production rate (section 3), polymer composition (section 4), molecular weight distribution (MWD) branching and crosslinking (section 5), particle size distribution, PSD, (section 6), and particle morphology (section 7) will be presented. In these sections feedback control strategies proposed and implemented in real-time experiments will be differentiated from those implemented only by computer simulation. Finally, a summary of the control strategies discussed in the previous sections and recent developments in both modeling and estimation techniques (artificial neural networks, ANN) and advanced control strategies (model predictive control) will be covered.

2. State Estimation in Emulsion Polymerization Reactors

Feedback control requires the process control variables to be determined on-line. In addition, for the implementation of some control techniques like model based control the knowledge of the current values of state variables is necessary. If there are process controlled outputs and state variables that cannot be measured on-line or the errors associated with their measurements are important they have to be estimated on-line. For this purpose state estimation techniques are currently used, being the most common ones Kalman filtering and Luenberger observers although non-linear optimization has also been used [1-14]. Recently, artificial neural networks (ANN) have also been used as estimators for chemical processes (see section 8). In general, state estimators used a reasonably accurate mathematical model of the emulsion polymerization process to estimate the states that are not measured, based on the values of those (or a measurement which is a combination of more than one state) which are available. Figure 2 presents a schematic of the Kalman filter estimation technique.

The key point of the Kalman filtering is the calculation of the filter gain that requires the knowledge of the covariance matrices for both the modeling and measurement errors. These are user-specified variances which are laborious to determine accurately (quite a few replicate experiments are required to have the covariance of the measurement errors and extensive comparison between model predictions and experimental results is needed to calculate the covariance of the modeling errors). If adequate values of these matrices are available the Kalman filter performs properly, otherwise the state estimates will be determined either by the measurements or the model predictions. Examples of the application of Kalman filters for state estimation in emulsion polymerization systems can be found in refs. [1, 6-12].

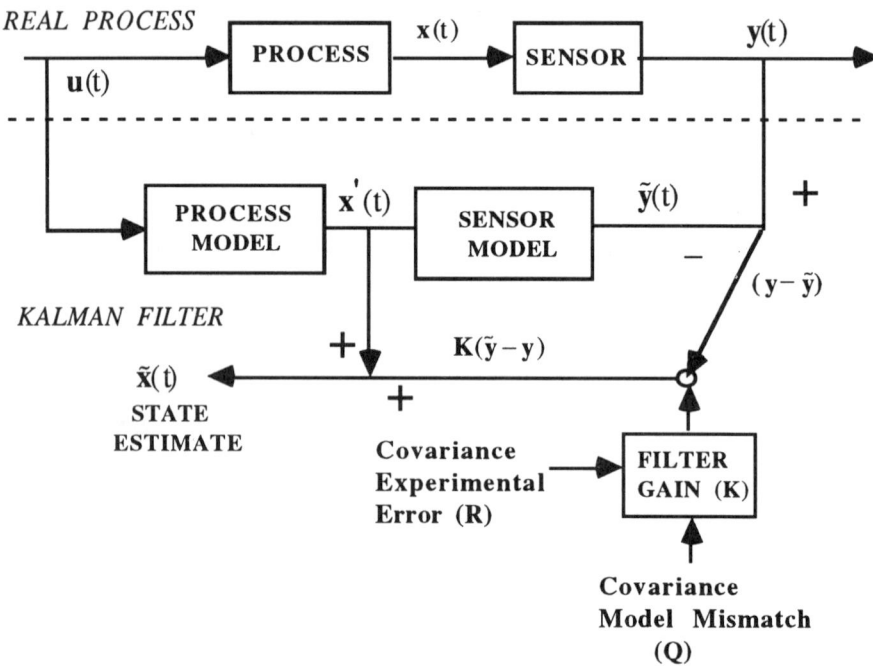

Figure 2. Schematic of the Kalman filter.

Non-linear optimization has also been used for state estimation by minimizing the following objective function:

$$J = \sum_{t=1}^{m} (y_{exp}(t) - y_{\text{mod}}(t))^{\text{T}} (y_{exp}(t) - y_{\text{mod}}(t)) \qquad (10)$$

where m is the number of experimental samples. Kozub and MacGregor [11] and Urretabizkaia *et al.* [13] applied this technique to semibatch emulsion polymerization processes. The main difference between the extended Kalman filter (EKF) and the non-linear optimization (NLO) is that the latter requires less tuning parameters than the EKF (the covariance matrices for the error in the process model and the measurements as well as for the initial error of estimation must be known or calculated). Nevertheless, the EKF requires less computational effort than NLO which can be an advantage estimating state variables of fast reacting emulsion polymerization processes. Kozub and MacGregor [11] compared both techniques by means of computer simulation and observed that the differences were not significant in the values of the estimated states.

An important issue in state estimation is that the observability criterion must be met in order for any state variable to be correctly estimated. This means that, based on the knowledge of the measured outputs and the inputs at time t, the entire state vector can be unambiguously determined (Ray [15]). Because of the limitations in on-line sensors [16], strict observability is difficult to achieve in emulsion polymerization. Thus, the molecular weight distribution and particle morphology are not observable using on-line measurements of monomer conversion and particle size. Similarly, copolymer composition is not strictly observable from calorimetric measurements. However, in practice, the two cases are completely different. Copolymer composition can be accurately estimated from calorimetric measurements because of the robust structure of the mathematical model describing the copolymerization and the availability of the parameters of this model (reactivity ratios and partition coefficients of the monomers). On the other hand, the estimation of the MWD from on-line measurements of monomer conversion and copolymer composition is much more uncertain because of both model mismatch and errors in the values of the model parameters.

3. Maximum Production Rate Under Safe Conditions

The maximum production rate can be achieved if the heat generation rate is held at the safe maximum heat removal capacity of the reactor. In semicontinuous isothermal processes the solids content of the reactor increases during the operation resulting in an increase of the latex viscosity that in turn causes a decrease of the overall heat transfer coefficient, U, and consequently, of the maximum heat removal capacity. If the effect of the solids content on U is known, the evolution of the maximum heat removal capacity can easily be calculated off-line and the control problem reduces to the tracking of a trajectory. Kozub and MacGregor [17] proposed the use of the monomer feed rate, initiator feed rate and in some cases reactor temperature as manipulated inputs to control the heat generation rate using non-linear inferential feedback strategies. Recently, Saenz de Buruaga *et al.* [18] showed in real-time experiments, that a simple proportional-integral control with dead-time compensation to account for the delay caused by mass transfer limitations was able to track the maximum heat removal capacity of a reactor using the monomer feed rate as the manipulated variable. They assessed the control strategy in the production of a VAc/VeoVa10 copolymer latex of 55 wt% solids content. Polymer quality was not taken into consideration in this study.

4. Polymer Composition

Copolymer composition has been by far the property of polymer latexes that has been more often controlled by using on-line feedback control strategies. To achieve this goal, the ratio of the comonomers in the polymerization loci should be controlled. Few of the strategies published in the literature were experimentally implemented and verified, and even less in industrial-like processes.

4.1. EXPERIMENTALLY IMPLEMENTED STRATEGIES

Guyot et al. [19] used a feedback control scheme to control the composition of a 20 wt% solids content styrene(St)/acrylonitrile(AN) latex. On-line gas chromatography (GC) was used to measure the unreacted amounts of monomer in the reactor and the feed rate of the more reactive monomer (St) necessary to keep constant the desired ratio of the monomers in the reactor was calculated. It was shown that although the desired monomer ratio was maintained constant into the reactor, the composition of the copolymer formed was richer in St than expected. Copolymer composition drift resulted from the different water solubilities of both monomers (acrylonitrile is much more water-soluble than styrene). Guyot et al. [20] and Rios and Guillot [21] using the same feedback strategy, produced homogeneous copolymers of butadiene-acrylonitrile and almost homogeneous terpolymers of acrylonitrile-styrene-methyl acrylate. A similar feedback approach but based in head space analysis was presented by Oliveres et al. [22] to minimize the compositional drift of an emulsion copolymerization of styrene/acrylonitrile.

After those pioneering works, advanced control strategies based on mathematical models of the process, appeared in the literature [1,13,17,23-28].

Dimitratos et al. [23] used a feedforward-feedback control to produce homogeneous VAc/BuA copolymers in latexes of low to medium solids contents. Leiza et al. [1] used a non-linear adaptive controller (NLA) combined with a classical proportional-integral controller (PI) to obtain homogeneous ethyl acrylate-methyl methacrylate latexes of 33 wt% solids content. The NLA controller was also used by Urretabizkaia et al. [13] and Asua et al. [26] for the polymerization of VAc/MMA/BuA of 55 wt% solids content and the copolymerization of VAc/BuA of 35 wt% solids content, respectively.

The control strategies used by Dimitratos et al. [23], Leiza et al. [1] and Urretabizkaia et al. [13] are summarized in Figure 3. The sample withdrawn from the reactor was analyzed by GC and a state estimation algorithm was used to calculate the values of the state variables. These values were updated to account for the time elapsed in sampling, analysis and state estimation. The updated states variables were used to calculate the control actions by combining a model based controller and a classical proportional-integral (PI) controller. The differences between the works are as follows:

i) Leiza et al. [1] and Urretabizkaia et al. [13] used a sampling device able to handle high solids content latexes (up to 55 wt%) whereas Dimitratos et al. [23] only were able to deal with low and medium ($\leq$ 30 wt%) solids contents, *ii*) Dimitratos *et al.* [23] and Leiza *et al.* [1] used an extended Kalman filter for state estimation whereas Urretabizkaia *et al.* [13] employed non-linear optimization,

iii) the model based controller used by Dimitratos *et al.* [13] was based on the solution of the following equation for the flow rate of the more reactive monomer

$$\frac{dp}{dt} = \frac{d\left(\frac{[A]_p}{[B]_p} \right)}{dt} = 0 \tag{11}$$

where $[i]_p$ is the concentration of monomer i in the polymer particles. On the other hand, Leiza *et al.* [1] and Urretabizkaia *et al.* [13] used a NLA controller based on the material balances of the monomers. The latter method allows to compensate for the errors in the composition of the initial charge as it is shown in Figure 4 that presents the evolution of the cumulative copolymer composition obtained in a simulated run in which the initial amount of the more reactive monomer was smaller than that required to produce a copolymer of the desired composition and only the model based controller was used. Figure 4 also shows that copolymer composition diverged from the desired one when the model based controller developed by Dimitratos *et al.* [23] was used,

$$\beta_i = \frac{[i]_p}{[A]_p}$$

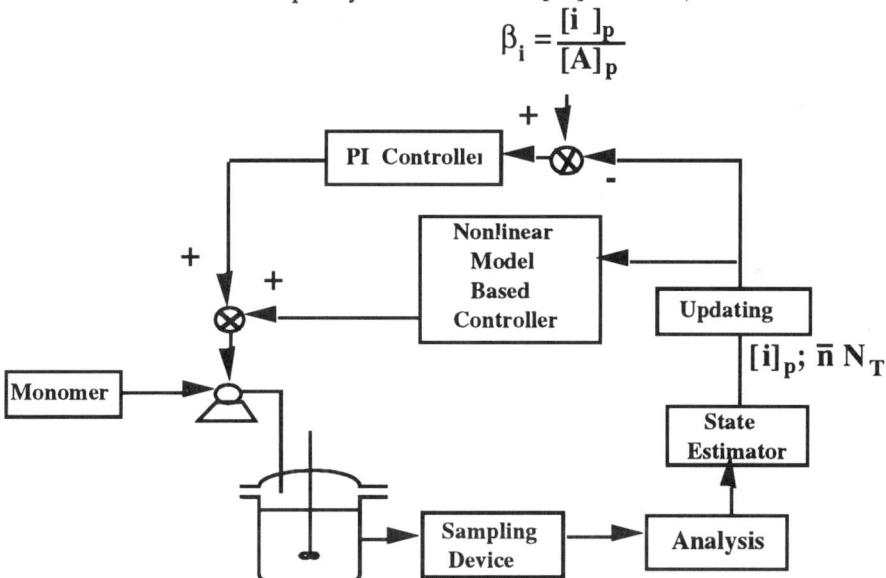

Figure 3. Schematic of the control strategies proposed by Leiza *et al.* [1], Dimitratos *et al.* [23] and Urretabizkaia *et al.* [13].

iv) Urretabizkaia *et al.* [13] used only the NLA controller because it was found by simulation that the PI controller did not provide any additional advantage. Figure 5 shows the evolution of the terpolymer composition produced for an experiment controlled by the NLA controller aiming at producing a 15/35/55 VAc/MMA/BuA terpolymer of 55 wt% solids. On the other hand, Asua *et al.* [26] used on-line reactor calorimetry to estimate the copolymer composition instead of on-line GC. This avoided the dead time associated with the GC measurements and allowed to control fast processes since the heat generated by polymerization can be obtained every 2 seconds.

A different approach to control the copolymer composition of emulsion polymerization processes have been proposed by Canu *et al.* [27] and Saenz de Buruaga *et al.* [18,28]. These approaches cannot be seen as a classical feedback control strategy, as depicted in Figure 1, since the measurement of the copolymer composition was not directly used in the feedback controller, but optimal monomer addition profiles were obtained as a function of conversion by solving off-line an optimization problem where one of the constraints was the homogeneous copolymer composition requirement. The profiles obtained a priori were implemented based on the on-line measurements of conversion obtained during the experiment. Canu *et al.* [27] used densitometry and sound velocity measurements to obtain an on-line estimation of conversion, and Saenz de Buruaga *et al.* [18, 28] used on-line calorimetry. From the proposed techniques, calorimetry seems to be the more robust and easiest to apply in an industrial environment.

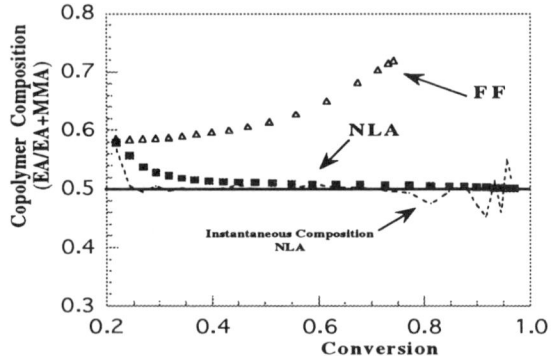

Figure 4. Simulation of the evolution of the cumulative copolymer composition during an emulsion polymerization of EA/MMA with an excess of EA in the initial charge.

The proposed control strategy was successfully applied by Canu *et al.* [27] to the emulsion copolymerizations of MMA/VAc, BuA/St and the terpolymerization of MMA/VAc/BuA. Saenz de Buruaga *et al.* [18,28] showed the performance of the strategy by producing BuA/VAc copolymers of both homogeneous composition and predefined composition profiles. Saenz de Buruaga *et al.* [18] also showed (Figure 6) that the control strategy devised lends itself to safe operation, since sudden inhibition (deliberately induced by the addition of an inhibitor during the exothermic phase of the polymerization) did not lead to monomer accumulation or thermal runaway once polymerization recommenced. Furthermore, the quality of the polymer was maintained at the target value.

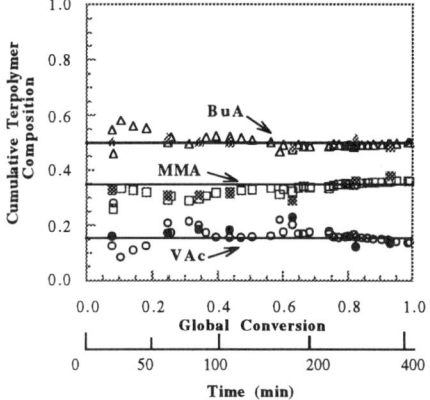 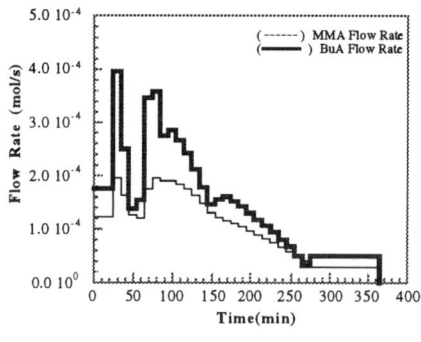

Figure 5. Evolution of cumulative terpolymer composition during the 55 wt% solids emulsion polymerization of VAc/MMA/BuA .

4.2. SIMULATION BASED STRATEGIES

Kozub and MacGregor [17] proposed a non-linear inferential feedback control strategy to control the copolymer composition, among others properties, in the emulsion polymerization of styrene/butadiene by means of computer simulation.

Their approach is similar to that of Dimitratos *et al.* [23] but modifying eq. (11) in order to develop a feedback configuration as follows

$$\frac{d\tilde{p}_i}{dt} = \frac{1}{\tau_i}(p_{i,set} - \tilde{p}_i) \tag{12}$$

$$p = \frac{[St]_p}{[St]_p + [Bu]_p} \quad \text{or} \quad p = \frac{R_{p,St}}{R_{p,St} + R_{p,Bu}} \tag{13}$$

where $\tilde{p}$ is the estimated controlled process output i, i.e. the copolymer composition expressed as in eq. (13), τ_i is the process control-loop response time constant, and, R_{pi} the polymerization rate of monomer i in the polymer particles.

Note that if eq. (12) is set equal to zero, the solution is completely equivalent to the feedforward part developed by Dimitratos *et al.* [23]. In agreement with the remarks made by Leiza *et al.* [1] on the feedforward controller developed by Dimitratos *et al.* [23], Kozub and MacGregor [17] pointed out that implementation of eq. (12) with the right hand side equal to zero (open-loop policy without feedback correction) for the case of errors in the initial charge leads to a very poor quality of the copolymer composition.

Vega [24] applied the input/output linearization technique (GLC) developed by Kravaris and Chung [25] to calculate the flow rate profile of the more reactive monomer

372

required to produce homogeneous composition copolymers in the simulated emulsion polymerization of acrylonitrile/butadiene.

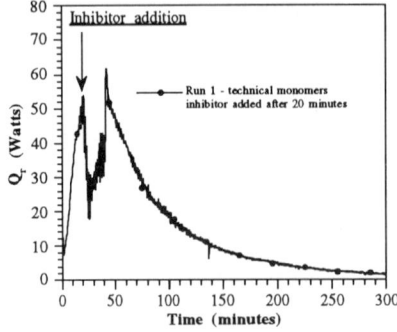 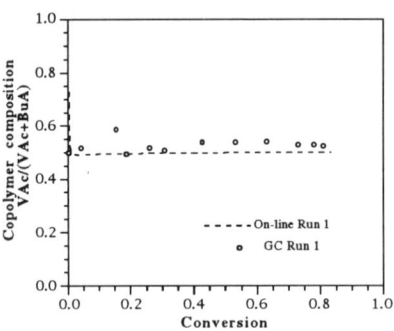

Figure 6. Evolution of the heat released by polymerization and the cumulative copolymer composition obtained in a controlled VAc/BuA emulsion polymerization.

It can be shown that the GLC controller reduces to

$$\frac{d\tilde{p}_i}{dt} = k_i\left[(\tilde{p}_i - p_{i,set}) + \frac{1}{\tau_{I,i}}\int (\tilde{p}_i - p_{i,set})dt \right] \qquad (14)$$

k_i and $\tau_{I,i}$ being the parameters of the PI controller for the propertie i, and which only differs from eq. (12) in the addition of an integral term in the feedback part of the controller. Kozub and MacGregor [17] suggested that if an state estimator is used to predict the states and then the controlled outputs, p, the addition of the integral part is not necessary.

5. Control of Molecular Weight Distribution (MWD) and Degree of Branching

Molecular weight distribution affects important end-use properties of the film such as elasticity, strength, toughness, and solvent resistance. Due to the difficulty of developing robust on-line sensors to measure the molecular weight and molecular weight distribution, few works appeared in the literature dealing with feedback control strategies to control the MWD.

On-line determination of MWD or chain length distribution (CLD) would be possible through the use of automated gel permeation chromatography (GPC), but to our knowledge no such application has been reported for an emulsion polymerization process. For solution polymerizations, experimental setups capable of performing this task reported by Ponnuswamy *et al.* [29], Budde and Reichert [30], and Ellis *et al.* [31].

Another possibility attempted by many authors has been the estimation of molecular weight from other available on-line measurements. Relative success was obtained in solution polymerization systems by Jo and Bankoff [2], Schuler and Suzhen [3], Schuler and Papadoulou [4], Papadopoulou [5], Ellis *et al.* [31] and Abedekun and Schorck [32]. The main problem encountered in the estimation being the non-observability of the molecular weight distribution from the available on-line measurements. Some of the proposed estimation schemes can be promising for emulsion processes.

5.1. EXPERIMENTALLY IMPLEMENTED STRATEGIES

An experimental application for an emulsion polymerization process was recently proposed by Canu *et al.* [27]. Their approach can be seen as an extension of the one presented for copolymer composition control. The problem of achieving a copolymer having a constant MWD was reduced to maintaining the instantaneous ratio of the monomer to the chain transfer agent constant over the polymerization time. Implicit is the assumption that the main mechanism prevailing in the termination reactions is transfer to chain transfer agent, CTA. Then, the control strategy reduces to calculate off-line the flow rates of monomer and chain transfer agent as a function of the conversion and its on-line implementation. Experimental implementation of such strategy was shown for the case of producing over the entire extension of reaction a homopolymer of styrene with an average number molecular weight $<M_n>$ of 1×10^5 g/mol using CCl_4 as chain stopper. They showed that their strategy produced a homopolymer of constant $<M_n>$ of $1,3 \times10^5$ g/mol over the entire extension of the reaction whilst in a batch process the molecular weight decreased considerably during the polymerization. Due to the low chain transfer rate constant to CTA, rather high concentration of CCl_4 was used (6 wt% on monomer). More efficient CTA such as n-dodecyl mercaptan may be used, but in this case mass transfer limitations might cause a time-delay. It has to be mentioned that this strategy cannot be considered as a true feedback control strategy but as an on-line conversion based trajectory tracking open-loop policy.

5.2. SIMULATION BASED STRATEGIES

There are also a limited number of simulation studies devoted to control the polymer molecular weight in emulsion polymerization. In most of the works, CTAs have been used as the manipulated variable to control the molecular weight of the polymer. Hamielec *et al.* [33] and Kanetakis *et al.* [34] have shown by computer simulation of a styrene-butadiene rubber latex train how to control the MWD and branching frequency by manipulating the CTA and splitting the monomers feeds between the first and subsequent reactors. More recently, Vega *et al.* [35] presented an approach to reduce off-spec product usually generated in the transient between steady states of an styrene/butadiene emulsion polymerization. They showed that their approach is superior to the "bang-bang" operations normally implemented in industry. In addition, Vega *et al.* [36] proposed also an optimized steady state operation in which monomers and CTA were fed along the reactor train and stated that $<M_n>$ can be fixed or forced to follow any prespecified profile.

Kozub and MacGregor [17] proposed a non-linear inferential feedback strategy to control the molecular weight and the degree of branching of an emulsion polymerization carried out in a semibatch reactor. In order to control the molecular weight they stated that to produce a desired cumulative property (number or weight average molecular weight and polydispersity) the corresponding instantaneous property should be kept constant throughout the process. For instance to control the number average molecular weight, $<M_n>$, the controlled variable, p, to be included in eq. (12) can be the following one

$$p = <M_n> = (F_{St}M_{ost} + F_{Bu}M_{oBu})\frac{d\nu_1/dt}{d\nu_0/dt} \tag{15}$$

where F_{St} and F_{Bu} are the instantaneous composition of the copolymer, ν_0 and ν_1 the first moments of the copolymer chain distribution, and M_{oi} the molecular weight of monomer i. The flow rate of the CTA was used as the manipulated variable. In order to control the degree of branching the controlled process variable is the instantaneous property defined as follows

$$p = Branching = \frac{d(\nu_0 B_{n3})/dt}{d\nu_1/dt} \tag{16}$$

being B_{n3} the frequency of tri-branching. A similar approach can also be applied for the degree of crosslinking. For the case of branching and crosslinking the more efficient manipulated variable was found to be the flow rate of monomer. The performance of the proposed non-linear inferential feedback strategy was shown in the simulated emulsion polymerization of styrene/butadiene. A multivariable objective having as goals the control of overall conversion, instantaneous copolymer composition and instantaneous weight average molecular weight was assessed. It was found that the inferential feedback controller provides a significant improvement relative to the open-loop policy (obtained solving eq. (12) with the right hand side equal to zero) when errors in the initial conditions were introduced.

6. Control of the Particle Size Distribution

The PSD of a polymer latex may be critical if the final product is going to be the latex itself, e.g. for paints and coatings. In other cases, the PSD will remain as an intermediate variable of less importance. An excellent review of the techniques available for on-line measurement of particle size and PSD is presented by Hergeth in this book [16]. One of the difficulties for the control of PSD using feedback strategies is related to the fast dynamics of the nucleation period itself. The fastest on-line measurement available right now will take no less than 5 to 10 minutes to be completed (due to the dilution step required) and the nucleation of a secondary population of particles by micellar mechanism would no take more than 5 minutes. Therefore, there is no time to influence the emerging population by manipulating emulsifier and initiator flow rate at least in a feedback fashion. Another difficulty arises from the limited knowledge of the mechanisms determining the PSD (particle nucleation and coagulation) available.

An attempt to control the PSD using a feedback strategy was presented by Gordon and Weidner [37] for the emulsion polymerization of vinyl chloride. Based on an energy balance of the reactor jacket (heat balance calorimetry) the conversion was inferred and its evolution (mainly slope changes) used as an indication of generation of new particle populations. This on-line information was used to modify the flow rate of emulsifier fed into the reactor. The type of controller used to manipulate the flow rate of emulsifier was not reported.

7. Control of Particle Morphology

Particle morphology can affect polymer end-use properties considerably, and hence controlling particle morphology is an important issue. However due to the difficulty of characterizing particle morphology, even by means of off-line techniques, the development of control strategies is still a challenging research field. Accordingly, most of the works are implementation of predefined strategies to achieve the desired polymers. Nevertheless, in a recent work, Echeverria [38] showed how to prepare core-shell particles by using an on-line feedback control strategy based on the NLA controller. They prepared particles having a homopolymer core of MMA or BuA and an interphase composed of a gradient composition from 100/0 to 0/100 as a function of the growing particle radius. Finally, a shell of the counterpart homopolymer (MMA or BuA) was incorporated.

8. Conclusions and Future Challenges

In the foregoing, the feedback control strategies aiming at controlling latex properties such as polymer composition, molecular weight distribution, particle size distribution and particle morphology are reviewed. These are only partial solutions to the global control problem posed in eqs (1) - (8). An overall solution of the global control problem has not been yet implemented. The main difficulties arise from both the lack of robust and fast enough sensors for on-line monitoring of the controlled process variables, the limited knowledge of the mechanisms involved in emulsion polymerization, and the complex models required to fit the highly non-linear behavior of emulsion polymerization. Recent developments in both on-line monitoring and kinetics and mechanisms in polymerization in dispersed media are reviewed in length in other sections of this book [16,39-43] and are not further discussed here, although the feedback control will benefit from any improvement in these fields. On the other hand, the complexity of the mathematical models based on first principles is inherent to the non-linear nature of emulsion polymerization. This complexity is a burden for the application of advanced non-linear controllers [44]. Recently, artificial neural networks (ANN) have successfully been used to model complex chemical process in which the fundamental mechanisms are poorly known [45]. ANNs are extremely flexible non-parametric models able to represent highly non-linear processes under both steady state (feedforward ANNs, [45]) and dynamic (recurrent ANNs[46]) conditions. One of the disadvantages of the ANNs is that extensive data is needed to estimate the values of the parameters of the ANN (to train the ANN). In addition, this data must be representative of the entire operation region because although ANNs can interpolate better than a first principles model, their extrapolation ability is very limited. ANNs can also be used in combination with first principles models [47,48]. In these arrangements the first principles model accounts for the main trends of the system and the ANNs takes care of the fine tuning. The advantages of these combined systems are that prior knowledge is incorporated into the model and that the training of the ANN usually requires less data. The disadvantage for on-line control is that the structure of the model remains complex and therefore rather demanding from the point of view of on-line computer time. ANNs have been used for both state estimation [46,49,50] and process control [45,51,52]. When ANNs are directly used as state estimators, in contrast to traditional methods, this technique does not utilize the measurement to estimate the state variables but uses the outputs of the model as state estimators [46]. ANNs have also been used for model based process control such as model predictive control [52]. An application to emulsion polymerization has been reported recently [51]. Irrespective of the mathematical model of the process, model predictive control seems to be the more promising control strategy to solve the global control problem posed by eqs (1)-(9).

9. References

1. Leiza, J.R., de la Cal, J.C., Meira, G.R., and Asua, J.M. (1992) On-line copolymer compostion control in the semicontinuous emulsion copolymerization of ethyl acrylate and methyl methacrylate, *Polymer Reaction Engineering* 1, 461-498.
2. Jo, J.M. and Bankoff, S.Q. (1976) Digital monitoring and estimation of polymerization reactors, *AIChE J.* 22, 361-369.
3. Schuler, H. and Suzhen, S. (1986) Real time estimation of the chain length distribution in a polymerization reactor, *Chem. Eng. Sci.* 40, 1891-1903.
4. Schuler, H. and Papadopoulou, S. (1986) Real time estimation of the chain length distribution in a polymerization reactor. II Comparison of estimated and measured distribution functions, *Chem. Eng. Sci.* 41, 2681-2683.
5. Papadopoulou, S. (1987) Gel permeation chromatography as a tool for the real-time estimation of the chain length distribution in a polymerization reactor, in Meeting of International Polymer Processing Society, Stuttgart.
6. Ahlberg, D.T. and Cheyne, J. (1979) Adaptive control of a polymerization reactor, *AIChE J. Symp. Ser., Chemical Process Control*, pp. 221-229.
7. Kiparissides, C., MacGregor, J.F., and Hamielec, A.E. (1981) Suboptimal stochastic control of a continuous latex reactor, *AIChE J.* 27, 13-20.
8. MacGregor, J.F. (1986) On-line reactor energy balances via Kalman filtering, IFAC Instrumentation and Automation in the Paper, Rubber, Plastics and Polymerization Industries, Ohio, pp. 35-39.
9. Dimitratos, J, Georgakis, C., El-Aasser, M.S., and Klein A. (1989) Dynamic modeling and state estimation for an emulsion copolymerization reactor, *Comp. Chem. Eng.* 13, 21-33.
10. Dimitratos, J., Georgakis, C., El-Aasser, M.S., and Klein A. (1991) An experimental study of adaptive Kalman filtering in emulsion copolymerization, *Comp. Chem. Eng.* 46, 3203-3218.
11. Kozub, D.J. and MacGregor, J.F. (1992) State estimation for semibatch emulsion polymerization reactors, *Chem. Eng. Sci.* 47, 1047-1062.
12. Eliçabe, G.E., Ozdeger, E.,Georgakis, C., and Cordeiro, C. (1995) On-line estimation of reaction rates in semicontinuous reactors, *Ind. Eng. Chem. Res.* 34, 1219-1227.

377

13. Urretabizkaia, A., Leiza, J.R., and Asua, J.M. (1994) On-line terpolymer composition control in semicontinuous emulsion polymerization, *AIChE J.* **40**, 1850-1864.
14. Appelhaus, P., and Engell, S. (1996) Design and implementation of an extended observer for the polymerization of polyethyleneterepthalate, *Chem. Eng. Sci.* **51**, 1919-1926.
15. Ray W.H., (1981) Advanced Process Control, Mac Graw Hill, New York.
16. Hergeth, W.D., (1996) On-line characterization methods I and II*in J.M. Asua (ed) Polymer Dispersions. Principles and Applications.* Kluwer Academic Publishers, Dordrectch.
17. Kozub, D.J. and MacGregor, J.F. (1992) Feedback control of polymer quality in semibatch reactors, *Chem. Eng. Sci.* **47**(4), 929-942.
18. Saenz de Buruaga, I., Echevarría, A., Armitage, P.D., de la Cal, J.C., Leiza, J.R., and Asua, J.M. (1997) On-line control of semibatch emulsion polymerization reactors based on calorimetry, *AIChE J.* (submitted).
19. Guyot, A., Guillot, J., Pichot, C., and Rios Guerrero, L. (1981) New design for producing constant-composition copolymers in emulsion polymerization, in , D.R. Basset and A.E. Hamielec (eds), *Emulsion Polymers and Emulsion Polymerization.*, ACS Symp. Series, pp 415-436.
20. Guyot, A., Guillot, J., Graillat, C., and Llauro, M.F. (1984) Controlled composition in emulsion copolymerization: Application to butadiene acrylonitrile copolymers, *J. Macromol. Sci.-Chem.* **A-21(687)**, 683-699.
21. Rios, L., and Guillot, L. (1989) Polymerisation acrylonitrile-styrene-acrylate de methyle, 2. Synthese de terpolymeres homogenes en composition par polymerisation en emulsion, *Macromol. Chem.* **183**, 531-548.
22. Oliveres, M., Recasens, F., and Puigjaner, L. (1988) Computer-aided operation of a fedbatch copolymer reactor: A control strategy based on an identified kinetic model and on-line vapor-phase observation, ISCRE 10, Basle.
23. Dimitratos, J., Georgakis, C. El-Aasser, M.S., and Klein, A. (1989) Control of product composition in emulsion copolymerization, in K.H. Reichert and W. Geiseler eds, *International Workshop on Polymer Reaction Engineering*, pp. 33-42.
24. Vega, J.R. (1993) Increase of the product quality in anionic homopolymerization and emulsion copolymerizations, PhD Thesis, Universidad Nacional del Litoral, Santa Fe, Argentina.
25. Kravaris, C. and Chung, C.B. (1987) Non-linear state feedback synthesis by global input/output linearization , *AIChE J.* **33**, 592-603.
26. Asua, J.M., Saenz de Buruaga, I., Arotçarena, M., Urretabizkaia, A., Armitage, P.D., Gugliotta, L.M., and Leiza, J.R. (1995) On-line control of emulsion polymerization reactors, in K.H. Reichert and H.U. Moritz (eds), *International Workshop on Polymer Reaction Engineering*, pp. 655-671.
27. Canu, P., Canegallo, S., Morbidelli, M., and Storti, G. (1994) Polymer quality control in latex semibatch reactors through on-line monitoring of conversion, *J.Appl. Polym. Sci.*, **54**, 1988.
28. Saenz de Buruaga, I., Arotçarena, M., Armitage, P.D., Gugliotta, L.M., Leiza, J., and Asua, J.M. (1995) On-line calorimetric control of emulsion polymerization reactors, *Chem. Eng. Sci.* **51**, 2781-2786.
29. Ponnusvamy, S., Shah, S.L., and Kiparissides, C. (1986) On-line monitoring of polymer quality in a batch polymerization reactor, *J. Appl. Polym. Sci.* **32**, 3239-3253.
30. Budde, U., and Reichert, K.M. (1988) Automatic polymerization reactor with on-line data measurement and reactor control, *Die Angewandte Makromolekulare Chemie* **161**, 195-204.
31. Ellis, M.I., Taylor, T.W., Gonzalez, V., and Jensen, K.F. (1988) Estimation of the molecular weight distribution in batch polymerization , *AIChE J.* **34**, 1341-1353.
32. Adebekun, D.K., and Schork, F.J. (1989) Continuous solution polymerization reactor control. 2. Estimation and nonlinear reference control during methyl methacrylate polymerization, *Ind. Eng. Chem. Res.* 28, 1846-1861.
33. Hamielec, A.E., MacGregor, J.F., Broadhead, T.O., Kanetakis, J., and Wong, F.Y.C. (1983) Dynamic steady state modeling of a latex reactor train-manufacture of cold SBR, ACS Rubber Technology Div. Meeting, Toronto.
34. Kanetakis, J., Wong, F.Y.C., Hamielec, A.E., and MacGregor, J.F. (1985) Steady state modeling of a latex reactor train for the production of styrene-butadiene rubber, *Chem. Eng. Sci. Comm.* **35**, 123.
35. Vega, J.R., Gugliotta, L.M., and Meira, G.R. (1995) Continuous emulsion polymerization of styrene and butadiene. Reduction of off-spec product between steady-states, *Latin American Applied Research* **25**, 77-82.
36. Vega, J.R., Gugliotta, L.M., Brandolini, M.C. and Meira, G.R. (1995) Steady-state optimization in a continuous emulsion copolymerization of styrene and butadiene, *Latin American Applied Research* **25**, 207-214.
37. Gordon, D.L., and Weidner, K.L. (1981) Control of particle size distribution through emulsifier metering based on rate of conversion, in D.R. Basset and A.E. Hamielec (eds.), *Emulsion Polymers and Emulsion Polymerization*, ACS Symp. Series n. 165, pp 515-532.
38. Echevarria, A. (1996) Ph.D. Dissertation, University of the Basque Country.
39. Gilbert, R.G. (1996) Mechanisms for radical entry and exit. Aqueous-phase influences on polymerization in *J.M. Asua (ed) Polymer Dispersions. Principles and Applications.* Kluwer Academic Publishers, Dordrectch.

378

40. van Herk, A.M., (1996) Particle growth in emulsion polymerization in *J.M. Asua (ed) Polymer Dispersions. Principles and Applications*. Kluwer Academic Publishers, Dordrectch.
41. Tauer, K., (1996) Particle nucleation in *J.M. Asua (ed) Polymer Dispersions. Principles and Applications*. Kluwer Academic Publishers, Dordrectch.
42. Gilbert, R.G. (1996) Particle size distributions in *J.M. Asua (ed) Polymer Dispersions. Principles and Applications*. Kluwer Academic Publishers, Dordrectch.
43. Charmot, D. (1996) Network formation in free radical emulsion polymerization in *J.M. Asua (ed) Polymer Dispersions. Principles and Applications*. Kluwer Academic Publishers, Dordrectch.
44. Bequette, B.N. (1991) Nonlinear control of chemical processes: A review, *Ind. Eng. Chem. Res.* **30**, 1391-1413.
45. Bhat, N., and McAvoy, T.J. (1990) Use of neural networks for dynamic modeling and control of chemical process systems, *Comp. Chem. Eng.* **14**, 573-583.
46. Karjala, T.W. and Himmelblau, D.M. (1994) Dynamic data rectification by recurrent neural networks vs traditional methods, *AIChE J.* **40**, 1865-1875.
47. Psichogios D.C. and Ungar, L.M. (1992) A hybrid neural network. First principles approach to process modeling, *AIChE J.* **38**, 1499-1511.
48. Thompson, M.L. and Kramer, M.A. (1994) Modeling chemical processes using prior knowledge and neural networks, *AIChE J.* **40**, 1328-1340.
49. Xu, L., Jiang, J.P., and Zhu, J. (1994) Supervised learning control of a non-linear polymerization reactor using CMAC neural network for knowledge storage, *IEE Process Control Theory* **141(1)**, 33-38.
50. Karjala, W.T. and Himmelblau, D.M. (1996). Dynamic Rectification of Data via Recurrent Neural Nets and the Extended Walman Filter, AIChE. J., **42** (8), 2225-2239.
51. Tsen, A.Y-D., Jang, S.S., Wong, D.S. and Joseph, B. (1996) Predictive control quality in batch polymerization using hybrid ANN models, *AIChE J.* **42**, 455-465.
52. Psichogios D.C. and Ungar, L.M. (1991) Direct and indirect model based control using artificial neural networks, *Ind. Eng. Chem. Res.* **30**, 2564-2573.

OVERVIEW OF USES OF POLYMER LATEXES

A. J. DeFUSCO, K. C. SEHGAL and D. R. BASSETT
Union Carbide Corporation
UCAR Emulsion Systems
Cary, North Carolina, USA

1. Introduction

Latex polymers derived from emulsion polymerization are used in a very large number of applications. The scope of this chapter is to review the more common uses and to emphasize the utility of emulsion polymerization to produce an extremely wide variety of polymers. The most visible use of latexes is in architectural paints, both interior and exterior, where a slow and steady replacement of alkyds and solvent-based paints has been under way for several decades. The differences in performance standards between interior and exterior uses have dictated the appropriate compositions, with vinyl-based polymers dominating the interior markets and acrylics generally prevalent in exteriors. Advances in adhesive technology, and the increasing use of branched vinyl esters, have produced cost-effective alternatives to acrylics, however. In non-architectural polymer applications, a major shift to latexes has occurred in the area of adhesives, caulks and sealants where the elastomeric properties obtainable with available monomers have been exploited. In the regard, the pressure polymers, using ethylene as a plasticizing co-monomer, have gained a sizable market share. Two remaining areas consuming large volumes of latexes include textile coatings and paper coatings. In textiles, acrylic polymers, modified for adhesion, flexibility and solvent resistance, account for most of the volume. In paper coatings styrene-butadiene copolymers have long dominated the market due to their low cost advantage. Finally, a fast growing area of latex consumption is industrial coatings where thermosetting latexes with excellent performance properties have been developed. Under the category of specialty emulsion polymers are included cement additives, rheology modifiers, flocculants, foams, and biomedical latexes. In each case, emulsion polymerization has proven to be an extremely versatile process for the production of polymers designed for specific application characteristics.

Several of the latex applications described above, including paper coatings, pressure sensitive adhesives and biomedical latexes, are extensively discussed elsewhere in the book. Consequently, the application areas covered in this chapter will

J. M. Asua (ed.), Polymeric Dispersions: Principles and Applications, 379–396.
© 1997 *Kluwer Academic Publishers. Printed in the Netherlands.*

be limited to architectural coatings, coatings for textiles, water-borne adhesives, caulks and sealants, and associative rheology modifiers.

2. Latexes In Architectural Coatings

2.1. INTRODUCTION

The scope of architectural coatings includes a variety of products which are geared to both professional contractors and individual homeowners. Most of the products that a consumer would find in a paint store or the paint section of a home center include products that can be classified as architectural coatings, including primers, enamels, sealers, and stains. The most common types of architectural coatings are latex house paints.

A latex paint can be defined as a dispersion of pigment in a latex polymer vehicle designed for application to a substrate in a thin film[1,2]. The primary purposes of latex paints are: a.) to decorate and b.) to protect. Latex paints have become extremely popular with professionals and consumers alike for a variety of reasons, most of which can be summarized below:

- Low odor
- Low toxicity
- Easy to apply (ideal for brush or roller application)
- Easy clean-up with soap and water
- Faster drying than oil-based paints
- Better exterior durability than oil-based paint
- Improved cost/performance

There are typically a number of ingredients that comprise a latex paint[3]. The latex polymer is, of course, the most important ingredient because it is the film-forming portion of the paint and the portion from which the paint derives most of its properties. But there are a number of other ingredients that comprise a typical house paint formulation. The following is a list of the most common ingredients in latex paints:

1. Primary Pigment -- This is the ingredient which contributes the most to
   the *color* and *hiding power* of a paint. In white paints, this pigment is
   titanium dioxide (TiO_2). In colored paints, this pigment can be one of
   a variety of different materials ranging from organic to inorganic
   compounds.
2. Extender Pigment -- This ingredient is much lower in cost than the
   primary pigment, but it can profoundly affect the performance of
   latex paints. Examples of the most common extender pigments include
   calcium carbonate, magnesium silicate (talc), aluminum silicate (clay),
   silica, and mica.

3. Coalescing Solvent & Plasticizer -- These ingredients are used to insure complete coalescence or fusing together of the latex polymer particles. Coalescing solvents are usually preferred over plasticizers because they will eventually escape from the film, while plasticizers will remain with the film indefinitely.

4. Dispersant & Surfactant -- These ingredients are used primarily to promote pigment wetting and ease of dispersion. Dispersants are generally anionic in nature, while the most common surfactants are typically nonionic.

5. Defoamer -- As the name suggests, this ingredient is used to aid removal of air that can be entrapped in a latex paint. There are generally two types of defoamers -- silicone and non-silicone types. Silicone types tend to be more effective, but these types can also contribute more readily to surface defects.

6. Preservative & Mildewcide -- These ingredients are used to inhibit microbial growth that is associated with any water-based system such as latex paints. Preservatives are used to inhibit microbial growth in the latex paint itself, while mildewcides are used to prevent growth of fungus and algae on the dried latex paint film.

7. Thickener & Rheology Modifier -- Other than the latex polymer and the primary pigment, the thickener or rheology modifier is arguably the most important ingredient in a latex paint because it has a profound affect on the application properties of the latex paint. The term "thickener" usually refers to an ingredient which only increases the low shear viscosity of latex paint; the term "rheology modifier" refers to an ingredient which affects both the low shear and high shear viscosities of latex paint, and impacts key application properties, such as roller spatter.

8. Miscellaneous Additives -- These ingredients include a host of specialty chemicals, which are used in very small amounts to significantly affect key latex film properties. Examples of these types of products include anti-mar, anti-slip, and anti-blocking additives.

2.2. LATEX POLYMER TYPES

There are a number of different latex polymers that are used in latex paints. The two most common latex polymers are acrylics and vinyl acrylics. Acrylic latexes are generally copolymers of methyl methacrylate and butyl acrylate and are used primarily in 1st line exterior flats and trims. Vinyl acrylics are copolymers of vinyl acetate and butyl acrylate, and -- unlike acrylic latexes -- are used primarily in interior flat and semi-gloss paints. Vinyl acetate homopolymers are used as vehicles for tape joint cement, and may still be used to a small degree in ceiling paints. Vinyl acetate is also co-polymerized with: a.) ethylene, to produce EVA latex polymers, which have a much smaller market niche than vinyl acrylics, and b.) VeoVa 10 branched vinyl ester, to produce more hydrophobic latex polymers, which are growing rapidly in popularity in exterior coatings. The final latex polymer type of note is styrene acrylic, which is a terpolymer of styrene, methyl methacrylate, and butyl acrylate. These polymers are used primarily in high gloss coatings.

There are several key attributes of latex polymers which determines their use in latex paints. These characteristics are listed below:

1.) <u>Monomer Composition</u> -- controls the overall durability and performance of the latex paint.
2.) <u>Glass Transition Temperature</u> -- reflects the hardness of the latex polymer and its suitability for use in trims or flats.
3.) <u>Particle Size</u> -- affects latex polymer rheology and adhesion properties.
4.) <u>Molecular Weight</u> -- affects latex polymer integrity, which becomes evident in scrub resistance and exterior durability.
5.) <u>Acid Content</u> -- affects adhesion to metal surfaces.
6.) <u>Adhesion Promoters</u> -- affects adhesion to difficult substrates, such as gloss alkyd and chalky surfaces.

2.3. PIGMENT BINDER RELATIONSHIP

An important aspect of latex paints is a concept known as "Pigment Volume Concentration." This term refers specifically to the following ratio:

$$\frac{\text{Pigment Volume, NVM Gals}}{\text{(Pigment Volume + Latex Polymer Volume, NVM Gals)}} \times 100$$

This term essentially describes the percentage of the dried latex film that is pigment. For example, a latex paint that is 50% PVC will dry to form a latex paint film which is 50% pigment (by volume).

The concept of PVC is a critical tool that is used by paint chemists when they develop new paint formulations[4]. PVC affects a number of paint performance properties including: hiding power, permeability, and durability. It also affects another key factor: economics of the latex paint formulation.

Of all of the properties affected by PVC, the most important is the gloss property. Gloss is a measurement of the light reflectance of a paint surface. The lower the PVC, the more "latex polymer rich" the paint film, and thus the higher is the gloss potential of that film. High Gloss and semi-gloss paints must be formulated at low PVC, while flat or eggshell paints are usually formulated at significantly higher PVCs. Latex paint systems are defined in large part by the gloss imparted by the paint film. It therefore follows that latex paint systems are defined in large part by the PVC of the latex paint formulation.

Most of what has been discussed thus far applies to all types of latex paints. However, latex paints can be further characterized as interior or exterior. The performance requirements of these two types of latex paints can be significantly different, depending on the end-use purpose.

2.4. INTERIOR LATEX PAINTS

To determine the key properties of interior paints, one needs to look no further than a Consumer Reports review of interior latex paints. Most paint companies which market a large percentage of their products to consumers want to be ranked favorably by this organization, and so the properties tested by the Consumer Reports Laboratory are often the properties on which most paint companies concentrate their product development efforts. These properties include scrub resistance, cleansability (which refers to the ease of removal of common household stains from the film), adhesion (which is actually a test of water resistance), blocking resistance (commonly referred to as "sticking"), hiding power, color properties, and application properties.

In addition to marketing paints for consumers or the do-it-yourself customers, most paint companies market a significant amount of material to professional paint contractors. The professional painters pay more attention to application and color properties than the average consumer, and far less attention to film properties such as scrub resistance and cleansability. The professional painters want to complete the paint job in the shortest amount of time. One of the few areas in which the paint contractors get very concerned is a property known as *touch-up*. This property refers to an instance in which the paint contractor sprays an initial coat of paint during one time of the day (typically early in the morning, when the temperature of the room is cold), and returns later in the day (when the temperature of the room has significantly increased) to "touch up" the areas that he missed during his initial application. If the paint applied in the second application does not match the initial coat, then the paint contractor knows that the paint job will not be accepted and he will need to re-paint the room.

Vinyl-based latex polymers, particularly vinyl acrylics, dominate the interior latex paint market. Other latex polymers cannot approach the cost effectiveness of vinyl acrylics, particularly in the area of scrub resistance and application properties. Vinyl-based latex polymers which use ethylene as the plasticizing co-monomer in place of butyl acrylate have exhibited particularly good performance in the area of low temperature touch-up properties. Although acrylics and styrene-acrylics are used infrequently in interior coatings, these latex polymers are used in certain small applications which require particularly good performance in a key property, such as high gloss, blocking resistance, or stain removal.

2.5. EXTERIOR LATEX COATINGS

Due primarily to the fact that exterior latex paints are exposed to the various elements of nature, namely, UV (sunlight), rain, and significant changes in temperature, exterior latex paints must withstand a far greater number of performance challenges than interior latex paints. Although systems are still defined to some degree by the gloss potential of the dried paint film, there are two other factors which play a key part in defining the requirements for exterior latex paints: 1.) geographical climate, and 2.) materials of construction.

Geographical climate plays an important role for a variety of reasons. First, the climate dictates the amount of freeze/thaw cycling that occurs in a given year. Freeze/thaw cycling is important when painting over dimensionally unstable woods, such as Southern Yellow Pine or fir plywood. The more the wood expands and contracts in response to the changes in temperature, the more stress is put on the coating to withstand this expansion and contraction. Ultimately, the latex coating will exhibit grain cracking, which can lead to flaking and loss of adhesion of the latex coating from the wood. Relative humidity and average sunlight also play a role in defining the requirements for exterior latex paint performance.

Materials of construction play an important role because the substrates that are most often used in new construction vary to a large degree and present different obstacles to performance. For example, Western Red Cedar (unlike Southern Yellow Pine) is considered a dimensionally stable wood, i.e. it does not undergo very much expansion and contraction with changes in temperature. However, this wood is dark-colored and can exude tannin stains through a paint film. These stains can cause severe discoloration of white paint films.

Another type of substrate, masonry (brick and concrete), can also cause problems for exterior latex paints. Certain types of latex polymers, such as vinyl acrylics, can hydrolyze (break down over time due to high pH) and lose adhesion. In addition to this, salts contained in the cement can permeate through a dark-colored latex paints and deposit at the surface, thus exhibiting unsightly white salt stains, a defect called efflorescence.

There are a number of other properties that are important to exterior latex paint performance. These include the following:

1.) Resistance To Dirt Pick-Up & Mildew
2.) Gloss Retention
3.) Chalk Resistance
4.) Chalk Adhesion
5.) Color Retention
6.) Low Temperature Application

Historically, while vinyl acrylics have dominated the interior latex paint market, acrylics have dominated the exterior latex paint market. There are many reasons for this, but essentially acrylics have proven to be the most durable latex polymer for exterior use, while exhibiting an excellent balance of grain-cracking resistance and dirt pick-up. Styrene acrylics have been used in masonry coatings due to their excellent efflorescence resistance, but these polymers have a tendency to chalk much faster than other latex polymer types. Use of vinyl-based latex polymers has increased over the past several years, due in part to use of branched vinyl esters as co-monomer with vinyl acetate and due in part to advancements in vinyl acrylic technology.

2.6. FUTURE TRENDS

Several trends are emerging in new latex paint technology. Low odor/low VOC paints have recently been introduced into the market. These paints have not as yet been widely accepted because they do not perform as well -- in certain key performance properties -- as conventional paints. Alkyds still occupy a relatively modest niche of the architectural coatings market, but "alkyd replacement" latex technology continues to close the gap between alkyd and latex paints. Part of this advancement involves rheology enhancement of the latex itself, which results in latex paints that exhibit excellent film build properties without the use of rheology modifiers. Yet another area of new technology lies in latexes which contain cross-linking functionality. All of these areas are sure to further advance the use of latex polymers in architectural coatings.

3. Latexes in Adhesives, Caulks and Sealants

3.1. INTRODUCTION

There are many instances at both home and work where two substances need to be joined together. This is accomplished by fastening the materials mechanically by bolts, rivets, staples, or bindings or by bonding them with an adhesive. Adhesive bonding, because of its versatility, ease of application, low cost and ever-improving performance, has become the preferred method of fastening in a growing number of application areas.

An adhesive can be defined in the simplest terms as a substance which can bond together and is capable of holding or preventing the separation of two similar or dissimilar materials (called adherends). Adhesives can be classified according to chemistry (Acrylic, PVAc, EVA, Rubber , Urethane, Epoxies, Phenolics, Starch/Dextrin, Cyanoacrylate etc.), polymer type (Thermoplastic, Thermoset) and application/end-use (Structural and Nonstructural). Structural adhesives referred to here are those which have relatively high Tg (> 30° C, Figure1) and which when applied to two substrates are designed to transfer load from one substrate to the other and result in a high shear strength of the order of 1000 psi. Nonstructural adhesives can be classified into specific categories such as pressure sensitive, contact, laminating and packaging etc. Caulks and sealants can also be grouped under the nonstructural type although there is sometimes tendency to refer to sealants as structural materials.

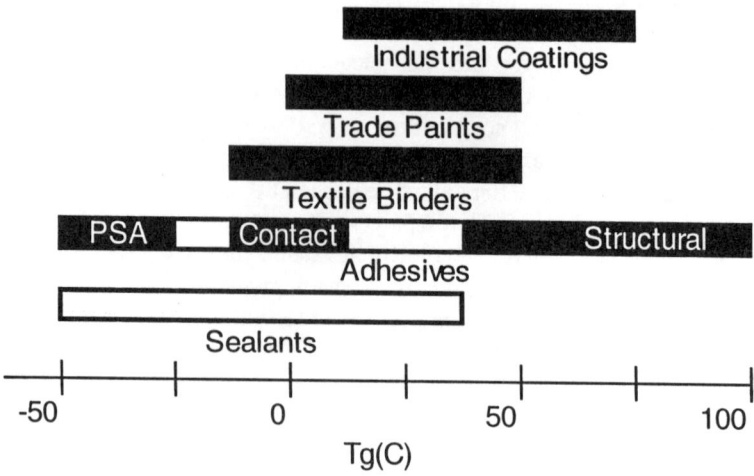

Figure 1. Latex adhesive/coating applications and glass transition temperature ranges

The following discussion is divided into three sections. The first section deals with latex adhesives, the second with caulks and sealants, and the third with polymer design for adhesives, caulks and sealants.

3.2. LATEX ADHESIVES

Latexes are finding increasing utility in nonstructural adhesives, caulks and sealants. They have several advantages over corresponding solution systems. Compared to solutions of resins in organic solvents, latex systems are non-flammable, non-polluting, low in toxicity, low in odor, higher in solids at lower viscosity, and often lower in cost. Government regulations have accelerated this movement towards water-borne systems. Producers of water-borne systems have responded to this opportunity through concurrent technological developments which have overcome most of the early limitations of such systems. In many instances, the change initiated only for economic or environmental reasons ultimately has led to improved performance.

Since latex adhesives have found extensive use in the nonstructural adhesives, the rest of the discussion will be confined to these type of adhesives. The use of some of the more commonly used latex adhesives for various applications is given in Table 1 below. Further details of the classification of adhesives and their uses are covered by Skeist[5].

Table1. Adhesive types used in various applications

Adhesive Type	Use
Acrylate	PSA(tapes, labels), Contact, Laminating, Textile, Paper
Vinyl Acetate homopolymer	Wood, White Glue
Ethylene Vinyl Acetate	Packaging, Film laminates
Styrene-Butadiene Rubber	Packaging, Carpet, Shoes, PSA
Neoprene	Contact, Laminating
Natural rubber	self adhesive envelops

General test methods for performance evaluation of different types of Latex adhesives can be grouped into three categories described below:

- Application properties: Coatability, dry time, open time/recoat time, green strength
- Adhesive properties: Strength of adhesion to various substrates in cleavage/peel and shear modes, peel strength, shear strength and tack in different geometries
- Mechanical/other properties: Shrinkage, tensile-elongation, stress relaxation/creep, compliance, water absorption, weathering.

Details of tests for a given application are provided in the ASTM manuals 15.06 and 15.09 and PSTC test methods [6]. Pressure Sensitive Adhesives (PSAs) and Contact adhesives are discussed in detail below.

3.2.1. Pressure Sensitive AdhesiveS (PSAs)

Pressure sensitive adhesives are soft ductile materials which in the dry state (i.e., free of solvent or water), are permanently tacky at room temperature and adhere to a variety of surfaces under only slight pressure. They have low glass transition temperature (Tg), less than -20°C (Fig.1) and low to medium molecular weight. This class of adhesives is being increasingly used in consumer, automotive and construction areas. PSAs are generally copolymers derived from acrylic, vinyl acetate, ethylene, styrene, butadiene and isoprene type of monomers. In many cases, depending on the nature of the base polymer, they are formulated with tackifiers, plasticizers and curing agents to enhance adhesive properties. Water-borne systems are modified with surfactants, defoamers and rheology modifiers to enhance application properties. A typical PSA end-use system consists of the adhesive, the carrier (polymeric or metallic film or paper backing) and, in many cases, silicone release liner. They find applications in tapes, labels, decals, floor tiles, wall coverings and wood grained films. Raw material supplier, coater and converter/printer form an integral part of the end use PSA product. A good monograph on PSA technology is available[7].

Peel, tack (quick stick) and shear (holding power) adhesion tests form the basis for evaluating the PSA performance. In general, while peel and tack go hand in hand with each other, shear follows an opposite direction: an adhesive with high peel and tack generally possesses low shear properties and vice versa. These properties are controlled by surface/rheological and bulk/viscoelastic contributions which in turn depend on the material and formulating components of the adhesive systems. An empirical relationship has been found to exist between peel and shear[8]. Efforts are always made to develop systems which have high peel-tack as well as shear.

3.2.2. Contact Adhesive

Contact Adhesive is a material which is applied to both of the substrates to be bonded. The adhesive is   allowed to partially dry on two surfaces, may have little residual tack but forms a strong bond when two coated surfaces are mated together under low and moderate pressure. Contact adhesives have also low Tg of the order of -10 to +10° C which is higher than that of the PSAs (Figure 1). This class of adhesives, especially the water-borne types, are based on neoprenes and acrylics. Some major use areas of these adhesives include:

Shoes:	Leather/Cloth
Automotive:	Headliners, Interior Upholstery, Vinyl Top
Furniture:	Plastic Laminate Table and Counter tops to Wood and Fiberboard
Trade Sales:	Wood/Wood, Cloth/Wood, Plastic/Wood, Cloth/Cloth, Metal/Vinyl etc.

These adhesives are characterized by good green (initial) strength, excellent ultimate bond strength, good high temperature bond strength and good metal adhesion. In addition they should have long open time (see below).

The most important performance parameter for a contact adhesive is the "open time-green strength " relationship. Open time is the time elapsed between the spreading of the last coat of adhesive and the assembly of the adhesive joint. Green strength can be defined as the strength of the adhesive joint shortly after assembly, e.g., less than one minute. Since the adhesive is normally only partially dry at the time of joint assembly, green strength is usually measured when the joint contains residual carrier like water or solvent. Green strength is important in several applications where it is desirable for the parts to remain in place without clamping, e. g., when bonding semi-rigid substrates which may not entirely be flat, such as formica and particle board. The development of good green strength requires that the adhesive build cohesive strength without its surface losing the ability to deform or exhibit auto tack under moderate pressure. Thus rapid development of green strength and extended open time are desirable features of a contact adhesive. In practice these conditions are hard to meet for water-based systems. New polymer design approaches are being pursued to meet this challenge for water-borne systems.

3.2.3. Areas for Future Improvement

Currently available water-based contact adhesives, both acrylics and neoprenes are in some ways, deficient when compared with the traditional neoprene solvent systems. An improvement in the open time-green strength relationship (longer open time and rapid build-up of green strength)is desired. Improvement in high temperature bond strength is also needed.

The latex pressure sensitive adhesives, because of the surfactants and some initiators used in their manufacture, suffer inherently from water sensitivity which adversely affects the adhesive, application and converting properties. A reduction in water sensitivity can make the latex adhesives suitable for use not only as replacement for solvent-borne systems but also in new applications. Some other areas of improvement include higher peel/tack-shear balance, adhesion to low energy plastic surfaces under different environmental conditions of relative humidity and temperature and application properties for high speed coating operation.

3.3. CAULKS AND SEALANTS

Caulks and sealants are materials used to fill gaps between similar and dissimilar surfaces which differ in surface energy and morphology. Their main function is to prevent the transmission of water, air and heat in the joint. The two terms, Caulks and Sealants are used interchangeably. However, there are differences in many performance characteristics[5,6]. Caulks are used on substrates where joint movement of the order of 10 % is expected. Generally they are relatively nonelastomeric, exhibit low elasticity and undergo appreciable plastic deformations under cyclic changes of heat and cold. Sealants, on the other hand, are relatively elastomeric materials and can undergo expansion or contraction of the order of 25 %. Thus ,in addition to meeting all the requirements of caulks, sealants are required to extend and compress much more under extreme environmental changes without losing their sealing properties. Caulks and sealants encompass a wide range of Tg (-20 to +10, Figure1) and find applications in architectural, maintenance, automotive and industrial areas. Water-borne systems, which are generally acrylics or vinyl acrylics, sometimes modified with small amounts of silanes, are widely used in homes for caulking, weather proofing and sealing leaks.

General test methods for caulks and sealants include extrusion rate, volume shrinkage, low-temperature flexibility, recovery, adhesion, slump resistance, tackiness-dirt pick up, cracking, hardness, flexibility, tensile/elongation, paintability, durability and package stability. These tests are done according to the ASTM manual-04-07 (C-920, C-570, C-834, C-836 and C-957) or and Federal specifications (TT-S-00227E, 00230C and 01543A)

Caulks and Sealants, like adhesives, can be classified according to chemistry, polymer type, application method, physical form and curing mechanism[9]. They are also classified in terms of low/medium and high performance [10]. Water-borne polymers find use in both the categories. The former are generally low cost with limited joint movement and low or no chemical cure while the latter are medium/high cost, are capable of joint movement with no or little shrinkage and have good adhesion to a wide range of substrates differing in surface energy and morphology. These substrates can include glass, masonry, concrete, wood, steel, aluminum, plastics etc. From a practical point of view they can be classified according to appearance as shown in the table below.

Table 2. General classification of acrylic latex caulks

Type	Clear	Translucent	Pigmented
Percent Latex	95	75-80	30-35
Polymer Tg (° C)	-10, +10	-25, -10	-35, 0
Pigment/Latex solids	0	0.02-0.06	2.5-3.2
Weight Solids (%)	59	60-64	80-85
Volume Solids (%)	55	55-59	70-75

3.3.1. Areas for future improvement

There is need for development of higher performance acrylic latex caulks and sealants systems through improvement in various areas to meet future challenges . One important area includes improvement in the elastomeric character to withstand large joint movements without losing sealing properties. Development of zero VOC systems, lower residual monomer and elimination of ethyl acrylate is also needed. Other areas of improvement include mildew growth, cracking of paints applied over uncured caulks, degassing and after flow during manufacture and application methods.

3.4. DESIGN OF LATEX ADHESIVES, CAULKS AND SEALANTS

It is clear from the above discussion that for optimum performance, PSA requirements require high peel-tack as well as high shear; those for contact adhesives require high green strength and long open time; and those for caulks and sealants need high flexibility-interfacial adhesion and toughness. It is not easy to achieve a given set of properties for a given application because there are conflicting polymer and formulation properties requirements. For example in PSAs, polymer viscoelastic properties required for high peel adhesion conflict with those required for high shear adhesion. Systematic studies are needed to identify parameters of polymer systems to design latex adhesive systems with optimum performance. The parameters that affect the properties of a latex adhesive system can be collected into three groups - polymer, emulsion and formulating parameters. The polymer parameters include the type of monomer (BA,EA,MMA,STY,ACN), Tg (-60 to + 10° C), functional groups (carboxyl, hydroxyl, amide) and molecular weight (medium to high); the emulsion parameters include stabilization system (anionic, nonionic), particle size (0.2 to 0.5 microns), initiation package (redox, thermal), feed method (uniform, core-shell); the formulating parameters include the use of post-additives like wetting agents (anionic), rheology modifiers (conventional, associative), tackifiers (rosin, terpenes, phenolics), adhesion promoters (silanes), defoamers (conventional), preservatives (conventional), plasticizers (benzoates, phthalates) and crosslinkers (external, ambient, ionic). Some aspects of polymer parameters are discussed here. Although these have been exemplified with PSA systems, the findings are applicable to adhesives and caulks and sealant systems in general.

3.4.1. *MonomerType/Tg*

Acrylic adhesives are based on higher chain monomers like ethyl, butyl, 2-ethylhexyl and octyl acrylates whose long, nonpolar side chain provide internal plasticization. Small amounts of higher Tg monomers , like styrene, methyl methacrylate, vinyl acetate and acrylonitrile impart cohesive strength; and functional monomers like those containing carboxyl, hydroxyl and nitrogen impart specific adhesion. For example calculated Tg of about - 45° C can be achieved through compositions such as n-butyl acrylate/methyl methacrylate in a ratio of 90/10 or 2- ethylhexyl acrylate/methyl methacrylate in a ratio of 80/20. A series of polymers encompassing a range of -54 to +8° C were prepared by varying the ratio of the major monomers while keeping other aspects constant and the effect on peel-tack and shear was studied [8]. The divergent influence of polymer Tg on tack-peel (adhesive or interfacial strength) and shear (cohesive strength) was seen. Tack and peel decreased and shear increased with increase in Tg. A high tack-peel value requires intimate contact between polymer segments (high mobility, low Tg) and the substrate. Shear adhesion on the other hand requires high cohesive strength, which unlike tack-peel is favored by stiff polymer segments, inter and intra-chain associations, chain entanglements and actual crosslinks (high Tg). These types of studies and observations help in the selection of the type and levels of monomers to achieve desired Tg and design proper systems for a given application. Lower Tg polymers will favor interfacial adhesion, long open time and flexibility while the higher Tg polymers will favor high cohesive strength, tensile strength and high bond strength at elevated temperatures.

3.4.2. *Molecular Weight and Distribution*

Molecular weight of latexes can be adjusted with initiator level, Chain Transfer Agent (CTA) and multifunctional monomers. In the present case, under constant monomer composition and polymerization conditions, the molecular weight was altered by changing the concentration of CTA from 0 (high molecular weight) to 1.0 % (low molecular weight) and the distribution was altered by process conditions [11]. The effect on peel strength and holding power (shear) was studied. The peel curve showed three distinct regions. In region 1, the peel force increased with increase in molecular weight (decrease in CTA concentration). The mode of failure was cohesive on a macroscopic scale. indicating sufficient mobility and deformation characteristics favorable for the formation of a good bond and insufficient cohesive strength to withstand the stresses applied during the debonding process. In region 2, the peel force attained a maximum value but the failure mode was still cohesive. The increase in peel strength appeared to be arising out of the increase in cohesive strength due to increase in the molecular weight. In region 3, peel strength decreased with increase in molecular weight, the failure mode shifted from cohesive to adhesive on a macroscopic scale. An increase in cohesive strength took place to a degree which markedly restricted the flow and deformation behavior of polymeric adhesive and inhibited the achievement of maximum number of contacts at the interface required for the formation of stronger interfacial bond. The shear adhesion showed a continuous increase with increase in the molecular weight which was due to the increase in the cohesive strength. The curve showed a steady increase initially followed by a sharp increase. Altering the molecular weight distribution through process variations considerably enhanced the adhesive properties balance (higher peel as well as shear for a given level of CTA). Thus proper control of molecular weight/distribution can result in designing new improved polymers for a given application.

3.4.3. *Functional Monomers*

Several polar functional groups such as carboxyl, hydroxyl, and nitrogen containing are incorporated into acrylic adhesive polymers to enhance specific interactions with organic and inorganic surfaces. In addition, functional groups like carboxyl can also enhance the stability of the latex. It has been observed[11] that the shear adhesion increased with increase in the carboxyl content. Dipole-dipole interactions due to carboxyl groups result in an increase in cohesive strength. Thus a modification in the rheological properties, as observed by an increase in T_g and elastic modulus with increase in carboxyl content takes place. In the low molecular weight range, the peel strength increases with increase in the carboxyl content. It is evident that a proper combination of the type of polar functional group, molecular weight/distribution and T_g can lead to optimization of interfacial and bulk properties essential for the development of high performance adhesives and caulks and sealants.

4. Latexes for Textile Applications

4.1. INTRODUCTION

The textile industry is a major user of emulsion polymers of various types in a wide variety of applications(12). The function of the latex is to serve as a coating, binder or adhesive in nonwovens, back coating, flocking, pigment printing, lamination, textile stiffening, and sizing applications. Latexes offer many advantages to the textile industry in common with those offered to the paint industry: minimum fire hazard and pollution as well as ease of clean-up and handling. In addition, high molecular weight at low viscosity, combined with available crosslinking chemistries, offer attractive performance and application properties.

A wide range of polymer types are used in the various textile applications. Because of the large differences in properties associated with each polymer class, it is important to match the appropriate polymer type with the intended application. Polymer latexes commonly used in textiles include vinyl acetate homopolymers, vinv-acrylics, all-acrylics, styrene-acrylics, styrene-butadiene and ethylene-vinyl acetate copolymers. Since there are significant cost differences among the polymer types, cost-performance factors are also important in choosing a polymer for a given end use.

The discussion here is intended to be an introductory outline of several major applications in the textile industry. Four applications have been chosen for illustration: nonwoven fabrics, direct coatings, back coatings and sizing [13].

4.2. NONWOVEN FABRICS

Nonwoven fabrics are produced by several techniques. In each case fabric is made by bonding random web or sheet structures with latex polymer using chemical or thermal means. Fabric bonding is used to produce sheets, gowns, masks and disposable articles. The polymer properties required are low tack (internal crosslinking), dry cleaning (external crosslinking or inclusion of acrylonitrile comonomer), good adhesion (functional polar groups such as hydroxyl or carboxyl), soft hand (control of polymer glass transition temperature,Tg), and good pot life (stable crosslinking chemistry). The most common post-crosslinking in textile applications involves the use of N-methylol acrylamide (NMA), although there is a concerted effort to avoid the presence of formaldehyde[14]. Novel low-temperature curing chemistries for this purpose are being developed in many laboratories. Latexes used for fabric bonding include acrylics, styrene-acrylics and poly(vinyl acetate).

Fiberfil end-uses include jackets, quilting, cushions and sleeping bags where high loft (air entrapment) is desired. Polymer requirements include good sprayability (rheology), wash resistance, dry cleaning and low tack. Most successful fiberfil binders are crosslinkable (NMA). Latex types include acrylics and styrene-acrylics.

Flocking is the application of short fibers to a substrate coated with a latex film to achieve a decorative or functional effect. Latexes, after curing, show excellent resistance to washing and dry-cleaning. In addition, they contribute the proper flexibility and porosity to fabrics with a wide choice of properties. In the flocking process the latex is applied to a fabric, and short fibers are allowed to fall onto the (wet) surface within an electrostatic field whose purpose is to orient the fibers in a vertical position. Polymer properties required include good adhesion to the substrate, abrasion resistance, heat stability, good laundering/dry cleaning, and high viscosity to avoid wicking into the fabric. In general, flocking polymers are crosslinkable to provide good anchoring to the fibers. Acrylic and styrene-acrylic latexes are preferred.

4.3. DIRECT COATINGS

Color coating of textiles involves the application of pigmented coatings directly to fabric as a finishing operation or as a printed pattern. In contrast to dyeing, the color pigments are bound to the fabric by the latex polymer. Pigment printing is especially useful on low energy surfaces such as polyester sheets and fabrics which do not readily accept dyes. Printing is accomplished by roller or screen printing techniques. Both methods require good rheology control and good mechanical stability. Desired polymer properties include good adhesion to polyester, flexibility (hand), crock resistance (blocking), laundering/dry cleaning and resistance to yellowing. Self-crosslinking is necessary to achieve the durability of the coating. Latex types include acrylics, styrene-acrylics and butadiene-acrylics.

4.4. BACK COATINGS

Fabrics are often backed with a polymer coating to impart dimensional stability, shrinkage control, improved handling properties and durability. These coatings are normally clear, but where hiding and cost reduction are desired the latex is often loaded with clay. The types of fabrics commonly back coated are carpets, upholstery, draperies, rainwear and bedding. In most cases high viscosity formulations are preferred to avoid strike-through. For carpets and upholstery fabrics, the performance requirements are minimal, and styrene-butadiene latexes are used. For higher valued applications, crosslinkable acrylics are used. Another type of back coating involves the use of foamed latex applied to drapery, upholstery and nonwoven fabrics to give high opacity, good flexibility and softness for specific applications. A modification of this process is to dry the foam coating, crush it into the fabric, then cure to give a coating of good durability. Desirable binder properties include durability, colloidal stability, adhesion, high pigment binding, foamability and self-crosslinking potential for washing/dry cleaning.

4.5. SIZING

Sizing is the application of a polymer film to yarns and fibers to be woven into fabric such that the yarn can withstand the weaving process with minimum breakage and fraying. Size coatings are usually removed after weaving. For natural fibers, starch, proteins, cellulosics and poly(vinyl alcohol) have been used. With synthetic fibers, however, acrylic sizing has taken over due to its better bonding and its ability to work at higher weaving speeds. In addition, they exhibit excellent removal properties, electrolyte compatibility and high binding strength.

5. Latex Thickeners and Rheology Modifiers

5.1. INTRODUCTION

Thickeners for latex paints have historically included natural and synthetic resins such as cellulosics and carboxylic polyacrylates [15]. These polymers generally have high molecular weights and thicken water-based formulations by virtue of having large hydrodynamic volumes in solution. The low-shear viscosities of paints can be effectively increased and thixotropic (shear-thinning) behavior achieved by deformation of the hydrated polymer under shear. The thickening behavior of these polymers is predictable and largely unaffected by paint components. Unfortunately, this thickening mechanism results in poor flow and leveling and promotes spattering of paint droplets on application. Recent developments in polymer design have eliminated these deficiencies and, at the same time, produced true rheology modifiers based on an entirely different mechanism, polymeric association[16].

5.2. ASSOCIATIVE POLYMERS

Associative polymers have relatively low molecular weights and are composed of a water-soluble backbone with two or more hydrophobic portions capable of associating through hydrophobic clusters linking the molecules together to form a network. This network can be designed to be more or less responsive to shear forces such that the network deforms, and then disintegrates, depending on the shear applied. In this way, the low- and high-shear viscosity responses can be affected independently so that good flow and leveling, as well as paint transfer, can be achieved simultaneously. Roller spatter is eliminated by the absence of high molecular weight polymeric reinforcement of paint filaments formed during film splitting. Moreover, a major advantage is the use of emulsion polymerization in the preparation of the leading class of associative polymers, hydrophobic alkali soluble emulsions (HASE).

LATEX PARTICLE SOLUBILIZATION

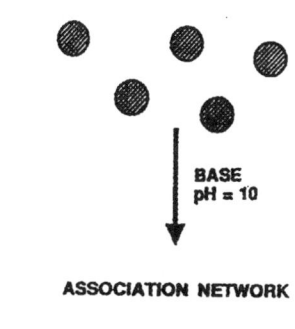

BASE
pH = 10

ASSOCIATION NETWORK

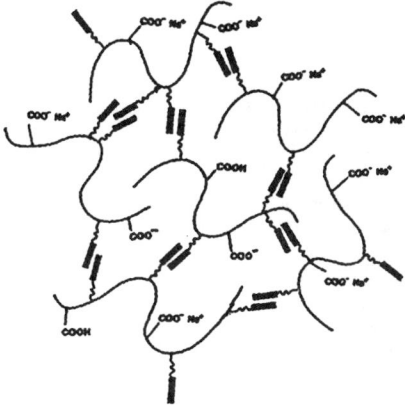

Figure 2. Latex particle solubilization upon neutralization to form an associative network

The composition of HASE polymers generally includes a carboxylic monomer, a flexibilizing monomer, and a macromonomer to which is attached a nonionic surfactant (ethylene oxide chain with terminal hydrophobe). Preparation is carried out using a traditional emulsion polymerization process. Solubilization is accomplished by neutralization of the incorporated carboxyl groups to produce enough electrostatic repulsion to break the particles into individual molecules which can then associate in a manner depicted in Figure 2. A major factor in the strength and extent of the association network is the size and structure of the hydrophobic clusters which, in turn, depend upon the nature of the individual hydrophobes [17].

The richness of emulsion polymerization lies in the capacity to produce a variety of polymeric designs capable of producing a wide range of useful rheological responses. Thus a family of rheology modifiers can be produced to satisfy the requirements of most aqueous paint systems. One problem with this technology is that it is based on interactions and, as a result, most surface active components in a given formulation can participate in the association network, sometimes leading to unpredictable behavior. These surfactant interactions are discussed in more detail elsewhere in this book.

6. References

1. Lambourne, R. (1987) *Paint and Surface Coating*, Ellis Horwood Publishers, Chichester.
2. Martens, C. R. (1981) *Waterborne Coatings, Emulsion and Water-Soluble Paints*, VanNostrand Reinhold Co., Publishers, New York.
3. Flick, E.W. (1994) *Water-Based Point Formulations*, vol. 3, Noyes Publications, Park Ridge, N.J.
4. Sherwin, M. A., Buttrick, G. W., Glancy, C. W. (1983) Understanding High-PVC Paints, American Paint and Coatings J., **68**, 44-52.
5. Skeist, I. (1990) *Handbook of Adhesives*, VanNostrand Reinhold Co., Publishers, New York.
6. Pressure Sensitive Tape Council (1992) *Test Method Manual*, 10th ed., Chicago.
7. Satas, D. (1989) *Handbook of Pressure Sensitive Adhesive Technology*, VanNostrand Reinhold Co., Publishers, New York.
8. Sehgal, K. C. (1993) Surface and bulk effects in the adhesion of emulsion pressure sensitive adhesives, 67th colloid and surface science symposium, Toronto.
9. Prane, J. W. (1989) *Sealants and Caulks*, Federation of Societies for Coatings Technology, Publishers, Blue Bell, PA.
10. Foster, V. R. (1987) Polymers in caulking and sealant materials, *J. Chem Ed.*, **64**, 861-868.
11. Sehgal, K. C. and Bassett, D. R. (1995) Structure-property-performance relationships in emulsion pressure sensitive adhesives, Proc. 18th Ann. Meeting, Adhesion Society, 292-304.
12. Clive, C. D. and Friddle, J. D. (1991) Specialty textile coatings- a formulation overview, *J. Coated Fabrics*, **21**, 32-41.
13. Underwood, S. M. private communication.
14. Bassett, D. R., Sherwin, M. A., Hager, S. L. (1979) Study of latex crosslinking by thermal evolution techniques, J. Coatings Tech. **51** (657), 65-72.
15. Schultz, D. N. and Glass, J. E. (1991) *Polymers as Rheology Modifiers*, ACS Symposium Series, **462**, American Chemical Society, Publishers, Washington.
16. Glass, J. E. (1989) *Polymers in Aqueous Solution*, ACS Advances in Chemistry Series, **223**, American Chemical Society, Publishers, Washington.
17. Jenkins, R. D., DeLong, L. M., and Bassett, D. R. (1996) Influence of alkali-soluble associative emulsion polymer architecture on rheology, in J. E. Glass (ed.) ACS Advances in Chemistry Series, **248**, American Chemical Society, Publishers, Washington.

FILM FORMATION

J. RICHARD
Centre de Recherches en Microencapsulation
8, rue André Boquel
Parc Scientifique des Capucins
49100 Angers, France

1. Introduction

One of the main features of a polymer latex to be used in a coating formulation, is its film forming behavior. The film formation process is of considerable industrial importance, since most of the applications of polymer latexes require the formation of a continuous film with high mechanical strength, toughness or scrub resistance. These applications, in which the polymer latex is thus used as a water-based binder, currently encompass various fields such as adhesives, paper coatings, paints, varnishes, carpet backing and textile sizing. Furthermore, the range of applications of these materials widens continuously as worldwide efforts are being made to reduce the volatile organic content of coating materials.

Upon drying, some latexes form transparent continuous films at room temperature, while others only give rise to a friable opaque material, i.e. a powder. From this observation, it appears that a minimum film-forming temperature (MFT) can be defined. It corresponds to the minimum temperature at which a latex cast on a substrate forms a continuous and clear film. Conversely, for a temperature lower than MFT, the dry latex remains opaque and powdery. For this reason, MFT corresponds to the practical temperature limit for applications of a latex in coatings and adhesives.

The mechanism of film formation from a latex is of both theoretical and practical interest, since it influences the final film structure and properties, and hence the ultimate performances of the resulting coatings. For instance, mechanical strength, adhesion behavior, as well as alteration of properties due to water, are strongly dependent on the final film structure.

From a phenomenological point of view, the film formation process has traditionally been considered to occur in three sequential stages. During the first stage (stage I), water evaporation takes place to the point where particles come into close contact and form a dense array. This stage is characterized by a constant rate of water evaporation, which is equal to the evaporation rate for an aqueous solution of electrolytes and emulsifiers with the same concentrations.

In the second stage (stage II), deformation of particles into polyhedra occurs to completely fill the space with a dense packing. It is induced by surface and osmotic forces associated with the presence of water in the intersticial spaces. The overall rate of water evaporation decreases as the intersticial voids become filled with polymer. This stage leads to a nascent film which displays a honeycomb-like cellular structure. This

J. M. Asua (ed.), Polymeric Dispersions: Principles and Applications, 397–419.
© 1997 *Kluwer Academic Publishers. Printed in the Netherlands.*

structure consists of deformed hydrophobic particle cores regularly separated by hydrophilic membranes which originate from the surface species stabilizing the initial colloidal dispersion. Only a residual quantity of water may remain in the film, which can evaporate through the polymer itself.

Finally, in the third stage (stage III), the film undergoes a ripening process which mainly concerns the interfacial membranes. In this process, often referred to as the further coalescence process, two events may occur : polymer diffusion across the particle boundaries and breaking up of the membranes, depending on the aging conditions and the features of the polymer particles. The classical sequence of stages for film formation is depicted in Figure 1.

This chapter begins with a short account of particle interactions and force balance which govern film formation. Then, it presents a critical review of the main theoretical models proposed in the literature for film formation. The characterization methods used to monitor stages I and II will be reviewed together with experimental results obtained recently. On this basis, driving forces and relevant mechanisms for film formation will be pointed out. In the same way, new investigation methods developed to characterize the ripening of film structure during the further coalescence process (stage III) will also be presented. Experimental results concerning the effect of various parameters on structure ripening will be discussed, mainly as regards aging conditions and latex features. It will make it possible to bring out the key parameters which govern stage III. Then, experimental results concerning the fate of the hydrophilic species which originate from the stabilizing system and build the membranes in the films, will be reviewed. The effect of structure modifications on film properties will be presented, mainly as regards mechanical and permeability properties.

2. Particle Interactions and Force Balance during Film Formation

From the standpoint of energy, film formation from a latex is favorable, since it results in a minimization of free energy through the large decrease of total surface. Film formation is not a spontaneous phenomenon though. A high activation energy related to the presence of stabilizing species at the surface of particles impedes the bringing together of particles and then prevents film formation from occuring spontaneously. This energy barrier originates either from electrostatic repulsions generated by surface ionic groups of particles, or from steric interactions due to non ionic chains grafted or adsorbed at the surface of particles. Thus, film formation consumes energy to bring particles together and then to deform the particles from spheres to dodecahedra. Brown first proposed to establish the balance of forces and interactions which act to achieve film formation and those which resist it [1].

At least, four forces are exerted which tend to favor stages I and II of the film formation process :
- the van der Waals forces between the spheres F_V, which are classical attractive forces leading to flocculation or coalescence of emulsions [2],
 - the capillary force F_C, resulting from the surface curvature of water present in the intersticial capillary system during water evaporation ; it should mainly act when particles are packed in an ordered array and begin to appear on the film surface,

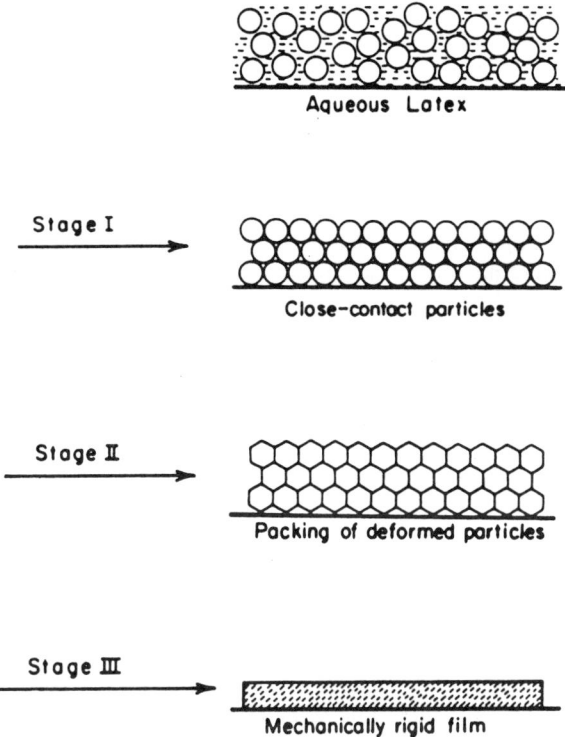

Figure 1. The phenomenological representation of the film formation process

- the force produced by the curvature of polymer surface F_S, which is related to interfacial tension ; it takes a significant value only after the particles have come into contact [1],
- the gravitational forces F_g, which leads to settling of dispersions.

Conversely, the forces which tend to resist film formation are :
- the repulsive interactions of the spheres F_R, which are of coulombic or steric origin in the case of electrostatically- or sterically-stabilized latexes, respectively ; these forces are responsible for the colloidal stability of latexes and for the energy barrier which prevents particles from being brought together [2, 3],
- the resistance of the spheres to deformation F_G, which is related to the elastic modulus of the polymer.

Then, for stages I and II to occur and film formation to take place, the following inequality must be fulfilled at any instant of these stages :

$$F_V + F_C + F_S + F_g > F_R + F_G \tag{1}$$

This force balance can be considered as a relevant starting point to discuss the many theoretical models proposed in the literature for film formation. Most of these models have been developed to bring out the driving forces, and to derive a criterion on latex features and drying conditions for film formation to occur.

3. Critical Review of Proposed Mechanisms for Film Formation

3.1. DRY SINTERING THEORY

This model is issued from the first investigation into the mechanism of film formation conducted by Dillon et al. [4]. These authors postulated that sintering of two particles which are in contact with each other occurs through viscous flow of polymer and particle deformation induced by a shearing stress. The shearing stress is generated by surface tension of polymer, i.e. polymer-air interfacial tension, which tends to minimize the surface area of the system. It was proposed that the relationship developed by Frenkel [5] for coalescence of spheres through purely viscous flow describes the film formation from latex dispersions (Figure 2).

According to Brown [1], the theory of latex particle fusion based on purely viscous flow cannot be considered as appropiate to describe the film formation process of latexes. Several criticisms of the dry sintering mechanism can be stated. This theory does not give any explanation about the force which puts particle into contact. Moreover, it does not take into account the major part played by water in film formation. Hence, Brown denoted that the surface tension could not provide the driving force, and would rather be replaced by the polymer-water interfacial tension. However, due to the presence of surface active species, the contribution of this interfacial tension would be too low (~ 10 mN/m) to be considered as responsible for film formation. Finally, this model cannot account for film formation from lightly crosslinked polymer latexes which may also form continuous films ; purely viscous flow is not possible in these systems.

For these reasons, Brown developed an alternative theory based on the capillary pressure as the main driving force instead of the polymer surface tension.

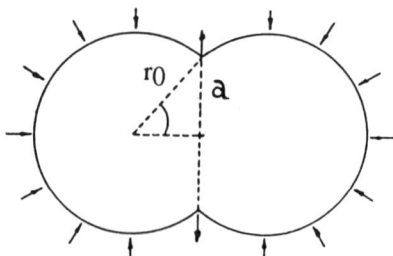

Figure 2. The dry sintering model : viscous flow of polymer and particle deformation under the action of surface tension of polymer.

3.2. CAPILLARY THEORY

Brown stated that the role of water in film formation is of utmost importance. He established the balance of forces which is presented in section 2 and the condition given in equation (1) to be fulfilled for film formation. Then, he argued that, among forces which tend to promote or resist film formation, F_V, F_S, F_g and F_R can be neglected, compared to capillary forces F_C and resistance to deformation F_G which bring the main contributions. It is worthwhile noticing that his is likely true when the particles are no longer mobile in the latex and are packed in an ordered array. At this point, polymer particles which are in contact, begin to appear on the surface, and water is confined in concave intersticial areas (Figure 3).

Then, particles may either resist deformation and water evaporates completely from the intersticial channels without film formation, or particles may deform causing film formation. Hence, the conditions for film formation can be described by the simple inequality :

$$F_C > F_G \qquad (2)$$

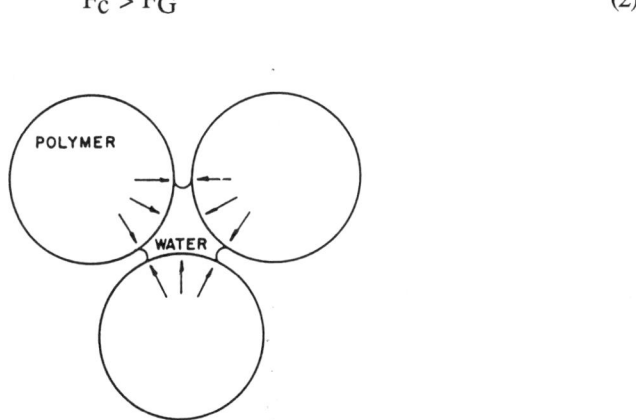

Figure 3. The capillary forces arising from water in intersticial areas.

Brown proposed a simplified quantitative development of his theory. The resistance of polymer spheres to deformation was derived using a model of two elastic spheres pressed together. Pressure required to bring the spheres together was calculated as a function of the elastic shear modulus G_t of the polymer. Capillary pressure was then derived applying the Laplace law in the narrow capillary which exists in the plane passing through the centers of three contiguous particles [1]. Assuming that pressures are exerted on the same surface area, Brown obtained the criterion for film formation as follows :

$$G_t < 35 \, \gamma / r_0 \qquad (3)$$

where γ is the surface tension of water in the capillaries and r_0 the radius of the polymer particles. G_t is the modulus value measured in a creep experiment over the period of time for water evaporation, i.e. duration of capillary forces. Very interestingly, Brown showed how MFT of the latex can be deduced from the above relationship and the creep modulus spectrum. For a latex with a given particle size r_0 and surface tension γ, MFT corresponds to the temperature for which the equality between the two terms of eq. (3) is fulfilled.

The model developed by Brown was one of the most commonly accepted models, but some points of his theory have needed adjustements and corrections leading to modified criterion for film formation.

3.3. MODIFICATIONS OF BROWN'S MECHANISM

Mason pointed out that the area over which the capillary pressure and the pressure to contact are exerted, are not identical. Therefore, the relationships between forces cannot be readily applied to pressures [6]. Furthermore, this author stated that capillary pressure is not constant throughout the second stage of drying and he derived the expression of capillary pressure as a function of particle deformation. Then, the resulting condition for film formation was found to be :

$$G_t < 266 \, \gamma \, / \, r_0 \qquad (4)$$

While Brown and Mason only considered elastic deformation, Lamprecht [7] and Eckersley et al. [8] pointed out that the viscoelastic relaxation of polymer should be taken into account to describe the deformation of spheres satisfactorily. Film formation can then occur either under fixed stress or with a deformation rate fixed by water evaporation. In the former case, film formation takes place if particle deformation after complete water evaporation is equal to that of dodecahedra in a close-packed arrangement. However, the latter situation, i.e. with deformation rate fixed by water evaporation, appears to be more realistic. A criterion for film formation, involving the time dependent creep compliance of the polymer $J_c(t)$, was derived based on this viscoelastic model [7, 8] :

$$1 \, / \, J_c(t) < \alpha \, \gamma \, / \, r_0 \qquad (5)$$

In this expression, α is a numerical factor which ranges between 34 [7] and 600 [8] , depending on the calculation conditions for the radius of the contatct area between the spheres.

Sheetz remarked that Brown description implicitly assumes that polymer-water contact angle is zero, which is seldom the case with real latexes [9]. Furthermore, the capillary force contains two contributions, normal (P_n) and parallel (P_p) to the film surface. This implies that the expression of capillary pressure P_n, which is assumed to be the driving force for film formation, should be modified to take into account the polymer-water contact angle θ :

$$P_n = 12.9 \, \gamma \, / \, r_0 \, \cos\theta \qquad (6)$$

Using Young equation and considerations about variations of polymer-water and polymer-air interfacial tensions, Sheetz concluded that the capillary force normal to the surface is approximately independent of the surface tension of water γ [9]. Variations of γ mainly affects the capillary force parallel to the surface. The criterion for film formation was then expressed as [9] :

$$G_t < 80 / r_0 \qquad (7)$$

3.4. ROLE OF POLYMER-WATER INTERFACIAL TENSION

Vanderhoff et al. [10, 11] remarked that film formation can also occur for latexes with diameter higher than 0.1 μm. However, capillary pressure is not great enough to be responsible for this process in these latexes. For this reason, an original model for film formation was developed, based on polymer spheres suspended in a droplet of water and surrounded by their double stabilizing layers. It was suggested that when two polymer spheres come into contact, two important radii of curvature r_1 and r_2 are set up (Figure 4) and must be taken into account to correctly derive the surface pressure P exerted on the particles.

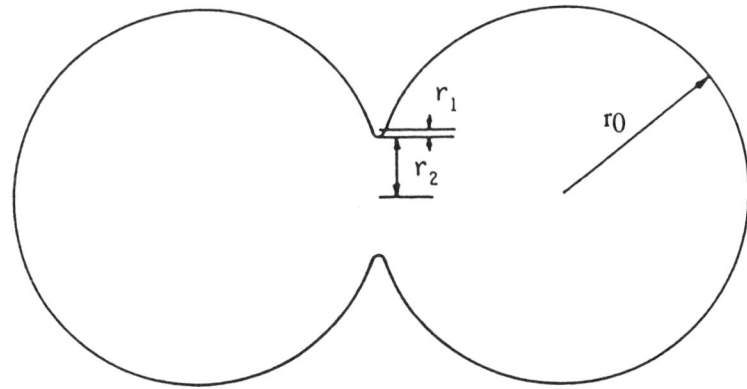

Figure 4 . Vanderhoff's model : radii of curvature to be taken into account.

Then, it turns out that the contribution of polymer-water interfacial tension γ is no longer negligible, since r_1 and r_2 are very small. Pressure on the particles is then increased by a term P_s expressed as follows :

$$P_s = \gamma_{pw} (1 / r_1 - 1 / r_2 + 2 / r_0) \qquad (8)$$

For γ_{pw} ranging ~ 10 mN/m, P_s becomes of the same order of magnitude as the capillary pressure, only for particle diameter R higher than 0.1 μm.

As regards the initial concentration step during which particles are brought into contact, Vanderhoff agreed that the capillary forces of Brown play a significant role to force particles together [11]. The pressure exerted by the water layer surrounding the particles was then calculated and stated to be responsible for bringing the particles together, the driving force being the water surface tension. Obviously, this model does not remain valid for stage II of the film formation process, since the water layer cannot be assumed to be preserved when particles begin to appear at the surface of the wet film.

The main criticism as regards the contribution of polymer-water interfacial tension to film formation concerns the dependence of pressure P_S upon particle size. The values of pressure P_S derived from Vanderhoff's model do not depend on the particle size. This implies that MFT of a particular latex should not be dependent upon particle size. This statement is in complete disagreement with Brown's theory which in contrast predicts that MFT increases as particle size increases [1]. This point appears to be very controversial. According to earlier work, it seemed to be confirmed that MFT is actually independent of particle size [12], while some authors have just recently found a significant variation [8].

3.5. COMBINATION OF CAPILLARY AND INTERFACIAL FORCES

Recently, Eckersley et al. [8] have shown that Vanderhoff's concept of capillary force and interfacial force acting in tandem to promote film formation is not inconsistent with the dependence of MFT with particle diameter, provided that the polymer is treated as a time-dependent viscoelastic material. A comprehensive model has been proposed, in which the capillary and interfacial forces are complementary [13]. The capillary force originates from water capillaries in the intersticial regions, while interfacial force arises from the tendency of spheres in contact to reduce surface area. The driving force is the tendency to reduce surface energy, both in the water capillaries and the contacting particles. The main assumptions are that deformations due to each force are additive and polymer behaves as a linear viscoelastic material. The radius of the contact region (a) between two particles has been calculated as a function of particle radius (r_0), ellapsed drying time (t), polymer modulus (G^*) and viscosity (η^*), surface tension of water in the capillaries (γ) and polymer-water interfacial tension (γ_{pw}), according to the expressions :

$$a = a_{capillary} + a_{interfacial} \tag{9}$$

$$a = (2.8 \, r_0^2 \, \gamma \, G^{*-1})^{1/3} + [3 \, \gamma_{pw} \, r_0 \, t \, (2 \, \pi \, \eta^*)^{-1}]^{1/2} \tag{10}$$

This model has been experimentally evaluated and validated by using an array of latexes with widely different physical properties. However, it is very sensitive to the values of the viscoelastic parameters G^* and η^*, which have to be measured under wet, hydroplasticized conditions [14]. Moreover, the time scales for deformation and flow processes are not exactly known, which makes it uneasy to determine the appropriate values of G^* and η^*.

3.6. WATER DIFFUSION THEORY

Sheetz [9] noted that the upper limit for compressive capillary pressure due to curvature of water at the surface of a capillary cannot exceed the tenacity of pure water. The order of magnitude of this parameter is 4 MPa, i.e. much less than the values reported by

Brown as required for film formation to occur [1]. Then, Sheetz ruled out the wet sintering of spheres under the only action of polymer-water interfacial tension as a possible mechanism for film formation. As a matter of fact, Sheetz showed experimentally that wet sintering was not able to completely compact the agglomerate of latex particles, although this mechanism plays a role confined to the early stages of film formation.

For these reasons, he proposed a new mechanism based on the diffusion of water through the polymer particles, which begin to emerge from the liquid at the surface of the film. Appearance of particles at film surface generates a compressive capillary force normal to the surface, which causes deformation of particles under the surface [9]. In the same time, the capillary force parallel to the surface brings a compressive contribution, which is strongly dependent on water surface tension. This compression accelerates the closing of the capillaries and results in a film surface sealed over with polymer which behaves as a permeable surface skin.

The energy for particle compaction and deformation within the film is assumed to be supplied as heat from the surroundings, and to be converted to useful work through water evaporation under isothermal conditions. This situation is similar to that of a vessel completely filled with water and covered with a frictionless piston, through which water vapor, but not to liquid water, can diffuse [9].

3.7. AUTOHESION THEORY

Voyutskii claimed that capillary forces and surface tension forces cannot account for the physical properties of latex-cast films, especially as regards mechanical strength [15]. He attributed the enhancement of film properties to autohesion, i.e. mutual interdiffusion of free polymer chains across particle-particle interfaces, which makes them more homogeneous.

This theory likely gives a good description of the further coalescence process, i.e. stage III of film formation, observed upon aging of the films. However, it cannot be considered as relevant for stages I and II of film formation.

4. Stages I and II : Characterization Methods and Recent Results

Since they were proposed, the above theories have been submitted to experimental test and evaluation so as to detect their inadequacies and to bring out the effective driving forces and relevant processes.

4.1. CUMULATIVE WATER LOSS

At first, experimental evaluation was performed using the measure of cumulative water loss with time to monitor the film formation process. This very simple method consists in weighing a latex sample placed in constant-temperature constant-humidity room at regular time intervals. The data obtained can be easily converted to the cumulative amount of water lost by evaporation as a function of time.

The curves classically display three well-defined regions which are correlated with the three steps of film formation, as depicted in Figure 1 [16] : (1) an initial step with constant evaporation rate, which can persist to ~ 60-75 wt% solids ; (2) an intermediate step in which the rate of water evaporation drops off rapidly to a low value ; (3) a final stage in which water evaporates again with a constant rate ~ 10-20 times smaller than that of the initial step.

Vanderhoff et al. [16] thoroughly investigated this process and they clearly showed that the evaporation rate in the first step is identical to that of pure water or dilute emulsifier solution. They also established that the end of this step corresponds to irreversible contact of particles with one another, the remaining water filling the interstices. Then, water evaporation results in the necking of these channels until polymer particles are completely deformed and densely packed.

Although this method can be considered as rather informative, it remains a macroscopic method which cannot easily provide useful information about the actual process, when used alone. For this reason, it must be combined with more local investigation methods, able to probe the ordering and morphology of particles during film formation. New powerful techniques have been recently developed so as to monitor film formation and probe the effect of various parameters on this process. These are environmental scanning electron microscopy (ESEM), small angle neutron scattering (SANS), and atomic force microscopy (AFM).

4.2. ESEM

Very recently, the advent of ESEM has made it possible to study the drying behavior of latexes at the microscopic level. The film formation process can be monitored visually under conditions that mimic drying under ambient conditions, in a specimen chamber where the sample is kept wet, since it is maintained at gas pressures above saturated water vapor pressure.

A successful method has been recently developed by Eckersley et al. [17] to approximate the actual drying of acrylic latexes under usage conditions. These authors have used ESEM to study the effect of molecular weight and cross-linking of the polymer chains on the film formation process. They have shown that uncross-linked particles with a very low molecular weight, referred to as the "highly fusible material", forms a surface skin, prior to complete evaporation of water. However, the nascent film resulting from fusion of particles remains porous enough to allow the passage of water, as evidenced from cumulative weight loss data [17]. It seems that the water flux is not hindered and water loss is not controlled by diffusion through the surface skin. In contrast, this behavior is not observed for another extreme case, i.e. a highly cross-linked sample ; in this latter case, particles protrude through the continuous water phase, while water level recedes through the film.

These observations concerning the formation of a skin of partly-coalesced latex near the surface above a reservoir of water have been confirmed by Keddie et al. [18]. These authors have used ESEM associated with phase-modulated ellipsometry to determine the rate-limiting step in the film formation process. They have shown that, for a "soft" latex, i.e. a latex with a low glass transition temperature (T_g), the rate-limitig step is water evaporation ; particles continuously deform and squeeze water to the surface. For a "hard" latex, i.e. a latex with a high Tg, there exists a drying front which recedes through the film and produces voids near the surface. Then, due to a high surface energy, these voids shrink upon viscous flow of polymers which appears to be the rate-limiting parameter. It is worthwhile noticing that, in this case, film formation only occurs after receding of the drying front, which indicates that surface tension of polymer plays a major part in film formation of "hard" latexes.

4.3. SANS

This powerful technique has been shown to be a well-adapted, direct method for the characterization of particle ordering and film formation [19]. As regards SANS, contrast is classically obtained by isotopic labelling, since the scattering centers are nuclei. Chevalier et al. [19] have used a simple method of labeling, which consists in adding deuterated water in the aqueous phase, hence providing good contrast between the continuous hydrophilic phase and the hydrophobic particle cores. They have obtained good quality SANS spectra for concentrated latexes and studied the occurence of film formation, following the evolution of the interference pattern as a function of volume fraction of polymer. Diffraction spots observed in the scattering experiments indicate that ordering of particles takes place within the concentrated dispersions with a face-centered cubic packing, for volume fractions of polymer ranging ~ 50 %. Then, due to strong repulsions between particles, ordering is retained during removal of water ; particles deform and are compressed to rhombic dodecahedra when volume fraction becomes higher than 74 %, to produce a cellular foam-like structure. This conclusion is only valid for latexes bearing thick polymeric stabilizing membranes grafted at their surface, which have a certain elasticity and connectivity.

When particles are only stabilized by thin membranes of surfactants which are rather mobile, a discontinuity appears in the ordering of particles ; fragmentation of the membranes separating the particle cores occurs before complete drying of the latex. Thus, the membranes appear to be unstable upon compression and deformation, so that they are suddenly expelled to large pools dispersed in the film or to the outer surface of the film [19]. The authors have interpreted this behavior in terms of a coalescence front, which is a boundary between a region of dry film and the liquid dispersion. The driving force for front moving is experimentally shown to be water evaporation ; it increases volume fraction of particles in the liquid dispersion which is compensated by sticking of particles to the front and coalescence with the dry film (Figure 5).

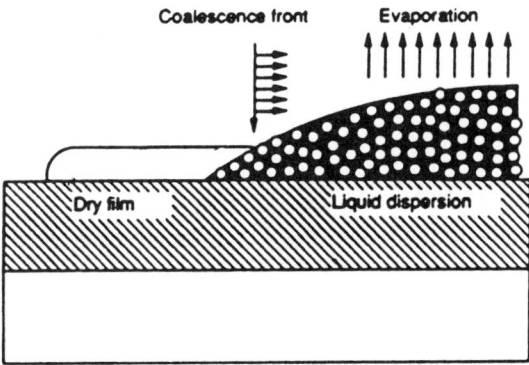

Figure 5 . Schematic representation of the coalescence front for surfactant-stabilized latexes.

408

This interpretation is consistent with a recent model of drying process proposed by Feng et al. [20].

Fragmentation of the membranes during film formation which leads to coalescence of the particles, has been described in terms of phase inversion and macroscopic segregation of hydrophilic material within a hydrophobic matrix [19]. These results have pointed out the utmost practical importance of the stabilizing membranes in the film formation process.

4.4. AFM

Meier et al. [21, 22] have recently started a debate about whether deformation of spheres during step II results in a reduction of their center-to-center spacing as predicted by most models. They have tried to bring an answer to this question and to evaluate a new theoretical model in which the top of the particle surface remains spherical during film formation, but with a changing radius of curvature. The deformation is thus assumed to be unisotropic. They have used AFM in the contact mode to determine the topography of film surface during film formation.

The kinetics of film formation at different temperatures has been followed, by measurement of surface profile to give the reported corrugation heights (i.e. the the peak to valley distance of latex particles, see Figure 3). They have worked with a poly(i-butylmethacrylate) latex, which is a high T_g material, spread as monolayer samples on a mica substrate, under either a dry or wet atmosphere. They have concluded that film formation is driven by polymer surface tension alone in the dry state, while in the wet case both capillary pressure and polymer-water interfacial tension play a significant role. Their results show that the rate of film formation is much faster under the wet condition than in the dry one, and that capillary pressures are high enough to cause deformation of polymer spheres even at temperatures ~ 15 °C below T_g.

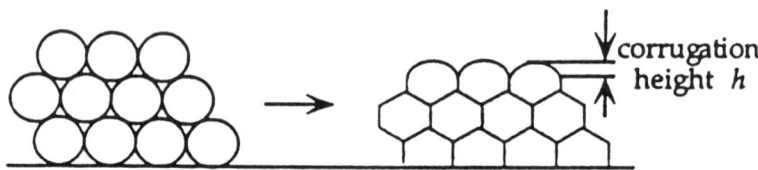

Figure 6. Corrugation heights measured by AFM.

A direct correlation of the kinetics of film formation and rheological properties of polymer has been established through the determination of the time-temperature superposition factor for the time-dependent relaxation modulus. Finally, the rate of film formation has been found to increase as particle size and molecular weight of polymers decrease, as expected from the driving stress for film formation and the viscoelastic response.

4.5. FILM FORMATION IN AN AQUEOUS ENVIRONMENT

A new investigation method has been developed by Dobler [23] to study the influence of polymer-water and polymer-air interfacial tensions on the deformation mechanism of particles during step II. Film formation has been studied for a dense packing of particles immersed in an excess of water, i.e. under conditions where polymer-water interfacial tension γ can be the only existing driving force. γ has been varied by copolymerizing various amounts of methacrylic acid comonomer at the surface of acrylic latexes [23]. Obviously, film formation in water is quite different from standard conditions, since water evaporation cannot play any part in the process then. However, it has been shown experimentally that polymer-water interfacial tension is able to cause film formation in water, even at temperatures much below T_g of the polymer, i.e. when the polymer modulus is as high as 100 MPa [23]. Moreover, the rate of film formation has been found to decrease when γ is decreased.

Very interestingly, kinetics of film formation in water has been compared to that observed when water is allowed to evaporate. It turns out that film formation of dense particle packings under conditions where water evaporation takes place is much faster than for the same packings in water. For this reason, it has been concluded that driving forces which cause particle deformation originates from water evaporation [23].

Comparison of kinetics with or without addition of surfactants in the latex has led to the conclusion that capillary forces do not seem to be involved in the process of particle deformation. This result rules out models proposed by Vanderhoff and Eckersley, and conversely brings a strong argument in favor of Sheetz's theory [9]. Observation of iridescent surface film formed at the very early stage of the process, i.e. for volume fraction ranging ~ 30-40 %, supports this conclusion [23].

5. Stage III : Characterization Methods for Structure Ripening and Recent Results

Until a few years ago, most of the attention in the literature about film formation was mainly devoted to stages I and II. The purpose was to develop an understanding of the specific driving forces responsible for particle deformation and the viscoelastic features which describe the polymer response to these forces. In addition, early studies by electron microscopy already demonstrated that, at the end of stage II, films have a honeycomb-like cellular structure [11], which is able either to disappear over time [16], or conversely to persist for months [24], depending on aging conditions and latex features.

In the past five years, more attention has been paid to stage III of film formation. The aim of the studies in this field has been to clarify the specific issues of whether, how and to what extent polymer chains within each particle are able to diffuse across the particle boundaries, i.e. the interfacial membranes. Another interesting purpose of the studies about further coalescence has been to investigate the ripening process of the film structure upon aging, and to determine the influence of latex structure and composition on this mechanism.

Transmission electron microscopy (TEM) of latex films, ultramicrotomed and appropriately stained, used to be the traditional method to examine film morphology and structure ripening [11, 16, 24, 25]. Alterations of film morphology upon aging and annealing were observed using this technique. Results suggested that polymer segment diffusion occurs across the interfaces when particle boundaries disappear. However, polymer diffusion needed to be quantified, so as to relate interdiffusion to the evolution of film morphology [26]. For this purpose, new experimental tools have been developed recently to examine stage III of film formation at the molecular level. These are direct non radiative energy transfer (DET) using fluorescence decay measurements, and small angle neutron scattering (SANS) using deuterated species to get contrast. The data obtained have been treated so as to derive chain diffusion coefficients and to relate further coalescence mechanism to polymer chain dynamics. Moreover, these techniques have been combined with a new imaging method, namely atomic force microscopy (AFM), and also spectroscopic techniques. The former method makes it possible to get topographic images of surfaces with high resolution. In addition, all these surface analysis techniques have been used to elucidate the fate of the hydrophilic stabilizing material, such as emulsifiers or polyacrylic acid chains for instance, during the further coalescence process.

5.1. DET

The Winnik group has been the first one to introduce DET technique to study the process of interparticle polymer diffusion in latex films [27]. They have used particles labeled with appropriate fluorescent groups, one with a dye such as phenanthrene which acts as an energy donor, and the other such as anthracene which acts as an energy acceptor [27-29]. The donor fluorescence decay profile is analyzed in terms of energy transfer through Förster mechanism. The perturbation introduced by the presence of acceptor dyes results in a strong deviation of the fluorescence decay from the natural unquenched purely exponential profile. It is then fitted and processed so as to quantify the state of mixing of donor- and acceptor-labeled chains [27].

The method has been applied either to 1 μm diameter poly(methyl methacrylate) particles or to 100 nm diameter poly(butyl methacrylate) (PBMA) particles. Results have demonstrated that polymer diffusion takes place across particle boundaries. Polymer diffusion coefficients have been derived using a planar sheet or a spherical model for Fickian diffusion of polymer chains, when diffusion is observed over sufficiently long times. In the case of PBMA latexes, a good agrement with the Fickian diffusion model has been found for polymer chains with relatively small average molecular weight, i.e. $M_W \sim 7.6 \times 10^4$ g mole^{-1}. A deviation from the theoretical prediction has been evidenced for chains with higher M_W values, which are likely to give rise to the formation of entanglements ; it is attributed to a larger M_W dependence of polymer diffusion for long entangled chains ($D \sim M_W^{-2}$) than for the short chains ($D \sim M_W^{-1}$) [28, 29]. The prominence of viscoelastic behavior of polymers has been established. It has been shown to result in a strong dependence of interdiffusion on factors such as temperature, molecular weight and distribution, and also the addition of low molecular weight diluents ("coalescing aids") which act as plasticizers to decrease the local constraints on the chains [29, 30]. An apparent activation energy for polymer interdiffusion has been derived for PBMA latexes and found to be nearly identical to that reported for the viscoelastic behavior of PBMA in bulk [28].

5.2. SANS

SANS has also been shown to be a very powerful tool to study the process of chain diffusion across interfaces in latex films, provided that deuterated species can be introduced in the films to get good contrast for neutron scattering. As a matter of fact, neutrons are scattered by nuclei of atoms and scattering is controlled by the density of scattering length. This parameter can be adjusted by isotopic substitution or labeling.

Two basic ideas have been exploited for this purpose. The former consists in forming films which contain a small amount of perdeuterated particles statistically distributed in a matrix of protonated latex particles. Then, the variations of the radius of gyration R_g of the deuterated chains can be investigated as a function of annealing time and temperature, and the diffusion coefficient of the chains across interfaces can be derived [31, 32]. The latter method consists in rehydrating dried latex films with D_2O in the vapor or liquid form. D_2O selectively labels the hydrophilic material which is mainly located in the interfacial membranes in the films [33]. Then, small angle scattering reflects the distribution of hydrophilic material within the hydrophobic medium made of particle cores. For this reason, it is possible to study the ripening of the ordered cellular structure of latex films, since interdiffusion strongly modifies the distribution of the hydrophilic material in the samples [33-35].

Using the former method, the BASF group [31, 32] has shown that the further coalescence process is actually related to a massive interdiffusion of chains of different particles. Furthermore, these authors have been able to derive the polymer diffusion coefficient D from the variations of R_g with time. The values of D obtained for latexes of PBMA polymer are very similar to that published by the Winnik group based on DET data. Hahn et al. [31, 32] have also investigated the effect of degree of cross-linking on the diffusion coefficient D. Unfortunately, their results have appeared to be fully inconsistent with polymer diffusion theory, since D has been found to increase with cross-linking. They have interpreted their data as follows : interdiffusion of the deuterated chains in the cross-linked rubber particles has not been possible ; diffusion of mobile chains could only occur on the surface of the cross-linked particles, in the interstices generated by an incomplete deformation of these particles.

Using the latter technique, the Rhône-Poulenc group [33-35] has performed a thorough investigation of the ripening process in films obtained from latexes stabilized by copolymerized carboxylic acid groups. As evidenced from TEM experiments, their films exhibit a cellular structure consisting of hydrophobic deformed particle cores regularly separated by a network of hydrophilic interfacial membranes (Figure 7). Their work has been focused on the evolution in the position and intensity of neutron diffraction peaks appearing in the neutron scattering pattern [35]. Position of the peaks is related to both spacings between planes of particles and particle diameters [34]. Vanishing of the peaks is considered as an evidence for destruction of interfacial membranes and hence interdiffusion of polymer chains.

Fragmentation of the membranes and their expulsion into large lumps dispersed in an hydrophobic matrix has been shown to occur upon annealing the films. The cross-linking density within the particle cores, together with the composition and neutralization of the membranes, have been varied so as to study the part played by both mobility of core chains and mobility of membranes in the further coalescence process [34, 35]. Then, key issues have been brought out from SANS experiments. It turns out

412

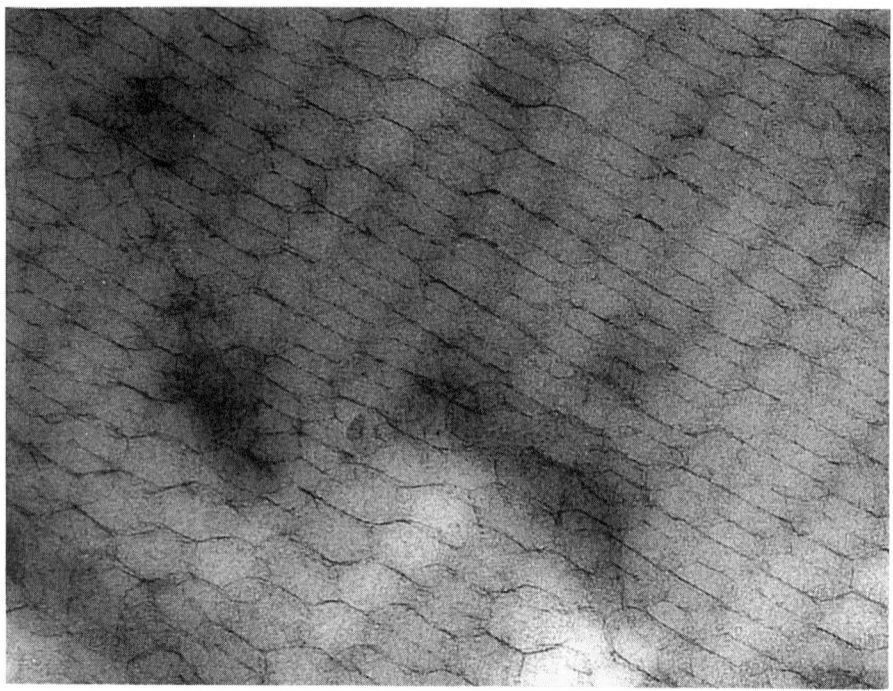

Figure 7. TEM photomicrograph of an ultramicrotomed stained film from a latex stabilized by copolymerized carboxylic acid groups (magnification : 70,000).

that highly cross-linked cores will never fuse together, even if the hydrophilic membranes are very mobile (surfactant membranes, for instance) or if temperature is raised : cross-linking of the cores completely prevents interdiffusion as evidenced from observation of SANS spectra (Figure 8).

Conversely, very mobile uncross-linked core chains cannot interdiffuse if the network of membranes remain rigid, for instance when temperature remains lower than their apparent glass transition temperature or when they have been physically cross-linked through neutralization of surface carboxylic groups [34].

Finally, the occurence of further coalescence has been shown to be related to the dynamical viscoelastic features of the polymer, derived from dynamic micromechanical analysis (DMA) spectra [35]. More precisely, local chain segment mobility has been calculated from DMA spectra and related to the translatory diffusion coefficient of chain segments, and then to the curvilinear diffusion coefficient of the chains. Semi-quantitative correlations have been established between these parameters and the occurrence of interdiffusion. These results show that the diffusion process is controlled by local molecular dynamics of the chains [35], in agreement with conclusions drawn from DET results [29].

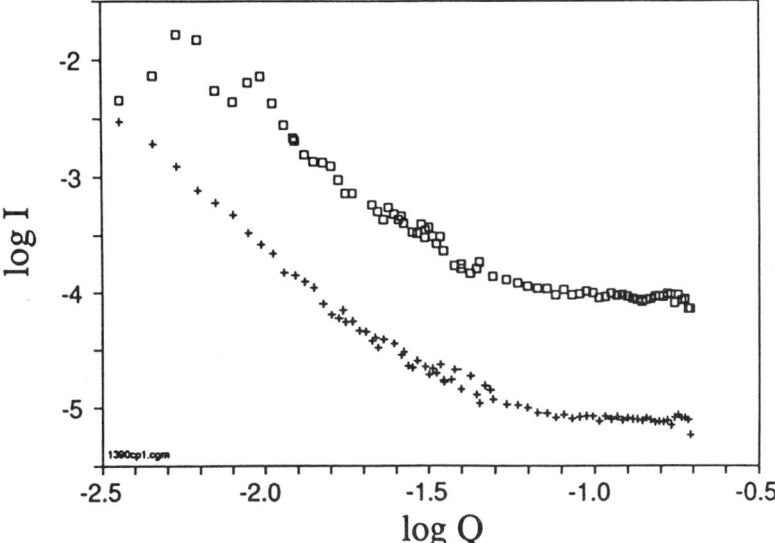

Figure 8. SANS spectra of latex films with highly cross-linked (☐) or uncross-linked (+) cores, stabilized by mobile surfactant membranes.

5.3. AFM

AFM has been mainly used to study the effect of surfactant post-added to latex dispersions, on surface topography of dried films. Both cases of anionic and nonionic surfactants derived from nonylphenol poly(glycol ether) have been considered. The roughness of film surface and the packing of particles have been investigated as a function of surfactant concentration [36, 37]. An optimal surfactant concentration corresponding to the full coverage of latex particles has been found to lead to a high packing order of particles at the film surface resulting in a low roughness. Results have been interpreted on the basis of stabilizing repulsions of either electrostatic [36] or steric [37] origin. In addition, it has been suggested that depletion interactions created by micelles, are responsible for increasing roughness when surfactant concentration becomes higher than the critical micellar concentration [37].

More recent work has dealt with the study of surfactant exudation in latex films. In the case of sodium dodecylsulfate (SDS) surfactant, it has been shown that annealing leads to a massive exudation of the surfactant towards the film surface when the latex is formulated with a coalescing aid. After exudation, SDS forms crystallites on the top of the film. The morphology of the exudates at the surface of the films, as well as their size and number have been investigated as a function of annealing time in different regions of film surface [38].

Finally, AFM has also been used to investigate the morphology of composite latex particles [39], core-shell particles [40], blends of particles with different glass transition temperatures [41] and films obtained from these latexes. Interestingly, microscopic surface topography has been related to macroscopic parameters like hardness and gloss [41].

5.4. SPECTROSCOPIC METHODS FOR SURFACE ANALYSIS

Spectroscopic methods have been developed to get insight into the composition of the surface layer of latex films. The evolution of this composition has been followed as further coalescence proceeds and low molecular weight species migrate throughout the film [42,43]. Zhao et al. [42] have performed surface analysis of latex films using three techniques for surface analysis, namely : attenuated total reflection with Fourier transform infrared spectroscopy (ATR-FTIR), X-ray photoelectron spectroscopy (XPS) and secondary ion mass spectroscopy (SIMS). The sampling depth of these techniques is ~ 3 μm, ~ 50 Å, and ~ 10 Å respectively. The concentration profile of anionic surfactants near the film interfaces has been characterized. An enrichment with surfactants has been shown to occur at both film-air and film-substrate interfaces, with a greater effect at the film-air interface. As further coalescence proceeds, concentration of surfactants at interfaces has been found to increase with aging time and to strongly affect the adhesion properties of latex films on a polypropylene substrate. It is worthwhile noticing that adhesion performance can be either enhanced or deteriorated depending on the nature of surfactant used for latex stabilization [42].

More recently, surface analysis methods such as XPS [44, 45] and contact angle measurements [44] have also been used to characterize the distribution of copolymerized methacrylic acid groups in latex films obtained from core-shell particles.

6. Effect of Structure Modifications on Film Properties

Structure of latex films has been very early recognized as a key feature governing film properties. This is the reason why the past forty years have witnessed so great a deal of research carried out in the field of film formation, with the intend to control film properties through control of film structure. Basically, two types of properties have been investigated as regards relationships with film structure. These are mechanical properties, including either viscoelastic behavior or ultimate strength, and permeability properties, including gas, water vapor and liquid water permeability.

6.1. MECHANICAL PROPERTIES

Dynamic viscoelastic properties of latex films obtained from particles bearing surface carboxylic groups have been shown to be strongly dependent on film structure [23, 46, 47]. In most of these films, when the carboxylic group content is higher than ~ 2 wt%, the cellular honeycomblike structure is preserved, unless the films are annealed at sufficiently high temperature or the core chains are very short and mobile. Then, the preservation of the hydrophilic interfacial membranes within the film has been associated with strong modifications in the viscoelastic spectra, which are large increases in both the storage (E') and loss (E") moduli in the rubbery region [46, 47]. These modification are attributed to specific interactions between carboxylic groups such as H bonding interactions, responsible for interfacial cross-linking within the membranes.

When carboxylic group content becomes higher than ~15 wt%, the rubbery modulus enhancement is accompanied by the appearence of a second transition, apart from the main glass transition of the films. This indicates that the membranes behave as a continuous segregated carboxylic phase with its own glass transition, and is macroscopically connected throughout all the film [46]. When the core chains are very mobile (short, uncross-linked polymer), neutralization of surface carboxylic groups leads to the preservation of cellular structure and results in similar effects, as regards viscoelastic behavior, i.e. an increase of the rubbery modulus and the appearence of a second transition [47] (Figure 9). Finally, a similar modification in the viscoelastic spectra has also been reported for latexes sterically-stabilized by partly grafted poly(vinyl alcohol) (PVA) chains, when annealing conditions enable PVA membranes to form a continuous network in the films [48].

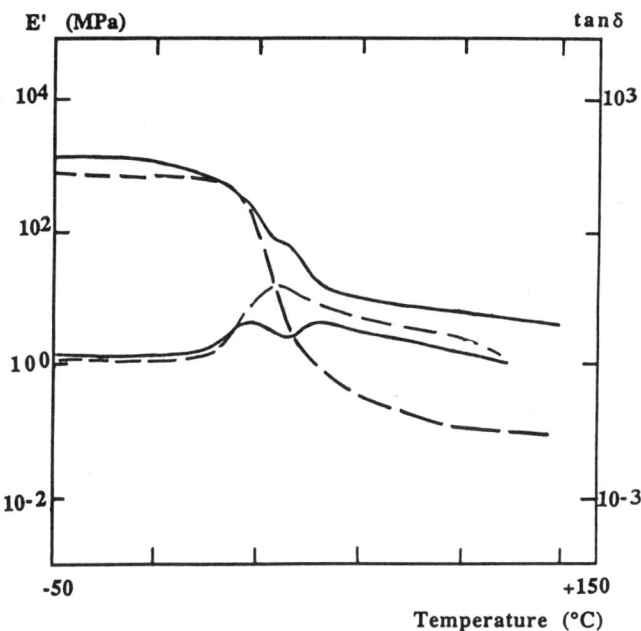

Figure 9. Effect of neutralization of surface carboxylic groups on isochronal DMA spectra (storage modulus E' and loss tangent tan δ = E''/E'). Full line : latex neutralized at pH 9 with NaOH ; dotted line : non-neutralized latex.

Hence, for all the above cases, relationships between structure modifications upon ripening and viscoelastic behavior have been definitely established [46-48]. Moreover, it is worthwhile noticing that the above effects are not observed for solvent-cast films of polymer latexes in which the network of membranes does not exist [23, 49].

As far as ultimate mechanical strength is concerned, the BASF group [50] has studied the effect of further coalescence. They have shown that annealing greatly enhances the mechanical strength of films obtained from uncross-linked latexes. According to these authors, this process takes place in two main steps after film formation. First, fast interdiffusion of chain ends and small chains in the interfacial membranes leads to a transition from a brittle to tough fracture behavior. Secondly, the final mechanical strength slowly develops by interdiffusion of long chains and formation of entanglements. This latter step has been identified as a process of critical importance for toughness development [50]. The interdiffusion process and formation of entanglements are highly hindered by cross-linking the chains, which tends to restrict ultimate mechanical strength of the films.

6.2. PERMEABILITY PROPERTIES

Chainey et al. [51] have thoroughly studied the helium permeability of latex films as a function of aging time, and compared it with helium permeability of films cast from solutions of the same polymers. Gas permeability of the two types of films have been shown to differ considerably. Just after film formation, permeability coefficients of latex films have been found to be considerably higher than that of the corresponding solvent-cast films. Moreover, they drop upon aging. This behavior is attributed to the occurence of the further coalescence process, i.e. interdiffusion of chains across particle boundaries (stage III of film formation), which reduces the diffusion of gas molecules in the interparticle regions. However, permeability coefficients of latex films have always been found to level off at a value which remains higher than that of the corresponding solvent-cast films. This result has been interpreted as showing that latex films never become completely homogeneous [51].

More recently, water vapor permeability properties of latex films have also been investigated as regards the effect of surfactants, cross-linking and particle surface functionalization [33, 52]. Diffusion of water molecules has been shown to be restricted to the interfacial network of hydrophilic membranes [33]. Once again, a reduction of permeability upon aging has been evidenced, which is attributed to further coalescence [52]. Permeablity has been shown to be mainly governed by the solubility of water vapor in the films, i.e. the affinity and the number of sorption sites for water in the film. When it is preserved, the network of membranes form a more permeable route through the films, which results in high water vapor permeabiliy values [33, 52]. Similar conclusions have been drawn for the permeation of liquid water, and electrolyte- or sucrose-containing aqueous solutions through latex films [53]. However, some authors earlier suggested that, after completion of further coalescence, diffusion of water should also take place through the particle core itself [16].

7. Concluding Remarks

A great deal of research has been carried out on both formation and structure-property relationships of latex films for about forty years. Most of the efforts have been devoted to stages I and II of the film formation process, and many different mechanisms and models have been proposed to describe this process. A lot of experimental evaluations have been performed, using sophisticated methods such as ESEM, SANS and AFM, so as to check the validity of these models and to bring out the actual driving forces for film formation.

However, until now, no unambiguous conclusion can be drawn from the experimental results, concerning the effective driving forces and relevant processes. The only definitely established issue which is admitted by most of the authors today, is that water evaporation plays a major part in particle gathering and deformation. Hence, driving forces for film formation should be associated to water evaporation. At this point, two main descriptions given in the literature seem to be in agreement with this statement. These are models based on the combination of capillary and interfacial forces [8, 13], derived from Brown's [1] and Vanderhoff's theories [10,11], on one hand, and Sheetz's theory of water diffusion through a permeable surface skin [9], on the other hand.

Many experimental results show the occurence of a non homogeneous drying of latex films with an accelerated film formation near the surface, and hence bring strong arguments in favor of Sheetz's model [17-19, 23]. However, a point of critical importance remains very controversial : is water diffusion actually hindered by the surface skin [23], or conversely does this thin nascent film remain very porous, so that it cannot control water diffusion [17]? This question is still open and experimental evidence is still required. In addition, at least the mechanism for formation of the surface skin is likely to be driven by capillary and interfacial forces, the deformation of particles being controlled by viscoelastic properties of particles [8, 13, 23]. From these remarks, it turns out that a single mechanism cannot be proposed yet for film formation. Behind the apparent simplicity of its phenomenological description, this process remains rather complex and dependent on many parameters, such as ambient conditions (relative humidity, temperature) [23] and type of polymer [18].

Although more recently investigated, the mechanism of stage III, i.e. further coalescence and structure ripening, appears to be rather well-elucidated. It is mainly based on massive polymer diffusion through particle interfaces and experimental evidence of this mechanism has been provided using DET and SANS [29, 31]. Moreover, it can be quantitatively related to polymer diffusion features, such as chain diffusion coefficient D [29, 31, 35]. In the special case where interfacial membranes made of hydrophilic material grafted on the particle cores are preserved in the films, ripening occurs through fragmentation of the membranes and expulsion of the hydrophilic material into large lumps dispersed in the hydrophobic matrix [34]. However, the question arises to know whether significant diffusion of short chains takes place in these films, prior to membrane fragmentation and expulsion.

Finally, the practical importance of the control of the film formation process has been emphasized by many authors. As a main issue, correlations between film properties, and final film structure resulting from further coalescence, have been established [33, 47, 50, 51].

From a more prospective viewpoint, it appears that considerable advances are being made nowadays in the design of two new types of polymer latexes. These are composite particles with controlled morphology [54, 55] and new reactive cross-linkable systems such as ambient-curable and heat-activated thermosetting polymer latexes [56, 57]. In these new systems, film formation process remains a critical step to be controlled and deserves to be thoroughly investigated. As a matter of fact, in composite systems, final morphology of the films will determine physical properties [54], and in reactive systems, diffusion kinetics of reactive chains during film formation will strongly influence the cross-linking mechanism together with the structure and performances of the final cross-linked films [57, 58].

418

8. References

1. Brown, G.L. (1956) Formation of films from polymer dispersions, *J. Polym. Sci.* **22**, 423-434.
2. Overbeek, J.Th.G. (1977) Recent developments in the understanding of colloid stability, *J. Colloid Interface Sci.* **58**, 408-422.
3. Napper, D.H. (1977) Steric stabilization, *J. Colloid Interface Sci.* **58**, 390-407.
4. Dillon, R.E., Matheson, L.A. and Bradford E.B. (1951) Sintering of latex particles, *J. Colloid Sci.* **6**, 108-117.
5. Frenkel, J. (1943) Viscous flow of crystalline bodies under the action of surface tension, *J. Phys. (U.S.S.R.)* **9**, 385-398.
6. Mason, G. (1973) Formation of films from latices. A theoretical treatment, *Br. Polym. J.* **5**, 101-108.
7. Lamprecht, J. (1980) Ein neues Filmbildungskriterium für wäβrige Polymer dispersionen, *Colloid Polym. Sci.* **258**, 960-967.
8. Eckersley, S.T. and Rudin, A. (1990) Mechanism of film formation from polymer latexes, *J. Coat. Technol.* **62** (780), 89-100.
9. Sheetz, D.P. (1965) Formation of films by drying of latex, *J. Appl. Polym. Sci.* **9**, 3759-3773.
10. Vanderhoff, J.W., Tarkowski, H.L., Jenkins, M.C. and Bradford, E.B. (1966) Theoretical considerations of the interfacial forces involved in the coalescence of latex particles, *J. Macromol. Chem.* **1**, 361-372.
11. Vanderhoff, J.W. (1970) Mechanism of film formation of latices, *Br. Polym. J.* **2**, 161-173.
12. Brodnyan, J.G. and Konen T. (1964) Experimental study of the mechanism of film formation, *J. Appl. Polym. Sci.* **8**, 687-697.
13. Eckersley, S.T. and Rudin, A. (1994) The film formation of acrylic latexes : a comprehensive model of film coalescence, *J. Appl. Polym. Sci.* **53**, 1139-1147.
14. Eckersley, S.T. and Rudin, A. (1993) The effect of plasticization and pH on film formaion of acrylic latexes, *J. Appl. Polym. Sci.* **48**, 1369-1381.
15. Voyutskii, S.S. (1963) *Autohesion and Adhesion of High Polymers*, Interscience Publisher, New York.
16. Vanderhoff, J.W., Bradford, E.B. and Carrington, W.K. (1973) The transport of water through latex films, *J. Polym. Sci. Symp.* **41**, 155-174.
17. Eckersley, S.T. and Rudin, A. (1994) Drying behavior of acrylic latexes, *Prog. Org. Coat.* **23**, 387-402.
18. Keddie, J.L., Meredith, P., Jones, R.A.L. and Donald A.M. (1995) Rate-limiting steps in the film formation of water-borne acrylic latices as elucidated with ellipsometry and environmental SEM, *Polym. Mater. Sci. Eng.* **73**, 144-145.
19. Chevalier, Y., Pichot, C., Graillat, C. Joanicot, M., Wong, K., Maquet, J., Lindner, P. and Cabane, B. (1992) Film formation with latex particles,*Colloid Polym. Sci.* **270**, 806-821.
20. Feng, J. and Winnik, M.A. (1995) Latex blends and kinetics of drying of latex dispersions, *Polym. Mater. Sci. Eng.* **73**, 90-91.
21. Meier, D.J. and Lin, F. (1995) Theoretical aspects of film formation, *Polym. Mater. Sci. Eng.* **73**, 84-85.
22. Lin, F. and Meier, D.J. (1995) AFM studies of latex film formation, *Polym. Mater. Sci. Eng.* **73**, 93-94.
23. Dobler, F. (1991) Mécanismes de Coalescence des Latex, *Thèse de Doctorat de l'Université Louis Pasteur de Strasbourg.*
24. Distler, D. and Kanig, G. (1978) Feinstruktur von Polymeren aus wäβriger Dispersion, *Colloid Polym. Sci.* **256**, 1052-1060.
25. Kast, H. (1985) Aspects of film formation with emulsion copolymers, *Makromol. Chem., Suppl.* **10/11**, 447-461.
26. Wang, Y., Kats, A., Juhué D., Winnik, M.A., Shivers, R.R. and Dinsdale, C.J. (1992) Freeze-fracture studies of latex films formed in the absence and presence of surfactant, *Langmuir* **8**, 1435-1442.
27. Pekcan, Ö, Winnik, M.A. and Croucher M.D. (1990) Fluorescence studies of coalescence and film formation in poly(methyl methacrylate) nonaqueous dispersion particles, *Macromolecules* **23**, 2673-2678.
28. Wang, Y., Zhao, C. and Winnik, M.A. (1991) Molecular diffusion and latex film formation : an analysis of direct non radiative energy transfer experiments, *J. Chem. Phys.* **95**, 2143-2153.
29. Wang, Y. and Winnik, M.A. (1993) Polymer diffusion across interfaces in latex films, *J. Phys. Chem.* **97**, 2507-2515.
30. Juhué, D. and Lang, J. (1994) Latex film formation in the presence of organic solvents, *Macromolecules* **27**, 695-701.
31. Hahn, K., Ley, G. Schuller, H. and Oberthür R (1986) On particle coalescence in latex films, *Colloid Polym. Sci.* **264**, 1092-1096.
32. Hahn, K., Ley, G. and Oberthür R (1988) On particle coalescence in latex films (II), *Colloid Polym. Sci.* **266**, 631-639.
33. Richard, J., Mignaud, C. and Wong, K. (1993) Water vapour permeability, diffusion and solubility in latex films, *Polym. Int.* **30**, 431-439.
34. Joanicot, M., Wong, K., Richard, J. Maquet, J. and Cabane, B. (1993) Ripening of cellular latex films, *Macromolecules* **26**, 3168-3175.

35. Richard, J. and Wong, K. (1995) Interdiffusion of polymer chains and molecular dynamics in dried latex films, *J. Polym. Sci. Part B Polym. Phys.* **33**, 1395-1407.
36. Juhué, D. and Lang, J. (1993) Effect of surfactant postadded to latex dispersion on film formation : a study by atomic force microscopy, *Langmuir* **9**, 792-796.
37. Juhué, D. and Lang, J. (1994) Latex film surface morphology studied by atomic force microscopy : effect of a non-ionic surfactant postadded to latex dispersion,*Colloids Surfaces A : Physicochem. Eng. Aspects* **87**, 177-185.
38. Juhué, D., Wang, Y., Lang, J. Leung, O.M., Goh, M.C. and Winnik M.A. (1995) Surfactant exudation in latex films, *Polym. Mater. Sci. Eng.* **73**, 86-87.
39. Butt, H.J. and Gerharz, B. (1995) Imaging homogeneous and composite latex particles with atomic force microscpe, *Langmuir* **11**, 4735-4741.
40. Sommer, F., Duc, T.M., Pirri, R., Meunier, G. and Quet, C. (1995) Surface morphology of poly(butyl acrylate)/poly(methyl methacrylate) core-shell latex by atomic force microscopy, *Langmuir* **11**, 440-448.
41. Butt, H.J. and Kuropka R. (1995) Surface structure of latex films, varnishes, and paint films studied with an atomic force microscope, *J. Coat. Technol.* **67** (848), 101-107.
42. Zhao, C.L., Holl, Y., Pith, T. and Lambla M. (1989) Surface analysis and adhesion properties of coalesced latex films, *Br. Polym. J.* **21**, 155-160.
43. Urban, M.W. (1995) Mobility of surfactants and latex film formation, *Polym. Mater. Sci. Eng.* **73**, 137-138.
44. Dobler, F., Affrossman, S. and Holl, Y. (1994) Surface analysis of model latex particles, *Colloids Surfaces A : Physicochem. Eng. Aspects* **89**, 23-35.
45. Arora, A., Daniels, E.S., El-Aasser, M.S., Simmons, G.W. and Miller A. (1995) Synthesis and characterization of core-shell ionomeric latexes. II. Surface analysis by X-ray photoelectron spectroscopy, *J. Appl. Polym. Sci.* **58**, 313-322.
46. Zosel, A., Heckmann, W., Ley, G. and Mächtle, W. (1987) Chemical heterogeneity in emulsion copolymers of carboxylic monomers,*Colloid Polym. Sci.* **265**, 113-125.
47. Richard, J. and Maquet, J. (1992) Dynamic micromechanical investigations into particle/particle interfaces in latex films, *Polymer* **33**, 4164-4173.
48. Richard, J. (1993) Thermomechanical behaviour of composite polymer films obtained from poly(vinyl acetate) latexes sterically stabilized by poly(vinyl alcohol), *Polymer* **34**, 3823-3831.
49. Charmeau, J.Y. Kientz, E. and Holl Y. (1995) Effects of film structure on mechanical and adhesion properties of latex films, *Polym. Mater. Sci. Eng.* **73**, 48-49.
50. Zosel, A. and Ley, G. (1993) Influence of cross-linking on structure, mechanical properties and strength of latex films, *Macromolecules* **26**, 2222-2227.
51. Chainey, M., Wilkinson, M.C. and Hearn, J. (1985) Permeation through homopolymer latex films, *J. Polym. Sci. Polym. Chem. Ed.* **23**, 2947-2972.
52. Roulstone, B.J., Wilkinson, M.C. and Hearn, J. (1992) Studies of polymer latex films: II. Effect of surfactants on the water vapour permeability of polymer latex films, *Polym. Int.* **27**, 43-50.
53. Steward, P.A., Hearn, J. and Wilkinson, M.C. (19995) Studies on permeation through polymer latex films, I. Films containing no or only low levels of additives, *Polym. Int.* **38**, 1-12.
54. Hidalgo, M., Cavaillé, J.Y., Guillot, J. Guyot, A., Pichot, C., Rios, L. and Vassoille, R. (1992) Polystyrene (1) -poly(butyl acrylate - methacrylic acid) (2) core-shell emulsion polymers. Part II : Thermomechanical properties of latex films, *Colloid Polym. Sci.* **270**, 1208-1221.
55. Vandezande, G.A. and Rudin, A. (1994) Novel composite latex particles for use in coatings, *J. Coat. Technol.* **66** (828), 99-108.
56. Craun, G.P. and Kimberley, D.S. (1993) Transesterification cure of thermosetting latex coatings, *U.S. Patent* **5,260,356.**
57. Inaba, Y., Daniels, E.S. and El-Aasser, M.S. (1994) Film formation from conventional and miniemulsion latex systems containing dimethyl meta-isopropenyl benzyl isocyanate (TMI)-- A functional monomer crosslinking agent, *J. Coat. Technol.* **66** (832) , 63-74.
58. Geurts, J.M., van Es J.J.G.S. and German A.L. (1995) Latexes with intrinsic crosslink activity, *Proceedings of the 21st International Conference in Organic Coatings Science & Technology (Athens)*, p. 221-235.

LATEX PAINT FORMULATIONS

JULIAN A. WATERS
University of Bristol
School of Chemistry
Cantock's Close
Bristol BS8 1TS
UK

1. Introduction

Latex paints have developed commercially over the last four decades or so to become the major products by volume in the coatings industry. Paints and coatings have been described in texts [1]. Nearly all latex paints are water-based and most are designed for the "decorative" or "architectural" paint market.

Non-aqueous latex paints are available commercially, but the output volume is small; the science and technology of the latices for these products has been considered elsewhere [2]. With a continuous phase of aliphatic hydrocarbon and the avoidance of water-sensitive components in their formulation, the paints offer a number of advantages especially for exterior use in terms of durability and application, but are obviously less acceptable on environmental grounds.

The formulations of latex paints have been developed partly to provide improving technical performance and partly to meet changing customer requirements. The features sought by customers vary in different countries, influenced by climatic conditions and areas of differing major use. The paint formulator has an understanding of how the paint properties, characteristics and aesthetics can be altered. Suppliers of components for the paints also benefit from knowing what the formulator has to do in order to move the properties and characteristics in a particular direction.

Aqueous latices are used in the manufacture of a wide range of decorative paint products, including primers and undercoats as well as the finishing or "top-coat" systems. For the latter, sheen levels vary from "matt" through various midsheens to semi-gloss (described as "silk" in some countries) to relatively high gloss. In addition to providing the correct sheen level, the paint formulation must be designed to give the required application and technical performance as a dry paint film. Latices continue to be developed for paints such as those required by car manufacturers and for "refinish" paints which are used following repairs to car bodies. Although such paints can be properly described as latex paints, they are of specialist composition and their polymer base is complex, and not confined to latex, and they will not be included here.

J. M. Asua (ed.), Polymeric Dispersions: Principles and Applications, 421–433.

2. Latex Paint Components

The latex paints are mixtures of two aqueous dispersions with a number of components minor by volume and often a co-solvent liquid (miscible with water) included. One of the dispersions is a high-solids latex and the other, the "mill-base", is produced by milling inorganic pigment and extender (filler) material in water with water-soluble polymers and other dispersants.

2.1. LATEX SELECTION

Almost all of the paint characteristics are influenced by the choice of the latex. Firstly, there is a choice between colloid-containing and colloid-free latex. In this context, the term "colloid" refers to a water-soluble polymer which is present in the early stage of the manufacture of the latex by emulsion polymerisation.

2.1.1 Colloid-containing Latices
Water-soluble polymers commonly employed as colloid include:
    substituted cellulose, e.g. hydroxyethyl cellulose
    partially-hydrolysed poly(vinyl acetate) (PVA)
    acrylic acid copolymer
    poly(vinyl pyrrolidone)
Usually the colloid is dissolved in the aqueous charge before adding monomer and commencing emulsion polymerisation. It is therefore subjected to attack by free radicals; this can result in scissioning of the polymer chains, especially with the celluloses, and to the formation of radicals pendant from the chains which consume some of the monomer to produce polymer tails or grafts. After completion of the polymerisation process, the colloid may be very different from its starting composition. It may have significantly enhanced surface-activity for example; in this event, attempts to manufacture the latex by addition of the colloid only after polymerisation, fail to achieve the same characteristics.

Characteristics which usually arise from the presence of colloid are shown below:

TABLE 1. Characteristics arising with colloid in latex

Latex	Derived Paint
enhanced colloidal stability - appears to be steric increased viscosity larger particle size	stability to freeze-thaw cycles, high electrolyte and high shear increase in pseudoplastic rheology [#]effective titanium complex thixotrope

[#]with cellulose and PVA

2.1.2 Colloid-free Latices
The presence of water-soluble polymer in the latex composition adversely affects some of the properties of the derived paints. In particular, paint characteristics involving performance under wet conditions such as scrub resistance and condensation staining can be impaired. These weaknesses can be avoided or reduced by selecting latex which has been manufactured without colloid, at the loss of the benefits above.

Colloid-free latex for paint use is manufactured by emulsion polymerisation processes, usually with a combination of anionic and non-ionic surfactants to provide colloidal stability for the particles. Ionic residues from the initiator which are attached at one end of polymer chains, contribute to the stability. The latices tend to be lower in viscosity and to have smaller particle sizes than colloid-containing systems.

2.1.3 Glass Transition Temperature

A number of factors influence the choice of the glass-transition-temperature (Tg) for the latex particles. It is necessary to ensure that good film integration occurs when the paint is applied over a range of temperatures down to some stipulated minimum. However, this is counter-balanced by a desire to maintain reasonable film-hardness for durability, resistance to dirt-pick-up and washability.

The important issue is the effective Tg, that is allowing for possible plasticisation of the polymer by other components in the paint, such as cosolvent or surfactants. A useful parameter here is "minimum film forming temperature" (MFT) which can be measured by spreading a sample of latex or modified latex over a metal platen over which a temperature gradient is maintained and identifying the minimum temperature at which adequate film integration has occurred [3]. Generally, with a higher loading of pigment and extender in the paint, a lower MFT is required (below):

TABLE 2. MFT requirements for different paints

Paint Type	Approximate MFT for Latex
Matt	5°C
Primer/Undercoat	7°C
Mid-sheen	15°C
Gloss	20°C

It would be expected that MFT was closely related to Tg. The relationship has been found to be linear for some colloid-containing latex types and non-linear for some colloid-free systems [4].

Latex particles for paints usually comprise copolymers and the Tg can be selected by changing the composition using the Fox equation [5] where the Tg for the copolymer (T_C) is given by

$$\frac{1}{T_C} = \frac{W_X}{Tg_X} + \frac{W_Y}{Tg_Y} \tag{1}$$

where the copolymer comprises a weight fraction (W) of the respective component which would have, as a homopolymer, a glass transition temperature (Tg).

2.1.4 Copolymer Particles

To provide the film polymer with a Tg in the desired range and with a selected balance of physical properties, the latex particles comprise copolymer with two or three constituents. The third and minor monomer constituent which may be present at only a few percent of the total composition, is often included in the copolymer to provide benefits to the paint or to the latex during storage prior to paint make-up. Some of the characteristics provided are listed (Table 3).

TABLE 3. Commonly used monomers

Principal Monomer	Plasticising Comonomer	Minor Monomer	Characteristics from Minor Monomer
methyl-methacrylate	butyl-acrylate 2-ethyl-hexyl-acrylate	acrylic acid or methacrylic acid	contributes to colloidal stability; increases paint viscosity.
styrene	butyl-acrylate 2-ethyl-hexyl-acrylate	acrylic acid or methacrylic acid	contributes to colloidal stability; increases paint viscosity.
vinyl acetate	vinyl-versatate butyl acrylate 2-ethyl-hexyl-acrylate	adhesion promoter	improves "wet adhesion"

Having selected a weight ratio of monomers to provide a given Tg and other characteristics, there is no guarantee that the particles prepared by emulsion polymerisation will have an uniform composition throughout their structure; the surface composition, affecting the MFT value, may be different from the bulk. Implications from this can be important to the paint formulator. This can be illustrated by considering a copolymer of butyl acrylate (BA) / methyl methacrylate (MMA) / acrylic acid (AA) which is of a type commonly used to prepare latex for paint. The acrylic acid proportion is a few percent of the total. The monomers have significantly different solubilities in water, respectively varying from low; from requiring a few percent to saturate water; to being fully miscible with water. Using published Q,e values [6], values for the reactivity ratios may be obtained as follows:

butyl acrylate / methyl methacrylate / acrylic acid

(a)   (b)   (c)

$$r_1(ab) = 0.34 \; r_2(ab) = 1.9$$
$$r_1(ac) = 0.32; \; r_2(ac) = 2.9$$
$$r_1(bc) = 0.75; \; r_2(bc) = 1.2$$

On the basis of the reactivity ratios and by considering the propagation rates from each of the three terminal radicals in turn, the composition of the growing polymers may be compared (Figure 1). It follows that with this copolymer composition the expectations would be:

(i) a tendency to produce AA-rich polymer
    - chains with sufficient AA content will be water-soluble
    (i.e. solution-polymer formation)

(ii) chains produced in the earlier stages of polymerisation will be richer in MMA / AA (higher Tg polymer and relatively hydrophilic). (But this can be offset by running the process at low free-monomer levels). Also the higher solubility of these monomers in the aqueous phase will in addition favour the formation of solution-phase oligomeric radicals richer in MMA / AA.

(iii) BA will be the slowest monomer to convert. At the end of the polymerisation, polymer chains rich in BA will be produced (low Tg and relatively hydrophobic).

In summary, the latex would be expected to comprise heterogeneous particles with some polymer (rich in AA) remaining in the continuous phase. With regard to the interest of the paint formulator, the solution polymer is likely to affect the paint rheology by giving an increase in low-shear viscosity and by reducing water-resistance properties.

Where latex particles comprise a mix of polymer chains of differing composition, the surface composition would normally be expected at equilibrium to be richer in the relatively more hydrophilic species but this effect does not appear to be significant when the particles are grown under normal conditions with low free-monomer levels. Also during emulsion polymerisation a wide spectrum of composition for the polymer chains will result, tending to mask this effect.

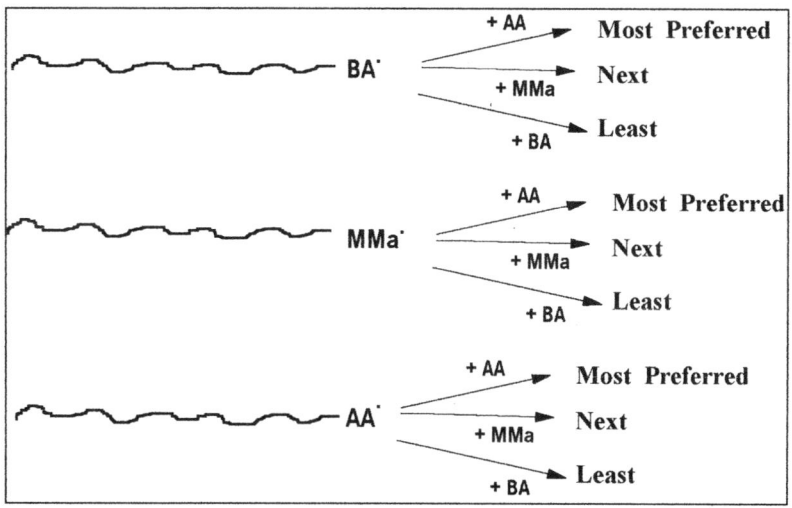

Figure 1. Propagation from the differing polymer end radicals.

2.1.5 *Composite Particles*

Composite latex particles, comprising at least two different (co)polymers within each particle, have been described widely, as is evident from the patent and open literature. Some preparation processes are likely to generate a spectrum of compositions [7] and most involve the polymerisation of a second set of monomer(s) in the presence of a first polymer [8-11]; this will usually lead to some graft polymer which has elements of both polymers and which will enhance the "internal blending" of the polymers within each particle. Recently alternative processes have been reported in which preformed polymer particles "engulf" second particles which may comprise inorganic material or another polymer [12,13].

Composite particles have been designed for use in specialist latex paints such as refinish for automobiles [14]. Use of composite particles in decorative paints does not yet appear to be wide-spread but this situation may be changing. Composite particles offer decorative paints an opportunity for different physical property balances and increased cost-effective performance, by controlled location of specific functional components (expensive) and by unobtrusive location of cheaper, poor performance components, for example by confining them to the centre of the particles. Wider use of composite particles may follow improved control of particle morphology during manufacture.

A number of theories, based on the interfacial energies involved, have been developed concerning the morphology of composite latex particles [12,15,16]. These theoretical considerations have included predictions for the equilibrium morphology when one of two polymer components increases in relative size by further polymerisation [17]; they have indicated the importance of the relative volumes of the components in determining the equilibrium morphology [12].

Internal re-arrangement of particle morphology, presumed toward an equilibrium structure, has been observed. In some cases this has been dramatic leading to particle disruption [18,19]. Rearrangement altering the minimum-film-forming temperature, has been induced by heat-treatment [4].

2.1.6 *Particle Blends*

Mixtures of different latex particles are of interest to the paint formulator because in addition to attaining desired physical property balances they offer potential for control of film morphology especially when differentiating film surface composition from bulk. Some of the properties of the paint film such as resistance to dirt-pick-up, gloss, scuffing and marking resistance are more associated with *surface* characteristics; for others, such as opacity and the physical properties governing durability, the *bulk* characteristics are more relevant. Most water-resistance properties such as scrub-resistance and swelling and softening when rewetted are thought to be governed mostly by bulk combined with the characteristics of the interface with the substrate. This is discussed below (4.2). It can be speculated that latex paint, fully optimised for technical performance might be formulated to include surface-migrating particles for surface properties and different latex particles for the bulk properties.

2.1.7 *Surfactant Selection*

The choice of surfactants used during the manufacture of the latex can affect the final paint properties. Because they remain in the final dry paint film, some of the properties will be determined by their location. A network of surfactant molecules throughout the film would be expected to "wick" moisture through the coating and reduce water-resistance properties; this would contrast with surfactant residues isolated at cusps within the film.

Surfactants can be selected from an extensive range of anionic and non-ionic materials. "Reactive" surfactants [20], carrying one or more copolymerisable double bonds are also available, as are reactive surfactant precursors which copolymerise with particle monomer to produce surface-active polymers or to become directly and covalently attached at the particle surfaces. The latter have been used as colloidal stabiliser to prepare non-ionic latex by emulsion polymerisation [21,22] or by dispersion polymerisation in water / alcohol mixtures [23]; excellent properties have been claimed for paints derived from these latices [24].

Surfactants which become covalently bound at the particle surfaces will not desorb, avoiding a problem which may otherwise occur during paint make-up.

2.1.8 *Removal of Residual Monomer*

Residues of monomer remaining at the end of latex manufacture give an odour to the derived paint and if they were present at a significant level, would pose a health risk to the operator during application and even possibly, in the case of interior paints, to subsequent room users.

It is usual for the latex manufacturer to add further free-radical initiator, at a suitable temperature, at the very end of the process and to allow sufficient time for the latex to be exposed to a significant free-radical flux. However, this treatment does not remove all traces, as discussed elsewhere [25]. The monomer traces may be removed chemically, usually involving saturation of the monomer double bond, or physically, for example by passing steam or gas through the latex [26]. It is possible to reduce levels to less than a few parts per million.

2.1.9 Biocide Addition

Latices, like other aqueous systems containing bio-degradable components, require protection from attack by micro-organisms. For aqueous latex and derived paint the principal sources of infection are by bacteria and fungi (including yeasts). When selecting biocide for use in the systems, it is necessary to test the candidate biocide's effectiveness against both classes of micro-organisms. Some biocides are less effective at different pH ranges and this qualification needs to be checked, mindful of the pH of the latex during storage and the pH of the paint.

Unsuccessful protection of either latex or paint may be manifest at first as a change in rheology or as a fall in pH arising from some decomposition of thickener (especially if cellulosic) or surfactant (although there may be alternative explanations for these changes). Increasing contamination leads to odour development and further physical deterioration and may lead to disastrous gassing.

2.2. MILL-BASE PREPARATION

The mill-base is a dispersion of inorganic pigment and extender particles. It is conveniently prepared with a high-speed disperser which comprises a disc fitted with small blades at its perimeter, rotated about a vertical axis within the aqueous phase held in a containing vessel. Firstly water and aqueous phase components are pumped to the vessel and with the disc rotating at high speed, the solid materials are metered in slowly. Water-soluble polymers, usually present in the aqueous phase, give an increase in viscosity and this appears to assit the comminution and dispersing process. After addition of all particulates the milling is continued for some time to ensure complete comminution and dispersal, giving a dispersion with a mobile, creamy consistency.

2.2.1 *Inorganic Pigment*

Rutile titanium dioxide is used to provide the paint with whiteness and to contribute, usually in a major way, to opacity. Many different grades of the pigment are available to the paint formulator. The rutile particles are supplied with differing surface layers such as silica, alumina, zinc oxide and zirconia. Selection from the range available is usually made on the basis of empirical testing and optimisation for paints of a given type and end-use. Optimum particle size for scatter is around 220nm. and pigment manufacturers produce material which can be dispersed down to this approximate range.

2.2.2 *Extender Particles*

Different minerals are used, including calcium carbonate, kaolin (china clay), talc and barytes. They contribute to opacity and colour and are used in large volumes in the paints with lower sheen. They are major components in "dry-hiding" systems where, after drying of the paint film, entrapped air contributes to the opacity by providing a matrix phase or occluded phase of lower refractive index.

2.2.3 *Dispersants*
Combinations of low molecular weight surfactants and water-soluble or dispersible polymers may be included in the mill-base formulation. These are required during the dispersing process to give colloidal stability; they also are important to help control the paint rheology. Highly efficient stabilisation giving near-Newtonian rheology is *not* the target here, as discussed later.

Because of the hydrophilic nature of the extender particle surfaces, efficient adsorption of low molecular weight hydrophobe moieties is not guaranteed. Partially soluble or interfacially active *polymers* have advantages arising from the possibility of multi-point (but weak) adsorption along their length.

2.3. OTHER COMPONENTS

Anti-foam agents, coalescing agent which reduces the effective Tg of the latex polymer and biocide are included in the aqueous phase when preparing the mill-base. Suitable biocides for protecting the paint in the can may be identified as described above (2.1.9). In addition, for some countries such as those with tropical climates, biocides may be required to protect the dry paint film. Other materials which are added to modify paint rheology or dry film opacity are described later.

3. Paint Preparation

The latex is added to the mill-base dispersion or the other way around, whilst stirring. The order-of-addition is chosen to avoid "shock" to any part of the system and is strictly adhered to during manufacture!. Some components may be added after the mixing operation and there may be adjustments to the pH with ammonia or alternative base; the water content may be adjusted. Some typical paint formulations are shown (Table 4) [27]. In these examples the latex comprised a medium-fine particle sized, cellulose ether stabilised, vinyl acetate-vinyl versatate copolymer ("Emultex VV573" - registered Trade Mark of the Harlow Chemical Co. Ltd).

A combination of several colour pigments may be used to give the required colour effect in the final film. Colour mixing is a sophisticated procedure and is carried out with great precision and reproducibility. There are two distinct operating procedures. Either the paints are fully pigmented before leaving the manufacturing site or they are delivered as base formulations for addition of tinting pigments at the retailing store premises to meet the individual customer's requirements.

3.1. PAINT CHARACTERISTICS

During the development of modified paint products and as a quality control check, a number of characterisations are made routinely. These usually include measurement of viscosity at low-shear and at high-shear for which purpose simple intruments are available commercially. Stability to added electrolyte and to repeated cycles of freezing and thawing may also be checked; paints based on latices with sterically-stabilised particles usually are more robust.

3.1.1 Particle Size Characterisation

A difficulty here is that the paint will contain particles of differing refractive index and density as well as size. However it may be possible to do this with a disc centrifuge photosediometer [28] because its operation involves physical fractionation of the particles by size and density. From the raw data (particles reaching the detector with time) it is necessary to be able to separate the peaks for the different particle types and to apply the correct density value to each.

TABLE 4. Examples of paint composition

Matt Paint			Semi-Gloss Paint		
Component	%age by Wt	%age by Vol	Component	%age by Wt	%age by Vol
water	21.1	27.6	water	15.1	18.4
polyphosphate	1.7	2.2	propylene glycol	3	3.8
maleate copolymer	0.4	0.5	acrylic acid copoly.	0.6	0.7
ammonia (0.91)	0.1	0.1	ammonia	0.1	0.1
biocide	0.2	0.3	biocide	0.2	0.2
anti-foam	0.2	0.3	anti-foam	0.2	0.2
hydroxyethyl cellulose (3%)	19.2	25.1	hydroxyethyl cellulose (3%)	14.5	19.7
titanium dioxide	9.7	3.1	titanium dioxide	19	5.7
calcium carbonate	21.4	12.7	talc	3	1.7
kaolin	5.1	3	titanate complex	0.3	0.4
coalescing agent	1	1.4	coalescing agent	1	1.3
latex	12.9	15.3	latex	36	40
opacifying polymer particles	7	8.4	opacifying polymer particles	7	7.8
		PVC=72%			PVC=35%

4. Paint Properties

The pigment volume concentration (PVC) may be defined as the volume of the inorganic particulates divided by the total dry film volume excluding any air in the film. In a "well-bound" paint there is sufficient polymer to fill the space between the inorganic particulates but above a critical value (C) for the PVC this is no longer the case and the dried paint film will include air. Because of the polydisperse nature of the particles and irregular shape of some, the CPVC is not easily predicted. It can be readily identified however from a range of paints of varying PVC due to a transition point when property performance is plotted against PVC; for example this can be done with opacity or washability assessment. In general, gloss, mid-sheen and exterior paints are formulated below CPVC, as are interior Matt paints which are designed to be more cleansable with improved resistance to marking and scuffing damage. Formulations above CPVC can offer higher opacity at lower cost and provide less expensive interior wall and ceiling paints.

430

4.1. RHEOLOGY

Decorative latex paints are applied by brush or roller and to a lesser degree by airless spray. Usually the paints are designed for both brush and roller but exceptions to this include the so-called "solid-emulsion" paints for which brush application is unsuitable. This duality for application leads to a need for some balancing in the in-built rheological characteristics.

For application by brush, the user requires (not always consciously) good pick-up of paint onto the brush, relative ease in avoiding drips from the brush and splash when applied to the substrate, easy unrestricted brushing and almost complete disappearance of brush marks before the paint dries.

Rheograms showing apparent viscosity plotted against shear-rate readily differentiate latex paints from fluid, oil-based gloss paints, which in the hands of an experienced person give unsurpassed brushing characteristics (Figure 2). The latex paints have a pronounced non-Newtonian character with steep shear-thinning behaviour and some hysteresis. At low shear rates there is a high sensitivity to disperse phase volume fraction (φ). The shear-thinning is believed to arise largely from weak, reversible flocculation in the paint which may be considered to give loose clusters of particles increasing the effective value for (φ); the cluster size is thought to decrease with increasing shear rate [29].

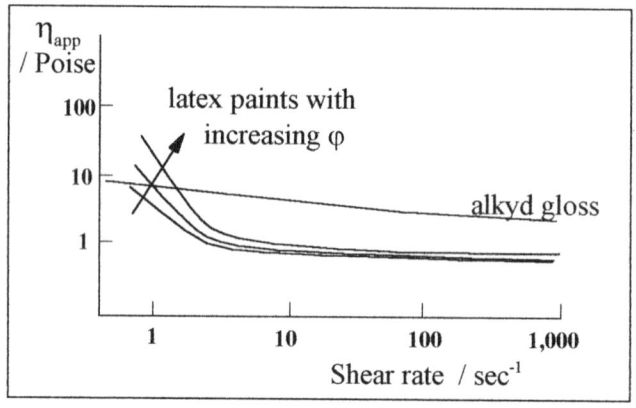

Figure 2. Rheograms comparing latex and oil-based paints
after breakdown of structure at high shear

During application by brush or roller, the wet paint is subjected to very high shear rate, in the order of $10,000 \text{ sec}^{-1}$, which is sufficient to break most of the paint structure. Because of the pseudoplastic behaviour, the viscosity at low shear is much higher. This is beneficial for improving resistance to sagging for the wet paint film on vertical surfaces but may hinder flow-out of brush-marks.

The flow-out of brush-marks was considered by Orchard and Smith [30]. Taking the brush-marks to be sinusoidal in cross-selection, they derived an equation for the amplitude (a) of the marks after time (t), where the initial amplitude is (a_0) and the wave lenght is (λ)

$$a = a_0 e^{-f} \qquad (2)$$

$$f = \frac{(2\pi)^4 \gamma h^3}{3\lambda^4} \int \frac{dt}{\eta} \qquad (3)$$

and where (η) is the viscosity. The mean paint thickness (h) and the surface tension (γ) are assumed to stay constant.

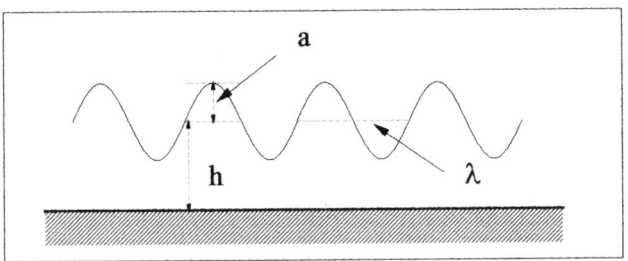

Figure 3. Assumed brush-mark cross-section

With oil-based paints, the equation is known to give quite good fit with experimental observation. Because decay of brush-marks is faster at larger values of (f), it will be faster at lower paint viscosity. The relevant shear-rate regime is very low and with the marked non-Newtonian behaviour of the latex paints and high sensitivity of viscosity to (φ), the relationship between decreasing brush-mark decay with increasing paint solids, which is observed in practice, is indicated.

Major changes in the paint rheology can be effected by selecting different polymer for the aqueous phase. A number of types of polymer are available. These include substituted celluloses (already discussed), "associative thickeners" (discussed elsewhere [31]) and (meth)acrylic acid copolymers. The latter may be supplied as latex with the particles dissolving when the pH is raised during paint preparation.

Different rheology is obtained depending on whether colloid-containing or colloid-free latex is selected and on the choice of minor monomers for the latex composition.

Rheology also involves aesthetics in that the appearance of the paint when its container is opened, and the appearance on a fully-laden brush, are often taken as indicators of "paint quality" by the user.

4.2. WATER RESISTANCE PROPERTIES

Depending on the paint type, a number of physical properties must be provided. These might include washability, resistance to house-hold stains and to scuffing and good adhesive performance to the substrate which it covers. Performance of the paint film when it is wet or damp can pose special difficulties. These arise largely because the paint was originally an aqueous composition and still contains a number of water-sensitive species. The problem is severe when the latex paint is applied over a well-sealed substrate, for example which is coated with an alkyd paint. For good performance under these conditions, modification to the formulation is required, usually effected by including a special monomer in the latex. Many such components ("adhesion promoters") are described in the patent literature; several are available commercially. Some include the heterocyclic ethylene urea (ureido) structure [32].

There are conflicting views on the possible mechanism by which "wet adhesion" is improved. One of these is that the components reduce the loss in modulus when the film becomes wet and consequently peel adhesive performance is retained; another is that they generate beneficial polar-polar interactions at the interface between the film and the substrate.

4.3. OPACITY

Comparison between latex paints (with disperse phase polymer) and oil-based gloss paints (with solution phase polymer) suggests that light scatter from the titanium dioxide pigment particles is more efficient in the latter systems. In a film derived from polymer particles and pigment particles at or near optimum size, a loss in scatter would be expected from any pigment-pigment contact. Ideally, all the pigment particles, at optimum size, would have only polymer particles as near-neighbours.

When drying a mixed dispersion of polymer and pigment particles this ideal arrangement will not result and random statistics will dominate the extent to which a pigment particle will include another pigment particle amongst its near-neighbours. This suggests that if modifications to the dispersions could be made such that pigment-pigment contacts could be reduced or eliminated, increases in opacity would be attained.

Much current research is devoted to finding cost-effective ways of improving opacity in latex paints, as is evident from the patent literature. Techniques which are being developed include complete or partial encapsulation of titanium dioxide particles with either film-forming or non-film-forming polymer [33]. Also special polymer particles have been developed such that when they dry as part of the paint film, an air-void is generated at the particle centre [34]. These particles are available commercially. A method to characterise such particles with a disc centrifuge, has been described [35].

As mentioned above, air is incorporated within paint films when formulated above the CPVC. The air reduces the refractive index of the matrix phase and increases the differential with the pigment and extender particles. However these paints have a number of performance limitations. Successful incorporation of air, at the correct location within the film and without loss of film properties and with full cleansability, will yield very significant technical advantages to the latex paint formulator.

5. References

1. Lambourne, R. (Ed) (1987). *Paint & Surface Coatings, Theory & Practice*, Ellis Horwood, Chichester.
2. Barret, K.E.J. (Ed) (1975). *Dispersion Polymerisation in Organic Media*, J. Wiley & Sons, New York.
3. Gordon, P.G., Davies, M.A.S. and Waters, J.A. (1984), *J. Oil Colour Chemists Assoc.* **67**, 197.
4. Hourston, D.J., Simpson, M.A., Waters, J.A. and Williams, M.J.P. (1994) Modified film from composite latex, Waterborne Coatings, UMIST, Manchester.
5. Fox, T.G. (1950) *Bull. Am. Phys. Soc. Series 2*, **1**.
6. Brandrup, J. and Immergut, E. (1989). *Polymer Handbook third Ed.*, Wiley-Interscience, New York.

7. Bassett, D.R. and Hoy, K.L. (1980), in D.R. Bassett and A.E. Hamielec (eds.), *Emulsion Polymers and Emulsion Polymerisation, ACS Symp. Ser.* **165**, 415.
8. Erickson, J.R. and Seidewand, R.J. (1981), in D.R. Bassett and A.E. Hamielec (eds.), *Emulsion Polymers and Emulsion Polymerization. ACS Symp. Ser.* **165**, 483.
9. Dimonie, V., El-Aasser, M.S., Klein, A. and Vanderhoff, J.W. (1984). *J. Polymer Sci., Polymer Chem. Edit.,* **22**, 2197.
10. Guillot, J., Guyot, A. and Pichot, C. (1990), in Candau, F. and Ottewill, R.H. (Eds) *An Introduction to Polymer Colloids,* Kluwer Academic Publishers, Dordrecht.
11. Greenhill, D.A., Hourston, D.J. and Waters, J.A. (1990). *ACS Ser.* **424**, 397.
12. Waters, J.A., to ICI plc (1989). *European Patent* 327,199.
13. Ottewill, R.H., Schofield, A.B. and Waters, J.A. (1996). Preparation of composite latex particles by engulfment, *Colloid Polym. Sci.* **274**, 763-771.
14. Backhouse, A.J. (1982) *J. Coatings Techn.* **54**, 83.
15. Berg, J., Sundberg, D.C. and Kronberg, B. (1986). *Polym. Mater. Sci. Eng.* **54**, 367.
16. Chen, Y.C., Dimonie, V. and El-Aasser, M.S. (1992). *J. Appl. Polym. Sci.* **45**, 487.
17. Waters, J.A. (1994)., *Coll. & Surf. A: Physicochem. & Eng. Aspects* **83**, 167.
18. Ugelstad, J. (1983). in *Science & Technology of Polymer Colloids.* Poehlein, G.W., Ottewill, R.H. and Goodwin, J.W. (eds), NATO ASI Series E, Applied Sciences **68**, Kluwer Academic Publishers, Dordrecht.
19. Keith, J.S., and Waters, J.A. (1993). Factors controlling the morphology of latex particles, First UK Polymer Colloids Forum, Bristol.
20. Allan, G.C., Aston, J.R., Grieser, F. and Healy, T.W. (1989), *J. Coll. Int. Sci.* **128**, 258.
21. Ottewill, R.H., Satgurunathan, R., Waite, F.A. and Westby, M.J. (1987), *Br. Polym. J.* **19**, 435-440.
22. Westby, M.J. (1988), *Coll. Polym. Sci.* **266**, 46.
23. Graetz, C., Thompson, M.W., Waite, F.A. and Waters, J.A. to Imperial Chemical Industries, Ltd., (1979). *British Patent* 2,039,497.
24. Palluel, A.J., Westby, M.J., Bromley, C.W.A., Davies, S.P. and Backhouse, A.J. (1986). Novel Aqueous Dispersion Polymers, *Proc. 12th. Int. Conf. Organic Coatings & Techn.,* Athens.
25. Kukulj, D. and Gilbert, R.G. (1997) Polymerization at high conversion, in J.M. Asua, (ed), *Polymeric Dispersions, Principles and Applications,* Kluwer Academic Publishers, Dordrecht.
26. Englund, S.M. (1981). Monomer Removal from Latex, *CEP Aug.* 55-59.
27. Harlow Chemical Co. Ltd. (1990). Polymers for Industry, Sales brochure.
28. McFadyen, P. and Fairhurst, D. (1993), *Clay Minerals,* **28**, 531.
29. Albers, W. and Overbeek, J.Th.G. (1960). *J. Coll. Sci.* **15**, 489
30. Orchard, S.E., Smith, N.D.P. et al. (1961). *JOCCA,* **44**, 618.
31. Jenkins, R.D. and Bassett, D.R., (1997) Synergistic interactions among associative polymers and surfactants, in J.M. Asua, (ed), *Polymeric Dispersions, Principles and Applications,* Kluwer Academic Publishers, Dordrecht.
32. Rhône-Poulenc Surfactants and Specialities (1993). Wet Adhesion Monomer for Latex Paints, Sales brochure.
33. Van Herk, A.M. (1997) Encapsulation of Inorganic Particles, in J.M. Asua (ed), *Polymeric Dispersions, Principles and Applications,* Kluwer Academic Publishers, Dordrecht.
34. Kowalski, A., Vogel, M. and Blankenship, R.M. to Rohm and Haas Co. (1981). *European Patent Application* 22,633.
35. Cooper, A.A., Devon, M.J. and Rudin, A. (1989). Use of a disk centrifuge to characterise voided latex particles, *J. Coatings Techn.* **61**, no 769, 25-29.

ENCAPSULATION OF INORGANIC PARTICLES

The Use of Emulsion Polymerization to Encapsulate Pigments and Fillers

A.M. VAN HERK

Department of Polymer Chemistry, Eindhoven University of Technology
P.O. Box 513, 5600 MB Eindhoven , The Netherlands

1. Application of Microencapsulated Particles

Microencapsulation is the process of obtaining small solid particulates, liquid droplets, or gas bubbles with a coating. In this review the process of encapsulating inorganic pigments and fillers with a polymer through suspension or emulsion polymerization is described.

The microencapsulation of pigment and filler particles is an important area of research, both in the academic world and in industrial laboratories. Much activities in the past decade have been aimed at obtaining inorganic powders, coated with an organic polymer layer. Such systems are expected to exhibit properties other than the sum of the properties of the individual components. In general, several benefits from this encapsulation step can be expected when the obtained particles will be applied in a polymeric matrix (e.g. plastics or emulsion paints):

- Better particle dispersion in the polymeric matrix
- Improved mechanical properties
- Improved effectiveness in light scattering in a paint film
- Protection of the filler or pigment from outside influences
- Protection of the matrix polymer from interaction with the pigment
- Improved barrier properties of a paint film

Amongst materials that have been encapsulated are $CaSO_3$, $CaCO_3$, $BaSO_4$, TiO_2, zeolites, talc, several clays like bentonite and kaolin, limestone, alumina, silica, $NiO.ZnO.Fe_2O_3$, Fe_2O_3, ZnO, Cr_2O_3, CdS, HgS, Cu, wollastonite, carbon black, graphite, disazo yellow, copper phtalocyanine, lackrot C and other azo pigments [1].

The applications of these encapsulated particles relate to the above mentioned benefits and can be found in filled plastics, paints, inks, paper coatings etc. of which in the next subsections examples will be given. The use of encapsulated particles as catalyst carriers has recently been reported [2].

J. M. Asua (ed.), Polymeric Dispersions: Principles and Applications, 435–450.
© 1997 *Kluwer Academic Publishers. Printed in the Netherlands.*

1.1. ENGINEERING PLASTICS

In engineering plastics the interaction between filler and plastic is very important for the mechanical properties like fracture toughness [3,4].

The treatment of inorganic particles with hydrophobicing or coupling agents like silanes, titanates, zirconates etc. is aimed at improving the compatibility with the matrix polymer and it is shown that indeed many rheological and mechanical properties can benefit from this step [5]. However it is also clear that this relatively simple treatment is not sufficient to produce a composite with properties close to that of the unfilled polymer. The search for even better properties has initiated the process of encapsulation of inorganic particles where an intermediate layer, which interacts strongly with both the filler surface and the matrix polymer, would provide the required improvement. Similar conclusions were reached by Dekkers and Heikens [6] on their study on the effect of interfacial adhesion on tensile behavior of polystyrene-glass-bead composites; they state that obviously a more drastic modification (than surface modification with silanes) near the glass beads' surface is required in order to obtain a composite both stiffer and tougher than the matrix material polystyrene, for instance, encapsulation of the glass beads within a layer of low modulus material.

Kolarik and coworkers [7] explored the properties of three-component composites consisting of a thermoplastic matrix (polypropylene) and an elastomer encapsulated filler (ethylene-propylene diene copolymer (EPDM) elastomer on $CaCO_3$).

The (relatively cheap) filler cores inside elastomer enhanced the apparent volume fraction of the incorporated elastomer. The adhesion at the matrix/filler interface turned out to be important and increases the yield stress of these ternary composites.

In many instances a thin polymer layer has been observed on inorganic surfaces, that have been in contact with a polymer, that is not extractable, even without the presence of covalent bonds between the surface and the polymer layer. This bound polymer does not necessarily lead to reinforcement of a filled polymer and in some cases this layer can degrade the tear strength of the material by allowing failure at a weak second interface between bound and matrix polymer. This degradation can be alleviated by cross-linking the polymer or by increasing the molecular weight to force either chemical or physical links across the weak second interface [8]. Processing these particles at higher temperatures and shear necessitates high molecular weights or crosslinking anyway and in that case covalent bonding with the surface is superfluous.

Ono [9] showed that carbon powders coated with polymethyl methacrylate were directly moldable into sheets which had excellent thermal properties and could also be used as electric conductive plastics. An expanded graphite-polymethyl methacrylate (PMMA) composite film turned out to form excellent diaphragm material for high fidelity loudspeakers.

1.2. COATINGS, INKS AND TONERS

One of the most important applications of encapsulated pigment and filler particles is in emulsion paints. One of the more expensive components of water-borne paints is the white pigment, usually titanium dioxide (rutile form). The pigment is added to obtain hiding power. The hiding power or opacity depends on the occurrence of light absorption and light scattering, for pigments with a high refractive index, like titanium dioxide, light scattering forms the main contribution to the hiding power. The light scattering effectiveness of the pigment particles depends on their particle size and on the interparticle distance. Agglomerates of pigment, already present in the wet paint film or formed by flocculation during the drying process, will reduce the scattering effectiveness of the dispersed pigment particles. By encapsulating the pigment particles it is expected that the chance of flocculation is reduced and that the dispersion in the final paint film is improved [10]. It has been suggested that the layer thickness could be optimized to obtain optimum spacing between titanium dioxide particles to achieve maximum light scattering [11].

In encapsulating the pigment particle an important adverse effect of the pigment could be influenced, that is the generation of radicals under the influence of UV light. These radicals can lead to degradation of the matrix polymer and thus leads to reduced durability. With the proper choice of the polymer layer the durability might be improved also. Other advantages are improved block resistance, less dirt pick up, better adhesion [3,12] and improved chemical resistance [3].

For the above mentioned reasons most commercial pigments already obtain inorganic and/or organic surface modifications. An additional benefit can be brought about by the formation of multilayers of polymer on inorganic particles [13] where for example rubber toughening effects can be introduced [7].

Besides inorganic pigment and filler particles also organic pigments have been encapsulated with polymer, e.g. copper-phtalocyanine and azo pigments [14].

Several fillers/extenders for paint applications have been encapsulated like $CaCO_3$, alumina, silica , wollastonite and clays like bentonite and kaolin [15].

Other applications of encapsulated pigments can be found in inks, paper coatings and electro-photographic toners.

1.3. MAGNETIC PARTICLES

When the inorganic particles are magnetically responsive this opens pathways to special applications like coupling of enzymes and antibodies to the surface of the magnetic particles after which drug targeting becomes possible. Also these particles can be used in biochemical separations [16]. Furthermore, these magnetic particles can be used in magnetic recording media, oil spill clean up and moldable magnetic powders [17].

Multilayer magnetic composite particles comprising of polymer latex particles with small (20 nm) $NiO.ZnO.Fe_2O_3$ particles heterocoagulated on the surface and these heterocoagulates encapsulated with polystyrene forms a recent interesting development where simple encapsulation with one layer of one type of polymer is extended to multilayer materials [18].

Huang [19] describes the preparation of magnetic latex particles through inverse emulsion polymerization. He encapsulates iron oxide with crosslinked hydrophilic polymer, these particles can be used as a seed to prepare aqueous hydrophobic magnetic latex particles. Cohen describes the precipitation of iron hydroxides in swollen polymer particles which are converted to the oxide by means of a heat treatment [20].

2. Encapsulation of Inorganic Particles

2.1. GENERAL PRINCIPLE

Emulsion polymerization is the technique that is used most often because of water-based coatings related applications of encapsulated pigment particles.

The inorganic particles (after hydrophobization) are dispersed with the normal surfactants and an emulsion polymerization is performed where the locus of polymerization is the hemi- or admicelle around the inorganic particle (Figure 1).

Usually 'maximum' properties are obtained when the inorganic particles are distributed evenly and as single (primary) particles in the matrix. This means that in the steps towards obtaining the final product keeping the particles well dispersed is of major importance. Initially the particles should be well dispersed in the aqueous phase and (partial) coagulation during the emulsion polymerization must be avoided because this leads to irreversible fixation of the coagulates.

On studying the kinetics of the emulsion polymerization of methyl methacrylate on TiO_2 [21,22] partial coagulation was observed with dark field microscopy which also showed up in the conversion time curves as temporarily rate retardation's (plateaus). By improving the mixing conditions this problem was diminished in later work [23].

To be able to disperse the inorganic particles in an aqueous medium, special stabilizing agents for the inorganic particles should be added or the surface should be hydrophobized in order to be able to use conventional emulsion polymerization surfactants. In order to improve dispersion of the pigment particles power ultrasound has been applied [11, 24]. Many pigment particles can be obtained with organic surface modifications commercially or alternatively they can be modified with silanes or titanates.

These so called coupling agents usually contain at least one smaller alkoxy group which can react with a hydroxyl group on the surface. The other groups can contain functionality's which can interact physically or chemically with the surrounding polymer matrix thereby aiding the dispersion of the filler.

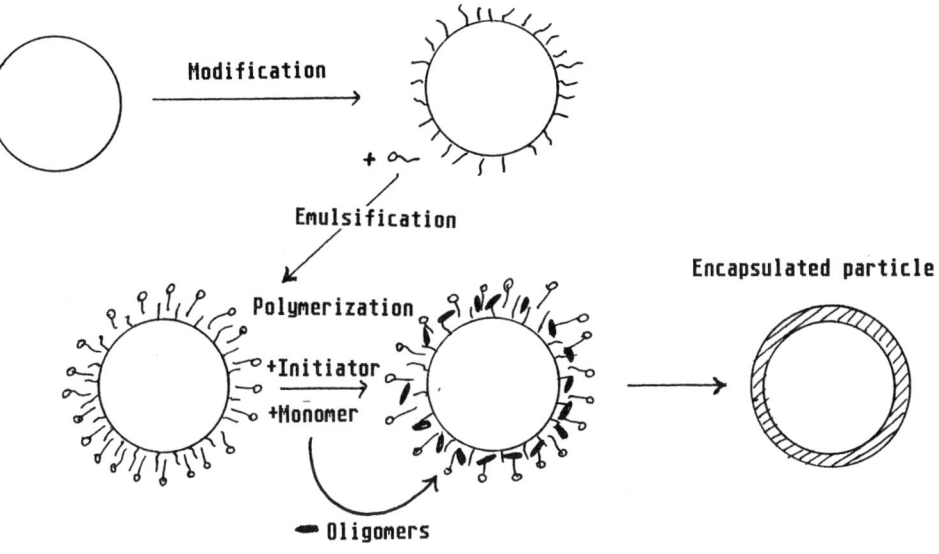

Figure 1. Schematic representation of encapsulation of inorganic (submicron) particles through an emulsion polymerization.

Water soluble titanates are also commercially available which can be used to stabilize aqueous pigment and filler dispersions. Usually these coupling agents are applied from organic solutions or by applying them directly to the dry powder, for example in a Henschel mixer.

Other additives which are used to make the surface hydrophobic are for example a combination of methacrylic acid and aluminum nitrate as a coupling agent [24] or groups like stearoic acid.

Polymerization on the surface of inorganic particles with water soluble monomers can also be regarded as a special case of emulsion polymerization. Aqueous solution polymerization's are reported by Chaimberg et al. [25] for the graft polymerization of polyvinylpyrrolidone onto silica . The nonporous silica particles were modified with vinyltriethoxysilane in xylene, isolated and dispersed in an aqueous solution of vinylpyrrolidone. The reaction was performed at 70 °C and initiated by hydrogen peroxide. Nagai et al [26] report on the solution polymerization of the quaternary salt of dimethylaminoethyl methacrylate with lauryl bromide, a surface active monomer, on silica gel. Although the aim was to polymerize only on the surface, also latex particles were formed.

2.2. THE EFFICIENCY OF ENCAPSULATION

Polymerization on the surface is in competition with the process of particle formation (Figure 2). Therefore the normal stabilization with micelle forming surfactants is not straightforward and the offered surface area of the inorganic particles is very important. So far mainly submicron particles have been encapsulated with this method.

The presence of conventional surfactants in encapsulation reactions introduces the problem that a delicate balance between stabilization of polymer and inorganic particles and formation of new particles has to be maintained [27,28].

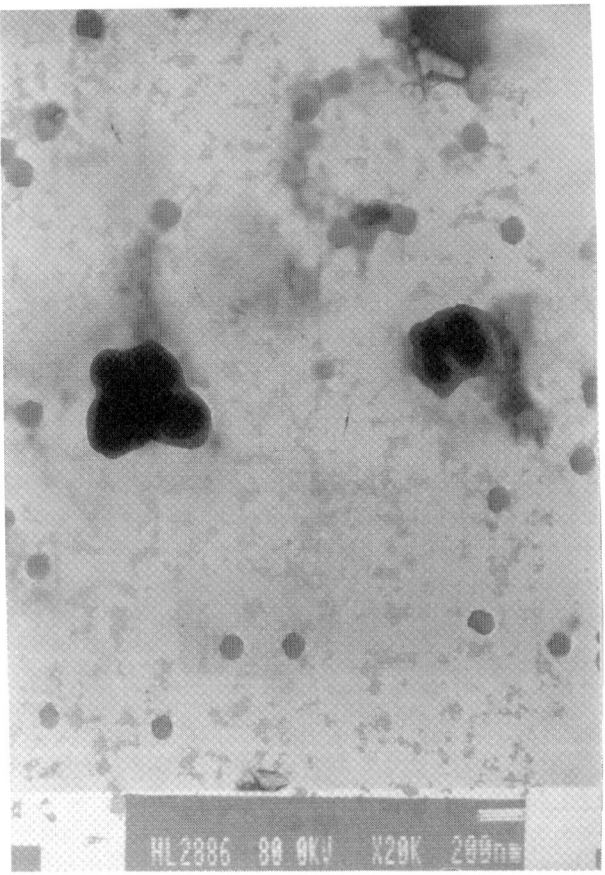

Figure 2. Transmission electron micrograph of the product of an encapsulation of TiO₂ with polymethyl methacrylate [27], showing encapsulated and free polymer particles.

This could be done by monitoring the surfactant concentration with on-line conductivity measurements and dynamically controlling the surfactant concentration [27]. Micellar nucleation is partly suppressed by adjusting the free surfactant concentration just below the critical micelle concentration (corrected for adsorption on TiO$_2$ [27,28]). However homogeneous nucleation can also occur, which will be more substantial with more water soluble monomers present. Therefore more free polymer will be formed and the efficiency will drop. For the encapsulation of TiO$_2$ in recipes with about 20 % solid contents, the highest efficiencies are usually found for styrene (25 %)[13,29]. The efficiency seems to depend on the water solubility of the monomers (see also section 2.3).

In the initial stages of the emulsion polymerization the electrostatic interaction between of the initiator fragment at the end of the chain and the charges on the surface can play a role in increasing the capturing efficiency of charged oligomers. Haga [30] used differently charged initiators and confirmed this electrostatic effect. However, even in the case that an electrostatic repulsion can be expected between the polymer produced and the TiO$_2$ surface, encapsulation proceeds rather well. Also in the latter case unextractable polymer was formed. Haga found that the molecular weight of the encapsulating polymer (non-extractable) was higher than that of the apparent encapsulating polymer (extractable) and attributed this effect to an increased monomer concentration on the TiO$_2$ surface. The encapsulating polymer also had a higher density, both for PMMA and polystyrene (PS) (Table 1).

TABLE 1. Densities of encapsulating polymer, apparent encapsulating polymer and polymer prepared by conventional emulsion polymerization (data taken from [30]).

	Density (g/cm3)	
	PMMA	PS
Conventional emulsion polymer	1.185	1.049
Apparent encapsulating polymer	1.187	1.034
Encapsulating polymer	1 .206	1.094

The data suggest that special structures with high density are formed of the polymer that is in direct contact with the surface. This is also in line with the increase in Tg that was observed in this case and other cases (Table 2).

Lee and Chiu [31] carried out emulsifier free emulsion polymerization of methyl methacrylate (MMA) in the presence of CaSO$_3$ particles with potassium persulphate as initiator. Besides encapsulation also secondary nucleation occurred. The tensile strength of PMMA/CaSO3 composite made from emulsion polymerization was higher than that of the composite made from mechanical blending.

According to Hergeth and coworkers [32] a minimum surface of the filler particles is needed to prevent secondary nucleation. To estimate this amount, a formula was derived for seeded emulsion polymerization with spherical particles and a water soluble initiator [32]. This formula was based on the observation that primary particles are produced by a collapse and micellization process of oligomeric chains. An upper limit for the particle size was estimated to be 100 nm for the encapsulation of silica with polyvinyl acetate. Because the surface area needed to prevent secondary nucleation is proportional to the monomer conversion per unit of time, the encapsulation efficiency can be improved by using monomer starved conditions.

The encapsulation of the larger filler particles (> 1 micron) is more difficult because the low surface area of the particles does not suffice to capture all the formed oligomers and therefore secondary nucleation is almost unavoidable in the normal emulsion polymerization approach. However, the process of heterocoagulation can also occur during an emulsion polymerization and in many instances is the main mechanism of encapsulation. This mechanism leads to non-uniform polymer layers but the resulting encapsulated particles can still improve the filler properties [33].

Using less water soluble monomers in combination with a nonionic initiator, the formation of surface active oligomers in the aqueous phase can be minimized thus increasing the efficiency. Janssen [27] used cumene hydroperoxide in combination with Iron (II) sulphate and a reductor aiming at initiation at the hydrophobic/hydrophilic interface of the modified TiO_2 particles. The efficiency was improved most for the polymerization of MMA whereas for styrene the differences were smaller. This can be explained by the fact that in the case of MMA more oligomers are formed in the aqueous phase per unit of time which leads to more secondary nucleation compared to styrene. Therefore more is gained in the case of MMA by using uncharged radicals.

Dekking used electrostatic interactions to adsorb a radical initiator at the surface of clays like kaolin and bentonite [15]. Laible et al. [10] and Caris et al. [34] used azo initiators that reacted to the surface of TiO_2 and Al_2O_3 . The initiation from the inorganic surface improved the encapsulation efficiency. Some forms of TiO_2 can generate radicals under the influence of UV-light that can initiate polymerization.

Furusawa et al. [35] adsorbed a thick layer of hydroxyl propyl cellulose (HPC) on silica particles and polymerized styrene in the presence of these particles with potassium persulphate as initiator. In the absence of surfactant polymerization of bare silica particles resulted in secondary nucleation. However, when the HPC coated particles were used, this resulted in well encapsulated materials. When surfactant (SDS) was added raspberry shaped particles were obtained.

Hasegawa et al. [36] carried out soapless emulsion polymerization of MMA on TiO_2 where a layer of surfactant was adsorbed. Besides sodium dodecyl sulphate also docecyltrimethyl ammonium bromide and polyoxyethylene sorbitan mono-oleate, a nonionic surfactant, was used. A minimum amount of surfactant adsorption was needed for uniform encapsulation. In this approach, ionic surfactants were more effective than nonionic surfactants because less nonionic surfactant adsorbed on the surface (also related to the low critical micelle concentration (CMC)).

As mentioned before [32] the oligomer production rate is important in relation to the prevention of secondary nucleation. Adding the monomer semicontinously aids to the prevention of formation of new polymer particles because the oligomer production rate is lowered.

So in principle, using small particles at a high concentration, a hydrophobic initiator, low surfactant concentrations and monomers with a low water solubility (added semicontinously) results in the highest efficiencies.

2.3. COPOLYMERIZATION

To improve compatibility between the polymer layer and the matrix it is interesting to vary the composition of the polymer layer by copolymerization. In that way the glass transition temperature (Tg) of that layer can also be adjusted. Several copolymerizations were performed on titanium dioxide pigment particles including the monomer combinations styrene (St)/ methyl acrylate (MA), styrene/methyl methacrylate and methyl methacrylate/butyl methacrylate (BMA) [13, 29]. When increasing the MA content in the S/MA copolymerizations (Table 2), the efficiency decreases.

When increasing the water solubility of the monomer in a homopolymerization or increasing the content of this water-soluble comonomer in a copolymerization the efficiency decreases (Table 2) [13,27,28,29].

The glass transition temperatures were measured for both the copolymer layer on the pigment particles and the newly formed copolymer particles. The glass transition temperatures are changed because of a change in chemical composition of the polymer layer and because of interaction with the surface.

TABLE 2. Efficiency and glass transition temperatures, data from [13,29]

Monomers	Ratio	Efficiency [a]	Polymer content [b]	Tg
	mol/mol	wt.%	wt.%	°C
MMA	-	7.0	5.7	
MMA/BMA	1.390	8.6	11.6	
S/MMA	0.920	8.6	6.9	
S	-	33.4	25.0	103.6[c]
MA	-	6.6	6.2	5.7
S/MA	0.082	4.8	8.7	17/65
S/MA	0.827	6.8	11.7	50
S/MA	1.000	7.7	12.6	65[d]
S/MA	1.654	7.8	12.8	75

[a] Efficiency: weight percentage of monomer reacted to the surface
[b] Polymer content: weight percentage polymer of the encapsulated particles
[c] Free polymer shows a Tg of 99 °C
[d] Free polymer shows two Tg's of 7 and 67 °C

2.4. SPECIAL APPROACHES

Besides emulsion polymerization with a separate monomer phase also emulsions consisting of dissolved monomer droplets together with inorganic particles were polymerized by the group of Ruckenstein [2, 37]. They created emulsions of decane, a monomer and silica in an aqueous solution of surfactant which were polymerized to latexes containing rather uniformly distributed inorganic particle clusters of submicrometer size. Besides silica also zeolite, TiO_2 , CuO and Cu were encapsulated in combination with fumed silica [2]. Some of these particles were functionalized with quaternary ammonium groups and in combination with $RuCl_3$ used as catalyst particles in the catalytic oxidation of toluene to benzoic acid.

Also inverse emulsions were prepared, there (the otherwise difficult) encapsulation with water soluble monomers like acrylamide was performed [37]. In a first step a colloidal dispersion was prepared by dispersing the silica particles in an aqueous solution of acrylamide containing a water soluble dispersant, a crosslinking agent like N,N-methylene bisacrylamide and an initiator. The colloidal system was dispersed in decane containing a suitable surfactant. The volume fraction of the continuous phase was only 0.1, the capsules therefore had a polyhedral shape. The capsules were 4-5 micron for alumina and 1-1.5 micron for silica.

Another interesting approach is the use of adsorbed surfactant bilayers as a locus of polymerization [38 and references therein, 39]. The first step is the formation of a so called admicelle, a bilayer of the surfactant molecules at the solid/aqueous interface (Figure 3). An admicelle may be considered to be the surface analogue of a micelle and therefore this approach can be compared to the emulsion polymerization approach. By adsolubilization the monomer solubilizes in this admicelles and can be polymerized by addition of persulfates or hydrophobic azo initiators. It was found that adsolubilization of styrene on alumina occurred at a nearly constant sodium dodecylsulphate (SDS) to styrene ratio of 2 : 1, suggesting a sandwich-type structure.

Another possibility which has not fully been explored yet but is under research in the group of German (Eindhoven University of Technology) is the use of vesicles (which intrinsically can form bilayers) as a locus of polymerization on the surface of inorganic particles (Figure 3C). Both the adsorption of vesicles (on for example glass beads) [40] as well as the polymerization within vesicle structures has been described [41].

In our group in several instances it has been observed that a double layer of SDS spontaneously forms on for example alumina or wollastonite particles. After modification of these inorganic particles with hydrophobic titanates the amount of SDS that adsorbs on the surface is halved, indicating that the hydrophobic parts of the titanates interact with a monolayer of surfactants.

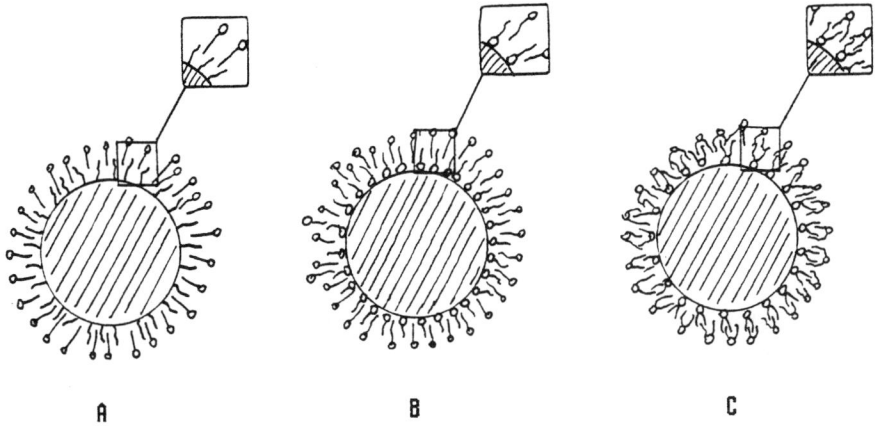

A   B   C

~ Hydrophobic tail

~o Ionic surfactant

⫍ Phospholipid

Figure 3. Schematic representation of the formation of bilayers (also called admicelles or hemimicelles).
(A) Adsorption of a monolayer of a conventional surfactant on a previously hydrophobized particle.(B)
Adsorption of a bilayer of surfactant molecules. (C) Adsorption of a vesicle bilayer.

3. Characterization of Microencapsulated Particles

Especially for microencapsulated particles the characterization of the particle morphology is very important because it is usually not trivial what type of structure is formed. For example, in the case of encapsulation through an emulsion polymerization step, the occurrence of partial coagulation and secondary nucleation may lead to products that have a high polymer content but can not be referred to as encapsulated particles. So in this respect scanning and transmission electron microscopy should be used additionally to the determination of the polymer content of the particles. The use of a relatively simple technique like dark field microscopy can also be very useful in monitoring e.g. coagulation phenomena and secondary nucleation.

Sometimes very simple tests like dispersabilities in water [38] or an organic solvent [36] or dyability with a hydrophilic dye like methylene blue [26] can give qualitative information on the success of the encapsulation procedure.

In the next subsections several techniques will be discussed in relation to their use in characterizing encapsulated particles.

3.1. POLYMER CONTENT

In literature the polymer content of a powder which is formed as a result of an encapsulation reaction is defined as: (polymer weight [g]/ powder weight [g]) *100 %.

The weight of polymer is usually determined by pyrolysis of this polymer and measuring the weight loss, either in a normal electric furnace [42] or in thermogravimetrical analytical equipment (TGA)[30, 33]. When the inorganic core material also loses weight on heating (like for example $Mg(OH)_2$), TGA is preferable because the separate contributions to the weight loss can be distinguished at different temperatures.

In case of emulsion polymerization the formed free polymer has to be removed first by repeated washing and centrifugation steps [33].

In many instances the polymer layer is extracted from the surface so that it remains possible to determine the molecular mass and/or the chemical composition distribution of the polymer [13, 30, 33, 42]. In some cases it is possible to dissolve the core material, like for example in the case of SiO_2 which dissolves easily in HF. Fukano [42] found that, after dissolving the SiO_2 with HF, the unextractable polystyrene had a molecular mass (distribution) that differed from the extractable polystyrene. The authors conclude that part of the surface polymer is chemically grafted to the surface and the amount of unextractable polymer varies with the type of inorganic substrate. In the case of encapsulation of $CaCO_3$ with polystyrene (initiator 4,4' azo-bis-(cyanopentanoic) acid, ABCA) [33] about 20 % of the surface polymer is tightly bound to the surface of $CaCO_3$.

3.2. PARTICLE SIZE AND MORPHOLOGY

One of the main problems in encapsulating filler and pigment particles is to prevent (partial) coagulation during the encapsulation process, because the coagulates will also become irreversibly encapsulated . The layer thickness and uniformity are other points of interest that are usually determined.

Particle size measurements are usually performed prior to the encapsulation with the usual techniques like dynamic light scattering (DLS), scanning electron microscopy (SEM) , transmission electron microscopy (TEM) or light microscopy (LM).

Only in those cases where the initial particles are spherical and the particle size distribution is narrow, particle size measurements with DLS can be used to determine the layer thickness. In general TEM is used to obtain information on particle size and layer thickness of the encapsulated particles. A nice example is given by Templeton-Knight [11] in the encapsulation of TiO_2 with polymethyl methacrylate where the use of ultrasound was studied in order to obtain uniform polymer coating of TiO_2.

The uniformity of the polymer coating improved on the use of ultrasound during the encapsulation process which could clearly be observed with TEM micrographs.

Sometimes artifacts ('diffraction halos') in TEM can be misinterpreted as polymer layers [28]. The presence of TiO_2 or other inorganic materials can prevent melting of the polymer ,which would otherwise occur because of the electron beam energy, because the inorganic materials absorb electrons [21]. With TEM the structure of the polymer layer can be observed which is not always uniformous but can also be raspberry shaped [35]. In emulsion polymerization sometimes heterocoagulation occurs which leads to polymer 'layers' which consists of latex particles [33, 36].

A technique which is very helpful in observing this phenomena is dark field microscopy.

SEM in many cases can also give a good impression of the encapsulation process [9, 37]. The use of both SEM and TEM on the same samples can give very clear evidence for the encapsulation [36].

The formation of homopolymer of styrene on the surface of calcium carbonate was detected by diffuse reflection infrared spectroscopy (DFIR) [43]. Functionalization of silica with silanes [44] and titanium dioxide with titanates [21] and the formation of copolymers of styrene and butyl acrylate on the surface of silicas was observed with DFIR measurements [44]. Energy dispersive spectroscopy performed on an emission electron microscope was used to determine the nature of areas (polymer or silica) in SEM micrographs [2].

One of the drawbacks to use SEM and TEM on the dried product is that it is not always clear whether coagulation has occurred during the encapsulation process or afterwards during the sample treatment.

3.3. MOLECULAR MICROSTRUCTURE OF ENCAPSULATING POLYMER

The molecular microstructure can be divided in intramolecular microstructural aspects like sequence distributions, tacticity and branching and intermolecular microstructure in terms of molecular weight and chemical composition.

Most examples of analysis of microstructure of encapsulating polymer relate to the intermolecular microstructure, mostly the molecular weight distribution. It is suggested that the monomer concentration (vinyl acetate) at the surface of silica is higher than in the aqueous phase as elucidated from ultrasonic velocity measurements [32]. Also the monomer ratio at the surface of TiO_2 particles in the copolymerization of styrene and methyl methacrylate was found to be different from that in free latex particles [13, 29]. The average lifetime of radicals on the surface of pigment/filler particles is influenced by the particle number through the entry and exit rate coefficients and the termination rate at the surface of the particles. It has been suggested that because TiO_2 cannot be penetrated by radicals the chance of termination is lowered and the average lifetime of radicals at the inorganic surface will be longer than that in normal polymer particles [22]. Both the local monomer concentration and the radical concentration will have an effect on the molecular weight distribution (Figure 4). Lee et al [45] observed in the emulsion polymerization of methyl methacrylate on $CaSO_3$ that the weight average molecular weight (Mw), obtained by gel permeation chromatography (GPC), initially decreased because at the initial stage of the reaction some coagulation occurred,

448

reducing the number of particles and thus increasing the termination rate. But when the gel effect became significant Mw increased again.

Electrostatic interaction between the positively charged surface and the anionic sulphate radicals increased the entry rate of radicals which led to lower molecular weight of polymethyl methacrylate formed on the surface of TiO_2 [36].

In general it is found that polymer on the surface of inorganic particles has a higher molecular weight than apparent encapsulating polymer or free polymer [30, 35, 42] (Figure 4).

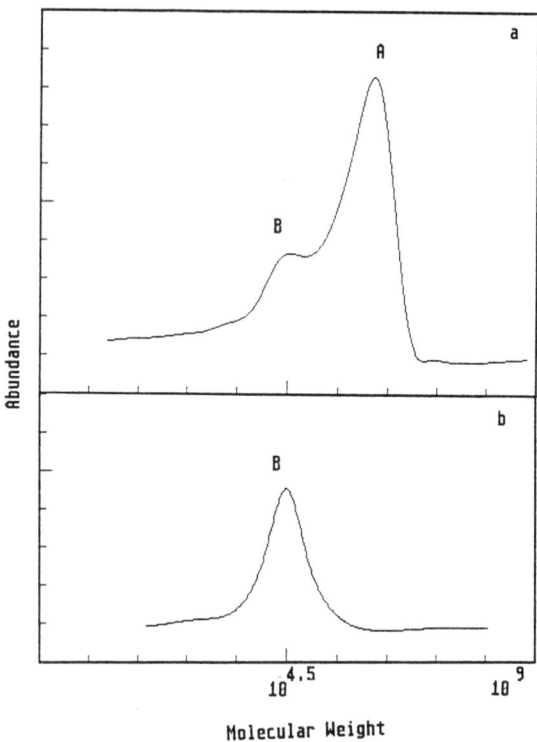

Figure 4. Gel permeation chromatogram of latex polymer separated from composite silica-polystyrene latex system. (a) Polymer formed on the surface of silica (marked A) (b) Polymer formed in the absence of silica (marked B) (Data from Furusawa et al.[35])

This effect is usually attributed to higher monomer concentrations at the surface [30]. Sometimes two peaks can be observed that can be attributed to the encapsulating polymer and the isolated latex particles formed by initiation in the aqueous phase [35]. The chemical composition distribution in the case of copolymerization of styrene and methyl methacrylate on TiO_2 was determined by high performance liquid chromatography [13]. The copolymer composition on the surface depends on many factors and interactions with the surface might change the monomer ratio, leading to different copolymers with a different Tg [29].

Also for homopolymers differences in Tg have been observed previously [29, 46].

For example for PMMA on silica particles an increase of more than 10 °C was found [46] which was explained by a reduced chain mobility of polymer chains directly in contact with the surface.

4. Conclusions and Future Outlooks

Encapsulation of organic and inorganic pigment and filler particles is possible although no universal method exists and for every system a particular method has to be optimized. For submicron particles the emulsion polymerization approach seems the most promising approach. For larger particles this method is less suitable and one has to resort to other techniques like suspension polymerization or heterocoagulation with small latex particles. The main problems are to obtain and maintain primary particles during the encapsulation reaction and at the same time have a high encapsulation efficiency. In that respect the use of semi-continuous addition of monomers and surfactant within the emulsion polymerization approach seems to be the most promising. To improve compatibility with the final matrix, copolymerization and core-shell polymerizations offer interesting possibilities.

Mechanical properties of plastics can benefit from the encapsulation of pigments and fillers. Also the quality of coatings can be improved, rendering more environmentally friendly coatings with a longer lifetime.

The large scale production of encapsulated pigment and filler particles has not fully developed yet, but with the increasing call for better properties and with the help of fundamental research in the near future major developments can be expected.

5. References

1. Hofman-Caris, C.H.M. (1995) *New J. Chem.* **18**,1087-1096.
2. Hong, L. and Ruckenstein, E. (1993) *J. Appl. Polym. Sci.* **48**,1773-1780.
3. Godard, P., Wertz, J.L., Biebuyck, J.J. and Mercier, J.P. (1989) *Pol. Eng. and Sci.* **29**,127-133.
4. Kendall, K. (1978), *Br. Polym.* J. **10**, 35-38.
5. Hawthorne, D.G., Hodgkin, J.H., Loft, B.C. and Solomon, D.H. (1974) *J. Macromol. Sci.-Chem.* **A8**, 649-657.
6. Dekkers, M.E.J. and Heikens D. (1983) *J. Appl. Polym. Sci.* **28**, 3809-3815.
7. Kolarik, J., Lednicky, F., Jancar, J. and Pukanszky, B. (1990) *Polymer Commun.* **31**, 201-204.
8. Kendall, K. and Sherliker, F.R. (1980) *Br. Polym. J.* **12**, 85-88.
9. Ono, T. (1986) *Org. Coat.* **18**, 279-296.
10. Laible, R. and Hamann, K. (1980) *Adv. in Coll. and Interf. Sci.* **13**, 65-99.
11. Templeton-Knight, R.L. (1990) *J. Oil Colour Chem. Assoc.* **1**, 459-464.
12. Hoy, K.L. and Smith, O.W. (1991) *Polym. Mater. Sci. & Eng.*, ACS Fall Meeting 1991, New York, **65**, 78-79.
13. Janssen, R.Q.F., Derks, G.J.W., van Herk, A.M. and German, A.L.

450

(1993) On-line monitoring and control of the (co-)polymer encapsulation of TiO2 in aqueous emulsion systems in '*Encapsulation and Controlled Release*', D.R. Karsa and R.A. Stephenson (eds.), Royal Society of Chemistry, Cambridge 102-116.

14. Kroker, R. and Hamann, K. (1978) *Angew. Makromol. Chemie* **13**, 1-22.
15. Dekking, H.G.G. (1965) *J. Appl. Polym. Sci.* **9**, 1641-1651.
16. Arshady, R. (1993) *Biomaterials* **14**, 5-15.
17. Buske, N. and Goetze, T. (1983) *Acta Polym.* **34**, 184-185.
18. Furusawa, K., Nagashima, K. and Anzai, C. (1993) *Kobunshi Ronbunshu* **50**, 337-342.
19. Huang, T.C.C. (1986) *The model encapsulation of iron oxides in latex particles by preparation and characterization of magnetic latex particles*, PhD thesis, Lehigh University
20. Cohen, B., Wong, T.K. and Hargitay, B. (1986) *Magnetically responsive reagent carrier*, Eur. Pat. 180384 A2, CA 105 P 75418n.
21. Caris, C.H.M., van Elven, L.P.M., van Herk, A.M. and German, A.L.(1989) *Br.Polymer J.* **21**, 133-140.
22. Caris, C.H.M., Kuijpers, R.P.M., van Herk, A.M. and German, A.L. (1990) Makromol. Chem., Makromol. Symp. **35/36**, 535-548.
23. Janssen, R.Q.F., van Herk, A.M. and German, A.L. (1993*) J. Oil Colour Chem. Assoc.* **11**, 455-461.
24. Lorimer, J.P., Mason, T.J., Kershaw, D., Livsey, I. and TempletonKnight, R. (1991) *Colloid Polym. Sci.* **269**, 392-397.
25. Chaimberg, M., Parnas, R. and Cohen, Y. (1989) *J. Appl. Polym. Sci.* **37**, 29212931.
26. Nagai, K., Ohishi, Y., Ishiyama, K. and Kuramoto, N. (1989) *J. Appl. Polym. Sci.* **38**, 21832189.
27. Janssen, R.Q.F. (1995) *Polymer encapsulation of titanium dioxide*, Ph D Thesis, Eindhoven ,The Netherlands
28. Caris, C.H.M. (1990) *Polymer encapsulation of inorganic submicron particles in aqueous dispersion*, Ph D Thesis, Eindhoven, The Netherlands
29. van Herk, A.M., Janssen, R.Q.F., Janssen, E.A.W.G. and German, A.L. (1993) *Polymer encapsulation of inorganic particles*, in Proceedings of the XIXth International conference in organic coatings science and technology, Athens Greece july 1993, 219-224
30. Haga, Y., Watanabe, T. and Yosomiya, R. (1991) *Angew . Makromol. Chem.* **189**, 2334.
31. Partch, R. (1993) *Mat Tech* **8**, 43-44.
32. Hergeth, W., Starre, P.,Schmutzler, K. and Wartewig, S. (1988) *Polymer* **29**, 1323-1328.
33. Janssen, E.A.W.G., van Herk, A.M. and German, A.L. (1993) *Polym. Prepr. (Am. Chem. Soc., Div. Polym. Chem.)* **34**, 532-533.
34. Caris, C.H.M. and German, A.L. (1989) *Preparation of polymers in emulsion, pigment particles and the modification of pigment particles*, Eur. Pat. Appl. 0 328 219 A1, CA 112 P 100763a.
35. Furusawa, K., Kimura, Y. and Tagawa, T. (1986) J. Colloid Interf. Sci. **109**, 69-76.
36. Hasegawa, M., Arai, K. and Saito, S. (1987) *J. Pol. Sci. Part A Pol. Chem. Ed.* **25**, 3231-3239.
37. Park, J.S. and Ruckenstein, E. (1990) *Polymer* **31**, 175-179.
38. Wu, J., Harwell, J.H., O'Rear, E.A. and Christian, S.D. (1988) *AIChE Journal* **34**, 1511-1518.
39. Meguro, K., Yabe, T., Ishioka, S., Kato, K. and Esumi, K. (1986) *Bull. Chem. Soc. Jpn.* **59**, 3019-3021.
40. Jackson, S., Reboiras, M.D., Lyle, I.G. and Jones, M.N. (1986) *Faraday Discuss. Chem. Soc.* **85**, 291-301.
41. Kurja, J., Nolte, R.J.M., Maxwell, I.A. and German, A.L. (1993), *Polymer* **34**, 2045-2049.
42. Fukano, K. and Kageyama, E. (1975) *J. Pol. Sci. Pol. Chem. Ed.* **13**,13091324 and 1325-1338.
43. Nakatsuka, T., Kawasaki, H., Yamashita, S. and Kohjiya, S. (1983) *J. of Colloid and Interface Sci.* **93**, 277-280.
44. Revillon, A., Espiard, P. and Guyot, A. (1991) *Double Liaison-Phys. Chimie Peint. Adh.* **431/432**, 29-42.
45. Lee, C-F. and Chiu, W-Y. (1993) *Polymer Int.* **30**,475-481.
46. Hergeth, W., Steinan, U., Bittrich, H., Simon, G. and Schmutzler, K. (1989) *Polymer* **30**, 254-258.

SOME TRENDS IN THE PREPARATION AND USE OF REACTIVE LATICES

J.J.G.S. VAN ES, J.M. GEURTS, J.M.G. VERSTEGEN
AND A.L. GERMAN
*Department of Polymer Chemistry, Eindhoven University of Technology,
P.O. Box 513, 5600 MB Eindhoven, The Netherlands*

1. Introduction

Reactive polymers constitute an important class of binder materials for application in, e.g., coatings and resins. Traditional solvent-borne systems consist of low molecular weight material dissolved in a suitable organic solvent. During application and evaporation of the solvent network formation occurs, resulting in an increase in the molecular weight, the viscosity and the glass transition temperature (T_g). Through the formation of a network these thermoset resins attain their desired properties: insolubility in most organic solvents, good water resistance, good hardness development, and good resistance against blocking.

Because of stricter regulations for the emission of volatile organic compounds, new coatings are being developed, such as high solids, powder coatings, UV-curable coatings, and, also, *water-borne* coatings [1].

Traditional latex (acrylic) paints consist of high molecular weight thermoplastic material without pendant functional groups. During film formation the T_g of the polymer needs to be lowered temporarily, which necessitates the use of cosolvents. Furthermore, the mechanical properties of these latex paints cannot match those of the traditional solvent-borne systems.

Thus, it would appear that by incorporating reactive groups in latex paints some of these problems might be solved. Therefore, thermoset or crosslinkable latices constitute an interesting class of dispersions that have a great potential for application in water-based coatings. However, as will be discussed in this paper, the translation from a solvent-borne system to a dispersed phase system is not that straightforward.

Nonetheless, many types of functional groups have been incorporated in polymer dispersions with the aim of improving properties through crosslinking during film formation. For a review of the older literature one is referred to the papers of Grawe and Bufkin [2-6]. More recent updates can be found in papers by Daniels and Klein [7] and by Ooka and Ozawa [8].

This article does not intend to present an overview of all types of functional groups that may be utilized in reactive latices, nor will it deal with emulsification of reactive polymers or post-polymerization modification as a means of introducing the desired functionality. Rather, it will focus on those aspects of emulsion polymerization, by which the incorporation and location of functional groups can be controlled, as well as by which the desired properties during film formation can be attained.

451

J. M. Asua (ed.), Polymeric Dispersions: Principles and Applications, 451–462.

452

After a brief introduction into crosslinking mechanisms during film formation, specific aspects of emulsion polymerization using reactive monomers will be illustrated, using results, recently obtained in our group, on amino-, epoxy-, and hydroxy-functional latices. Attention will be especially directed towards the incorporation of the functional monomer, the prevention of premature crosslinking, and the control of the location. Next, some aspects of film formation when working with a reactive latex (blend) will be discussed. The paper will end with a personal view with regard to the future developments of reactive latices.

2. Crosslink Types and Mechanisms during Film Formation

The most commonly made distinction between types of crosslinking is that between interstitial, homogeneous and interfacial crosslinking (Figure 1) [2,7].

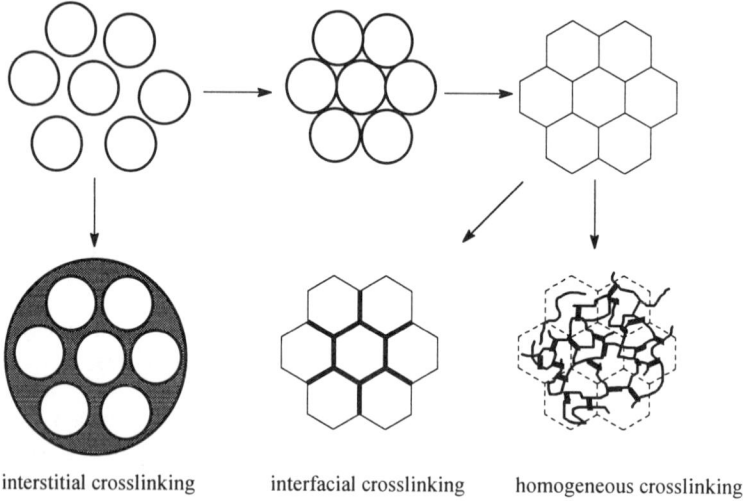

interstitial crosslinking   interfacial crosslinking   homogeneous crosslinking

Figure 1. Schematic representation of the different types of crosslinking during film formation

In interstitially crosslinked latex systems thermoplastic latex particles are imbedded in a polymer network. This network is formed by water-soluble functionalized resins or water-dispersable functionalized particles that fill the voids of the matrix of thermoplastic material during film formation.

Homogeneous crosslinking requires a random distribution of reactive groups in the latices and, additionally for two-pack systems, of the external crosslinker during film formation. This will ensure that crosslinking will occur homogeneously, resulting in a film with no local differences in crosslink density. Generally, these systems are capable of self-condensation or autoxidation.

Self-condensing latices contain monomers that are capable of selfcondensation, for instance [2-5]: N-methylol (meth)acrylamide and derivates thereof, hydroxymethyl-ated diacetone acrylamide, and (meth)acryloyloxypropyltrialkoxysilanes. Crosslinking during film formation is induced thermally or by addition of a catalyst. The main problem when working with these latices is premature crosslinking due to hydrolysis.

Autoxidizable latices are air drying through allylic oxidation of mono- or poly-unsaturated side chains. Monomers used to prepare these latices include cyclohexenyl and allyl acrylate, and vinyl condensation products of drying oils, amongst others. Crosslinking is induced thermally or by addition of a suitable metal ion catalyst.

Interfacial crosslinking [5,7-9] is encountered when working with particles that do not have a random distribution of functional groups or when an external crosslinker is used that does not provide a homogeneous crosslinking, because of, for instance, diffusion problems into the former particles during film formation. The crosslinking is usually concentrated at or close to the surface of the former particles.

Latices that show interfacial crosslinking usually contain functional groups that can be crosslinked with external crosslinkers. Important examples include:
* Carboxy-functional latices that can be crosslinked with metal ions, such as Mg^{2+}, Al^{3+} and Zn^{2+}, with melamine or urea-formaldehyde resins, and with epoxy-containing curing agents, amongst others. These latices are produced by incorporating carboxy-functional monomers, such as (meth)acrylic, maleic or fumaric acid.
* Hydroxy-functional latices with, e.g., melamines, urea-formaldehyde, isocyanates, and epoxies. These latices are generally produced by copolymerization of hydroxy-functional monomers, such as hydroxyethyl and 2-hydroxy-1-propyl (meth)acrylate. - These latices will be discussed more extensively in the next section.
* Epoxy-functional latices, produced by copolymerizing epoxy-functional monomers such as glycidyl (meth)acrylate and allyl glycidyl ether, can be crosslinked with amines and carboxylic acids, amongst others. These latices will also be discussed more extensively below.
* Acetoacetoxy-functional latices, produced by copolymerizing acetoacetoxyethyl (meth)acrylate, can be crosslinked with amines and carboxylic acids, amongst others.

Beside loss of functionality due to hydrolysis, a major disadvantage of these two-pack systems may be a limited application time after addition of the crosslinking agent. A solution to this problem may be the use of a blend of complementary reactive latices [10]. In this approach complementary groups that are reactive at ambient temperature are incorporated in separate latices. Upon mixing a stable blend can be obtained because the electrostatic and/or steric repulsion between the latex particles will also prevent premature reaction. Only during film formation and the resultant interdiffusion of polymer chains, crosslinking can occur for these so-called two-pack-in-one-pot systems (Figure 2). The systems presently under investigation comprise the amino-epoxy and the amino-acetoacetoxy systems.

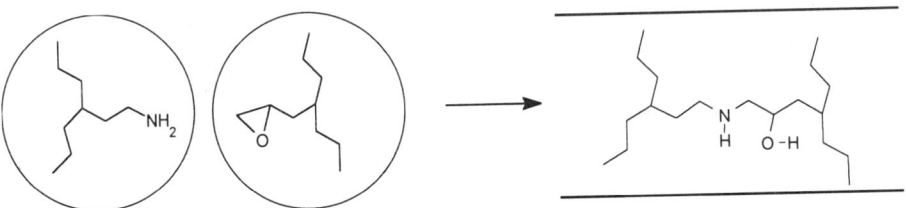

Figure 2. Principle of two-pack-in-one-pot systems

When investigating the crosslink behaviour of reactive latices and their blends during film formation, it is essential to be able to exert control over the amount and location of the functional groups in the latex particles. The study of the emulsion copolymerization behaviour of reactive monomers is therefore of utmost importance.

3. Emulsion Copolymerization Using Reactive Comonomers

Translation from a solvent-borne to a water-based reactive system is not that straight-forward. Evidently, many reactive groups are too sensitive towards hydrolysis to permit their use in an aqueous environment. Also, the incorporation of functional monomers during emulsion copolymerization may give rise to specific problems.

Compared with the other monomer(s) in the recipe, the reactive monomer usually is the more water soluble one or it can even be water miscible. This may result in competitive, or even dominant, polymerization of the reactive monomer in the water phase and, consequently, in a low degree of incorporation inside the latex particles.

The reactive functionality, generally a polar or charged group located closely to the vinylic double bond, may affect the copolymerization behaviour, i.e., a strong composition drift may occur. The resultant inhomogeneity in the chemical composition distribution of reactive functionality in the polymer chains and even throughout the latex particles may adversely affect the final properties after crosslinking the film.

Also, molecular weights obtained in conventional emulsion polymerization are much higher than those in solution polymerization. This implies that divinylic impurities present in the reactive monomer or side reactions of the functional group leading to grafting and crosslinking will influence the properties of the final latex more prominently and may lead to premature network formation inside the latex particles. This will adversely affect the polymer interdiffusion and crosslinking behaviour during the final stages of the film formation.

Using results, recently obtained in our group, on amino-, epoxy- and hydroxy-functional latices some of these problems and possible solutions will now be discussed in more detail.

3.1. AMINO-FUNCTIONAL LATICES

The preparation of amino-functional latices seems highly desirable, as they may function in many crosslinking systems (e.g., with epoxy- or acetoacetoxy groups). For these types of crosslinking reactions a primary amino group is highly desirable.

Primary amino functionality can be introduced in latices by emulsion copolymerization using methacrylic- [11] or vinylbenzylic [12] ammonium salts or, else, by post-polymerization modification, e.g., by treatment of carboxy-functional latices with aziridine (ethylene imine) [13]. Whereas the latter method is undesirable because of the toxicity of residual aziridine, the former methods can only be used to incorporate amino-functional monomer in a low amount [11] under special conditions [12] at the surface. A good method to incorporate primary amino-functional monomers randomly inside film-forming acrylic latex particles is still lacking.

Therefore, and also because of their limited commercial availability, it was decided to prepare a series of homologous w-amino-1-alkyl methacrylates and to systematically investigate their (emulsion) copolymerization behaviour [14]. Methacrylic monomers were chosen, because of their decreased reactivity in Michael addition and their decreased sensitivity towards amine-induced polymerization, as compared with their acrylic counterparts.

Amino-functional methacrylates can be prepared in good to excellent yields by reacting methacryloyl chloride with the HCl-, or p-toluenesulfonic acid (pTsOH-), salt of the corresponding w-aminoalcohol (Figure 3; cf. Table 1).

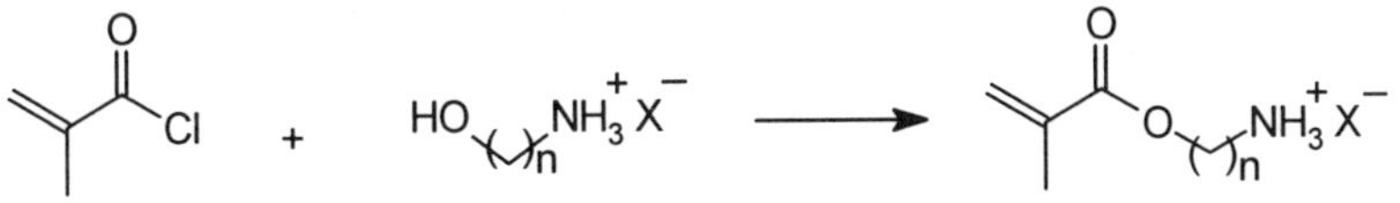

Figure 3. Synthesis of w-amino-1-alkyl methacrylate acid salts (HX = HCl or pTsOH)

As can be expected (cf. Table 1), the water solubility of the w-amino-1-alkyl methacrylate salt is strongly reduced by elongating the carbon chain connecting the double bond and the ammonium group. This may be beneficial for their intended application in emulsion polymerization. The 11-amino-1-undecyl methacrylate.HCl (AUMA.HCl) forms micelles.

Also, the pK_a of the monomer shows a marked increase in this series, implying that the monomer becomes more nucleophilic and, thus, more reactive in the crosslinking. This also implies that the pH during an emulsion polymerization of monomers

TABLE 1. Amino-functional methacrylate(acid salt)s: synthesis yields, water solubility and pK_a

monomer			yield (%)	$[M.HX]_{aq}$ (mM)[a]	pK_a	$[M]_{aq}$ (mM)[a]
name	X	n				
AEMA	Cl	2	50	1200	8.1	-
APrMA	OTs	3	92	466	9.3	-
APMA	OTs	5	88	195	10.2	53
AUMA	Cl	11	51	69	10.3[b]	<2

[a] Measured at 60 °C. [b] Determined below the CMC.

containing a free primary amino group will be high: for 5-amino-1-pentyl methacrylate (APMA) and for AUMA the pH should be around 12. At this high pH and at the intended temperature of emulsion polymerization (60 °C) hydrolysis of ester groups present in the functional monomer and in the backbone monomer might be expected, however control experiments have indicated that the extent of hydrolysis is not significant.

Upon deprotonation aminoethyl methacylate (AEMA) and 3-amino-1-propyl methacrylate (APrMA) show a fast intramolecular rearrangement to give N-hydroxyethyl- and N-3-hydroxypropylmethacrylamide, respectively (Figure 4) [15]. APMA and its homologues no longer give this rearrangement, because of unfavourable transition states. However, they are sensitive to Michael addition and deprotonation should be performed shortly before polymerization. Michael addition can not be avoided when copolymerizing with good Michael acceptors: for this reason, and also in order to obtain random copolymers, only methacrylates can be used as backbone monomers in batch emulsion polymerizations with these monomers.

Short chain w-aminoalkyl methacrylate salts, such as AEMA.HCl, can be (co)polymerized in solution. As was also observed earlier [15], upon deprotonation the aminoalkyl ester groups in these polymers no longer undergo rearrangement to N-hydroxyalkylamide groups because of increased steric hindrance by the polymer backbone. Homologous w-aminoalkyl methacrylates, e.g. APMA, can be copolymerized in solution, both

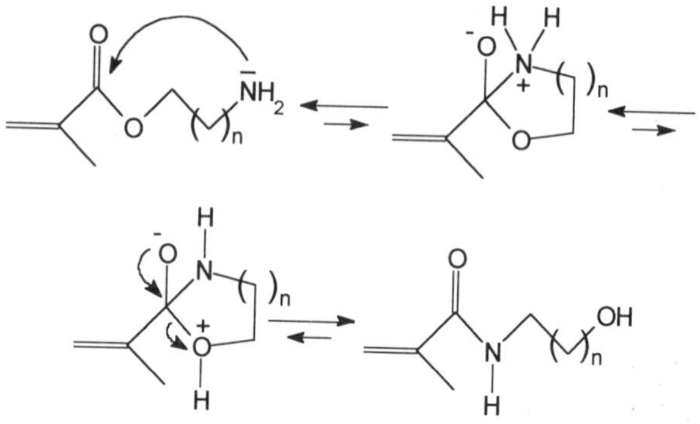

Figure 4. Intramolecular rearrangement of w-amino-1-alkyl methacrylates AEMA (n=1) and APrMA (n=2)

in their free amine form and as their acid salt, in the latter case followed by deprotonation. When using w-amino-1-alkyl methacrylate salts in copolymerization with BMA a strong composition drift occurs: due to charge repulsion the incorporation of the amino-functional methacrylate salt in an already charged polymer chain will be hampered. Use of APMA and AUMA (i.e., with a free amino group) results in a significantly enhanced incorporation. Solution copolymers can be emulsified to yield stable artificial amino-functionalized latices, containing polymer of compara-tively low molecular weight, but with a broad particle size distribution.

When using short chain w-amino-1-alkyl methacrylate salts, such as AEMA.HCl and APMA.pTsOH, under standard batch emulsion copolymerization conditions, the hydrophobic comonomer is almost exclusively found in the latex particles, while the amino-functional monomer predominantly polymerizes in the aqueous phase. Because of their limited water solubility, it is possible to incorporate to some extent comparatively long chain w-aminoalkyl methacrylate salts, such as the HCl-salt of 11-aminoundecyl methacrylate (AUMA).

Aminoalkyl methacrylate salts, such as AEMA.HCl, can be incorporated in emulsifier-free copolymerizatons using a nonionic redox initiator system (see also [11]). Upon titration with a nonionic surfactant and deprotonation, stable latices with a narrow particle size distribution are obtainable with maximally 0.5 % by weight of primary amino-functional monomer incorporated at the surface.

When neutralized, aminoalkyl methacrylates, such as APMA and AUMA, can be incorporated in batch emulsion copolymerizations. However, the extent to which they become incorporated, is still low. This cannot be attributed to their water solubilities, which are sufficiently low (cf. Table 1). Control of the pH is essential: if no base is added during the experiment the pH will drop quickly and the free amino group will become protonated, so that the amino-functional monomer predominantly polymerizes in the water phase. But even when the pH is kept at ca. 12, the extent of incorporation is low. Possibly, Michael addition in the monomer droplets at the elevated tem-perature of polymerization results in the formation of products so hydrophobic that they can no longer be transported through the water phase.

In order to enhance the fraction of amino-functional groups in the latex, emulsion copolymerizations need to be performed under semi-continuous, or monomer-starved, conditions. When the monomers are added as a cooled pre-emulsion and the pH in-side the reactor is kept at ca. 12, most of the primary amino-functional monomer is incorporated and monodisperse, sterically stabilized latex particles are obtained with a narrow particle size distribution. Incorporation of other primary amino-functional monomers in semicontinuous emulsion polymerization is presently under investigation in our laboratories.

3.2. EPOXY-FUNCTIONAL LATICES [16]

Glycidyl methacrylate (GMA) is the monomer commonly used to incorporate epoxy groups in a latex during emulsion polymerization. However, GMA is somewhat sensitive to hydrolysis and the byproduct formed, 2,3-dihydroxypropyl methacrylate, will influence the polymerization process. Also, crosslinking inside the particles is observed. The latter will have an adverse effect on the interdiffusion of polymer chains required at a certain stage during film formation [17].

Internal crosslinking of GMA-functional latex particles can be prevented by controlling the molecular weight by means of a chain transfer agent (CTA), such as n-dodecyl mercaptan or carbon tetrabromide. Although the CTA will not prevent the occurrence of grafting and of some crosslinking, it effectively prevents the formation of an infinite network.

Crosslinking inside the latex particles cannot be prevented by performing emulsion polymerizations at ambient temperature. Again, application of a CTA at only moderate concentration leads to the desired reduction in molecular weights and properly prevents the formation of an infinite network. In addition, the extent of hydrolysis of epoxy groups is significantly reduced. When the CTA is used in higher concentrations at room temperature, it also retards the polymerization, probably by influencing the entry and exit rates (cf. [18]).

3.3. HYDROXY-FUNCTIONAL LATICES

For the incorporation of the commonly used monomers hydroxyethyl methacrylate (HEMA) and -acrylate (HEA), emulsion polymerizations are generally performed semicontinuously in the presence of large amounts of non-ionic soaps in order to prevent gelation. Both the amount of soap and the heterogeneity of the copolymer are undesirable, when studying film formation of reactive latices. Control over particle morphology is dubious in view of the extent of water phase polymerization and of physical adsorption of polymer rich in hydroxy-functional monomer, produced in the water phase.

In order to allow the preparation of a more homogeneous hydroxy-functional latex under batch emulsion polymerization conditions with an anionic soap (SDS) and an anionic initiator (SPS), the use of more hydrophobic, i.e. less water-soluble w-hydroxy-functional monomers 3-hydroxy-1-propyl (HPMA), 4-hydroxy-1-butyl (HBMA), 6-hydroxy-1-hexyl (HHMA) and 8-hydroxy-1-octyl (HOMA) methacrylate has been investigated [19].

These monomers can be prepared by reaction of an excess diol with methacryloyl chloride (route 1 in Figure 5) or by reaction of an w-haloalcohol with potassium methacrylate under phase-transfer conditions [20] (route 2) in generally good yields (see Table 2). As could be expected, upon elongating the carbon chain connecting the hydroxy and methacrylate moieties, the water solubility of these monomers shows a strong decrease. Also, reactivity ratios in the solution copolymerization with styrene (cf. Table 2) indicate a trend towards the formation of alternating copolymers.

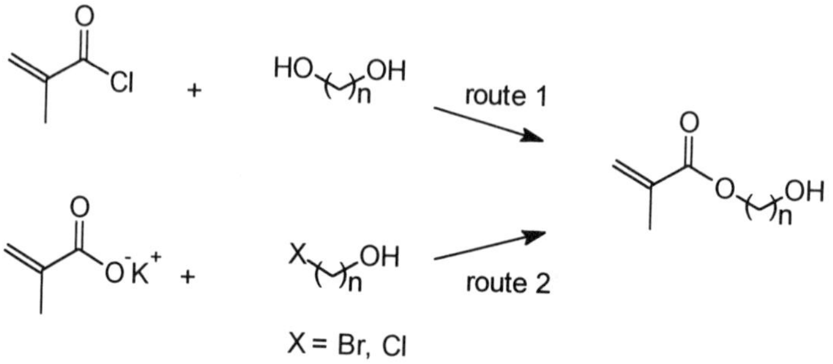

Figure 5. Synthesis of w-hydroxy-1-alkyl methacrylates

TABLE 2. Hydroxy-functional methacrylates (1): synthesis yields, water solubility, and reactivity ratios with styrene (2)

monomer		yield (%)		$[M]_{aq}$ (mM)[a]	react. ratio	
name	n	route 1	route 2		r_1	r_2
HEMA	2	-	-	∞	0.48	0.27
HPMA	3	60	70	382	0.26	0.16
HBMA	4	64	-	170	0.16	0.10
HHMA	6	70	75	37	-	-
HOMA	8	30	-	5	-	-

[a] Measured at 50 °C.

Batch emulsion copolymers with styrene were analyzed with gradient polymer elution chromatography (GPEC) and NMR spectroscopy. Irrespective of the monomer ratio, formation of hydroxy-functional homopolymer can effectively be prevented, the sole exception being recipes rich in HPMA. In the latter case this can be overcome by reduction of the initiator concentration and, thereby, of the extent of termination in the aqueous phase.

4. Film Formation of Reactive Latices

Generally, the mechanism of film formation of latices is divided into three phases: (a) evaporation of the water or drying, (b) coalescence and deformation of the latex par- ti-cles, and (c) cohesive strength development by further coalescence of adjacent par-ticles and interdiffusion of polymer chains. Throughout film formation the mechanical strength of the film will increase, but in particular so during the last stage of the film formation, when as a result of polymer interdiffusion a critical degree of entanglement is attained [7].

Additionally, in the case of reactive systems mechanical strength can be attained through crosslinking. In general, crosslinking will not significantly alter the glass transition temperature because of the high molecular weights of the emulsion copoly-mers involved.

However, the incorporation of reactivity does not necessarily have to lead to impro-ved properties or the latter may be less than expected. In principle, the presence of the crosslinker (often a polar compound present in the water phase) during film formation will lead to some intraparticle grafting and crosslinking, especially in the outer shell of the reactive particles. Undoubtedly, this will retard the polymer interdiffusion in the final stage of film formation, possibly to the extent that the maximally attainable degree of entanglement is reduced. Beside this, the diffusion of the crosslinker into the interior of the former latex particles may become hampered so that non-homogeneous crosslinking may be the result.

A solution to this problem may be a significant reduction of the molecular weight: the extent of grafting and crosslinking will influence the polymer interdiffusion to a lesser extent, because polymer diffusion will no longer occur through reptation. Additionally, the glass transition temperature may now show a strong increase.

An alternative solution could be to separate the two complementary reactivities by incorporating them into separate particles, so that only in the final stage of film formation crosslinking can occur.

4.1 FILM FORMATION OF TWO-PACK-IN-ONE-POT SYSTEMS

When a film is prepared from a blend of latices, containing complementary reactive functional groups, two major issues have to be considered additionally. In general, the particle size distribution of a blend will be extremely broad or, in the case of narrowly distributed constituent latices, bimodal. This may affect the minimum film formation temperature (MFT) and the packing in the early stages of film formation. The latter effect determines the surface area, across which the complementary reactive chains can diffuse. Additionally, there may be a competition between polymer-polymer interdiffu-sion and polymer-polymer crosslinking.

The effect of bimodality of the particle size distribution on film formation has been investigated by measuring the MFT and by examining the surface of films dried below their MFT by means of Atomic Force Microscopy (AFM). Monodisperse model latices, prepared by emulsifier-free emulsion polymerization of butyl methacrylate, are use᷾ ᵢ order to exclude any effect of surfactants migrating during film formation. In specific cases highly ordered packings can be observed [21].

From this, two strategies emerge to make use of the effect of the particle size distribution in crosslinking during film formation for blends of reactive latices. On the one hand, functional monomers of complementary reactivity can each be built into small latex particles of exactly the same size and stabilized in exactly the same fashion. This will ensure a random packing of latex particles, so that the contact area established in the early stages of film formation will be maximized, while the path length for interdiffusion will be limited by their small size. Recipes have been developed that allow the incorporation of amino-, epoxy- and acetoacetoxy functionality in such latices.

On the other hand, bimodal particle size distributions may be used, wherein large particles, consisting of a hard core and a soft shell with reactive functional groups become embedded in a matrix of significantly smaller soft particles of complementary reactivity. Crosslinking during film formation will result in a film that has hard domains randomly distributed inside a flexible matrix, which should be able to attain sufficient mechanical stability through the crosslinking.

The progress of the crosslinking reaction during film formation may be monitored by various techniques, such as FT-IR and DMTA. However, the maximally attainable degree of crosslinking may be restricted because of increasing diffusion limitations during polymer-polymer interdiffusion, caused by the increasing degree of crosslinking. Thus, the all important question will be: how do the rates of chemical reactivity and of polymer-polymer interdiffusion compare? Now that the preparation of functional latices appears under control, the effect of these and other important factors can be evaluated.

What is the effect of the average molecular weight? For the preparation of non-internally crosslinked particles, emulsion copolymerization in the presence of a chain transfer agent can be used, but usually gives material of still rather high molecular weight. For reactive latices, containing polymers of intermediate or relatively low molecular weight, solution copolymerization, followed by emulsification, may be used, but will afford relatively broad particle size distributions. A better method would be the use of cobaloximes for catalytic chain-transfer [22] in the emulsion copolymerization of functional monomers. From these latices blends can be prepared with different combinations of average molecular weights, i.e., low vs. low, high vs. high, low vs. high.

The intrinsic reactivity of the system under investigation will be another factor. For example, it is well known that at ambient temperature reactivity in the amino-acetoacetoxy system is significantly higher than that in the amino-epoxy system. This may imply that for each of these systems, a different range of molecular weights will give optimal results.

The effects of these factors on chemical reactivity are being evaluated using FT-IR, sol-gel determinations and DMTA. For crosslinking in blends of complementary reactive latices significantly higher plateau values in the DMTA curves are found than for the corresponding solution-type model crosslinking systems. These first results indicate that two-pack-in-one-pot systems present an interesting opportunity for crosslinking reactive latices.

5. Conclusions and Prospects for Further Research

In this paper a number of problems, that one frequently encounters in the preparation of reactive latices using functional monomers, have been discussed; and for a number of important types of reactive latices, solutions have been offered. Preferential, or exclusive, water phase polymerization of the reactive monomer, as well as inhomogeneity of the chemical composition distribution in the polymer chains and throughout the particles can be prevented through modification of the functional monomer. Premature crosslinking inside the latex particles, which would adversely affect the polymer interdiffusion and crosslinking behaviour during film formation, and loss of functionality due to hydrolysis can be circumvented by developing suitable polymerization conditions. With these results in hand the control of the distribution of functional groups throughout the latex particles (morphology control) can now be attained. Using these well defined reactive latices a number of fundamental questions encountered in the film formation of reactive latices can now be addressed.

A few focal points for future research are evident. For the crosslinking systems the rates of the polymer-polymer interdiffusion will have to be determined, e.g., using the fluorescence quenching technique [23], as developed by Winnik et al. [24], or small angle neutron scattering [25]. Another method to obtain information on polymer-polymer interdiffusion from DMTA measurements, as suggested by Richard and Wong [26], is presently being applied to crosslinkable systems in our laboratories [27].

Control of molecular weight remains a focal point for future research. The development of yet more active chain transfer agents, the application of new functionalized chain transfer agents [28] in emulsion to obtain telechelic polymers, as is presently studied in our group, and the (further) investigation of catalytic chain transfer in the preparation of low molecular weight reactive latices, will be of interest.

The application of pseudo-living, or controlled, radical polymerization [29,30] in emulsion has in principle become possible [31] and will be further developed. In combination with fluorescence quenching this may allow the quantification of the effect of molecular mass on polymer interdiffusion and in crosslinking systems. Also, reactive block copolymers and gradient copolymers may become available, which will offer a variety of new opportunities to control the crosslinking behaviour of reactive latices.

6. References

1. Padget, J. (1994) Polymers for water-based coatings - a systematic overview, *Journal of Coating Technology* **66 (839)**, 89-105.
2. Bufkin, B.G. and Grawe, J.R. (1978) Survey of the applications, properties, and technology of crosslinking emulsions Part I, *Journal of Coating Technology* **50 (641)**, 41-55.
3. Grawe, J.R. and Bufkin, B.G. (1978) Survey of the applications, properties, and technology of crosslinking emulsions Part II, *Journal of Coating Technology* **50 (643)**, 67-83.
4. Bufkin, B.G. and Grawe, J.R. (1978) Survey of the applications, properties, and technology of crosslinking emulsions Part III, *Journal of Coating Technology* **50 (644)**, 83-108.
5. Grawe, J.R. and Bufkin, B.G. (1978) Survey of the applications, properties, and technology of crosslinking emulsions Part IV, *Journal of Coating Technology* **50 (645)**, 70-100.
6. Bufkin, B.G. and Grawe, J.R. (1978) Survey of the applications, properties, and technology of crosslinking emulsions Part V, *Journal of Coating Technology* **50 (647)**, 65-96.
7. Daniels, E.S. and Klein, A. (1991) Development of cohesive strength in polymer films from latices: effect of polymer chain interdiffusion and crosslinking, *Progress in Organic Coatings* **19**, 359-378.
8. Ooka, M. and Ozawa, H. (1994) Recent developments in crosslinking technology for coating resins, *Progress in Organic Coatings* **23**, 325-338.

462

9. Noomen, A. (1989) The chemistry and physics of low-emission coatings. Part 2. Waterborne two-pack coatings, *Progress in Organic Coatings* **17**, 27-39.
10. Geurts, J.M., Van Es, J.J.G.S., and German, A.L. (in press) Latices with intrinsic crosslink activity, *Progress in Organic Coatings* .
11. Le Fevre, W.J. and Sheetz, D.P. (1963) Emulsion Polymerization with amino alcohol esters as cationic comonomers, U.S. Patent 3,108,979.
12. Ganachaud, F., Mouterde, G., Delair, Th., Elassari, A., and Pichot, C. (1995) Preparation and characterization of cationic polystyrene latex particles of different aminated surface charges, *Polymers for Advanced Technologies* **6**, 480-488.
13. Simms, J.A. (1971) German Patent 1,520,590.
14. Geurts, J.M., Welland, R.W.A., Göttgens, C.M., Van Es, J.J.G.S., and German, A.L. (in prep).
15. Smith, D.A., Cunningham, R.H., and Coulter, B. (1969) Anomalous behaviour of a polymeric amino ester, *Journal of Polymer Science: Part A-1* **8**, 783-784.
16. Geurts, J.M., Jacobs, P.E., Muijs, J.G., Van Es, J.J.G.S., and German, A.L. (1996) Molecular mass control in methacrylic copolymer latexes containing glycidyl methacrylate, *Journal of Applied Polymer Science* **61**, 9-19.
17. Richard, J. (1997) Film formation In J.M. Asua (Ed.) *Polymeric Dispersions. Principles and Applications*, Kluwer, Dordrecht.
18. Lichti, G., Sangster, D.F., Whang, B.C.Y., Napper, D.H., and Gilbert, R.G. (1982) Effect of chain-transfer agents on the kinetics of the seeded emulsion polymerisation of styrene, *Journal of the Chemical Society, Perkin Transactions 1* **82**, 2129-2145.
19. Verstegen, J.M.G., Schoenmakers, G.J., Urlings, D.M., Van Es, J.J.G.S., and German, A.L. (in prep).
20. Curci, M., Mieloszynski, J.-L., and Paquer, D. (1993) Synthesis of functionalized acrylates, *Orga-nic Preparations and Procedures International* **25**, 649-657.
21. Geurts, J.M., Lammers, M., and German, A.L. (1996) The effect of bimodality of the particle size distribution on film formation of latices, *Colloids and Surfaces A: Physicochemical and Engineering Aspects* **108**, 295-303.
22. Sanayei, R.A. and O'Driscoll, K.F. (1989) Catalytic chain-transfer in polymerization of methyl methacrylate. i chain-length dependence of chain-transfer coefficient, *Journal of Macromolecular Science - Chemistry* **A26**, 1137-1149, and references cited therein.
23. Feng, J., Pham, H., Macdonald, P., Winnik, M.A., Geurts, J.M., Zirkzee, H.F., Van Es, J.J.G.S., and German, A.L. (in prep).
24. Pekcan, Ö., Winnik, M.A, and Croucher, M.D. (1990) Fluorescence studies of coalescence and film formation in poly(methyl methacrylate) nonaqueous dispersion particles, *Macro molecules* **23**, 2673-2678.
25. Hahn, K., Ley, G., Schuller, H., and Oberthur, R. (1986) On particle coalescence in latex films, *Colloid and Polymer Science* **264**, 1092-1096.
26. Richard, J. and Wong, K. (1995) Interdiffusion of polymer chains and molecular dynamics in dried latex films, *Journal of Polymer Science, Part B: Polymer Physics* **33**, 1395-1407.
27. Geurts, J.M., Peerlings, C.C.L., Van Es, J.J.G.S., and German, A.L. (in prep).
28. Meijs, G.F., Morton, T.C., Rizzardo, E., and Thang, S.H. (1991) Use of substituted allylic sulfides to prepare end-functional polymers of controlled molecular weight by free-radical polymerization, *Macromolecules* **24**, 3689-3695.
29. Georges, M.K., Veregin, R.P.N., Kazmaier, P.M., and Hamer, G.K. (1993) Narrow molecular weight resins by a free radical polymerization process, *Macromolecules* **26**, 2987-2988.
30. Matyjazewski, K. (1993) From "living" carbocationic to "living" radical polymerization, *Journal of Macromolecular Science - Pure and Applied Chemistry* **A31**, 989-1000.
31. Bon, S.A.F., Bosveld, M., Klumperman, B., and German, A.L. (1996) Controlled radical po lymerization in emulsion, *Macromolecules* submitted for publication.

REACTIVE SURFACTANTS

K. TAUER
MPI für Kolloid- und Grenzflächenforschung
Kantstraße 55
D-14513 Teltow, Germany

1. Introduction

A reactive surfactant is an amphiphilic molecule with an additional functionality that provides it with chemical reactivity. An important class of reactive surfactants are those that act in radical polymerizations either as comonomer (SURFMERS), or as initiator (INISURFS), or as chain transfer agent (TRANSURFS) [1, 2].

Because the stability of polymer dispersions is a property of practical importance the search for an optimized strategy to equip polymer dispersions with a sufficient stability is a matter of continuous research during the last thirty years. The kernel is to reach a sufficient stability as long as it is required (polymerization, conditioning, storage), and to allow coagulation or coalescence of particles when it is needed (separation of polymer from latex, film formation). Conventional surfactants could lead to problems during final applications of polymers if there is no more need for particle stability [3].

Especially for SURFMERS, it is to point out that other interesting applications are known and also a matter of intensive research as for instance polymerization in organized states (micelles, vesicles, adsorption layers). As there are interrelations between the application of SURFMERS in heterophase polymerization and their homopolymerization behavior the following references are given for the interested reader [4-13]. The most important results with interrelations to heterophase polymerization are: (1) the proof that the rate of polymerization goes up enormously if the SURFMER concentration is above the critical micelle concentration (CMC) [8]; (2) that a topological polymerization of only one micelle is extremely unlikely [6, 11] although in some special cases of SURFMERS with slow micellar dynamics molecular weight data indicate a confinement of polymerization to one micelle [9, 14]; (3) that it seems also to be impossible to confine a radical polymerization to a monomer swollen micelle [4]; and (4) that surface modification due to grafting of adsorbed SURFMERS is possible [10] as well as interfacial polymerization [7].

However, in the following this contribution is confined to applications of reactive surfactants in heterophase polymerizations. As basic ideas, first results and obvious problems with respect to the application of reactive surfactants have been reviewed recently [2, 15, 16] this contribution is an attempt to feature the topic from another angle and to present some new results as well.

J. M. Asua (ed.), Polymeric Dispersions: Principles and Applications, 463–476.
© *1997 Kluwer Academic Publishers. Printed in the Netherlands.*

2. General Features

The main reason for the application of reactive surfactants in heterophase polymerizations is the hope to improve final product properties and/or to provide polymer dispersions with new properties. The particular way to try this is a covalent binding of surfactants to polymer particles. This is another approach than the well known way to incorporate stabilizing groups that are not surface-active like polymeric stabilizers [17] and ionic groups arising from initiators or ionic comonomers [18].

It is a basic principle that the surface-active properties of any kind of reactive surfactant will be changed during polymerization [19-22]. Therefore, it could be possible that during heterophase polymerizations with reactive surfactants stability problems occur and a higher amount of coagulum is produced compared with non-reactive surfactants. And indeed, stability problems have been observed in several cases for INISURFS, SURFMERS; and TRANSURFS [23]. For instance, it turned out that acrylated alkyl ethoxylate surfactants are significantly less effective as stabilizers in styrene, vinyl acetate (VAC), and methyl methacrylate (MMA) emulsion polymerization than the non-reactive analogues [24]. With respect to polymerization phenomena, the self-aggregation (micellization and/or adsorption) leads to a non-isotropic distribution of the reactive surfactants as well as to a spatial fixation. This behavior is harmless for common surfactants but may have drastical consequences for reactive surfactants in emulsion polymerization as they will form radicals. These consequences are in case of INISURFS compared to common water-soluble initiators a drastically reduced primary radical efficiency due to an enhanced cage effect [25-27] and a higher overall activation energy for the mean degree of polymerization [28]. For SURFMERS one can expect peculiarities in the copolymerization behavior as the main monomer and the SURFMER are separated from each other and not isotropically mixed. Furthermore, one should expect an influence on particle nucleation due to the polymerizability of the surfactant and its high concentration in the water phase compared to the main monomer at the beginning of the polymerization. Another effect that could become crucial is the possibility of a spontaneous polymerization in an ordered state even in the absence of an initiator. For instance, such a behavior is known for sodium alkyl 2-hydroxy-3-methacryloyloxypropyl phosphates-SURFMERS [29], other amphiphilic vinyl monomers [30], , but also for MMA and styrene in micellar solutions of anionic (sodium alkylbenzenesulfonates) and non-ionic (alkylated ethoxylate) surfactants [31, 32]. If a spontaneous polymerization occurs it might lead to complications during the preparatory work for the polymerization as well as in case of a semi-batch or continuous polymerization during feeding as polymerization can start at quiet low temperatures in a range from 5 °C to 60 °C [33].

Once particles are formed, reactive surfactants are not any longer located only in the dispersion medium as they will adsorb at the particle water interface. Consequently, the polymerization behavior of reactive surfactants may change. The polymerization may take place mainly in an adsorption layer consisting of admicelles formed by the reactive surfactant containing adsolubilized main monomer. For example, results are known concerning the admicellar copolymerization of sodium 10-undecen-1-yl sulfate with styrene [34]. Furthermore, these authors reported that for solution copolymerization and admicellar copolymerization the copolymer composition is significantly different in a way that the admicellar copolymer contains much more SURFMER repeat units. This may be due to the local increase in the concentration of reactive surfactants in an adsorption layer compared to solution. For SURFMERS as well as for TRANSURFS this means an enhanced reaction probability with entering radicals. Whereas in the case of INISURFS the situation is different as radicals are formed in the adsorption layer. The consequences are (1) an enhanced primary recombination rate as pointed out above and (2) due to the higher concentration an enhanced decomposition rate that offsets to a certain extend the higher primary recombination rate and is the reason that even though with INISURFS high polymerizations rates are obtainable [25].

Furthermore, one really has to be aware that with the application of reactive surfactants the control possibilities with respect to polymerization process and product properties are changed. If one will be able to tailor the copolymerization behavior of SURFMERS in dependence on the composition of the main monomers an additional control possibility will be gained. On the other hand, in case of INISURFS a control possibility is lost as surfactant and initiator are combined and two former independent actions depend now on each other. Whereas, the application of a TRANSURF links molecular weight distribution and colloid stability.

Finally, some remarks concerning the design of reactive surfactants. Firstly, one should design structures that are stable against hydrolysis to preserve the advantage of covalent binding of surfactants to the particle surface. Secondly, the combination of at least two tasks in one molecule makes demands with respect to the chemical purity. This is more important for reactive surfactants than for simple surfactants as even minor amounts of impurities may significantly disturb polymerization but may have no effect on stability.

These features of the application of reactive surfactants in heterophase polymerization clearly underline the peculiarities compared to conventional systems. The problems are obvious as well, but also the attraction of this topic as there is today still more unknown than known.

3. SURFMERS

The development of a SURFMER was successful if : (1) it is able to stabilize the polymer dispersion through the entire polymerization and storage at a high solid content level (at least 50 %), (2) it is covalently bound to the particle surface at the end of the polymerization, (3) the improvements of final application properties justify the higher costs.

To the best of the author's knowledge the first scientific paper with respect to a polymerizable surfactant was published in 1956 [19]. This is a queer story as the first paper dealing with a man-made polymeric surfactant was already published 5 years earlier [35]. This first polysoap was a cationic one prepared through the reaction of n-dodecyl bromide with poly-2-vinylpyridine. Contrary, the first polymerizable surfactants were anionic namely sodium allyl α-sulfopalmitate and stearate, respectively, prepared by esterification of the corresponding α-sulfo fatty acids with allyl alcohol [19].

It was in 1970 that for the first time the application of a SURFMER in emulsion polymerization was reported [36, 37]. The system investigated was a styrene/butadiene emulsion polymerization in the presence of sodium 9-(and 10)-acrylamido stearate. The maximum latex surface coverage attained by polymerization in an adsorption layer of that SURFMER was around 80 % [36]. The mechanical stability of the latexes stabilized with in situ polymerized SURFMER was found to be higher than that stabilized with monomeric SURFMER [37].

Undoubtedly, both homopolymerization behavior and copolymerization behavior of the polymerizable surfactant with the main monomers are crucial for the success of a SURFMER. However, its copolymerization behavior is only hardly to investigate as the situation is different from an ordinary isotropic copolymerization for at least two reason. These are: (1) the ratio of the monomers is around 2% SURFMER based on main monomer and (2) the anisotropic nature of the reaction system with different reaction loci during the course of the polymerization. Nevertheless, one can get some idea about the behavior of a SURFMER starting with „ordinary" copolymerization parameters for the different polymerizable groups (acrylic, vinylic, maleic, ...) (cf. Table 1).

TABLE 1. Range of copolymerization parameters (r_1 / r_2) for styrene, MMA, and VAC (monomer 2, row 1) with comonomers representing different polymerizing groups of SURFMERS (monomer 1, column 1) [38]

	Styrene		MMA		VAC	
	r_1	r_2	r_1	r_2	r_1	r_2
Maleic anhydride	0-0.05	0-0.097	0-0.08	0.46-6.4	0	0.019
Di ethyl maleate	0-0.7	6.07-8.0	0	20-341	0.04	0.171
Di ethyl fumarate	0.02-0.11	0.29-0.39	0.04-0.05	2.1-40.3	0.33-0.43	0.011-0.09
1-Acrylamido-1-deoxy-D-glucitol	0.03	2.42	0.05	3.75	0.98	0.03
Methyl acrylate	0.8	0.192	0.4	2.15	2.58	0.405
Butyl acrylate	0-0.29	0.44-1.23	0.11-0.43	0.92-2.86	3.48	0.018
Allyl acetate	0	90	0	23-99.2	0.45-0.7	0.6-1.0
1-Vinyl imidazol	0.071	9.94	0.014	4.36	1.9	0
2-Vinyl pyridine	0.75-1.81	0.46-0.55	0.64-1.1	0.27-0.42	13.65-30	0
N-Vinyl succinimide	0.01-7	0.05-10.7	0-0.048	0.01-9.94	1.99-5.68	0.072-0.229

The value of these data for heterophase copolymerizations with SURFMERS may be doubtful and unfortunately, the values span a wide range for each monomer combination, but other data are not yet available. Nevertheless, one can draw some conclusions for designing a SURFMER with a proper polymerizing group for a given main monomer (cf. Table 2). The main criterion for the efficiency of a SURFMER is its influence on latex stability and only in second place the degree of covalent binding. The basic assumption for the interpretation of the r-values with respect to SURFMER efficiency is that the SURFMER is only able to stabilize as long as it is present at the particle water interface. If the SURFMER has a high tendency to polymerize from the start the chance is high that a considerable amount will be buried inside the particles. This scenario is bad with respect to stability but good with respect to covalent binding. Case 1 of Table 2. should be worse than case 2. For case 2 of Table 2 ($r_1 > 1$ and $r_2 < 1$) copolymerization leads to a copolymer enriched with SURFMER units. It depends on the particular conditions if the copolymers will act as stabilizer or flocculant for the dispersion. Case 3 of Table 2. is very similar to a classical emulsion polymerization as the SURFMER acts more as stabilizer than as comonomer. This is true as long as the conversion of the main monomer is low and a free monomer phase still exists but, during the third stage the SURFMER should copolymerize more and more.

TABLE 2. Possible relations between r-values and SURFMER efficiency with respect to latex stability and covalent binding

#	Relation	Meaning	Example	Stability	Binding
1	$r_1, r_2 < 1$	Cross propagation	Diethyl maleate / VAC	sad or good	good or sad
2	$r_1 > 1, r_2 < 1$	Both radicals prefer SURFMER	Butyl acrylate / VAC	good or sad	good or sad
3	$r_1 < 1, r_2 > 1$	Both radicals prefer the main monomer	Diethyl maleate / styrene	good	good

It is surprising that based on these crude data some reasonable predictions are possible. So for instance, it has been experimentally observed that sodium sulfopropyl alkyl maleates show a different behavior than the corresponding fumarates in styrene emulsion polymerization [39]. At 20 % conversion all monomeric fumarate SURFMER is used up whereas, the maleates start to react in a larger extend at 80 % conversion. From the r-values (Table 1) one can predict that the maleate SURFMERS represent case 3 whereas the fumarates more likely represent case 1. The copolymer consisting of fumarate SURFMER and styrene act as stabilizer in that particular case [39].

Another example for a case 1 system is VAC and a maleate. One would expect for this particular system some problems with the stability during the polymerization as the reactivity of the maleate group is enhanced compared to styrene. Indeed, the sulfopropyl alkyl maleates lead in that case to coagulation at higher solid contents [40]. Whereas, for butyl acrylate and styrene as monomer mixture maleate SURFMERS work very well up to a solid content of 50 % [41].

The data in Table 1 illustrate that the design of a suited SURFMER for VAC is more difficult than for MMA and styrene. Only one combination indicates a possible case 3 namely, if the SURFMER has an allylic polymerizable group. But, the problems with allylic monomers in radical polymerization due to degradative chain transfer to monomer are well known [42].

Now, as the polymerizable group is chosen and if the amphiphilic part of the SURFMER is given as well, the next step is the arrangement of these group in the molecule either near the hydrophic group, or near the hydrophilic group, or among them. Examples will be given below for each type and their application in emulsion polymerization. The first example is a sodium sulfo dodecylstyryl ether SURFMER where the polymerizable group is located in the hydrophobic part (cf. Figure 1) [43].

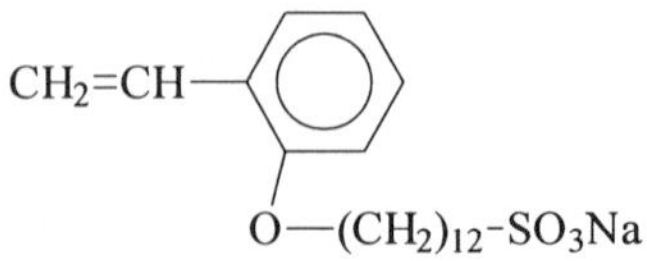

$$CH_2=CH-\bigcirc$$
$$O-(CH_2)_{12}-SO_3Na$$

Figure 1. Dodecyl styryl ether SURFMER [43]

The authors investigated the yield of the SURFMER bound to the surface in a styrene seeded emulsion polymerization (solid content around 11 %) in dependence on the type of initiator. It turned out that the surface yield (the amount of SURFMER bound to the particle surface relative to the total amount of SURFMER) was higher in case of water soluble initiators (potassium peroxidisulfate 54,6 % and biacetyl 52,3 %) compared to water insoluble systems (benzoyl peroxide 15,5 % and lauroyl peroxide 10,0 %). This result is surprising as one could expect the reverse order due to the location of the double bond in the hydrophobic part of the SURFMER.

A SURFMER where the double bond is located near to the hydrophilic part is the sodium dodecyl allyl sulfosuccinate (TREM LF-40 Henkel AG, cf. Figure 2).

$$C(O)O-(CH_2)_{11}-CH_3$$
$$CH_2$$
$$CH-SO_3Na$$
$$C(O)O-CH_2-CH=CH_2$$

Figure 2. Dodecyl allyl sulfosuccinate SURFMER [44 - 46]

In a series of papers results were reported concerning the kinetics [44] and the polymerization mechanism [45] of VAC emulsion polymerization with that SURFMER and its non-polymerizable analogue. The authors have found water soluble oligomers in the aqueous phase at the end of the polymerization only when the SURFMER was used. An increase in the initial SURFMER concentration leads to an increase in the surface yield, to a decrease in the rate of polymerization, and to smaller particles. It turned out that compared to the non-polymerizable analogue the degradative chain transfer to the SURFMER due to the allyl group is responsible for the peculiarities observed [46].

The third example is a siloxane sugar surfactant where the polymerizable methacrylate group is located between the hydrophobic and hydrophilic part (Figure 3).

The synthesis of this SURFMER is described in [47]. The combination of a siloxane and a sugar is a rather unusual for an emulsion polymerization surfactant. The main reason for its application in emulsion polymerization is that this compound forms multilamellar vesicles in aqueous solutions indicated by the appearance of streaming birefringence. A polymerization of these structures is possible if an uncharged PEGA200 initiator [17] is used [48]. This is a different situation compared to the other SURFMERS as the chance that a topological polymerization takes place is much higher.

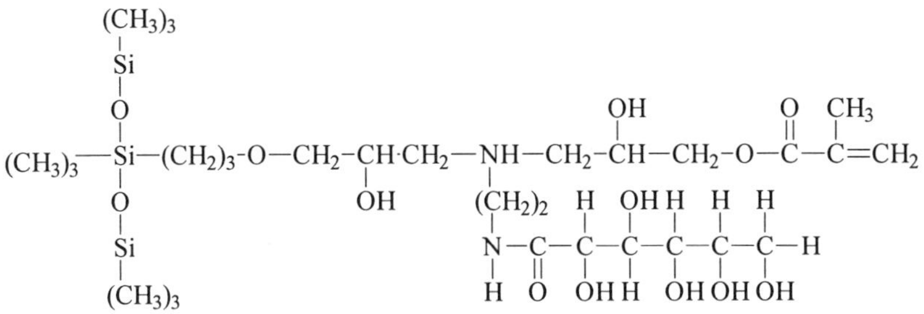

$$(CH_3)_3$$
$$|$$
$$Si$$
$$|$$
$$O$$

$$(CH_3)_3-Si-(CH_2)_3-O-CH_2-CH \cdot CH_2-NH-CH_2-CH-CH_2-O-C-C=CH_2$$

with branches:

- On the left Si: O below, then Si, then $(CH_3)_3$
- $CH \cdot$ bears OH
- CH bears OH (upper) and the group $(CH_2)_2$
- To the right: O (double bond to C) and CH_3

$$(CH_2)_2 \quad H \ OH H \ H \ H$$
$$| \qquad\qquad | \ | \ | \ | \ | \ |$$
$$N-C-C-C-C-C-H$$
$$| \ || \ | \ | \ | \ |$$
$$H \ O \ OH H \ OH OH OH$$

Figure 3. Siloxane gluconamide methacrylate SURFMER [47]

The emulsion polymerizations were carried out with styrene as main monomer and PEGA200 as initiator at 60 °C. In a first series the solid content of the only sterically stabilized dispersions was limited to around 5 % depending on the amount of SURF-MER. The influence of the SURFMER concentration on the average particle size is very pronounced (cf. Table 3). All polymerizations went off without the formation of coagulum.

TABLE 3. Effect of SURFMER concentration on average particle size

Run	[S] (w-%) [1]	d_p (10^{-9} m) [2]
A1	2,0	116,2
A2	1,0	184,3
A3	0,6	242,2
A4	0,3	361,6
A5	0,1	945,0
A6	0,05	1098

1) - weight percent SURFMER based on styrene
2) - volume average diameter from dynamic light scattering

Transmission electron microphotographs of these latexes reveal a surprising result. It is clearly seen that the particle size distribution is extremely broad. In this range of SURFMER concentration very large particles of up to several microns in diameter are formed beside very tiny particles with a diameter of only a few nanometers (Figure 4).

A further increase in SURFMER concentration leads to the disappearance of the large particle fraction. Figure 5 shows two examples of particle size distributions obtained when the SURFMER concentration is drastically increased. In both cases the starting aqueous solution of 4.5 w-% SURFMER is streaming birefringent. Figure 5.A results from a semibatch polymerization with monomer feed at 70 °C. The polymerization was started with an initial charge of 100 g of water, 4.5 g of SURFMER, 10 g of styrene, and 2.06 g of PEGA200. After 30 minutes another 42 g of styrene were fed over a period of 3 hours. The final latex was obtained after a total reaction time of 9 hours with a solid content of around 35 %. The volume weighted average particle diameter estimated by counting 2,900 particles on the electron

470

microphotographs is 65.3 nm whereas dynamic light scattering gives a volume weighted diameter of 67.8 nm.

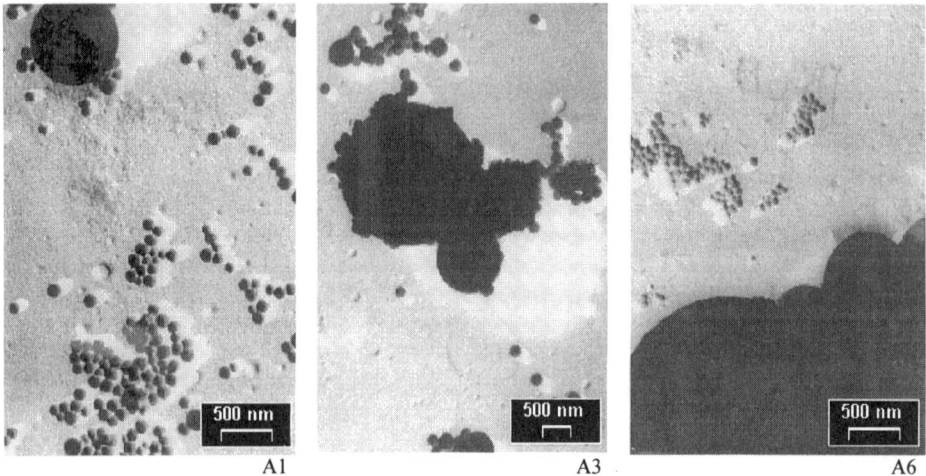

Figure 4. Transmission electron moicrophotographs of latexes A1, A3, and A6

The dispersion depicted in Figure 5.B corresponds to a polymerization in vesicles. A solution of 4.5 g of SURFMER, 2.06 g of PEGA200, and 0.653 g of styrene in 100 g of water (also with that amount of styrene and PEGA200 the solution retains streaming birefringence) was placed in a thermostated water bath and allowed to polymerize at 70 °C for 6 hours.

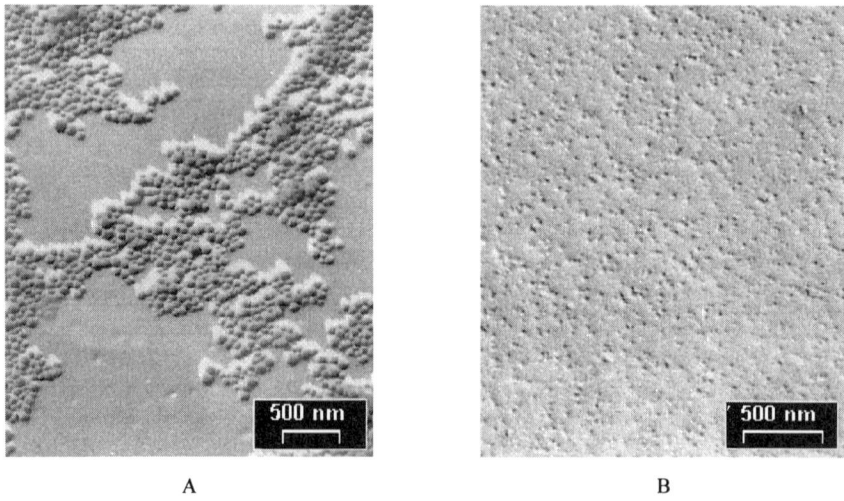

Figure 5. Transmission electron microphotographs of polystyrene latexes prepared with
   the siloxanyl gluconamide SURFMER
   A - 8,6 w- % SURFMER based on styrene
   B - 689 w- % SURFMER based on styrene (vesicle polymerization)

After polymerization the solution is no longer streaming birefringent. The volume weighted average particle diameter estimated with dynamic light scattering is 24.3 nm. It is not to decide whether or not the vesicular structure is retained after polymerizations.

The results reveal that only at high SURFMER concentrations emulsion polymerization of styrene results in latexes with a monomodal particle size distribution. At SURFMER concentrations of 2 w % and lower based on monomer the latexes always show multimodal particle size distributions. For latexes with a solid content over 30 % the SURFMER concentration must be higher than 5 w- % based on monomer. It is likely that due to a topological polymerization of that SURFMER a certain amount is used up and not available for particle stabilization.

4. INISURFS

Surface active initiators may by symmetrical or non-symmetrical with respect to the radical generating group. The symmetrical INISURFS are examples of a special class of so called dimeric or gemini surfactants. These surfactants possess dynamic properties and form structures that are drastically different from those of common single-chain surfactants [49]. The development of INISURFS represents a considerable extension of the classes of gemini surfactants as until today mainly cationic dimeric surfactants have been investigated.

If for both INISURFS and SURFMERS the same criteria of success are valid one important difference with respect to polymerization strategy is evident. A complete covalent binding of the stabilizing groups arising from an INISURF is only possible if during the last period of the polymerization the temperature is raised to achieve the highest possible decomposition and reaction. This is of course not possible as it takes around 32 hours at 100 °C and $k_I = 6 \ 10^{-5} \ s^{-1}$ that the initiator concentration drops to 1/1000 of its initial value. Another reason why the goal of a complete covalent binding is unrealistic is the fact that the radical efficiency is extremely low. Efficiency values lower than 0.1 % have been observed for symmetrical azo-INISURFS [27]. But, even if it could be increased to usual values the goal remains unattainable as any value less than 1 prevents to be successful.

Since the surface activity of the INISURFS is the main reason for the low efficiency there seems to be no way out of this dilemma. Even the half way back i.e., non-symmetrical INISURFS that lead to one surface active and to one non-surface active radical turned out to be not successful. The radical efficiency increases only marginal [25, 26].

In spite of the low radical efficiencies the overall polymerization rates are high. The reason for that is an increased decomposition rate due to a much higher local concentration of the INISURF in an ordered state (micelles or adsorption layer) compared to an isotropic solution. The concentration in a micelle can reach values up to $10^3 \ M \ m^{-3}$ based on volume of ordered state. However, this opens up another way to higher efficiencies: dilution with surfactants that form with the INISURF mixed ordered states. That this way is successful was very recently shown with a combination of an non-symmetrical INISURF (Figure 6) and an anionic surfactants [50].

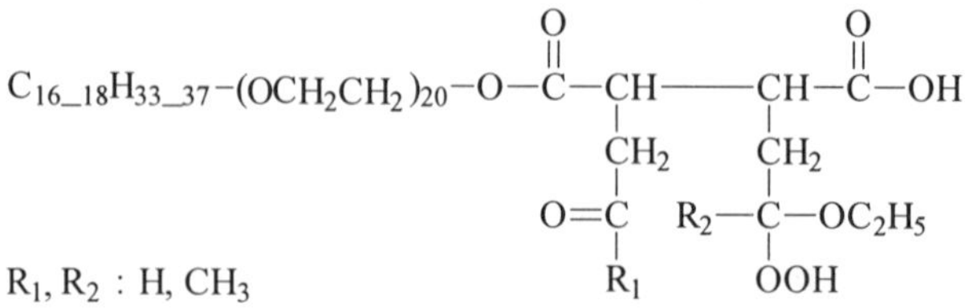

$$C_{16\_18}H_{33\_37}-(OCH_2CH_2)_{20}-O-\overset{\overset{O}{\|}}{C}-\underset{\underset{\underset{R_1}{O=C}}{CH_2}}{CH}———\underset{\underset{\underset{OOH}{R_2-C-OC_2H_5}}{CH_2}}{CH}-\overset{\overset{O}{\|}}{C}-OH$$

$R_1, R_2 : H, CH_3$

Figure 6. Non-symmetrical hydroperoxide INISURF [50]

Furthermore, these authors observed under the same conditions an increase in the radical efficiency with increasing temperature. A temperature increase from 60 °C to 80 °C caused an increase in the radical efficiency from 0.04 to 0.1 under the particular experimental conditions. The explanation is, that with increasing temperature the decomposition outside the ordered states contributes more and more as the activation energy for the decomposition in homogeneous solution is higher than that in micelles or in adsorption layers. The validity of this dilution concept for other INISURFS still has to be proved.

The INISURFS known so far always contain an ester linkage that makes them susceptible to hydrolysis. The anionic PEGAS INISURFS contain even two ester linkages per radical [17, 28]. The synthesis of INISURFS with a higher chemical stability was one topic over the last years. Figure 7 shows an anionic surface active initiator with only one ester group per radical [51]. This bis[2-(4'sulfophenyl)ethylene]-2,2'-azodi-isobutyrate (PEAS) starts to decompose markedly at temperatures above 70 °C. Emulsion polymerizations of styrene with variable amounts of PEAS at 90 °C lead to monodisperse latexes with a ratio weight average to number average particle size of less than 1.01 (estimated from TEM micrographs) if the amount of initiator is less than 8 weight-% based on monomer.

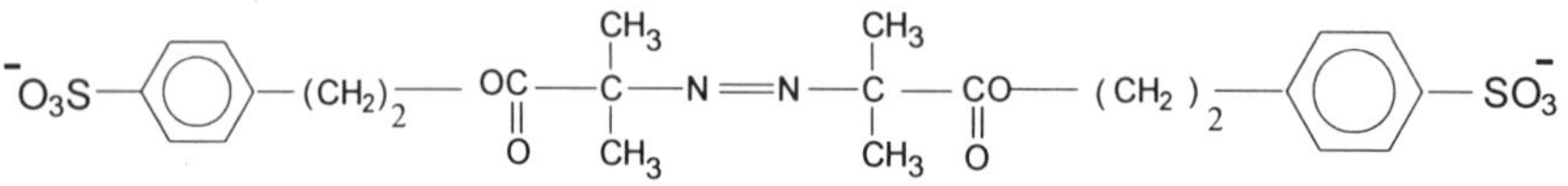

$$^-O_3S-\langle\bigcirc\rangle-(CH_2)_2-\underset{\underset{O}{\|}}{OC}-\underset{\underset{CH_3}{|}}{\overset{\overset{CH_3}{|}}{C}}-N=N-\underset{\underset{CH_3}{|}}{\overset{\overset{CH_3}{|}}{C}}-\underset{\underset{O}{\|}}{CO}-(CH_2)_2-\langle\bigcirc\rangle-SO_3^-$$

Figure 7. Sulfonated INISURF (PEAS) [51]

Another new class of INISURFS without any easily hydrolysable bonds are 2,2'-azobis(N-2'-methylpropanoyl-2-amino-alkyl-1)-sulfonates (AAS) (Figure 8) [51].

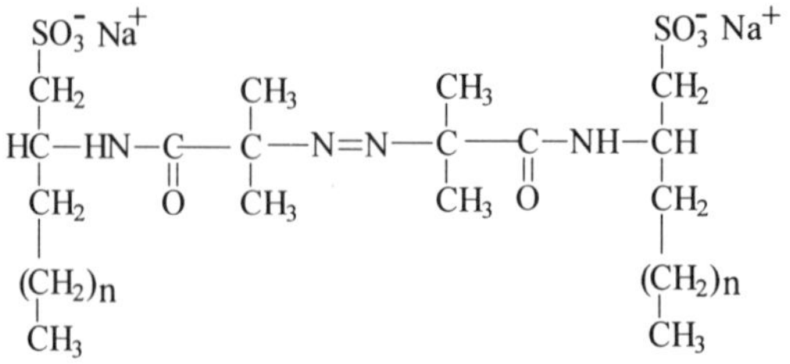

$$\underset{\underset{\underset{CH_3}{(CH_2)_n}}{\underset{CH_2}{HC}}}{\overset{\overset{SO_3^- Na^+}{CH_2}}{}}-HN-\underset{\underset{O}{\|}}{C}-\underset{\underset{CH_3}{|}}{\overset{\overset{CH_3}{|}}{C}}-N=N-\underset{\underset{CH_3}{|}}{\overset{\overset{CH_3}{|}}{C}}-\underset{\underset{O}{\|}}{C}-NH-\underset{\underset{\underset{CH_3}{(CH_2)_n}}{\underset{CH_2}{CH}}}{\overset{\overset{SO_3^- Na^+}{CH_2}}{}}$$

Figure 8. INISURF without any hydrolysable groups (AAS) [51]

These INISURFS have been prepared by a modified RITTER-reaction in one step starting from 2,2'-azobisisobutyronitrile, oleum, and the corresponding long chain α-olefin. A second step is the neutralization of the acid. Although, a small amount of coagulum forms during the polymerization the final latexes are stable up to a solid content of 50 %. Contrary to PEAS and also PEGAS [15] AAS's lead to non-monodispers latexes. Figure 9 shows that the particle size distributions consist of a small and a large fraction. The diameter of the large fraction depends on the alkyl chain length in a way that the longer the alkyl chain the smaller these particles. Whereas, the mean diameter of the small fraction is around 50 nm and nearly independent of the alkyl chain length.

These two contrary examples with respect to particle size distribution reveal a little bit the potential that is given with the application of INISURFS. However, there is still a lack concerning systematic kinetic investigations. The main reason for this situation is that the purity of INISURFS available today is not sufficient.

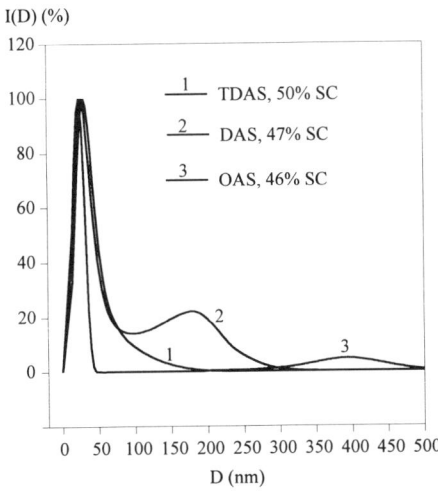

Figure 9. Influence of the alkyl chain length of AAS-INISURFS on the number weighted particle size distribution of polystyrene latexes;
TDAS - n = 10; DAS - n = 6; OAS - n = 4;
Particle size distribution determined with capillary hydrodynamic fractionation
Polymerisation: semibatch, monomer and INISURF solution feed, 90°C,
Initial charge: 0.5g INISURF, 10g styrene, 40g water,
Feed: 40g monomer, 10g water, 2g INISURF

A common property of SURFMERS and INISURFS is their capability to spontaneous polymerizations. In case of TDAS a polymerization at temperatures lower than 40 °C was observed if an emulsion was prepared consisting of 100 g of water, 167 g of monomers (86,85 g of butyl acrylate, 75,15 g of styrene, 5,0 g of acrylic acid), and 11,1 g of TDAS. For example, the conversion was 50 % at 30 °C after 50 minutes polymerization time.

5. TRANSURFS

The number of studies devoted to TRANSURFS is very limited, cf. [2]. To the authors knowledge there are only two papers both dealing with surface active thiols as transfer agents in styrene emulsion polymerization. In [52] is shown that ω-thio-decylsulfonate is covalently bound to the particle surface. The application of an ethoxylated undecyl-thiol with a variable number of ethylene glycol units in styrene emulsion polymerization initiated with 2,2'-azobis(2-methyl-N-2-hydroxyethyl)-propionamide) leads to sterically stabilized monodispers latexes up to solid content of around 10 % [53]. The amount of coagulum formed is around 10 % if the number of ethylene glycol units is larger than 50. On the other hand the particle size is the smaller the shorter the ethylene glycol chain.

6. Conclusions

As frequently occurs, things do not turn out so simply as thought when started a work. A lot of different reactive surfactants have been prepared and tested in heterophase polymerizations but, until today no results are known where the goals (high solid content, no surface active compounds in latex serum, no coagulum formation during polymerization, improvement of product properties) have been completely reached. But, improvements of latex and/or polymer properties are known [2, 15, 16].

However there is also no reason to despair of going on this way. So for instance, it seems to be useful to start for practical applications with a partial replacement of common surfactants by SURFMERS to avoid instabilities during conditioning. Tailoring of properties of reactive surfactants with respect to chemical structure and constitution (position of reactive groups relative to hydrophobic or hydrophilic molecule part) might be a way to reach future success. Another possibility to increase the degree of binding could be the application of multi-functional reactive surfactants as for instance INISURFS with an additional transfer functionality, or SURFMERS with an additional radical generating functionality, or TRANSURFS with an additional double bond, or any other possible combination.

7. References

1. Kusters, J.M.H. and German, A.L. (1990) Emulsions polymerization with INISURFS: emulsifiers with initiating properties, *33rd IUPAC International Symposium on Macromolecules*, July 8-13, **1990** Montreal Book of Abstracts, Session 2.5.6
2. Guyot, A. and Tauer, K. (1994) Reactive surfactants in emulsion polymerization, *Adv.Polym.Sci.* **111**, 43-65
3. Dickstein, J. (1986) Relationship of chemical structure to applications of specialty monomers, *Polym.Prepr.* **27**, 427-428
4. Hyde, A.J. and Robb, D.J.M. (1963) The apparent molecular weights of polymerized soap micelles, *J.Amer.Chem.Soc.* **67**, 2089-2092
5. Kammer, U. and Elias, H.-G. (1972) Zur Assoziation von Seifen V. Synthese und Polymerisation von Seifen aus N-Mathacryloyl-11-alkylamino-undecansäure-Typs, *Kolloid-Z. u. Z. Polymere* **250**, 344-351
6. Hamid, S. and Sherrington D. (1986) Polymerized micelles: fact or fancy?, *J.Chem.Soc., Chem.Commun.* **12**, 936-938
7. Law, T.K., Florence, A.T. and Whateley, T.L. (1986) Stabilization of emulsions by interfacial polymerisation of poloxamer surfactants derivatives, *Coll.Polym.Sci.* **262**, 167-170
8. Egorov, V.V. and Subov, V.P. (1987) Radical polymerisation in aggregates of ionic surface active monomers in water, *Usp.Chim.* **56**, 2076-2097
9. Nagai, K. and Elias, H.-G. (1987) Polymerizable amphiphiles, 3 Polymerization of micellized 1-O-3-(4-vinylphenyl)propyl-b-D-glucopyranose, *Makromol.Chem.* **188**, 1095-1127
10. Torstensson, M. and Hult, A. (1992) Surface modification of polymers using surfactant monomers, *Polym.Bull.* **29**, 549-556

11. Cochin, D., Zana, R. and Candau, F. (1992) Photopolymerization of micelle-forming monomers. 2. Kinetic study and mechanism, *Macromol.* **26**, 5765-5771

12. Anton, P., Köberle, P. and Laschewsky, A. (1993) Recent developments in the field of micellar polymers, *Makromol.Chem.* **194**, 1-27

13. Nagai, K. (1994) Polymerization of surface active monomers and applications, *Macromol.Symp.* **84**, 29-36

14. Pucci, B., Polidori, A., Rakotomanomana, N.Chorro, M., and Pavia, A.A. (1993) Synthèse de Glycolipides Polymérisables Dérivés du Tris(hydroxyméthyl)aminométhane: Préparation de Micelles Polymérisées, *Tetrahedron Lett.* **34**, 4185-4188

15. Tauer, K. (1995) Emulsion polymerization with reactive surfactants, *Polymer News* **20**, 342-347

16. Tauer, K., Goebel, K.-H., Kosmella, S., Neelsen, J., and Stähler, K. (1988) Neuere Entwicklungen bei der Synthese von Polymerdispersionen, *Plaste Kautschuk* **35**, 373-378

17. Tauer, K. (1994) Block copolymers prepared by emulsion polymerization with poly(ethylene oxide)-azo-initiators, *Polym.Adv.Technol.* **6**, 435-440

18. Liu, L-J. and Krieger, I.M. (1977) Control of surface charge on polymer latex particles in P. Becher and M.N. Yudenfreund (eds.), *Emulsions, Latices, and Dispersions*, Marcel Dekker, Inc., New York and Basel, pp. 41-69

19. Bistline, R.G., Stirton, A.J., Weil, J.K., and Port, W.S. (1956) Synthetic detergents from animal fats. VI. Polymerizable esters of alpha-sulfonated fatty acids, *J.Amer.Oil Chem.Soc.* **33**, 44-45

20. Paleos, C.M., Stassinopoulou, C.I., and Malliaris, A. (1983) Comparative studies between monomeric and polymeric sodium 10-undecenoate, *J.Phys.Chem.* **87**, 251-254

21. Richtering, W., Löffler, R., and Burchard, W. (1992) Comparison between monomeric and polymeric surfactants. 2. Properties of polysurfactants in aqueous and nonaqueous solution, *Macromol.* **25**, 3642-3650

22. Boyer, B., Lamaty, G., Leydet, A., Roque, J.-P., and Sama P. (1992) Polymerized surfactants. I. The influence of polymerization on the solubilization of hydrophobic substrates by ionic surfactants, *New J.Chem.* **16**, 883-886

23. Guyot, A., Asua, J.M., Sherrington, D.C., German, A.L., and Tauer, K. (1994/1995) Results obtained within a European network, Reactive surfactants in heterophase polymerization for high performance polymers, CT 93 - 0159 (DG DSCS)

24. Ferguson, P., Sherrington, D.C., and Gough, A. (1993) Preparation, characterization and use in emulsion polymerization of acrylated alkyl ethoxylate surface-active monomers, *Polymer* **34**, 3281-3292

25. Ivanchev, S.S., Pavljuchenko, V.N., and Byrdina, N.A. (1987) Elementary reactions of the emulsion polymerization of styrene with the localization of radical formation acts at the interface, *J.Polym.Sci.: Part A: Polym.Chem.*, **25**, 47-62

26. Kusters, J.H.M. (1994) *INISURFS: surface-active initiators*, PhD thesis Technical University Eindhoven, The Netherlands

27. Kusters, J.M.H., Napper, D.H., Gilbert, R.G., and German, A.L. (1992) Kinetics and particle growth in emulsion polymerization systems with surface-active initiators, *Macromol.* **25**, 7043-7050

28. Tauer, K. and Kosmella, S. (1993) Synthesis, characterization and application of surface active initiators, *Polym.Intern.* **30**, 253-258

29. Yasuda, Y., Rindo, K., Tsushima, R., and Aoki, S. (1993) Spontaneous polymerization of amphiphilic vinyl monomers 2. Spontaneous polymerization of sodium alkyl 2-hydroxy-3-methacryloyloxypropyl phosphates in micellar and inverse micellar systems, *Makromol.Chem.* **194**, 485-491

30. Yasuda, Y., Rindo, K., Tsushima, R., and Aoki, S. (1993) Spontaneous polymerization of amphiphilic vinyl polymers, 4. Spontaneous polymerization of methacrylic derivatives of quarternary ammonium bromides with a long alkyl chain, *Makromol.Chem.* **194**, 1893-1899

31. Aoki, S. (1995) Spontaneous polymerization in micelles as reaction loci, *J.Synth.Org.Chem.Jpn.* **53**, 423-431

32. Asahara, T., Seno, M., Shiraishi, S., and Arita, Y. (1970) The polymerization of vinyl monomers in the presence of surface active agents. I. The polymerization of methyl methacrylate, *Bull.Chem.Soc.Jpn.* **43**, 3895-3898

33. Yasuda, Y., Rindo, K., and Aoki, S. (1992) Spontaneous polymerization of amphiphilic vinyl monomers. 1. Spontaneous polymerization of sodium dodecyl 2-hydroxy-3-methacryloyloxypropyl phosphate, *Makromol.Chem.* **193**, 2875-2882

476

34. Glatzhofer, D.T., Cho, G., Lai, C.L., O'Rear, E.A., and Fung, B.M. (1993) Polymerization and copolymerization of sodium 10-undecen-1-yl sulfate in micelles and in admicelles on the surface of alumina, *Langmuir* **9**, 2949-2954
35. Strauss, U.P. and Jackson, E.G. (1951) Polysoaps. I. Viscosity and solubilization studies on an n-dodecyl bromide addition compound of poly-2-vinylpyridine, *J.Polym.Sci.* **VI**, 649-659
36. Greene, B.W., Sheetz, D.P., and Filer, T.D. (1970) In situ polymerization of surface-active agents on latex particles I. Preparation and characterization of styrene/butadiene latexes, *J.Coll.Interf.Sci.* **32**, 90-95
37. Green, B.W. and Sheetz, D.P. (1970) In situ polymerization of surface-active agents on latex particles II. The mechanical stability of styrene/butadiene latexes, *J.Coll.Interf.Sci.* **32**, 96-100
38. Greenley, R.Z. (1989) Free radical copolymerization reactivity ratios, in J. Brandrup and E.H. Immergut (eds.), *Polymer Handbook*, John Wiley & Sons, New York, pp. II/153-II/266
39. Goebel, K.-H. and Stähler, K. (1994) Emulsion copolymerization of styrene with monomeric emulsifiers, *Polym.Adv.Technol.* **6**, 452-454
40. Schoonbrood, H.A.S., and Asua, J.M. (1996), to be published
41. Goebel, K.-H., and Tauer, K. (1990), unpublished results
42. Odian, G. (1991) *Principles of Polymerization*, J. Wiley & Sons, Inc., New York
43. Tsaur, S-L. and Fitch, R.M. (1987) Preparation and properties of polystyrene model colloids I. Preparation of surface-active monomer and model colloids derived therefrom, *J.Coll.Interf.Sci.* **115**, 450-462
44. Urquiola, M.B., Dimonie, V.L., Sudol, E.D., and El-Aasser, M.S. (1992) Emulsion polymerization of vinyl acetate using a polymerizable surfactant. 1. Kinetic studies, *J.Polym.Sci.: Part A: Polym.Chem.* **30**, 2619-2629
45. Urquiola, M.B., Dimonie, V.L., Sudol, E.D., and El-Aasser, M.S. (1992) Emulsion polymerization of vinyl acetate using a polymerizable surfactant. II. Polymerization mechanism, *J.Polym.Chem.:Part A: Polym.Chem.* **30**, 2631-2644
46. Urquiola, M.B., Sudol, E.D., Dimonie, V.L., and El-Aasser, M.S. (1993) Emulsion polymerization of vinyl acetate using a polymerizable surfactant. III. Mathematical model, *J.Polym.Chem.: Part A. Polym.Chem.* **31**, 1403-1415
47. Wagner, R., Richter, L., Weiland, D., Reiners, J., and Weißmüller, J. (1996), Silicon modified carbohydrate surfactants (II). Siloxanoyl containing branched structures, *Appl.Organaomet.Chem.* in press
48. Tauer, K. (1995) unpublished results
49. Karaborni, S., Esselink, K., Hilbers, P.A.J., Smit, B., Karthäuser, J., van Os, N.M., and Zana, R. (1994) Simulating the self-assembly of gemini (dimeric) surfactants, *Science* **266**, 254-256
50. Pavljuchenko, V.N., Lesnikova, N.N., Byrdina, N.A., and Ivanchev, S.S. (1995) Kinetic pecularities of emulsion polymerization of styrene with a surface-active initiator (russ.), *Dokl.Akad.Nauk* **342**, 70-72
51. Tauer, K. and Sedlak, M. (1995), to be published
52. Fifield, C.C. (1985) Phd thesis University of Connecticut, University Microfilms International 5869 (1992), Ann Arbor, Michigan USA
53. Vidal, F., Guillot, J., and Guyot, A. (1995) Surfactants with transfer agent properties (transurfs) in styrene emulsion polymerization, *Colloid.Polym.Sci.* **273**, 999-1007

SYNERGISTIC INTERACTIONS AMONG ASSOCIATIVE POLYMERS AND SURFACTANTS

RICHARD D. JENKINS* AND DAVID R. BASSETT†
Union Carbide Corporation
* Union Carbide Asia Pacific Technical Center, The Pasteur,*
† UCAR Emulsion Systems, Cary, North Carolina, 27511, USA

1. Introduction

The need to improve the performance of formulated aqueous systems containing associative polymers makes elucidating the interactions between associative polymers and surfactants an important contemporary problem in associative polymer technology. For example, adding surfactant to a latex formulation may cause its viscosity to either increase or decrease, depending on the outcome of competitive adsorption of thickener and surfactant at particle interfaces and at network junctions in the aqueous phase [1]. Even though this subject has been widely studied by both academic and industrial laboratories, the capricious nature of these interactions is not understood on a molecular level. Recent studies have examined the interaction of nonionic, anionic, cationic, or zwitterionic surfactants with a variety of associative polymer types, such as those based on backbones of poly(ethylene oxide) [2-13], cellulose [14-16], copolymers of acrylic acid, methacrylic acid, or other carboxylated monomers [6,17-19], and copolymers of acrylamide [20,21]. The charter of the NATO Advanced Study Institute is to "point out unsolved problems and speculate about future research directions" [22]. In the spirit of this charter, this chapter describes the phenomenology that applies to a variety of associative polymer types resulting from interactions between associative polymers and surface active agents to highlight the need for more fundamental research.

1.1. ADSORPTION OF SURFACTANTS TO POLYMERS

Surfactants adsorb to polymers because the adsorbed state of the surfactant on the polymer is energetically more favorable than in regular micelles [23,24]. Among the several modes of interaction are adsorption of the surfactant to the polymer backbone and clustering of the surfactant on the hydrophobe of the associative polymer. The clustering of surfactant molecules along the associative polymer forms a pseudo-micellar structure that shields the hydrophobic groups of the associative polymer from contact with water. In addition, associative polymers have hydrophobic and hydrophilic segments that form segregated hydrophobic and hydrophilic regions in solution. Surfactant molecules interact with these interfaces by binding the hydrophobic portion of the surfactant with the hydrophobic block of polymer (i.e., Nagarajan's "Type 5" topology) [25]. The more hydrophobic the polymer, the greater the adsorption of surfactant to it [26]. The preferred mode of interaction depends on the concentrations and chemical natures of the surfactant (e.g., its solubility, the structure of the

477

J. M. Asua (ed.), Polymeric Dispersions: Principles and Applications, 477–495.
© 1997 Kluwer Academic Publishers. Printed in the Netherlands.

hydrophobe of the surfactant) and polymers (e.g., its molecular weight, solubility, size of the hydrophobes and placement within the polymer [10].

Polymer / surfactant systems generally show two critical concentrations: the first occurs at the concentration where the surfactant first interacts with, and adsorbs to, the polymer, and the second occurs at the concentration where the adsorption sites on polymer are saturated [27]. The concentration of surfactant at the first transition is usually less than critical micellar concentration (cmc) of the pure surfactant in solution, and the second transition coincides with the cmc of the surfactant in solution. For surfactant concentrations between the two transitions, the interaction between polymer and surfactant is stoichiometric [27,28]. The stoichiometry of interaction for the formation of mixed aggregates of polymer and surfactant depends on the polymer alkyl group / surfactant molar ratio [10,29,30].

The presence of hydrophobes in the associative polymer causes adsorption of the surfactant to the polymer at concentrations that are lower than occur with an equivalent homopolymer to reduce the onset concentration of network formation in solution. Although interactions between nonionic polymers and surfactants are traditionally considered weak, they can be substantially strong and attractive when hydrophobic groups are on the polymer [14]. Depending on the stoichiometry of the surfactant and associative polymer, adsorption of surfactant can increase or decrease the extent of polymer / polymer aggregation, and the extent of intermolecular association. Solution rheology is highly sensitive to the manner in which the surfactant binds to the polymer, and is complicated due to competition between effects that reinforce the network and effects that reduce polymer / polymer interaction [31]. Therefore, a "systems" approach maximizes economic and performance benefits when using associative polymers, as contrasted to an "additive" approach where the components of the formulation are selected individually without regard to the presence of other ingredients [32].

1.2. SURFACTANT CO-THICKENING AND VISCOSITY MAXIMUM

Interactions among associative polymers and surfactants can produce a "co-thickening" effect, where the rheological properties of systems containing associative polymer increase with added surfactant. Often the properties increase to a maximum as surfactant concentration increases and subsequently decrease as surfactant concentration increases. This phenomenon also occurs with conventional water soluble polymers as well (especially the well - studied poly(oxyethylene) / sodium dodecyl sulfate system) [33-35], although the viscosity enhancement with surfactant for conventional polymers is considerably less than with associative polymers. The decrease in viscosity is expected from a law of mass action that usually governs the equilibrium process of adsorption of the surfactant to the polymer: statistically, it is more likely that an associative polymer's hydrophobe will encounter a surfactant hydrophobe as compared to another associative polymer hydrophobe when the number of surfactant molecules in solution is larger than the number of associative polymer molecules, screening the interactions between associative polymer hydrophobes. Although the rheological consequences of these interactions are system dependent, they seem general across the many types of anionic and nonionic associative polymers and surfactants.

The literature is divided on whether surfactant micelles are needed to produce the viscosity maximum, and whether the viscosity maximum occurs at the cmc of the surfactant. Some studies found that the viscosity of an associative polymer solution increased to a maximum at a surfactant concentration near the cmc of the surfactant, and then decreased as the concentration of surfactant in solution increased. For example, a maximum in solution viscosity of a hydrophobically modified hydroxyethylcellulose occurred at a concentration of sodium dodecyl sulfate just below its cmc [14]. Similarly, the solution viscosity of hydrophobically modified poly(acrylic acid) reached a maximum at the cmc of SDS, whose magnitude increased with increasing molecular weight and hydrophobe concentration [17].

In contrast, other studies showed that the viscosity maximum occurred at surfactant concentrations much less or much greater than the cmc of the surfactant. The viscosity of a 2.5% solution of associative polymer based on linear poly(ethylene oxide) with hexadecyl end-groups [36] increased to a maximum, and then subsequently decreased as the concentration of sodium dodecyl sulfate increased [5]. The maximum decreased as the molecular weight of the associative polymer increased, and disappeared altogether for the largest two molecular weight polymers of the study. Rather than occurring at the cmc of SDS, the viscosity maximum occurred when the concentration of hydrophobes in the associative polymer stoichiometrically equaled the concentration of sodium dodecyl sulfate hydrophobes in solution. Blank experiments in which sodium chloride was added to model associative polymer solutions showed that at concentrations where sodium dodecyl sulfate maximized the viscosity of the associative polymer solution, sodium chloride actually decreased the viscosity of the solution. These results indicated that the influence of sodium dodecyl sulfate on the viscosity of associative polymer solutions results from an interaction between the alkane chain of sodium dodecyl sulfate and the associative polymer. In a similar study, Hulden found that the viscosity maximum for solutions of model nonionic telechelic associative polymers based on poly(oxyethylene) with various hydrophobic end-groups shifted to higher surfactant concentration with decreasing thickener molecular weight or decreasing thickener hydrophobicity [29,30]. The ratio of the number of surfactant hydrophobes to the number of thickener hydrophobes at the viscosity maximum for the mixed aggregate was roughly constant at 1.7 for SDS and 5.2 for a ten mole ethoxylate of nonylphenol, indicating that the viscosity maximum was related to the size of the mixed aggregates, and that the surfactant had increased the strength or lifetime of the association junction. Likewise, the concentration of either an anionic or nonionic surfactant that was required to produce the maximum in the viscosity of four commercial nonionic comb associative polymers based on poly(oxyethylene) was either near the cmc of the surfactant or much greater, depending on the particular surfactant being studied [4]. With SDS, the viscosity maximum occurred at 0.5 cmc; with nonionic 15-S-9 (i.e., $b\text{-}C_{13}H_{27}(OCH_2OCH_2)_9OH$), the viscosity maximum occurred at concentrations ranging from approximately 0.03% to 5%, depending of the commercial associative polymer. Similarly, a maximum in solution viscosity with respect to 15-S-9 concentration with a hydrophobically modified styrene - maleic anhydride terpolymer polymers occurred at a concentration of over one thousand times greater than the cmc of the surfactant [7]. In a study for hydrophobically modified acrylamide, the concentration of SDS that strongly maximized solution viscosity was well below the cmc of SDS [20].

1.3. MECHANISMS FOR VISCOSITY MAXIMUM

Early explanations for the co-thickening phenomenon and resultant viscosity maximum invoked the notion of "cross-linked micelles" or a "micellar bridge" [2,7,37]. Due to its high molecular weight and multiple hydrophobes on a given backbone, the associative polymer can participate in multiple surfactant micelles, effectively bridging the micelles or acting as a physical "cross-linker" in solution. By this hypothesis, the viscosity increases near the cmc of the neat surfactant, and decreases at larger concentrations due to the stoichiometry of the surfactant relative to the polymer. Since this hypothesis depends on the presence of surfactant micelles to explain co-thickening, it does not explain the co-thickening that takes place when the surfactant concentration is below the cmc of the surfactant. It also fails to account for the influence of the surfactant on the concentration at which the onset of associative networking begins with the associative polymer (i.e., the "cmc" for the polymer / surfactant complex, which may be different from the cmc of the surfactant in solution).

An alternate view considers how the surfactant influences the junctions of the associative network, which are reflected in the viscoelastic properties of the solutions. The storage modulus and relaxation time constant passed through a maximum with respect to surfactant addition, which indicates that the increase in the shear moduli results from an increase in the elasticity of the network, and not just from the enhancement of the low shear viscosity of the solution. [5,6,11]. The maximum in storage modulus reflects a change in the number or functionality of network junctions. The change in the relaxation time constant reflects a change in the binding energy of the junction, its strength, or in the average residence time of the polymer hydrophobe in a dynamic associative junction [6].

Annable et al. explain surfactant induced maxima in rheological properties of associative polymer solutions as an entropically driven change from intramolecular associations (i.e., self-loops) in the network that do not contribute to modulus to intermolecular associations that do contribute to modulus [11]. The concept of added surfactant converting intramolecular associations into intermolecular associations to explain the viscosity maximum was independently suggested by Biggs et al. [20]. This is consistent with the "flower" or "rosette" model advocated by Yekta et al. [38]. Since self-loops force the functionality of the association cluster to be small due to geometric packing constraints, the conversion from intra- to inter- molecular associations increases the potential functionality of the associative junction. These effects increase viscosity, shear modulus, and relaxation time constant as surfactant concentration increases. Once all of the intramolecular associations have been converted to intermolecular associations, additional surfactant dilutes the number of associative polymer hydrophobes in a junction thereby reducing rheological properties.

2. Experimental

2.1. SURFACTANTS

Two or more conventional surfactant hydrophobes can be chemically joined to form one larger composite Complex hydrophobe with large molar volume [39]. This Complex hydrophobe behaves as a "pre-associated" version of the conventional hydrophobes from which it is made to increase the strength of association and thereby improve the thickening efficiency of an associative polymer [40]. A homogeneous Complex hydrophobe results when the conventional hydrophobes from which it is composed are all identical, while a heterogeneous Complex hydrophobe results when the conventional hydrophobes from which it is composed are different. In this work, the homogeneous Complex hydrophobe was composed of two nonylphenyl substituents, and the heterogeneous Complex hydrophobe was composed of an octylphenyl substituent and a nonylphenyl substituent. Complex hydrophobes were prepared and were subsequently ethoxylated to 20, 40, and 80 moles in a pressure autoclave following the procedure described in reference [41]. The number average molecular weights of the surfactants were determined by end-group analysis (hydroxyl number) and by gel permeation chromatography.

TERGITOL® NP series of ethoxylated nonylphenol surfactants, TERGITOL® 15-S series of ethoxylated of linear secondary alcohols of 12 to 15 carbon units, and TRITON® ethoxylated octylphenol surfactants were used as supplied as 100% active ingredient from Union Carbide Corporation.

2.2. ALKALI - SOLUBLE ASSOCIATIVE EMULSION POLYMERS

Associative macromonomers were prepared by reacting a surfactant with alpha, alpha, dimethyl meta-isopropenyl benzyl isocyanate (TMI® (meta), American Cyanamid) using the procedure described in reference [42]. Alkali-soluble associative emulsion copolymers were prepared by the conventional semi-continuous emulsion polymerization of various weight fractions of methacrylic acid, ethyl acrylate, and associative macromonomer [43]. The resulting latexes were diluted to the desired concentration and neutralized to a pH of 9 with 2-amino 2-methyl 1-propanol (AMP-95, Angus Chemical Company) to convert the resulting latexes into solutions.

2.3. NONIONIC ASSOCIATIVE POLYMERS BASED ON POLY(OXYETHYLENE)

Nonionic telechelic associative polymers were made by chain extending linear poly(oxyethylene) (CARBOWAX® 8500, Union Carbide Corporation) with isophorone diisocyanate (Aldrich) in a toluene solution that had been dried by azeotropic distillation. The chain extended polymer was subsequently terminated with conventional (i.e., hexadecanol, Aldrich; nonylphenol, Union Carbide Corporation) and Complex hydrophobic groups [36]. Nonionic associative polymers with comb architecture were synthesized following the method of reference [44]. The viscosity average molecular weights of the nonionic associative polymers were determined from intrinsic viscosity measurements in 40/60 solvent mixture of BUTYL CARBITOL® (Union Carbide Corporation) and water by weight at 25°C, as calibrated with poly(oxyethylene) standards (Pressure Chemical Company).

3. Results and Discussion

3.1. ALKALI - SOLUBLE ASSOCIATIVE EMULSION POLYMERS

Since alkali-soluble associative polymers carry an anionic charge, electrostatic interactions are superimposed onto the interactions due to adsorption of surfactants to associative polymer. Since the interactions are a balance of hydrophobic and ionic interactions, the behavior of hydrophobically modified poly(electrolytes) depends on concentration, temperature, and ionic strength. The polymers interact strongly with oppositely charged surfactants. Surfactant adsorption can take place even when the polymer and surfactant have like-charges when hydrophobic interactions are strong enough to overcome repulsion [18]. The counterion from anionic surfactants contracts

482

the chain dimensions and simultaneously enhances aggregation so that viscosity can either increase or decrease. The latter effect can also increase viscosity in addition to the aggregation induced by the added surfactant [19]. The rheology depends in a complicated way on polymer concentration, electrolyte concentration, amount and type of hydrophobe substitution, pH, and surfactant concentration. As described in forthcoming papers, the co-thickening effect can be used to enhance the shear-thinning viscosity profile, and to add time-dependent or shear history-dependent properties by selecting the relative concentrations of the appropriate surfactant / associative polymer pair [45,46]. This section investigates the impact of the structure of the surfactant (i.e., HLB, hydrophobe structure, moles of ethoxylation) and the composition of alkali - soluble associative emulsion polymers (i.e., carboxyl monomer level, associative macromonomer level) on alkaline solution viscosity.

3.1.1. Influence of Surfactant Structure

Whether or not the viscosity of an associative polymer solution increases or decreases when surfactant is added depends on the structure of the surfactant, and its concentration in solution. Figure 1 shows the effect of various nonionic surfactants having 20 moles of ethoxylation on the viscosity of 0.5 wt. % solutions of an alkali - soluble associative polymer composed of 40% MAA, 50% EA, and 10% associative macromonomer that carried a 40 mole ethoxylate chain terminated with a homogenous Complex hydrophobe.

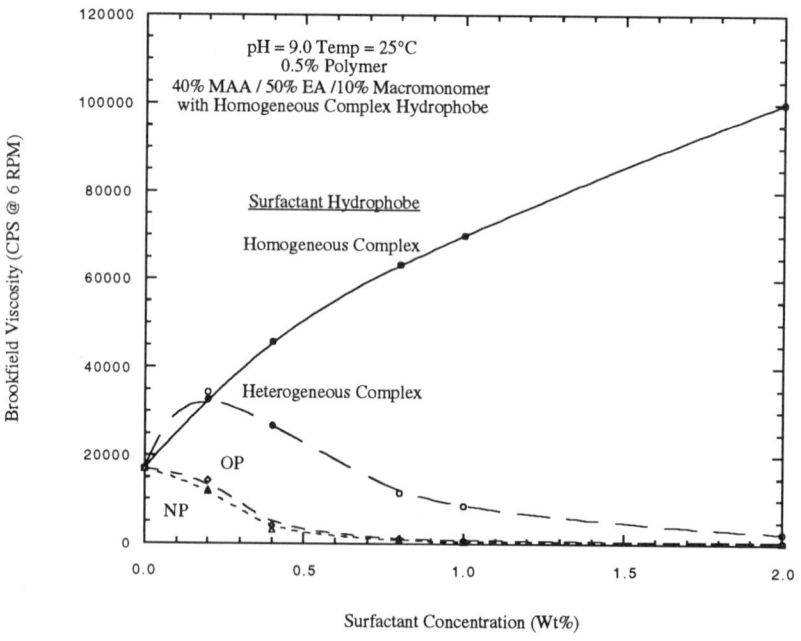

Figure 1. Influence of 20 Mole Ethoxylate Nonionic Surfactant Hydrophobe Structure on the Viscosity of Solutions of an Alkali - Soluble Associative Polymer.

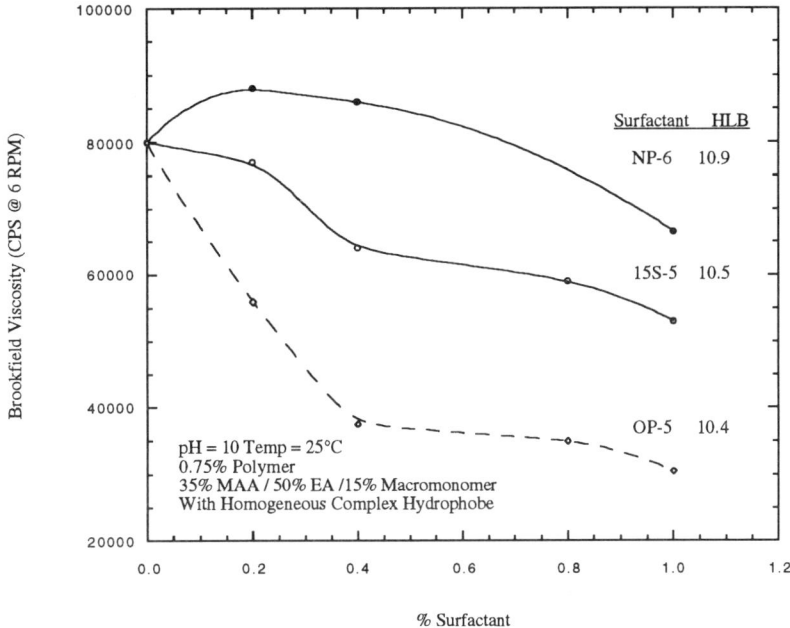

Figure 2. Influence of Surfactant Hydrophobe Structure at Constant HLB on the Alkaline Solution Viscosity of an Alkali - Soluble Associative Polymer.

The surfactants had either a nonylphenyl, octylphenyl, homogeneous Complex, or heterogeneous Complex hydrophobe. The viscosity of the solution increased as the concentration of the surfactant with the homogeneous Complex hydrophobe increased. The viscosity of the solution increased to a maximum and subsequently decreased to less than the solution without surfactant as the concentration of the surfactant with the heterogeneous Complex hydrophobe increased. The viscosity of the solution decreased as the concentration of either the conventional octylphenyl or nonylphenyl surfactant increased. In these experiments the best co-thickening occurred when the structure of the hydrophobe of the surfactants matched the structure of the structure of the hydrophobe in the associative polymer. Thus, whether or not the surfactant co-thickened with the polymer depended on the structure of the hydrophobe of the surfactant.

Figure 2 shows the influence of the concentration of three different surfactants that have 5-6 moles of ethoxylation (and are therefore of nearly the same HLB), but vary in hydrophobe structure (i.e., octylphenyl, nonylphenyl, and linear secondary alcohol of 12-15 carbon units) on the solution viscosity of an associative polymer composed of 35% MAA , 50% EA, and 15% associative macromonomer that carried a 20 mole ethoxylate chain terminated with a homogenous Complex hydrophobe. The viscosity of the solution containing the nonylphenyl ethoxylate increased to a maximum and subsequently decreased as surfactant concentration increased, whereas the viscosity of the solution containing the octylphenyl ethoxylate decreased as surfactant concentration increased. The solution with the linear secondary alkyl hydrophobe (i.e., 15-S-5) exhibited a response in-between those of the octylphenyl and nonylphenyl based surfactants. Surfactants that have marginal solubility in water (i.e., detergents), such as NP-6, often produced synergistic co-thickening, whereas surfactants with larger water solubility, such as 15-S-5, generally lowered solution viscosity. Nonetheless, that the response depended on hydrophobe structure shows that the interaction between associative polymers and surfactants depended on more than just the HLB of the surfactant.

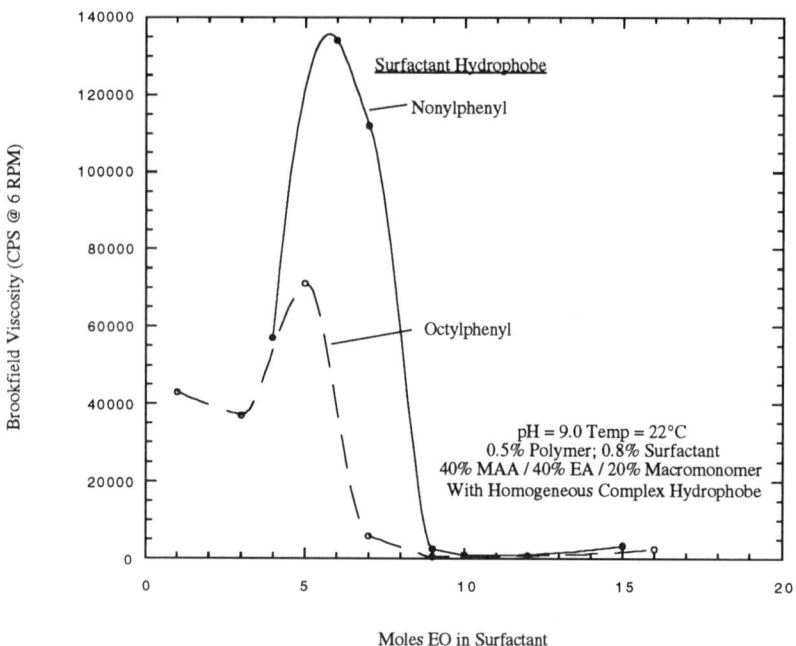

Figure 3. Influence of Nonionic Surfactant Hydrophobe Structure and Moles of Ethoxylation on the Viscosity of Solutions of an Alkali - Soluble Associative Polymer.

The solution viscosities of associative polymers were dramatically sensitive to the degree of ethoxylation of nonionic surfactants. Figure 3 shows the influence of the moles of ethoxylation of a nonylphenyl ethoxylate and an octylphenyl surfactant on the viscosity of an alkali - soluble associative polymer composed of 40% MAA, 40% EA, and 20% associative macromonomer that carried a poly(oxyethylene) chain of 40 moles terminated with the homogenous Complex hydrophobe. The concentration of the polymer was 0.5 wt. %, and the concentration of the surfactants was 0.8 wt. %, which was above the published values of the cmcs of these surfactants [47]. With either nonylphenyl or octylphenyl based surfactants, solution viscosity dramatically increased and subsequently decreased as the moles of ethoxylation in the surfactant increased. The viscosity maximum occurred with 6 moles of EO for both the octylphenyl and nonylphenyl based surfactants. For example, the viscosity decreased almost three orders of magnitude between NP-7 versus NP-9, highlighting the formulation sensitivity often encountered when associative polymers are used. The viscosity increase was larger with the nonylphenyl based surfactant than with the octylphenyl based surfactant. This shows the value of using a "systems approach" when formulating with associative polymers. The figure below compares the idealized structures of the nonylphenyl and octylphenyl hydrophobes in the surfactant.

Nonylphenyl Structure   Octylphenyl Structure

The nonylphenyl hydrophobe has a benzyllic hydrogen, while the octylphenyl hydrophobe does not. The octyl substituent in the octylphenyl hydrophobe is symmetric, while the nonyl substituent in the nonylphenyl hydrophobe is not. Since the molecular weights of the hydrophobes differ by only one carbon atom, but differ in the structure of the alkyl substituent, the selectivity shown in Figure 3 must have been related to entropic effects, analogous to a key fitting into a lock. Commercial samples of nonylphenol and octylphenol both contain a crude mixture of isomers differing in the structure of the alkyl substituent and in where the alkyl group is substituted on the benzyllic ring (i.e., ortho, meta, and para). Octylphenol usually contains a smaller fraction of impurities than nonyphenol. Yet even with this crude mixture of hydrophobic structures in the surfactant and despite the polydispersity inherent in the polymers and the surfactants, the shape of the curve approached that of a delta function to imply a level of selectivity in the association analogous to that usually only encountered in biological systems. This dramatic sensitivity suggests that entropic effects related to structure (i.e., as a key fits into a lock) were more important than enthalpic effects.

3.1.2. Influence Of Alkali - Soluble Associative Polymer Structure

Figure 4 shows the influence of the moles of ethoxylation in a nonylphenyl ethoxylate on the solution viscosity of an alkali - soluble associative polymer composed of 35% MAA and various levels of associative macromonomer (as stated in the Figure) that carried an 80 mole ethoxylate chain terminated in a homogenous Complex hydrophobe, with the balance being ethyl acrylate. The polymer concentration was 0.75 percent by weight, and the surfactant concentration was 0.8 percent by weight. The viscosity increased to a maximum and subsequently decreased as the moles of ethoxylation in the surfactant increased. The moles of ethoxylation that maximized solution viscosity did not depend strongly on the amount of macromonomer. The magnitude of the co-thickening at the viscosity maximum increased as the concentration of macromonomer increased. Synergistic co-thickening was largest with low mole ethoxylates. Therefore, co-thickening depended on the interaction between the surfactant and the associative macromonomer.

Figure 5 shows the influence of the moles of ethoxylation in a nonylphenyl ethoxylate on the solution viscosity of an alkali - soluble associative polymer composed of 15% of associative macromonomer that carried a 80 mole ethoxylate chain terminated in a homogenous Complex hydrophobe and various levels of methacrylic acid (as stated in the Figure), with the balance being ethyl acrylate. The polymer concentration was 0.75 percent by weight, and the surfactant concentration was 0.8 percent by weight.

486

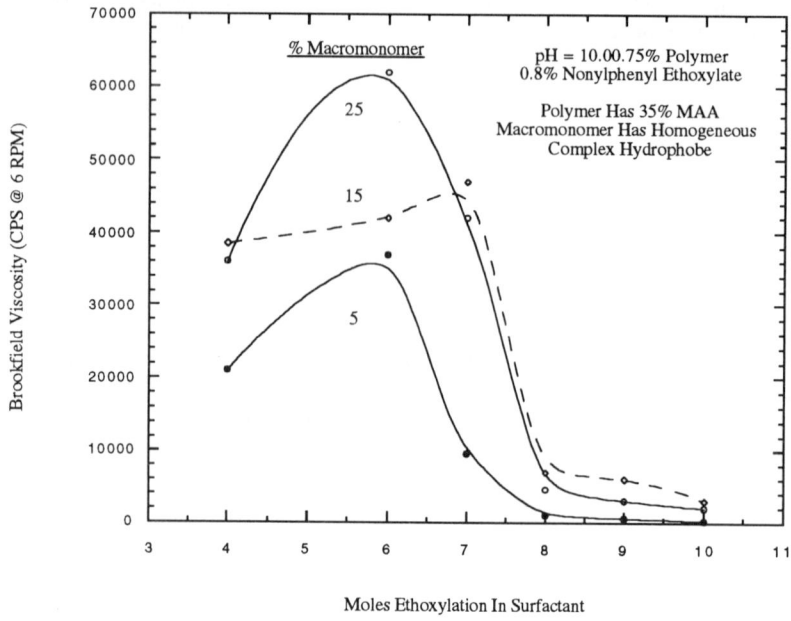

Figure 4. Influence of Associative Macromonomer Concentration on Surfactant Co-Thickening Effect.

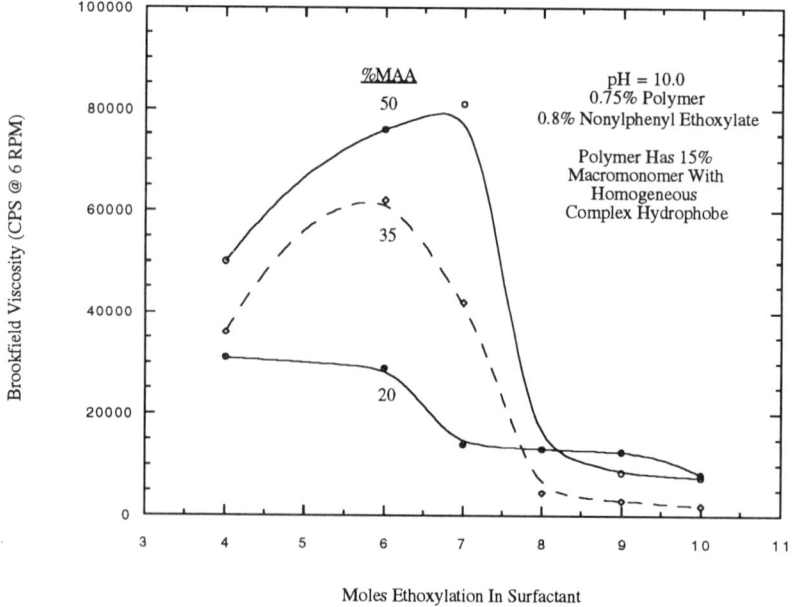

Figure 5: Influence of Methacrylic Acid Concentration on Surfactant Co-Thickening.

The viscosity increased to a maximum and subsequently decreased as the moles of ethoxylation in the surfactant increased. The magnitude of the viscosity at maximum increased as the carboxyl concentration in the polymer increased when the surfactant had low moles of ethoxylation. This increased interaction could be due to adsorption of the surfactant onto the backbone of the polymer through the ethoxylated portion of the surfactant to increase the effective associative functionality of the polymer.

3.2. NONIONIC ASSOCIATIVE POLYMERS

This section investigates the impact of the surfactant (i.e., HLB, hydrophobe structure, moles of ethoxylation), and the structure of the polymer (i.e., hydrophobe structure, linear telechelic and comb architecture) on the solution viscosity nonionic (i.e., poly(oxyethylene) based) associative polymers.

3.2.1. Linear Telechelic Associative Polymers
Figure 6 shows the effect of various nonionic surfactants with 20 moles of ethoxylation on the viscosity 2.5% solutions of a linear poly(oxyethylene) polymer of viscosity average molecular weight of approximately 90,000 with homogenous Complex hydrophobic end-groups.

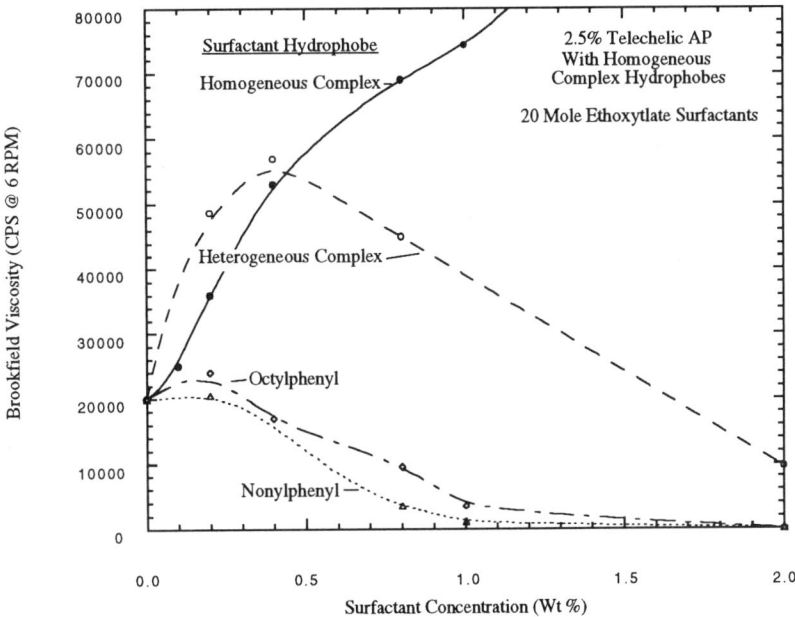

Figure 6. Influence of 20 Mole Ethoxylate Nonionic Surfactant Hydrophobe Structure on the Viscosity of Solutions of Linear Nonionic Associative Polymer With Homogeneous Complex Hydrophobic End-Groups.

The surfactants used in Figure 6 were the same as those used in Figure 1, and the polymers used the same homogeneous Complex hydrophobes. For the polymer of Figure 6, solution viscosity increased as the concentration of the surfactant based on the homogenous Complex hydrophobe increased in a manner similar to that demonstrated for the alkali - soluble associative polymer of Figure 1. Solution viscosity increased to a maximum and subsequently decreased as the concentration of the surfactant based on the heterogeneous Complex hydrophobe increased. The viscosity of the solution containing the surfactant based on the heterogeneous hydrophobe was the largest at concentrations when the surfactant concentration was less than that required to produce the maximum in solution viscosity. At surfactant concentrations beyond that required to produce the viscosity maximum, the viscosity of the solution containing the homogenous Complex hydrophobe was the largest. As a class, surfactants based on the Complex hydrophobes produced greater levels of co-thickening as compared to nonylphenyl or octylphenyl, especially at higher levels of ethoxylation.

Figure 7 shows the influence of the moles of ethoxylation of a nonylphenyl ethoxylate and an octylphenyl ethoxylate on the viscosity of 2.5 wt. % solutions of linear poly(oxyethylene) polymer of viscosity-average molecular weight of approximately 90,000 that had homogenous Complex hydrophobic end-groups (Figure 7a) or n-hexadecyl end-groups (Figure 7b). The viscosity of solutions for both polymers followed similar trends. Solution viscosity increased as the concentration of the 20 mole ethoxylate of the homogeneous Complex hydrophobe increased. Solution viscosity increased to a maximum and subsequently decreased as the concentration of the 40 and 80 mole ethoxylates of the homogeneous Complex hydrophobe increased. The magnitude of co-thickening was larger for the polymer capped with the homogeneous Complex hydrophobe. In general, the magnitude of the co-thickening at the viscosity maximum decreased as the length of the hydrophile in the surfactants increased, and the trend was reversed at large surfactant concentrations.

The three panels of Figure 8 present the influence of the number of moles of ethylene oxide in the surfactant hydrophile on the viscosity of 1% aqueous solutions of nonionic telechelic associative polymers of similar molecular weight (all have viscosity -average molecular weight of approximately 90,000). The only difference among the thickeners was the hydrophobic end-groups: homogeneous Complex hydrophobe for Figure 8a; heterogeneous Complex hydrophobe for Figure 8b; and n-hexadecyl hydrophobes for Figure 8c. The solution viscosities of all three polymers exhibited the same qualitative response with nonylphenyl or octylphenyl ethoxylates, although the magnitude of the co-thickening response depended on the structure of the polymer hydrophobic end-group. Solution viscosity showed a maximum at 5-6 moles of ethoxylation in the surfactant, and the associative polymer using the homogeneous Complex hydrophobe showed a larger relative increase in the viscosity maximum as compared to that of the polymer using a n-hexadecyl hydrophobe. This was another example of a specific entropic interaction between associative polymers and surfactants was generically applicable across various associative polymer types and structures.

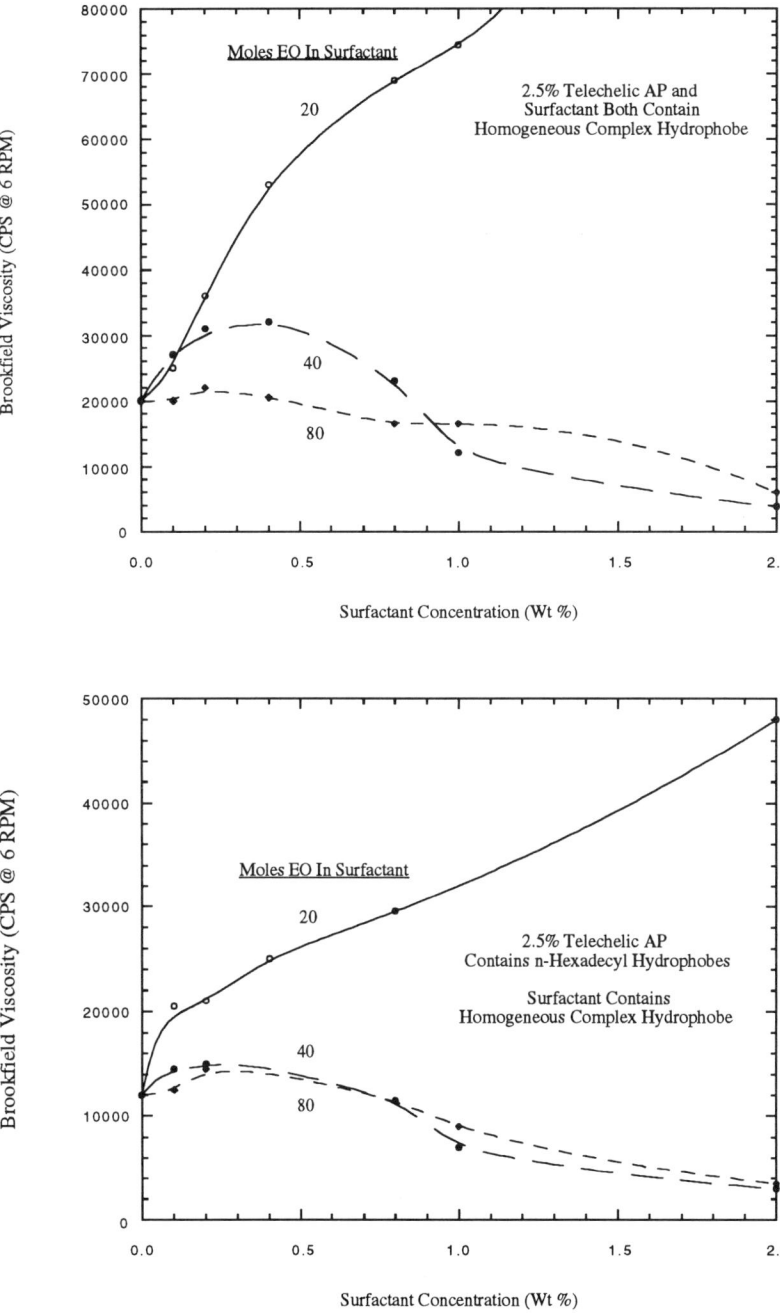

Figure 7. Influence of Concentration and Moles of Ethoxylation in Surfactants Based on the Homogeneous Complex Hydrophobe on the Aqueous Solution Viscosity of Linear Nonionic Associative Polymers End-Capped With: Top) Homogeneous Complex Hydrophobes; And Bottom) n-Hexadecyl Hydrophobes.

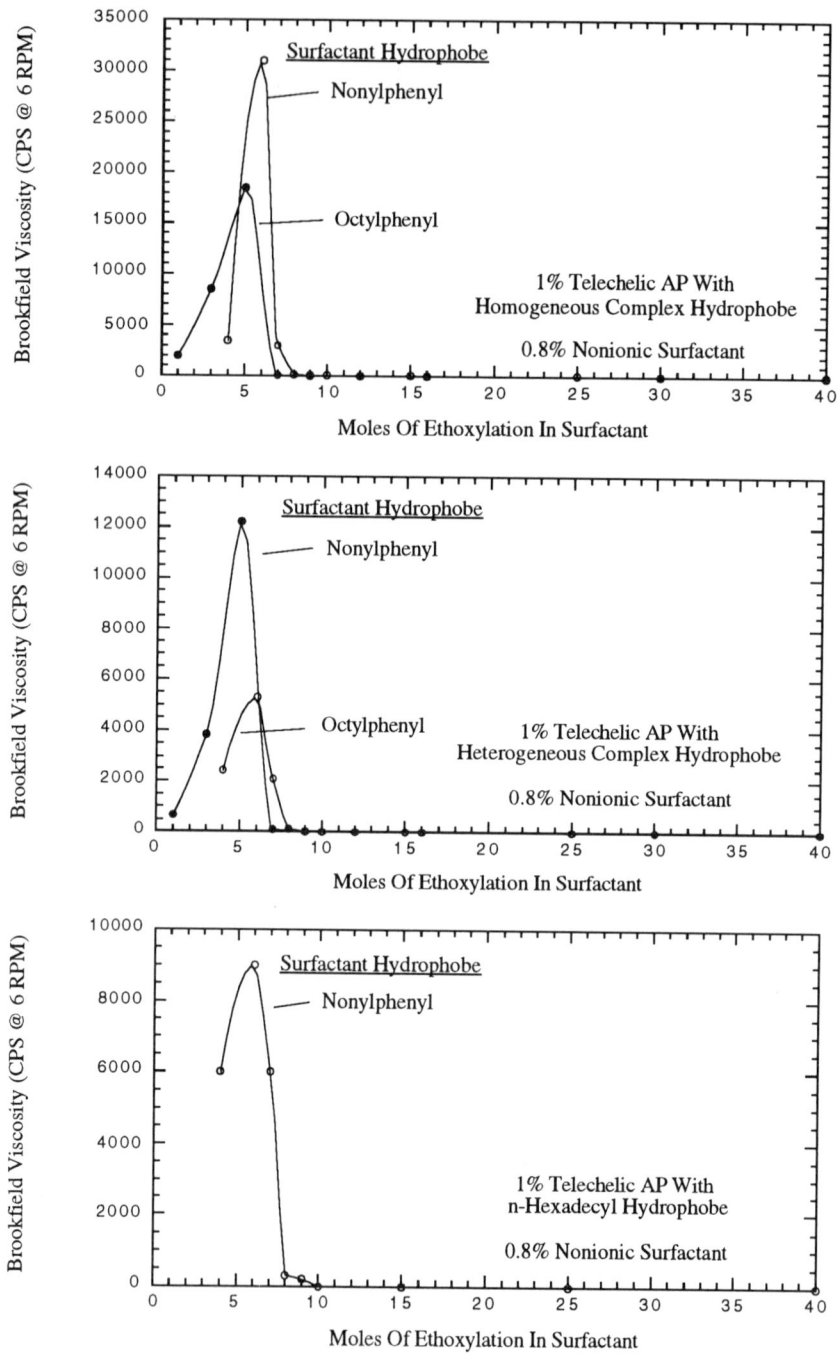

Figure 8. Influence of Nonionic Surfactant Hydrophobe Structure and Moles of Ethoxylation on the Solution Viscosity of Nonionic Linear Associative Polymers Capped with Top) Homogeneous Complex Hydrophobes; Middle) Heterogeneous Complex Hydrophobes; and Bottom) n-Hexadecyl Hydrophobes.

3.2.2. Associative Polymers With Comb Architecture

Figure 9 presents the influence of the number of moles of ethylene oxide in the surfactant hydrophile on the viscosity of 1% aqueous solutions of nonionic associative polymers with comb architecture. These comb polymers differed only in hydrophobe structure, and had similar number average molecular weights of approximately 110,000-120,000 and had similar placement or "bunching" of the hydrophobes along the polymer backbone because they were made using identical processes [44]. Like the telechelic polymers of Figure 8, these polymers were nonionic. Like the polymers of Figure 3, these polymers had a comb structure, although the hydrophobes of the comb polymers of Figure 9 were attached closely to the polymer backbone, while the hydrophobes of the polymers of Figure 3 were extended away from the backbone on a flexible poly(oxyethylene) chain. The polymers of Figures 3, 8a, and 9b had the homogeneous Complex hydrophobe in common. The co-thickening phenomenology of the comb polymers shown in Figure 9 paralleled the behavior of the telechelic polymers of Figure 8 and the alkali - soluble polymers of Figure 9, although the magnitude of the co-thickening effect depended on the hydrophobe used in the comb polymers and on the concentration of hydrophobic groups in the polymer. This was yet another example of a specific phenomenon that is generally qualitatively applicable across the various classes of associative polymers of various backbone and hydrophobe structures.

4. Summary And Conclusions

Associative polymers interact in a capricious manner with surfactants to produce system-dependent rheological responses. This chapter showed several examples of interactions that both built and destroyed associative structure in solution, and how this behavior depended in a complicated manner on the structures and concentrations of the surfactant and associative polymer. And yet, qualitative similarities in the co-thickening behavior of certain surfactants that were sparingly soluble in water (i.e. detergents) applied in general across the chemical classes of associative polymers and to the various polymer architectures within these classes. This entropic "lock-in-key" behavior seems analogous to selectivity often seen in biological systems.

Although the interactions between associative polymers and surfactants are not yet completely understood from a fundamental perspective, they are technologically useful in formulating high performance waterborne systems, or in reducing formulation cost by driving thickener efficiency. To arrive at a scientific understanding of the interactions, rheology studies should be expanded to include extensional viscosity, linear and non-linear viscoelastic properties, the time dependent responses, and the shear - history dependent responses of solutions formulated with surfactants, because these tests probe and alter the associative structure in solution. Surfactant-induced changes in the associative polymer network structure that can either increase or decrease solution viscosity and viscoelasticity must come either from change in the functionality of the network junctions, their average lifetime, or from an increase in the number of hydrophobes that are in the network by converting intramolecular associations into intermolecular associations [6,45,46]. Mapping how surfactant and associative polymer structures influence this behavior with well designed and characterized polymers and surfactants will facilitate deducing a mechanism.

Since rheology characterizes the association mechanism indirectly, fundamental understanding would be facilitated by techniques that characterize the polymer /

492

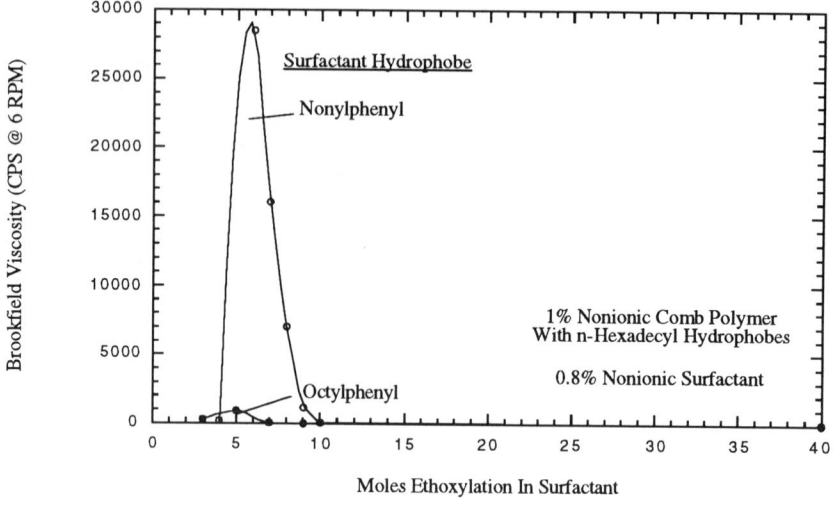

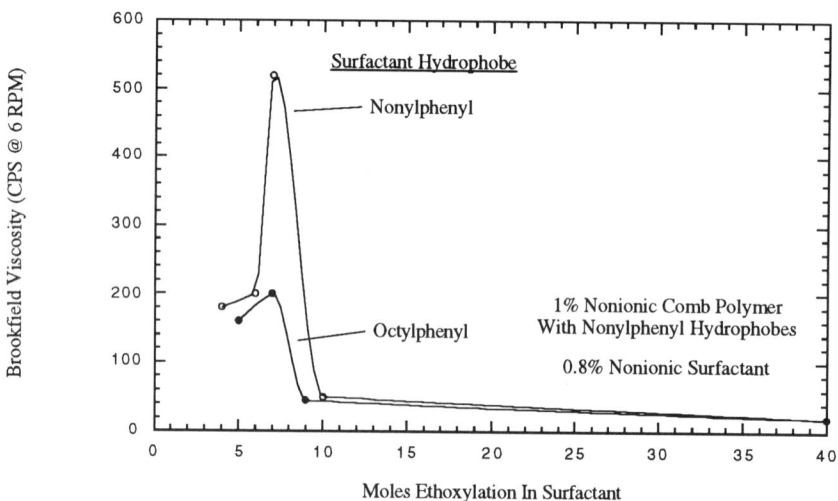

Figure 9. Influence of Nonionic Surfactant Hydrophobe Structure and Moles of Ethoxylation on the Solution Viscosity of Nonionic Comb Associative Polymers With: Top) n-Hexadecyl Hydrophobes; and Bottom) Nonylphenyl Hydrophobes.

surfactant complex in solution directly, such as light-scattering, fluorescence, SANS, SAXS, etc. [48-53], and performing these tests under various shear conditions where possible.

An analogy may exist between surfactant / associative polymer mixtures and the phase behavior of conventional polymer / surfactant systems [31] or of the phase behavior of mixed surfactant systems, where the formation of ordered phases in the latter is well known in the literature. Recent work that has begun to view associative polymer systems as consisting of liquid crystalline micro-domains can be applied to systems that contain both associative polymers and surfactants [53].

5. References

1. Thibeault, J. C., Sperry, P. R., and Schaller, E. J. (1986)Effect of surfactants and cosolvents on the behavior of associative thickeners in latex systems, in J. E. Glass (ed.) *ACS Advances in Chemistry Series* **213**, American Chemical Society, p. 375.
2. Glancy, C. W., and Bassett, D. R., (1984) Effect of Latex Properties on the Behavior of Nonionic Associative Thickeners in Paint, Proceedings of the PMSE Division of the ACS, **51**, p. 348.
3. Bassett, D. R., and Glancy, C. W. , (1989) Interactions of Associative Polymers with Surfactants in Aqueous Media, Symposium on Polymeric Surfactants, ACS National Meeting, Miami Beach.
4. Karunasena, A., Brown, R. G., and Glass, J. E., (1989) Hydrophobically Modified Ethoxylated Urethane Architecture. Importance for Aqueous and Dispersed Phase Properties, in J. E. Glass (ed.) *ACS Advances in Chemistry Series* **223**, American Chemical Society, p. 495.
5. Jenkins, R. D., (1990), Ph.D. Dissertation, Lehigh University.
6. Jenkins, R. D., Sinha, B. R., and Bassett, D. R., (1991) Associative Polymers with Novel Hydrophobe Structures, Proceedings of the PMSE Division of the ACS, **65**, p. 72.
7. Lundberg, D. J. , Ma, Z., Alahapperuna, K., Glass, J. E., (1991) Surfactant Influences on Hydrophobically Modified Thickener Rheology, in D. N. Schultz and J. E. Glass (eds.) *ACS Symposium Series* **462**, American Chemical Society, p.234.
8. Lundberg, D. J., Glass, J. E., and Eley, R. R., (1991) Viscoelastic Behavior Among HEUR Thickeners, *Journal of Rheology*, **35**(**6**), p. 1255.
9. Francois, J., (1994) Association in Water of Model Hydrophobically End - Capped Poly(ethylene oxide), *Progress in Organic Coatings* **24**, p. 67.
10. Binana-Limbele, W., Clouet, F., and Francois, J., (1993) Hydrophobically End - Capped Poly(ethylene oxide) Urethanes. Part 3: Effect of Sodium Dodecyl Sulfate on Their Association in Aqueous Solution, *Colloid and Polymer Science* **271**, p.748.
11. Annable, T., Buscall, R., Ettelaie, R., Shepherd, P., and Whittlestone, D., (1994) The Influence of Surfactants on the Rheology of Associating Polymers in Solution, *Langmuir* , **10**, p. 1060.
12. Tarng, M. R., Kaczmarski, J. P., Lundberg, D. J., and Glass, J. E., (1996) Comparative Flow Properties of Model Associative Thickener Aqueous Solutions , in J. E. Glass, (ed.) *ACS Advances in Chemistry Series* **248** , American Chemical Society, p. 305.
13. Zhou, L.,. (1990) Ph. D. Dissertation, Lehigh University.
14. Sau, A. C., and Landoll, L. M., (1989) Synthesis and Solution Properties of Hydrophobically Modified (Hydroxyethyl) Cellulose, in J. E. Glass (ed.) *ACS Advances in Chemistry Series* **223**, American Chemical Society, p. 343.
15. Gelman, R. A., (1987) TAPPI p. 159, as reviewed in reference [13].
16. Hoffmann, H., Hofmann, S., Kastner, U., (1996) Viscoelastic Surfactant Systems Under Shear, in J. E. Glass, (ed.) *ACS Advances in Chemistry Series* **248** , American Chemical Society, p. 219.
17. Iliopoulos, I., Wang, T. K., and Audebert, R., (1991) Viscometric Evidence of Interactions Between Hydrophobically Modified Poly(Sodium Acrylate) and Sodium Dodecyl Sulfate, *Langmuir,* **7**, p. 617.
18. Magny, B., Iliopoulos, I., Audebert, R., Piculell, L., and Lindman, B., (1992) Interactions Between Hydrophobically Modified Polymers and Surfactants, *Progress in Colloid and Polymer Science,* **89**, p. 118.
19. Wang, T. K., Iliopoulos, I., and Audebert, R., (1991) Aqueous Solution Behavior of Hydrophobically Modified Poly(Acrylic Acid) in S. W. Shalaby, C. L. McCormick, and G. B. Butler (eds.) *ACS Symposium Series* **467**, American Chemical Society, p. 218.
20. Biggs, S., Selb, J., and Candau, F., (1992) Effect of Surfactant on the Solution Properties of Hydrophobically Modified Polyacrylamide *Langmuir* **8**, p.838.
21. Amis, E. J., Hu, N., Seery, T. A. P., Hogen-Esch, T. E., Yassini, M., and Hwang, F., (1996) Associating Polymers Containing Fluorocarbon Hydrophobic Units, in J. E. Glass, (ed.) *ACS Advances in Chemistry Series* **248** , American Chemical Society, p. 279.

22. NATO Advanced Study Institute on Recent Advances in Polymeric Dispersions, (1996) in Elizondo, Spain.
23. Goddard, E.D., (1986) Polymer - Surfactant Interaction. Part 1: Uncharged Water - Soluble Polymers and Charged Surfactants, *Colloids and Surfaces* **19** p.255.
24. Goddard, E. D., (1993) Applications of Polymer - Surfactant Systems in E. D. Goddard and K. P. Ananthapadmanabhan (eds.) *Interactions of Surfactants with Polymers and Proteins*, CRC Press, p. 395.
25. Nagarajan, R. (1982) On the Nature of Interactions Between Polymers and Surfactants in Dilute Aqueous Solutions , *Polymer Preprints* , **22(2)** , p. 30.
26. Breuer, M. M., and Robb, I. D., (1972) Interaction Between Macromolecules and Detergents, *Chemistry and Industry* **1**, p. 530.
27. Jones, M.N., (1967) The Interaction of Sodium Dodecyl Sulfate with Polyethylene Oxide, *Journal of Colloid and Interface Science* , **23** , p.36.
28. Cabane, B., (1982) Organization of Surfactant Micelles Absorbed on a Polymer Molecule in Water: A Neutron Scattering Study, *Journal de Physique*, **43** , p. 1529.
29. Hulden, M., (1994) Hydrophobically Modified Urethane - Ethoxylate (HEUR) Associative Thickeners. 1. Rheology of Aqueous Solutions and Interactions With Surfactants, *Colloids and Surfaces A: Physicochemical and Engineering Aspects*, **82**, p. 263.
30. Hulden, M., (1994), Ph. D. Dissertation, Abo Akademi University, Finland.
31. Lindmann, B., and Thalberg, K., (1993) Polymer - Surfactant Interactions - Recent Developments, in E. D. Goddard and K. P. Ananthapadmanabhan (eds.) *Interactions of Surfactants with Polymers and Proteins*, CRC Press, p. 203.
32. Howard, P. R., Leasure, E. L., Rosier, S. T., and Schaller, E. J., (1991) Systems Approach to Rheology Control, in D. N. Schultz and J. E. Glass (eds.) *ACS Symposium Series 462*, American Chemical Society, p.207.
33. Tam, K. C., (1996) Rheological Properties of Poly(Ethylene Oxide) in Anionic Surfactant Solutions, in J. E. Glass, (ed.) *ACS Advances in Chemistry Series* **248** , American Chemical Society, p.205.
34. Brackman, J. C., (1991) Sodium Dodecyl Sulfate Induced Enhancement of the Viscosity and Viscoelasticity of Aqueous Solutions of Poly(Ethylene Oxide). A Rheological Study on Polymer - Micelle Interaction, *Langmuir* **7**, p. 469.
35. Brown, W., Fundin, J., de Graca Miguel, M., (1992) Poly(Ethylene Oxide) - Sodium Dodecyl Sulfate Interactions Studied Using Static and Dynamic Light Scattering, *Macromolecules* **25** , p. 7192.
36. Jenkins, R. D., Bassett, D. R., Silebi, C. A., and El-Aasser, M. S., (1995) Synthesis and Characterization of Model Associative Polymers, *Journal of Applied Polymer Science* **58** p. 209.
37. Bieleman, J. H., Riestuis, F. J. J., and Van Der Velden, P. M., (1986) The Application of Urethane Based Polymeric Thickeners in Aqueous Coating Systems, *Polymers Paint Colours Journal*, **176(4169)** p. 450.
38. Yekta, A., Duhamel, J., Adiwidjaja, H., Brochard, P., and Winnik, M. A., (1993) Association Structure of Telechelic Associative Thickeners in Water, *Langmuir* **9**, p. 881.
39. Jenkins, R. D., Bassett, D. R., and Shay, G. D., (1994) Polymers Containing Complex Hydrophobic Groups, US Patents 5,292,828; 5,352,734; and 5,401,802.
40. Jenkins, R. D., DeLong, L. M., and Bassett, D. R., (1996) Influence of Alkali - Soluble Associative Emulsion Polymer Architecture on Rheology, in J. E. Glass, (ed.) *ACS Advances in Chemistry Series* **248** , American Chemical Society, p.425.
41. Jenkins, R. D., Bassett, D. R., Smith, D. E., Argyropoulos, J. N., Loftus, J. E., and Shay, G. D., (1996) Complex Hydrophobe Compounds, US Patent 5,488,180.
42. Jenkins, R. D., Bassett, D. R., and Shay, G. D., (1994) Polymers Containing Macromonomers, US Patents 5,292,843; 5,342,883; and 5,405,900.
43. Jenkins, R. D., Bassett, D. R., Sterlen, R. A., and Daniels, W. B., (1995) Processes for Preparing Aqueous Polymer Emulsions, US Patents 5,399,618; and 5,436,292.
44. Hoy, R.C., and Hoy, K. L., (1984) US Patent 4,426,485.
45. English, R. J., Gulati, H. S., Smith, S. W., Kahn, S. A., and Jenkins, R. D., (1996) Alkali - Soluble Associative Polymers: Solution Rheology and Interactions With Nonionic Surfactants, *Proceedings of the XIIth International Congress on Rheology*.
46. English, R. J., Gulati, H. S. , Kahn, S. A., and Jenkins, R. D., (1996) Solution Rheology of a Hydrophobically Modified Alkali - Soluble Associative Polymer, *Journal of Rheology*, submitted.
47. TRITON and TERGITOL Surfactants Product Literature, Union Carbide Corporation.
48. Alami, E., Rawiso, M., Isel, F., Beinert, G., Binana-Limbele, W., and Francois, J., (1996) Model Hydrophobically End - Capped Poly(Ethylene Oxide) In Water, in J. E. Glass, (ed.) *ACS Advances in Chemistry Series* **248** , American Chemical Society, p. 343.
49. Yekta, A., Nivaggioli, T., Kanagakingam, S., Xu, B., Masoumi, Z., and Winnik, M.A., (1996) Urethane - Coupled Poly(Ethylene Glycol) Polymers Containing Hydrophobic End-Groups. NMR Characterization as a Step Toward Determining Aggregation Numbers in Aqueous Solutions, in J. E. Glass, (ed.) *ACS Advances in Chemistry Series* **248** , American Chemical Society, p. 364.
50. Macdonald, P. M., Uemura, Y., Dyke, L., and Zhu, X., (1996) Self- Diffusion Coefficients of Associating Polymers From Pulsed - Gradient Spin - Echo Nuclear Magnetic Resonance Spectroscopy, in J. E. Glass, (ed.) *ACS Advances in Chemistry Series* **248** , American Chemical Society, p. 377.

51. Uemura, Y., and Macdonald, P. M., (1996) NMR Diffusion and Relaxation Time Studies of HEUR Associating Polymer Binding to Polystyrene Latex, *Macromolecules* **29**, p. 63.

52. Rao, B., Uemura, Y., Dyke, L., and Macdonald, P. M., (1995) Self- Diffusion Coefficients of Hydrophobic Ethoxylated Urethane Associating Polymers Using Pulsed - Gradient Spin - Echo Nuclear Magnetic Resonance, *Macromolecules* **28**, p. 531.

53. Abrahansen-Alami, S., Alami, E., Francois, J., (1996) The Lyotropic Cubic Phase of Model Associative Polymers: Small Angle X-Ray Scattering (SAXS), Differential Scanning Calorimetry (DSC), and Turbidity Measurements, *Journal of Colloid and Interface Science* **179** p. 20.

LATEX APPLICATIONS IN PAPER COATING

DO IK LEE
Emulsion Polymers Research
The Dow Chemical Company
Midland, Michigan 48674
USA

1. Introduction

The main objectives of coating papers and paperboards are to improve their aesthetic appearance and printability. Coatings impart smoothness, gloss, brightness, and opacity to the base sheets for improved appearance, and provide them with enhanced printability which requires resistance to ink film-splitting forces, smoothness, ink holdout and gloss, sharp halftone reproduction, etc. The properties and printability of coated papers are affected by the base sheets (fiber types, sheet formation, internal sizing, and base weight), coating materials (pigment types, binder types, rheology modifiers, water-retention aids, lubricants, defoamers, etc.), coating formulations (ratios of coating components, solids and pH's), coating process (coating application types and speed), coat weights, drying conditions (dryer types, drying temperature, drying time, and final moisture level), etc. In other words, many factors affect the properties and printability of coated papers, and they are inter-dependent with each other.

Various types of soft latexes, such as styrene-butadiene (S/B), styrene-butyl acrylate (S/BA), and polyvinyl acetate (PVAc) latexes, are widely used as binders in paper coatings. Latex particles not only bond pigment particles and adhere them to the base sheets, but also significantly affect coating formulation rheology, coated paper properties, and printability. Also, polystyrene latexes of various particle sizes are used as plastic pigments to improve coating smoothness and gloss. Recently, hollow sphere pigments have been introduced as easier finishing plastic pigments. Overall, latexes play a major role in paper coatings.

This paper will briefly review papermaking, coating, and printing processes and describe paper coating formulations, coated paper properties, and printability. Then, the paper will extensively review and discuss the effects of latexes on the rheology of paper coating formulations, coating properties, and printability in terms of the physico-chemical and colloidal properties of latex particles and the viscoelastic properties of latex polymers. Finally, some future research challenges will be presented.

J. M. Asua (ed.), Polymeric Dispersions: Principles and Applications, 497–513.

2. Brief Review of Papermaking, Coating, and Printing [1-4]

Woods (soft and hard) are reduced to cellulose fibers forming wood pulps, and lignin is separated from cellulose by chemical, semi-chemical or mechanical pulping process. Stocks of wood pulps are formed into sheets on fourdrinier machine to produce papers and on fourdrinier or cylinder machine to make paperboards. Although base sheets can never be smooth and uniform, the main objective of papermaking is to achieve better uniformity, smoothness, and strength. Especially for those base sheets to be subsequently coated, their uniformity is one of the most important factors to achieve smooth coated papers and paperboards. Papers and paperboards are coated with various coating formulations by a variety of coating processes: size-press, air-knife, rod, and blade coating process. Among these coating processes, the blade coating process is most widely used, and there are several different types of blade coaters: inverted blade, puddle blade, tube-loaded blade, and short-dwell blade coater. Figure 1 shows uncoated and coated papers, respectively. Figure 2 shows the difference in printability between uncoated and coated papers. The coated paper shows sharper halftone dots. There are three major printing processes: letterpress, offset, and rotogravure. These different printing processes demand different coated paper properties for printability.

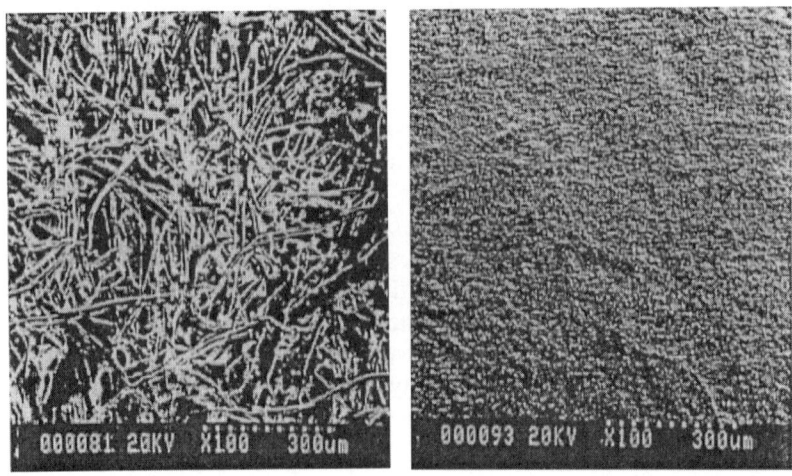

Figure 1. Uncoated and coated papers.

3. The Structure of Paper Coatings and Coated Papers: Pigment-Binder-Air and Fiber-reinforced Pigment-Binder-Air Composites

Paper coatings are highly pigmented, porous coatings. In terms of the pigment volume concentration relationships used in paints, the pigment volume concentrations of paper

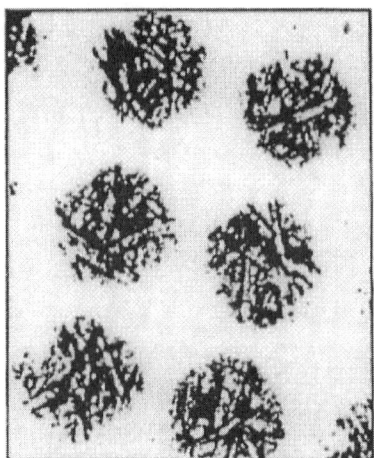

Figure 2. Reproduction of halftone dots on uncoated and coated papers, respectively.

coatings range from 60% (80/20 by weight) to greater than 90% (96/4 by weight). In other words, they are well above the critical pigment volume concentrations (CPVC's) of various pigments used in paper coatings. Figure 3 shows the surface profiles of No. 1 Clay/styrene-butadiene (S/B) latex coatings on polyester films at the ratios ranging from a pure latex coating (0/100) to an almost pure pigment coating (99/1 by weight). Figure 4 shows the coating opacity as a function of pigment concentration. Figure 5 shows the coating gloss vs. PVC. The opacity, brightness, and gloss of coating increase with increasing pigment concentrations above the CPVC which is about 50 to 55% by volume, that is, 72/28 to 76/24 by weight. It is not surprising to see that the opacity and brightness increase, as coating porosity increases with increasing pigment concentration, but it is very unique for paper coatings that the gloss increases with increasing pigment concentration above the CPVC, that is, the gloss increases with decreasing latex level. This binder level-coating gloss relationship has been a topic of intense research in paper coatings. As mentioned already, the pigment concentrations of paper coatings are well above the critical volume concentration where the coating porosity starts to become part of the coatings, the opacity suddenly increases, and the gloss turns around and starts to increase. Therefore, paper coatings are unquestionably pigment-binder-air composites. Figure 6 shows the cross-section of a coated paper. It is quite obvious from the figure that the structure of coated paper is a fiber-reinforced pigment-binder-air composite.

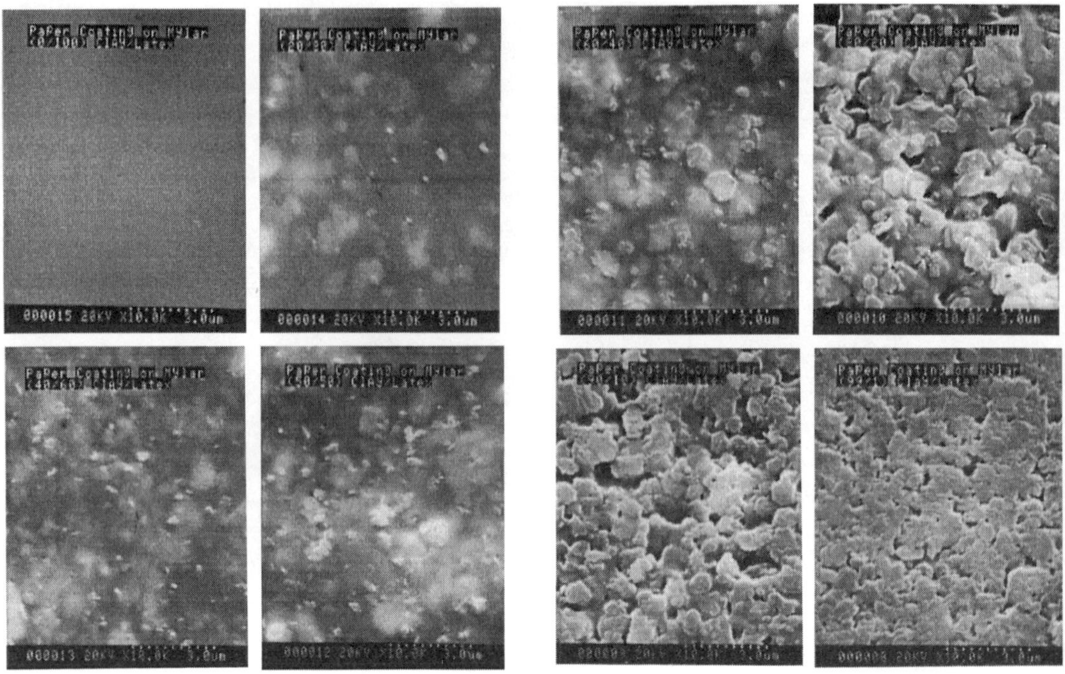

500

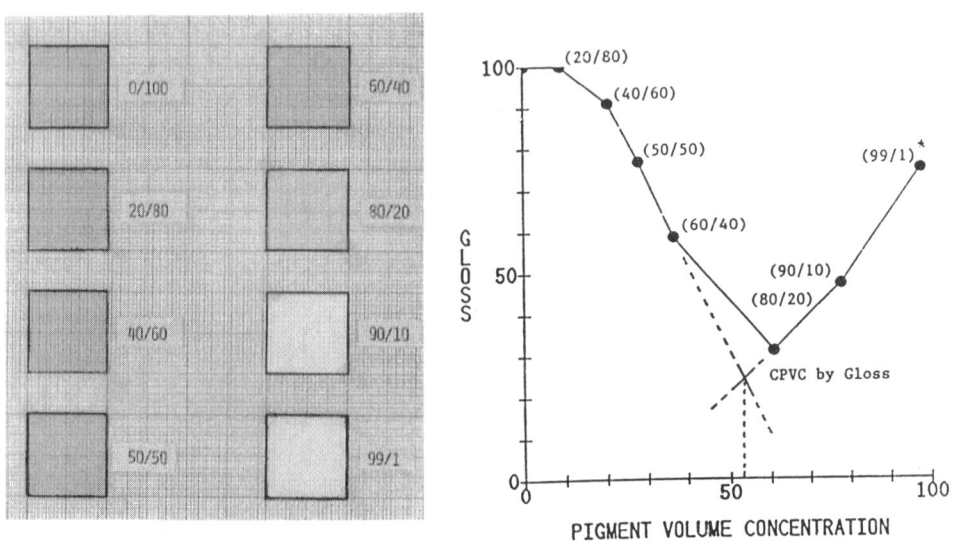

Figure 3. The surface profiles of No. 1 Clay/S/B latex coatings on polyester films: 0/100 @ Top-1, 20/80 @ T-2, 40/60 @ Bottom-1, 50/50 @ B-2, 60/40 @ T-3, 80/20 @ T-4, 90/10 @ B-3, and 99/1 @ B-4.

Figure 4. The opacity of No. 1 Clay/S/B latex coatings on a graph paper as a function of the pigment to latex ratios by weight.

Figure 5. Gloss vs. PVC
* by weight

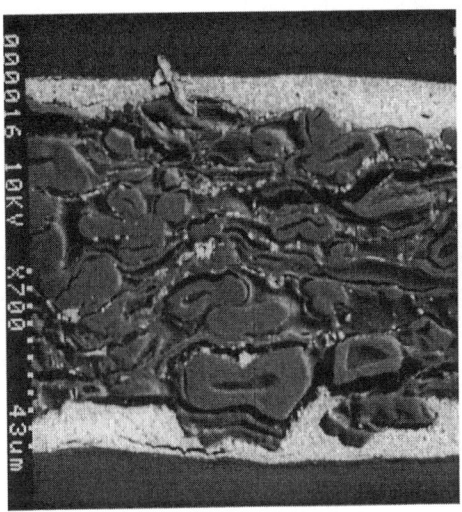

Figure 6. The cross-section of a coated paper

4. Paper Coating Formulations and Wet Coating Properties [3,5,6]

Basic materials used in paper coatings are pigments, binders, and some additives, such as viscosity modifiers, water-retention aids, and lubricants, and water is a carrier. Coating grade pigments are premium clay, No. 1 clay, No. 2 clay, precipitated calcium carbonate, fine and coarse ground calcium carbonates, and titanium dioxide. These pigments are dispersed in water with proper types and amounts of dispersants, such as tetra-sodium pyrophosphate (TSPP) and sodium polyacrylates. The optimum dispersant level corresponds to the minimum low-shear viscosity of a slurry system, at which the slurry is fully dispersed or deflocculated. Recently, a variety of structured pigments, such as calcined clays and chemically flocculated clays have been introduced to the paper coating industry. Because pigments are the major components of paper coatings, the convention of specifying their composition is to express all components other than pigments by weight based on 100 parts pigments, e.g., 100 Pigments/15 Binders/0.5 Thickener/0.5 Lubricant/0.1 Defoamer. In some cases, all components including the pigments are normalized to 100% by weight.

Both natural and synthetic binders are used in paper coatings. Natural binders are starch, soy-protein, and casein. Synthetic binders are styrene-butadiene (S/B) latexes [7], styrene-acrylate (S/A) latexes, polyvinyl acetate (PVAc) latexes [8], acrylic latexes [9], vinyl-acrylic latexes, and polyvinyl alcohols (PVA) [10]. The binder level varies depending on the type of printing methods: 8-15 parts/100 parts pigments for letterpress, 15-20 parts for sheet offset, 10-16 for web offset, and 4-10 parts for rotogravure printing. Also, the binder level depends on the type of binders used. The lower binder level is needed for the stronger binder.

502

In addition to dispersants already mentioned, various additives, such as viscosity modifiers, water-retention aids, insolubilizers and crosslinking agents for natural binders and latexes, foam control agents, lubricants, and various bases, are added to the coating formulations, as needed.

Coating formulation solids vary depending on the type of coating processes. Air-knife coating requires low solids (35 to 50%), while blade coating solids range from 50% to 70% or higher. The current trend is to increase coating solids as high as possible, pushing the blade coating solids above 70%.

Paper coating formulations are designed to produce the best possible coated papers and paperboards at acceptable costs without problems associated with coating operations, coater runnability, etc. Viscosity modifiers, water-retention aids, anti-foamers/defoamers, lubricants, etc. are judiciously used for the above-stated purposes.

5. The Important Properties of Coated Papers and Paperboards

The following properties of coated papers and paperboards are important for their appearance or printability, and some are also important for special applications:

a. Surface Strength or Pick Strength (IGT Dry Pick, Prufbau Dry Pick, Vandercook Proofing Press, etc.)
b. Water Resistance or Wet Pick Strength (Finger Rub, IGT Wet Pick, Prufbau Wet Pick, Adams Wet Rub, Taber Wet Rub, etc.)
c. Gloss (a measure of the specular reflection at the angle of incident angle)
d. Brightness (the reflection of blue light peaking at a wavelength of 475 nm)
e. Opacity (the amount of light transmitted by paper)
f. Smoothness (levelness)
g. Porosity
h. Compressibility
i. Stiffness
j. Ink Receptivity (K&N ink and Croda ink tests)
k. Mottle (non-uniform print appearance)
l. Water Repellency (the ability of coated paper to remove surface moisture and thereby allow ink transfer).
m. Back Trap Mottle (mottle caused by the transfer of printed ink from the paper back to the subsequent blankets)
n. Web Offset Blister Resistance (the ability of printed paper to withstant blistering during ink drying)
o. Water Absorption
p. Glueability (water-based glue, hot-melt glue, etc.)

Some of the above-listed properties are routinely tested for quality control and the others are sometimes evaluated for specific applications.

6. The Effects of Latexes on the Properties of Wet Coating Formulations and Coated Papers and Paperboards

6.1. THE COLLOIDAL PROPERTIES OF LATEXES AND THEIR INTERACTIONS WITH VARIOUS COATING COMPONENTS [11-18]

Since carboxylated latexes were introduced to the paper coating industry in the early 1960's, they have been most widely used as synthetic binders in paper coatings. The carboxylated latexes are not only compatible with various paper coating pigments and natural binders, but also improve the overall colloidal stability of coating formulations. Polyvinyl acetate latexes (PVAc) are very stable by themselves, but they interact with clay particles in the coating formulations via hydrogen bonding between alcohol groups on PVAc latex particle surfaces and silanol groups on clay particle faces. Because of these interactions, the viscosity of coating formulations containing clays and PVAc latexes is higher than that of those coating formulations containing non-interacting latexes, such as carboxylated latexes. Such pigment-binder interacting coating formulations result in more open coatings. In fact, coating rheology, immobilization concentrations, holdout, fiber coverage, structure, brightness, opacity, gloss, etc. are strongly affected by both the colloidal properties of latex particles and their interactions with the pigment particles and other additives, such as co-binders and thickener molecules, in the coating formulations. For this reason, controlling the colloidal behaviors of various coating components and their colloidal interactions in the coating formulations is one of the most important paper coating formulation technologies controlling the properties of coating formulations and coated properties.

Controlled destabilization or wet coating structure formation of paper coating formulations by the use of electrolytes or polymeric flocculants alters their rheological properties, but may lead to better fiber coverage and more porous, bulkier coatings. Figure 7 shows the transmission electron micrographs of the cryo-microtomed cross-sections of the coating formulations containing different amounts of electrolytes, respectively: control vs. additional electrolyte added. These micrographs clearly show the effect of electrolyte addition on the state of the coating dispersions. While the control dispersion is fully dispersed, the dispersion containing an added electrolyte caused clay particles to be destabilized and flocculated without latex particles involved. The reason for this situation was that the latex particles studied were much more stable than the clay particles. However, ideally speaking, we would prefer forming wet coating structures involving both clay and latex particles. Either pigment-interacting latex particles or polymeric species interacting with both latex and pigment particles would provide wet coating structures containing both latex and pigment particles. Figure 8 shows the effect of electrolyte addition on the fiber coverage and smoothness of coated papers. As can be seen from the pictures, the addition of electrolytes improves both fiber coverage and coating smoothness.

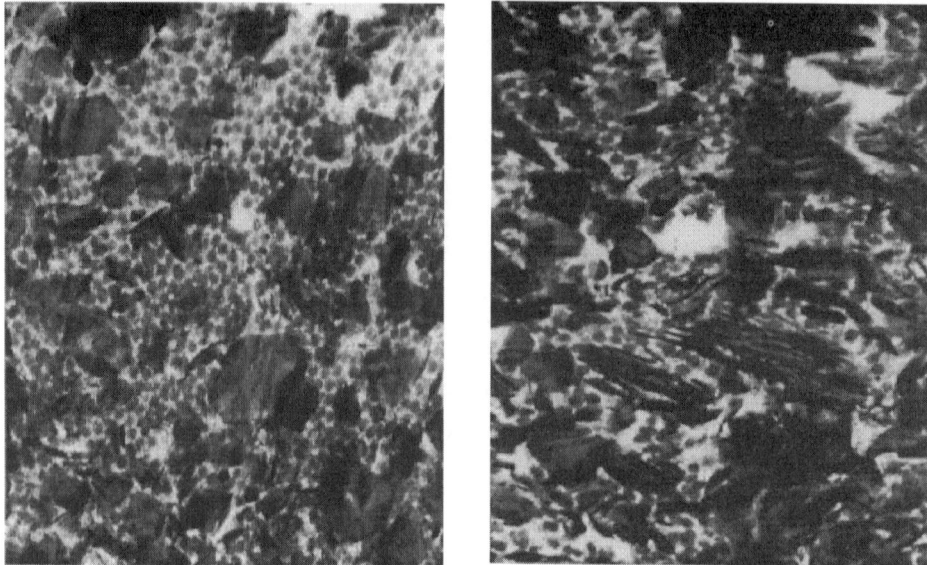

Figure 7. Transmission electron micrographs showing the cryo-microtomed cross-sections of the coating formulations containing different amounts of electrolytes, respectively: control vs. additional electrolyte.

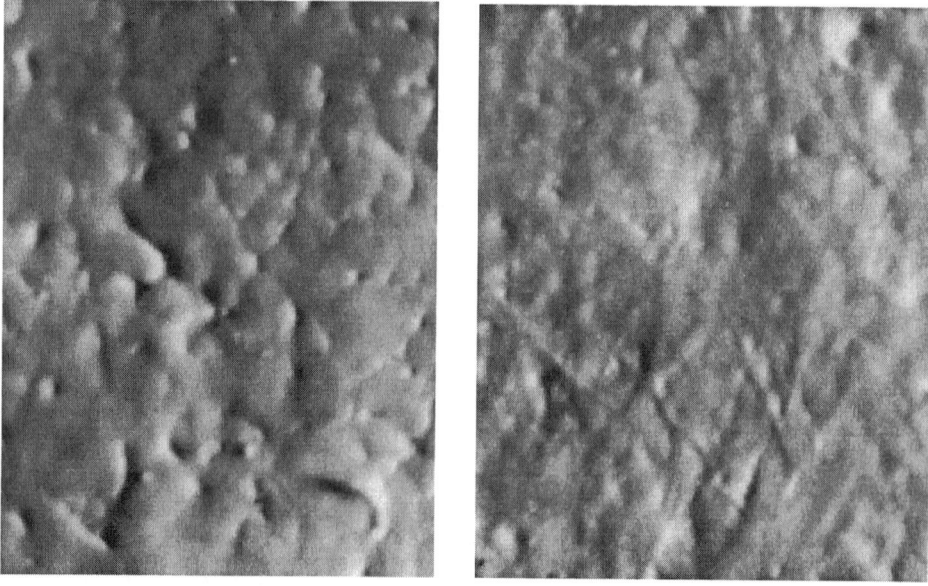

Figure 8. Scanning electron micrographs showing the effect of electrolyte addition to paper coating formulations on fiber coverage and coating smoothness: control vs. additional electrolyte.

6.2. THE EFFECT OF LATEXES ON THE RHEOLOGY AND RUNNABILITY OF PAPER COATING FORMULATIONS [19-22]

The rheology of paper coating formulations is affected by the physico-chemical and colloidal properties of latexes, such as colloidal stability, mechanical stability, particle size and size distribution, vinyl acid type and amount, surface functional groups and amounts, pH, particle deformability, etc. Figure 9 shows the effect of the colloidal properties of latexes and paper coating formulations on high-shear rheology. Figures 10 and 11 show the effects of latex particle size and size distribution, respectively, on the high-shear rheology of paper coating formulations. The high-shear rheology also improves with increasing latex carboxylations and pH. It has been found that the high-speed blade runnability of paper coating formulations improves as their high-shear viscosity decreases. Therefore, the high-speed blade runnability of paper coating formulations improves with increasing colloidal stability of latexes, decreasing latex particle size, increasing small particle size fractions of bimodal latexes, increasing latex carboxylation and pH, etc.

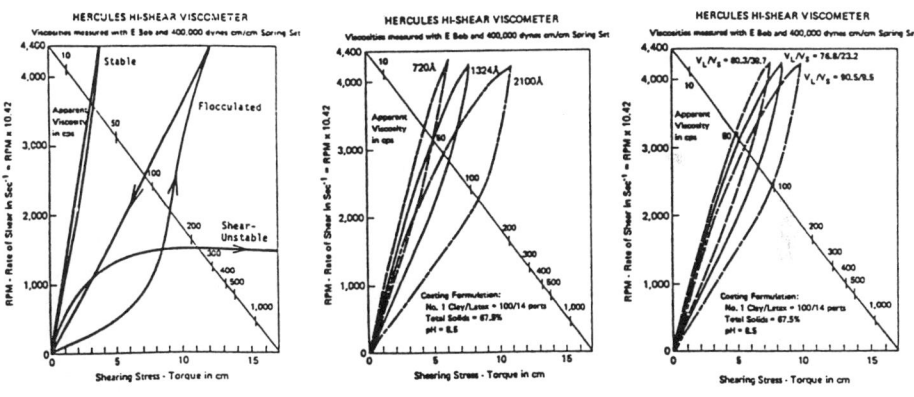

Figure 9. The effect of the colloidal properties of coating formulations on high-shear rheology [19]

Figure 10. The effect of latex particle size on high-shear rheology [20]

Figure 11. The effect of the small to large particle size ratios of bimodal latexes on high-shear rheology [20]

6.3. PAPER COATING PROPERTIES VS. S/B LATEX COMPOSITION

Figure 12 shows the wet rub, pick rating, varnish holdout, and gloss of coated paper vs. S/B latex composition [23]. Although this study was done more than 40 years ago, the results are still very valuable. If the S/B composition is converted to the corresponding Tg's and various coating properties are re-plotted against Tg, the property-Tg relations

506

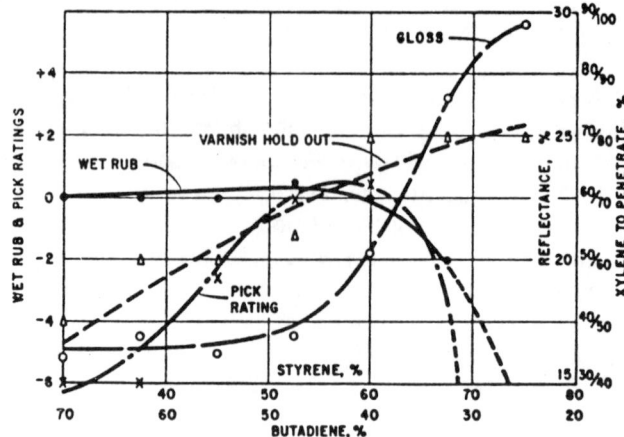

Figure 12. The properties of coated paper vs. S/B latex composition: wet and dry pick, gloss and varnish
holdout vs. S/B ratios [23]

will become universal, that is, they will be applicable to all types of latex polymers as a
function of Tg's. The pick strength is optimum over a narrow range of Tg's (0-20°C),
whereas the wet strength is generally favored by soft latexes. Coating gloss vs.
composition or Tg indicates that there is a transition composition or Tg for gloss. This
relationship is understood in terms of coating shrinkage due to latex particle deformation.

6.4. THE VISCOELASTIC PROPERTIES OF PAPER COATINGS [24,25]

The viscoelastic properties of paper coatings are directly dependent on both those of latex
polymers used as binders and the structure of pigment-binder-air composite coatings. The
shear modulus G' of paper coatings below the Tg of the latex polymer is mainly
dependent on their air-void volume, while the coating modulus above the latex polymer
Tg is affected by both the amount and modulus of the latex polymer and the composite
coating structure, as shown in Figure 13 [25]. The characterization of paper coatings by
dynamic mechanical spectroscopy not only provides information on the properties of
paper coatings, such as calenderability, stiffness, blister resistance, rotogravure
printability, etc., but also is capable to assess their composite coating structure.

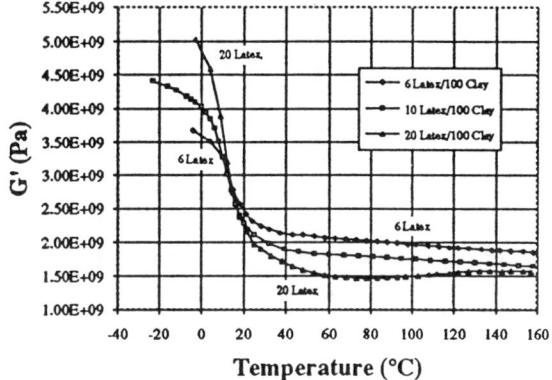

Figure 13. Dynamic mechanical spectra (G') of paper coatings containing different amounts of a latex: 6,10, and 20 parts per 100 parts No. 1 Clay [25]

6.5. BINDING STRENGTH VS. LATEX PARTICLE SIZE, PARTICLE SURFACE FUNCTIONALITY, AND MOLECULAR ARCHITECTURE

It is well known that dry pick strength increases with decreasing particle size of latexes and increasing latex carboxylation, as shown in Figures 14 and 15. Dry and wet pick strengths of paper coatings are differently affected by a crosslinking density of S/B latex polymers [26]. Figure 16 shows that dry pick increases with increasing gel content of S/B latex polymers, but wet pick peaks at lower gel contents.

The binding strength of latexes in paper coatings is influenced by their chemical and structural aspects in the fiber-reinforced pigment-binder-air composites. The chemical aspect can be controlled by the surface functionality of lat ϲ particles which can promote

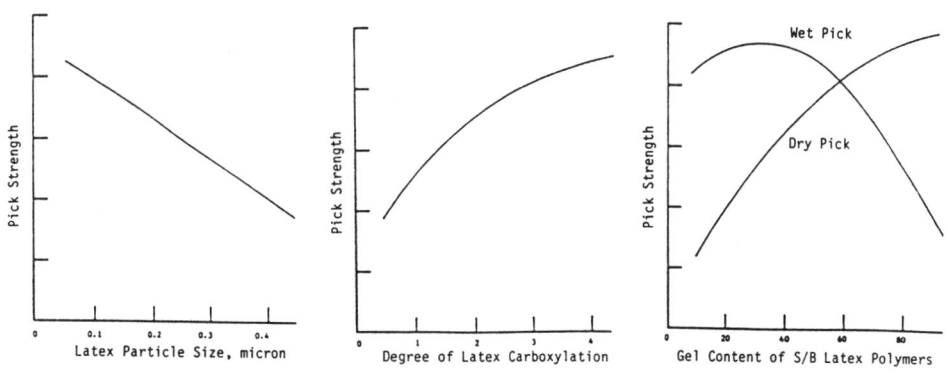

Figure 14. Pick strength vs. latex particle size.

Figure 15. Pick strength vs. latex carboxylation.

Figure 16. Dry and wet pick vs. latex polymer gel content.

508

broadly-defined acid-base interactions, while the structural aspect can be affected by both the viscoelastic properties of latex polymers and the properties of wet paper coating formulations. Undoubtedly, this area will be continuously investigated in search of better and better binding latexes for paper coatings.

6.6. COATING GLOSS VS. LATEX PROPERTIES [23,27-35]

The optical properties of paper coatings have been already discussed in section 3. As stated already, the opacity and brightness of paper coatings are strongly affected by the pigment volume concentrations: the higher the pigment volume concentrations, the greater the porosity (air-voids) and, in turn, the higher the opacity and brightness. Although opacifying pigments such as titanium dioxide can be used to enhance paper coating opacity, the light-scattering from the air-voids within the coatings is the main source of the paper coating opacity. Since coating gloss is one of the most important properties of coated papers, the subject will be discussed here in more details.

The gloss of coated papers is largely dependent on their surface smoothness: the smoother the surface, the higher the gloss. Paper coating gloss decreases with increasing binder level to the CPVC, as shown in Figure 5. Also, coating gloss depends on latex polymer Tg and drying temperature. Figure 17 explains why the coating gloss is affected by binder level, latex Tg, and drying conditions.

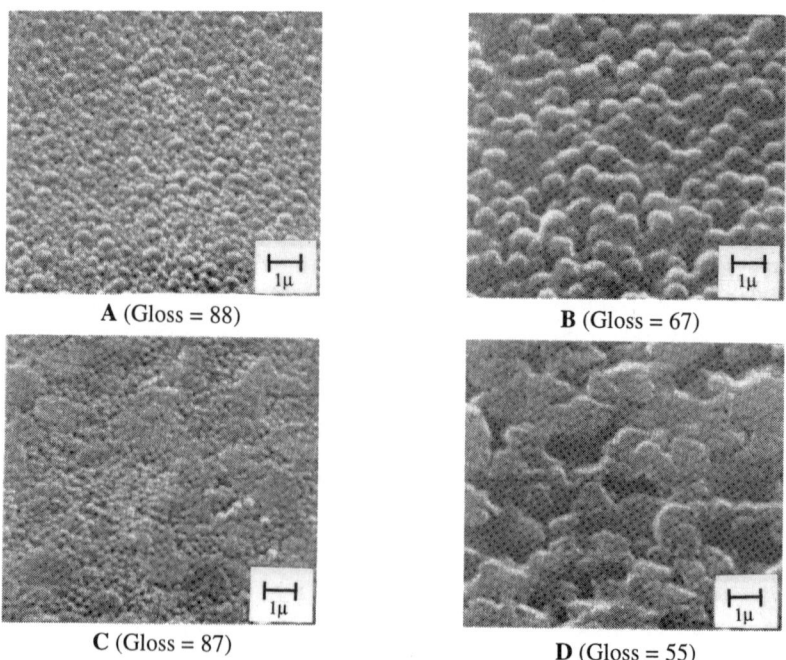

A (Gloss = 88) B (Gloss = 67)

C (Gloss = 87) D (Gloss = 55)

Figure 17. Scanning electron micrographs showing the effect of latex film formation on coating surface roughening: plastic pigment/non-film forming and film forming latexes, respectively, (top left and right) and clay/non-film forming and film forming latexes, respectively, (bottom left and right) [29].

Since the coating gloss is a measure of surface smoothness, anything which causes surface roughening or smoothing will affect the coating gloss. Flocculation or destabilization of coating formulations may cause a reduction in gloss. Plastic pigments improve coating gloss by reducing coating shrinkage and making coatings easier to finish. For this reason, hollow plastic pigments further improve coating gloss over that of the conventional plastic pigments. Coated papers are mostly supercalendered, while coated paperboards are gloss-calendered. Soft-nip thermofinishing [36] is also used.

6.7. PRINTABILITY VS. LATEX PROPERTIES [37]

Printability can be defined by the combination of press runnability and print quality. Although press runnability and print quality are closely interdependent, the former is more concerned with how fast and how long a printing press can be run trouble-free, while the latter is more concerned with the quality of reproduction, print gloss, print mottle, back trap mottle, water repellency, missing dots, etc. Some of the important properties affecting print quality are discussed below.

6.7.1. *Ink Gloss [26]*
Ink gloss depends not only on the unprinted initial sheet gloss, but also on the surface porosity and ink-binder interactions. Generally, the higher the initial sheet gloss, the higher the ink gloss. Ink gloss increases with decreasing latex particle size, suggesting that the ink gloss is higher with tighter coatings. Ink gloss increases with increasing gel content. This result indicates that the ink gloss improves with decreasing interactions between latex polymer and ink solvent. This finding is also supported by the effect of acrylonitrile content of latexes on ink gloss.

6.7.2. *Ink Receptivity and Ink Absorption*
Ink receptivity and ink absorption are two important properties of coated papers to be printed. There are a number of factors affecting the ink receptivity and ink absorption of coated surfaces, but the most important factor is the binder level in the coatings: the lower the binder level, the higher the receptivity and absorption of ink. Among different types of latexes, PVAc latexes are more ink receptive than either S/B latexes or acrylic binders [8]. Generally, ink receptivity increases with increasing particle size and increasing hardness of latex polymers. Also, it can be increased by limiting the coalescence of latex particles [30].

6.7.3. *Web Offset Blister Resistance [38-41]*
Web offset printing requires a blister resistance of coated papers during drying the printed papers. PVAc latexes are more blister resistant that either S/B or acrylic latexes. Although PVAc latexes produce more porous coatings, the porosity is not the only reason for their excellent blister resistance. Their high thermal flow behaviors are responsible for blister resistance. Based on this finding, blister resistant S/B latexes were developed by reducing their crosslinking density. Many studies have been made on the effect of the gel content of S/B latex polymers on blister resistance. The blister resistance of S/B latexes improves with their decreasing gel content. However, as shown in Figure 16, their binding strength also decreases with decreasing gel. Therefore, blister

510

resistance and binding strength are compromised for the development of blister resistant S/B latexes. Figure 18 shows the effect of gel content on the blister resistance and binding strength of S/B latexes.

6.7.4. *Rotogravure Printability [11,42,43]*

Rotogravure printability depends on both the transfer of ink from recessed cells to paper at the printing nip and the reproduction of the transferred ink dots. Therefore, rotogravure printability requires good fiber coverage and compressibility of coated papers. As mentioned earlier, good fiber coverage and coating smoothness can be achieved by controlling the interaction of latexes with pigments or the colloidal stability of coating formulations, thus lowering their immobilization points [11-18]. Figure 19 shows the effect of latex polymer softness on missing dots: the softer the latex polymer, the better the rotogravure printability in terms of missing dots.

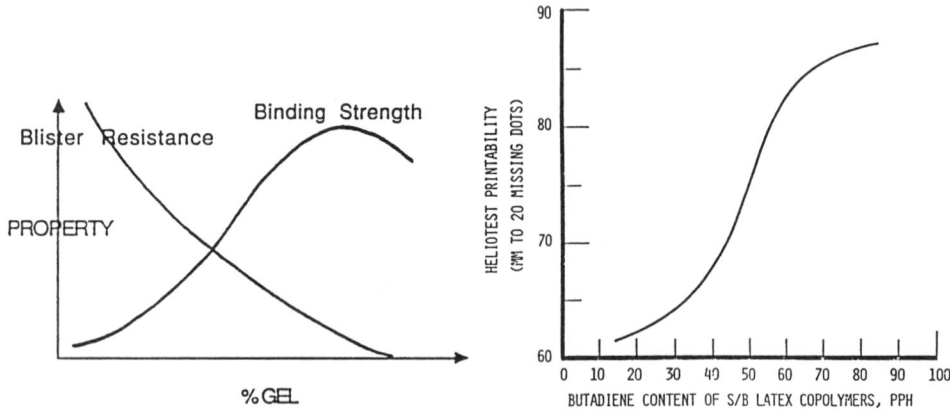

Figure 18. The effect of gel content on blister resistance and binding strength

Figure 19. Rotogravure printability vs. butadiene content of S/B latex copolymers [43]

7. Plastic Pigments [44-47]

Conventional plastic pigments [44,45] are carboxylated polystyrene latexes of various particle sizes ranging from 100 nm to 600 nm. They are widely used as partial replacement of inorganic pigments to achieve high gloss, brightness, ink receptivity, and blister resistance. Small particle size plastic pigments are better for both sheet and ink gloss, while large particle size plastic pigments are better for brightness, opacity, and ink receptivity. Recently, hollow sphere plastic pigments [46,47] have been introduced into the paper coating industry. They improve the opacity and finishability of paper coatings beyond those achieved by the conventional solid plastic pigments. Figure 20 shows a freeze-fractured surface of paper coating containing hollow plastic pigment particles.

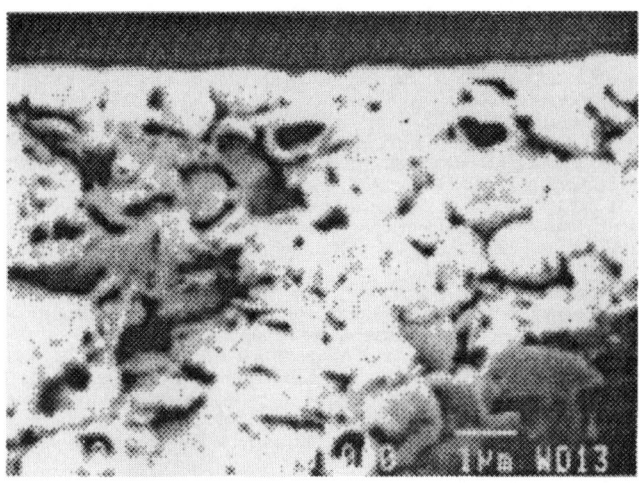

Figure 20. Scanning electron micrograph showing a freeze-fractured surface of paper coating containing hollow plastic pigment particles [47]

8. Futute Research Challenges

As already discussed, the use of latexes in paper coatings markedly improves coating formulation properties, coated paper and paperboard properties, and printability. Since they were first introduced in the late 1940's, latexes have been continuously upgraded in quality and performance and have kept pace with the needs of the paper coating industry. In the past, latex producers have successfully resolved may challenges: pigment compatibility in the 1950's, casein and protein compatibility in 1960's, blade runnability and web offset blister resistance in 1970's, high solids paper coatings in 1980's, double coatings, etc. Now, latex researchers have new challenges on hand: super-binder, super-runnability binder, high-glossing and easy-finishing binder, ink solvent-resistant binder, high ink gloss binder, print mottle-free binder, sole binder, binder migration control, coating structure control, better fundamental understandings of coating, drying, finishing, and printing processes, etc. The paper coating industry is a dynamic industry always moving forward to the cutting edge of science and technology. Consequently, latex researchers must be more creative and innovative than ever to meet the challenges.

512

9. References

1. Casey, J.P (ed) (1980) Pulp and Paper, 3rd. ed., Vol. 1 and 2, John Wiley & Sons, New York.
2. Booth, G.L. (1970) *Coating Equipment and Processes*, Lockwood, New York.
3. Hagemeyer, R.W. (1980) Pigment coating (Chapter 22), in J.P. Casey (ed.), *Pulp and Paper*, 3rd ed., Vol. 4, John Wiley & Sons, New York, pp. 2013-2187.
4. Bruno, M.H. and Walker, W.C. (1980) Printing (Chapter 23), in J.P. Casey (ed.), *Pulp and Paper*, 3rd ed., Vol. 4. John Wiley & Sons, New York, pp. 2191-2275.
5. Hagemeyer, R.W. (ed.) (1976) *Paper Coating Pigments*, 4th ed., TAPPI Monograph Series 38.
6. Sinclair, A.R. (ed.) (1975) *Synthetic Binders in Paper Coatings*, TAPPI Monograph Series 37.
7. Heiser, E.J. and Kaulakis, F. (1975) Styrene-butadiene latex (Chapter 3), in Sinclair, A.R. (ed.), *Synthetic Binders in Paper Coatings*, TAPPI Monographs Series 37, pp. 22-63.
8. Walsh, T.F. and Gaspar, L.A. (1975) Polyvinyl acetate latex (Chapter 5), in Sinclair, A.R. (ed.), *Synthetic Binders in Paper Coatings*, TAPPI Monographs Series 37, pp. 98-119.
9. Latimer, J.J. and DeGroot, H.S. (1975) Acrylic binders (Chapter 6), in Sinclair, A.R. (ed.), *Synthetic Binders in Paper Coatings*, TAPPI Monographs Series 37, pp. 120-136.
10. Jerzec, R.C. and Cogan, G.P. (1975) Polyvinyl alcohol (Chapter 4), in Sinclair, A.R. (ed.), *Synthetic Binders in Paper Coatings*, TAPPI Monographs Series 37, pp. 64-97.
11. Lee, D.I., Louman, H., and Moss, M.H. (1975) Development of total synthetic binder for rotogravure paper, *Tappi J.* **58** (9), pp. 79-82.
12. Baumeister, M. and Kraft, K. (1981) *Tappi J.* **64** (1), p. 85.
13. Lee, D.I. (1981) Coating structure modification and coating hold-out mechanisms, TAPPI Coating Conference Proceedings, TAPPI PRESS, Atlanta, pp. 143-153.
14. Beck, U., Gossens, J.W.S., Rahlwes, D., and Wallpott, G. (1983) Coating color structure and water retention, TAPPI Coating Conference Proceedings, TAPPI PRESS, Atlanta, pp. 47-54.
15. Whalen-Shaw, M. (1989) Coating structure - Part I: A mechanistic view to the development of wet coating structure, TAPPI Coating Conference Proceedings, TAPPI PRESS, Atlanta, pp. 9-19.
16. Lepoutre, P. and Lord, D. (1990) Destabilized clay suspensions: Flow curves and dry film properties, *J. Colloid Interface Sci.* **134**, pp. 66-73.
17. Chonde, Y., Roper, J.A., and Lee, D.I. (1991) The importance of electrokinetic measurements for understanding the colloidal phenomena occurring in paper coating formulations, Symposium on Paper Coating Fundamentals, TAPPI PRESS, Atlanta, pp. 21-33.
18. Salminen, P. and Fors, S. (1992) Fundamental approaches for optimizing fibre coverage in blade coating, TAPPI Coating Conference Proceedings, TAPPI PRESS, Atlanta, pp. 7-22.
19. Lee, D.I. (1983) The effect of latexes on the rheology of high solids paper coatings, presented at the Panel Discussion, 1983 TAPPI Coating Conference and TAPPI Korea **15** (2), pp. 7-23.
20. Van Gilder, R., Lee, D.I., Purfeerst, R., and Allswede, J. (1983) High solids latexes for paper coatings, *Tappi J.* **66** (11), pp. 49-53.
21. Roper, J.A. and Attal, J.F. (1993) Evaluation of coating high speed runnability using pilot coater, rheological measurements, and computer modeling, *Tappi J.* **76** (5), p. 55.
22. Roper, J.A. (1995) An introduction to the rheology of paper coating, TAPPI Coating Binders Short Course.
23. Taber, D.A. and Stein, R.C. (1957) Effect of latex variables on properties of coating colors and coated papers, *Tappi J.* **40** (2), pp. 107-117.
24. Yamaguchi, Y., Ishikawa, O., Yamashita, T., and Tsuji, A. (1993) A study of viscoelastic properties of coated layer in paper coating, *Advanced Coating Fundamentals*, TAPPI PRESS, Atlanta, pp. 51-58.
25. Kan, C.S., Kim, L.H., Lee, D.I. and Van Gilder, R.L. (1996) Viscoelastic properties of paper coatings: Structure/property relationship to end use performance, TAPPI Coating Conference Proceedings, TAPPI PRESS, Atlanta, pp. 49-60.
26. Uchida, A. (1984) *Kami Pa Gikyoshi* **38** (12), p. 1196.
27. Pinder, J.A. (1972) TAPPI CA Report No. 36.
28. Gate, L., Windle, W., and Hine, M. (1973) *Tappi J.* **56** (3), p. 61.
29. Lee, D.I. (1974) A fundamental study on coating gloss, TAPPI Coating Conference Proceedings, TAPPI PRESS, Atlanta, pp. 97-114.
30. Lee, D.I. (1982) Development of high-gloss paper coating latexes, TAPPI Coating Conference Proceedings, TAPPI PRESS, Atlanta, pp. 125-135.
31. Watanabe, J. and Lepoutre, P. (1982) A mechanism for the consolidation of the structure of clay-latex coatings, *J. Appl. Polym. Sci.* **27** (11), pp. 4207-4219.
32. Lee, D.I. and Hendershot, R.E. (1986) Development of low glossing paper coating latexes: Theories and concept, TAPPI Coating Conference Proceedings, TAPPI PRESS Atlanta, pp. 31-34.
33. Groves, R., Penson, J.E. and Ruggles, C. (1993) Styrene-butadiene latex binders and coating structure, TAPPI Coating Conference Proceedings, TAPPI PRESS Atlanta, pp. 187-205.
34. Erikson, U. and Rigdahl, M. (1993) Difference in consolidation and properties of kaolin-based coating layers induced by CMC and starch, *Advanced Coating Fundamentals*, TAPPI PRESS Atlanta, pp. 19-30.
35. Stanislawska, A. and Lepoutre, P. (1996) Consolidation of pigmented coatings: Development of porous structure, *Tappi J.* **79** (5), pp. 117-125.
36. Vreeland, J.H. (1975) U.S. Patent 3,873,345 and (1978) U.S. Patent 4,112,192.

37. Van Gilder, R.L. and Purfeerst, R.D. (1994) Commercial six-color press runnability and the rate of ink-tack build as related to the latex polymer solubility parameter, TAPPI Coating Conference Proceedings, TAPPI PRESS Atlanta, pp. 229-241.
38. Hagymassy, J., Lee, D.I., Schmitt, J.A., Givens, S.P., and Haynes, L.U. (1978) An investigation of the web offset blister problem, *Tappi J.* **61** (1), p. 49.
39. Sekiguchi, S., Matsumoto, K., Izaki, N., and Uchida, A. (1977) TAPPI Coating Conference Proceedings, TAPPI PRESS, Atlanta, p. 17.
40. Yamawaki, K., Sasagawa, Y. and Tsuji, A. (1991) TAPPI Coating Conference Proceedings, TAPPI PRESS, Atlanta, p. 199.
41. Schowob, J.M., Guyot, C. and Richard, J. (1991) TAPPI Coating Conference Proceedings, TAPPI PRESS, Atlanta, p. 207.
42. Fernandez, J.M., Petersen, W.F., and Koval, J.G. (1983) TAPPI Coating Conference Proceedings, TAPPI PRESS, Atlanta, p. 137.
43. Lee, D.I. (1984) The influence of latexes on rotogravure printability, presented at *the Panel Discussion*, 1984 TAPPI Coating Conference and TAPPI Korea **16** (2), p. 10.
44. Heiser, E.J. and Shand, A. (1973) *Tappi J.* **56** (1), p. 70.
45. Heiser, E.J. and Shand, A. (1973) *Tappi J.* **56** (2), p. 101.
46. Hemenway, C.P., Latimer, J.J., and Young, J.E. (1985) *Tappi J.* **68** (5), p. 102.
47. Brown, J.T. (1991) TAPPI Coating Conference Proceedings, TAPPI PRESS, PRESS, Atlanta, p. 113.

POLYMER COLLOIDS FOR BIOMEDICAL AND PHARMACEUTICAL

APPLICATIONS

C. PICHOT, T. DELAIR, A. ELAÏSSARI
Unité Mixte CNRS-bioMérieux
Ecole Normale Supérieure de Lyon
46, allée d'Italie
69364 Lyon Cedex 07, France

1. Introduction

Polymer colloids can be used either as models in academic research dealing with colloid phenomena or as dispersed material in a wide variety of industrial applications. One major reason for such a development is related to the basic understanding in the production of these latexes through an increasing number of free-radical heterogeneous polymerization processes as well as in their complete characterization in terms of molecular or colloidal properties. Two major reasons motived the attractive interest for such polymer colloids in biological, medical and pharmaceutical applications: i) their unique and versatile properties with regard to the control of the particle shape and morphology, size and size distribution, surface chemistry, polymer nature, etc., and ii) the huge progress in molecular biology during the last two decades resulting in a vast number of available biomolecules (monoclonal antibodies, peptides, DNA probes, enzymes, recombinant proteins etc.) and sophisticated methodologies (amplification techniques, protein engineering, capillary electrophoresis, etc.).

It should be reminded that these dispersed polymer materials can be obtained by conventional emulsion, emulsifier-free, micro or miniemulsion polymerization providing particles in the submicron size range (ca<1 μm) [1]. Other heterogeneous processes such as dispersion, precipitation polymerization or emulsification techniques (solvent extraction/solvent evaporation) usually lead to polymer particles mainly in the supermicron size range.

For several reasons, polystyrene latexes have been first and largely used in the biomedical field (especially because of the hydrophobic nature of the polymer favoring irreversible adsorption of antibodies). Earlier applications of polystyrene latexes in immunoassays date back to 1956 by Singer et al. [2]. Extensive reviews on the subject have been reported dealing with the advantages and drawbacks of such dispersed material as well as in their performances [3]. At the same time, numerous systematic studies have been reported on the control of the particle size and surface properties of polystyrene latexes whether they were produced in the presence or in the absence of emulsifiers [4, 5]. Since 1970, extensive research has been directed towards the

J. M. Asua (ed.), Polymeric Dispersions: Principles and Applications, 515–539.

516

preparation of various kinds of particles better adapted for a specific application as well as the understanding of the many complex phenomena related to the use of synthetic particles for in-vitro or in-vivo applications.

Regardless of the preparation technique, an adequate selection of the formulation recipe (upon varying the nature and concentration of a main monomer, initiator, surface-active agent, functional or reactive monomers, ionic strength and pH of the continuous medium) allows one to tailor different particle morphologies as illustrated in Figure 1.

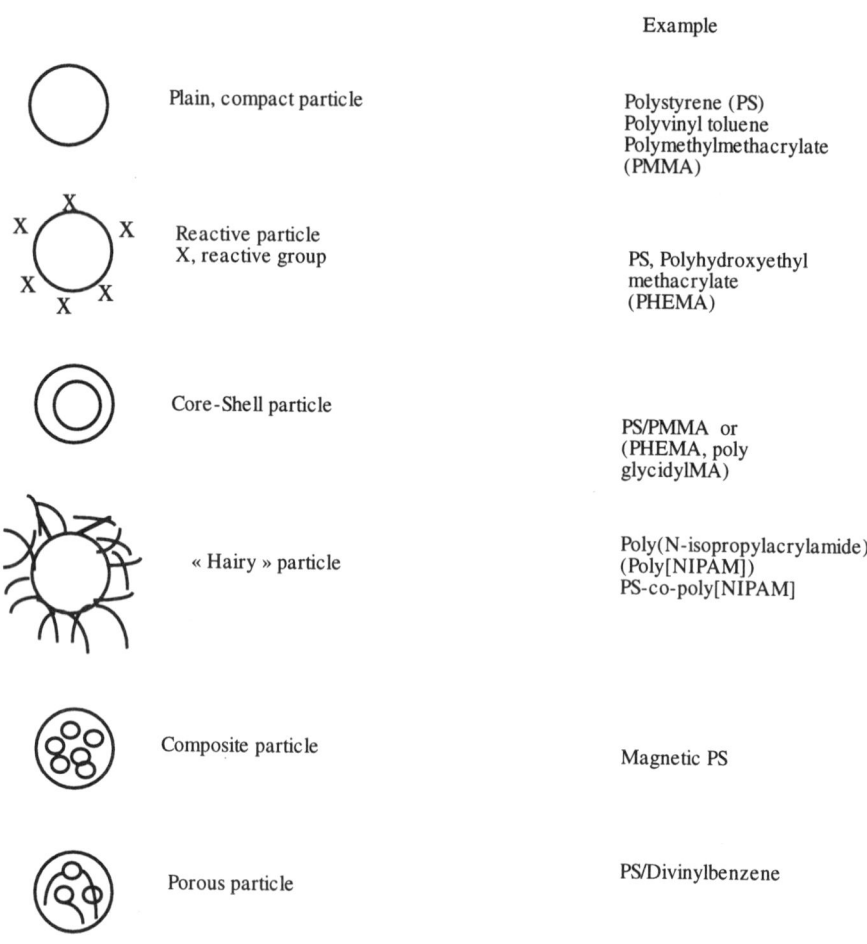

Example

Plain, compact particle — Polystyrene (PS) Polyvinyl toluene Polymethylmethacrylate (PMMA)

Reactive particle X, reactive group — PS, Polyhydroxyethyl methacrylate (PHEMA)

Core-Shell particle — PS/PMMA or (PHEMA, poly glycidylMA)

« Hairy » particle — Poly(N-isopropylacrylamide) (Poly[NIPAM]) PS-co-poly[NIPAM]

Composite particle — Magnetic PS

Porous particle — PS/Divinylbenzene

Figure 1: Different particle morphologies

Plain (or compact) microparticles have been extensively used, (particularly with polystyrene and derivatives), in a broad size range in between 20 nm to 100 μm. The development of underlined functionalized (or reactive) particles is of great importance for controlling the hydrophilic-hydrophobic balance of the water-polymer interface and also to introduce reactive surface groups able to covalently attach biomolecules. It appears that "hairy" particles (among them stimuli-responsive particles) can be produced mainly from poly(acrylamide) derivatives, providing novel applications in medicine and biotechnology. A particular attention has also been paid to the preparation of composite particles containing inorganic colloids, for instance ferrites, leading to magnetic particles offering numerous applications. Porous particles are also of interest, especially for chromatographic analysis. Finally, the preparation of biodegradable particles has become an active field of research since many criteria are required as regards to the compatibility of such colloids for in-vivo applications.

Concerning application of polymer colloids in the biological and pharmaceutical domains, a schematic categorization can be proposed as listed in Table 1. Four main domains may be considered using these polymer colloids as calibration standards, tracers, solid-phase supports or targets.

TABLE 1 : Survey of the main applications of polymer colloids
(adapted from ref. [6])

Main Use	Current Application	Nature of particles
Calibration Standards	Identification and enumeration of lymphocytes, virus, bacteria	Hydrophobic (R, F)
Tracer/Marker	Diagnostic assays NMR imaging	Hydrophobic (M, F, R)
Solid-phase support	Immunoassays / Cell separation Chromatography Haemoperfusion Cell activation Cell motility Enzyme immobilization	Hydrophobic (M, F, R) Hydrophilic (R)
Targeting	Drug delivery Gene therapy	Biodegradable (R, M)

R = reactive groups ; F = Label group ; M = magnetic

The present review aims at describing several aspects dealing with the use of polymer colloids in the biological and pharmaceutical fields :

- first, the main features which need to be taken into account, mostly as regards to colloidal and surface properties.

- second, the preparation of different kinds of monosized polymer particles: functionalized (including reactive), hydrophilic, biodegradable, labelled.

- finally, the immobilization of biological molecules onto these various particles.

This review will mostly cover the case of polymer particles in the colloidal range (i.e, nanoparticles) as produced by some of the above-mentioned heterogeneous polymerization processes. A great number of extensive papers already described the preparation and the applications of particles in the supermicron range especially those produced by the dynamic swelling method as reviewed by Ugelstad and co-workers [7,8,9].

2. Structural Requirements Concerning Polymer Colloids

When dealing with polymer colloids for biological applications, a large number of variables should be considered with respect to the molecular, surface and colloidal properties of the particles. For a given application, any selected latex must be characterized as completely as possible. Table 2 gives a list of several properties which need a special attention, together with the main characterization techniques.

However, prior to particle characterization, latex cleaning is often a preliminary task in order to get free (and sometimes to analyse) residual monomer, surfactants, water-soluble oligomers and polymers, and others impurities (such as initiator traces, electrolytes, additives, etc.). An extensive review of the subject can be found in the literature with emphasis on specific advantages and drawbacks of each cleaning technique [10].

Concerning the chemical structure of the particle, glassy polymers (at room temperature) such as polystyrene, polymethylmethacrylate, polyvinyltoluene and various copolymers of styrene with butadiene, hydroxyethylmethacrylate, glycidyl methacrylate are preferred over soft polymers. In order to avoid sedimentation upon storage, the choice of a polymer should take into account the particle density relative to that of the buffer continuous phase.

The control of particle size and size distribution is a very important requirement since it defines the available surface area for the attachment of biologically-active macromolecules. In the nanometric size range, batch polymerizations are currently used for producing particles between 0.08-0.4 µm. For larger particles (up to 1 µm) seeded polymerizations are more suitable since, in addition, a self sharpening effect causes the size distribution to be narrower.

Monodispersity criteria is often required whether the latex particles are used as calibration standards or as solid-phase supports. Indeed, for sake of reproducibility, when using latexes in immunoassays (especially for agglutination tests) it is crucial that each particle bears the same number of antibodies (or antigens). Polymerization recipes should therefore allow a nucleation stage as short as possible.

The presence of surface charges (in the absence of polymer stabilizers) is necessary for ensuring efficient colloidal stability of the latex particles during synthesis, upon storage and when they are mixed with biological fluids (exhibiting often significant ionic strength). Surface charges are usually originated either from initiator fragments or from ionic monomers (carboxylic acid monomers, for instance) introduced into the formulation recipe. The nature and the density of surface charges must be considered, as a function of the physicochemical properties of the biomolecules as well as of the kind of biological application. Amphoteric latex particles can be adequately used for giving, depending on pH, negative or positive surface charges.

Reactive groups on the latex surface are often required, especially for the immobilization of biomolecules containing mainly COOH, NH_2 and SH groups. They can be introduced either during the latex synthesis or later, by functionalization of preformed particles. For the first route, many functional monomers are now available (or can be synthesized) bearing carboxyl, activated esters, sulfonate, hydroxy, amide, aldehyde, quaternary amine or polyoxyethylene groups. The characterization of such latexes can be performed using complementary techniques providing the precise number

TABLE 2 : Variables to be considered in the design of polymer colloids

Properties	Characterization Techniques
Shape of particle Particle size and size distribution	Electron microscopy Atomic force microscopy Quasielastic light scattering
Surface charge density	Conductometry Potentiometry
Surface potential	Electrophoresis Dielectric spectroscopy
Number of reactive groups	Colorimetry Fluorescence
Interface polarity	Contact angle X-Ray Photoelectron Spectroscopy (XPS)
Surface morphology	XPS, Secondary Ion Mass Spectroscopy (SIMS), NMR
Stability against electrolyte, temperature	Coagulation kinetics
Composition of latex serum	NMR, SEC, HPLC

of reactive groups per particle. Several problems are not yet solved, concerning: the surface morphology, for instance in "hairy"-like particles; the location and the distribution (homogeneous or patchy) of polar or charged end- groups; the deprotection of some reactive groups; etc.

3. Reactive Particles

In biomedical applications, latex particles are used as carriers of biological species such as enzymes, antibodies, DNA fragments and peptides. In some cases, for particular purposes, it may be required to tether the biomolecule to the polymer support via the formation of a covalent bond, rather than by a mere physical adsorption. In this respect, functional groups should be present at the surface of the microspheres and synthetic methods of preparation of functionalized particles have to be developed along with characterization techniques.

Most of the coupling methods of biological molecules to synthetic supports rely on a chemical reaction involving the ε-lysine amino groups of the former since these groups are most often available, (historically this method has been developed for proteins). However, thio, hydroxy and carboxy groups can possibly be used as well, but at a much lower extend than their aminated counterparts. Thus, many kinds of particles have been produced bearing functional groups susceptible of reacting with a primary amine, either directly, such as aldehyde, epoxy or chloromethyl groups, or indirectly, by forming a covalent bond after a so-called activation step, as it is the case for carboxyl and hydroxyl groups.

Carboxylated particles, the most popular ones, are commercially available at different particle sizes, functional group densities and even coloured. They can be synthesized by copolymerization of (meth)acrylic acids with a great variety of monomers like styrene or acrylates; a review by Blackley [11] dealt with the main issues in handling ionogenic monomers in emulsion polymerization.

Two main processes are usually employed for the syntheses of functional particles by copolymerization of a "basic" monomer, styrene for instance, with a functional monomer. The simpler process is the batch one, in which all the comonomers are loaded and polymerized in one single step, as we described for the syntheses of thio-functionalized latexes [12] and aminated latexes [13, 14]. A two stage process, the so-called core-shell process, can also be used to obtain functional particles: it consists in the polymerization of a mixture of the functional monomer and the basic monomer at the surface of either preformed particles (seed polymerization) or by adding the monomer mixture at high conversion during the synthesis of the particles (shot polymerization) [15, 16].

Actually any kind of functional particles can be obtained by copolymerization of an appropriate monomer as reported in Table 3.

The major problem, when dealing with functional comonomers, is to assess the amount of functional groups actually anchored at the surface of the beads. Thus, specific titration techniques have to be developed on cleaned latexes (i.e, free of polyelectrolytes) [27]. Conductometric or potentiometric titrations are well known methods to quantify available carboxyl groups. In the author's laboratory, chemical methods have been developed to quantify amino groups [13, 14], and radiochemical techniques for aldehyde and thio groups [21, 12]. The principle is to use a product capable of reacting specifically with the functional groups to be titrated on the particle surface. Then, quantification is achieved either by measuring the optical density at a proper wavelength in the aqueous phase (or directly on the latex), or by measuring of the radioactivity bound onto the latex sample after reaction of the nanosphere with a radiolabelled specific counterpart.

From our results, a two-step functionalization process was more efficient to obtain a high incorporation yield of the functional monomer, which can be introduced in the second step, when the formation and growth of the particles are almost complete.

Some non specific characterization methods, mainly spectroscopic or spectrometric, can also lead to the determination of the number of functional groups at the surface of the particles. NMR can prove very useful : for instance, ^{1}H NMR spectroscopy of dried polystyrene latex particles dissolved in deuterochloroform allows

TABLE 3: List of various monomers with their corresponding functional group.

Nature of monomer	Name	Functional group	Reference
Ionic	N-trimethyl-N-ethylmethacrylate ammonium.	$N(CH_3)_3{}^+Cl^-$	[17]
	3-methacrylamidinopropyltrimethyl ammonium chloride.	"	[18]
	vinyl benzyltrimethyl ammonium chloride.	$-N(CH_3)_3{}^+Cl^-$	[19]
	vinylbenzyl isothiouronium chloride	$-S-C(NH_2)_2{}^+Cl^-$	[12]
	4 - vinyl benzyl amine hydrochloride.	$-NH_3{}^+,Cl^-$	[13, 14]
Nonionic Hydrophobic	4-Vinyl benzylchloride	-Cl	[20]
	4-Vinyl benzaldehyde	-CHO	[21]
	Vinyl benzylamine	-NH2	[22]
	Glycidyl methacrylate	$-CH-CH_2$ $\diagdown \!\! O \!\! \diagup$	[16]
	4-Vinylbenzyltrifluoroacetamide	$-NHCOCF_3$	[22]
Nonionic Water-Soluble	Hydroxyethylmethacrylate	OH	[23]
	Acrolein	-CHO	[24, 25]
	6(-Methacryloxyloxy)hexyl β-D-cellobioside	saccharidic moiety	[26]

the determination of the overall amount of functional monomer incorporated in the particles, whereas ^{1}H NMR of dried polystyrene latex particles redispersed in d$_6$-DMSO, a poor solvent for polystyrene, gives access the amount of surface bound functional groups [22]. It is worth noting that NMR analyses are less sensitive than colorimetric titrations as observed by Brooks et al., who compared the chemical, radiochemical and ^{1}H NMR analytical techniques in the determination of surface aldehyde groups on surfactant free polystyrene/polyacrolein latexes [28]. XPS (X-Ray Photo Electron spectroscopy) is another technique to demonstrate the incorporation of a functional monomer, and to quantify it, if the latter bears an atom not present in the basic monomer, which is quite often the case [22]. New characterization techniques are emerging, in particular Time-Of-Flight/Secondary Ion Mass Spectrometry (TOF-SIMS),which has been used successfully by Davies et al for the surface chemical analyses of a series of colloids. These authors showed that XPS and TOF-SIMS provide complementary information about functional groups on the particle surfaces which, in addition, are correlated with electrophoretic mobility and particle size data [29].

4. Hydrophilic Latex Particles

A great deal of effort has been devoted to the preparation of latex particles bearing a hydrophilic layer, especially in the field of diagnostic assays, for several reasons: (i) emulsion copolymerization with appropriate polar monomers can be performed using batch, seeded or shot growth procedures; (ii) a higher colloidal stability against ionic strength is usually conferred to hydrophilic particles in comparison with hydrophobic ones, (iii) decrease in the physical adsorption of most proteins or biological entities is obtained as well as a better biocompatibility (then, denaturation of proteins and enzymes can be avoided).

Numerous papers have been devoted to the synthesis of latex particles with a hydrophilic surface. It should be mentioned the pioneer work of Okubo et al. [30] who prepared polystyrene latexes by emulsifier-free emulsion copolymerization of styrene with various hydrophilic monomers such as 2-hydroxyethyl methacrylate (HEMA), ethyl methacrylate (EMA), methyl acrylate (MA). Seeded emulsion polymerization can be alternatively performed to control the incorporated amount of HEMA on the surface, therefore to adjust the hydrophilic-lipophilic balance.

Hydrophilic particles can also be produced by polymerization of polyethylene oxide containing macromonomers or amphiphilic monomers [31]. In the recent years, several papers reported on the incorporation of carbohydrate moieties onto latex surfaces. Davies et al. [32] reported the emulsion copolymerization of a galactose derivative with styrene to yield latex particles containing galactose groups. The functionalization has been evidenced through surface chemical analysis (SIMS, XPS). In the author's laboratory, much attention has been paid to design suitable structures bearing saccharide moieties allowing to ensure high yield of surface incorporation. Charreyre et al. [33] first prepared a surface-active hexylmethacrylate-terminated disaccharide monomer, the 6-(2-methylpropenoyloxy) hexyl β-D-cellobioside (CHMA) and performed a seed copolymerization technique using polystyrene seed particles. Formulation recipes were optimized so as to favor surface incorporation of CHMA as revealed by NMR and XPS analysis (6 to 66 % of the introduced amount). Revilla et al. [34] described the synthesis and characterization of 11-(N-p-vinyl benzyl)amido undecanoyl maltobionamide monomer (LIMA); it was then copolymerized in the presence of styrene either by batch or seeded or techniques. The location of the carbohydrate moieties at the surface of the latex particles has been determined using various methods (^{1}H-NMR, XPS) and yield of surface incorporation were found quite high (60 to 90 %), especially using the batch process. It was suggested that significant incorporation of the hydrophilic saccharide can be ensured through a good control of both the surface-activity and the reactivity of the selected amphiphilic monomers [35].

Recently, much attention has been focused on the preparation of hydrophilic latex particles exhibiting stimuli-responsive properties. Latexes based on poly[N-isopropylacrylamide] (NIPAM) or other acrylamide derivatives are good examples since such polymers are thermally-sensitive. They exhibit a lower critical solution temperature (LCST) property or cloud point (i.e, 32°C for Poly[NIPAM]).

This LCST property characterizes the transition from expanded hydrophilic particles (below the LCST) to hydrophobic shrunk ones (above the LCST) due to the release of hydrophobically-bound water. More details on the preparation techniques, properties and applications on poly[NIPAM] can be found in a recent extensive review [36]. One may take advantage of such property for several applications: i) to prevent hydrophobic adsorption of proteins and to favor the covalent binding when grafting is preferred; (ii) to adsorb proteins above the LCST and to desorb them below, many innovating applications were developed as recently described by Kawaguchi et al. [37].

The preparation and properties of such thermosensitive latex particles have been first reported by Pelton et al. [38] and Wu et al. [39] using NIPAM monomer. Both authors showed that small and monodisperse particles can be obtained through a radical-initiated precipitation process provided a crosslinker (N,N'-methylenebisacrylamide (MBA)) be added in the recipe in order to prevent redissolution of the polymer below the LCST; microgel (or hydrogel) latex particles are then produced. In addition, the LCST of the resulting particles can be changed through the introduction of an hydrosoluble monomer like acrylamide [36]. Recent studies gave more details on polymerization mechanism and colloidal properties of these poly[NIPAM] latexes as a function of temperature, pH and ionic strength.

A recent work of Meunier at al. [40] dealt with the preparation and characterization of cationic poly(N-isopropylacrylamide) copolymer bearing amino groups. It was synthesized using NIPAM, MBA, a cationic monomer (1,5-aminoethylmethacrylate hydrochloride (AEM)), and N,N'-azobis(amidinopropane) dihydrochloride as initiator. It is worth mentioning the dramatic effect of the functional monomer (even at very small amounts of AEM, 0.1-1.0 mole % related to NIPAM) on the polymerization kinetics which results in the formation of monosized particles. However for AEM concentrations higher than 2 mole %, polydisperse latexes were obtained together with the production of large amounts of polyelectrolytes. The presence of amine groups at the particle surface was directly evidenced and quantified by colorimetric titration and NMR analysis, and indirectly by electrophoretic mobility.

5. Biodegradable Particles

In the last decade, there has been much increasing interest in the development of drug targeting systems as an alternative to the natural delivery of a drug, with the purpose to increase the efficacy through a better tissue distribution. For the structural design of colloidal polymer carriers intended for use in drug targeting, several critical requirements must be fulfilled, among which : biocompatibility, biodegradability, bioresorbability, non toxicity, stability on storage, and appropriate size. In addition, three main objectives have to be considered concerning the properties of such systems : i) to carry the drug (which can be encapsulated, solubilized, bound, attached on or into the particle) ; ii) to reach the target site ; and iii) to deliver the drug at a controlled release rate) [41].

Various colloidal systems have been tailored in order to achieve controlled drug delivery such as: liposomes, polymeric vesicles, macromolecular prodrugs, micro and nanoparticles. A great deal of effort has been devoted to the preparation of the latter ones and Table 4 describes the main strategies which can be used for producing well defined nanoparticles.

Before giving more details, two main remarks should be emphasized :

- as proposed by Kraux et al., in a recent review [53], nanoparticles encompass nanocapsules in which the drug is confined inside a cavity surrounded by a polymeric membrane and nanospheres in which the drug is dispersed throughout the particles.

- contrary to in-vitro applications, for drug targeting, very few synthetic polymers fulfilled the above enumerated criteria; three major kinds were generally considered : poly(meth)acrylics, polyalkylesters and polycyanoacrylates, mostly since their

524

degradation leads to bioresorbable polymer chains.

It is worthy to mention the polyalkylcyanoacrylate (PACA) nanoparticles. First, they were found to fulfil several major requirements for drug targeting: good drug-loading capacity, biodegradability, low toxicity. Second, they can be easily produced in water with good reproducibility through an anionic dispersion polymerization process as pioneered by Couvreur et al. [45]. The presence of a nonionic polymer (dextran) or surfactant (polyethylene oxide based, poloxamer or pluronic) is needed for ensuring steric stabilization of the particles. In addition, pH of the aqueous medium should be kept low for controlling the formation of particles. Monosized particles down to 30 nm can be produced upon increasing the concentration of polymeric surfactant or by adding SO_2 to the monomer phase [47]. Molecular weight analysis showed that polycyanoacrylate particles are built by an entanglement of a large number of small oligomeric species (50 to 1000). Studies about the loading capacity of such nanoparticles were carried out in the case of hydrophobic or hydrophilic drugs [46]. In addition, antisens oligonucleotides were found to be protected from the degradation by exonuclease when adsorbed onto cationic PACA nanospheres [54].

TABLE 4 : Preparation and characteristics of biodegradable nanoparticles

Method	Nature of Polymer	Particle size range (nm)	Reference
Emulsion and dispersion polymerization	Poly[Acrylic and methacrylic esters] , Poly[Acrylamide]	100-1000 (M)	[42, 43, 44]
Anionic dispersion polymerization in water	Poly[alkylcyanoacrylate]	50-200 (M)	[45, 46, 47]
(Pseudo) anionic dispersion polymerization in non polar media	Poly[esters]	<1000 (M-P)	[48]
Anionic interfacial polymerization	Poly[alkylcyanoacrylate]	200-300 (M)	[49]
Emulsification	Poly[esters] natural polymers	100-1000 (P)	[50]
Desolvation	Poly[esters]	100-300 (M-P)	[51]
Shearing of lamellar systems	-	<1000 (M)	[52]

M = monodisperse ; P = polydisperse

Polyalkylester nanoparticles, especially those made of lactic and glycolic acids derivatives, can be produced as nanospheres or as nanocapsules using emulsification approaches (using various solvent extraction procedures). The protocol is as follows [50]: i) dissolution of the preformed polyester in a low boiling point organic solvent ; ii) dispersion of this polymer solution in a continuous media (generally water) in the presence of surface-active agents (or with serum albumin as a colloidal stabilizer) ; iii) extraction or evaporation of the organic solvent ; iv) recovering of the resulting nanospheres by filtration, decantation or centrifugation.

Most of examples reported in the literature on the preparation of microparticles generally led to broad size distribution. To synthesize nanoparticles, it is necessary to adjust manufacturing parameters, especially the ratio of the polymer solution to the continuous phase, which needs to be decreased and the stirring speed has to be significantly increased during emulsification [50].

Such procedures may present several drawbacks (presence of surface-active agents, size polydispersity, etc.) and attempts were recently disclosed for preparing monodisperse polyester nanoparticles. A new and simple method, so-called desolvation technique suitable for isolating macromolecules from liquid media, was proposed by Stainmesse et al. [51]. The principle consists in the introduction of the polymer dissolved in a liquid L_1, into a precipitation medium, a liquid L_2 which is a non solvent of the polymer. Building of the phase diagrams (L_1, L_2, polymer) is a perquisite stage to determine the domains of nanoparticle formation (i.e, corresponding to a low polymer concentration (0.5 to 1.5 %). Stable nanoparticles could be obtained by this method without adding any surfactant and with a polydispersity of nearly 10 %. The understanding of the nanoprecipitation mechanism and the optimization of an experimental model is thought to be very convenient for prediction of the nanoparticle formation.

It should be also mentioned the recent work of Slomkowski [48] who prepared poly(esters) particles by pseudoanionic polymerization of ester monomer (ε-caprolactone-lactic acid) in hydrocarbon media. Control of stability was achieved by using poly(dodecylacrylate)-g-poly(ε-caprolactone) amphiphilic copolymer, and monosized monosized latex particles were obtained, making then good candidates as bioadsorbable protein carriers.

Finally, it is worth mentioning the original work of Diat et al.[52] who investigated the shearing effect on lyotropic lamellar phases. The authors showed that monodisperse multilamellae vesicles with high encapsulation ratio can be obtained with a size range depending upon the shear rate. Observation of such oignon-like particles by cryofracture seems to reveal a polyhedral structure instead of a spherical one.

When dealing with nanoparticles for in vivo applications, many steps should be examined in many details so as to evaluate the performances of these colloidal systems in drug targeting. A large body of studies has been addressed to this very important domain and it is out of scope of this review to deal with this aspect.

It can be just mentioned that two important criteria have to be considered : i) the loading capacity of the nanoparticles which is strongly dependent upon the physicochemical properties of the carrier and of the drug (there are two main loci for loading the drug, either by inclusion inside the particle or by surface adsorption through physical forces or covalent immobilization); and ii) the drug release which implies kinetic studies [41, 44, 53].

6. Labelled Particles

In the biomedical field, most particles are used as carriers and therefore are most often plain, i.e. have no particular physical properties besides being particles. The first application of plain latex particles as markers in the biomedical domain was in agglutination tests where the microspheres were used as support as well as reporter: the formation of aggregates, easily detected with naked eyes, was a physical proof that the immunological reaction between particle-bound antibodies and antigens had occurred. The following section is devoted to beads exhibiting particular physical properties such as coloured, fluorescent and magnetic particles, with an overview of their preparation modes and applications, specially as reporter groups.

6.1. COLOURED AND FLUORESCENT PARTICLES

Coloured particles can be obtained by using an oil soluble dye. The beads are temporarily swollen with the dye/solvent mixture in order to let the dye get inside the colloid. Then, the solvent is distilled off, leaving the dye stranded inside the particles [55]. Polymerization of pyrrole is one very simple means of making coloured particles which leads to polypyrrole latexes, which are intensely black. Furthermore these particles can easily be functionalized via many different routes [56].

The preparation of fluorescent microspheres has recently been reviewed [6]. As for coloured particles, the fluorescent dye can be entrapped inside the beads but it can, as well, be directly bound to the surface of functionalized latexes [57] either directly or via the use of a funtionalized linear polymer. In this latter case, poly(2-vinylnaphtalene-alt-maleic anhydride) was bound to amino-terminated polystyrene linear chains and used as a surfactant in emulsion polymerization of styrene, producing fluorescent monodisperse latex particles [58].

Coloured particles have recently been used in a rapid thin-layer chromatographic method for quantification of C-reactive protein in sera, in order to differentiate between viral and bacterial infections. In this test, based on a so-called sandwich format, a first antibody is bound to defined zones on a thin-layer immunoaffinity membrane, while the second antibody is covalently tethered to deeply coloured blue latex particles. When the antigen is present in the sample, a blue line is observed on the membrane. Quantitation is achieved by scanning reflectometry or with a modified bar code reader [59].

Fluorescent particles are widely used as markers in many different biomedical applications such as immuno and genetic fluorescence [60], in neurosciences for brain cell labelling in vivo [61], and in flow cytometry, as a reference for the enumeration of platelets [62].

6.2. MAGNETIC PARTICLES

A lot of scientific papers, as well as patents, exist on the preparation of magnetic particles. Briefly, two main processes can be used, either encapsulation of ferrite by polymers during the polymerisation reaction or impregnation of preformed particles with magnetic material; both techniques have been reviewed in [6]. In the encapsulation procedure, the main issue is to polymerize the monomers around the metallic particles without letting any metal susceptible of being in contact with the biomolecules, a fact that might result in a loss of the biological properties. Ugelstad et al. developed an elegant and efficient, impregnation procedure: oxidative chemical groups, such as nitro moieties, are covalently anchored inside the whole volume of monodisperse polymer particles; then, an aqueous solution of Iron (II) is made to diffuse from the outer phase into the core of the microspheres. Ferrous salts get subsequently oxidized to iron oxy-hydroxy derivatives, and, on heating, these oxides precipitate under the Fe_3O_4 or γ-Fe_2O_3 form, as evenly distributed magnetic particles inside the polymer beads. With this technique, particles can be obtained with a fair variety of surface functional groups [8].

The main use of magnetic particles consists in the removal or isolation of specific species from complex mixtures. For instance, ex-vivo removal of cancer cells from body fluids can be achieved. Many other separation applications have been described with, in particular involving DNAs or antibodies, in [63].

Magnetic particles can be also used as reporter groups, just like fluorescent or coloured latexes, for the detection of an immunological reaction in a sandwich format for instance. To the bottom of a titration well, an antibody (specific of the antigen to detect) is immobilized. Then, a serum is added in the well; if antigens are present, they bind to the immobilized antibodies which, if present in the sera to be analyzed, can bind to the bottom of the well. The detection occurs by incubating a second antigen-specific antibody bound to magnetic particles. The amount of immobilized magnetic particles depends on the quantity of antigen present in the serum and is quantified by weighing the amount of unbound particles which are removed from the bottom of the well using a magnet. This detection is very sensitive and doesn't require any separation steps [64].

7. Immobilization of Biomolecules onto Latex Particles

The immobilization of biomolecules (proteins, nucleic acids etc.) onto latex particles can be achieved either via physical adsorption or via chemical attachement. In this section we will separately deal with physical adsorption and then with covalent binding procedures, though for the latter far less theoretical work has been published than on adsorption.

7.1. PHYSICAL ADSORPTION

The adsorption of biomolecules at solid/liquid interfaces is of considerable importance in biological technology and natural processes, so physical adsorption was first studied and the subject is well documented. The next chapter by Norde reports a detailed study related to this aspect.

To understand the interactions governing such phenomena, a number of techniques have been developed, providing information regarding :
- Adsorption/ desorption kinetics at the solid wall.
- Exchange kinetics between adsorbed and bulk molecules.
- Conformation and reorganization of the adsorbed biomolecules.
- Alteration of physical and biological properties.

The adsorption of proteins (γ-globulins, fibrinogen, albumin, enzymes, antibodies, etc.) on hydrophobic, hydrophilic or charged latexes has been extensively studied by several authors [65-70] in order to understand the mechanism of such process. This research work was motivated by the importance of protein adsorption in blood compatibility or incompatibility of polymeric materials, drug delivery, diagnostics and fundamental studies. The adsorption on hydrophobic and charged (PS) latexes was first investigated using Bovine Serum Albumin (BSA) and Human Serum Albumin (HSA)

528

as model proteins; the adsorption is generally a very rapid and irreversible process and principally governed by hydrophobic interactions rather than by electrostatic ones as pointed out by Norde and Lyklema [71]. Nevertheless, the adsorbed amounts of proteins were dependent upon experimental conditions (pH, ionic strength and surface charge density of the latex) as well as physicochemical properties of the biomolecules [72, 73]:

The effect of pH on the adsorbed amounts of proteins has been largely studied and discussed based on adsorption isotherms [68, 73], which usually exhibit two domains : (i) a rapid initial increase of the interfacial concentration for low solution concentrations, (ii) a plateau value reached for higher bulk concentrations. The protein adsorption isotherms were interpreted using a Langmuirian model [74, 75] which provides the affinity constant and the maximum adsorbed amount for a given protein/latex system. The validity of the Langmuir representation was established by plotting the linear variation of Ns^{-1} versus C^{-1} (where Ns is the amount of adsorbed protein and C, the bulk protein concentration). Another approach (Langmuir-Freudlich isotherm [69]) consists in plotting the adsorbed amount (Ns) as a function of bulk protein concentration (C), on a log-log scale. The slope was interpreted in terms of interaction type between protein and the interface. In most of cases, the adsorption plateau ranged between the adsorbed amounts for closely packed side-on and end-on monolayers. The maximum protein adsorption on hydrophobic and charged latex particles occurred for pH close to the isoelectric point (IEP) of the protein (5.5 for IgG and 4.5 for BSA etc.) [72, 73]. The presence of such maximum has been interpreted in term of conformation size and stability of the protein which decreases at the IEP [73, 76]. Smaller adsorbed amounts were observed for pH values far from the IEP, a result which was interpreted as a consequence of the lateral electrostatic repulsions between adsorbed macromolecules. Another interpretation proposed by Elgersma et al. [72] is that the IEP corresponds to the protein-covered latex particles instead of to the protein only.

The effect of ionic strength on the protein adsorption isotherms has been systematically investigated in order to assess the magnitude of the electrostatic interactions. For oppositely charged systems (positively charged latexes and negatively charged protein or vice-versa), the protein adsorbed amounts decrease as the ionic strength of the medium increases due to the screening of the charges and consequently the decreased of the electrostatic attractions. Such a behavior has been reported in many works [76, 67, 77], dealing with BSA, HSA, IgG and latexes bearing positive or negative charges. The effect of salinity reflects that electrostatic forces, in addition to hydrophobic ones, also contribute to the interactions between proteins and charged surfaces.

Additionally, the dependence of the adsorption upon the surface charge density has been reported by many authors, Burns et al. [78] (HSA adsorption on amphoteric latex surface), Hidalgo-Alvarez et al. [77, 79] (IgG adsorption onto anionic-charged PS latexes), Elgersma et al. [72] and Duinhoven et al. [80, 81] (Enzyme adsorption on PS latexes). The adsorbed amount of proteins on hydrophobic and cationic charged latexes increases upon increasing the surface charge density above the IEP of the protein, whereas for anionic charged latexes it is below the IEP when the adsorbed amount increases upon increasing the surface charge density.

Since the adsorption of proteins onto polymer particles was principally governed by hydrophobic interactions, as observed for various protein/polystyrene systems, the best method to reduce this adsorption phenomena is to change the HLB of the surface. Recently, several authors reported that core-shell polystyrene latexes bearing various hydrophilic shells, (2-hydroxyethyl methacrylate (HEMA) [76, 82, 83], ethyl methacrylate (EMA) [83], methyl acrylate (MA) [30], acrylamide [76], methyl methacrylate (MMA) [30], 11-(N-p-vinyl benzyl)amido undecanoyl maltobionamide monomer (LIMA) [84], hexylmethacrylate-terminated oligosaccharide hexyl β-D-cellobioside (CHMA) [85], N-isopropylacrylamide (NIPAM) [37, 86], NIPAM/glycidyl methacrylate (NIPAM/GMA) [87] exhibit low protein adsorption. As an illustration, the effect of the amount of LIMA on the BSA adsorption onto functionalized polystyrene latexes is presented in Figure 2. The BSA adsorbed amount decreases as the surface functionalization with LIMA monomer increases. Under such conditions, it is possible to carry out covalent binding of proteins onto reactive latex surfaces with very little simultaneous adsorption.

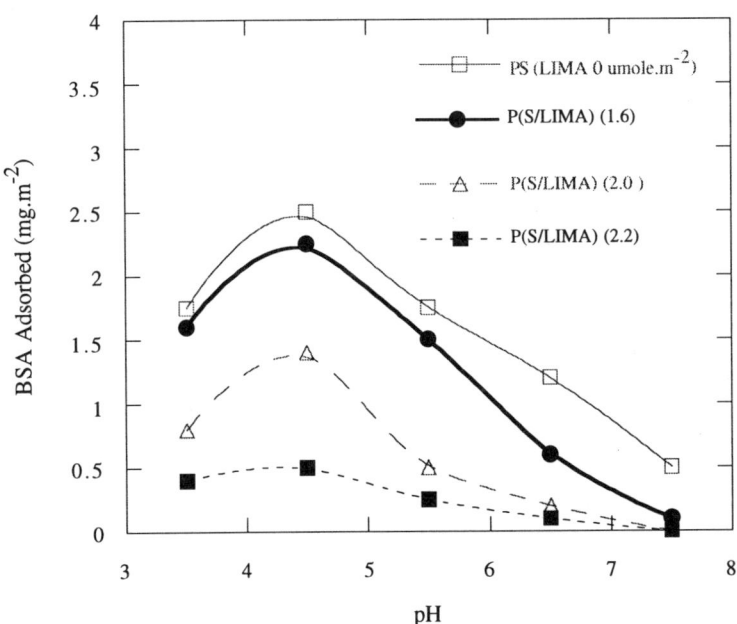

Figure 2. pH dependence of BSA adsorption onto PS and P(S/LIMA) copolymer latexes (0, 1.6, 2.0 and 2.2 μmole.m^{-2} of LIMA fixed on the latex particles) (25°C, ionic strength 0.01M) [84].

In the field of thermally-sensitive nanoparticles, Kawaguchi et al. [37] reported the temperature dependency adsorption of proteins on polyN-isopropylacrylamide hydrogel microspheres. The adsorption of human gamma globulin (HGG) below the lower critical solution temperature was found to be lower compared to that above the LCST. The desorption was controlled by the temperature of the aqueous medium and the incubation time. These results were interpreted with respect to latex surface morphology : below the LCST, the particle surfaces is hydrophilic, whereas, above the LCST, the particle surface become hydrophobic and hence can accommodate a higher adsorbed amount of proteins. These results have been confirmed by Yoshioka et al. [86] who investigated the

530

adsorption of IgG and BSA onto poly[N-isopropylacrylamide] grafted silica gel.

The adsorption of large deoxyribonucleic acids onto sulfate polystyrene latex particles has recently been reported by Yamaoka et al. [88]; native-DNA adsorbed amount was found to increase as ionic strength increases contrary to heat-denatured DNA. A systematic investigation on the adsorption behavior of short single-stranded oligonucleotides onto monodisperse polystyrene latexes differing in nature and surface charge density is currently performed in the author's laboratory. [89, 90]. Adsorption isotherms were established for various pH, surface charge density, ionic strength of the aqueous medium, the chain length of the oligonucleotide. The experimental findings suggested that electrostatic interactions between oligonucleotides and latex particles bearing cationic charges controlled the adsorption phenomena. The adsorbed amount of oligonucleotide increases when decreasing the pH of the medium or the surface charge density, irrespective of oligonucleotide sequence as shown in Figure 3.

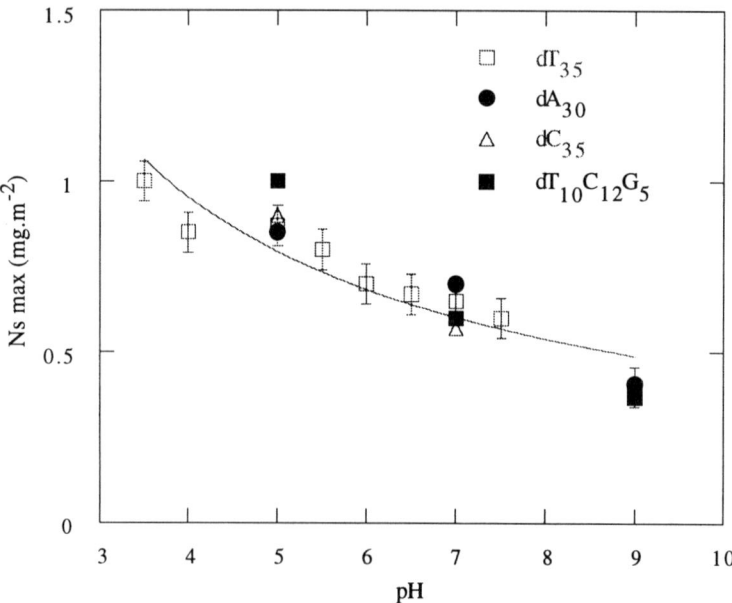

Figure 3 . Maximal adsorbed amount of dT_{35}, dA_{30} , dC_{35} and $dT_{10}C_{12}G_5$ on latex particles bearing amino groups as a function of pH at 10^{-2} ionic strength [91].

On the contrary, the effect of ionic strength on the maximal adsorbed amount of oligonucleotides was not significant. The contribution of hydrophobic interactions was investigated by studying the adsorption of oligonucleotides onto anionic latex particles (sulfate groups). It was pointed out that, in this case, the pH of the medium was not a determinant factor, since the amounts of adsorbed oligonucleotides were low and in the same range. These values were close to the extrapolated one for cationic latex at zero surface charge density (or similarly corresponding to ζ-potential equal to zero).

The conformation of adsorbed single-stranded DNA fragments onto cationic and anionic latexes has been reported by Walker et al. [92] using hydroxyl radical foot printing method. When the particles are positively charged all sugar residues are in direct contact points with the latex surface. However, only a few contacts are found between single-stranded DNA fragments and an anionic surface.

7.2. COVALENT COUPLING ON FUNCTIONALIZED PARTICLES

7.2.1 Direct coupling

Functional groups like aldehyde, chloromethyl and epoxide readily react with the primary amine of a biological molecule. The reaction occurs by nucleophilic displacement of the chlorine atom of chloromethyl styrene groups [93] or by ring opening of the oxirane of glycidyl methacrylate functionalized latex particles [94], to yield a secondary amine. As a result, the biological molecule is covalently linked to the functional colloid. Aldehyde groups react with amino moieties by forming an imine derivative with concomitant water elimination. Most authors stabilize the imine bond by performing, with a borohydride derivative, the reduction of the carbon-nitrogen double bond to the corresponding more stable alkylamine carbon-nitrogen single bond. Bale Oenick et al., for instance, showed that immobilization of proteins was more efficient in the presence of a reducing agent [93], whereas, according to Slomkowsky's results, the multidentate immobilization involving Schiff bases is very stable, and no protein leakage was observed on storing, even after many days [95].

Other chemical groups have specific coupling procedures, in particular, sulfhydryl groups react with activated carbon-carbon double bonds to yield a mercapto ether, as used in [96], or with other thio groups to form disulfide bonds [12].

7.2.2 Indirect coupling

In the following section, the immobilization procedure of biological molecules requires an extra step prior to coupling: the functional groups borne by the latexes are unreactive as such and need to be 'activated' as many biochemists call it. The most popular technique is the activation of carboxylated latexes by carbodiimide [97]. The intermediate obtained on reaction of a carboxylic acid group with a carbodiimide is fairly unstable, specially in an aqueous environment, and has to be quickly reacted with the protein to be immobilized. This is the major drawback of any immobilization protocol based on the use of highly reactive and unstable intermediates; thus, optimal experimental conditions will be the best compromise between hydrolysis of the reactive species by water molecules, and the actual bond formation reaction that links the biomolecules to the solid phase.

The activation and coupling steps seldom happen simultaneously because of the possible reticulation of the proteins, which very often bear both amino and carboxyl groups. In some cases, it is worth adding an extra step: first, reaction in one pot of the carboxylated latex, carbodiimide and N-hydroxysuccinimide to form the N-hydroxysuccinimide ester as an intermediate and then, after removal of excess chemicals, introduction of the desired biomolecule. Though this procedure seems more lengthy than the direct reaction with carbodiimide, the coupling of the protein is often more efficient by this way, in terms of conservation of the biological specific properties of the natural macromolecule.

TABLE 5: Various covalent immobilization procedures

Solid Support Functional Groups	Biological Molecule Functional Group (Reaction Type)	Comments	Reference
CHO	NH$_2$ (Addition)	No activation step required. Borohydride reduction may prove useful	[93, 95]
ClCH$_2$-	NH$_2$ (Nucleophilic substitution)	No activation step required. Sluggish reaction as Cl- is a poor leaving group.	[93]
CH-CH$_2$ \ / O	NH$_2$ (Nucleophilic substitution)	No activation step required. A high pH coupling medium is often needed.	[94]
SH	SH (Reduction)	No activation step required. Disulfice bond formation. Very susceptible to oxidation and displacement by other SH groups.	[12]
SH	Maleimide Iodoacetyl (Addition to activated C=C bond)	No activation step required. Requires modification of the biomolecule	[96]
COOH	NH$_2$ (Nucleophilic substitution)	An activation step is required. Activated esters sensitive to hydrolysis by coupling buffer	[96]
NH$_2$	NH$_2$	An activation step with a bifunctional reagent is required.	[55]
OH	NH$_2$ (Nucleophilic substitution)	An activation step of the support is needed. This step often requires an organic medium	[55]

This last point is very important when one wants to immobilize on a synthetic support macromolecules exhibiting biological activity. Coupling on latexes high loads of a biological molecule is of no use, if most of it is denatured (i.e. lost its properties), so the binding procedures have to be adapted, and a variety of conditions need to be tested for each case.

The activation of hydroxy functionalized particles can be achieved by reaction with tosyl chloride or carbonyl di-imidazole in order to introduce on the particle surface a good leaving group for subsequent covalent immobilization of the natural macromolecule by nucleophilic substitution [55]. However, for the sake of efficiency, the activation step should take place in an organic solvent, therefore only crosslinked particles can be used, otherwise they might dissolve in the organic medium.

An overview of the different available covalent immobilization procedures is given in Table 5, with a comment on the main features of each technique

Perspectives : New immobilization procedures are emerging and, though probably still at an early stage, they are worth mentioning. A very promising method relies on the use of imprinted polymers reviewed by Wulff [97]. In 1995, Mosbach et al reported the synthesis of an adsorbent with high selectivity for the enzyme RNase A, by a surface imprinting procedure based on metal coordination. A metal chelating monomer was polymerized onto methacrylate particles in the presence of metal ions and RNase A. The 'shape' of the metal complexed enzyme was thus imprinted at the surface of the beads which were capable of separating RNase A, from a mixture with Lysosyme [98].

Phenyl boronic acid derivatives can form stable complexes with cis-diols in basic conditions [99]; in 1994, Takagi et al. prepared by emulsion polymerization of styrene, butyl acrylate and m-acrylamidophenylboronic acid microspheres having phenyl boronic groups at the surfaces [100]. Then, they could bind D-Glucose via complex formation between a phenylboronic acid moiety and a cis-diol group of D-Glucose.

8. Conclusions

There is an increasing demand of well-defined polymer nanoparticles suitable for numerous applications in medicine and biology and this review showed that many reliable preparation procedures are now available. Most of these techniques rely on communition methods, namely based on free-radical heterogeneous polymerizations which are quite versatile for controlling the particle size, the surface properties and the particle morphology. Much effort has also been devoted in recent years on the production of various monosized biodegradable nanoparticles either from heterogeneous polymerizations or from emulsification and precipitation processes.

As regards to polymer chemistry, there is still a need of new kinds of nanoparticles with innovative properties; a nonexhaustive list is suggested below: i) polymer particles with specific structure for mimicking various phenomena involved in living systems (molecular recognition at the surface of cells, membranes, bacteria, virus, etc); ii) reactive particles with a better control of the accessibility, reactivity and surface density of the functional groups; iii) very small nanoparticles (below 60 nm) as produced by microemulsion or micellar polymerization processes, taking into account of the extensive understanding in the physicochemistry of these disperse systems (particularly suitable for drug targeting); iv) hydrophilic and stimuli-responsive particles ("smart material") for various biotechnology (sensors) and biology purposes.

The characterization of these polymer colloids is also a crucial task in view of the many problems encountered when such supports interact with biologically-active macromolecules or drugs. Fundamental and systematic studies are still necessary to better understand the mechanisms of interactions and the consequences on the performances of the biomolecule activity (whether they are used for in-vitro or in-vivo applications). The complexity of the many reported phenomena should stimulate coordinated multidisciplinary research involving chemists, physicists, biologists, etc.

534

9. References

1. Daniels, E.S., Sudol, E.D., El-Aasser, M.S. (eds.) (1992) Preparation, characterization and applications of polymer latexes, *ACS Symp. Series*, **492**, Washington, D.C.

2. Singer, J.M., Plotz, C.M. (1996), The latex formation test.1-Application to the serological diagnosis of rheumatoid arthritis, *Am. J. Med.*, **31**, 888.

3. Singer, J.M. (1987) The use of polystyrene latexes in medicine, in M.S. El-Aasser, R.M. Fitch (eds.), *Future Directions in Polymer Colloids*, Martinus Nijhoff Publishers, Dordretch, pp. 372-394.

4. Van den Hul, H.J., Vanderhoff, J.W. (1971) "Clean monodisperse latexes as model colloids", in R.M. Fitch (ed.) *Polymer Colloids*, Plenum Press, NY, 1-26.

5. Goodwin, A.R., Ottewill, R M, Pelton, R H, Vianello, G., Yates, D.E. (1978) Control of particle size in the   formation of polystyrene latices , *Br. Polym.J.*, **10**, 173-180.

6. Arshady, R. (1993) Microspheres for biomedical applications : preparation of reactive and labelled microspheres, *Biomaterial*, **14**, 5-15.

7. Rembaum, A., Margel S, (1978), Design of polymeric immunospheres for cell labeling and separation, *Br. Polym. J*, **10**, 275-280.

8. Ugelstad, J., Berge, A., Ellingsen, T., Schmid, R., Nilson, T.N., Mork, P.C., Stenstad, P., Hornes, E., Olswik, Ø (1992) Preparation and applications of new monosized polymer particles, *Prog. Polym.*, **17**, 17.

9. Ugelstad, J., Stenstad, P., Kilaas, L., Prestvik, W.S., Rian, A., Nustad, K., Huge, R., Berge, A. (1996) Biochemical and biomedical applications of monodispersed polymer particles, *Macromol. Symp.*, **101**, 491-500.

10. El-Aasser, M.S., Methods of latex cleaning, in G.W. Poehlein, R.H. Ottewill, J.W. Goodwin (eds.) *Science and Technology of Polymer Colloids*, NATO ASI Series, E68, Ed. Nijhoff, The Hague, 1983, 422-448.

11. Blackley, D.C., (1983) Preparation of carboxylated latices by emulsion polymerization, in G.W. Poehlein, R.H. Goodwill, J.W. Goodwin (eds.), *Science and Technology of Polymer Colloids*,   NATO ASI Series, E68, Nijhoff, The Hague, 203-219.

12. Delair, Th., Pichot, C., Mandrand, B. (1994) Synthesis and characterization of cationic latex particles bearing sulfhydryl groups and their use in the immobilization of Fab antibody fragments, *Colloid Polym. Sci.* , **272**, 72-81.

13. Delair, Th., Marguet, V., Pichot, C., and Mandrand, B. (1994) Synthesis and characterization of cationic amino functionalized latexes, *Colloid Polym. Sci.*, **272**, 962-970.

14. Ganachaud, F., Mouterde, G., Delair, Th., Elaïssari, A., and Pichot, C. (1994) Preparation and characterization of cationic polystyrene latex particles of different aminated surface charges, *Polym. for Adv.Tech.*, **6**, 480-488.

15. Sarobe J., and Forcada J. (1996) Synthesis of core-shell type polystyrene monodisperse particles with chloromethyl groups, *Colloid Polym. Sci* , **274**, 8-13.

16. Kling, J.A., and Ploehn, H.J. (1995) Synthesis and characterization of epoxy-functional polystyrene particles, *J. Polym. Sci. Part A: Polym. Chem.*, **33**, 1107-1118.

17. Brouwer, W.M., Van Der Vegt, M., and Van Haaren, P. (1990) Particle surface characteristics of permanently charged poly(styrene-cationic monomer) latices, *Europ. Polym. J.*, **26**, 35-39.

18. Hassanein, M., Abdel-Hay, F.I., El-Hefnawy, T., and El-Esawy, A.(1994) Autooxidation of 2-6, di-t-butyl phenol catalyzed by Cobalt (II)-Schiff-base complex bound to cationic latexes, *Europ. Polym. J.*, **30**, 335-337.

19. Ford, N.T., Yu, H., Lee, J.J., and El-Hamshary H. (1993) Synthesis of monodisperse cross-linked

polystyrene latexes containing vinylbenzyl)trimethyl-ammoniumchloride units, *Langmuir,* **9**, 1698-1703.

20. Verrier-Charleux, B., Graillat, C., Chevalier, Y., Pichot, C., Revillon, A. (1991) Synthesis and characterization of emulsifier-free quaternarized vinylbenzylchloride latexes, *Colloid. Polym. Sci.,* **269**, 398-405.

21. Charleux, B., Fanget, P., Pichot, C. (1992) Radical-initiated copolymers of styrene and p-formylstyrene, 2) Preparation and characterization of emulsifier-free copolymer latices, *Macromol. Chem. Phys.,* **193**,,205-220.

22. Charreyre, M.T., Razafindrakoto, V., Véron, L., Delair, Th., and Pichot, C. (1994) Radically-initiated copolymers of styrene with 4-vinylbenzylamine and its trifluoroacetaminde derivative, 2) Preparation of latex particles bearing amino groups, *Macromol. Chem. Phys.* , **195**, 2153-2167.

23. Tamai, H., Fuji, I., and Suzawa, T. (1987) Surface characterization of hydrophilic functional polymer latex particles, *J. Colloid. Interface. Sci.* , **118**, 176-181.

24. Yan, C., Zhang, X., and Sun, Z. (1990) Poly(styrene-co-acrolein) latex particles : copolymerization and characteristics, *J. Appl.. Polym. Sci.,* **40**, 89-98.

25. Basinska, T., Slomkowsky, S., and Delamar, M. (1993) Synthesis and characterization of polystyrene core/polyacrolein shell latexes, *J. Bioactive Compatible Polymers* , **8**, 205-219.

26. Charreyre, M.T., Boullanger, P., Delair, Th., Mandrand, B., LLauro M.F and Pichot, C. (1993) Preparation and polymerization studies with an hexylmethacrylate cellobioside monomer, *Macromol. Chem. Phys.,* **194**., 117-130.

27. Vanderhoff, J.W., Van den Hul, H.J., Tausk, R.J.M, and Overbeck, J.T.G. (1970) *Clean Surfaces : Their preparation and Characterization for interfacial surfaces,* M. Dekker, New York.

28. Le Dissez, C., Wong, P.C., Mitchell, K.A.R., and Brooks, D.E. (1996) Analysis of surface aldehyde functions on surfactant-free polystyrene/polyacrolein latex, *Macromolecules,* **29**, 953-959.

29. Davies, M.C., Lynn, R.A.P., Hearn, J., Paul, A.J., Vickerman, J.C.,and Watts, J.F. (1995) Surface chemical characterization using XPS and TOF-SIMS of latex particles prepared by the emulsion copolymerization of functional monomers with methyl methacrylate and 4-vinylpyridine, *Langmuir* , **11**, 4313-4322.

30. Okubo, M., Azuma, I., Hattori, H. (1992) Preferential adsorption of bovine fibrinogen dimer onto polymer microspheres having heterogeneous surfaces consisting of hydrophobic and hydrophilic parts, *J. Appl. Polym. Sci.,* **45**, 245-251.

31. Ito, K. (1994), Polyethylene oxide macromonomers., in M.K. Mishra (ed.), *Macromolecular Design : Concept and Practice.*,Polymer Frontiers international Inc, 129-160.

32. Davies, R.C., Lynn, R.A.P., Davis, S.S., Hearn, J., Watts, J.F., Vickerman, J.C., Paul, A.J. (1993), Preparation of polymer latex particles with immobilized sugar residues and their surface characterization by X-ray photoelectron spectroscopy and time-of-flight secondary ion mass spectroscopy, *Langmuir,* **9**, 1637-1645.

33. Charreyre, M.T., Boullanger, P., Delair, Th., Mandrand, B., and Pichot, C. (1993) Preparation and characterization of polystyrene latexes bearing disaccharide surface groups, *Colloid Polym. Sci.,* **271**, 668-679.

34. Revilla, J., Elaïssari, A., Pichot, C., Gallot, B. (1995) Surface functionalization of polystyrene latex particles with a liposaccharide monomer, *Polym. for Adv.Tech,* **6**, 455-464.

35. Demharter, S., Richtering, W. and Mulhaupt, R. (1995) Emulsion polymerization of styrene in the presence of carbohydrate-based amphiphiles, *Polym. Bull.,* **34**, 271-277.

36. Schild, H. G. (1992) Poly(N-isopropylacrylamide): experiment, theory and application, *Prog, Polym. Sci.,***17**, 163-249.

37. Kawaguchi, H., Fujimoto, K., Mizuhara, Y. (1992) Hydrogel microspheres III. Temperature-dependent

536

adsorption of proteins on poly-N-isopropylacrylamide hydrogel microspheres, *Colloid Polym. Sci.* **270**, 53-57.

38. Pelton, R. H., Chibante, P. (1986) Preparation of aqueous lateces with N-isopropylacrylamide, *Colloids Surfaces*, **20**, 247-257.

39. Wu, X., Pelton, R. H., Hamielec, A. E., Woods, D. R., McPhee, W. (1994) The kinetics of poly(N-isopropylacrylamide) microgel latex formation *Colloid Polymer Science*, **272**, 467-477.

40. Meunier, F., Elaïssari, A., Pichot, C. (1995) Preparation and characterization of cationic poly(n-isopropylacrylamide) copolymer latexes, *Polym. for Adv.Tech*, **6**, 489-496.

41. Garny, R. (1986) Controlled drug delivery with colloidal polymeric systems, *NATO ASI Series E, Applied Sciences*, **106**, 195-211.

42. Lukowski, G., Müller, R.H., Müller, B.W., Dittgen, M. (1993) Acrylic acid copolymer nanoparticles for drug delivery. Part II characterization of nanoparticles surface-modified by adsorption of ethoxylated surfactants, *Colloid. Polym. Sci.*, **271**, 100-105.

43. Rolland, A., Gibassier, D., Saclo, P., Leverge, R. (1986) Purification et propriétés physicochimiques de suspensions de nanoparticules de polymères acryliques, *J. Pharm. Belg.*, **41**, 94-105.

44. Couvreur, P., Couarraze, G., Devissagnet, J.P., Puisieux, F. (1996) Nanoparticles : Preparation and Characterization in S. Benita (ed.) *Microencapsulation, Methods and Industrial Applications*, Marcel Dekker, New-York, 183-211.

45. Couvreur, P., Vautier, C. (1991) Polyalkylcryanoacrylate nanoparticles as drug carrier : present state and perpectives, *J. Cont. Release*, **17**, 187-199.

46. Seijo, B., Fattal, E., Roblot-Treupel, L., Couvreur, P. (1990) Design of nanoparticles of less than 50nm diameter : preparation characterization and drug loading, *Int. Jal. of Pharmaceutics*, **62**, 1-7.

47. Lescure, F., Zimmer, C., Roy, D., Couvreur, P. (1992) Optimisation of polyalkylcyanoacrylate nanoparticle preparation : influence of sulfur dioxide and pH on particle characteristics, *J. Colloid Interface Sci.*, **254**, 77-96.

48. Slomkowski, S. (1996) Controlled polymerization of ε-caprolactone from macromolecules to microspheres, *Macromol. Symp.*, **103**, 213-228.

49. Couvreur, P., Dubernet, C., Puisieux, F. (1995) "Controlled drug delivery with nanoparticles : Current possiblity and future trends", *Eur. J. Pharm.*, **41 (1)**, 2-13.

50. Arshady, R. (1991) Preparation of biodegradable microspheres and microcapsules, *J. Control. Release*, **17**, 1-21.

51. Stainmesse, S., Orecchioni, A.M., Nakache, E., Puisieux, F., Fessi, H. (1995) Formation and stabilization of a biodegradeble polymeric colloidal suspension of nanopaticles, *Colloid. Polym. Sci.*, **273**, 505-511.

52. Diat, O., Roux, D., Nallet, F. (1993) Effect of shear an a lyotropic lamellar phase, *J. Phys. II*, **3**, 1427-1452.

53. Kraux, H.J., Schwartz, A., Rohdewald, P. (1985) Polylactic and nanoparticles, a colloidal drug delivery system for lipophilic drugs, *Int. J. Pharm.*, **27**, 145-155.

54. Chavany, C., Le Doan, T., Couvreur, P., Hélène, C. (1992) "Polyalkylcyanoacrylate nanopartcles as polymeric carriers for antisense oligonucleotides", *Pharm. Res.*, **9**, 441.

55. Bangs, L.B. (1984), *Uniform Latex Particles,* Seragen Diagnostics Inc.

56. Tarcha, P.J., Misun, D., Finley, D., Wong, M., and Donovan, J.J. (1992) Synthesis, analysis, and immunodiagnostic applications of polypyrrole latex and its derivatives, in E.S. Daniels, E.D. Sudol, M.S. El-Aasser (eds.), *Polymer Latexes - Preparation, Characterization, and Applications*, ACS Symposium series , **492**, 347-367.

57. Charreyre, M.T., Yekta, A., Winnik, M.A., Delair, Th., and Pichot, C. (1995) Fluorescent energy

transfert   from fluorescein to tetramethylrhodamine covalently bound to the surface of polystyrene particles, *Langmuir*, **11**, 2423-2428.

58. Cao, T., Weiping, Y., and Webber, S.E. (1994) Poly(2-vinylnaphtalene-alt-maleic acid)-graft-polystyrene as a photoactive polymer micelle and stabilizer for polystyrene latexes, *Macromolecules*, **27**, 7459-7464.

59. Nilsson, S., Lager, C., Laurell, T., and Birnbaum, S. (1995) Thin-layer immunoaffinity chromatography with bar code quantitation of C-reactive protein, *Analytical Chemistry*, **67**, 3051-3056.

60. Cheung, S.W. (1993) Methods for making fluorescent microspheres, United State Patent, 5,194,300

61. Paulke, B.R., Hartig, W.,and Bruckner, G. (1992) Synthesis of nanoparticles for brain cell labelling in vivo, *Acta Polymer.*, **43**, 288-291.

62. Dickerhoff, R., and Von Ruecker, A. (1995) Enumeration of platelets by multiparameter flow cytometry using platelet-specific antibodies and fluorescent reference particles, *Clinical and Laboratory Haematology* **17**, 163-172.

63. Advances in Biomagnetic Separation (1994) Uhlen,M., Hornes, M., and Olsvik, Ø. (eds), Eaton Publishing Co., Natik.

64. Rohr, T.E. (1995) United State Patent 5, **445**, 971.

65. Andrade, J. D. (1985) Surface and Interfacial Aspects of Biomedical Polymers, Volume 2, Protein Adsorption, Plenum Press, New York.

66. Andrade, J. D., Hlady, V., Wei, A. P. (1992) Adsorption of complex proteins at interfaces, *Pure and Appl. Chem.*, **64**, 1777-1781.

67. Hondo, A., Higashitani, K. (1992) Adsorption model proteins with wide variation in molecular properties on colloidal particles,.*J. Colloid and Interface Sci.*, **150**, 344-351.

68. Betton, F., Theretz, A., Elaïssari, A., Pichot, C. (1993) Adsorption of bovine serum albumin onto amphiphilic acrylic acid copolymer-stabilized polystyrene latex particles, *Colloids and Surfaces B: Biointerfaces*, **1**, 97-106.

69. Yoon, J., Y., Park, H., Y., Kim, J., H., Kim, W., S. (1996) Adsorption of BSA on highly carboxylatex microspheres-quantitative effect of surface functional groups and interaction forces, *J. Colloid and Interface Sci.* **177**, 613-620.

70. Ball, V., Huetz, P., Elaïssari, A., Cazenave, J. P., Voegel, J. C., Schaaf, P. (1994) Kinetics of exchange processes in the adsorption of proteins on solid surfaces, *Proc. Natl. Acad. Sci USA*, **91**, 7330-7334.

71. Norde, W., and Lyklema, J. (1979) Thermodynamics of protein adsorption : Theory with special reference the adsorption of human plasma albumin and bovine pancreas ribonuclease at polystytene surfaces, *J. Colloid and Interface Sci.* **71**, *350-366.*

72. Elgersma, A., Zsom, R. L., J., Norde, W., Lyklema, J. (1991) The adsorption of different types of monoclonal immunoglubin on positively and negatively charged polystyrene latices, *Colloids and Surfaces*, **54**, 89-101.

73. Elgersma, A., Zsom, R., L., J., Norde, W., Lyklema, J. (1990) The adsorption of bovine serum albumin on positively and negatively charged polystyrene latices, *J. Colloid and Interface Sci.*, **138**, 145-156.

74. Aptel, J. D., Thomann, J. M., Schmitt, A., Bres, E. F. (1988) Adsorption of Human albumin onto hydroxyapatite. Static and dynamic studies, *Colloids and Surfaces*, **32**, 159-171.

75. De Baillou, N., Voegel, J. C., Schmitt, A. (1985) Adsorption of human albumin and fibrinogen onto heparin-like materials. I. Adsorption isotherms, *Colloids and Surfaces*, **16**, 271-288.

76. Suzawa, T., Shirahama, H. (1991) Adsorption of plasma proteins onto polymer latices, *Advances in Colloid and Interface Science*, **35**, 139-172.

77. Martin, A., Puig, J., Galisteo, F., Serra, J., Hidalgo-Alvarez, R. (1992) Biocolloids and biosurfaces: On some aspects of the adsorption of immunoglobulin-g molecules on polystyrene microspheres, *J.*

Dispersion Science and Tehnology, **13**, 399-416.

78. Burns, N., Holmberg, K., Brink, C. (1996) Influence of surface charge on protein adsorption at an amphoteric surface: Effects of varying acid to base ratio, *J. Colloid and Interface Sci.*, **178**, 116-122.

79. Hidalgo-Alvarez, R., Galisteo-Gonzalez, F. (1995) The adsorption characteristics of immunoglobulins, *Heterogeneous Chemistry Reviews*, **2**, 249-268.

80. Duinhoven, S., Poort, R., Van der Voet, G., Agterof, W. G. M., Norde, W., Lyklema, J. (1995) Driving forces for enzyme adsorption at solid-liquid interfaces: 1. The serine protease savinase, *J. Colloid and Interface Sci.*, **170**, 340-350.

81. Duinhoven, S., Poort, R., Van der Voet, G., Agterof, W. G. M., Norde, W., Lyklema, J. (1995) Driving forces for enzyme adsorption at solid-liquid interfaces: 2. The fungal lipase lipolase, *J. Colloid and Interface Sci.*, **170**, 351-357.

82. Kamei, S., Okubo, M., Matsumoto, T., Yamamoto, M. (1986) Adsorption of trypsin onto styrene-2-hydroxyethyl methacrylate copolymer microspheres and its enzymatic activity, *Colloid Polym. Sci.*, **264**, 743-747.

83. Walker, D. S., Garrison, M. D., Reichert, W. M. (1993) Protein adsorption to HEMA/EMA copolymers studied by integrated optical techniques, *J. Colloid and Interface Science*, **157**, 41-49.

84. Revilla, J., Elaïssari, A., Carriere, O., Pichot, C. (1996) Adsorption of bovine serum Albumin onto polystyrene latex particles bearing saccharidic moieties, *J. Colloid and Interface Sci.*, **180**, 405-412.

85. Charreyre, M. T., Revilla, J., Elaïssari, A., Pichot, C., Gallot, B. (1996) Surface-functionalized polystyrene latexes with liposaccharide monomers : preparation, characterization and applications, (in press). in R.J. Spontak (ed), ACS Meeting *Advances in polysaccharides characterization and applications*, M. Dekker.

86. Yoshioka, H., Mikami, M., Nakai, T., Mori, Y. (1994) Preparation of poly(N-isopropylacrylamide)-grafted silica gel and its temperature-dependent interaction with proteins, *Polym. for Adv. Tech.* **6**, 418-420.

87. Kondo, A., Kaneko, T., Higashitani, K. (1994) Developement and application of thermo-sensitive immunomicrospheres for antibody purification, *Biotechnology and Bioengineeiring*, **44**, 1-6.

88. Yamaoka, K., Fukudome, K., Mukaiyama, N., Shirahama, H., Suzawa, T. (1990) Adsorption of deoxyribonucleic acid andpoly(cytidylic acid)-poly(inosinic acid) onto poly(styrene) latex, *J. Colloid and Interface Sci.*, **136**, 519-526.

89. Elaïssari, A., Cros, P., Laurent, V., Mandrand, B. (1994) Adsorption of oligonucleotides onto negatively and positively charged latex particles, *Colloid and Surfaces A: Physicochemical and Engineering Aspect*, **83**, 25-31.

90. Elaïssari, A., Pichot, C., Delair, Th., Cros, Ph., Kürfurst, R. (1995) Adsorption and desorption studies of polyadenylic acid onto positively charged latex particles, *Langmuir*, **11**, 1261-1267.

91. Ganachaud, F., Elaïssari, A., Laayoun, A., Pichot, C., Cros, Ph (1997), Adsorption of single stranded DNA fragment onto aminated latex particles, (*Langmuir*, submitted).

92. Walker, H. W., Grant, S. B., (1995) Conformation of DNA block copolymer molecules adsorbed on latex particles as revealed by hydroxyl radical footprinting *Langmuir*, **11**, 3772-3777.

93. Bale Oenick, M.D., and Warshawsky, A. (1991) Protein immobilization on surface modified latices bearing aldehyde groups, *Colloid Polym. Sci.*, **269**, 139-145.

94. Inomata, Y., Wada, T., Fujimoto, K., and Kawaguchi, H. (1994) Preparation of DNA-carrying affinity latex and purification of transcription factors with the latex, *J. Biomater. Sci. Polymer Edn.*, **5**, 293-302.

95. Basinska, T., Kowalczyk, D., Miska, B., and Slomkowsky, S. (1995) Interaction of proteins with polymeric latexes, *Polym. Adv. Technol.*, **6**, 526-533.

96. Pelton, R. (1990) Chemical reaction at the latex-solution interface, in F. Candau and R.H. Ottewill (eds.), *Scientific Methods for the Study of Polymer Colloids and Their Applications,* Kluwer Academic Publishers, Dordrecht, 493-516.

97. Wulff, G. (1991) in U.K. Pandit and F.C. Alderweurekt, (eds) *Bioorganic Chemistry in Healthcare and Technology* ,NATO ASI Series A, **207**, Plenum Press, 55-68.

98. Kempe, M., Glad, M., Mosbach, K. (1995) An approach towards surface imprinting using the Enzyme Ribonuclease A, *J.of Molecular Recognition* **8**, 35-39.

99. Ferrier, R.J. (1972) Applications of phenylboronic acid in carbohydrate chemistry, in R.L. Whistler and J.N. BeMiller *Methods in Carbohydrate Chemistry*, Academic Predss, 419-426.

100. Tsukagoshi, K., Kawasaki, R., Maeda, M., Takagi, M. (1994) Preparation and sugar binding property of microspheres having surface-anchored phenylboronic acid groups, *Chemistry Letters*, 681-684.

INTERACTION OF PROTEINS WITH POLYMERIC AND OTHER COLLOIDS

WILLEM NORDE
Wageningen Agricultural University
Dreijenplein 6
6708 HB Wageningen
The Netherlands

1. Introduction

In various applications finely dispersed particles, among which polymer colloids, are used as carriers for bio-macromolecules, proteins in particular. Examples of such applications can be found in e.g. drug targeting and drug delivery systems [1], diagnostic tests (immunolatices) [2], immobilization of enzymes in biocatalysis [3] and in chromatographic separation and purification procedures [4]. The main advantage of using sorbent material of colloidal dimensions is the large surface to volume ratio and, consequently, the relative large amount of adsorbed material in a given volume.

When a protein adsorbs from an (aqueous) solution onto a (solid) surface it changes its environment, which, in turn, may affect its three-dimensional structure or, at least, its structural stability. In view of the structure-function relation for proteins, the biological activity may as well be affected by the adsorption process. In essentially all applications of adsorbed proteins this matter is of crucial importance. In many cases uncontrolled, non-specific and irreversible adsorption hampers the preservation of maximum biological activity of the proteins.

Adsorption of proteins is the net result of an interplay of various interactions, among which electrostatic, hydrophobic and steric interactions are the most important ones. Structural rearrangements in the protein molecule may affect the conformational entropy of the protein which, by itself, influences the affinity between the protein and the surface. Furthermore, the altered flexibility of the protein changes the ability of the molecule to form optimum bonds with the sorbent surface.

The overall adsorption process could be imagined as occurring in three consecutive steps:
(a) protein transport from the bulk solution to the sorbent surface;
(b) interaction and attachment of the protein molecule with the surface, possibly involving perturbation of the protein structure;
(c) relaxation of the adsorbed protein molecule into its final conformation.

In this article I will focus on steps (b) and (c), and in particular I will discuss how physical-chemical properties of the sorbent surface and of the protein molecule are expected to influence the mode of adsorption. The discussion is illustrated with experimental data obtained for well-defined systems and the paper is concluded by giving some examples that are of more practical relevance.

J. M. Asua (ed.), Polymeric Dispersions: Principles and Applications, 541–555.
© 1997 *Kluwer Academic Publishers. Printed in the Netherlands.*

2. Physical-Chemical Characteristics Affecting Protein Adsorption

2.1. CHARGE ON THE SORBENT SURFACE AND THE PROTEIN MOLECULE

In an aqueous environment both the protein molecule and the sorbent surface are, as a rule, electrically charged. The charge originates from the presence of surface groups and from specific adsorption of ions from solution. For proteins, acidic and basic groups, primarily located at the aqueous periphery of the molecule, associate or dissociate with protons. Depending on the type of sorbent material, its surface charge is determined by strong acidic or basic groups (e.g. sulphate or quaternary ammonium groups originating from the polymerization initiator on polymeric latices), by dissociation equilibria of weak acidic or basic groups (e.g. oxide surfaces, carboxylated polymer latices, surfaces of various biological materials), or by the uptake of ions other than protons (e.g. silver halides). Other surfaces may not contain charged groups, such as those of polyoxymethylene oxide, polystyrene, teflon, etc.). Adsorption of ions from solution also leads to charging the surface provided that this adsorption is specific (which means that adsorption forces are partly of non-electrical nature so that the adsorbing ions can overcome an opposing electrical potential).

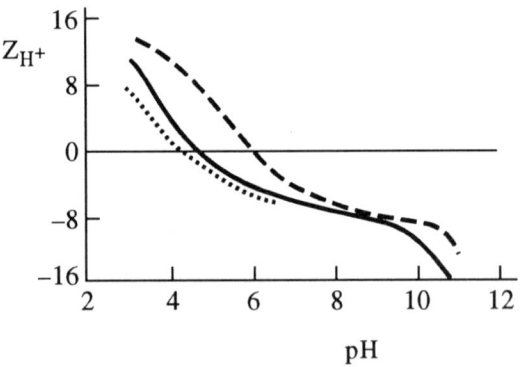

Figure 1. Proton titration curve for α-lactalbumin in solution (—) and adsorbed on positively (····) and negatively (- - -) charged polystyrene surfaces. Ionic strength 0.05 M; T = 25°C.

Charged protein molecules and surfaces, including the specifically adsorbed ions, are surrounded by counterions that are diffusely distributed in the solution. The surface charge and the counter charge together form the so-called electrical double layer.

When the protein molecule and the sorbent surface approach each other their electrical double layers overlap. This causes a redistribution of charges and a change in the polarity of the interfacial region. These environmental alterations may induce shifts in the dissociation behavior of weak acidic and basic groups (including those of the amino acid residues of the protein) in the sorbent-protein contact zone.

In Figure 1 the proton titration curves for α-lactalbumin (αLA) from bovine milk before and after adsorption on positively and negatively charged polystyrene (PS) latices are compared. This comparison is only valid in the pH region where the sorbent material does not contribute to the titration. The nature of the charged groups on the positive latex is $= {}^{\oplus}NH -$ and that on the negative latex $-OSO_3^{\ominus}$. In both systems αLA was adsorbed at pH 7 to 80% of its saturation values. Details on the experimental procedure and the data analysis are given in reference [5]. It is clear that the proton titration curve of the protein is affected by adsorption and the reversal of the shift on changing the sign of the latex surface charge indicates that the shift occurs primarily for electrostatic reasons.

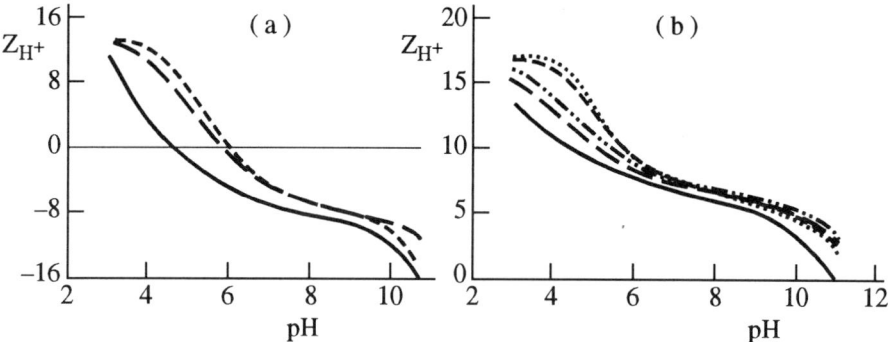

Figure 2. Proton titration curves for (a) α-lactalbumin and (b) lysozyme in solution (—) and adsorbed on negatively charged polystyrenesulphonate surfaces for various degrees of coverage of the sorbent surface by the protein: - - 80%; - ·· - 60%; - - - 40%; ··· 20%. Ionic strength 0.05 M; T = 25°C.

In Figure 2 the influence of the degree of coverage of the sorbent surface by the protein is displayed. In these examples the proteins are lysozyme (LSZ) from hen's egg and αLA; the sorbent is negatively charged polystyrene sulphonate (PSS) latex. For αLA the shift in the titration curve is essentially independent of the surface coverage. For LSZ the change in the titration behaviour increases with decreasing surface coverage, suggesting more extensive structural perturbations in the LSZ molecules at lower surface coverage.

Adsorption of globular proteins to solid surfaces leads to relatively compactly structured adsorbed layers [6]. Hence, the region of contact between the protein layer and the sorbent surface has a low dielectric constant relative to that of the aqueous medium. The presence of a net amount of charge in that region would result in an extremely high potential which is energetically very unfavorable. In addition to charge adjustments in the protein (i.e., protonation or deprotonation) accumulation of charge in the contact zone can be prevented by the incorporation of low-molecular-weight ions concomitant with the formation of the adsorbed protein layer.

Experimental evidence for the uptake of ions in the adsorbed layer is scarce. Using γ-spectrometry and electron paramagnetic resonance the uptakes of Na^+, Ba^{2+} and Mn^{2+} in layers of human serum albumin on negatively charged PS latices have been directly assessed (see Figure 3). It was found that the number of cations incorporated tend to increase with increasing pH, i.e., with increasing charge antagonism between the protein and the sorbent surface. Unlike the electrostatic effect, the chemical effect of this ion incorporation is unfavorable, simply because the non-aqueous, proteinaceous environment is a poorer "solvent" for those ions as compared to water. Indeed, maximum affinity for protein adsorption has been found under conditions where the protein charge just compensates the charge on the sorbent surface so that ion incorporation is not needed to eliminate the charge antagonism.

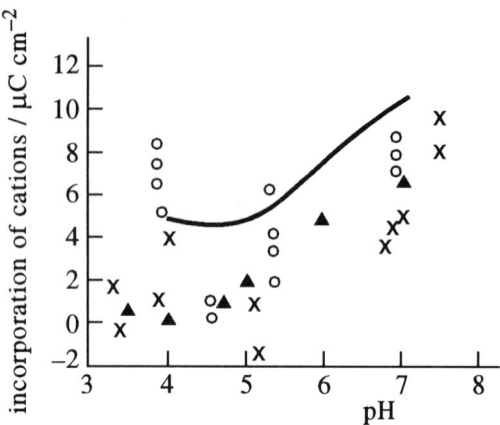

Figure 3. Incorporation of cations in a monolayer of human serum albumin adsorbed on negatively charged polystyrene surfaces, as predicted from a model (—) and as measured from 50 mM solutions of NaCl (o), $BaCl_2$ (x) and $MnCl_2$ (▲). T = 25°C.

2.2. PROTEIN AND SORBENT HYDROPHOBICITY

The notion "hydrophobicity" is related to the observation that the solubility in water decreases with decreasing polarity of the solute.

Protein molecules contain both polar and apolar parts. In globular proteins in an aqueous environment the apolar residues tend to be buried in the interior of the molecule where they are shielded from contact with water. However, due to other interactions and to geometrical constraints it is, as a rule, not possible to hide all the apolar parts in the interior and to expose all the polar parts at the aqueous periphery of the protein molecule. For the relatively small proteins (such as LSZ, αLA, ribonuclease, cytochrome c, etc.), having a molar mass of ca. 15,000 Da, apolar atoms occupy about 40%-50% of the water-accessible surface area. For larger proteins, which have a lower surface/volume ratio, the fraction of apolar atoms at the surface is usually less. Furthermore, water-soluble, non-aggregating proteins show a more or less even distribution of the polar and apolar residues over their surface so that no pronounced hydrophobic regions are present.

Sorbent surfaces may range between extremely hydrophilic (e.g. glass, silica and other metal oxides) to extremely hydrophobic (e.g. polystyrene, teflon).

When the surfaces of the protein and the sorbent are primarily hydrophilic dehydration would oppose adsorption. If adsorption still occurs some hydration water may be retained in the protein-sorbent contact zone. When (one of) the contacting surfaces (is) are hydrophobic dehydration of (that) those surface(s) would promote protein adsorption.

Results from hydrophobic interaction chromatography studies [7] confirm that the polarity of the protein exterior influences its adsorption. However, not only the hydrophobicity of the protein surface, but the *overall* hydrophobicity of the protein molecule may be relevant for the adsorption behavior. The overall hydrophobicity influences the protein structure stability which, in turn, affects the adsorption affinity. This matter will be discussed in more detail in Section 2.4.

It is intrinsically difficult to trace the effect of the hydrophobicity of the sorbent surface because variation in the hydrophobicity involves variation in the chemical composition and, often, in the electrical charge density. Nevertheless, as a trend, proteins adsorb in larger amounts at more hydrophobic surfaces [8, 9, 10]. The desorbability at hydrophobic surfaces is found to decrease with time, probably due to progressing structural rearrangements in the protein allowing increased hydrophobic bonding between the protein and the sorbent surface [11]. There are some indications in literature [12] that proteins, being amphipolar themselves, have a maximum affinity for surfaces of intermediate polarity.

2.3. SORBENT SURFACE MORPHOLOGY

Most surfaces are not completely smooth and rigid, i.e. parts of the polymer molecules may protrude into the aqueous environment, giving the surface a "hairy" character. The morphology of the surface is determined by the balance of the energetic and entropic contributions to the mutual interactions among monomeric units in the polymer and water molecules. If polymer parts extend into the solution the surface will respond dynamically to protein adsorption. On the one hand this would offer the possibility to optimize contact by conforming to the protein shape; on the other hand deposition of a layer of protein molecules would reduce the conformational entropy of those sorbent surface protuberances.

Grafting or pre-adsorbing water-soluble oligomers or polymers have been used to tune protein adsorption. In particular, polyethylene oxide (PEO) has been proven to be rather successful in producing protein-resistant surfaces [13, 14]. The protein repellency

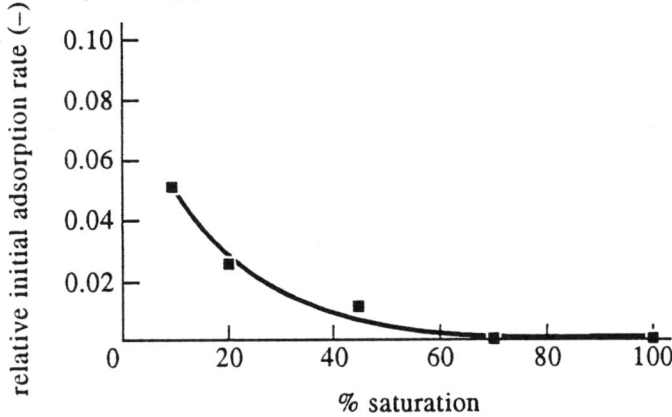

Figure 4. Relative adsorption rate of lipase at a hydrophobic silanized silica surface on which PEO is preadsorbed. For details is referred to the text.

can be attributed to steric repulsion, i.e. osmotic pressure and elastic restoring forces in the PEO chains [15]. The effectiveness of PEO to reduce protein adsorption is determined by its density on the sorbent surface and, to a minor extent, the chain length of the PEO [15]. As an example, in Figure 4 the relative adsorption rate of a lipase (Lipase B from Candida rugosa) onto a hydrophobic surface is given as a function of the degree of coverage of the sorbent by PEO chains containing 127 ethylene oxide monomers [16]. The figure shows a steep retardation of lipase adsorption and essentially complex suppression at surface coverages beyond 50%.

2.4. PROTEIN STRUCTURE STABILITY

The structure of a globular protein molecule in aqueous solution is only marginally stable. Hence, the protein structure may be readily perturbed by changing the environmental conditions, such as pH, temperature or addition of other solutes. Protein structure stability may also be affected by introducing a foreign interface to the system. Nevertheless, as a rule, globular protein molecules retain a compact structure after adsorption [6]. It is likely that proteins tend to rearrange their conformation to allow their hydrophobic residues to contact hydrophobic regions on the sorbent surface. It is, therefore, to be expected that the native-state structural stability of the protein and the sorbent hydrophobicity are the main factors determining the degree of structure perturbation in adsorbing protein molecules.

The mere observation that many water-soluble proteins spontaneously adsorb on electrostatically repelling surfaces even if they are hydrophilic and that the tendency to adsorb increases with decreasing stability of the structure in solution suggest that there is a driving force for adsorption that is related to protein structure stability. For instance, hydrophobic parts of the protein that are buried in the interior of the dissolved molecule may, after adsorption, be exposed to the sorbent surface where they are still shielded from water. Such structural changes involve a decrease in *intra*molecular hydrophobic bonding which leads to destabilization of secondary structures as α-helices and β-sheets. Thus, adsorption could very well lead to rupture of ordered secondary structure and, hence, to an increased conformational entropy of the protein. If the conformational entropy gain overcompensates unfavorable effects of electrostatic repulsion and hydrophilic dehydration the protein could still spontaneously adsorb on a hydrophilic, like-charged surface. Indeed, spontaneous adsorption under such conditions was found to be endothermic [17] and must therefore be driven by entropy increase.

The most direct way to detect structural rearrangements upon adsorption is by comparing spectral properties of the protein in the dissolved and the adsorbed states. However, the complex nature of the solid-water interface often hinders straightforward interpretation of the spectroscopic signals from the protein. For that reason, various authors [18, 19, 20] have compared the secondary structure of desorbed protein in solution with that of native protein before adsorption. Only a few studies have been reported [21, 22, 23, 24] where the structure in the adsorbed state has been successfully assessed. In these studies the degree of ordered secondary structure is lower in the adsorbed state. After displacement from the sorbent surface some proteins do refold into their native structure, whereas others do not [22].

More recently, micro-differential scanning calorimetry (DSC) experiments have provided complementary evidence for adsorption-induced structural perturbation [25, 26,

27]. These experiments showed that the thermal stability of the protein structure in the adsorbed state is less stable than in the native state in solution. The difference is larger for more hydrophobic sorbent surfaces and also for those proteins of which the native structure in solution is less stable.

In summary, protein adsorption is a complex process that is controlled by a number of subprocesses, the major ones being (a) changes in the state of hydration of the sorbent surface and parts of the protein molecule, (b) electrostatic interactions between the protein and the sorbent material, and (c) structural rearrangements in the adsorbing protein molecules. These subprocesses interplay among each other. For instance, favorable dehydration of a hydrophobic surface requires close contact between the sorbent surface and the adsorbed protein layer, which, for compact globular proteins, may be optimized by structural rearrangements that increase the molecular flexibility. If such structural rearrangements include a reduction of secondary structure the rotational mobility of the polypeptide backbone and, hence, its conformational entropy will be enhanced. Furthermore, an increased backbone flexibility improves the ability of the protein to form ion-pairs with oppositely charged groups on the sorbent surface and to reduce lateral electrostatic repulsion between like-charged proximal adsorbed protein molecules.

Based on these considerations it is to be expected that all proteins adsorb on hydrophobic surfaces even under electrostatically adverse conditions. With respect to their adsorption behavior on hydrophilic surfaces distinction may be made between structurally stable ("hard") and labile ("soft") proteins. The hard proteins adsorb on hydrophilic surfaces only if they are electrostatically attracted. The soft proteins are more liable to undergo structural changes when they adsorb and the ensuing conformational entropy gain may be sufficiently large to cause spontaneous adsorption on a hydrophilic, like-charged surface.

3. Protein Adsorption in Model-Systems

The model-systems contain one type of well-defined protein and one type of well-characterized solid surface in an aqueous medium containing one type of low-molecular-weight electrolyte. Table 1 summarizes some relevant properties of the proteins. LSZ, ribonuclease (RNase) from bovine pancreas and αLA are relatively small and they have almost identical sizes and shapes. They have different isoelectrical points, so that at a given pH these proteins contain different numbers of charged groups. The values for the denaturation temperature, and the Gibbs energy of denaturation indicate that the stability of the native structure in solution decreases in the order LSZ > RNase > αLA. The structural stability of αLA further decreases by removing the Ca^{2+}-ion from the protein. Bovine serum albumin (BSA) is about five times larger than the other proteins. Its isoelectric point is also at low pH and the structural stability is comparable to that of αLA.

The sorbent materials are supplied as finely dispersed particles (colloids). Some of their properties are collected in Table 2. The sorbents cover all different combinations of hydrophobicity and sign of surface charge. Thus, these model systems allow systematic investigation of the influences of hydrophobicity, electrical charge and protein structure stability on protein adsorption.

For all these systems, the adsorption isotherms, where the amount of protein per unit area of sorbent surface is plotted against the protein concentration in solution after

TABLE 1. Some properties of the proteins that are relevant for their adsorption behavior

protein	LSZ	RNase	αLA	αLA(–Ca^{2+})	BSA
molar mass (Da)	14600	13680	14200	14200	67000
isoelectric point (pH units)	11.1	9.4	4.3	4.1	4.6
denaturation temperature (°C) (at pH of maximum stability)	76	70	63	41	65
Gibbs energy of denaturation (J g^{-1})					
heat	4.1	3.2	1.5	...	...
denaturant	4.0	3.9	1.9	0.7	...

LSZ: lysozyme; RNase: ribonuclease; αLA: α-lactalbumin; αLA(–Ca^{2+}): Ca^{2+} depleted αLA; BSA: bovine serum albumin.

TABLE 2. Some properties of the sorbent particles

	0.05 M electrolyte				
	phosphate buffer pH 7.0			Acetate buffer pH 5.5	Borate buffer pH 9.5
	PS$^+$	PS$^-$	SiO$_2^-$	αFe$_2$O$_3^+$	αFe$_2$O$_3^-$
Nature of charged groups	=$^+$NH–	–OSO$_3^-$	–O$^-$	–OH$_2^+$	–O$^-$
Charge density (mC m^{-2})	+27	–23	...	...	...
Zeta-potential (mV)	+32	–69	–39	+20	–47
Hydrophobicity (contact angle with water)	82°	82°	0°	hydrophilic	
Surface area (m^2 g^{-1})	12.4	10.0	100	36.0	36.0

adsorption, show well-developed plateau values, Γ_{pl}. In Figure 5 these plateau values are given. The charge of the proteins in qualitatively indicated by "+" and "–" signs. At the hydrophobic PS surfaces all proteins adsorb, irrespective of the electrostatic interaction. However, the influence of electrostatic interaction is reflected in the value for Γ_{pl}. At the hydrophilic hematite (αFe$_2$O$_3$) surface, where dehydration is unfavorable, the structurally most stable proteins LSZ and RNase adsorb only if electrostatically attracted, but the less stable αLA and BSA adsorb even when electrostatically repelled. A similar behavior is observed at the hydrophilic silica (SiO$_2$) surface; here the influence of protein structure stability is clearly demonstrated by the difference between the Γ_{pl}-values for αLA and αLA(–Ca^{2+}).

The influence of electrostatics is further investigated by varying the charge on (one of) the components. Figure 6 shows Γ_{pl}-values for RNase and BSA on various surfaces as a function of pH, i.e. as a function of charge on the protein and, in case of αFe$_2$O$_3$, on the sorbent. If global electrostatic forces between the protein and the sorbent dominate the adsorption, Γ_{pl} should be a monotonic function of pH. This is indeed observed for RNase on αFe$_2$O$_3$. However, at both positively and negatively charged PS surfaces Γ_{pl} for RNase is almost invariant with the charge contrast between the protein

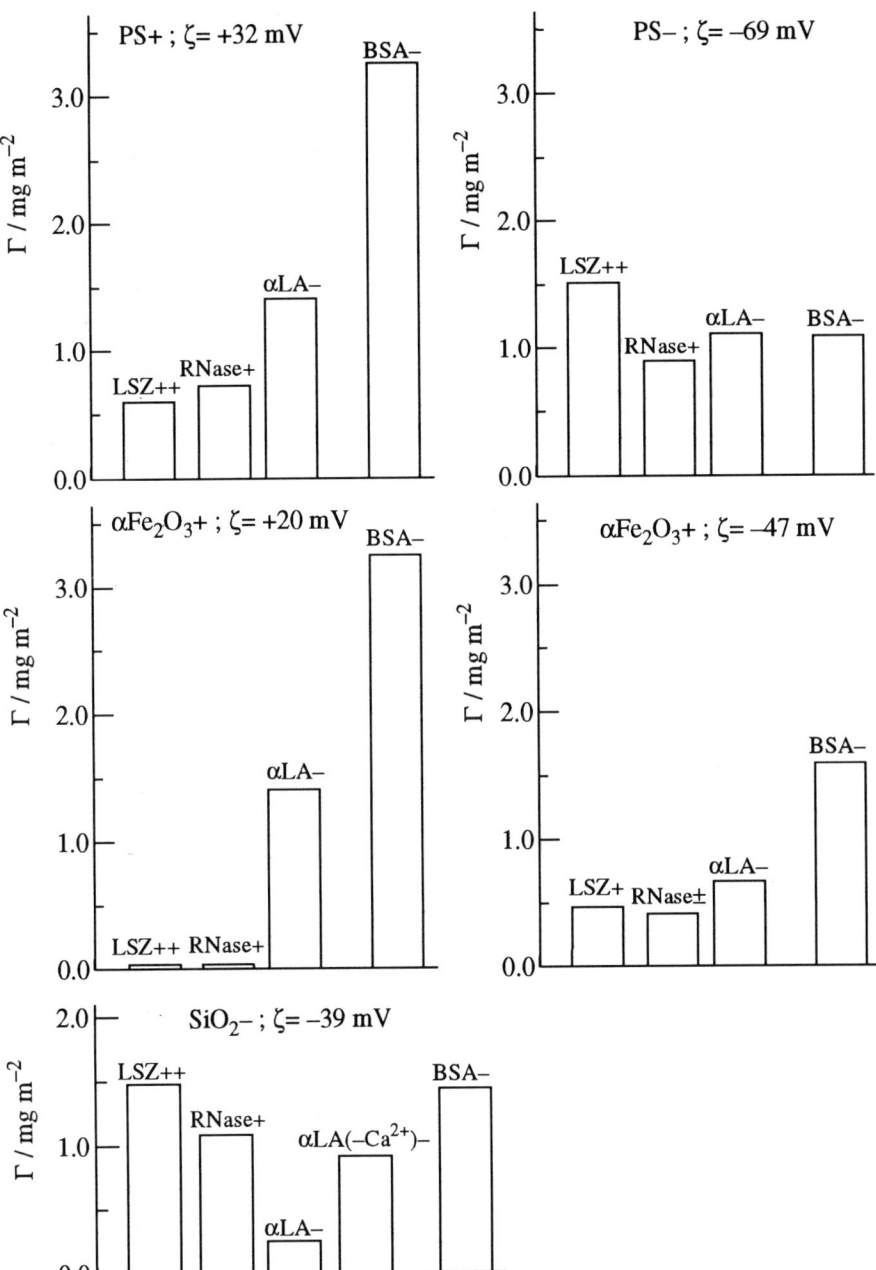

Figure 5. Plateau values of adsorption isotherms of lysozyme (LSZ), ribonuclease (RNase), α-lactalbumin (αLA), Ca^{2+}-depleted αLA (αLA($-Ca^{2+}$)) and serum albumin (BSA) on hydrophobic polystyrene (PS) and hydrophilic hematite (αFE$_2$O$_3$) and silica (SiO$_2$) surfaces. An indication of the sorbent surface charge is given through the ζ-potential of the bare surface and of the proteins by the + and − signs. Ionic strength 0.05 M; T = 25°C.

550

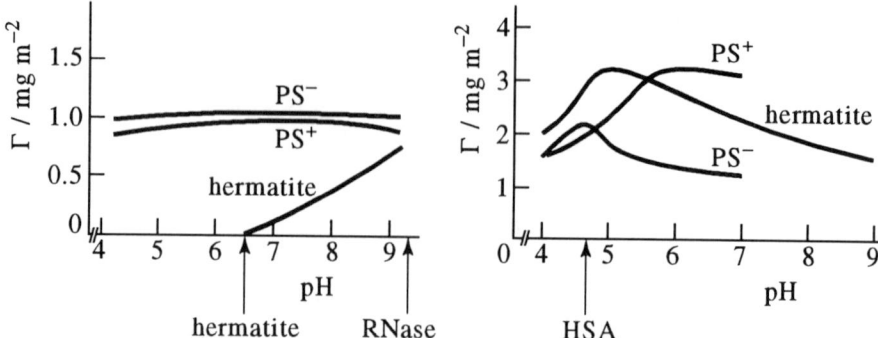

Figure 6. Plateau-values of adsorption of ribonuclease (left) and human serum albumin (right) on polystyrene and hematite surfaces. The charge of the polystyrene is indicated by the + and – signs and the isoelectric points of hematite and the proteins is indicated by the arrows. Ionic strength 0.01 M and T = 25°C.

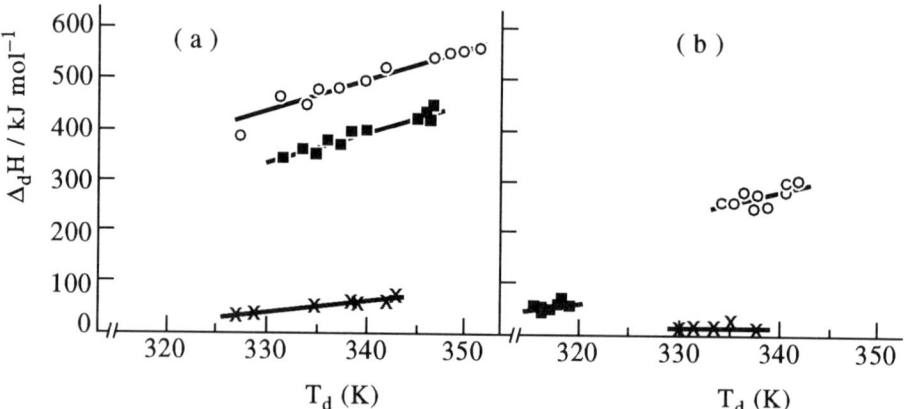

Figure 7. Temperature-induced denaturation enthalpy for (a) lysozyme and (b) α-lactalbumin in solution (o) and adsorbed on hematite (■) and polystyrene (x). Ionic strength 0.05 M.

and the sorbent, suggesting that hydrophobic dehydration effects overrule electrostatics. With BSA all curves show a maximum value near the isoelectric point of the protein-covered sorbent particle. Such a Γ_{pl} (pH) profile is quite often observed for proteins at surfaces of different nature [5, 28, 29, 30]. It suggests that the Γ_{pl} (pH) dependency reflects a characteristic of the protein rather than protein-sorbent interaction. Indeed, a variety of experimental data indicated that for BSA on PS the reduction in Γ_{pl} at either side of the isoelectric point is due to progressive rearrangements in the structurally labile protein molecule [31].

Heat-induced denaturation enthalpies for LSZ and αLA in the dissolved and the adsorbed states are given in Figure 7. It can be inferred that on the hydrophobic PS surface LSZ and, even more so, αLA have lost most of their ordered structure. At the hydrophilic αFe_2O_3 surface LSZ retains most of its structure, whereas the less stable αLA loses it almost completely. Again, this explains why αLA does and LSZ does not adsorb on a hydrophilic like-charged surface.

TABLE 3. Percentage of α-helix in proteins before and after adsorption on silica particles.

protein			before adsorption	adsorbed state (plateau-adsorption)
LSZ				
	pH	4.0	32	25
		4.7	33	30
		7.0	32	n.m.
BSA				
	pH	4.0	69	n.m.
		4.7	70	n.m.
		7.0	74	38
n.m.: not measurable, because of flocculation of the colloidal dispersion.				

More direct evidence for differences in structural perturbations between hard and soft proteins is presented in Table 3, where the influence of a hydrophilic SiO_2 surface on the α-helix content in LSZ and BSA is summarized. The helix reduction is larger for the least stable protein. The helix reduction from 74% to 38% , as observed for BSA at pH 7, involves 212 peptide units and, hence, an entropy increase of the polypeptide backbone of $2R \ln 2^{212} = 2442$ J K^{-1} mol^{-1}. At 298 K this gives a contribution to the Gibbs energy of adsorption of 728 kJ mol^{-1}. Even if this is an overestimate (the rotational freedom along the polypeptide chain will be restricted by its attachment to the sorbent surface through the anchoring of various amino acid residues), it may still be sufficiently large to outweigh opposing effects from hydrophilic dehydration and electrostatic repulsion.

In summary, the adsorption data for the model-systems confirm the expectations with respect to the roles of electrical charge, hydrophobicity and protein structure stability, as stated in section 2

4. Two Cases of Practical Relevance

4.1. ADSORPTION-INDUCED CHANGES IN PROTEOLYTIC ACTIVITY

Proteolytic enzymes (proteases) are applied in e.g. detergents to degrade proteinaceous stains from soiled cloth. In detergent industry the protease Subtilisin 309 from Bacillus lentus (trademark Sarinase®) is mainly used. Savinase has a molar mass of 28,000 Da; its isoelectric point is at pH 10. The interaction of this protein with the various interfaces it encounters in liquid detergents may affect its enzymatic activity. Hence, we investigated how adsorption of different types of surfaces affect the autolytic activity of Savinase. It is generally assumed that globular protein molecules must unfold to some extent prior to proteolytic attack. Adsorption-induced structural changes could therefore make Savinase more susceptible for autodigestion. On the other hand, adsorption at a surface could lead to a decreased accessibility of the peptide bonds to be cleaved and, hence, to a protection against autolysis.

552

The sorbents used in this study were suspensions of PS, teflon and SiO_2. All three sorbents are negatively charged; the surfaces of teflon and PS are hydrophobic and that of SiO_2 is hydrophilic. Further details on these sorbents are given in the references [32] and [33].

As judged from the initial parts of the isotherms [33], the affinity of Savinase is higher for PS and teflon than for SiO_2. The Γ_{pl}-values for PS and teflon are essentially the same, i.e. 2.1 mg m^{-2}, whereas it is 1.8 mg m^{-2} for SiO_2.

The influence of adsorption on the proteolytic stability is illustrated in Figure 8. It is observed that hydrophobic surfaces stimulate autolysis of Savinase, whereas, especially at prolonged incubation times, the hydrophilic SiO_2 surface seems to have a protecting effect. This result is in line with the conclusion, made in section 2.4, that a hydrophobic surface stimulates advanced structural perturbation in proteins, which, in the example given here, causes a higher susceptibility towards autolytic attack.

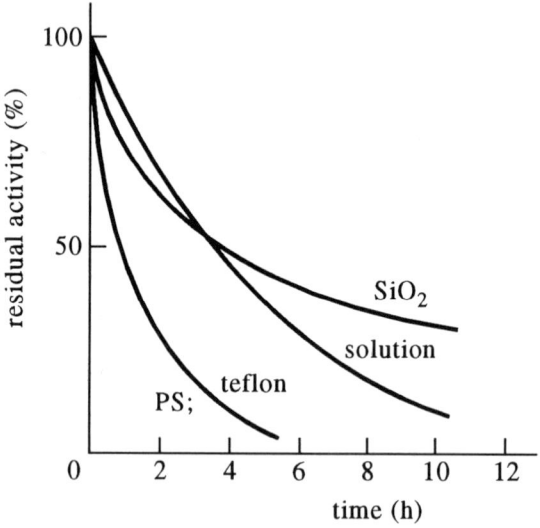

Figure 8. Inactivation of Savinase in solution and adsorbed on different surfaces. For details see text. Ionic strength 0.01 M; pH 8; T = 20°C.

4.2. THE IMMUNOLOGICAL ACTIVITY OF ADSORBED IgG

Over the past decades immuno gamma globulin (IgG) adsorbed on polymer latices (so-called immunolatices) has been extensively used in medical diagnostics. IgG molecules in solution have a Y-shaped conformation. The antigen binding sites are located at the far ends of the identical "arms" of the Y, the so-called F(ab)-parts. The "leg" of the Y is called the Fc-part. The use of immunolatices requires preservation of the immunological activity of IgG after adsorption. However, the adsorbed IgG molecules may have lost their ability to bind antigens. This could be caused by a decreased accessibility of the antigen binding sites when the F(ab)-parts are in close contact with the sorbent surface.

Furthermore, adsorption-induced structural perturbations may reduce the antigen binding capacity.

To obtain insight in how protein-sorbent interactions influence the orientation and conformation of IgG molecules, the adsorption behavior of two monoclonal IgGs (directed towards the human pregnancy hormone hCG), and their constituent $F(ab')_2$ and Fc fragments are studied [34].

TABLE 4. Molar mass and isoelectric points of two monoclonal IgG's (a-hCG) and their $F(ab')_2$ and Fc fragments.

protein	#1			#2		
	IgG	$F(ab')_2$	Fc	IgG	$F(ab')_2$	Fc
molar mass (kDa)	150	100	50	150	100	50
isoelectric point (pH units)	5.8	5.9	6.0	6.9	8.5	6.1

Information on these proteins is given in Table 4. The isoelectric points of the $F(ab')_2$ and Fc fragments of IgG#1 are rather similar, which suggests that the charge is evenly distributed over the F(ab) and Fc-parts. The different parts of IgG#2 have different isoelectric points, so that, at pH 7, the F(ab)-parts are positively and the Fc-part is negatively charged. The Fc-parts of the two IgGs are more or less identical.

Figure 9 shows the adsorption behavior of the IgGs and their corresponding $F(ab')_2$ and Fc fragments on negatively charged silica surfaces. The charge on the protein is

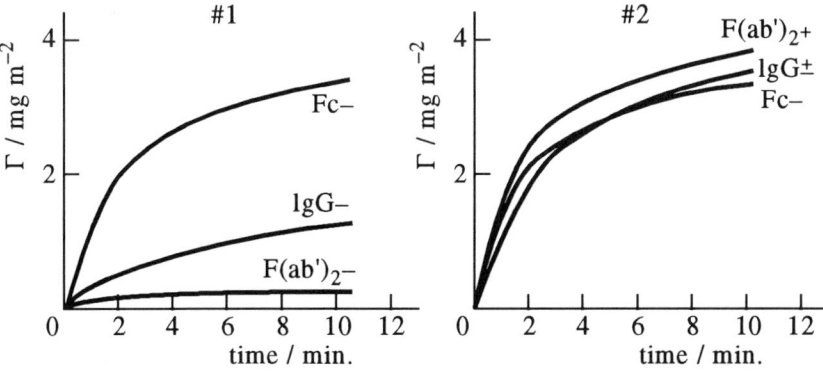

Figure 9. Adsorption of immuno gamma globulins and their corresponding fragments on negatively charged, hydrophilic silica surfaces. The charges on the proteins are indicated by the + and − signs. Ionic strength 0.005 M; pH = 7; T = 20°C.

qualitatively indicated by "+" and "−" signs. FTIR spectroscopy revealed that the structural stability with respect to adsorption of the Fc fragment is much less than that of the $F(ab')_2$ fragment [34]. In line with this, the adsorption behavior of the Fc fragment is typically that of a "soft" protein, i.e.. it adsorbs on a hydrophilic, electrostatically repelling surface, whereas the "hard" $F(ab')_2$ fragments only adsorb at the hydrophilic silica when it is electrostatically attracted (cf. the Figures 9a and 9b).

Thus, both the susceptibility to structural adaptation upon adsorption and electrostatic interaction may be utilized to orient the IgG molecule on the surface. The

554

resulting variation of the hCG binding capacity of both IgGs as a function of pH is shown in Figure 10. For IgG#1 at pH < 6, where both the F(ab) and Fc-parts are electrostatically attracted, the orientation of the adsorbed IgG molecules may be more or less random. However, when the F(ab) and Fc-parts are electrostatically repelled (pH > 6) the "soft" Fc-part still has the tendency to adsorb, causing a preferential orientation of the IgG molecule on the surface such that the F(ab)-parts are oriented towards the solution being available to bind hCG. With IgG#2 the F(ab)-parts are more positively charged than the Fc-parts, so that the IgG molecule preferentially adsorbs with its F(ab)-parts down to the negatively charged SiO_2 surface. Only at higher pH, where the F(ab)-parts are negatively charged as well, the preference seems to be less pronounced, as inferred from the rise in the hCG binding capacity.

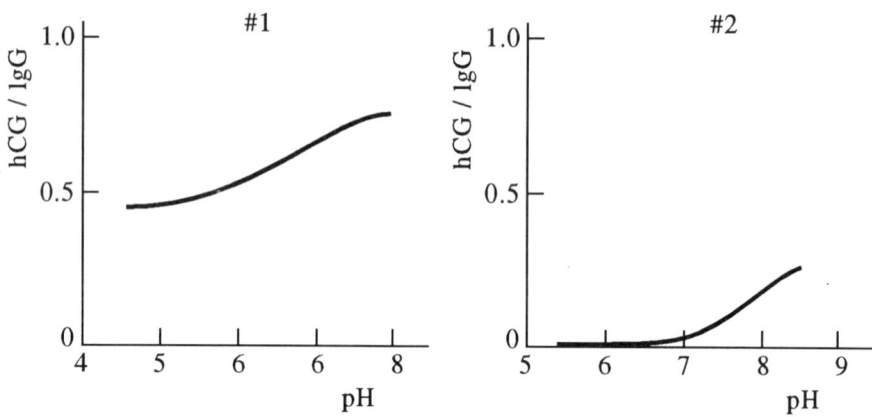

Figure 10. Antigen (hCG) binding capacity of immuno gamma globulins adsorbed on negatively charged hydrophilic silica surfaces. For details *see* text.
Ionic strength 0.01 M; T = 20°C.

5. References

1. Bae, Y.H., Okano, T. and Kim, S.W. (1990) Temperature dependence of swelling of crosslinked poly (N,N-alkyl substituted acrylamides) in water, *J. Polym. Sci. Polym. Phys.* **28**, 923-936.
2. Klein, F., Bronsveld, W., Norde, W., van Romunde, L.K.J. and Singer, J.M. (1979) A modified latex-fixation test for the detection of rheumatic factors, *J. Chem. Pathol.* **32**, 90-92.
3. Malcata, F.X., Reyes, H.R., Garcia, H.S., Hill Jr., C.G. and Amundson, C.H. (1990) Immobilized lipase reactors for modification of fats and oils, *J. Am. Oil Chem. Soc.* **67**, 870-910.
4. Chaiken, I.M., Shai, Y., Fassina, G. and Calicetti, P. (1988). Current development of analytical affinity chromatography: design and biotechnological uses of molecular recognition surfaces, *Makromol. Chem. Macromol. Symp.* **17**, 269-280.
5. Haynes, C.A., Sliwinski, E. and Norde, W. (1994) Structural and electrostatic properties of globular proteins at a polystyrene-water interface, *J. Colloid Interface Sci.* **164**, 394-409.
6. Haynes, C.A. and Norde, W. (1994) Globular proteins at solid-liquid interfaces, *Colloids and Surfaces B: Biointerfaces* **2**, 517-566.
7. Regnier, F.W. (1987) The role of protein structure in chromatographic behavior, *Science* **238**, 319-323.
8. Elwing, H., Welin, S., Askendal, A., Nilssou and Lundström, I. (1987) A wettability gradient method for studies of macromolecular interactions at the liquid-solid interface, *J. Colloid Interface Sci.* **91**, 248-255.
9. Gölander, C.-G., Lin, Y.-S., Hlady, V. and Andrade, J.A. (1990) Wetting and plasma-protein adsorption studies using surfaces with a hydrophobicity gradient, *Colloids and Surfaces* **49**, 289-302.

555

10. Van den Berg, E., Elwing, H., Askendal, A. and Lundström, I. (1991) Protein immobilization to 3-aminopropyl triethoxy silane/glutaraldehyde surfaces: characterization by detergent washing, *J. Colloid Interface Sci.* **143**, 327-335.

11. Elwing, H., Askendal, A. and Lunström, I. (1989) Desorption of fibrinogen and γ-globulin from solid surfaces induced by a nonionic detergent, *J. Colloid Interface Sci.* **128**, 296-300.

12. Baszkin, A. and Lyman, D.J. (1980) The interaction of plasma proteins with polymers, *J. Biomed. Mater. Res.* **14**, 393-403.

13. Lee, J., Martic, P.M. and Tan, J.S. (1989) Protein adsorption on pluronic copolymer-coated polystyrene particles, *J. Colloid Interface Sci.* **131**, 252-266.

14. Tan, J.S. and Martic, P.M. (1990) Protein adsorption and conformational change on small polymer particles, *J. Colloid Interface Sci.* **136**, 415-431.

15. Jeon, S.I., Lee, H.J., Andrade, J.D. and de Gennes, P.G. (1991) Protein-surface interactions in the presence of polyethylene oxide, *J. Colloid Interface Sci.* **142**, 149-166.

16. Schroën, C.G.P.H., Van der Voort maarschalk, K., Cohen Stuart, M.A., van der Padt, A. and Van 't Riet, K. (12995) Influence of pre-adsorbed block copolymers on protein adsorption: surface properties, layer thickness and surface coverage, *Langmuir* **11**, 3068-3074.

17. Norde, W. (1992) Energy and entropy of protein adsorption, *J. Dispersion Sci. Tech.* **13**, 363-377.

18. Walton, A.G. and Soderquist, M.E. (1980) Behavior of proteins at interfaces, *Croat. Chem. Acta 53*, 363-372.

19. Chan, B.M.C. and Brash, J.L. (1981) Adsorption of fibrinogen on glass: reversibility aspects, *J. Colloid Interface Sci.* **82**, 217-225.

20. Norde, W., MacRitchie, F., Nowicka, G. and Lyklema, J. (1986) Protein adsorption at liquid-solid interfaces: reversibility and conformational aspects, *J. Colloid Interface Sci.* **112**, 447-456.

21. Kondo, A., Oku, S. and Higashitani, K. (1991) Structural changes in protein molecules adsorbed on ultra fine silica particles, *J. Colloid Interface Sci.* **143**, 214-221.

22. Norde, W. and Favier, J.P. (1992) Structure of adsorbed and desorbed proteins, *Colloids Surfaces* **64**, 87-93.

23. Kondo, A., Oku, S., Murakami, F. and Higashitani, K. (1993) Conformational changes in protein molecules upon adsorption on ultrafine particles, *Colloids Surfaces B: Biointerfaces* **1**, 197-201.

24. Barbucci, R., Casolaro, A. and Magnani, A. (1992) Characterisation of biomaterial surfaces: ATR-FTIR, potentiometric and calorimetric analysis, *Clinical Mater.* **11**, 37-51.

25. Steadman, B.L., Thompson, K.C., Middaugh, C.R., Matsuno, K., Vrona, S., Lawson, E.Q. and Lewis, R.V. (1992) The effects of surface adsorption on the thermal stabilities of proteins, *Biotechn. Bioengin.* **40**, 8-15.

26. Haynes, C.A. and Norde W. (1995) Structures and stabilities of adsorbed proteins, *J. Colloid Interface Sci.* **169**, 313-328.

27. Yan, G., Li, J., Huang, S.-C. and Caldwell, K. (1995) Calorimetric observations of protein conformation at solid/liquid interfaces, *ACS Symp. Ser.* **602**, 256-268.

28. Bagchi, P. and Birnbaum, S.M. (1981) Effect of pH on the adsorption of immunoglobulin G on anionic poly (vinyl toluene) model latex particles, *J. Colloid Interface Sci.* **83**, 460-478.

29. Morrissey, B.W. and Stromberg, R.R. (1974) The conformation of adsorbed blood proteins by infrared bound fraction measurements, *J. Colloid Interface Sci.* **46**, 152.

30. Koutsoukos, P.G., Mumme-Young, C.A., Norde, W. and Lyklema, J. (1982) Effect on the nature of the substrate on the adsorption of human plasma albumin, *Colloids and Surfaces* **5**, 93-104.

31. Norde, W. and Lyklema, J. (1978) The adsorption of human plasma albumin and bovine pancreas ribonuclease at negatively charged polystyrene surfaces, *J. Colloid Interface Sci.* **66**, 257-265; 266-276.

32. Maste, M.C.L., Van Velthoven, A.P.C.M., Norde, W. and Lyklema, J. (1994) Synthesis and characterization of a short-haired poly(ethylene oxide)-grafted polystyrene latex, *Colloids and Surfaces* **A83**, 255-260.

33. Maste, M.C.L., Rinia, H.A., Brands, C.M.J., Egmond, M.R. and Norde, W. (1995) Inactivation of a subtilisin in colloidal systems, *Biochim. Biophys. Acta* **1252**, 261-268.

34. Buijs, J. (1995) Immunoglobulins and their fragments on solid surfaces. Ph.D. thesis Wageningen Agricultural University, Wageningen, The Netherlands.

INDEX

558